T0134984

Advances in Intelligent Systems and Computing

Volume 755

Series editor

Janusz Kacprzyk, Polish Academy of Sciences, Warsaw, Poland
e-mail: kacprzyk@ibspan.waw.pl

The series "Advances in Intelligent Systems and Computing" contains publications on theory, applications, and design methods of Intelligent Systems and Intelligent Computing. Virtually all disciplines such as engineering, natural sciences, computer and information science, ICT, economics, business, e-commerce, environment, healthcare, life science are covered. The list of topics spans all the areas of modern intelligent systems and computing such as: computational intelligence, soft computing including neural networks, fuzzy systems, evolutionary computing and the fusion of these paradigms, social intelligence, ambient intelligence, computational neuroscience, artificial life, virtual worlds and society, cognitive science and systems, Perception and Vision, DNA and immune based systems, self-organizing and adaptive systems, e-Learning and teaching, human-centered and human-centric computing, recommender systems, intelligent control, robotics and mechatronics including human-machine teaming, knowledge-based paradigms, learning paradigms, machine ethics, intelligent data analysis, knowledge management, intelligent agents, intelligent decision making and support, intelligent network security, trust management, interactive entertainment, Web intelligence and multimedia.

The publications within "Advances in Intelligent Systems and Computing" are primarily proceedings of important conferences, symposia and congresses. They cover significant recent developments in the field, both of a foundational and applicable character. An important characteristic feature of the series is the short publication time and world-wide distribution. This permits a rapid and broad dissemination of research results.

More information about this series at http://www.springer.com/series/11156

Ajith Abraham · Paramartha Dutta
Jyotsna Kumar Mandal · Abhishek Bhattacharya
Soumi Dutta
Editors

Emerging Technologies in Data Mining and Information Security

Proceedings of IEMIS 2018, Volume 1

 Springer

Editors
Ajith Abraham
Machine Intelligence Research Labs
Auburn, WA, USA

Abhishek Bhattacharya
Institute of Engineering and Management
Kolkata, West Bengal, India

Paramartha Dutta
Department of Computer and Systems
  Sciences
Visva-Bharati University
Santiniketan, West Bengal, India

Soumi Dutta
Institute of Engineering and Management
Kolkata, West Bengal, India

Jyotsna Kumar Mandal
Department of Computer Science
  and Engineering
University of Kalyani
Kalyani, India

ISSN 2194-5357   ISSN 2194-5365   (electronic)
Advances in Intelligent Systems and Computing
ISBN 978-981-13-1950-1   ISBN 978-981-13-1951-8   (eBook)
https://doi.org/10.1007/978-981-13-1951-8

Library of Congress Control Number: 2018950802

This Springer imprint is published by the registered company Springer Nature Singapore Pte Ltd.
The registered company address is: 152 Beach Road, #21-01/04 Gateway East, Singapore 189721, Singapore

Organizing Committee

Patron

Prof. (Dr.) Satyajit Chakrabarti, Institute of Engineering and Management, India

Conference General Chair

Dr. Bimal Kumar Roy, Indian Statistical Institute, Kolkata, India

Convener

Dr. Subrata Saha, Institute of Engineering and Management, India
Abhishek Bhattacharya, Institute of Engineering and Management, India

Co-convener

Sukalyan Goswami, University of Engineering and Management, India
Krishnendu Rarhi, Institute of Engineering and Management, India
Soumi Dutta, Institute of Engineering and Management, India
Sujata Ghatak, Institute of Engineering and Management, India
Dr. Abir Chatterjee, University of Engineering and Management, India

Key Note Speakers

Dr. Ajith Abraham, Machine Intelligence Research Labs (MIR Labs), USA
Dr. Fredric M. Ham, IEEE Life Fellow, SPIE Fellow, and INNS Fellow, USA
Dr. Sheng-Lung Peng, National Dong Hwa University, Hualien, Taiwan
Dr. Shaikh Fattah, Editor, IEEE Access and CSSP (Springer), Bangladesh
Dr. Detlef Streitferdt, Technische Universität Ilmenau, Germany
Dr. Swagatam Das, Indian Statistical Institute, Kolkata, India
Dr. Niloy Ganguly, IIT Kharagpur, India

Dr. K. K. Shukla, IIT (BHU.), Varanasi, India
Dr. Nilanjan Dey, Techno India College of Technology, Kolkata, India

Technical Program Committee Chair

Dr. J. K. Mondal, University of Kalyani, India
Dr. Paramartha Dutta, Visva-Bharati University, India
Abhishek Bhattacharya, Institute of Engineering and Management, India

Technical Program Committee Co-chair

Dr. Satyajit Chakrabarti, Institute of Engineering and Management, India
Dr. Subrata Saha, Institute of Engineering and Management, India
Dr. Kamakhya Prasad Ghatak, University of Engineering and Management, India
Dr. Asit Kumar Das, IIEST, Shibpur, India

Editorial Board

Dr. Ajith Abraham, Machine Intelligence Research Labs (MIR Labs), USA
Dr. J. K. Mondal, University of Kalyani, India
Dr. Paramartha Dutta, Visva-Bharati University, India
Abhishek Bhattacharya, Institute of Engineering and Management, India
Soumi Dutta, Institute of Engineering and Management, India

Advisory Committee

Dr. Mahmoud Shafik, University of Derby
Dr. Mohd Nazri Ismail, National Defence University of Malaysia
Dr. Bhaba R. Sarker, Louisiana State University
Dr. Tushar Kanti Bera, University of Arizona, USA
Dr. Shirley Devapriya Dewasurendra, University of Peradeniya, Sri Lanka
Dr. Goutam Chakraborty, Professor and Head of the Intelligent Informatics Lab, Iwate Prefectural University, Japan
Dr. Basabi Chakraborty, Iwate Prefectural University, Japan
Dr. Kalyanmoy Deb, Michigan State University, East Lansing, USA
Dr. Vincenzo Piuri, University of Milan, Italy
Dr. Biswajit Sarkar, Hanyang University, Korea
Dr. Raj Kumar Buyya, The University of Melbourne
Dr. Anurag Dasgupta, Valdosta State University, Georgia
Dr. Prasenjit Mitra, The Pennsylvania State University
Dr. Esteban Alfaro-Cortés, University of Castilla-La Mancha, Spain
Dr. Ilkyeong Moon, Seoul National University, South Korea
Dr. Izabela Nielsen, Aalborg University, Denmark

Dr. Prasanta K. Jana, IEEE Senior Member, Indian Institute of Technology (ISM), Dhanbad

Dr. Gautam Paul, Indian Statistical Institute, Kolkata

Dr. Malay Bhattacharyya, IIEST, Shibpur

Dr. Sipra Das Bit, IIEST, Shibpur

Dr. Jaya Sil, IIEST, Shibpur

Dr. Asit Kumar Das, IIEST, Shibpur

Dr. Saptarshi Ghosh, IIEST, Shibpur, IIT KGP

Dr. Prof. Hafizur Rahman, IIEST, Shibpur

Dr. C. K. Chanda, IIEST, Shibpur

Dr. Asif Ekbal, Associate Dean, IIT Patna

Dr. Sitangshu Bhattacharya, IIIT Allahabad

Dr. Ujjwal Bhattacharya, CVPR Unit, Indian Statistical Institute

Dr. Prashant R. Nair, Amrita Vishwa Vidyapeetham (University), Coimbatore

Dr. Tanushyam Chattopadhyay, TCS Innovation Lab, Kolkata

Dr. A. K. Nayak, Fellow and Hony. Secretarý, CSI

Dr. B. K. Tripathy, VIT University

Dr. K. Srujan Raju, CMR Technical Campus

Dr. Dakshina Ranjan Kisku, National Institute of Technology, Durgapur

Dr. A. K. Pujari, University of Hyderabad

Dr. Partha Pratim Sahu, Tezpur University

Dr. Anuradha Banerjee, Kalyani Government Engineering College

Dr. Amiya Kumar Rath, Veer Surendra Sai University of Technology

Dr. Kandarpa Kumar Sharma, Gauhati University

Dr. Amlan Chakrabarti, University of Calcutta

Dr. Sankhayan Choudhury, University of Calcutta

Dr. Anjana Kakoti Mahanta, Gauhati University

Dr. Subhankar Bandyopadhyay, Jadavpur University

Dr. Debabrata Ghosh, Calcutta University

Dr. Rajat Kr. Pal, University of Calcutta

Dr. Ujjwal Maulik, Jadavpur University

Dr. Himadri Dutta, Kalyani Government Engineering College, Kalyani

Dr. Brojo Kishore Mishra, C. V. Raman College of Engineering (Autonomous), Bhubaneswar

Dr. S. Vijayakumar Bharathi, Symbiosis Centre for Information Technology (SCIT)

Dr. Govinda K., VIT University, Vellore

Dr. Ajanta Das, University of Engineering and Management, India

Technical Committee

Dr. Vincenzo Piuri, University of Milan, Italy

Dr. Mahmoud Shafik, University of Derby

Dr. Bhaba Sarker, Louisiana State University

Dr. Mohd Nazri Ismail, National Defence University of Malaysia

Dr. Tushar Kanti Bera, Yonsei University, Seoul
Dr. Birjodh Tiwana, LinkedIn, San Francisco, California
Dr. Saptarshi Ghosh, IIEST, Shibpur, IIT KGP
Dr. Srimanta Bhattacharya, Indian Statistical Institute, Kolkata
Dr. Loingtam Surajkumar Singh, NIT Manipur
Dr. Sitangshu Bhattacharya, IIIT Allahabad
Dr. Sudhakar Tripathi, NIT Patna
Dr. Chandan K. Chanda, IIEST Shibpur
Dr. Dakshina Ranjan Kisku, NIT Durgapur
Dr. Asif Ekbal, IIT Patna
Dr. Prasant Bharadwaj, NIT Agartala
Dr. Somnath Mukhopadhyay, Hijli College Kharagpur, India, and Regional Student
Coordinator, Region II, Computer Society of India
Dr. G. Suseendran, Vels University, Chennai, India
Dr. Sumanta Sarkar, Department of Computer Science of University of Calgary
Dr. Manik Sharma, Assistant Professor, DAV University, Jalandhar
Dr. Rita Choudhury, Gauhati University
Dr. Kuntala Patra, Gauhati University
Dr. Helen K. Saikia, Gauhati University
Dr. Debasish Bhattacharjee, Gauhati University
Dr. Somenath Sarkar, University of Calcutta
Dr. Sankhayan Choudhury, University of Calcutta
Dr. Debasish De, Maulana Abul Kalam Azad University of Technology
Dr. Buddha Deb Pradhan, National Institute of Technology Durgapur
Dr. Shankar Chakraborty, Jadavpur University
Dr. Durgesh Kumar Mishra, Sri Aurobindo Institute of Technology, Indore,
Madhya Pradesh
Dr. Angsuman Sarkar, Secretary, IEEE EDS Kolkata Chapter and Kalyani
Government Engineering College
Dr. A. M. Sudhakara, University of Mysore
Dr. Indrajit Saha, National Institute of Technical Teachers' Training and Research,
Kolkata
Dr. Bikash Santra, Indian Statistical Institute (ISI)
Dr. Ram Sarkar, Jadavpur University
Dr. Priya Ranjan Sinha Mahapatra, Kalyani University
Dr. Avishek Adhikari, University of Calcutta
Dr. Jyotsna Kumar Mandal, Kalyani University
Dr. Manas Kumar Sanyal, Kalyani University
Dr. Atanu Kundu, Chairman, IEEE EDS, Heritage Institute of Technology, Kolkata
Dr. Chintan Kumar Mandal, Jadavpur University
Dr. Kartick Chandra Mondal, Jadavpur University
Mr. Debraj Chatterjee, Manager, Capgemini
Mr. Gourav Dutta, Cognizant Technology Solutions
Dr. Soumya Sen, University of Calcutta
Dr. Soumen Kumar Pati, St. Thomas College of Engineering and Technology

Mrs. Sunanda Das, Neotia Institute of Technology Management and Science, India
Mrs. Shampa Sengupta, MCKV Institute of Engineering, India
Dr. Brojo Kishore Mishra, C. V. Raman College of Engineering (Autonomous), Bhubaneswar
Dr. S. Vijayakumar Bharathi, Symbiosis Centre For Information Technology, Pune
Dr. Govinda K., Vellore Institute of technology
Dr. Prashant R. Nair, Amrita University
Dr. Hemanta Dey, IEEE Senior Member
Dr. Ajanta Das, University of Engineering and Management, India
Dr. Samir Malakar, MCKV Institute of Engineering, Howrah
Dr. Tanushyam Chattopadhyay TCS Innovation Lab, Kolkata
Dr. A. K. Nayak Indian Institute of Business Management, Patna
Dr. B. K. Tripathy, Vellore Institute of Technology, Vellore
Dr. K. Srujan Raju, CMR Technical Campus, Hyderabad
Dr. Partha Pratim Sahu, Tezpur University
Dr. Anuradha Banerjee, Kalyani Government Engineering College
Dr. Amiya Kumar Rath, Veer Surendra Sai University of Technology, Odisha
Dr. S. D. Dewasurendra,
Dr. Arnab K. Laha, IIM Ahmedabad
Dr. Kandarpa Kumar Sarma, Gauhati University
Dr. Ambar Dutta, BIT Mesra & Treasurer—CSI Kolkata
Dr. Arindam Pal, TCS Innovation Labs, Kolkata
Dr. Himadri Dutta, Kalyani Government Engineering College, Kalyani
Dr. Tanupriya Choudhury, Amity University, Noida, India
Dr. Praveen Kumar, Amity University, Noida, India

Organizing Chairs

Krishnendu Rarhi, Institute of Engineering and Management, India
Sujata Ghatak, Institute of Engineering and Management, India
Dr. Apurba Sarkar, IIEST, Shibpur, India

Organizing Co-chairs

Sukalyan Goswami, University of Engineering and Management, India
Rupam Bhattacharya, Institute of Engineering and Management, India

Organizing Committee Convener

Dr. Sajal Dasgupta, Vice-Chancellor University of Engineering and Management

Organizing Committee

Subrata Basak, Institute of Engineering and Management, India
Anshuman Ray, Institute of Engineering and Management, India
Rupam Bhattacharya, Institute of Engineering and Management, India
Abhijit Sarkar, Institute of Engineering and Management, India
Ankan Bhowmik, Institute of Engineering and Management, India
Manjima Saha, Institute of Engineering and Management, India
Biswajit Maity, Institute of Engineering and Management, India
Soumik Das, Institute of Engineering and Management, India
Sreelekha Biswas, Institute of Engineering and Management, India
Nayantara Mitra, Institute of Engineering and Management, India
Amitava Chatterjee, Institute of Engineering and Management, India
Ankita Mondal, Institute of Engineering and Management, India

Registration Chairs

Abhijit Sarkar, Institute of Engineering and Management, India
Ankan Bhowmik, Institute of Engineering and Management, India
Ankita Mondal, Institute of Engineering and Management, India
Ratna Mondol, Institute of Engineering and Management, India

Publication Chairs

Dr. Debashis De, Maulana Abul Kalam Azad University of Technology, India
Dr. Kuntala Patra, Gauhati University, India

Publicity and Sponsorship Chair

Dr. J. K. Mondal, University of Kalyani, India
Dr. Paramartha Dutta, Visva-Bharati University, India
Abhishek Bhattacharya, Institute of Engineering and Management, India
Biswajit Maity, Institute of Engineering and Management, India
Soumik Das, Institute of Engineering and Management, India

Treasurer and Conference Secretary

Dr. Subrata Saha, Institute of Engineering and Management, India
Rupam Bhattacharya, Institute of Engineering and Management, India
Krishnendu Rarhi, Institute of Engineering and Management, India

Hospitality and Transport Chair

Soumik Das, Institute of Engineering and Management, India
Nayantara Mitra, Institute of Engineering and Management, India
Manjima Saha, Institute of Engineering and Management, India
Sreelekha Biswas, Institute of Engineering and Management, India
Amitava Chatterjee, Institute of Engineering and Management, India

Web Chair

Samrat Goswami
Samrat Dey

Foreword

Welcome to the Springer International Conference on Emerging Technologies in Data Mining and Information Security (IEMIS 2018) held on February 23–25, 2018, in Kolkata, India. As a premier conference in the field, IEMIS 2018 provides a highly competitive forum for reporting the latest developments in the research and application of information security and data mining. We are pleased to present the proceedings of the conference as its published record. The theme of this year is Crossroad of Data Mining and Information Security, a topic that is quickly gaining traction in both academic and industrial discussions because of the relevance of privacy-preserving data mining (PPDM) model.

IEMIS is a young conference for research in the areas of information and network security, data sciences, big data, and data mining. Although 2018 is the debut year for IEMIS, it has already witnessed a significant growth. As an evidence of that, IEMIS received a record of 532 submissions. The authors of the submitted papers are from 35 countries around the world. Authors of the accepted papers are from 11 countries.

We hope that this program will further stimulate research in information security and data mining and provide practitioners with better techniques, algorithms, and tools for the deployment. We feel honored and privileged to serve the best recent developments in the field of WSDM to you through this exciting program.

Kolkata, India

Bimal Kumar Roy
General Chair, IEMIS 2018

Preface

This volume presents the proceedings of the International Conference on Emerging Technologies in Data Mining and Information Security (IEMIS 2018), which took place in the University of Engineering and Management in Kolkata, India, on February 23–25, 2018. The volume appears in the series "Advances in Intelligent Systems and Computing" (AISC), which is one of the fastest growing book series, published by Springer Nature, one of the largest and most prestigious scientific publishers,. AISC is meant to include various high-quality and timely publications, primarily conference proceedings of relevant conference, congresses, and symposia but also monographs, on the theory, applications, and implementations of broadly perceived modern intelligent systems and intelligent computing, in their modern understanding, i.e., tools and techniques of artificial intelligence (AI), computational intelligence (CI)—which include data mining, information security, neural networks, fuzzy systems, evolutionary computing, as well as hybrid approaches that synergistically combine these areas—but also topics such as multiagent systems, social intelligence, ambient intelligence, Web intelligence, computational neuroscience, artificial life, virtual worlds and societies, cognitive science and systems, perception and vision, DNA- and immune-based systems, self-organizing and adaptive systems, e-learning and teaching, human-centered and human-centric computing, autonomous robotics, knowledge-based paradigms, learning paradigms, machine ethics, intelligent data analysis, and various issues related to "big data," security, and trust management. These areas are at the forefront of science and technology and have been found useful and powerful in a wide variety of disciplines such as engineering, natural sciences, computer, computation and information sciences, ICT, economics, business, e-commerce, environment, health care, life science, and social sciences. The AISC book series is submitted for indexing in ISI Conference Proceedings Citation Index (now run by Clarivate), EI Compendex, DBLP, SCOPUS, Google Scholar, and SpringerLink and many other indexing services around the world. IEMIS 2018 is a debut annual conference series organized at the School of Information Technology, under the aegis of the Institute of Engineering and Management. This idea came from the heritage of the other two cycles of events: IEMCON and UEMCON, which were organized by the Institute of Engineering and Management under the leadership of Prof. Dr. Satyajit Chakraborty.

In this volume of "Advances in Intelligent Systems and Computing," we would like to present the results of studies on selected problems of data mining and information security. Security implementation is the contemporary answer to new challenges in the threat evaluation of complex systems. Security approach in theory and engineering of complex systems (not only computer systems and networks) is based on the multidisciplinary attitude to information theory, technology, and maintenance of the systems working in real (and very often unfriendly) environments. Such a transformation has shaped natural evolution in the topical range of subsequent IEMIS conferences, which can be seen over the recent years. Human factors likewise infest the best digital dangers. Workforce administration and digital mindfulness are fundamental for accomplishing all-encompassing cybersecurity. This book will be of extraordinary incentive to a huge assortment of experts, scientists, and understudies concentrating on the human part of the Internet and for the compelling assessment of safety efforts, interfaces, client-focused outline, and plan for unique populaces, especially the elderly. We trust this book is instructive yet much more than it is provocative. We trust it moves, driving per user to examine different inquiries, applications, and potential arrangements in making sheltered and secure plans for all.

The Programme Committee of the IEMIS 2018 Conference, its organizers, and the editors of these proceedings would like to gratefully acknowledge participation of all the reviewers who helped to refine contents of this volume and evaluated conference submissions. Our thanks go to, in the alphabetic order, Prof. Bimal Kumar Roy, Dr. Ajith Abraham, Dr. Sheng Lung peng, Dr. Detlef Streitferdt, Dr. Shaikh Fattah, Dr. Celia Shahnaz, Dr. Swagatam Das, Dr. Niloy Ganguly, Dr. K. K. Shukla, Dr. Nilanjan Dey, Dr. Florin PopentiuVladicescu, Dr. Dewan Md. Farid, Dr. Saptarshi Ghosh, Dr. Rita Choudhury, Dr. Asit Kumar Das, Prof. Tanupriya Choudhury, Prof. Arijit Ghosal, Prof. Rahul Saxena, Prof. Monika Jain, Dr. Aakanksha Sharaff, Prof. Dr. Sajal Dasgupta, Prof. Rajiv Ganguly, Prof. Sukalyan Goswami.

Thanking all the authors who have chosen IEMIS 2018 as the publication platform for their research, we would like to express our hope that their papers will help in further developments in design and analysis of engineering aspects of complex systems, being a valuable source material for scientists, researchers, practitioners, and students who work in these areas.

Auburn, USA	Ajith Abraham
Santiniketan, India	Paramartha Dutta
Kalyani, India	Jyotsna Kumar Mandal
Kolkata, India	Abhishek Bhattacharya
Kolkata, India	Soumi Dutta

Contents

Part II   Cloud Computing

Part III   Computational Mathematics

Part VII Expert System

About the Editors

Dr. Ajith Abraham received Ph.D. from Monash University, Melbourne, Australia, and M.Sc. from Nanyang Technological University, Singapore. His research and development experience includes over 25 years in the industry and academia spanning different continents like Australia, America, Asia, and Europe. He works in a multidisciplinary environment involving computational intelligence, network security, sensor networks, e-commerce, Web intelligence, Web services, computational grids, data mining, and applications to various real-world problems. He has authored/co-authored over 350 refereed journal/conference papers and chapters, and some of the papers have also won the best paper awards at international conferences and also received several citations. Some of the articles are available in the ScienceDirect Top 25 Hottest Articles—http://top25.sciencedirect.com/index. php?cat_id=6&subject_area_id=7.

He has given more than 20 plenary lectures and conference tutorials in these areas. He serves the editorial board of several reputed International journals and has also guest-edited 26 special issues on various topics. He is actively involved in the Hybrid Intelligent Systems (HIS); Intelligent Systems Design and Applications (ISDA); and Information Assurance and Security (IAS) series of international conferences. He was General Co-chair of the Tenth International Conference on Computer Modeling and Simulation (UKSIM'08), Cambridge, UK; Second Asia International Conference on Modeling and Simulation (AMS 2008), Malaysia; Eighth International Conference on Intelligent Systems Design and Applications (ISDA'08), Taiwan; Fourth International Symposium on Information Assurance and Security (IAS'07), Italy; Eighth International Conference on Hybrid Intelligent Systems (HIS'08), Spain; and Fifth IEEE International Conference on Soft Computing as Transdisciplinary Science and Technology (CSTST'08), Cergy-Pontoise, France, and Program Chair/Co-chair of the Third International Conference on Digital Information Management (ICDIM'08), UK, and Second European Conference on Data Mining (ECDM 2008), the Netherlands.

He is Senior Member of IEEE, IEEE Computer Society, IEE (UK), ACM, etc. More information at: http://www.softcomputing.net.

Dr. Paramartha Dutta was born in 1966 and did his bachelors and masters in statistics from the Indian Statistical Institute, Calcutta, in the years 1988 and 1990, respectively. He completed his M.Tech. in computer science from the same institute in the year 1993 and Ph.D. in engineering from the Bengal Engineering and Science University, Shibpur, in 2005. He has served in the capacity of research personnel in various projects funded by the Government of India, which include Defence Research Development Organization, Council of Scientific and Industrial Research, Indian Statistical Institute. He is now Professor in the Department of Computer and System Sciences of the Visva-Bharati University, West Bengal, India. Prior to this, he served Kalyani Government Engineering College and College of Engineering in West Bengal as full-time faculty member. He remained associated as visiting/guest faculty of several universities/institutes such as West Bengal University of Technology, Kalyani University, Tripura University.

He has co-authored eight books and has also seven edited books to his credit. He has published more than two hundred technical papers in various peer-reviewed journals and conference proceedings, both international and national, as well as several chapters in edited volumes of reputed international publishing houses like Elsevier, Springer-Verlag, CRC Press, John Wiley. He has guided six scholars who already had been awarded their Ph.D. apart from one who has submitted her thesis. Presently, he is supervising six scholars for their Ph.D. program.

He is the co-inventor of ten Indian patents and one international patent which are all published apart from five international patents which are filed but not yet to be published.

He, as an investigator, could implement successfully projects funded by All India Council for Technical Education, Department of Science and Technology, of the Government of India. He has served/serves in the capacity of external member of Boards of Studies of relevant departments of various universities encompassing West Bengal University of Technology, Kalyani University, Tripura University, Assam University. He had the opportunity to serve as the expert of several interview boards conducted by West Bengal Public Service Commission; Assam University, Silchar; National Institute of Technology, Arunachal Pradesh; Sambalpur University; etc.

He is Life Fellow of the Optical Society of India (FOSI), Institution of Electronics and Telecommunication Engineers (FIETE), and Institute of Engineering (FIE); Life Member of Computer Society of India (LMCSI), Indian Science Congress Association (LMISCA), Indian Society for Technical Education (LMISTE), and Indian Unit of Pattern Recognition and Artificial Intelligence (LMIUPRAI); the Indian affiliate of the International Association for Pattern Recognition (IAPR); Senior Member of Associated Computing Machinery (SMACM) and Institution of Electronics and Electrical Engineers (SMIEEE), USA.

Dr. Jyotsna Kumar Mandal completed his M.Tech. in computer science from the University of Calcutta in 1987, was awarded Ph.D. (engineering) in computer science and engineering by Jadavpur University in 2000, was Former Dean, Faculty of Engineering, Technology and Management, for two consecutive terms since 2008, and is presently working as Professor of Computer Science and Engineering, Kalyani University, Kalyani, Nadia, Government of West Bengal. He is Ex-Director, IQAC, Kalyani University, and Chairman, CIRM, Kalyani University. He was appointed as Professor in Kalyani Government Engineering College through Public Service Commission under Government of West Bengal. He started his career as Lecturer at NERIST, under MHRD Government of India, Arunachal Pradesh, in September 1988. He has teaching and research experience of 30 years. His areas of research are coding theory, data and network security, remote sensing and GIS-based applications, data compression, error correction, visual cryptography, and steganography. He has produced 21 Ph.Ds., two scholars have submitted, and eight are pursuing. He has supervised 3 M.Phil. graduates and more than 50 M.Tech. dissertations and more than 100 M.C.A. dissertations. He is Chief Editor of CSI Journal of Computing and Guest Editor of MST Journal (SCI indexed) of Springer. He has published more than 400 research articles out of which 154 articles in various international journals. He has published five books from LAP Germany and one from IGI Global. He was awarded A. M. Bose Memorial Silver Medal and Kali Prasanna Dasgupta Memorial Silver Medal in M.Sc. from Jadavpur University. India International Friendship Society (IIFS), New Delhi, conferred "Bharat Jyoti Award" for meritorious service, outstanding performance, and remarkable role in the field of computer science and engineering on August 29, 2012. He received "Chief Patron" Award from CSI India in 2014. International Society for Science, Technology and Management conferred "Vidyasagar Award" in the Fifth International Conference on Computing, Communication and Sensor Network on December 25, 2016. ISDA conferred Rastriya Pratibha Award in 2017.

Abhishek Bhattacharya is Assistant Professor in the Department of Computer Application at the Institute of Engineering and Management, India. He did his M.Sc. from the Biju Patnaik University of Technology and completed his M.Tech. in computer science from BIT, Mesra. He remained associated as visiting/guest faculty of several universities/institutes in India. He has three books to his credit. He has published 20 technical papers in various peer-reviewed journals and conference proceedings, both international and national, as well as chapters in edited volumes of reputed International publishing houses. He has teaching and research experience of 13 years. His areas of research are data mining, network security, mobile computing, and distributed computing. He is the reviewer of couple of journals of IGI Global, Inderscience Publications, and Journal of Information Science Theory and Practice, South Korea.

He is Member of International Association of Computer Science and Information Technology (IACSIT), Universal Association of Computer and Electronics Engineers (UACEE), International Association of Engineers (IAENG), Internet Society as a Global Member (ISOC), the Society of Digital Information and Wireless Communications (SDIWC), and International Computer Science and Engineering Society (ICSES); Technical Committee Member of CICBA 2017, 52nd Annual Convention of Computer Society of India (CSI 2017), International

Conference on Futuristic Trends in Network and Communication Technologies (FTNCT-2018), ICIoTCT 2018, ICCIDS 2018, and Innovative Computing and Communication (ICICC-2018); and Advisory Board Member of ISETIST 2017.

Soumi Dutta is Assistant Professor at the Institute of Engineering and Management, India. She is also pursuing her Ph.D. in the Department of Computer Science and Technology, Indian Institute of Engineering Science and Technology, Shibpur. She received her B.Tech. in information technology and her M.Tech. in computer science securing first position (gold medalist), both from Techno India Group, India. His research interests include social network analysis, data mining, and information retrieval. She is the member of several technical functional bodies such as the Michigan Association for Computer Users in Learning (MACUL), the Society of Digital Information and Wireless Communications (SDIWC), Internet Society as a Global Member (ISOC), and International Computer Science and Engineering Society (ICSES). She has published several papers in reputed journals and conferences.

Part I
Artificial Intelligence

Application of Common ANN for Similar Datatypes in On-line Monitoring and Security Estimation of Power System

Shubhranshu Kumar Tiwary, Jagadish Pal and Chandan Kumar Chanda

Abstract Monitoring a power network is an important task which is very complicated. To monitor the power systems, the developed nations are relying and shifting
more and more towards soft-computing and pattern recognition techniques with the
rapid improvements in the computation. In the work elaborated here, a report on the
employment of a common multilayer feed-forward net, to the security estimation of
a power network has been reported. The model built, is a 5-bus system, developed
on the Simulink environment of a MATLAB R2013a version 32-bit software. The
outcome was confirmed on a Hardware-In-Loop (HIL) device, on RT Lab Simulator
of OPAL RT Technologies. The analysis is presented in this work for the perusal of
the readers.

1 Preface

Applying the ANN in field is very confusing for any type of power network, especially making the choice for the type of neural net. In this work, a report on the
use of a multilayer feed-forward neural net with error backpropagation, for on-line
monitoring of MW (Active power) and MVAR (Reactive power) flowing on the lines
of a modified 5 bus power system has been elaborated. The work is particular in the
sense, that a multilayer neural net model with seven inputs, is used to monitor the
MW power flow on all seven lines of the power network, also a second neural net

S. K. Tiwary (✉) · J. Pal · C. K. Chanda
Department of EE, IIEST, Shibpur, Howrah 711103, India
e-mail: mail_shubh2005@yahoo.co.in

J. Pal
e-mail: jagadish_pal@hotmail.com

C. K. Chanda
e-mail: ckc_math@yahoo.co.in

© Springer Nature Singapore Pte Ltd. 2019
A. Abraham et al. (eds.), *Emerging Technologies in Data Mining and Information
Security*, Advances in Intelligent Systems and Computing 755,
https://doi.org/10.1007/978-981-13-1951-8_1

3

model is used to monitor the MVAR power on all seven lines as well [1]. Initially, a modified power network is developed on the Simulink environment of MATLAB, then it is simulated on RT Lab. Then the MW and MVAR flow of all lines are recorded. Then, these real-time data are tabulated in an Excel sheet and their corresponding states are classified in a separate Excel sheet. The state classification of the power network at each instant is done by comparing power flowing, with the thermal limits of the transmission lines. We must include equal numbers of each state classification (namely secure and insecure) for good convergence of ANN training. After that, we train the ANN in off-line mode. When the training has converged, and gives accurate classification results, we use it on RT lab simulation and observe the performance [2, 3].

2 Power System Model Evolution

The 1-line diagram of a 5-bus power system is as given below in Fig. 1.

In this power system, there is no load attached to Bus 2. The Simulink model for the above network is shown in Fig. 2.

In this model, there are two generators. The source at Bus 1 and Bus 2 is 100 MVA at base voltage of 25 kV, each. The load at Bus 3, Bus 4 and Bus 5 are 20 MW, 15 MW and 15 MW also at 25 kV, respectively. There are 7 distributed parameters line block of 100 km length with similar capacitance, inductance and resistance [5, 6]. As shown in Fig. 2, each line contains a V-I mask, to monitor Voltage and Current at each instant. Then we run the simulation and record the power flows. Figure. 3, shows the model.

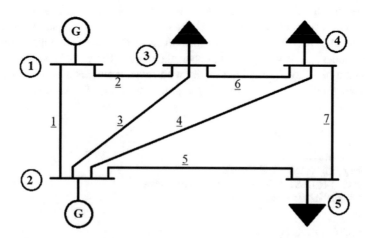

Fig. 1 Modified 5-bus network

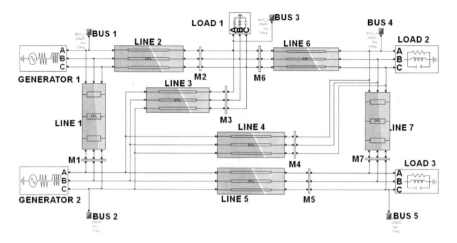

Fig. 2 Simulink model of the modified 5-bus network

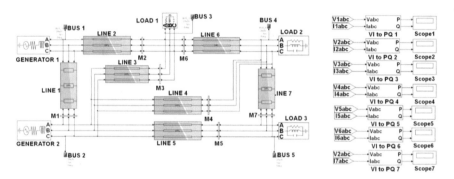

Fig. 3 The Simulink model with scope monitoring MW and MVAR power on all lines

After initializing the model, we run the Simulation and record the power flows for developing ANN models. The power flowing on line 1, as seen in Scope is shown below in Fig. 4.

Figure 4 shows the Scope output, of power flowing on line 1, both W and VAR as well. The scope reading of Scope 2, 3, 4, 5, 6, and 7 give the readings of power flows on lines 2, 3, 4, 5, 6, and 7, respectively. The plot in the upper half of the graph is the MW power flowing on the line while that in the lower half is the MVAR power. The Y-axis gives the magnitude of power flowing while X-axis gives the time elapsed.

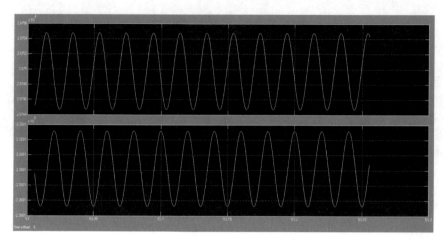

Fig. 4 Scope reading of active power (W) and reactive power (VAR) on line no. 1

Table 1 Range of Active power flowing on the lines in Watt (W)

Line no.	AP_{max}	AP_{min}
Line 1	$2.8754 * 10^4$ W	$2.8742 * 10^4$ W
Line 2	$8.2634 * 10^4$ W	$8.2618 * 10^4$ W
Line 3	$6.5556 * 10^4$ W	$6.5543 * 10^4$ W
Line 4	$2.9021 * 10^4$ W	$2.9013 * 10^4$ W
Line 5	$2.9013 * 10^4$ W	$2.9006 * 10^4$ W
Line 6	6.34 W	5.84 W
Line 7	−0.394 W	−0.454 W

3   Neural Net (FFBPN) Model Development

From the last segment, we note the power flowing on all lines in the form of a sinusoidal curve [7–9]. For developing the Neural Net model, the range between the crest and trough (Fig. 4 for line 1) is noted, as shown in Table 1, where AP_{max} is the maximum MW power flowing on the line while AP_{min} is the minimum MW power flowing. Table 1 shows the range of MW power flow on all the lines.

Similarly, Table 2 gives the range between the VAR power flowing on the lines. Here, RP_{max} is the maximum reactive power flowing on the lines while RP_{min} is the minimum reactive power flowing.

As mentioned previously, we need to develop an ANN model, that can monitor the MW as well as MVAR power flowing on all the lines together. The Feed-forward backpropagation neural net is a supervised learning type of model, for which the classification must be user defined [4, 5]. And we also need training data to train the network. To ensure proper training of ANN, we used 300 datasets, out of which 150 datasets are for Secure state classification and 150 datasets for Insecure state classi-

Table 2 Range of reactive power flowing on the lines in VAR

Line no.	RP_{max}	RP_{min}
Line 1	$-2.26813 * 10^6$ VAR	$-2.268124 * 10^6$ VAR
Line 2	649 VAR	631 VAR
Line 3	−352 VAR	−367 VAR
Line 4	−121.9 VAR	−129.4 VAR
Line 5	3.02 VAR	−3.98 VAR
Line 6	125.3 VAR	124.85 VAR
Line 7	0.64 VAR	0.585 VAR

fication. Secure state is numerically classified as 0 and insecure state is numerically classified as 1 [10–12]. When the active power flow on all the lines are between AP_{max} and AP_{min} the system is secure, numerically 0, otherwise it is insecure, numerically 1. Similarly, when reactive power flow on the lines are between RP_{max} and RP_{min} the system is secure, numerically 0 otherwise it is insecure, numerically 1. For training datasets, we calculate the data in an arithmetic series starting from the minimum values of active or reactive power, by adding a step size which is calculated by the formula mentioned below in (1) and (2).

$$SS_{MW} = (AP_{max} - AP_{min})/150 \tag{1}$$

$$SS_{MVAR} = (RP_{max} - RP_{min})/150, \tag{2}$$

where, SS_{MW} = step size to calculate active power array for training ANN.

And, SS_{MVAR} = step size to calculate reactive power array for training ANN.

Then, we can calculate the 300 training datasets using simple C programming codes and tabulate them in a Microsoft (MS) Excel sheet in a matrix of dimension 7×300. For these training data, classification has to be given in a separate Excel sheet (namely 0 or 1). Then, we initialize a feed-forward neural net with seven input nodes, three hidden nodes, and one output node. After that we start training the ANN, until it converges to a good accuracy [13, 14]. After the training converges, we develop a Simulink model for the ANN and incorporate them in the power network as in Fig. 5.

In Fig. 5, of the two ANN models, one represents active power monitor and another represents reactive power monitor [15]. Thereafter, the model, excluding the two scope, is converted into a subsystem (named Master Subsystem, SM_in) and the two scopes are also converted into a separate subsystem (named Console Subsystem, SC_out) and loaded onto the HIL simulator. This model separation is done for appropriate simulator core allocation. The model is shown below Fig. 6.

After uploading the model, we run the simulation in real-time on RT Lab at a high device frequency of 50 μs for a time of 30 s. The observation is discussed below.

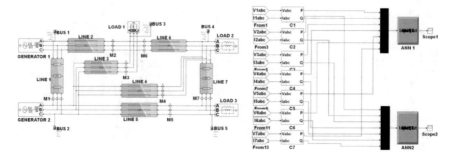

Fig. 5 ANN models monitoring the MW and MVAR power

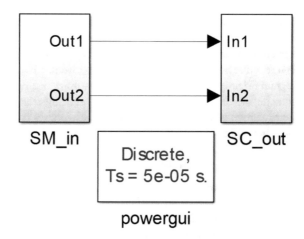

Fig. 6 Model uploaded to HIL simulator for real-time simulation

4   Observations

Neural net gives the very good results of monitoring in real-time on RT Lab. When the power network is simulated in a steady state with neural net monitor, the state of the power network is classified as 0 or a value very close to 0, which means the system is in a secure state of operation [16]. Figure 7 shows the secure power flow classification by ANN, classified as 0 or very close to zero, which means the power flow is in a steady state [17]. The upper graph represents output of ANN monitoring MW power flow while lower graph is for MVAR power flow.

To verify the results of ANN performance in the event of a tripping, we induce a fault [17] after five seconds and get the results as shown in the Scope output above in Fig. 8.s

As can be seen from Fig. 8 above, when a fault is induced on a line after 5 s for a duration of 40 ms the neural net output goes from secure state (0) to insecure state (1). It means that, in the event of a fault, the ANN will warn the operator about the tripping immediately. The above results prove that a common ANN can prove to be a very good monitoring tool for the whole power network [17–19].

5   Concluding Remarks

From the work elaborated in this paper, its observed that an ANN model gives optimum outputs for state estimation and monitoring in real time, if properly educated and examined. From the above study, two important points were noted.

5.1. *It is simpler and less cumbersome to train and use a multiple input multiple layer ANN model to monitor the power system*

5.2. *Although easier to develop, the used ANN model here cannot identify the line on which the tripping took place*

For future work, neural net may be tested using Python or C programming in a real power system network/utility.

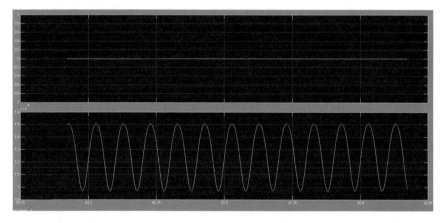

Fig. 7   Steady-state power flow classification for MW (upper curve) and MVAR (lower curve)

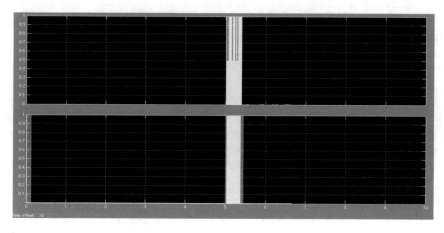

Fig. 8 Scope showing ANN output as 0 in secure state and 1 in insecure state after inducing fault at 5–5.4 s

References

1. Khazaei, J., Piyasinghe, L., Miao, Z., Fan, L.: A real-time digital simulation modelling of single-phase PV in RT-LAB. In: Proceedings of PES General Meeting, Conference and Exposition 2014, pp. 1–5, National Harbor, MD, USA, IEEE (2014)
2. Wu, W., Wu, X., Jing, L., Li, J.: Design of modular multilevel converter hardware-in-loop platform based on RT lab. In: Proceedings of International Power Electronics and Motion Control Conference, (IPEMC-ECCE Asia), pp. 1–6. IEEE (2016)
3. Huang, H., Liu, X., Wang, H., Wu, H.: Fault analysis of RT-Lab semi-physical simulation for east fast control power supply. In: 3rd International Future Energy Electronics Conference and ECCE Asia 2017, Kaohsiung, Taiwan, pp. 1369–1374. IEEE (2017)
4. Hsu, Y.Y., Chen, C.R.: Tuning of power system stabilizers using an artificial neural network. IEEE Trans. Energy Convers. **6**(4), 612–619. IEEE (1991)
5. Srinivasan, D., Chang, C.S., Liew, A.C., Leong, K.C.: Power system security assessment and enhancement using artificial neural network. In: Proceedings of EMPD '98, 1998 International Conference on Energy Management and Power Delivery, Singapore, vol. 2, pp. 582–589. IEEE (1998)
6. Aggoune, M.E., Atlas, L.E., Cohn, D.A., Damborg, M.J., El-Sharkawi, M.A., Marks, R.J.: Artificial neural network for power system static security assessment. In: IEEE International Symposium on Circuits and Systems, vol. 1, pp. 490–494. IEEE (1989)
7. Sobajic, D.J., Pao, Y.H.: Artificial neural net based dynamic security assessment for electric power systems. IEEE Trans. Power Syst. **4**(1), 220–228. IEEE (1989)
8. Pao, Y.H., Sobajic, D.J.: Combined use of unsupervised and supervised learning for dynamic security assessment. In: IEEE Proceedings of PICA '91, pp 278–284, Baltimore, MD. IEEE (1991)
9. Aggourne, M., El-Sharkawi, M.A., Park, D.C., Damborg, M.J., Marks, R.J.: Preliminary results on using artificial neural networks for security assessment. IEEE Trans. Power Syst. **6**(2), 890–896. IEEE (1991)
10. Zhou, Q., Davidson, J., Fouad, A.A. Application of artificial neural networks in power system security and vulnerability assessment. In: IEEE PES 1993 Winter Meeting, Paper No. 93, vol. 9, issue 1, pp. 525–532. IEEE (1993)

11. Jain, T., Srivastava, L., Singh, S.N., Erlich, I.: New parallel radial basis function neural network for voltage security analysis. In: Proceedings of the 13th International Conference on, Intelligent Systems Application to Power Systems, pp. 320–326. IEEE (2005)

12. Nakawiro, W., Erlich, I.: Online voltage stability monitoring using artificial neural network. In: 2008 Third International Conference on Electric Utility Deregulation and Restructuring and Power Technologies, pp. 941–947. IEEE (2008)

13. Saeh, I.S., Khairuddin, A.: Static security assessment using artificial neural network. In: 2nd IEEE International Conference on Power and Energy (PECon 08), pp. 1172–1178. IEEE (2008)

14. Ding, N., Benoit, C., Foggia, G., Bessanger, Y., Wurtz, F.: Neural network-based model design for short-term load forecast in distribution systems. IEEE Trans. Power Syst. **31**(1), 72–81. IEEE (2015)

15. Bulac, C., Tristiu, I., Mandis, A., Toma, L.: Online power system voltage stability monitoring using artificial neural networks. In: The 9th International Symposium on Advanced Topic in Electrical Engineering, Bucharest, Romania, pp. 622–625. IEEE (2015)

16. Basnet, S.M.S., Aburub, H., Jewell, W.: An artificial neural network-based peak demand and system loss forecasting system and its effect on demand response programs. In: 2016 Clemson University Power Systems Conference (PSC), pp. 1–5. IEEE (2016)

17. Tiwary, S.K., Pal, J.: ANN application for voltage security assessment of a large test bus system. In: Proceedings of 6th IEEE International Conference on Computer applications in Electrical Engineering—Recent Advances (CERA-2017), EE Department, IIT Roorkee, India, pp. 344–346. IEEE. https://doi.org/10.1109/CERA.2017.8343350

18. Tiwary, S.K., Pal, J.: ANN application for MW security assessment of a large test bus system. In: Proceedings of 3rd IEEE International Conference on Advances in Computing, Communication & Automation (ICACCA-2017), Tula's Institute, The Engineering & Management College, Dehradun, pp. 59–62. IEEE. https://doi.org/10.1109/ICACCAF.2017.8344661

19. Tiwary, S.K., Pal, J., Chanda, C.K.: Mimicking on-line monitoring and security estimation of power system using ANN on RT lab. In: Proceedings of 2017 IEEE Calcutta Conference (CALCON), Kolkata, pp. 100–104. https://doi.org/10.1109/calcon.2017.8280704

Crop Prediction Models—A Review

Supreeth S. Avadhani, Aashrith B. Arun, Varun Govinda
and Juyin Shafaq Imtiaz Inamdar

Abstract Crop yield is directly dependent on climatic and weather conditions. A lot of research has been done studying the dependency of weather on crop yield. Crop prediction models have proven to be successful in increasing the crop yield. Soil parameters and atmospheric parameters are used by the models to predict the suitable crop. Parameters such as type of soil, pH, phosphate, potassium, organic carbon, sulphur, manganese, copper, iron, depth, temperature, rainfall, humidity have shown to influence the yield of crop. In this paper, we review the research conducted by several researchers in this direction with a logical conclusion.

Keywords Artificial neural network (ANN) · Weather · Crop prediction models

1 Introduction

The most important goal of agricultural production is to achieve maximum crop yield at a minimum cost. Weather and soil conditions play a vital role in crop yield. Unfavourable conditions are a major reason for loss of crop. Predicting these unfavourable conditions will result in maximum crop production.

2 Background

The paper is supported by a thorough research of the most recent papers in the relevant field. This paper is classified based on two parameters: soil-based and weather-based.

S. S. Avadhani (✉) · A. B. Arun · V. Govinda · J. S. I. Inamdar
The National Institute of Engineering, Mananthavadi road,
Mysuru 570008, Karnataka, India
e-mail: supreethavadhani@gmail.com

A. B. Arun
e-mail: aashrith97@gmail.com

© Springer Nature Singapore Pte Ltd. 2019

13

A. Abraham et al. (eds.), *Emerging Technologies in Data Mining and Information Security*, Advances in Intelligent Systems and Computing 755,
https://doi.org/10.1007/978-981-13-1951-8_2

2.1   *Soil*

Dahikar et al. [1] have discussed the advantage of using Artificial Neural Network (ANN) in crop prediction taking into consideration quality of soil, acidic content of soil, organic chemical components, depth of soil and atmospheric conditions. The use of the supervised learning algorithm, feed forward backward propagation algorithm has reduced the errors in predictions as the algorithm is given both the example input and the anticipated output. Furthermore, these systems can be used to provide information to the farmers about the expected yield so that they can plan accordingly for the coming season.

Shahane et al. [2] discuss about developing a computer-based soil crop analysis system that analyses results of soil tests and underlying conditions in order to recommend ideal crops to be grown, possible outcome yields and recommend the right fertilizers to be applied to the soil. The Portal consists of tools that are as follows: Soil Crop Analysis, Soil Crop Matching Tool, Crop Yield Calculation, Fertilizer Recommendation, User Results, Crop Calendar, the formulae to Calculate and Analyze Nutrients in Soil, the Formula to Match Crops with Soil and an yield calculator.

Prof. Honawad et al. [3] investigate the development of digital image analysis approach for estimation of physical properties of soil in lieu of conventional laboratory approach which eliminates drawbacks such as manual involvement, time consumption, chances of human errors, and uncertain prediction. The signal processing method, improved the quality of the original image using filters and computing the features in the enhanced images. The algorithm proposed in the paper uses feature extraction methods like colour quantization and Texture-based feature extraction which is done by applying Gabor filter and Law mask. Then the matching is achieved by applying statistical measurements like Mean, standard deviation, skew, and kurtosis.

Hiremath et al. [4] propose a feature extraction algorithm using wavelet decomposed images of a soil image and its complementary image for texture classification. The retrieved sub-band images construct the features. The results obtained provide a merger of detail sub-bands with approximation sub-band. The rate of classification is improved at minimized computational cost.

Ramana Reddy et al. [5] computed statistical parameters derived from grey level co-occurrence parameters on sequential window (SW) and random window (RW). Pre-processing methods are applied on RW and SW. The RW on both pre-processed and non-pre-processed methods exhibits same percentage of classification as in the case of normal SW method. The classification rate after pre-processing by mean shows a better result.

Kanjana Devi et al. [6] have quantised the impacts of various factors like soil, weather, rain, fertilizers and pesticides on crop by using statistical methodologies on historical yield of crops. A comparison of various data mining techniques was applied to find out the most accurate technique. One of the major parameter considered is soil. It focuses on classification of soil fertility rate using K-means, Random tree. They have used association rule mining algorithm which proved to be efficient for analysing the agricultural soil data for crop yield production.

2.2 Weather

Jones et al. [7] have re-designed the DSSAT crop model [DSSAT.net] which uses DSS (Decision Support System) algorithm as it had become increasingly difficult to maintain the DSSAT crop models because there were different sets of code for different crops. The new design contains a modular approach. It consists of multiple modules such as soil module, a crop template, a weather module and a module for competition for light and water among the crops, soil and atmosphere. Further the author describes the essential data and the techniques deployed for the crop models. The re-designed DSSAT model provided considerable opportunity to its developers and other scientific community for greater co-operation in interdisciplinary research and in solving certain problems at field and farm levels.

Fernando et al. [8] assess the economic impact on annual coconut production data from 1971 to 2001 in the region. Their research showed that the loss in economy in crop shortage was around US$50 million.

Dempewolf et al. [9] provide a premise for designing and developing a practical wheat yield prediction model for Punjab Province, Pakistan. This analysis was restricted to a certain number of training years and on a single wheat growing region. This method made use of previous 6 years satellite data including the present year and also yield statistics of the past 6 years. The number of training years will be dependent on the region. The limitation of this methodology is that the MODIS satellite observations are at comparatively coarse spatial resolution relative to the field size distribution. In addition to this the optical satellite data are vulnerable to negative effects of cloud cover.

Ji et al. [10] developed accurate estimation techniques to predict rice yields in the preparation process. The reasons for the study were: (1) identify the effectiveness of Artificial Neural Networks in predicting the rice yield in mountainous region, (2) evaluate the performance of the ANN relative to developmental variation parameters (3) compare the effectiveness of multiple linear regression models with ANN models (Table 1).

3 Conclusion

A lot of research has been done in the field of crop prediction. Parameters such as atmospheric composition, organic content, etc. have been used to improve the prediction. Experimental implementations have shown positive results. On the other hand, practical implementation can cause overproduction and underproduction of

Table 1 Tabulation of the parameters used in various crop prediction models

	[1]	[2]	[3]	[7]	[11]	[6]	[12]
pH	✓	✓	✗	✓	✓	✓	✓
Carbon	✓	✗	✗	✓	✓	✓	✓
Magnesium	✓	✗	✗	✓	✓	✗	✗
Phosphate	✓	✗	✗	✓	✓	✗	✗
Potassium	✓	✓	✗	✓	✓	✗	✗
Nitrogen	✗	✓	✗	✓	✗	✗	✗
Rainfall	✓	✗	✗	✓	✗	✗	✗
Humidity	✓	✗	✗	✓	✗	✗	✓
Temperature	✓	✗	✗	✓	✗	✗	✗
Electrical conductivity	✗	✗	✗	✗	✓	✗	✓
Crop requirement	✗	✗	✗	✗	✗	✗	✗
Effect on environment	✗	✗	✗	✗	✗	✗	✗
Proposed model	ANN	Custom models	Mean, standard deviation, skew and kurtosis	DSSAT was model	Least median squares regression	Association rule mining algorithm	Agro-algorithm

certain crops in the same season which has been the main reason behind over pricing of crops and excessive wastage. Furthermore, none of the crop prediction models have taken into consideration the adverse effects of excessive agricultural cultivation on the environment. In order to avoid the above mentioned problems, a couple of parameters must be added to the crop prediction models.

They are:

1. Assess the present environmental condition of that particular region and predict a suitable crop such that the environmental conditions are improved as well as maximum yield is obtained.
2. Assess the requirement of each crop depending on the previous data so that neither crop shortage nor wastage occurs due to repeated prediction of the same crop.

Acknowledgements The authors are grateful to Dr. Anitha C, Assistant Professor, Department of Computer Science & Engineering, The National Institute of Engineering, Mysuru for her guidance and helpful comments. The authors also would like to thank the HOD, Department of CSE and the Principal for their continuous support and encouragement.

References

1. Dahikar, S.S., Rode, S.V.: Agricultural crop yield production using artificial neural networks. Int. J. Innov. Res. Electr. Electron. Instrum. Control Eng. **2**(1), 683–686 (2014)
2. Shahane, S.K., Tawale, P.V.: Prediction on crop cultivation. In: Int. J. Adv. Res. Comput. Sci. Electron. Eng. (IJARCSEE) **5**(10) (2016)
3. Honawad, S.K., Chinchali, S.S., Pawar, K., Deshpande, P.: Soil classification and suitable crop prediction. In: National Conference on Advances in Computational Biology, Communication, and Data Analytics, pp. 25–29 (2017)
4. Hiremath, P.S., Shivashankar, S.: Wavelet based features for texture classification. GVIP J. **6**(3) (2006)
5. Ramana Reddy, B.V., Suresh, A., Radhika Mani, M., Vijaya Kumar, V.: Classification of textures based on features extracted from preprocessing images on random windows. Int. J. Adv. Sci. Technol. **9** (2009)
6. Kanjana Devi, P., Shenbagavadivu, S.: Enhanced crop yield prediction and soil data analysis using data mining. Int. J. Modern Comput. Sci. **4**(6) (2016)
7. Jones, J.W., Hoogenboom, G., Porter, C.H., Boote, K.J., Batchelor, W.D., Hunt, L.A., Wilkens, P.W., Singh, U., Gijsman, A.J., Ritchie, J.T.: The DSSAT cropping system model. Eur. J. Agron. **18**(3), 235–265 (2003)
8. Fernando, M.T.N., Zubair, L., Peiris, T.S.G., Ranasinghe, C.S., Ratnasiri, J.: Economic value of climate variability impacts on coconut production in Sri Lanka. In: AIACC Working Papers, Working Paper No. 45 (2007)
9. Dempewolf, J., Adusei, B., Becker-Reshef, I., Hansen, M., Potapov, P., Khan, A., Barker, B.: Wheat yield forecasting for Punjab province from vegetation index time series and historic crop statistics. Remote Sens. **6** (2014)
10. Ji, B., Sun, Y., Yang, S., Wan, J.: Artificial neural networks for rice yield prediction in mountainous regions. J. Agric. Sci. **145**, 249–261 (2007)
11. Gholap, J., Ingole, A., Gohil, J., Gargade, S., Attar, V.: Soil data analysis using classification techniques and soil attribute prediction. Int. J. Comput. Sci. Issues **9**(3) (2012)
12. Kushwaha, A.K., Bhattachrya, S.: Crop yield prediction using agro algorithm in Hadoop. Int. J. Comput. Inf. Technol. Secur. **5**(2) (2015)

Rainfall Prediction: A Comparative Study of Neural Network Architectures

Kaushik D. Sardeshpande and Vijaya R. Thool

Abstract Artificial neural networks have wide range of application, one of which is time series prediction. This paper represents a case study on time series prediction as an application of neural networks. The case study was done for the rainfall prediction using the local database in India. The results were obtained by the comparative study of neural network architectures like back propagation (BPNN), generalized regression (GRNN), and radial basis function (RBFNN).

Keywords Artificial neural networks (ANN) · Big data · Time series prediction Rainfall prediction · BPNN · GRNN · RBFNN

1 Introduction

India is a developing country where the majority of its population has depended on agriculture as their primary occupation and source of income. In such countries, the rainfall plays a vital role in country's overall growth. So in such conditions, rainfall prediction is very important [1]. For rainfall prediction, typically, the parameters that are considered previously are rainfall, vapour content of that region, percentage of cloud cover, and the average temperature in the region where the study is being carried out.

Rainfall prediction in India has a good history. In earlier days, people used to predict the rainfall just by analysing the cloud situations. In earlier days, the weather is forecasted with respect to atmospheric changes and signs from the planetary astral alterations and the signs of rain were based on observation of the lunar phases and weather forecasts were based on the movement of winds. Ancient weather fore-

K. D. Sardeshpande (✉) · V. R. Thool
Department of Instrumentation Engineering, SGGS Institute of Engineering and Technology, Nanded 431606, Maharashtra, India
e-mail: kds.sirdeshpande@gmail.com

V. R. Thool
e-mail: vrthool@sggs.ac.in

© Springer Nature Singapore Pte Ltd. 2019
A. Abraham et al. (eds.), *Emerging Technologies in Data Mining and Information Security*, Advances in Intelligent Systems and Computing 755, https://doi.org/10.1007/978-981-13-1951-8_3

casting methods usually relied on observed patterns of events, also termed pattern recognition.

Rainfall is a natural process and has a high degree of variability in every way [2]. The new era of rainfall forecasting is the numerical model forecasting. The basic idea of this technique is to create a model on the basis of observed data. These forecasting models have reduced the human interference as a forecaster. So that the accuracy and sustainability in the rainfall prediction get amplified [3].

In proper rainfall forecasting, the factors listed above are taken into consideration, which includes too much complexity. To handle such complexity, we need an intelligent system to forecast such complex time series [4]. And also, we know ANN are inspired from the human biological neural system, which are attributed to the parallel and distributed processing nature [2]. So ANN can handle such complexity and are better at the forecasting of such time series.

Artificial neural networks are the mathematical models inspired from the human way of learning, which consist of learning from examples, fault tolerant and parallel processing of data. Basically, the neural network maps the input with output vectors by adjusting the weights typically by the gradient descent rule. It can be done by either in supervised or unsupervised way [5].

The ANN approach to the time series prediction has gained too much popularity due to its capacity to handle the noisy data and to predict or to classify them very efficiently [1]. Due to the features like adaptivity and sustainability in any condition, ANN has a wide range of applications, like pattern recognition, time series prediction, intelligent process modelling, etc. [6]. Time series prediction is one of the major applications of the artificial neural networks.

The multilayered neural network is more beneficial for the complex systems. Basically, the forecasting models must exhibit proper approximation [7]. Many forecasting models fail to give the correct approximation, when there are nonlinearities in the system [8]. Whereas neural networks are better at optimization and approximation as it requires the larger data set, and it gives best fit and the nonlinear mapping as well [4]. That is why ANN obtained too much popularity in the field of time series prediction [2].

In this paper, we have implemented three different neural network architectures, they are backpropagation, generalized regression, and radial basis function. The results are obtained by the comparative study of these architectures.

2   Architectures of ANN

2.1   Radial Basis Function

It is a special kind of neural network architecture. The basic idea behind the radial basis function is Cover's theorem on separability of patterns [9], which states that the linearly non-separable pattern is linearly separable in the higher dimensional

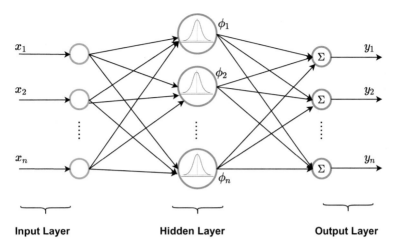

Fig. 1 Radial basis function architecture

space than a lower dimensional space. For a particular classification problem, we determine a plane between the two classes. But when we consider a linearly non-separable pattern, we have to go by the way stated by Cover [9].

In higher dimensional space, we will determine a hyper-surface that will classify the classes. So the determination of such hyper-surface is nothing but the application of radial basis function in neural networks [10]. For this, we need a hidden layer which will provide a set of basis functions which forms a basis for mapping of input layer with the output layer. Thus, RBF is nothing but the mapping of linearly non-separable input space to the linearly separable hidden space. The Euclidean distance between the input vector and the centres is calculated and the basis function is applied to it [1]. For such mapping, the most commonly used basis functions are the Gaussian basis functions. Figure 1 shows the architecture of radial basis function neural networks.

$$\phi(r) = exp\left[\frac{-r^2}{2\sigma^2}\right] \qquad (1)$$

where $r = ||x - c||$, the Euclidean distance between the input vector x and the centres c and $\phi(r)$ is the output of the hidden layers. Whereas the output of the network is the summation of products of weights and outputs of hidden layer, given as

$$y = \sum_{i=1}^{n} w_i \phi_i \qquad (2)$$

2.2 Generalized Regression

The GRNN is proposed by Specht in 1991 [11]. It is a variation in RBFNN [1]. It does not require an iterative training procedure like in backpropagation. It can capture any function between the input and output vectors. This type of neural network is an efficient estimator as the estimation error reaches to zero if the size of training set becomes large. Thus, it gives better results in the forecasting applications. It is based on the kernel regression techniques. In the conventional regression method, the mean square error (MSE) gets minimized while the generalized regression method predicts the probability density function (PDF) for the provided set of training data. The PDF is acquired without any previous knowledge of the data and that is why it is known as generalized regression [11]. It consists of four layer input, hidden, summation, and the output layer. The architecture of GRNN is shown in Fig. 2.

The mathematical expression for the estimation results of the generalized regression method using Gaussian kernel can be stated as [11]

$$\hat{Y}(X) = \frac{\sum_{i=1}^{n} Y^i \, exp(-\frac{r_i}{\sigma})}{\sum_{i=1}^{n} exp(-\frac{r_i}{\sigma})} \tag{3}$$

where

$$r_i = \sum_{j=1}^{p} |X_j - C_j^i| \tag{4}$$

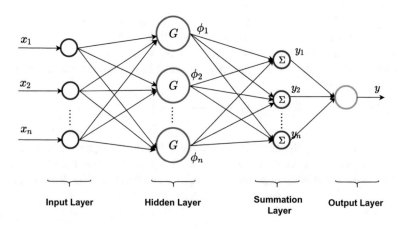

Fig. 2 Generalized regression architecture

2.3 Backpropagation Algorithm

The backpropagation algorithm is used for the training of the multilayered feed-forward neural network (MLFF). The backpropagation algorithm is presented in 1987 by Rumelheart and McClelland [12]. This is the basic algorithm used for the learning of MLFF. The BPNN is a MLFF neural network consisting of input layer, at least one hidden layer and the output layer. This algorithm is quite similar to feedback mechanism but not the same. In backpropagation, there is no such bidirectional path for the outputs to go back at the inputs.

The three interconnected layers are having synaptic weight connections on them. These weights get updated during the training. The error terms are calculated at the output layer by comparing the outputs with the desired values, and these error terms are propagated to the previous layers to it and accordingly the weights of that layer get updated [13]. This process is carried out till the error terms are propagated up to the input layer. This is why back propagation algorithm is an iterative training procedure. The training is done as the error function becomes sufficiently small. The error generated gets minimized by the gradient descent rule. The error terms get minimized by achieving the global minima. This operation is done by applying partial differentiation to the error terms. The architecture of the BPNN is given in Fig. 3.

The mathematical expressions involved in the backpropagation algorithm can be stated as follows. The mean square error is given as

$$E = \frac{1}{2} \sum_{k=1}^{K} (d_k - O_k)^2 \tag{5}$$

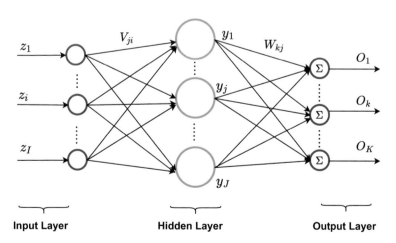

Fig. 3 Backpropagation architecture

where d_k is the desired output and O_k is actual output of neuron. Now the change in weights is obtained by applying partial differentiation to Eq. 5. Now the weight updating at hidden layer is given as the rule stated below

$$V_{ji}^{new} = V_{ji}^{old} - \eta \frac{\partial E}{\partial V_{ji}}$$
$$V_{ji}^{new} = V_{ji}^{old} - \eta \Delta V_{ji} \tag{6}$$

where V_{ji} is the synaptic weight connection between input layer and hidden layer and ΔV_{ji} is the change in weights of hidden layer. The change in weights is expressed as the partial differentiation of the error terms w.r.t the weight vectors.

By putting value of change in weights, we get the result as

$$V_{ji}^{new} = V_{ji}^{old} + \eta f'(net_j) \sum_{k=1}^{K} (d_k - O_k) f'(net_k) W_{kj} z_i \tag{7}$$

where η is the learning rate, $f net_k$ is the output of neuron from output layer, $f(net_j)$ is the output of neuron from hidden layer, W_{kj} is the weight connection between hidden layer and output layer and z_i is the input vector. Here, the term $(f'(net_j)(d_k - O_k) f'(net_k) W_{kj})$ is referred as error signal from hidden layer δ_{yj}.

The weight updating rule at output layer is

$$W_{kj}^{new} = W_{kj}^{old} - \eta \frac{\partial E}{\partial W_{kj}}$$
$$W_{kj}^{new} = W_{kj}^{old} - \eta \Delta W_{kj} \tag{8}$$

here ΔW_{kj} is the change in weights. By putting value of change in weights, we get the final result as

$$W_{kj}^{new} = W_{kj}^{old} + \sum_{k=1}^{K} \eta (d_k - O_k) f'(net_k) y_j \tag{9}$$

where y_j is the output of the hidden layer and input to the output layer. Here, the term $((d_k - O_k) f'(net_k))$ is referred as the error signal from output layer δ_{ok}. As the error signal from hidden layer also consists error signal from output layer (refer Eqs. 7 and 9), the error is backpropagated.

3 Methodology

The main area of study was the *Nanded district* from *Maharashtra State* of India. As per conventional rainfall prediction methods, we have collected the monthly average data for the area we have chosen to study. The dataset consists of previous rainfall

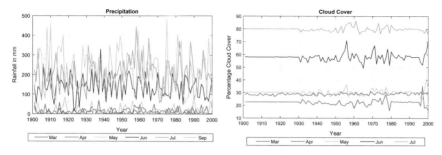

Fig. 4 Precipitation and cloud cover

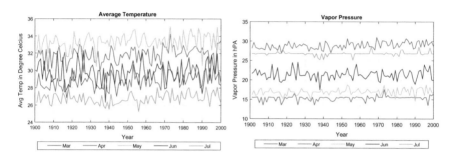

Fig. 5 Average temperature and vapour pressure

(Precipitation), percentage cloud cover, average temperature and vapour pressure. The dataset is pictured in Figs. 4 and 5. The dataset is obtained from the website, India water portal [14].

The actual methodology is pictured in Fig. 6. The inputs to the neural network are normalized between 0 and 1. The three neural network architectures were created in MATLAB and fed with the proper inputs and the targets. The inputs to the network were precipitation, average temperature, cloud cover and the vapour pressure from March to June, while the targets to the network were precipitation from September. The tabulated and the pictured results are given in Sect. 4.

4 Results and Simulation

The dataset consisted of the past hundred years data from 1901 to 2000 for the climate features listed in Sect. 3. Out of these datasets, 90 years' data were used for the training of the neural network architectures, while the remaining 10 years' data used for the prediction. Figure 7 shows the training results while Fig. 8 shows the results of prediction, that is, rainfall observed and the predicted rainfall from 1991 to 2000.

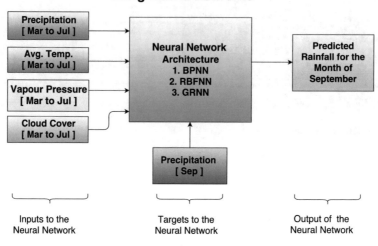

Fig. 6 Actual methodology

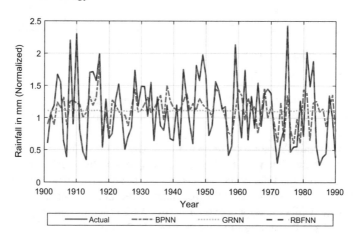

Fig. 7 Training results

The results are tabulated in the Table 1. RBFNN gave splendid results while train-ing, but while testing it fails to give proper output, i.e. it gave overfitting. BPNN gave good response while training as well as testing. GRNN failed to give good response in both training as well as testing, as it failed to handle the complexities from the data. It is found that de-normalized RMSE are 14.2682, 9.9748 and 6.9340 for RBFNN, GRNN and BPNN, respectively, for testing.

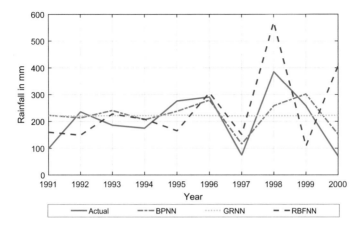

Fig. 8 Results of prediction

Table 1 Performance indices

Network	Training		Testing	
	MSE	Regression	MSE	Regression
BPNN	0.2322	0.4016	0.1202	0.7656
RBFNN	1.6204×10^{-22}	1.00	0.5090	0.3623
GRNN	0.2752	0.1826	0.2847	−0.1020

5 Conclusion

The application-oriented study of the different neural network architectures is presented in this paper. For the simulation purpose, we have used the different neural network architectures based upon their capabilities to handle the data. Out of these architectures, we got better results from BPNN and RBFNN, while GRNN failed to handle the data with such nonlinearities and has resulted in poor performance.

As we know the neural networks with more hidden layers are known as deep neural networks, which is a new emerging technology in the field of artificial intelligence. This same application can be driven by the deep neural network. The results with more accuracy and more efficient neural network model can be obtained by the application of deep learning.

References

1. Manek, A.H., Singh, P.K.: Comparative study of neural network architectures for rainfall prediction. In: Proceedings of Technological Innovations in ICT for Agriculture and Rural Development (TIAR), 2016 IEEE. IEEE (2016)
2. Luk, K.C., Ball, J.E., Sharma, A.: Study of optimal model lag and spatial inputs to artificial neural network for rainfall forecasting. J. Hydrol. **227**(1), 56–65 (2000)
3. Mitra, P., et al.: Flood forecasting using internet of things and artificial neural networks. In: 2016 IEEE 7th Annual Information Technology, Electronics and Mobile Communication Conference (IEMCON). IEEE (2016)
4. Toth, E., Brath, A., Montanari, A.: Comparison of short-term rainfall prediction models for real-time flood forecasting. J. Hydrol. **239**(1), 132–147 (2000)
5. Debenham, R.M., Garth, S.C.J.: Neural networks for artificial intelligence? In: IEE Colloquium on Current Issues in Neural Network Research, IET (1989)
6. Illingworth, W.T.: Beginner's guide to neural networks. In: Proceedings of the IEEE 1989 National Aerospace and Electronics Conference, 1989, NAECON 1989. IEEE (1989)
7. Feng, L-H., Lu, J.: The practical research on flood forecasting based on artificial neural networks. Expert Syst. Appl. **37**(4), 2974–2977 (2010)
8. French, M.N., Krajewski, W.F., Cuykendall, R.R.: Rainfall forecasting in space and time using a neural network. J. Hydrol. **137**(1-4), 1–31 (1992)
9. Cover, T.M.: Geometrical and statistical properties of systems of linear inequalities with applications in pattern recognition. IEEE Trans. Electron. Comput. **3**, 326–334 (1965)
10. Haykin, S.S., et al.: Neural Networks and Learning Machines, vol. 3. Pearson, USA (2009)
11. Specht, D.F.: A general regression neural network. IEEE Trans. Neural Netw. **2**(6), 568–576 (1991)
12. Rumelhart, D.E., McClelland, J.L., PDP Research Group: Parallel Distributed Processing, vol. 1. MIT Press, Cambridge, MA, USA (1987)
13. Zurada, J.M.: Introduction to Artificial Neural Systems, vol. 8. West, St. Paul (1992)
14. http://www.indiawaterportal.org/

Prediction of Bacteriophage Protein Locations Using Deep Neural Networks

Muhammad Ali, Farzana Afrin Taniza, Arefeen Rahman Niloy, Sanjay Saha and Swakkhar Shatabda

Abstract In phage therapy, bacteriophage proteins are used to kill bacteria that cause infection. The knowledge of the location of the bacteriophage proteins plays an important role here. In this paper, we propose a supervised learning based method to predict the locations of bacteriophage proteins. First, we address the problem of predicting whether a bacteriophage is extracellular or located in the host cell. Second, we also address the subcellular location prediction problem of the phage proteins. For the host located proteins, the proteins could either be located in cell membrane or in the cytoplasm. We have successfully used deep feed-forward neural network on a standard training dataset and achieved good results for both of the prediction problems. Our method uses an optimal set of features for classification and achieves 87.7% and 98.5% accuracy for two of the prediction problems which is 3.5% and 6.3% improved than the previous state-of-the-art results achieved for these problems, respectively.

Keywords Supervised learning · Deep neural networks · Feature selection
Protein subcellular localization

M. Ali · F. A. Taniza · A. R. Niloy · S. Saha · S. Shatabda (✉)
Department of Computer Science and Engineering, United International University,
Madani Avenue, Satarkul, Badda, Dhaka, Bangladesh
e-mail: swakkhar@cse.uiu.ac.bd

M. Ali
e-mail: raju4uiu@gmail.com

F. A. Taniza
e-mail: taniza19@gmail.com

A. R. Niloy
e-mail: arefeenniloy@gmail.com

S. Saha
e-mail: sanjay@cse.uiu.ac.bd

© Springer Nature Singapore Pte Ltd. 2019
A. Abraham et al. (eds.), *Emerging Technologies in Data Mining and Information Security*, Advances in Intelligent Systems and Computing 755,
https://doi.org/10.1007/978-981-13-1951-8_4

1 Introduction

Due to the abuse of antibiotics and the wide spectrum that leads killing beneficial bacteria being killed by the antibiotics, phage therapy is gaining popularity due to the precise nature of their ability to kill specific bacteria [1]. Knowledge of the subcellular locations of the phage proteins plays an important role in viral phage therapy and the underlying mechanism [2]. However, most of the experimental methods are time consuming and expensive.

A good number of computational methods have been proposed in the literature to address the protein subcellular localization problem and for phage proteins in particular [3–5]. In this paper, we address two problems related to phage protein locations modeled as supervised learning problem of classification. Throughout this paper, we will refer to them as **PH versus non-PH** and **PHM versus PHC** problems.

1. **PH versus non-PH**: Classification of phage proteins as host located and nonhost located or extracellular phage proteins.
2. **PHM versus PHC**: Classification of the phage proteins whether they are located on cell membrane or cell cytoplasm in a host cell. This problem is the subcellular localization problem of host located proteins.

In this paper, we present a novel prediction method for both of these problems. Our predictor is based on a deep neural network trained on standard benchmark training datasets. We have used features extracted from sequence, PSSM profile, and secondary structure information of a given phage protein. Different feature ranking techniques were deployed to reduce the number of features and with the optimal set of features, our method was able to predict maximum classification accuracy of 87.7% on the PH versus no-PH problem and 98.5% on the PHM versus PHC problem. These are significantly improved over the previous state-of-the-art results achieved so far on the same datasets by other methods. The main motivation for this research was to test the effectiveness of deep neural networks in solving the challenging classification problems like protein subcellular localization and the promising results motivate us to further investigate in future.

Rest of the paper is organized as follows: Sect. 2 presents a brief overview of the related work; Sect. 3 presents our prediction method and the selected datasets and evaluation strategy; Sect. 4 describes the experimental results and analysis; the paper concludes with a comment on the future work in Sect. 5.

2 Related Work

A number of tools and prediction methods have been developed in the literature for subcellular localization of proteins [6] and particularly for phage protein localization [3]. A webtool named PHAST was proposed in [7]. PHAST is a graphical tool capable of identifying phage sequences. It also annotates and visualizes phage sequences in bacterial genomes or plasmids. PhiSpy tool proposed in [8] used different features based on the sequence and its similarity. An enhanced version of PHAST webserver was proposed in [9]. It was called PHASTER which was at least four times faster than PHAST.

Many state-of-the-art supervised learning algorithms like support vector machines (SVM) [4], random forest [10], etc., have been used in the literature for predicting and identifying phage proteins or their locations in the literature. In a very recent work, ANOVA-based incremental forward search based feature selection techniques were used to predict phage protein locations in a host cell by Ding et al. [3]. They used SVM as a classifier.

To the best of our knowledge, multilayer neural network or deep neural networks have not been exploited in the literature. However, there have been use of neural networks in some related problems, with single hidden layer only in [11].

3   Methods and Materials

In this section, we describe the materials and methods used in this model. Rest of the section is organized in the following manner as suggested in [12]: a description of the datasets, protein representation and feature extraction, feature selection methods, description of the classification algorithm, and the parameters used for performance evaluation.

3.1   Dataset

For the two problems addressed in this paper, phage protein instances were taken from Uniprot Database [13] for which their locations were experimentally determined. This is a highly reliable data and first proposed in [3]. Among these selected phage proteins, those were removed from the dataset that are fragments of other proteins or contain nonstandard amino acids. Furthermore, any protein sequence with 30% or more similarity in the dataset were removed. In total, the number of protein instances was 278. For any binary classification problem, the dataset consists of two parts: positive dataset and negative dataset.

$$\mathbb{S} = \mathbb{S}^+ \cup \mathbb{S}^-$$

For the first problem of PH versus non-PH problem, 144 are host located problems and labeled as positive instances in our study and the rest are extracellular proteins labeled as negative instances. Among the 144 host located proteins, we label the cell membrane located proteins as positive and cell cytoplasm located proteins as negative in the second problem of PHM versus PHC classification. A brief summary of the datasets is given in Table 1.

Table 1 Summary of the datasets used

Problem	Total instances	Positive instances	Negative instances
PH versus non-PH	278	Host located	Extracellular
		144	134
PHM versus PHC	144	Cell membrane	Cell cytoplasm
		68	76

3.2   Feature Generation

A large number of features are used in this paper. The proposed features are divided into three groups: sequence based, position specific scoring matrix (PSSM) based, and structure based.

To extract the features for a given phage protein instance, we need three information: the protein sequence, a PSSM matrix generated from PSI-BLAST [14], and an SPD file generated by SPIDER2 [15]. PSSM matrix contains the evolutionary information and SPD files contains information about the secondary structure of the protein such as secondary motifs, torsional angles, and accessible surface area (ASA) for each amino acid residue. For each of the feature groups, different types of features were generated. A small description of the features is given below:

1. Amino acid sequence based features: This group includes composition (20) and physicochemical properties based features (105). Total number of features in this group is 125 [16].
2. PSSM-based features: This feature group includes monogram (20), bigram (400), 1-lead bigram (400), auto-covariance (200), and segmented distribution (200) based features. Total number of features in this group is 1220.
3. Structure-based features: This feature group includes motif occurrence (3) and composition (3), ASA composition (1), torsional angles sine and cosine composition (8), bigram (64) and auto-covariance (80) and structural probabilities composition (3), bigram (9), and auto-covariance (30). Total number of features in this group is 201.

In total, there were 1546 features extracted for each of the protein sequences. These features were previously explored in [2]. Here, we provide a small description of the features extracted. The details of the method can be found in [2].

3.3   Feature Selection

It is to be noted that the total number of features generated is higher than the number of instances for both of the problems. Selection and reduction of features often lead to better classification accuracy and reduce the curse of dimensionality. There are many techniques in the literature for reducing features from a dataset like genetic algorithms, sparse elimination technique, principal component analysis, incremental forward search, greedy best first search, etc. In our work in this paper, we investigated two feature selection methods: ANOVA-based ranking and recursive feature selection.

1. **ANOVA Based Ranking**: Analysis of variance (ANOVA) can be used to measure the significance of a feature group. First, proportion of the variance for a particular feature is calculated using a simple ratio of treatment sum of squares to the total sum of squares. This ratio is an indicator of the proportion of variance explained by that feature. The higher the ratio is the better. Thus, a simple ranking technique is to select top and significant features according to rank with respect to this ratio.

2. **Recursive Feature Elimination (RFE)**: Recursive feature elimination (RFE) [17] in each iteration of the algorithm ranks the features on a given training dataset using simple criteria like correlation coefficient and removes a single feature with lowest significance. Iteratively, it continues to remove features and thus generates optimal set of features for a given dataset.

3.4   Deep Neural Network

We have employed deep neural network as our classification tool. Architecture of deep learning network is shown in Fig. 1. Deep neural networks are multilayer perceptron models with features fed into the input layer and the desired decision variable is sought in the output layer. Deep neural networks (DNN) have gained popularity in recent years for solving classification problems and machine learning problems. This has gained momentum due to fascinating discoveries [18] in the field like shared parameters, rectilinear output units, dropout, etc. We have used a multilayer feedforward network with multiple hidden layers in our work for both of the prediction problems.

Fig. 1 A schematic diagram for deep neural network with two hidden layers

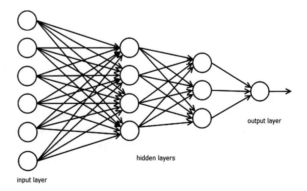

3.4.1 DNN Library Selection

There are a number of deep neural network libraries, tools or packages available in the literature. The most popular among them are Theano, Torch, Caffe, DeepLearning4J,Keras and TensorFlow. In this paper, we chose TensorFlow library because it is simple and easy to use. Furthermore TensorFlow provides support for multiple CPU and GPU. TensorFlow does numerical computation using data flow graphs known as stateful dataflow graphs. The edges of the graphs are the multidimensional data arrays communicated between the nodes in the graphs which are the mathematical operations. Tensorflow is python based and publicly available as free to use.

3.5 Performance Evaluation

To ensure the robustness of classification or prediction methods often it is necessary to sample the dataset in order to validate the results statistically. Two popular methods in sampling is train–test percentage split or use of independent test set. In absence of independent set more robust sampling methods are k-fold cross-validation and jackknife tests. In k-fold cross-validation, the original training data is randomly sample in each iteration into equal sized k partitions. Each partition is called a fold. In each fold, a single partition is used as test data and the rest of the training set is used to train a model. To remove the bias, the test dataset in randomly shuffled and the whole process is performed multiple times and average or aggregate performance measures are reported only. In this work, we have employed tenfold cross-validation.

Another popular sampling method often used for biological data or medical data is jack knife test. It is a leave-one-out test where each time only a single instance is tested over the trained model on the rest of the data set. To make a comprehensive comparison with the methods in the literature, we have also used jack knife tests on both of the datasets.

We have used accuracy of the binary classifiers as the main performance measure in this paper. Accuracy is defined as in the following equation:

$$Accuracy = \frac{TP + TN}{TP + TN + FP + FN} \tag{1}$$

For any binary classification problem, one class is regarded as the positive class and the other as negative class. We have introduced them at the beginning of this section. Here, TP is the number of true positives or the number of correctly classified positive samples. Similarly, TN denotes the number of true negatives or the number of correctly classified negative samples. FN and FP are respectively number of false negatives and false positives, i.e., these are the incorrectly classified samples. Thus, the maximum accuracy in percentage could be 1 and minimum 0. The higher the accuracy the better the prediction algorithm performs.

4 Experimental Results

In this section, we present the results of the experiments that found in this study. All the methods were implemented in Python language using Python3.4 version and TensorFlow [19]. Each of the experiments was carried 10 times. We have performed several experiments to find the best parameters and settings for our proposed method. The best setting for our method was found a deep feed-forward neural network with three hidden layers and with 99 optimal features selected using recursive feature elimination.

4.1 Comparison with Other Methods

In this section, we compare the performance of our proposed method with other state-of-the-art methods. We have selected PHPred [3] the previous state-of-the-art method to compare with our method. The maximum accuracy achieved by our method and PHPred are given in Table 2. From the values reported in Table 2, we could notice that our proposed method can achieve significantly improved accuracy for both of the problems compared to PHPred in both of the sampling tests.

4.2 Comparison of ANOVA and RFE

In this section, we present the results to show the comparison between two feature selection methods ANOVA-based feature ranking (AFR) and recursive feature elimination (RFE). This analysis is only shown on the first problem, which is PH versus non-PH problem. For this experiment, a single hidden layer feed-forward neural network was used with varying the number of neurons in the hidden layer from 10 to 30. The results using tenfold cross-validation are shown in Table 3. We could notice from the bold faced values in the table that RFE features are performing significantly better in comparison to other feature sets. From this experiments, we decided to use the 99 features selected by RFE.

Table 2 Comparison of results achieved by our method with other predictors

Method name	PH versus non-PH (%)	PHM versus PHC (%)
PHPred	84.2	92.4
Our method (tenfold)	93.0	100
Our method (Jack Knife)	87.7	98.5

Table 3 Results achieved for various feature elimination techniques on a single hidden layer feed-forward network on PH versus non-PH problem

Feature selection method	Hidden layer size					
	$h_1 = 10$		$h_1 = 20$		$h_1 = 30$	
	Max	Avg	Max	Avg	Max	Avg
RFE	**0.92**	0.61	**0.92**	**0.75**	0.87	0.69
AFR	**0.92**	0.55	0.85	0.56	0.78	0.59
ALL	0.89	**0.71**	**0.92**	0.68	**0.89**	0.69

Fig. 2 Comparison of maximum and minimum accuracy achieved by different number of hidden layers using the feed-forward deep neural network

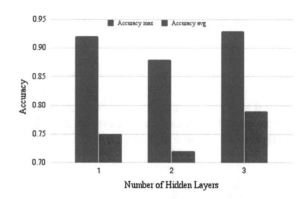

4.3   Number of Hidden Layers

Using the optimal set of features generate by the RFE feature selection technique, we further experimented on the number of hidden layers in the feed-forward deep neural network. Generally, it is assumed that more the hidden layers are, higher the training times are. However, we varied the number of hidden layers $k = 1, 2, 3$. For, $k = 1$, the number of neurons in the hidden layers was varied $n = 10, 20, 30$. For two hidden layers, we varied the number of neurons in first hidden layer, $h_1 = 10, 20, 30$ and number of neurons in the second hidden layer, $h_2 = 10, 20, 30$. For three hidden layers, we varied the number of neurons in each hidden layer with the values $10, 15, 25$. The best values were found using three hidden layers with the number of neurons in each layers as: $h_1 = 20$, $h_1 = 15$ and $h_3 = 10$. Comparison of the results in terms of maximum and average accuracy for different numbers of hidden layers in depicted in Fig. 2. For this experiment too, we used tenfold cross-validation.

4.4   Discussion

The experiments done in this study show how the reduced feature set and the application of deep feed-forward neural network were able to achieve better prediction accuracy for two phage bacterial protein subcellular localization problem: PH versus non-PH and PHM versus PHM. Initial experiments were carried out to justify the reduction of features and number of hidden layers. Best parameters were used to generate results using two standard robust sampling methods and compared to the previous state-of-the-art methods, our proposed method was able to produce results with significantly improved accuracy.

5   Conclusion

In this paper, we have investigated the applicability and effectiveness of deep neural networks for two important problems in computational biology. We have used a large number of features and used recursive feature elimination to reduce the number of features and used a three-layer feed-forward network to train the datasets for both of the problems. Our proposed method effectively discriminated host located phage proteins from extracellular phage proteins and was able to predict the subcellular locations of host located proteins, cell cytoplasm, or cell membrane. Initial promising results encourage us to further investigate the effectiveness of recent advances in deep learning network in general to other more sophisticated and complex problems in biological data. In future, we will also try to enrich the dataset of phage proteins and establish e webserver as a usable tool for the bioinformaticians.

References

1. Deresinski, S.: Bacteriophage therapy: exploiting smaller fleas. Clin. Infect. Dis. **48**(8), 1096–1101 (2009)
2. Shatabda, S., Saha, S., Sharma, A., Dehzangi, A.: iphloc-es: identification of bacteriophage protein locations using evolutionary and structural features. J. Theor. Biol. **435**, 229–237 (2017)
3. Ding, H., Liang, Z.Y., Guo, F.B., Huang, J., Chen, W., Lin, H.: Predicting bacteriophage proteins located in host cell with feature selection technique. Comput. Biol. Med. **71**, 156–161 (2016)
4. Ding, H., Yang, W., Tang, H., Feng, P.M., Huang, J., Chen, W., Lin, H.: Phypred: a tool for identifying bacteriophage enzymes and hydrolases. Virologica Sinica **31**(4), 350 (2016)
5. Ding, H., Feng, P.M., Chen, W., Lin, H.: Identification of bacteriophage virion proteins by the anova feature selection and analysis. Mol. BioSyst. **10**(8), 2229–2235 (2014)
6. Sharma, R., Dehzangi, A., Lyons, J., Paliwal, K., Tsunoda, T., Sharma, A.: Predict gram-positive and gram-negative subcellular localization via incorporating evolutionary information and physicochemical features into chou's general pseaac. IEEE Trans. NanoBiosci. **14**(8), 915–926 (2015)
7. Zhou, Y., Liang, Y., Lynch, K.H., Dennis, J.J., Wishart, D.S.: Phast: a fast phage search tool. Nucleic Acids Res. (2011). gkr485

8. Akhter, S., Aziz, R.K., Edwards, R.A.: Phispy: a novel algorithm for finding prophages in bacterial genomes that combines similarity-and composition-based strategies. Nucleic Acids Res. **40**(16), e126–e126 (2012)
9. Arndt, D., Grant, J.R., Marcu, A., Sajed, T., Pon, A., Liang, Y., Wishart, D.S.: Phaster: a better, faster version of the phast phage search tool. Nucleic Acids Res. **44**(W1), W16–W21 (2016)
10. McNair, K., Bailey, B.A., Edwards, R.A.: Phacts, a computational approach to classifying the lifestyle of phages. Bioinformatics **28**(5), 614–618 (2012)
11. Galiez, C., Magnan, C., Coste, F., Baldi, P.: ViRALpro: a new suite for identifying viral capsid and tail sequences (2015)
12. Chou, K.C.: Some remarks on protein attribute prediction and pseudo amino acid composition. J. Theor. Biol. **273**(1), 236–247 (2011)
13. Consortium, U., et al.: UniProt: a hub for protein information. Nucleic Acids Res. (2014). gku989
14. Altschul, S.F., Madden, T.L., Schäffer, A.A., Zhang, J., Zhang, Z., Miller, W., Lipman, D.J.: Gapped BLAST and PSI-BLAST: a new generation of protein database search programs. Nucleic Acids Res. **25**(17), 3389–3402 (1997)
15. Yang, Y., Heffernan, R., Paliwal, K., Lyons, J., Dehzangi, A., Sharma, A., Wang, J., Sattar, A., Zhou, Y.: Spider2: a package to predict secondary structure, accessible surface area, and main-chain torsional angles by deep neural networks. Prediction of Protein Secondary Structure, pp. 55–63 (2017)
16. Dubchak, I., Muchnik, I., Mayor, C., Dralyuk, I., Kim, S.H.: Recognition of a protein fold in the context of the scop classification. Proteins Struct. Funct. Bioinform. **35**(4), 401–407 (1999)
17. Guyon, I., Weston, J., Barnhill, S., Vapnik, V.: Gene selection for cancer classification using support vector machines. Mach. Learn. **46**(1), 389–422 (2002)
18. LeCun, Y., Bengio, Y., Hinton, G.: Deep learning. Nature **521**(7553), 436–444 (2015)
19. Abadi, M., Agarwal, A., Barham, P., Brevdo, E., Chen, Z., Citro, C., Corrado, G.S., Davis, A., Dean, J., Devin, M., et al.: Tensorflow: Large-scale machine learning on heterogeneous distributed systems (2016). arXiv:1603.04467

Classification of Phishing Websites Using Moth-Flame Optimized Neural Network

Santosh Kumar Majhi and Pragati Mahapatra

Abstract Phishing websites are taking a toll in today's Internet-infused world. These types of websites try to attack the classified information of the user on the Internet database, masquerading as the trusted website. They even use the logo and the website address of the original website to come off as the original one to the user. In this project, we deal with the classification of such websites from the real ones using the standards set by W3C. The Moth-flame algorithm is used as a learning algorithm to optimize the feedforward neural network and to classify the websites.

1 Introduction

There has always been a delicate discussion encircling the advent of technology and with that, cyber security comes into the limelight. The users are very much vulnerable to the process of acquiring sensitive information by feign techniques. This might cause financial damages, loss of information and identity, maligning of the e-commerce authenticity and the customer's confidence in the brand gets debilitated. This automated form of identity theft creates an air of suspicion regarding the use of the internet for commercial transactions.

Phishing is the act of acquiring confidential information such as passwords, user-names, security codes and credit card numbers, etc., by impersonating a genuine website in an electronic transmission. This fraudulent attempt incorporates use of fake E-mail IDs but unbelievable similarity with the original websites. The detection of such websites has been classified into two categories List-based and Heuristic-

S. K. Majhi (✉) · P. Mahapatra
Veer Surendra Sai University of Technology, Burla 768018, Odisha, India
e-mail: smajhi_cse@vssut.ac.in

P. Mahapatra
e-mail: pragatimahapatra96@gmail.com

© Springer Nature Singapore Pte Ltd. 2019
A. Abraham et al. (eds.), *Emerging Technologies in Data Mining and Information Security*, Advances in Intelligent Systems and Computing 755,
https://doi.org/10.1007/978-981-13-1951-8_5

based. List-based approach uses two lists, such as, blacklist and whitelist. Obviously, the blacklist consists of phishing URLs and the whitelist consists of legitimate URLs. Whenever a page is loaded by a browser, the browser checks if the URL is contained in the blacklist and if it is found in it, appropriate measures are taken and if not, then the URL is considered to be legitimate. The blacklist may be stored at the client or at the central server. The basic idea is that the frequently visited sites are added to the whitelist and if not so, the user is either denied access or asked to add it to the whitelist. The pros of this type of approach is that it is efficient in speed and simplicity but the limitation is that a lot of time is consumed to add the phishing websites to blacklist and the newly created phishing URLs go undetected. The second approach, the heuristic-based approach uses data mining algorithms to check for phishing websites. It checks one or more characteristics like the hypertext markup language (HTML), Uniform Resource Locator (URL), or the page content itself. The main advantage of this approach over the former is that it can also distinguish newly created phishing websites. It is effective in the true sense of the term since it works by selecting a set of distinctive features in the URLs that tell the legitimate ones from the fake. Hence, this is also called as the feature-based methods [6].

This paper proposes to deal with the classification of phishing websites using confusion matrix and calculating the accuracy for various training data sets. Since the task of phishing detection is a type of classification task, and it aims at assigning each data set to one of the predefined classes and since phishing detection requires two sets of data (also called binary classification problem), the two possible values can be very well represented in a confusion matrix. The mentioned characteristics of the web page are then tallied with the two sets of URLs to distinguish the two kinds of websites. More the accuracy of the confusion matrix, more effectively the detection is done.

1.1 Literature Review

The features depicting the phishing websites are collected and database query and mining concepts are used to classify the websites [1]. In another contribution, Random Forest machine learning algorithm is used to classify URL of phishing websites [6]. Neural networks and acceptance of noisy data are structured to classify the websites [4]. Different rule-induction algorithms are used, using a software tool to predict phishing websites [3]. Phishing websites are classified using the features that distinguish trusted websites from the phishing ones [5]. The minimization of error rate and increment of classification rate accuracy has been explained clearly [7].

2 Materials and Methods

2.1 Overview of Feedforward Neural Network

In this paper, we have taken three layers of neural network: the input layer, the hidden layer, and the output layer. The input values from each are fed to the input nodes and they are multiplied by their respective weights assigned to them and then it is then fed into the hidden layer which sends the summed up outputs of the input layer to an activation function. The same process is iterated for the output layer as well. Finally, an output is obtained. We use the Moth-flame optimization algorithm as the learning algorithm here.

2.2 Moth-Flame Optimization Algorithm

To travel in a straight line at night, moths reference the moon such that it is always to their left. Since the moon is far away, so referencing it helps to move in a straight line [2]. But they are not able to distinguish between the moon and artificial lights such as flames. So if a moth is nearer to a flame, it becomes its reference point. This leads to the moth spiraling into the flame. This particular behavior of the moth is mathematically modeled to devise the Moth-Flame Optimization algorithm.

Since, moths go towards the flame in a spiral manner, so a spiral function is used to move a moth towards a single flame. In every iteration, one flame is assigned to each moth so that they move towards it only. This is done so as to avoid getting stuck at the local optima. Here, a logarithmic spiral given in Eq. (1) is used.

$$D \times ebt \times \cos(2\pi t) + F, \tag{1}$$

where b is a constant, t is a random number in range $[-1, 1]$, F is a flame and D is distance between a moth M and its flame F. The distance is calculated by Eq. (2).

$$D = |F - M| \tag{2}$$

Due to these equations, the next position of a moth is always with respect to its corresponding flame. "t" defines the closeness of the next position of the moth to the flame. If $t = 1$, the moth is farthest from its flame and if $t = -1$, the moth is closest to its flame. Thus, the moth flies around the flame and of course, not necessarily towards the flame. Here, there is exploration as well as exploitation. To make exploitation increase with every iteration, t can be any random number within the range $[r, 1]$ where r linearly changes from -1 to -2 with multiple iterations.

The F matrix always has the n best solutions. The OF matrix and OM matrix is calculated after each iteration. The F matrix is sorted according to OF matrix. If a better solution is found by a moth at any iteration (known from OM), then that position replaces the last row of the F matrix. The moths are assigned flames in a chronological order. The last moth takes the last flame (worst) and first moth takes the first flame (best). This ensures that there is no trapping in local optima.

To increase exploitation in the later iterations, Eq. (3) is used to reduce the number of flames.

$$No. of\ flames = n - (iteration\ number) * (n - 1)/Total\ iterations \quad (3)$$

Due to this, in the last iterations, the position of the moths is updated only with respect to the best few possible flames.

The time complexity of this algorithm is given in Eqs. (4), (5) and (6).

$$O(t \times (O(sorting) + O(Position\ update))) \quad (4)$$

$$= O(t \times (n \times \log(n)) + n \times d) \quad (5)$$

$$= O(t \times n \times \log(n) + t \times n \times d), \quad (6)$$

where t is total number of iterations and d is number of dimensions.

2.3 Proposed Algorithm

Building the networks and adjusting the weights Here, the dataset has 29 attributes. So, in the feedforward artificial neural network, 29 nodes are there in the input layer. There is only one output with values -1 and 1, indication suspicious, non-phishing and phishing websites. So, only one node is there in the output layer. By trial-and-error method, it has been found that only one hidden layer with four nodes is sufficient to produce satisfactory results. Since some values in the dataset are missing, so these need to be handled. There are different techniques to do so. Some of them are using mean of all the values for that attribute or using the mean of the values of that attribute that belongs to the same class or deleting that tuple, etc. Here the whole tuple is deleted. After deletion, the dataset is updated and this dataset is used. Once the network is set up, the weights and biases need to be balanced. To do this, search agents are formed. Each search agent is an array of 45 floating point values in range $[-1, 1]$. The range is set arbitrarily. Here 100 search agents are used. The determination of 45 search agents is determined as follows

- There are 29 input nodes to be connected to four hidden nodes. So total number of weights needed to connect them is $29 \times 4 = 116$.
- There is only one output node which is connected to the four hidden nodes. So number of weight needed is $4 \times 1 = 4$.
- Each of the nodes in hidden and output layer have a bias value. Thus there are $4 + 1 = 5$ bias values.
- Thus each search agent has $116 + 4 + 5 = 125$ values.

Fig. 1 Flowchart of MFO
algorithm

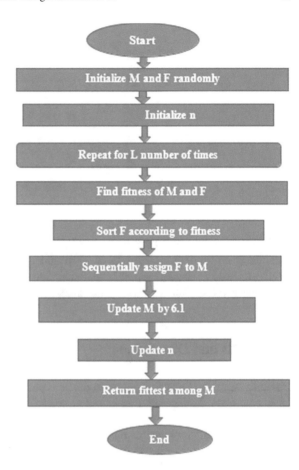

Fig. 1 Flowchart of MFO algorithm

Initially the 100 search agents are initialized with random values in range $[-1, 1]$. The values of each search agent are assigned as weights and biases of the feedforward artificial neural network. The neural network is run to get the output. The error for each search agent is calculated by comparing the target with the output. This error is the objective function and needs to be minimized by changing the values of the search agents. To optimize the values of the search agents, optimization algorithms are used which are described in the next sections.

After optimization, the fittest search agent (the search agent with least error) is got. The values of the best search agent are used as the weights and biases of the network and the network is tested.

Here a total of 11050 instances are used. The first 8500 instances are used for training and next 2550 instances are used for testing. In the end a confusion matrix is created for training, testing and total dataset. The flowchart given in Fig. 1 shows the flow of the algorithm used.

Fig. 2 Convergence curve

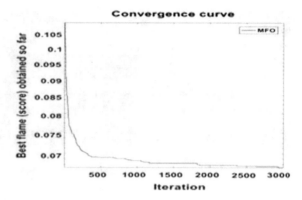

There are a total of 29 attributes that has been taken into account to test for a phishing website collected from the UCI repository. They are given a range of values from $-1, 0$ and 1. The dataset has been divided into testing and training sets. Each of the values given in Table 1 describes the dataset and their respective value ranges [5].

3 Results and Discussions

Basically, trial and error is required to reach to building a fine neural network with basic parameter description. Bad neural networks may produce unfit models. Error values may be large for over-fitted models. This can be avoided by adding more neurons to the hidden layer, or by adding a completely new layer in all. The main aim is to correct the error in weight vector of the neural network by adapting a training or learning algorithm to it. In this we use Moth-flame optimization as the learning algorithm to adjust the weight vectors. Our used model is capable of the following advantages:

1. MFO has better robustness i.e. it works well under conditions where other algorithms do not give satisfactory results.
2. MFO has better global optimization and the convergence accuracy is high (Tables 2, 3 and 4).

Evaluation Parameters After calculating the accuracy, specificity, sensitivity, error rate, prevalence and mean squared error for all the results, we finalize the best solutions of accuracy in Table 5 for the three sets of dataset (Fig. 2).

A few experiments have been conducted on classification of phishing websites using different techniques. One such technique was developed using Neural Network and then passed to an activation function of log-sigmoid [5]. The accuracy found in the method was 90.35% for testing data set. The accuracy is found to be less than that found by the Moth-flame Optimization algorithm.

Table 1 Attributes and their respective value ranges of data set

Attributes	Description	Range
Having IP address [1]	The URL has an IP address, it implies it is phishing	$\{-1, 1\}$
URL length	Length of URL >75, it is phishing, <54 but >75, it is suspicious	$\{-1, 0, 1\}$
Shortening_Service	Tiny URL, it is phishing Long URL, it is legitimate	$\{1, -1\}$
Having_At_Symbol [1]	If URL has @ symbol, it is phishing	$\{1, -1\}$
Double_slash_redirecting	If the position of the last occurrence of "//" in the URL >7, it is phishing	$\{-1, 1\}$
Prefix_Suffix	If domain part includes $(-)$, this is phishing	$\{-1, 1\}$
Having_Sub_domain	If 1 dot is there in domain part, it is legitimate If 2 dots, it is suspicious Otherwise, it is phishing	$\{-1, 0, 1\}$
SSLfinal_state		$\{-1, 1, 0\}$
Domain registration length	If domain expires in less than 1 year, it is phishing. Otherwise, legitimate	$\{-1, 1\}$
Favicon	If icon is loaded from external domain, it is phishing. Else, legitimate	$\{1, -1\}$
Port	If port # is of the preferred status, it is phishing. Else, legitimate	$\{1, -1\}$
HTTPS_token	If HTTP token is used in domain part, it is phishing. Else, legitimate	$\{-1, 1\}$
Request_URL2	If the percentage is <22%, it is legitimate. If it lies between 22% and 61%, it is suspicious. Else, it is phishing	$\{1, -1\}$
URL_of_Anchor	If it is <31%, it is legitimate. If it is >31% and <67%, it is suspicious. Else, it is phishing	$\{-1, 0, 1\}$
Links_in_tags	If it is <17%, it is legitimate. If it is $\geq$17% but $\leq$81%, it is suspicious. Else, phishing	$\{1, -1, 0\}$
SFH (Server Form Handler)	If it is "about: blank" or empty, it is phishing. If it redirects to a different domain, it is suspicious. Else, it is legitimate	$\{-1, 1, 0\}$
Submitting_to_email	If it uses mail (), it is phishing. Else, legitimate	$\{-1, 1\}$
Abnormal_URL	If host name is not present, it is phishing. Else, it is legitimate	$\{-1, 1\}$
Redirect	If the website has been redirected $\leq$1 times, it is legitimate. If $\geq$2 but <4 times, it is suspicious. Else, phishing	$\{0, 1\}$

(continued)

Table 1 (continued)

Attributes	Description	Range
On_mouseover	If there is no change of status bar on "mouseover" event, it is legitimate else, it is phishing	{1, −1}
RightClick	If right click doesn't work, it is phishing. Else, it is legitimate	{1, −1}
PopUpWindow	If the window contains text fields, it is phishing. Else, legitimate	{1, −1}
Iframe [1]	If it uses Iframes, it is phishing. Else, legitimate	{1, −1}
Age_of_domain	If it is ≥6 months, it is legitimate. Else, phishing	{−1, 1}
DNSRecord	If there is no such record, it is phishing. Else, it is legitimate	{−1, 1}
Web_traffic	If the rank of the website is<1, 00,000, it is legitimate. If it is>1, 00,000, it is suspicious. Else, it is phishing	{−1, 0, 1}
Page_Rank	If it is <0.2, it is phishing. Else, legitimate	{−1, 1}
Google_Index	If the webpage has been indexed by Google, it is legitimate. Else, phishing	{1, −1}
Links_pointing_to_page	If the number of links pointing to page is 0, it is phishing. If it is>0 but ≤2, it is suspicious. Else, it is legitimate	{1, 0, −1}
Statistical_report	If host belongs to the registered phishing addresses, it is phishing. Else, it is legitimate	{−1, 1}
Result		{−1, 1}

Table 2 Training data

	Target 0	Target 1
Observed 0	7462	520
Observed 1	38	480
Correct classification	0.99493	0.52000
Observed 0	7462	536
Observed 1	38	464
Correct classification	0.99493	0.53600
Observed 0	7451	504
Observed 1	49	657
Correct classification	0.99346	0.43410
Observed 0	7458	526
Observed 1	42	474
Correct classification	0.99440	0.52600

Table 3 Testing data

	Target 0	Target 1
Observed 0	2271	145
Observed 1	1	133
Correct classification	0.99955	0.52158
Observed 0	2265	85
Observed 1	7	193
Correct classification	0.99691	0.30575
Observed 0	2259	93
Observed 1	13	185
Correct classification	0.99427	0.33453
Observed 0	2261	77
Observed 1	11	201
Correct classification	0.99515	0.27697

Table 4 Total data

	Target 0	Target 1
Observed 0	9763	665
Observed 1	39	613
Correct classification	0.99602	0.52034
Observed 0	9727	621
Observed 1	45	657
Correct classification	0.99539	0.48591
Observed 0	9710	597
Observed 1	62	681
Correct classification	0.99365	0.46713
Observed 0	9719	603
Observed 1	53	675
Correct classification	0.99457	0.47183

Table 5 Evaluation parameters calculation

Training data		Testing data	Total data
Accuracy (%)	93.49	96.54	94.06
Specificity (%)	49.60	72.30	52.00
Sensitivity (%)	99.34	99.51	99.45
Error rate (%)	6.51	3.46	5.94
Prevalence (%)	88.23	89.09	88.43
Mean squared error (%)	30.16	2.37	33.16

4 Conclusion

Effective anti-phishing tools are essential for restricting access of phishing websites. Not just good classification but with good accuracy, we need to develop algorithms to implement this. Constant improvements are necessary to be made nevertheless for more accuracy in determining the phishing websites.

References

1. Alkhozae, M.G., Batarfi, O.A.: Phishing websites detection based on phishing characteristics in the webpage source code. Int. J. Inf. Commun. Technol. Res. **1**(6) (2011)
2. Mirjalili, S.: Moth-flame optimization algorithm: a novel nature-inspired heuristic paradigm. Knowl. Based Syst. **89**, 228–249 (2015)
3. Mohammad, R.M., Thabtah, F., McCluskey, L.: Intelligent rule-based phishing websites classi-fication. IET Inf. Secur. **8**(3), 153–160 (2014)
4. Mohammad, R.M., Thabtah, F., McCluskey, L.: Predicting phishing websites based on self-structuring neural network. Neural Comput. Appl. **25**(2), 443–458 (2014)
5. Prajapati, U., Sangal, N., Patole, D.: Fraud website detection using data mining. Int. J. Comput. Appl. **141**(3) (2016)
6. Sananse, B.E., Sarode, T.K.: Phishing URL detection: a machine learning and web mining-based approach. Int. J. Comput. Appl. **123**(13) (2015)
7. Yamany, W., Fawzy, M., Tharwat, A., Hassanien, A.E.: Moth-flame optimization for train-ing multi-layer perceptrons. In: 2015 11th International Computer Engineering Conference (ICENCO), pp. 267–272. IEEE, Dec 2015

Design and Analysis of Intrusion Detection System via Neural Network, SVM, and Neuro-Fuzzy

Abhishek Tiwari and Sanjeev Kumar Ojha

Abstract An intrusion detection system is continuous observation of system or over the network assessment of an intruder or any other attacks. In this paper, design, and analysis of intrusion detection system via neuro-fuzzy, neural network and SVM technique for the improvement misuse detection system. The proposed approachable to enhancement anomaly detection and improve these techniques for anomaly detection.

Keywords Intrusion detection · Neural network · Neuro-fuzzy · Fuzzy system SVM

1 Introduction

An intrusion detection system is able to identify intrusion attacks before the actually happened. The working strategy of intrusion detection system, analysis the user profile or system and find out the pattern and train these pattern to a system for the identification of intrusion [1]. It is persistently assessment of monitor or over the network for identification of networks such as malicious activities or policy violations [2]. If any unseen malicious or intrusion activity arrive then add these feature to our intrusion detection system for better accuracy into future over the identification of anomalies in network [3].

A. Tiwari (✉)
Department of Computer Science & Engineering, Galgotias Educational
Institutions, Greater Noida, India
e-mail: abhishektiwari.122101.cse@gmail.com

S. K. Ojha
Department of Radio Engineering & Cybernetic, Moscow Institute
of Physics and Technology, Zhukovsky, Russia
e-mail: kumarojhasanjeev@gmail.com

© Springer Nature Singapore Pte Ltd. 2019   49
A. Abraham et al. (eds.), *Emerging Technologies in Data Mining and Information
Security*, Advances in Intelligent Systems and Computing 755,
https://doi.org/10.1007/978-981-13-1951-8_6

2 Intrusion Detection System Architecture

The proposed approach for intrusion detection system categories into three phases: Data preprocessing, training, and testing [4, 5]. These are the main working architecture of any intrusion detection model.

Step 1: First Data preprocessing of the dataset for the purpose of raw data to machine readable form and remove the dummy information into datasets.
Step 2: In this phase training the intrusion detection model for the purpose of identification of intrusion.
Step 3: Persistently assessment of intrusion and track the activity and behavior of the intrusion (Fig. 1).

There are several architecture modules as:

Network Data Monitoring: In this module, monitor the packet whatever have available into network source and track the feature [6, 7].

Preprocessing: In this phase, remove the dummy information and ready to use the data for intrusion detection model. Feature Extraction: In this phase, track the packets and identify the feature and after extracting the feature submit into classifier module. In classifier model, similar kind of feature categories into clustering due to easily predict the feature of more intrude [8]. *Classifier:* Classification of the feature is a very important factor for identity the more intrude based on the feature. Classifier classify the feature of the profile. The neural network is able to handle both classification and pattern recognition.

Training: In this phase based on feature trained to the intrusion detection system for the purpose of easily identifying more intrude [9]; Testing: if intrusion is found out then message is sent to the user; *Knowledgebase:* This phase is very important for training purpose due to whatever data or feature observe, this phase gives the ability to take decision intrusion available or not [10, 11].

Fig. 1 Working architecture of intrusion detection system [15]

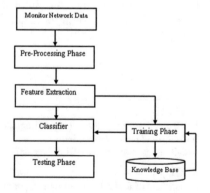

3   Intrusion Detection System Life Cycle

There are several phases of intrusion detection system life cycle such as development, evaluation, and selection, maintenance [12], continuous persistance [13] (Fig. 2).

3.1   Evaluation Performance

The evaluation parameters of intrusion detection system identify the available resource during system operation [13]. For the purpose of identification of intruder [14], evaluation parameters such as the behavior of intruder [1], utilization of resource and accuracy are used [15].

3.2   Deployment

In this phase, training the system regarding an identification of intruder behavior, deploy and acceptable if any new activities arrive.

3.3   Operation and Working Strategy

During the overall working strategy, intrusion detection system focus on the monitor and whatever activities going on over the system. Some time new strong intrusion attack sends the false message to the administrator.

Fig. 2 IDS life cycle [1]

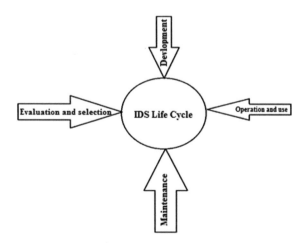

4 Proposed Approach

4.1 IDS Using Neural Network

Based on the profile of intruder behavior, intrusion detection model predicts the intrusion attacks over the system or network. The address of solution of this issues artificial neural network play very important role due to based on a characteristic of attackers its draw a curve and minimize the error corresponding to the input, find out the best-fit solution of the problem. The basic unit of the neural network is neurons, it communicates the information node to node. The arrangement of neurons that is topology based on the security or feature parameters. In which neurons take into the data and multiple with weight factor and summaries overall conclusion produce the output corresponding to one neuron. The fully connected topology of neurons indicates that computation process is simultaneous. The adjustment of magnitude weight for the optimize the solution, define as training rules, for example, define into basic neural network model as defined in Fig. 3.

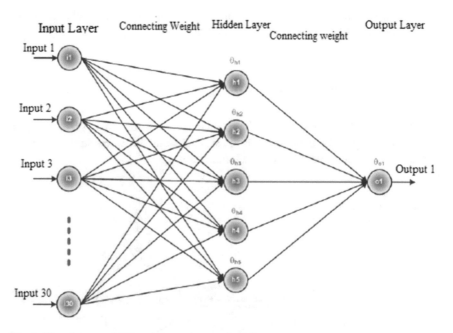

Fig. 3 Three layers architecture of neural networks [15]

4.2 Intrusion Detection System Via Neuro-Fuzzy

Neuro-fuzzy allocates the membership value of all the arguments based on alpha-cut value and select data as per requirement. It gives the real-time results and checks all the possible valuable combination.

$$\text{IF} < \text{P} > \text{THEN} \ < \text{C} >$$

Here, P and C are linguistics variables. P denotes the premise and C denote consequent.

The advantages of the fuzzy systems are

- Ability to represent linguistic variables;
- Interpretation of the outcome easily due to allocating the membership value of all the premises.
- Rule-Based selection of premises.

  Disadvantages are

- Unable to generalised
- Based on the requirement, select the premise via alpha-cut.

4.3 IDS Using Support Vector Machine

Support vector machine is a supervised learning for classification or regression problems. When classification is done by SVM then as output we get training input and whatever outline we have got that is feature space [15, 16].

4.4 Fundamental Working Strategy of SVM

Pattern recognition and classification process are two measure concepts whatever we have used into intrusion detection process. For l training data apply into two class classification process as

$$\left(\vec{x_1}, y_1 \right)..\left(\vec{x_i}, y_i \right) \ldots \left(\vec{x_l}, y_l \right)$$

discriminate function $f\left(\underset{x}{\rightarrow} \right)$ as $y = f\left(\underset{x}{\rightarrow} \right),$

where $yi = f + 1, j1$ g is the class label for sample data point $\underset{xi}{\rightarrow}$. If data is linear separation then function optimal hyperplane consider as

$$\vec{w} \cdot \vec{x} - b = 0$$

4.5 Kernel Functions of SVM

SVM solve the problem using "kernel trick" to solve. These kernel function K define for nonlinear function.

$$K\left(x_i^{\rightarrow}, x_j^{\rightarrow}\right) = K\left(\vec{x_i}\right).K\left(\vec{x_j}\right)$$

At that time we required only K during SVM process and useless for calculate Φ. When Mercer's Condition hold only then kernel function K exists.

$$K\left(x_i^{\rightarrow}, x_j^{\rightarrow}\right) = K\left(\vec{x_i}\right).K\left(\vec{x_j}\right) \text{iff}, \quad \text{for any } g(\vec{x})$$
$$\int g(\vec{x})^2 \, d(\vec{x}) \text{ is finite},$$

Then $\int K\left(\vec{x_i}, \vec{x}\right)g\left(\vec{x_i}\right)g\left(\vec{x_j}\right)d\left(\vec{x_i}\right)g\left(\vec{x_j}\right) \geq 0$

The nonlinear function of SVM is define as

$$f(\vec{x}) = sgn\left(\sum_{i=1}^{1} \alpha_i \gamma_i K\left(x_i^{\rightarrow}, x_j^{\rightarrow}\right) - b\right)$$

5 Experimental Outcomes

The ROC curve concludes the summarized form of the accuracy of the model. In the proposed approach, retrieval rate of ROC Curve of AUS $= 0.97712$ (Table 1).

5.1 ROC Analysis

In ROC analysis, check the intrusion detection model accuracy based on comparing the i sample class and predefine storage class of j. The size of the similarity matrix is a number of test samples by a number of classes (Fig. 4).

Table 1 ROC curve data

S. No.	Cut-off	Sensitivity	Specificity	Efficiency	PLR	NLR
1	0.00	0.0009	0.9916	0.4963	0.1063	1.0075
2	0.00	0.0023	0.9787	0.4905	0.1062	1.0195
3	0.00	0.0165	0.9787	0.4976	0.775	1.0049
4	0.00	0.0317	0.9787	0.5052	1.487	0.9894
5	0.00	0.0475	0.9787	0.5131	2.228	0.9733
6	0.00	0.0648	0.9787	0.5217	3.0409	0.9556
7	0.00	0.0825	0.9787	0.5306	3.8717	0.9375
8	0.00	0.1005	0.9787	0.5396	4.7158	0.9191
9	0.00	0.1168	0.9787	0.5478	5.4832	0.9024
10	0.00	0.1322	0.9787	0.5554	6.2021	0.8867
11	0.00	0.5952	0.9787	0.787	27.909	0.4136
12	0.00	0.6105	0.9786	0.7946	28.529	0.398
13	0.00	0.6248	0.9785	0.8016	29.0081	0.3834
14	0.00	0.6381	0.9782	0.8082	29.2819	0.3699
15	0.00	0.6506	0.9765	0.8135	27.7045	0.3578
16	0.00	0.7344	0.9751	0.8547	29.4629	0.2724
17	0.00	0.7635	0.975	0.8693	30.5642	0.2425
18	0.00	0.7697	0.975	0.8724	30.8065	0.2362
19	0.00	0.7755	0.975	0.8753	31.036	0.2302
20	0.00	0.7806	0.975	0.8778	31.2385	0.2250
21	0.00	0.7856	0.975	0.8803	31.4346	0.2199
22	0.00	0.7906	0.975	0.8828	31.6349	0.2148
23	0.00	0.7956	0.975	0.8853	31.8353	0.2096
24	0.00	0.8006	0.975	0.8878	32.0356	0.2045
25	0.00	0.8056	0.975	0.8903	32.2359	0.1994
26	0.00	0.8107	0.975	0.8929	32.44	0.1941
27	0.00	0.8159	0.975	0.8954	32.6465	0.1888
28	0.00	0.8209	0.975	0.8980	32.8472	0.1837
29	0.00	0.826	0.975	0.9005	33.0512	0.1785
30	0.00	0.831	0.975	0.9030	33.2516	0.1733
31	0.00	0.8361	0.975	0.9055	33.4535	0.1682
32	0.00	0.8411	0.975	0.9080	33.6543	0.163
33	0.00	0.8461	0.975	0.9106	33.8563	0.1578
34	0.00	0.8512	0.975	0.9131	34.0583	0.1526
35	0.00	0.8562	0.975	0.9156	34.2606	0.1475
36	0.00	0.8613	0.975	0.9181	34.463	0.1423
37	0.00	0.8663	0.975	0.9207	34.6654	0.1371

(continued)

Table 1 (continued)

S. No.	Cut-off	Sensitivity	Specificity	Efficiency	PLR	NLR
38	0.00	0.8714	0.975	0.9232	34.8675	0.1319
39	0.00	0.8764	0.975	0.9257	35.0698	0.1267
40	0.00	0.8815	0.975	0.9283	35.273	0.1215
41	0.00	0.8866	0.975	0.9308	35.4717	0.1163
42	0.00	0.8931	0.975	0.9341	35.7331	0.1096
43	0.00	0.8992	0.975	0.9371	35.9737	0.1034
44	0.00	0.9128	0.975	0.9439	36.5196	0.0894
45	0.00	0.9274	0.975	0.9512	37.1039	0.0745
46	0.00	0.9462	0.975	0.9606	37.8546	0.0552
47	0.00	0.9598	0.975	0.9674	38.3984	0.0413
48	0.00	0.9738	0.975	0.9744	38.9591	0.0269
49	0.00	0.979	0.975	0.977	39.1673	0.0216
50	0.01	0.9841	0.975	0.9795	39.3545	0.0163
51	0.01	0.9904	0.9749	0.9827	39.5061	0.0098
52	0.08	1.0000	0.9744	0.9872	39.0097	0.0000
53	0.94	1.0000	0.9681	0.9841	31.3795	0.0000
54	0.97	1.0000	0.9493	0.9746	19.7167	0.0000
55	0.98	1.0000	0.9182	0.9591	12.226	0.0000
56	0.99	1.0000	0.8962	0.9481	9.6383	0.0000
57	0.99	1.0000	0.8891	0.9445	9.0155	0.0000
58	0.99	1.0000	0.884	0.942	8.6189	0.0000
59	0.99	1.0000	0.8789	0.9394	8.2557	0.0000
60	0.99	1.0000	0.8718	0.9359	7.8032	0.0000
61	0.99	1.0000	0.8666	0.9333	7.4989	0.0000
62	0.99	1.0000	0.8596	0.9298	7.1202	0.0000
63	0.99	1.0000	0.8545	0.9272	6.8726	0.0000
64	0.99	1.0000	0.8495	0.9247	6.6424	0.0000
65	0.99	1.0000	0.8423	0.9212	6.3418	0.0000
66	0.99	1.0000	0.8332	0.9166	5.9952	0.0000
67	0.99	1.0000	0.8238	0.9119	5.6739	0.0000
68	0.99	1.0000	0.8139	0.907	5.3745	0.0000
69	0.99	1.0000	0.8041	0.9021	5.1056	0.0000
70	0.99	1.0000	0.794	0.897	4.8544	0.0000
71	0.99	1.0000	0.7842	0.8921	4.6334	0.0000
72	0.99	1.0000	0.7756	0.8878	4.4555	0.0000
73	0.99	1.0000	0.7672	0.8836	4.2959	0.0000
74	1.00	1.0000	0.7589	0.8794	4.1476	0.0000
75	1.00	1.0000	0.7511	0.8755	4.0174	0.0000

(continued)

Table 1 (continued)

S. No.	Cut-off	Sensitivity	Specificity	Efficiency	PLR	NLR
76	1.00	1.0000	0.7449	0.8725	3.9204	0.0000
77	1.00	1.0000	0.7367	0.8683	3.7974	0.0000
78	1.00	1.0000	0.7290	0.8645	3.6896	0.0000
79	1.00	1.0000	0.7217	0.8609	3.5935	0.0000
80	1.00	1.0000	0.7162	0.8581	3.524	0.0000
81	1.00	1.0000	0.7075	0.8537	3.4184	0.0000
82	1.00	1.0000	0.6843	0.8422	3.1677	0.0000
83	1.00	1.0000	0.6776	0.8388	3.102	0.0000
84	1.00	1.0000	0.6523	0.8261	2.876	0.0000
85	1.00	1.0000	0.6387	0.8193	2.7678	0.0000
86	1.00	1.0000	0.6288	0.8144	2.694	0.0000
87	1.00	1.0000	0.6171	0.8085	2.6116	0.0000
88	1.00	1.0000	0.6065	0.8032	2.5412	0.0000
89	1.00	1.0000	0.5935	0.7967	2.4599	0.0000
90	1.00	1.0000	0.5799	0.79	2.3805	0.0000
91	1.00	1.0000	0.5632	0.7816	2.2891	0.0000
92	1.00	1.0000	0.5483	0.7741	2.2138	0.0000
93	1.00	1.0000	0.5320	0.766	2.1368	0.0000
94	1.00	1.0000	0.5179	0.7589	2.0741	0.0000
95	1.00	1.0000	0.5038	0.7519	2.0154	0.0000
96	1.00	1.0000	0.4923	0.7462	1.9697	0.0000
97	1.00	1.0000	0.4789	0.7395	1.9191	0.0000
98	1.00	1.0000	0.0184	0.5092	1.0188	0.0000
99	1.00	1.0000	0.0061	0.5031	1.0061	0.0000
100	1.40	1.0000	0.0000	0.5000	1.0000	0.0000
101	2.40	1.0000	0.0000	0.5000	1.0000	Inf

5.2 False Acceptance Rate (FAR)

In which incorrectly accept the data that are not authorized to access it. This phenomenon is known as false acceptance rate (Figs. 5 and 6).

5.3 False Recognition Rate

The false recognition rate is defined ratio of the number of false recognitions and the number of identification attempts (Fig. 7).

False Reject Rate can be calculated as (Fig. 8):

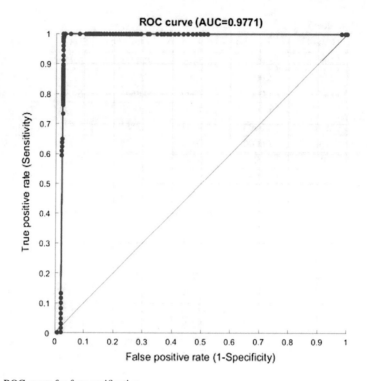

Fig. 4 ROC curve for face verification

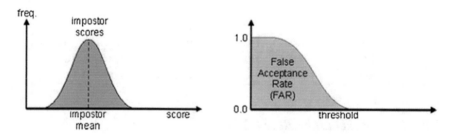

Fig. 5 False acceptance rate (FAR)

$$\text{FRR(n)} = \frac{\text{Number of rejected verification attempts for a qualified individual n}}{\text{Total number of verification attempts for that qualified individual n}}$$

$$\text{and} \quad \text{FRR} = \frac{1}{N} \sum_{n=1}^{N} FRR(n)$$

where "n" is the total number of enrolments.

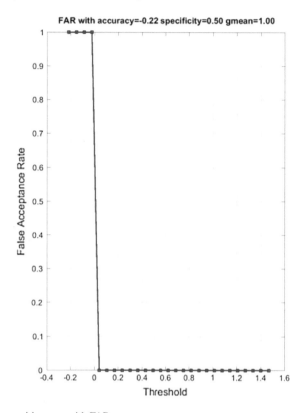

Fig. 6 Face recognition rate with FAR

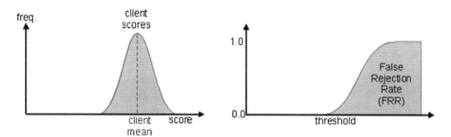

Fig. 7 False recognition rate (FRR)

5.4 *Comparison Graph*

In this Fig. 9 shows an accuracy comparison between neuro-fuzzy, neural network, and SVM technique.

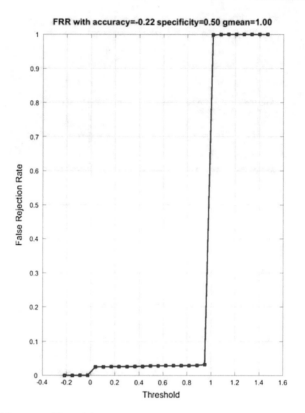

Fig. 8 FRR (False recognition rate)

Table 2 Comparison table between using technique

SN	Technique	Accuracy
1	Neuro-fuzzy [5]	79.079121
2	Neural network [15]	91.350967
3	SVM [12]	67.128467

The accuracy of three techniques with their graphs is calculated as follows.

In this Fig. 10 shows an accuracy comparison between neuro-fuzzy, neural network, and SVM technique and find accuracy value is neuro-fuzzy accuracy is 79.079121, neural network accuracy is 91.350967 and SVM accuracy is 67.128467 (Table 2).

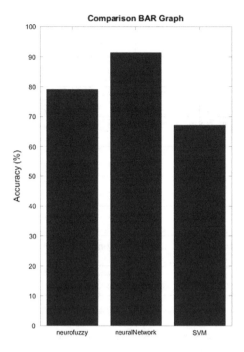

Fig. 9 Accuracy comparison graph

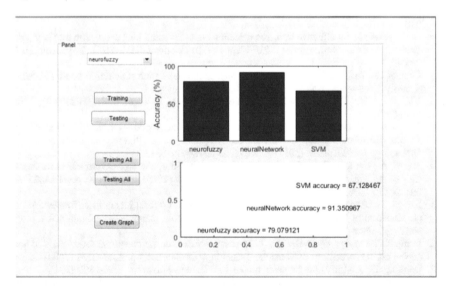

Fig. 10 Accuracy graph with comparison value

6   Conclusion

In the conclusion of this paper, we focused and analysis the current existing challenges and proposed schema regarding minimizing the issues regarding same. It address the solution of anomaly detection system, in which scope for training data set again train/update if found any unseen behavior of the user as input profile. Training rules able to deal the weight magnitude for corresponding input value and optimize the solution.

References

1. Kumar, I., Virmani, J., Bhadauria, H.S.: A review of breast density classification methods. In: Proceeding of 2nd International Conference on Computing for Sustainable Global Development INDIACom, pp. 1960–1967 (2015)
2. Kumar, I., Virmani, J., Bhadauria, H.S.: Wavelet packet texture descriptors based four-class BIRADS breast tissue density classification. Procedia Comput. Sci. **70**, 76–84 (2015)
3. Kumar, I., Bhadauria, H.S., Virmani, J., Thakur, S.: A classification framework for prediction of breast density using an ensemble of neural network classifiers. Biocybern. Biomed. Eng. **37**, 217–228 (2017)
4. Sabahi, F.: Intrusion detection: a survey. In: The Third International Conference on Systems and Networks Communications. IEEE Computer Society (2008)
5. Chandrasekhar, A.M. et al.: Intrusion detection technique by using k-means, fuzzy neural network and SVM classifiers. In: IEEE Xplore (2013). https://doi.org/10.1109/iccci201.6466 310
6. Le, T.-H. et al.: An effective intrusion detection classifier using long short-term memory with gradient descent optimization. In: IEEE Xplore (2017). https://doi.org/10.1109/platcon.2017. 7883684
7. Cannady, J.: Artificial neural networks for misuse detection. In: National Information Systems Security Conference (2006)
8. Vladimir, V.N.: The Nature of Statistical Learning Theory. Springer, Berlin, Heidelberg, New York (2005)
9. Tiwari, A. et al.: An effective approach for secure video watermarking based on H.264 standard. In: 3rd IEEE International Conference on Computational Intelligence and Communication Technology IEEE Xplorer (2017). 978-1-5090-6218-8/17/$31.00
10. Tiwari, A., Kamlesh K.Gupta, An effective approach of digital image watermarking for copyright protection. Int. J. Big Data Secur. Intell. **2**(1), 7–17 (2015). http://dx.doi.org/10.14257/ij bdsi.2015.2.1.02, ISSN: 2383-7047SERSC
11. Kumar, I., Bhadauria, H.S., Virmani, J., Thakur, S.A.: A hybrid hierarchical framework for classification of breast density using digitized film screen mammograms Multimedia Tools and Applications pp. 1–25 (2017)
12. Deng, H., Zeng, Q., Agrawal, D.P.: SVM-based intrusion detection system for wireless ad hoc networks. In Proceedings of Vehicular Technology Conference, pp. 2147–2151 (2003)
13. Denning, D.: An intrusion-detection model. IEEE Trans. Softw. Eng. **SE-13**(2) (2016)
14. Abhishek Tiwari, Neelesh Kumar Jain and Devraj Tomar, Analysis of multiscale transform based digital image watermarking for multimedia files, Int. J. Sci. Res. Devel. ISSN: 2321-0613 vol 2(2) pp. 177–182 (2014)
15. Tiwari, A.: Real time intrusion detection system using computational intelligence and neural network: review, analysis and anticipated solution of machine learning. Springer Book Series: Information Technology and Applied Mathematics (2017). ISBN: 978-981-10-7590-2, ISSN 2194-5357

16. Yan, H. et al.: ANN-based multi classifier for identification of perimeter events. In: IEEE Xplore (2011). https://doi.org/10.1109/iscid.2011.141

17. Malhotra, S. et al.: Genetic programming and K-nearest neighbour classifier based intrusion detection model. In: IEEE Xplore, ISBN (2017). https://doi.org/10.1109/confluence.2017.794 3121

18. Ghosh, A.K.: Learning program behavior profiles for intrusion detection. In: USENIX (1999)

Grammar-Based White-Box Testing via Automated Constraint Path Generation

Bijoy Rahman Arif

Abstract Software testing is an indispensable procedure for assuring software quality and test case generation is one of the major stages of software testing odyssey. In this paper, the author presents a grammar-based white-box testing which integrates some source code analysis, grammar-based test generation, and constraint solving as a whole. A preliminary implementation of grammar-based white-box testing, Java White-box Unit Tester (JWBUT), has been developed. JWBUT can generate well-distributed authentic test cases using the grammar-based white-box testing scheme, which starts from static analysis of Java source code, transforms control flow into context-free grammar (CFG), applies to grammar-based test generator to produce a set of constraint paths, and finally, generates test cases by solving the constraint paths. The author has experimented JWBUT on a set of Java methods to demonstrate the effectiveness of grammar-based white-box testing.

1 Overview

Software testing is a formal approach to assure two aspects of the software: quality and requirement [1]. It is performed on different phases of software development life cycle to ensure the quality by finding bugs and code errors. Then, it fixes those problems in the next phase of software testing. This way it improves efficiency and compactness of existing code so that it follows the requirements of software. Test case generation is considered an important part of software testing because test data helps to find the behaviors of software: correctness, resource allocation, time consumption, prime paths, and so on. These behaviors are important parameters to software quality assurance and that is why the quality of test case generation is considered vital to advance the state of the art in software testing [2]. Because of its big impact on software behaviors, test case generation is accomplished as a major software testing step even though it is one of the most labor-intensive tasks [3].

B. R. Arif (✉)
Independent University, Dhaka 1229, Bangladesh
e-mail: bijoyarif@iub.edu.bd

© Springer Nature Singapore Pte Ltd. 2019
A. Abraham et al. (eds.), *Emerging Technologies in Data Mining and Information Security*, Advances in Intelligent Systems and Computing 755,
https://doi.org/10.1007/978-981-13-1951-8_7

65

White-box testing is comparatively a new technique for automatic test case generation. The main difference between white box and other boxes (black box or gray box) is about the knowledge of source (structure) being perceived or not during software testing [4]. We must note that the black box and white-box ideas are not limited in the test case generation only [5] but also the way we can achieve it. White box, as the name suggested, has partial or full idea of source during software testing.

As white box does have knowledge of the source, that is why it is also called design-based testing [1]; it creates the inputs which are relevant and authentic for particular source, eventually, it gets rid of exhaustiveness required by other boxes to find bugs. If it can create the outputs associated with inputs, it ensures the bug nature of source [6].

1.1 Problem Statement and Our Approach

Researchers believe white-box testing is an effective technique for generating test cases based on knowledge of the source code. It generates a set of test inputs which are relevant and has better coverage on testing execution paths. However, there are three fundamental problems on the white-box testing approach:

Path Explosion Since a program usually involves branch structures and loop structures, there are typically infinite number of execution paths. Thus, one problem is how to generate a small yet sufficient set of test cases to cover all paths. Even with the target of generating a small set of test cases, due to nested loop structures, generating a single test case often invokes significant recursion and non-termination issues.

Coverage Issue Real programs may contain nested loops and branch structures with direct or indirect recursive methods. It is another challenge problem to generate a finite set of test cases to cover all complex execution paths in a well-distributed manner.

Complex Constraints Practical programs may involve complex constraints. Sometimes even though we obtain a concrete execution path, the constraints along the path may beyond the computational power of any available constraint solver.

On the problem for solving complex constraints, many new powerful constraint solvers [7–9] have developed during the last decade. Using those constraint solvers has enabled much more broader applications of constraint solving for test case generation. However, in this paper, we will focus on addressing the first two problems: *path explosion* and *coverage issues*.

Many techniques have been proposed to get around the path explosion problem. One main strategy lies on narrowing down the testing parts of a program by using constraints or assertions to direct exploration of execution paths [10–13], using grammar-bounded inputs [14, 15], or other heuristic approaches [16, 17]. Another strategy is exhaustive path generation with explicit annotation controls [18, 19]. For example, the lava tool [18] takes a seed, which consists of a high-level description that guides the production process, to generate effective test suites for Java virtual machine. YouGen [19] supports many extra-grammatical annotations, e.g., a depth bound, which guide or constraint effective test generation. All these techniques typically alleviate the path explosion problem by sacrificing the balanced coverage feature. These techniques may fail to discover program behaviors that a systematic test generation technique can discover.

1.2 Contribution and Significance of Research

The main objective of this paper is to propose a new grammar-based white-box testing approach which explores source code in a systematic manner to generate constraint paths, yet handles the path explosion and the coverage issues properly and automatically. Our new approach combines three techniques of source code analysis, grammar-based test generation, and constraint solving as an integrated whole, and generates authentic test cases which lead to find possible bugs from given source codes in a fully automatic way. To demonstrate the effectiveness of our grammar-based white-box test generation approach, we developed a preliminary tool, Java White-Box Unit Tester (JWBUT), which can take a wide range of Java methods and create authentic test cases as outputs. JWBUT consists of three main utility programs: (1) a generator from Java source code to context-free grammar, (2) a grammar-based test generator, and (3) a logic-based constraint solver.

1.3 Organization of Report

Our paper is divided to six parts. In Sect. 1, we introduced the concept of white-box testing with major ideas. In Sect. 2, we are going to discuss about the symbolic execution, a way to achieve white-box testing. We will talk about our tool in Sect. 3. In Sect. 4, we will discuss about JWBUT with different parameters of white-box testing and how our tool differs from other tools. In Sect. 5, a detail experimental result is given about the tool's different metrics and performance. Finally, we are going conclude by discussing the constraints and the ways to improve our work in Sect. 6.

2 White-Box Testing via Symbolic Execution

White-box test case generation explores the source or the binary structure, which makes it a better choice for reliability and authenticity. Among different techniques of white-box testing, symbolic execution [20], a formal and systematic exploration using symbolic abstraction of variables at each point of program to generate test cases has recently got much attention of researchers [3].

2.1   Symbolic Execution Tree (SET)

We present the concept of a symbolic execution tree (SET) to illustrate how the white-box testing works using symbolic execution. A symbolic execution tree (SET) is a symbolic abstract which simulates different possible execution paths of a system. Each node in a SET contains a tuple of three components: (1) the program counter, which is typically the line number for each statement in a source code, indicating the next statement to execute, (2) the status of input variables, represented in a symbolic way, and (3) a path constraint (PC), which is a cumulation of all constraints among symbolic variables from the beginning node (root node) to current node of code. The root node is labeled by $(1, \{x_1 = X_1, x_2 = X_2, \ldots\}, PC : true)$, where 1 means the beginning line of code, each x_i is an actual input variable and each X_i is a constraint variable which represents the possible state of actual variable x_i, and the initial PC is always true. A transition from a parent node to a child node simulates advance of execution. A transition may be labeled by a conditional Boolean value, either *true* or *false*, or a selection value (e.g., a value from a switch case). A *successful* path in SET is a finite path from the root to terminal leave nodes, in which the program counter is corresponding to the end of an execution scenario (such as a return or exit statement). However, a successful path may not be feasible in practice due to possible unsatisfiability of PC. A *feasible* path in SET is a successful path and the PC in its leaf node is *satisfiable*. It is worthy to be mentioned that a path in SET can possibly be infinite, which simulates an execution scenario of infinite loop. Thus, test case generation in the white-box testing essentially generates test cases corresponding to each feasible path.

2.2   SET Example

Example 1   Consider the GCD method which takes two positive integer inputs a and b, and returns their greatest common divisor.

Figure 1 shows the SET tree for the code fragment in Example 1. The GCD method contains a while loop which potentially involves a large number of different symbolic execution paths, since the while loop can be executed by any finite number of times and then followed by an exit. If the condition at statement 1 satisfies (True), PC is updated with constraint A != B and enters the loop to statement 2. If the condition is not satisfied (False) then the symbolic execution terminates at statement 7 with constraint $A == B$ added in PC. Everytime execution reaches the end of loop (line 6), the symbolic execution starts again from statement 1 with all previous constraints

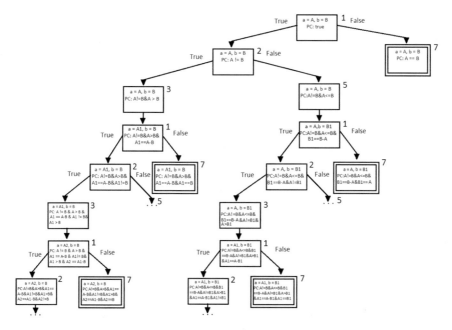

Fig. 1 Symbolic execution tree for the GCD method

Input: Two Positive Integers: a, b.
Output: the GCD of a and b.

```
    int GCDMethod(int a, int b)
1:      while (a != b) {
2:          if (a > b)
3:              a = a - b;
4:          else
5:              b = b - a;
6:      }
7:      return b;
```

---

in PC adding with an extra constraint at statement 1. Every PC terminated with constraint at statement 7 is a successful program path. If the PC associated with a successful program path is satisfiable, it will generate a test case; on the other hand, if the PC is unsatisfiable, the program path is infeasible and does not lead to a test case. In our example, the program path $(1, 2, 3, 1, 2, 3, 1, 7)$ may generate test case $A = 15, B = 5$; again program path $(1, 2, 5, 1, 2, 3, 1, 7)$ may generate test case $A = 6, B = 9$. Table 1 shows the PCs and their solutions to corresponding program paths. As we generate more PCs, we will get more test inputs.

Table 1 Test data and path constraints of the GCD method

Path	PC	Program input
1,7	A == B	A = 5 B = 5
1,2,3,1,7	A != B &	A = 10 B = 5
	A > B &	
	A1 == A − B &	
	A1 == B	
1,2,5,1,7	A != B &	A = 5 B = 10
	A ≤ B &	
	B1 == B − A &	
	B1 == A	
1,2,3,1,2,3,1,7	A != B &	A = 15 B = 5
	A > B &	
	A1 == A − B &	
	A1 != B &	
	A1 > B &	
	A2 == A1 − B &	
	A2 == B	
1,2,5,1,2,3,1,7	A != B &	A = 6 B = 9
	A ≤ B &	
	B1 == B − A &	
	A != B1 &	
	A > B1 &	
	A1 == A − B1 &	
	A1 == B1	
...	...	...
...	...	...

3 Java White-Box Unit Tester (JWBUT)

As we proposed in Sects. 1 and 2, our new tool is going to use symbolic execution of source code in a systematic manner. The tool combines three techniques of source code analysis, grammar-based test generation, and constraint solving as an integrated whole and generates authentic test cases which lead to find possible bugs from given source codes in a fully automatic way, yet it handles two important performance metric the path explosion and the coverage issues properly and automatically. A generic abstraction of the proposed tool is given in Fig. 2.

We have developed Java White-box Unit Tester (JWBUT) on the proposed abstraction of grammar-based white-box testing. JWBUT involves three major stages of execution. They are given below:

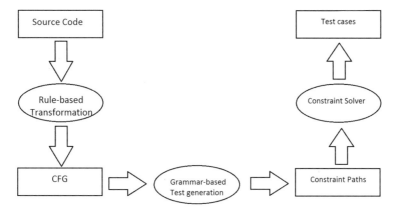

Fig. 2 Generic abstraction of the states and steps of JWBUT

Source Code ⇒ Context-free Grammar First, static analysis of a given source code
is performed to parse the program into a derivation tree, from which a control
flow graph will be extracted by rule-based transformation, and then a subsequent
conversion procedure is invoked to map the control flow graph to an equivalent
context-free grammar (CFG).

Context-free grammar ⇒ Constraint Paths Second, given a CFG, an automated
grammar-based test generator (GTG) is adopted to generate a set of well-
distributed path constraints, each of which contains a sequence of constraints
along an execution path.

Constraint Paths ⇒ Test Inputs Third, generating test inputs (or test cases) from constraint paths through symbolic execution using a constraint solver.

In next following sections, we are going to briefly describe each step.

3.1   Source Code⇒ Context-Free Grammar

Source Code ⇒ Token First, syntactic analysis of a given source code is performed
to transfer the program into tokens.

Token ⇒ Parse Tree Second, using the Java Backus–Naur Form (BNF), the stream
of tokens are transferred to an unambiguous well-defined parse tree.

Parse Tree ⇒ Control Flow Third, the parse tree is transferred to the control flow,
the possible generic execution paths of source code.

Control Flow ⇒ CFG Finally, the control flow is transferred to an unambiguous and
terminating CFG.

Table 2 Few random test inputs generated for the GCD method

No.	A	B	no.	A	B
1	4	7	6	14	5
2	9	4	7	5	5
3	5	4	8	8	11
4	5	3	9	11	7
5	7	2	10	4	9
...	...	...	...	...	...
...	...	...	...	...	...

3.2   Context-Free Grammar ⇒ Constraint Path

We have used a Java-based tool, Gena [21], for Grammar-based Test Generation
(GTG). The input of Gena is any consistent and terminating context-free grammar
(CFG) and using that CFG, it generates the constraint paths. Here, the constraint path
means any of the possible route to traverse from the beginning to the end of CFG,
expressed as the constraints on inputs.

Gena is well-developed automatic tool for generating constraint paths from CFG.
As it is automatic, it requires zero control inputs from user and generates well-
distributed constraint paths. It utilizes a novel dynamic stochastic model where each
variable (terms of rule in CFG) is associated with tuple of probability distribution
adjusted dynamically during the derivation. The adjustment is based on tabling strategy to keep track of recursion.

3.3   Constraint Path ⇒ Test Input

We create the test inputs from constraint paths compatible with Constraint Level
Programming over Finite Domain (clpfd) library of SWI-Prolog [22]. Library clpfd
has its own syntax and we create instruction level file accordingly. We should keep in
mind that the scope of clpfd is limited and hence, we are forced to bound to limited
scope. But the performance of JWBUT in this part can be significantly improved
using more powerful constraint solver mentioned in Sect. 1 (Table 2).

4   JWBUT—A Fully Automatic White-Box Unit Tester

Our tool, JWBUT, creates authentic test cases with zero external annotation from
the outside world in a fully automatic fashion. Test cases, generated by JWBUT,
are authentic because of mainly two reasons: each test case represents a particu-

lar program path from the start to the end of a program and test cases, associated with particular program paths, are getting more and more compound as the process explores more deeper and wider of a particular program. Moreover, JWBUT does not require any kind of manual labor from tester's side—a fully automatic tool. White-box testing can be judged using three merits with respect to the source code:

Statement Coverage—ensure every single line of code is tested.
Branch Coverage—ensure every branch (e.g., true or false) is tested.
Path Coverage—ensure all paths are tested.

4.1 Statement Coverage

Among all three merits of white-box testing, statement coverage is the easiest one to ensure. Invoking a handful of test cases, using JWBUT, can guarantee the whole statement coverage. Gena, an integral part of JWBUT, works with a doubling strategy— each time the specific branch has been chosen, its probability to choose next time getting half than the branches not being chosen. So complex program, even with a lot of branch statements, can be covered with the highest $nlogn$ invocation of test cases where n is the number of total statements.

4.2 Branch Coverage

Because of doubling strategy of Gena, all branches would be covered within maximum $nlogn$ invocation of test cases where n is the number of total statements. White-box strategies of narrowing the testing parts by using constraints [10–13], using grammar-bounded inputs [14, 15], other heuristic approaches [16, 17], or with explicit annotation controls [18, 19] would take much higher number of test cases invocation, yet there is no guarantee. For example, the lava tool [18] takes a seed, and the number of test cases invocation needed to full branch coverage depends on mercy of seed which is not usually balanced and is often random.

4.3 Path Coverage

Each test case, generated by JWBUT, is a complete program path from the start to the end of a program. Full code coverage of a program with loops depends on the number of times the loops are going to be executed. Things are getting worse with branch statements within loops. Hence, full path coverage of a moderate length program, having a few loops with branch statements is even quite absurd. JWBUT ensures to generate finite yet well-covered authentic test cases of the program.

5 Performance of JWBUT

Java White-box Unit Tester (JWBUT) is the tool we accomplished for white-box testing with some constraints imposed by the associated tools like SWI-Prolog, Gena, etc. The performance measurement of JWBUT performs on different kind of test suites collected from various sources. The suites, used for performance measurement of JWBUT, are tiny to moderately large snippets of real-world Java codes. These Java snippets are composed of different loops, conditions, and composition of both so that the different facets of JWBUT can be explored.

5.1 Results and Findings

We have evaluated performance of JWBUT on 14 suites. We know actually what these methods do and their outputs. The performance of JWBUT on these suites is satisfactory. For example, Fibonacci method is bug free and should create Fibonacci numbers; while passing in our tool, it creates Fibonacci numbers as test cases.

On the other hand, if GCD method has bugs; while passing in our tool, it creates bugs as test cases. Hence, a tester can check the test cases, the tester has the details of constraint paths, and corresponding test inputs created. Using these files, the tester can readily find out where the code broken and associated faults in code.

The detail results of all the suites are given in Table 3.

Table 3 Number of test cases invoked and created for the test suites

Test suite	Invoke	Constraint path	Test sase
1. Square	50	50	50
2. Cubic	50	50	50
3. Sum	50	50	50
4. Factorial	50	50	50
5. Factorial1	50	50	50
6. Fibonacci	100	100	100
7. Fibonacci1	100	100	100
8. Prime	500	500	9
9. Prime1	500	500	9
10. Prime2	500	500	9
11. LCM	1000	1000	50
12. GCD	1000	1000	16
13. Divisor	1000	1000	11
14. PrimeorDivisor	1000	1000	14

6 Conclusion

Our preliminary implementation of JWBUT has three major steps: rule-based transformation, grammar-based test generation, and constraint solver. To accomplish each step, we have six major modules: preprocessor, control flow generator, grammar generator, constraint path generator, postprocessor, and constraint solver. First four modules are written in declarative logic programming style; next module is written in Java because of the associative grammar-based test generator (GTG), Gena, finally, we have used constraint solver facility of SWI-Prolog. We believe JWBUT can produce well-distributed and authentic test cases for wide variety of Java methods in fully automatic fashion. It alleviates path explosion and coverage issues in a greater extent. But we would like to address two major bottlenecks of our tool. One is related with constraint path generation and another one is related with constraint solving.

First, the performance of constraint path generation is limited by the tool, Gena. Overall performance of JWBUT is highly dependent on Gena. Gena is written in Java using a double strategy. It creates balanced, well-distributed and homogeneous constraint paths. But this nature creates a big problem on generating number of test cases. For example, if there is a conditional statement within a loop, not all constraint paths created by JWBUT might be *satisfiable*. In worst case, for initial round, it might create one test case out of two paths; for second round, it might create one test case out of four paths; for third round, it might create one test case out of eight paths; and this way, it might create only thirteen test cases out of one thousand paths. Hence, Gena can be easily blown away before it creates enough numbers of test cases. This is one of the major problems using Gena and also for our tool. There is always a solution to this problem. If we can assign a probability or bias to particular condition more than its negation, the number of test cases will increase rapidly. This is a vital important future work we can do for further development of our tool. Subsequent version of JWBUT, Gena would be replaced by Aroj (a proposed tool by author) which will use the biasing capability of the conditional statements.

Second, we generate test cases using instruction level file generated by our postprocessor to constraint solver. We have used Constraint Level Programming over Finite Domain (clpfd) library of SWI-Prolog. The merit of test cases depends on how well the constraint solver is and how well we have tuned the constraint solver. The better way we can use constraint solver, the authenticity of test cases increases and the possibility of finding bugs also increases rapidly. Sometimes it happens a constraint path is *satisfiable*, but constraint solver is not well tuned so that we do not get test case. A complex constraint is always another issue to get nontrivial test cases. A well-chosen constraint solver with nicely tuned is the best way to alleviate these problems.

Acknowledgements I am really thankful and grateful to Dr. Hai-Feng Guo (University of Nebraska—Omaha), Dr. Qiuming Zhu (University of Nebraska—Omaha), and Dr. Yaoqing Yang (University of Nebraska—Lincoln) for their kind help.

References

1. Hetzel, C.W.: The Complete Guide to Software Testing. QED Information Sciences, vol. 2. Wellesley, MA (1988)
2. Ould, M.: Testing—a challenge to method and tool developers. Softw. Eng. J. **6**(2) (1991)
3. Anand, S., Harrold, M.J.: An orchestrated survey on automated software test case generation: test data generation by symbolic execution. J. Syst. Softw. (2013)
4. Patton, R.: Software Testing, vol. 2. ACM (2005)
5. Pan, J., Koopman, P., and Siewiorek, D.: A dimensionality model approach to testing and improving software robustness. In: Proceedings of AUTOTESTCON (1999)
6. Parrington, N., Roper, M.: Understanding Software Testing. Willey (1989)
7. de Moura, L.M., Bjørner, N.: Z3: an efficient SMT solver. In: Proceedings of the 14th International Conference on Tools and Algorithms for the Construction and Analysis of Systems (TACAS'08), pp. 337–340 (2008)
8. Dutertre, B., de Moura, L.M.: A fast linear-arithmetic solver for DPLL(T). In: Proceedings of the 18th International Conference on Computer Aided Verification (CAV'06), pp. 81–94 (2006)
9. Ganesh, V., Dill, D.L.: A decision procedure for bit-vectors and arrays. In: Proceedings of the 19th International Conference on Computer Aided Verification (CAV'07), pp. 519–531 (2007)
10. Anand, S., Godefroid, P., Tillmann, N.: Demand-driven compositional symbolic execution. In: Proceedings of the 14th International Conference on Tools and Algorithms for the Construction and Analysis of Systems, pp. 367–381. Springer (2008)
11. Godefroid, P.: Compositional dynamic test generation. In: Proceedings of the 34th Annual ACM SIGPLAN-SIGACT Symposium on Principles of Programming Languages (POPL'07), pp. 47–54 (2007)
12. Bjørner, N., Tillmann, N., Voronkov, A.: Path feasibility analysis for string-manipulating programs. In: Proceedings of the International Conference on Tools and Algorithms for the Construction and Analysis of Systems, pp. 307–321 (2009)
13. Veanes, M., Bjørner, N.: Alternating simulation and IOCO. In: Petrenko, A., da Silva Simão, A., Maldonado, J.C. (eds.), Proceedings of the 22nd IFIP WG 6.1 International Conference on Testing Software and Systems (ICTSS'10), pp. 47–62. Springer (2010)
14. Godefroid, P., Kiezun, A., Levin, M.Y.: Grammar-Based whitebox fuzzing. In: Proceedings of the ACM SIGPLAN 2008 Conference on Programming Language Design and Implementation (PLDI'08), pp. 206–215 (2008)
15. Majumdar, R., Xu, R.G.: Reducing test inputs using information partitions. In: Proceedings of the 21st International Conference on Computer Aided Verification (CAV'09), pp. 555–569 (2009)
16. Anand, S., Pasareanu, C.S., Visser, W.: Symbolic execution with abstraction. Int. J. Softw. Tools Technol. Transf. **11**, 53–67 (2009)
17. Chipounov, V., Kuznetsov, V., Candea, G.: S2e: a platform for in-vivo multi-path analysis of software systems. In: Proceedings of the 16th International Conference on Architectural Support for Programming Languages and Operating Systems (ASPLOS'11), pp. 265–278 (2011)
18. Sirer, E.G., Bershad, B.N.: Using poduction Grammars in software testing. In: Proceedings of the 2nd Conference on Domain-Specific Languages, pp. 1–13. ACM (1999)
19. Hoffman, D.M., Ly-Gagnon, D., Strooper, P., Wang, H.Y.: Grammar-based test generation with YouGen. Softw. Pract. Exp. **41**(4), 427–447 (2011)
20. King, J.C.: A new approach to program testing. Programm. Methodol. LNCS **23**, 278–290 (1975)
21. Guo, H., Qiu, Z.: A dynamic stochastic model for automatic grammar based test generation. Softw. Pract. Exp. (2014)
22. SWI-Prolog. http://www.swi-prolog.org
23. Hopcroft, J.R., Motwani. R., Ullman, J.D.: Introduction to Automata Theory, Languages, and Computation. Addison Wesley (2001)

24. Sen, K., Marinov, D., Agha, G.: Cute: a concolic unit testing engine for C. In: Proceedings of FSE (2005)
25. Godefroid, P., Klarlund, N., Sen, K.: DART: directed automated random testing. ACM Sigplan Notices, vol. 40(6). ACM (2005)
26. Majumdar, R., Xu, R.: Directed test generation using symbolic Grammar. In: Proceedings of ASE. ACM (2007)
27. Csallaner, C., Samaragdakis, Y.: JCrasher: an automatic robustness tester for Java. Softw. Pract. Exp. (2004)
28. Fraser, G., Arcuri, A.: Whole test suite generation. IEEE Trans. Softw. Eng. **39**(2) (2013)
29. Zhang, S., Saff, D., Bu, Y., Ernst, M.D.: Combined static and dynamic automated test generation. In: Proceedings of ISSTA. ACM (2011)
30. Kugler, H., Harel, D., Pnueli, A., Lu, Y., Bontemps, Y.: A dimentionality model approach to testing and improving software robustness. In: Proceedings of AUTOTESTCON (1999)
31. AT&T Research: Graphviz. http://www.graphviz.org

Music Playing and Wallpaper Changing System Based on Emotion from Facial Expression

Protik Hosen, Nuruzzaman Himel, Md. Asaduzzaman Adil, Ms. Nazmun Nessa Moon and Ms. Fernaz Narin Nur

Abstract Our faces contain a huge amount of information and it plays an important role in finding out of an individual's behavior. Facial expression analysis is the most effective way to detect human emotion. In this paper, we propose an approach for playing music and change the wallpaper for personal computer based on recognizing emotions through facial expressions displayed in live video streams. Facial expression recognition refers to the classification of facial features such as happiness, sadness, fear, disgust, surprise, and anger. To extract this type of feature we detect faces in front of a webcam we build own dataset according to facial expression and corresponding Haar classifiers are used. Then we make a playlist based on emotion and user interest and also build a model of the changing wallpaper in such a way that it will play songs fitting to our mood and changing our desktop wallpaper to happy images if we are sad. If we are angry it may change it to calming images.

1 Introduction

Face is one of another medium for expressing the feelings of a human mind. Generally, we express our emotions by voice but these emotions mainly bloom on our face. Different emotions lead different facial expressions like anger, joy, surprise,

P. Hosen · N. Himel · Md. A. Adil (✉) · Ms. N. N. Moon · Ms. Fernaz Narin Nur
Computer Science and Engineering, Daffodil International University (DIU), 02 Mirpur Rd, Dhanmondi, Dhaka 1207, Bangladesh
e-mail: adil3214@diu.edu.bd; Zamanadil1995@gmail.com

P. Hosen
e-mail: protik3380@diu.edu.bd

N. Himel
e-mail: himel3375@diu.edu.bd

Ms. N. N. Moon
e-mail: moon@daffodilvarsity.edu.bd

Ms. Fernaz Narin Nur
e-mail: narin@daffodilvarsity.edu.bd

© Springer Nature Singapore Pte Ltd. 2019 79
A. Abraham et al. (eds.), *Emerging Technologies in Data Mining and Information Security*, Advances in Intelligent Systems and Computing 755,
https://doi.org/10.1007/978-981-13-1951-8_8

sad, and excitement, etc. If we could recognize these facial expressions we easily detect on which mood we are. According to this theme, we try to build a system which automatically plays music and change wallpaper based on our mood.

Music is a kind of audible art which is able to create entertainment in the minds of the people with consistent sound and is called a form of refreshment in our mechanical life. Generally, if we want to listen music we have to search or think what kind of music our mind want to listen, but in different situation this process is very annoying, so the main objective is to design a music player which easily generate a playlist of songs and change wallpaper based on current emotional state.

2 Literature Survey

There have been several methods of face detection and recognition and audio features from an audio signal. But there have very few methods that can make playlist easily based on facial expression. Some of which try to describe below.

Brain Computer Interface (BCI) proposed [1] mobile multimedia controller. It consists of EEG hardware which monitoring the cognitive state of mind. This system requires the active manpower which can continuously to control multimedia. BCI systems require very costly EEG machines that are not available and in this machine, there is a limited number of feasibility. This system contains a cognitive detection algorithm which built in the mobiles that continuously monitor the EEG signal from EEG machines and it can recognize the user's mind. But our system can reduce such kinds of drawbacks. Any hardware doesn't need in our system because our system is software base that's why it is not costly. Our method can reduce manpower and gives us better performance.

Voice/speech for emotion recognition [2] is the environment in which system is set up. If the system is set up in an open field then surrounding noise of the environment will create an effect on the system's performance. A voice activity detection algorithm has been attached to this system to reduce the noise. It detects the speech that comes from user's input signal. Using user's input signal it can recognize emotion rate. Emotion recognition rate in this system is not good. But our system can recognize the emotion correctly and can play music based on emotion.

Electroencephalography (EEG) signals [3] to detect human emotions. EEG records the activity of the neurons in our brain. EEG signals mainly detect emotion from our mind and ignoring all others external characteristics like facial expressions. There have two types of classifiers such as Support Vector Machine (SVM) and Linear Discriminant Analysis (LDA). EEG signal work with seven emotion.

The problem with this system is that it is very complex and computational time is very high. On the other hand our system so much comfortable and user- friendly. Our system can also reduce this drawback which gives us better performance.

To extract facial features and audio features from an audio signal numerous approaches have been built but for emotion based music playlist are in less number and this system are very complex and memory requirement high for using the sensor, which gives a lesser accuracy in generation of a playlist.

Our system resolves the drawbacks by avoiding the employment of any additional hardware, design an automated emotion-based music player for the generation of customized playlist which gives high efficiency.

A very useful two-dimensional (Stress v/s energy) model plotted on two axes with emotions which is depicted by a two-dimensional co-ordinate system and that is lying on either two axes or the four quadrants formed by the two-dimensional plot is proposed by Thayer [4].

Jung Hyun Kim [5] tested and analyzed tags of music mood and A-V values from a total 20 subjects, the analysis results classified the A-V plane into eight regions using k-means clustering algorithm.

Londhe et al. [6] proposed an approach which analyzes extracted facial expression features based on an accurate and efficient statistical. Which mainly focused on the changes in curvatures on the face.

The author used Artificial Neural Networks (ANN) for extracting features into six major universal emotions like anger, disgust, fear, happy, sad, and surprise.

The author [7] proposed correlation method which knows as the nearest neighbor method. Which returns the nearest score between two images base at an angle. Two types of images are used in this method one is training images and another test images which are converted into column vectors.

The author used PCA algorithm [8, 9] for facial expression and to classify this expression for an individual face from [10, 11]. Using this facial expression we can play a song from the music playlist.

Most of the facial expressions depend on size and shape of the facial parts such as eyes, lip, mouth, eyebrows, nose, etc. Geometric and Appearance Based Methods [12, 16] also follow in this process. This method also uses characteristics point on the face for classifying the expressions some shape models [13, 14]. Differ between two facial landmarks makes the system reliable.

3 Proposed System

Our proposed system framework is as shown in Fig. 1. This system is based on face recognition algorithm. When a person starts using his/her personal computer and if our system process is running then Detecting his/her face on the webcam then the face image is captured and pre-processed for further processing. The system takes some images of user's image and updates the emotion model over time. The user must be given facial expression according to particular emotions to create the model. As not more than one person can enter in the system if more than one person in front of webcam then the system can find emotion but the main problem is the system can't play songs for multiple emotions. The user must build another model playlist of songs and wallpaper model and trained all the model.

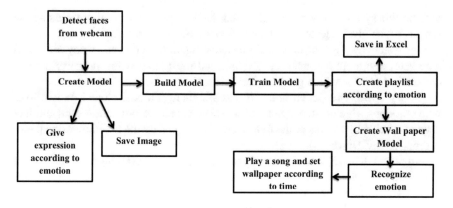

Fig. 1 Proposed diagram model for emotion awareness music player and changing wallpaper based on emotion

The system detects emotion on user's face then it will play songs fitting to user's mood and changing the desktop wallpaper. In term of changing the wallpaper, it takes a value from the user that indicates the wallpaper changing time interval.

3.1 Detecting User's Face on the Webcam and Processing the Face for Building Emotion Dataset

Face detection is the most important part of this project. For detecting human face Haar classifier cascades first be trained. To train this classifier, PCA algorithm and Haar feature must be implemented. We will use trained HAAR cascade classifier with openCV.

We can get it from OpenCV [15] library called "haarcascade_frontalface_default.xml". This XML produce the best result from testing. It can detect faces in a certain condition such as facing camera directly. It gives better accuracy of the recognizer and requires less training data.

Using this classifier, face is detected from the webcam stream and crop the face and convert the image to grayscale to improve detection speed and accuracy. Figure 2 shows the procedure of face detection and processing the face image.

Figure 3 shows the demonstration of detect face and crop the face.

Using this principle, we build emotion dataset and detect particular emotion. Figure 4 shows the block diagram of creating emotion dataset.

To create dataset user must give facial expression in webcam streaming according to the particular emotion. Then train the model and system will generate a trained XML file. Figure 5 shows this procedure.

Fig. 2 Block diagram of
face detection and
processing face image

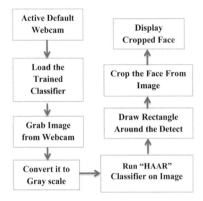

Fig. 3 Face detect and crop

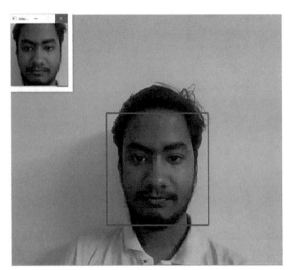

3.2 Create Playlist and Emotion Links

Organize particular emotion based songs to play create a playlist in. m3u file format
and set the emotion based playlist on an excel sheet. Figure 6 shows the playlist. m3u
file format.

Figure 7 shows the emotions links file on an excel sheet.

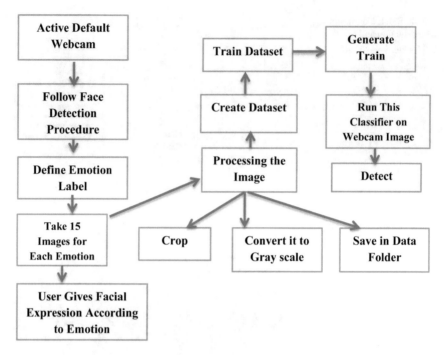

Fig. 4 Block diagram of creating emotion dataset

3.3 Emotion Extraction Module

Using emotion model actual emotion on user's face can be detected. The model trained for a single person so this work much better when it used to that same person. Emotion can be detected using facial features comparison.

3.4 Select Proper Music and Wallpaper Based on Detected Emotion

When the system detects the emotion then it picks a proper playlist for the user and executes the playlist. In order to change the wallpaper, it adds timer function that indicates changing wallpaper time interval.

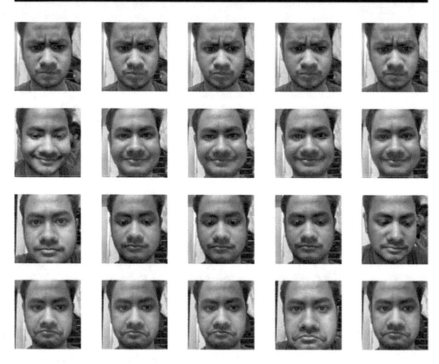

Fig. 5 Emotion dataset

Fig. 6 Playlist file

```
#EXTM3U
F:\music\1.mp3
F:\music\2.mp3
F:\music\3.mp3
F:\music\4.mp3
F:\music\5.mp3
F:\music\6.mp3
F:\music\7.mp3
F:\music\music.mp3
```

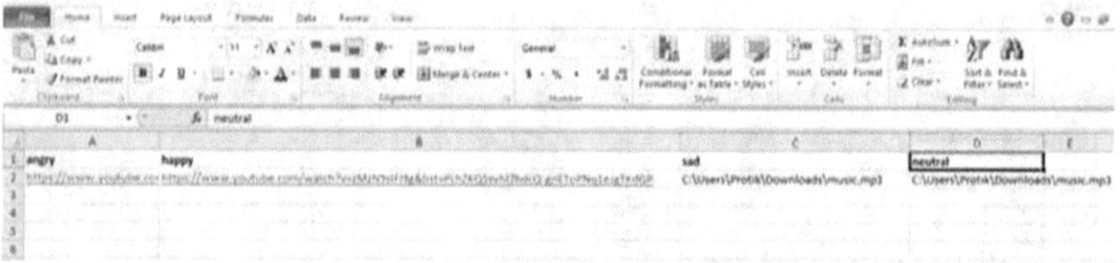

Fig. 7 Excel sheet

4 Conclusion and Future Plan

This system is a great opportunity for music listeners which reduce the searching time and unnecessary computational time, with better accuracy and efficiency. Our system can work even in poor light condition and low camera resolution. It also acts as the therapy system when our mood is in adversely situation. Our system is now only working on windows-based operating system. The future plan of our system is to work on different platform.

References

1. Tseng, K.C., Wang, Y.-T., Lin, B.-S., Hsieh, P.H.: Brain computer interface-based multimedia. In: 8th International Conference on Intelligent Information Hiding and Multimedia Signal Processing, Piraeus,pp. 277–280 (2012)
2. Chew, L.W., Seng, K.P., Ang, L.-M., Ramakonar, V., Gnanasegaran, A.: Audio-Emotion recognition system using parallel classifiers and audio feature analyzer. In: 2011 3rd International Conference on Computational Intelligence, Modelling Simulation, pp. 210–215 (2011)
3. Bhardwaj, A., Gupta, A., Jain, P., Rani, A., Yadav, J.: Classification of human emotions from EEG signals using SVM and LDA classifiers. In: 2015 2nd International Conference on Signal Processing and Integrated Networks (SPIN), pp. 180–185 (2015)
4. Thayer: The biopsychology of mood & arousal. Oxford University Press (1989)
5. Wong, J.-J., Cho, S.-Y.: Facial emotion recognition by adaptive processing of tree structures
6. Londhe, R.R., Pawar, D.V.P.: Analysis of facial expression and recognition based on statistical approach. Int. J. Soft Comput. Eng. (IJSCE) 2 (2012)

7. Larkin, T.K.: Face Recognition using Kernel. New Jersey Institute of Technology, Newark, NJ (2003)
8. Chakrabartia, D., Duttab, D.: Facial expression recognition using Eigenspaces. In: International Conference on Computational Intelligence Modeling Techniques and Applications, pp. 755–761 (2013)
9. Çarikçi, M., Özen, F.: A face recognition system based on eigenfaces method. In: INSODE, pp. 118–123 (2011)
10. Kabani, H., Khan, S., Khan, O., Tadvi, S.: Emotion based music player. Int. J. of Eng. Res. Gen. Sci. 3(1), 750–756 (2015)
11. Zaware, N., Rajgure, T., Bhadang, A., Sapkal, D.D.: Emotion based music player. Int. J. Innov. Res. & Dev. 3(3), 182–186 (2014)
12. Happy, S.L., Routray, A.: Automatic facial expression recognition using features of salient facial patches. IEEE Trans. Affect. Comput. 1–12 (2015)
13. Pantic, M., Patras, I.: Dynamics of facial expression: recognition of facial actions and their temporal segments from face profile image sequences. IEEE Trans. Syst. Man Cybern. 36(2), 433–449 (2006)
14. Pantic, M., Rothkrantz, L.J.M.: Facial action recognition for facial expression analysis from static face images. IEEE Trans. Syst. Man Cybern. 34(3), 1449–1461 (2004)
15. https://docs.opencv.org/2.4/doc/tutorials/objdetect/cascade_classifier/cascade_classifier.html
16. Youssif, A.A.A., Asker, W.A.A.: Automatic facial expression recognition system based on geometric and appearance features. Comput. Inform. Sci. 4(2), 115–124 (2011)

Gender Recognition Inclusive with Transgender from Speech Classification

Ghazaala Yasmin, Omkar Mullick, Arijit Ghosal and Asit K. Das

Abstract Automatic gender classification system has prompted a pertinent of increasing amount of applications, particularly the rise of social platforms and criminal investigation. Focus of substantial past researches was limited towards the discrimination of male and female gender only. Recently transgender has achieved legal recognition. So, any gender classification system should consider this third gender also. But unfortunately there is a lack of good gender classification system which can discriminate all the three types of gender well. This proposed work uses judiciously chosen acoustic features for classification of three classes of genders from their solo voice. The proposed system has been pursued with the sampled audio data extracted from audio signal. From the sampled data, acoustic features like tempo, pitch and spectral flux have been extracted using the idea of pattern recognition. The extracted feature set has been served for classification to predict the gender of a given unknown voice.

Keywords Pitch · Tempo · Spectral flux · Speech recognition · Gender classification

G. Yasmin (✉)
Department of Computer Science & Engineering, St. Thomas' College
of Engineering and Technology, Kolkata, West Bengal, India
e-mail: me.ghazaalayasmin@gmail.com

O. Mullick
Department of Electronics and Communication Engineering, St. Thomas' College
of Engineering and Technology, Kolkata, West Bengal, India
e-mail: omkarmullick@gmail.com

A. Ghosal
Department of Information Technology, St. Thomas' College of Engineering
and Technology, Kolkata, West Bengal, India
e-mail: ghosal.arijit@yahoo.com

A. K. Das
Department of Computer Science and Technology, Indian Institute
of Engineering Science and Technology, Shibpur, Howrah, India
e-mail: akdas@cs.iiests.ac.in

© Springer Nature Singapore Pte Ltd. 2019   89
A. Abraham et al. (eds.), *Emerging Technologies in Data Mining and Information Security*, Advances in Intelligent Systems and Computing 755,
https://doi.org/10.1007/978-981-13-1951-8_9

1 Introduction

Highlighting the statement in Webster's dictionary, "Speech" is the "communication or expression of thoughts in spoken words" [1]. Speech signal acts not only as a communication interface among people but also carries the information regarding the speaker. Any linguistic and non-linguistic characteristics of a speaker can be a measurable parameter to discriminate gender. Audio signal carries features based on three major domains—frequency, time, and perceptual domain. The features, computed based on these three domains, are the effective measures for identifying whether a speaker is male, female, or transgender. With the current concern of world-wide research and development, gender classification has received a great deal of attention among speech researchers. Gender identification has gained its importance because of legal inclusion of third type of gender, i.e., transgender apart from male and female. The objective of the proposed work is to introduce an efficient scheme to classify male, female as well as transgender through their speech.

1.1 Literature Survey

There are many research works performed by the researchers for gender classification. Past approaches for estimate gender classification has shown convincing results. Ali et al. [1] introduced a methodology for gender classification from speech by extracting feature using First Fourier Transform (FFT). Alías, Socoró and Sevillano [2] proposed perceptual features extracted from music, speech, and environmental sound. Subramanian, Rao and Roy [3] performed classification of different audio sound by extracting features like pitch, timbre and rhythmic. Both the works presented in [4, 5] proposed the scheme for speech detection by extracting perceptual features. Harb and Chen [6] classified gender using Mel Frequency Cepstral Coefficients (MFCC). Ghosal and Dutta [7] invoked a combination of frequency domain feature and time domain feature named as zero-crossing rate (ZCR) and spectral flux for the identification for male and female voice. Pahwa and Aggarwal [8] extracted mel co-efficient feature for gender recognition using Support Vector Machine (SVM). Kumar et al. [9] introduced multichannel gender classification from speech in movie audio. Taskeed et al. [10] suggested an efficient approach of classification using local directional pattern (LDP). Lartillot et al. [11] presented MIRtoolbox, which contains the efficient features related to time, frequency, and perceptual domain. This novel scheme was further explored based on these domain features by Müller and Ewert [12]. Grosche and Müller [13] revealed the predominant local pulse from music recording. Srivastava [14] proposed Weka-Tool for different pattern recognition purposes including classification. Malhi and Gao [15] proposed Principal Component Analysis (PCA) based feature selection scheme. Grosche and Müller [16] introduced a MATLAB toolbox for extracting and analyzing tempo-based features.

1.2 Summary

The proposed work is organized as follows: Proposed methodology is described in Sect. 2. Some preliminary notions which are used for feature extraction are explained in this section. Section 3 depicts experimental results and performance analysis and finally, the conclusion is made in Sect. 4.

2 Proposed Methodology

Speech can be categorized into different groups based on the genders of the speakers. These categories have their individual discriminative nature in frequency and perceptual domain. Moreover, transgender acquires its own distinctive factors which are clearly reflected in these domains. While observing the characteristics of male, female, and transgender speech, it is noticed that these three genders mainly differs in perceptual and frequency domain. The proposed work aims to extract the attributes based on perceptual features such as tempo and pitch with a combination of statistical calculations like mean, standard deviation for highlighting these features to get a promising result. From the past work it is learnt that spectral flux is a good frequency domain feature for male and female gender classification. Inspired from this, spectral flux based features are also employed in this work. Figure 1 depicts the preview of the proposed work.

2.1 Feature Extraction

The prior attempts have highlighted the utility of temporal features which includes zero-crossing rate, linear prediction co-efficient, spectral shape-based features like fast Fourier transform (FFT) and MFCCs. These features have been loaded with the combination of perceptual features like loudness or sharpness. To make the result more precise, the presented work is incorporating a fusion of perceptual features with a frequency domain feature called spectral flux.

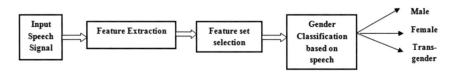

Fig. 1 Schematic diagram for gender recognition system

2.1.1 Pitch-Based Features

Pitch is presumed to be the perceived form of frequency. While noticing the perceptual differences among the talking of human genders, pitch is found as an essential feature as it varies widely for male, female, and transgender. Pitch is the measure of shrillness of speech. Being a numerical figure, absolute quantity pitch is a subjective measure for persons. In this work speech signals for male, female, and transgender have been decomposed into 108 frequency bands. But, there are total 88 effective frequency bands as all of the first 20 frequency bands are zeros. Each of these 88 bands are divided into frames of short duration. For each of these frequency bands, Short Time Mean Square Power (STMSP) is calculated. Therefore, mean STMSP has been calculated which leads generation of 88 pitch-based features as there are total 88 effective frequency bands.

Figure 2a–c depict the STMSP distribution spread through the 88 frequency bands for male, female and transgender respectively. To understand the differences of STMSP distribution clearly, only mean value of STMSP for frequency bands are considered. From these figures, it can be clearly understood that the STMSP distribution for male, female and transgender varies widely. The Music Information Retrieval (MIR) Toolbox for Matlab [9] created in the University of Jyvaskyla, Finland has been used for pulling out feature extraction. The chromagram toolbox [15] has also been used here for extracting pitch-based features.

2.1.2 Tempo-Based Feature

The rhythm of speech while speaking bears an important perceptual description. Tempo signifies how fast the speech is being delivered. It is observed that fastness of delivering speech varies for male, female, and transgender.

To highlight this difference among the speeches of male, female, and transgender, tempo is also considered along with pitch-based features. To generate the tempo-based features, novelty curve has to be computed from the speech signal. Novelty curve acts as an onset detection variance. In order to extract the features, this novelty curve has been split into non-overlapping tempo windows. Furthermore, the absolute values of first 30 fourier coefficients have been calculated from each window. For each of these 30 coefficients mean and standard deviation has been considered. Consequently a 60 dimensional tempo feature was generated.

The tempogram ("tempogram is a representation of tempo in the form of time vs beats per minute plot"—it is required to be specified) plot for male, female, and transgender are depicted in Fig. 3(a), (b), and (c) respectively. Tempo is denoted in terms of beats per minute (BPM). The diagrams represents that features based on tempo are able to capture the rhythmic aspect well and it contributes in discriminating the genders. To obtain tempo-based features tempogram toolbox [16] has been used.

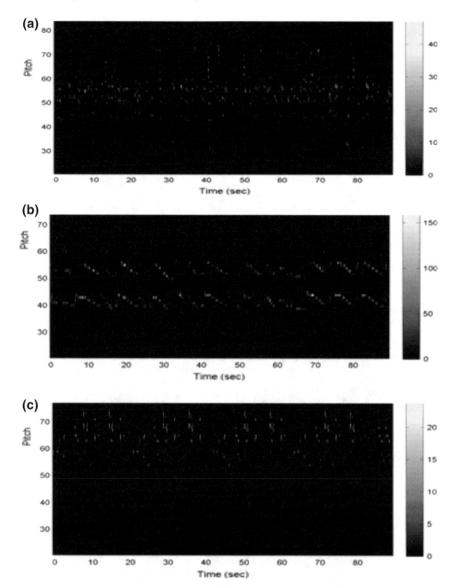

Fig. 2 a, b, c STMSP distribution for male, female and transgender speech respectively

2.1.3 Spectral Flux Based Feature

In frequency domain for every spectrum, the difference between current spectrum and the spectrum preceding to it is calculated. For every spectrum of the sample window, Spectral flux measures how rapidly the power spectrum of a signal differs.

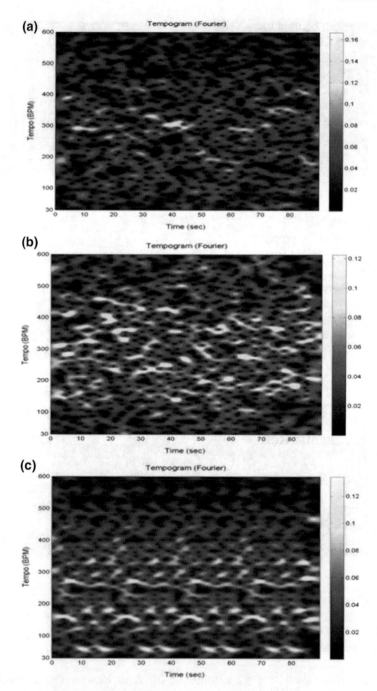

Fig. 3 **a**, **b**, **c** Tempogram plot for male, female and transgender speech respectively

This feature can be used to assess the timbre of a certain audio signal. Let, $SF(j)$ represents the spectral flux of jth spectrum. $S(j, i)$ corresponds to the value of the ith bin for the jth spectrum and $S(j-1, i)$ is equivalent for the spectrum before j. The value of each bin of the $(j-1)$th spectrum is subtracted from the values of the jth spectrum for each bin and then sum of the computed differences is taken as the flux of jth spectrum using Eq. (1). For n spectrum, the mean and standard deviation has been considered.

$$SF(j) = \sum_{i=0}^{n-1} s(j, i) - s(j - 1, i) \tag{1}$$

From spectral flux two sets of feature values has been calculated—mean and standard deviation.

Henceforth in the proposed method, 150 values of features (88 pitch-based features, 60 tempo-based features and 2 spectral flux based features) as a whole has been used.

In the domain of pattern recognition, a system is said to be more efficient if the feature set is of small dimension. This fact has motivated to impose a feature selection technique for the proposed system. For this purpose, feature selection algorithm called principal component analysis (PCA) [15] has been carried out. This algorithm evaluates significance of the feature components by taking number of Eigen-vectors. In the proposed system, 150 features have been carried out for PCA and the feature set is reduced to a 35-dimensional feature vector.

3 Experimental Results

For classification, some standard well-known supervised classifiers of different types have been employed. The goal of these classifiers is to build a model by taking known set of input data and known set of classes. The mentioned classifiers in Table 1 has been taken into account with 10-folds cross validation.

Table 1 Accuracy (%) of male, female and transgender speech discrimination

Classifer type	Classific scheme	Male	Female	Transgender
Bayes'	Naïve bayes	94.6	93.6	92.0
Function	Neural network	89.7	88.1	87.3
Tree	Random forest	93.3	92.1	90.4
Lazy	Adabooster	94.5	93.9	90.2
Meta	MCC (Multiclass Classifier)	91.7	91.2	89.8

Table 2 Comparison of performance in terms of classification accuracy

Precedent approach	Male (%)	Female (%)	Transgender (%)
Pahwa and Aggarwal (Considering MFC1)	89.8	78.5	85.1
Pahwa and Aggarwala (2016) (Ignoring MFC1)	85.3	88.3	84.8
Ali, Islam and Hossain	98.7	74.5	73.1
Proposed work	93.4	93	91

The proposed experiment has been carried out with a collection of 150 speech files, 50 for each of male, female, and transgender. The duration of each of the speech file is 90 s. The data set has been collected from various websites through Internet. The samples are of type mono quantized with 16-bit. For classification, weka-tool [14] have used. The dataset has been tested by using well known classifiers and the result has been tabulated in Table 1.

The strength of discriminating power of the proposed feature set has been compared with some past research works. The feature set proposed by Pahwa and Aggarwal in paper [8] and that obtained by Ali et al. in paper [1] has been applied on the present dataset. The comparative result has been reflected in Table 2.

The comparative analysis reveals that the current proposed feature set outperforms those two works. Here, the result has been tabulated by performing the average of the classification accuracy achieved for individual classifier mentioned in Table 1. The same calculation has been done for the proposed work.

3.1 Test Case Study on Ambiguous Audio Data

Apart from 150 sample audio data, the proposed idea has been applied on some of the audio data collected by recording the voice of common personality, where the identification of the gender is even difficult by people too. The methodology has been tested for the male voices, which are sounds similar to female and vice versa. Five such speakers' real data has been taken into consideration and the result has been summarized in Table 3. M, F, and T are symbolized for male and female and Transgender respectively. The score value represents 0/1, where 1 stands for correct identification and zero denotes incorrect identification. From Table 3, it is observed that all five speakers are recognized correctly, which demonstrates the effectiveness of the proposed feature extraction technique.

Table 3 Result score by testing the proposed method on ambiguous data

Precedent approach	Result	Score
Speaker 1 (M)	M	1
Speaker 2 (M)	M	1
Speaker 3 (F)	F	1
Speaker 4 (F)	F	1
Speaker 5 (T)	T	1

4 Conclusion

The prime goal of the proffered scheme is to develop a robust system for gender recognition. The scheme has made an attempt to explore the area of gender classification through speech by considering another class of gender, i.e., transgender. From the study of preliminary research, it has been observed that perceptual and frequency domain features are distinctly occupying the strength of differentiating the audio files. The recommended approach has been come up with an encouraging outcome by importing the combination of these two domains. However, it is impartial to notify that the proffered system may explore to be dealt with many other problems and requirements such as sub division among male speech identification with respect to age and same as for female and transgender. In future, this is expectant that the trial for the mentioned problem would come up with an encouraging result by applying it on some benchmark data. The presented procedure is likely to be explored in a hierarchical way for recognizing speaker. The speaker will be recognized by discriminating their gender and age hierarchically.

Acknowledgements "This chapter does not contain any studies with human participants or animals performed by any of the authors."

References

1. Ali, Md.S., Islam, Md.S., Hossain, Md.A.: Gender recognition system using speech signal. Int. J. Comput. Sci. Eng. Inf. Technol. (IJCSEIT) **2.1**, 1–9 (2012)
2. Alías, F., Socoró, J. C., Sevillano, X.: A review of physical and perceptual feature extraction techniques for speech, music and environmental sounds. Appl. Sci. **6.5**, 143 (2016)
3. Subramanian, H., Rao, P., Roy, S.D.: Audio signal classification. In: EE Dept, IIT Bombay, pp. 1–5 (2004)
4. Bach, J.H., Anemüller, J., Kollmeier, B.: Robust speech detection in real acoustic backgrounds with perceptually motivated features. Speech Commun. **53**(5), 690–706 (2011)
5. Richard, G., Sundaram, S., Narayanan, S.: An overview on perceptually motivated audio indexing and classification. Proc. IEEE **101**(9), 1939–1954 (2013)
6. Harb, H., Chen, L.: Gender identification using a general audio classifier. In: Proceedings of 2003 International Conference on Multimedia and Expo, 2003. ICME'03, vol. 2, pp. II–733. IEEE (2003)

7. Ghosal, A., Dutta S.: Automatic male-female voice discrimination. In: Issues and Challenges in Intelligent Computing Techniques (ICICT), pp. 731–735. IEEE (2014)
8. Pahwa, A., Aggarwal, G.: Speech feature extraction for gender recognition. Int. J. Image Graph. Signal Process. **8**(9), 17–25 (2016)
9. Kumar, N., et al.: Robust multichannel gender classification from speech in movie audio. Interspeech **2016**, 2233–2237 (2016)
10. Jabid, T., Kabir, Md.H., Chae, O.: Gender classification using local directional pattern (LDP). In: 2010 20th International Conference on Pattern Recognition (ICPR), pp. 2162–2165. IEEE (2010)
11. Lartillot, O., Toiviainen, P., Eerola, T.: A matlab toolbox for music information retrieval. In: Data Analysis, Machine Learning and Applications, pp. 261–268. Springer, Berlin, Heidelberg (2008)
12. Müller, M., Ewert, S.: Chroma Toolbox: MATLAB implementations for extracting variants of chroma-based audio features. In: Proceedings of the 12th International Conference on Music Information Retrieval (ISMIR) (2012)
13. Grosche, P., Müller, M.: Extracting predominant local pulse information from music recordings. IEEE Trans. Audio Speech Lang. Process. **19**(6), 1688–1701 (2011)
14. Srivastava, S.: Weka: a tool for data preprocessing, classification, ensemble, clustering and association rule mining. Int. J. Comput. Appl. **88**, 10 (2014)
15. Malhi, A., Gao, R.X.: PCA-based feature selection scheme for machine defect classification. IEEE Trans. Instrum. Meas. **53**(6), 1517–1525 (2004)
16. Grosche, P., Müller, M.: Tempogram toolbox: matlab implementations for tempo and pulse analysis of music recordings. In: Proceedings of the 12th International Conference on Music Information Retrieval (ISMIR), Miami, FL, USA (2011)

A Survey on Artificial Intelligence Techniques in Cognitive Radio Networks

R. Ganesh Babu and V. Amudha

Abstract Cognitive radio (CR) is the solution for the current spectral underutilized problems, Context awareness and environment awareness are the key functions of CR nowadays. A software radio with reconfiguration capacity will become Cognitive Radio by imparting intelligence to SDR using Artificial Intelligence Techniques. There are processes in CR such as spectrum sensing, monitoring, and management involves the use of AI techniques. Artificial Intelligence (AI) is directed along with "cognitive" functions for better learning and classification. Recently deep learning involves more "self-learning" algorithms without any supervision. Hence it is important to discuss the various AI techniques for better resource optimization in CR networks. Optimization parameters are different for different techniques depend upon the radio environment. The type of AI technique for a particular application is random.

Keywords Cognitive radio · Artificial intelligence techniques
Artificial neural networks (ANN) · Genetic algorithms (GA)

1 Introduction

Cognitive radio is the solution to Spectrum scarcity in communication networks without any dedicated licensed spectrum. The cognitive Radio operations are majorly spectrum sensing, Spectrum mobility and spectrum management. In Fig. 1, CR operations depend upon the five state cognition cycle: Observe, Orient, Plan, Decide and Act (OOPDA). In observe phase it listens from the environment by different sensors, the orient phase establish priority for actions, then in "decide" it selects the best

R. Ganesh Babu
St. Peter's Institute of Higher Education and Research, Chennai 600054, Tamil Nadu, India
e-mail: ganeshbaburajendran@gmail.com

V. Amudha (✉)
St. Peter's College of Engineering and Technology, Chennai 600054, Tamil Nadu, India
e-mail: ammukitcha@gmail.com

© Springer Nature Singapore Pte Ltd. 2019   99
A. Abraham et al. (eds.), *Emerging Technologies in Data Mining and Information Security*, Advances in Intelligent Systems and Computing 755,
https://doi.org/10.1007/978-981-13-1951-8_10

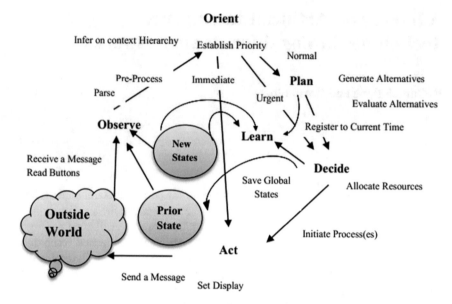

Fig. 1 Simplified cognition cycle

to act. The cycle will be repeated for old as well new stated of input to learn, CR cycle states are different for different situations. In order to learn and act Artificial Intelligence (AI) algorithms are used in literature. The type of learning and decision depends either as supervised learning or unsupervised learning. For classification with less false rate simple learning algorithms are used.

Channel selection, power and topology control, adaptive modulation and coding are taking place in the "act" phase of the cognition cycle. In this paper, we focus on the 'orient' and 'plan' aspects of cognition. Mainly used for spectrum sensing, which depends on game theory, artificial intelligence, multi-objective reasoning, and deep learning systems without any supervision. Dynamic spectrum sensing can be implemented by many methods.

Neural Networks (NNs) are imposed multiple combination of node points which is organic neurons in human brain. Each neurons are joined by multiple link connections and easily interact with each other [1]. All node points receives multiple input data at the input node and produce multiple classes of output. In Hidden Markov Model (HMM) the states are hidden but there is transition probability which decides to remain in the same state or to transition. If the decision is not binary then Fuzzy Logic [2] is used to perform different levels of possibility in output levels. In cognitive radio upper layer issues fuzzy Logic is used to Improve TCP layer Error Detection. Evolutionary algorithm like Genetic Algorithms (GA) is also used for cognitive radio spectrum sensing to provide maximum capability of detection of spectrum holes. In this paper Case Based Reasoning (CBR) is discussed in section which offers continuously learning process and autonomously improving the classification performance. All these AI techniques involve self optimization, self monitoring, self repair, self protection, self adaptation and self healing of ad hoc cognitive radio networks used in emergency ad hoc conditions. In the following chapters various AI techniques are presented.

2   Neural Networks (NN)

In general human brain is composed of 100 billion nerve cells called neurons. Each of them attached to another thousand cells by Axons [3]. These sensory inputs produce electric urges and they can move quickly through neural network. Each neuron is responsible to produce a single output value by accumulating inputs from different neurons through a activation function. The number of input nodes depends upon parameters we considered for training and the output depends upon the classes we want to classify. The hidden nodes are random which are still in research. Each link connection is specified with particular weight. The weights are updated during training by Back propagation algorithm. Neural networks provide signal detection, modulation identification and classification in cognitive radio networks. The classifications of signal processing perform simple multi operations on the data module. A cognitive radio is a reconfiguration network where it accepts any type of modulation in the signal received and reconfigures itself accordingly. There are different types of network topology used in ANN called feed forward and feed backward.

There are different ANN topologies that depends upon the learning process as either supervised or unsupervised in Fig. 2. Multilayer Perceptron is a supervised network which could be used for cognitive radio spectrum sensing. The number of input nodes depends on the number of parameters such as BW, Power, BER, and Modulation. The output nodes are two (primary and secondary). The networks is trained by back propagation algorithm, the weights are adjusted accordingly. Sigmoid function is used as activation function $f(x)$. The diagram is shown as in Fig. 2. The number of hidden nodes are randomly varied and found to be five hidden nodes performance is better for the classification.

There are different learning or training methods that are used such as: Supervised Learning, Reinforcement Learning, Stochastic Learning and Unsupervised Learning as shown in Fig. 3.

3   Hidden Markov Model (HMM)

The concept of HMM arises from the well-known Markov chain, the states emit physically observable symbols. In Fig. 4, only the input and output states are observable directly where the intermediate states are hidden (observed in Bayesian sense) that is why the name is HMM. Hidden Markov models are significantly having more applications in speech processing and in machine vision compared to observable

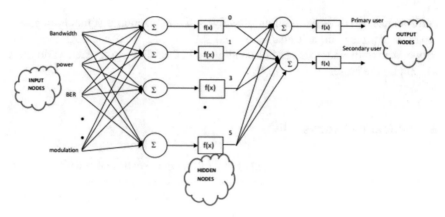

Fig. 2 Multilayer perceptron ANNs

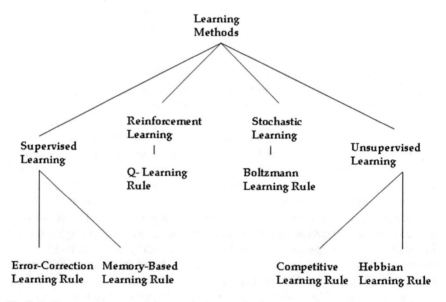

Fig. 3 Different learning or training methods

Markov models. Already there are papers presented the HMM for speech recognition algorithms [4].

In cognitive radio networks, information is processed and the channel condition statistics are studied to yield whether channel is occupied by primary or secondary user. It can capable to improve classification modules according to the classification models to measure particular distance between different states in Fig. 5.

N "hidden" States $S = \{S_1, S_2, \ldots S_N\}$

Series of states $Q = \{Q_1, Q_2, \ldots\}$

Observations Sequence $O = \{O_1, O_2, \ldots\}$

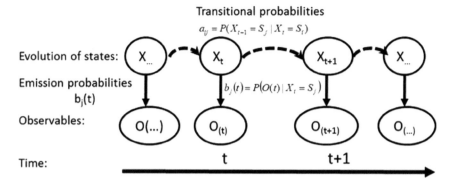

Fig. 4 Basic HMM model

Fig. 5 Basic channel model
of HMM

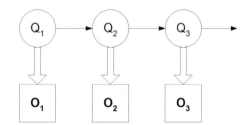

3.1 HMM-Based Cognitive Radio Spectrum Sensing

HMM is used to classify the state of the primary user by using the spectrum sensing techniques. However, most of these techniques make instantaneous decisions based on current measurements, and they do not consider the past status of the primary user. The prediction of primary user activity allows from all available data and choose a CR to better utilization of spectrum. The observation probability is considered for Power, BER, Bandwidth, and Modulation. The particular parameters for a channel statistics is mapped to a codebook index. Baum Welch algorithm is used for Training the opportunistic parameters for primary and secondary. Hence all the channel statistics are averaged finally to yield only two models for primary and secondary. The trained codebook indices are used testing phase. Viterbi algorithm is used to find the optimum path in the model to select the current channel condition from the past sequences of previous channel states as shown in Fig. 6.

Each spectrum band corresponding, the A (Transition), B (Observation) and π (initial) model parameters are computed for primary and secondary users. The A, B and π matrices of all bands is averaged to get a generalized reference model λ. as shown in Fig. 7, the input random spectrum parameters constituting the feature vector. Next, the nearest codebook index vector is found for feature vector. HMM models worked on the codebook indices to produce and the output probabilities. The output probabilities of each model are calculated by viterbi algorithm.

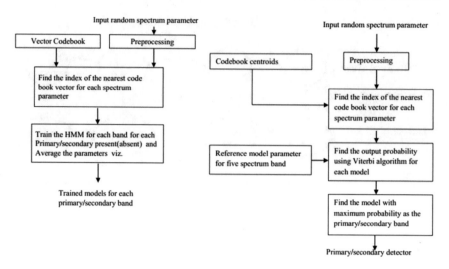

Fig. 6 Block diagram of HMM training and recognition

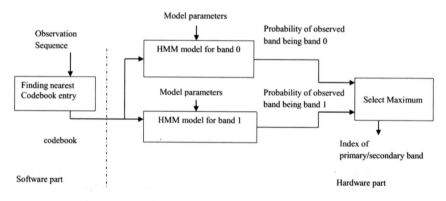

Fig. 7 HMM recognition system

Out of two output probabilities, the model which gives the maximum probability will be said as "primary" or "secondary".

4 Fuzzy Logic (FL)

Fuzzy logic is based on mathematical theory of fuzzy sets, which is a classical set theory. Instead of binary classification, fuzzy logic introduces degree of closeness to output which is a more human behavior classification than binary. The fuzzy classification is based on reasoning flexibility which accounts uncertainties and inaccuracies. For context aware reasoning Fuzzy logic is the best choice. Fuzzy logic used for

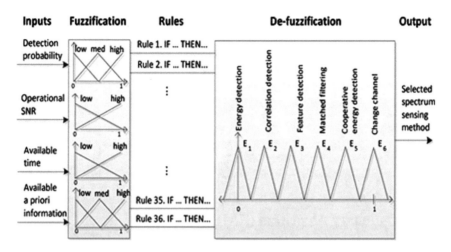

Fig. 8 spectrum decision of rule based selection system

spectrum sensing techniques, where compromise decision-making is required. Fuzzy operators NOT, AND, and OR enables to combine different conditions to make a decision. FL takes input from different types of information to make a model which is simple and human understandable way.

There are already channel classification using FL is already discussed. In [5] channel classification based on new rule is presented. The block diagram is shown in Fig. 8. The spectrum sensing selection consists of four different input parameters are: probability, SNR, available time duration, and priori information. Other inputs are also can be added. The first process is input fuzzification which converts measurable values to fuzzy set variables by using Membership Functions (MBFs). The second step is applying fuzzy rule consisting of IF THEN clauses are applied on the fuzzified values which maps inputs and outputs of the decision-making. As the reverse process the output from the fuzzy reasoning is next mapped to real-world data for spectrum sensing by using output MBFs. Different spectrum sensing methods such as energy detection, cyclostationary, and clustering methods can be selected based on the result of the rules. In [5] general, triangular input and output membership functions (MBFs) are typically used. The membership functions are assigned for the two inputs: SNR and available time as low and high. The inputs available a priori information of three functions such as low, medium, or high. The output is considered the classes as in [5].

In Fig. 9, Decision-making and learning system incorporate spectrum sensing selection methods, adaptive transformation, finding the performance and depends up on it the FL will be updated with new rules and inputs.

Fig. 9 Spectrum availability
detection by learning system

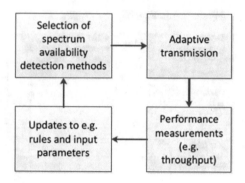

5 Evolutionary Algorithm (EA)

The GA is developed from the CR model and contributes the self-learning and evolution abilities in terms of the cognitive human improvement process.

The general flow of the GA algorithm is presented in Fig. 10. Evolutionary algorithms are more random in nature, that's why they are very close to understand the spectrum situations to classify primary or secondary in an efficient way compared to HMM. Genetic algorithm approach begins by defining structure of a chromosome with important genes. Genes considered for CR are frequency, modulation, power, and BER. Each gene is contributing for the decision-making process. The selected chromosomes and their ranges are shown in Table 1. The implementation details can be found in [5]. In this paper, assumption has been made that fitness function that is equally dependent on all the four parameters. The fitness function uses weighted sum approach in GA. All the four parameters set as an equal weight each. The input is given by Secondary User (SU) or application for QoS requirements are compared against the population of chromosomes. GA has capability of multi-objective optimization, so that overall fitness is computed as the cumulative sum of the individual fitness of each parameter (gene).

Table 1 Selected chromosomes and their ranges

Gene Parameter	Values
Frequency (MHz)	40–910
Power (dbm)	−90 to −40
BER	10^{-1} to 10^{-9}
Modulation	BPSK, QPSK, 8-QAM, 16-QAM

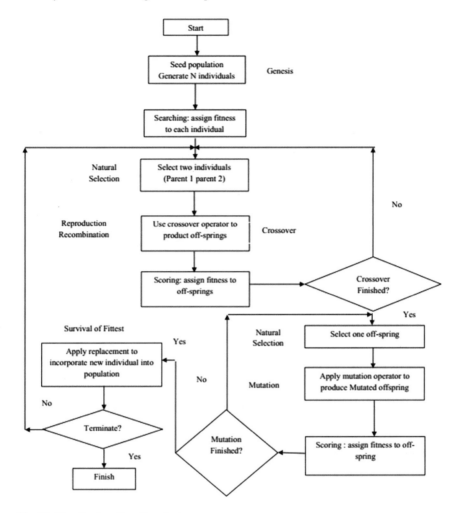

Fig. 10 Genetic algorithm-flowchart

6 Case-Based Reasoning (CBR)

It is one of essential AI technique based on past knowledge decision selection. The case-based reasoning chosen by the past cases and it is best applicable to the current problem. The optimization adopts to practice the each and every problem with less time consumption. The basic case-based reasoning is shown in Fig. 11. Since Cognitive radio works in the principle context reasoning. i.e., the environment awareness applications mostly depends up on the past situations of climatical conditions sensed. If the condition is new it will be stored as a database for the future cases.

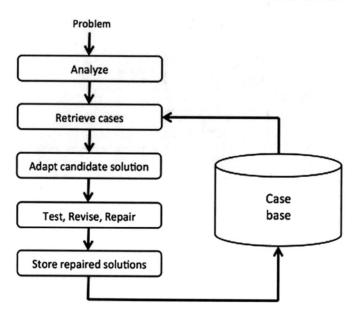

Fig. 11 Basic case-based reasoning

7 Deep Learning and Cognitive Computing

Deep learning is the art of research in the recent days in the field of machine learning and pattern recognition. Deep learning is a reinforcement learning of ANN with more hidden layers and nodes, useful in the field of computer vision (CV) and Natural Language Processing (NLP). It is not a new technique in the field of AI, but it requires high-speed computing and large memory to store the meta-data. It is an unsupervised learning with large data sets. Classification accuracy is high for many classes. In cognition, learning is necessary for the classification and prediction. Classification performance mainly depends on learning algorithm's used. Deep Belief Network (DBN) model are applied to improve accuracy.

Cognitive computing involves he collection of all AI techniques discussed in the paper. It depends on cognitive thinking not on perception, the detailed discussion is given in [6]. In cognitive computing, selection of input factors is more important because a wrong input factor leads to an unwanted decision by CR networks.

8 Conclusion

AI technique selection provides the best realization of cognitive engine design with spectrum sensing, monitoring, and management. Summarized several AI methods are in this paper. In the radio environment, AI shows best attainment to solve a complexity problem. Improving the choice of AI techniques, learning and preserving the past capabilities are essential for CR environment such as deep learning methods.

References

1. Clancy, T.C.: Dynamic spectrum access in cognitive radio networks. http://128.8.127.3/~jkat z/THESES/clancy.pdf (2006)
2. Goldberg, D.E.: A book on Genetic Algorithms in Search, Optimization, and Machine Learning. Addison-Wesley Pub Co (1989)
3. Rabiner, L.R., Juang, B.-H.: A Book on Fundamentals of Speech Recognition (2004)
4. Amudha, V., Venkataramani, B., Vinoth Kumar, R., Ravishankar, S.: SOC Implementation of HMM Based Speaker Independent Isolated Digit Recognition System. https://doi.org/10.110 9/vlsid.2007.144 (2007)
5. Amudha, V., Ramesh, G.P.: Dynamic Spectrum Allocation for Cognitive Radio Using Genetic Algorithm. https://pdfs.semanticscholar.org/6e28/dafb1d44e3cf870933993fe7846f3e9e93be. pdf (2013)
6. Matinmikko, M., Ser, J.D., Rauma, T., Mustonen, M.: Fuzzy-Logic Based Framework for Spectrum Availability Assessment in Cognitive Radio Systems. https://doi.org/10.1109/jsac.2 013.131117 (2013)
7. Singh, S.K., Singh, G., Pathak, V., Roy, K.C.: Spectrum management for cognitive radio based on genetic algorithm. Int. J. Adv. Res. Comput. Sci. (2011). http://dx.doi.org/10.26483/ijarcs. v2i1.295
8. Mitola, J., Maguare, J.Q.: Cogn. Radio Mak. Softw. Radio More Pers. (1999). https://doi.org/ 10.1109/98.788210
9. Zhao, Z., Peng, Z., Zheng, S., Shang, J.: Cogn. Radio Spectr. Alloc. Evol. Algorithm (2009). https://doi.org/10.1109/TWC.2009.080939
10. Rieser, C.J., Rondeau, T., Bostian, C.W., Gallagher, T.: Cognitive Radio Testbed: Further Details and Testing of a Distributed Genetic Algorithm Based Cognitive Engine for Programmable. https://doi.org/10.1109/milcom.2004.1495152 (2004)
11. Meziane, F., Vadera, S.: Artificial Intelligence Applications for Improved Software Engineering Development: New Prospects. https://doi.org/10.4018/978-1-60566-758-4.ch014 (2010)
12. Tewari, J., Arya, S., Singh, P.N.: Approach of intelligent software agents in future development. Int. J. Adv. Res. Comput. Sci. Softw. Eng. http://ijarcsse.com/Before_August_2017/docs/pap ers/Volume_3/5_May2013/V3I4-0319.pdf (2013)
13. Ammar, H.H., Abdelmoez, W., Hamdi, M.S.: Software Engineering Using Artificial Intelligence Techniques: Current State and Open Problems. https://pdfs.semanticscholar.org/0010/5 a98161f98000cefd1880f39fc005319ec33.pdf (2012)
14. Harman, M.: The role of artificial intelligence. Softw. Eng. (2011). https://doi.org/10.1109/R AISE.2012.6227961
15. Jain, P.: Interaction Between Software Engineering and Artificial Intelligence-A Review, www. enggjournals.com/ijcse/doc/IJCSE11-03-12-072.pdf (2011)
16. Nachamai, M., Vadivu, M.S., Tapaskar, V.: Enacted Software Development Process Based on Agent methodologies. https://pdfs.semanticscholar.org/2bc8/e776687cc749d0bd5b1bfbc2f28 f22cc6a3a.pdf (2011)

17. Ganesh Babu, R., Amudha, V.: Spectrum Sensing Techniques in Cognitive Radio Networks: A Survey. https://doi.org/10.1016/j.procs.2016.05.158
18. Ganesh Babu, R., Amudha, V.: Analysis of Distributed Coordinated Spectrum Sensing in Cognitive Radio Networks. https://www.ripublication.com/ijaer_spl/ijaerv10n6spl_114.pdf (2015)
19. Ganesh Babu, R., Amudha, V.: Performance Analysis of Distributed Coordinated Spectrum Sensing in Cognitive Radio. Networks (2015). https://doi.org/10.5829/idosi.mejsr.2015.23.ss ps.13
20. Ganesh Babu, R., Amudha, V.: Cluster Technique Based Channel Sensing in Cognitive Radio Networks. www.serialsjournals.com/serialjournalmanager/pdf/1463980714.pdf (2016)

Bangla Handwritten Digit Recognition Using Convolutional Neural Network

AKM Shahariar Azad Rabby, Sheikh Abujar, Sadeka Haque
and Syed Akhter Hossain

Abstract Handwritten digit recognition has always a big challenge due to its variation of shape, size, and writing style. Accurate handwritten recognition is becoming
more thoughtful to the researchers for its educational and economic values. There
had several works been already done on the Bangla Handwritten Recognition, but
still there is no robust model developed yet. Therefore, this paper states development
and implementation of a lightweight CNN model for classifying Bangla Handwriting Digits. The proposed model outperforms any previous implemented method with
fewer epochs and faster execution time. This Model was trained and tested with ISI
handwritten character database Bhattacharya and Chaudhuri (IEEE Trans Pattern
Anal Mach Intell 31:444–457, 2009, [1], BanglaLekha Isolated Biswas et al. (Data
Brief 12, 103–107, 2017, [2]) and CAMTERDB 3.1.1 Sarkar et al. (Int J Doc Anal
Recogn (IJDAR) 15(1):71–83, 2012, [3]). As a result, it was successfully achieved
validation accuracy of 99.74% on ISI handwritten character database, 98.93% on
BanglaLekha Isolated, 99.42% on CAMTERDB 3.1.1 dataset and lastly 99.43% on
a mixed (combination of BanglaLekha Isolated, CAMTERDB 3.1.1 and ISI handwritten character dataset) dataset. This model achieved the best performance on
different datasets and found very lightweight, it can be used on a low processing
device like-mobile phone.The pre-train model and code for all these datasets can be
found on this link https://github.com/shahariarrabby/Bangla_Digit_Recognition_C
NN.

A. S. A. Rabby · S. Abujar (✉) · S. Haque · S. A. Hossain
Department of Computer Science and Engineering, Daffodil International University,
Dhanmondi, Dhaka 1205, Bangladesh
e-mail: sheikh.cse@diu.edu.bd

A. S. A. Rabby
e-mail: azad15-5424@diu.edu.bd

S. Haque
e-mail: sadeka15-5210@diu.edu.bd

S. A. Hossain
e-mail: aktarhossain@daffodilvarsity.edu.bd

© Springer Nature Singapore Pte Ltd. 2019   111
A. Abraham et al. (eds.), *Emerging Technologies in Data Mining and Information
Security*, Advances in Intelligent Systems and Computing 755,
https://doi.org/10.1007/978-981-13-1951-8_11

Keywords Bangla handwritten recognition · Convolutional neural network Pattern recognition · Deep learning · Computer vision · Machine learning

1 Introduction

Handwritten digit recognition development research is rapidly evolving and reshaping the must automation fields like—automatic check reading, automatic number plate reading, digital postal service, Optical Image Recognizing (OCR), etc. Due to its various aspects of uses, computer vision researchers verily feel to work on it and improve—quality and performance indeed. But handwritten recognizing is more challenging compared to the typed letter. Because different people write in a different way and which creates a higher degree of variance in written style. Also, there are some similarities between different characters shape. The situation of overwriting makes it more challenging for accurately classifying the handwritten digit.

Nowadays, deep learning, especially the Convolutional neural network is working better in the purpose of classifying these types of recognition work rather other machine learning methods.

The Bangla alphabet incorporates the writing system for the Bangla language together with the Assamese alphabets. Bangla is the fifth most widely writing language in the world. With 250 million speakers, Bangla is the seventh most spoken language in the world by population. Also, it is one of the major languages in the Indian subcontinent and is the first language in Bangladesh. But the research held on Bangla handwriting recognition is very few compared with other languages like English, Arabic, Hindi, Chinese, etc.

Every language has its own Wscript. Bangla came from Sanskrit script which is completely different from English or other popular scripts. There are 50 characters, 10 numerical digits, more than 200 compound characters and modifier are available in Bangla language. Also, it has many similar characters, some of them are different considering small dot and line. And compound characters can be made by joining two characters. Almost all Bangla consonant can be used to make a new compound character. This makes it more difficult to achieve a good result and better performance with Bangla Handwritten character recognition. There are many other applications which may use this Bangla digit recognition system. Such as Bangla Handwritten character base OCR (Optical Character Recognition), Picture to text to speech, Bangla ID card reading, Number plate reading, vehicle tracking, Post office automation, etc.

2 Related Work

Several studies show that CNN can be successfully applied in the complex image classification work. For English MNIST [4] CNN got a remarkable result.

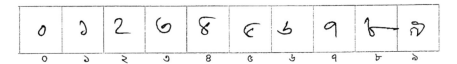

Fig. 1 Example of Bangla digit

There are few works available on Bangla digit recognition who reaches remarkable result in Bangla Handwritten digit recognition. *"Handwritten numeral databases of Indian scripts and multistage recognition"* works on mixed numerals [1]. The main feature of their model includes matra (the upper part of the character), vertical line and double vertical line and for the MLF classifier, the feature is constructed from the stroke feature of the characters (Fig. 1).

"Automatic Recognition of Unconstrained Off-Line Bangla Handwritten Numerals" [5] is also concluded with some good research work. The proposed method mentioned, used to extract features from a concept called water reservoir. It was implemented and continued further use in postal automation sector. Another work is *"Handwritten Bangla Digit Recognition Using Deep Learning"* [6]. Where a CNN model was used, that achieves 98.78% accuracy on CMATERDB 3.1.1 datasets.

3 Proposed Methodology

The proposed method uses a Convolutional Neural Network which has many phases, such as Dataset preparation, model training, etc., which is described below.

3.1 Datasets

This method used three datasets. ISI handwritten character database, CMATERDB 3.1.1 and BanglaLekha-Isolated datasets. Those datasets have Bangla characters and digits.

ISI handwritten character database has total 23392 numerical images where 19392 is for train and 4000 for the test. The image has different pixel size and images edge look smooth.

The CMATERDB 3.1.1 has the total 6000-digit image. Each class contains 600 BMP format 3 channel image. Most of the images are noise-free 32 × 32 pixels and almost correct labeling and no overwriting characters images. But the image edge looks blockier (Fig. 2).

BanglaLekha-Isolated datasets contain 19748 image each class contains average 1974 image each in 1 channel grayscale png format. Each image has different pixel size. Some image has incorrect labeling and some image is overwritten with others.

But the image ages are looking smooth. Then also join these two datasets and make a new dataset with 25748 images.

3.2 Dataset Preparation

For ISI handwritten character database and CMATERDB 3.1.1, we first converted it to one channel grayscale to reduce computational expense and invert the color. Black for background and white for digit. For BanglaLekha-Isolated, first fixed some incorrect labeling and delete some incorrect image. This dataset already containing inverted images. Than resized all images to 28 × 28 pixel.

Then converted both dataset's 28 × 28-pixel image into a 784 + (1 label) D matrix and store all the image pixels into a CSV file to reduce hard disk read and fast computation. Also, perform a Minmax (1) normalization to reduce the effect of illumination differences. Moreover, the CNN converge faster on [0…0.1] data than on [0…0.255].

$$Z_i = \frac{X_i - \text{minmum}(X)}{\text{maximum}(X) - \text{minmum}(X)} \tag{1}$$

Then convert the 10 labels into one hot encoding.

[1, 0, 0, 0, 0, 0, 0, 0, 0, 0] = ০   [0, 0, 0, 0, 0, 1, 0, 0, 0, 0] = ৫

[0, 1, 0, 0, 0, 0, 0, 0, 0, 0] = ১   [0, 0, 0, 0, 0, 0, 1, 0, 0, 0] = ৬

[0, 0, 1, 0, 0, 0, 0, 0, 0, 0] = ২   [0, 0, 0, 0, 0, 0, 0, 1, 0, 0] = ৭

[0, 0, 0, 1, 0, 0, 0, 0, 0, 0] = ৩   [0, 0, 0, 0, 0, 0, 0, 0, 1, 0] = ৮

[0, 0, 0, 0, 1, 0, 0, 0, 0, 0] = 8   [0, 0, 0, 0, 0, 0, 0, 0, 0, 1] = ৯

Then convert the 784 D into 28 × 28 image matrix.

Fig. 2 Example image of dataset

(a) ISI   (b) CMATERDB   (c) BanglaLekha

3.3 Proposed Model

Algorithm 1:

  1: ADAM (Learning Rate)
  2: For 30 iterations in all batch do:
  3:    Convolution 1 (Filter, Kernel Size, Stride, Padding, Activation)
  4:    Convolution 2 (Filter, Kernel Size, Stride, Padding, Activation)
  5:    MaxPool 1 (Pool Size)
  6:    Dropout (Rate)
  7:    Convolution 3 (Filter, Kernel Size, Stride, Padding, Activation)
  8:    Convolution 4 (Filter, Kernel Size, Stride, Padding, Activation)
  9:    MaxPool 2 (Pool Size)
  10:   Dropout (Rate)
  11:   Dense (Units, Activation, Kernel initializer, Bias Initializer)
  12:   Dropout (Rate)
  13:   Dense (Units, Activation, Kernel initializer, Bias Initializer)
  14: end for

Proposed Model in this paper use ADAM [7] optimizer with a learning rate of 0.001. The model has 9-layer CNN. For convolution 1 and 2, where filter size is 32, kernel size is (5 × 5), Stride is (1 × 1), "same" padding with ReLU (2) activation. Followed by a 5 × 5 max-pooling layer. Then used 25% dropout [8] to reduce overfitting.

$$ReLU(X) = MAX(0, X) \qquad (2)$$

For convolution 3 and 4, the filter is 64, kernel size is (3 × 3), Stride is (1 × 1), "same" padding with ReLU activation. Followed by a 2 × 2 max-pooling layer. Then used 25% dropout [8]. Then flatten the layer and use a Dense layer with 256 units with ReLU activation and 50% dropout. At final output layer, used 10 units with SoftMax (3) activation. Figure 3 is showing the neural network architecture.

$$\sigma(z)_j = \frac{e^{z_j}}{\sum_{k=1}^{K} e^{z_k}} \, for \, j = 1, \ldots k \qquad (3)$$

3.4 Optimizer and Learning Rate

The choice of optimization algorithm can make a sufficient change for the result in Deep Learning and computer vision work. The Adam paper says, "…many objective functions are composed of a sum of sub-functions evaluated at different subsamples of data; in this case, optimization can be made more efficient by taking gradient steps w.r.t. individual sub-functions…" [8]. The Adam optimization algorithm is an

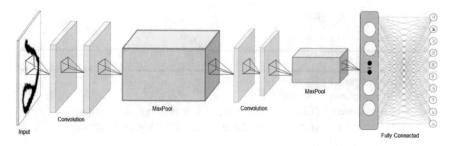

Fig. 3 Proposed CNN model

extension to stochastic gradient descent that recently adopting most of the computer vision and natural language processing application. The method computes individual adaptive learning rates for different parameters from estimates of first and second moments of the gradients.

Proposed method used ADAM (4) Optimizer [8] with learning rate = 0.001.

$$v_t = (1 - \beta_2) \sum_{i=1}^{t} \beta_2^{t-i} \cdot g_i^2 \tag{4}$$

When using a neural network to perform classification and prediction task. A recent study shows that cross-entropy function performs better than classification error and mean square error [9]. Cross-entropy error, the weight changes do not get smaller and smaller and so training is not s likely to stall out. Proposed method used categorical cross entropy (4) as loss function.

$$L_i = - \sum_j t_{i,j} \log(p_{i,j}) \tag{5}$$

To make the optimizer converge faster and closer to the global minimum of the loss function, using an automatic Learning Rate reduction method [10]. Learning rate is the step by which walks through the minimum loss. If higher learning rate use it will quickly converge and stuck in a local minimum instead of global minima. To keep the advantage of the fast computation time with a high Learning Rate, after each epoch model dynamically decreases the learning rate by monitoring the validation accuracy.

After some epochs manually checked the accuracy and decrease the learning rate to reach the global minima.

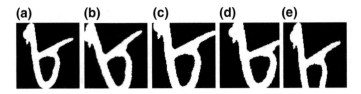

Fig. 4 **a** Original Image, **b** Rotated image, **c** Zoomed image, **d** Width shifted image, **e** Height shifted images

3.5 Data Augmentation

To avoid overfitting, artificially expand the handwritten dataset. This data transformation will create some variance that can occur when someone else writing the digits. For Data augmentation, several methods are chosen:

- Randomly shifting height and width 10% of the images.
- Randomly rotate our training image 10°
- Randomly 10% zoom the training image (Fig. 4).

3.6 Training the Model

Trained the model with different training set and validation set with a batch size of 86. In training time, the Learning Rate reduction formula will monitor the validation accuracy and reduce the learning rate. After 30 epochs, manually monitored the accuracy result and reduce the learning rate and train again 5–10 epochs and again manually set the learning rate couple of times.

The pre-train model and code for all these datasets can be found on this link https://github.com/shahariarrabby/Bangla_Digit_Recognition_CNN.

4 Evaluate the Model

The proposed model is applied to different datasets and get a pretty good result on train, test and validation sets which is shown below.

4.1 Train, Test, Validation Sets

For ISI handwritten character database used 15513 images as training set and 3879 images (20%) in the validation set. And the test set with ISI's 4000 images. For

BanglaLekha-Isolated Datasets used 15766 images as training set and 3942 images (20%) in the validation set. And make the test set with CMATERDB 3.1.1 datasets 6000 images. For CMATERDB 3.1.1 datasets used 4800 images as training set and 1200 image (20%) in the validation set. And make the test set with BanglaLekha-Isolated Dataset's 19708 images. For Mixed dataset used 80% from all of three dataset which is 39279 images on the training set, and 10% (4911) images on validation set and remaining 10% (4910) images in the test set.

4.2 Model Performance

For ISI handwritten character database, after 30 epoch model gets 99.35% accuracy on the training set and 99.74% accuracy on the validation set. Then tested the model with testing set and got 99.58% accuracy. Figure 5 shows the loss value and accuracy of the training set and the validation set and the Confusion Matrix (Table 1).

For BanglaLekha-Isolated dataset, after 50 epoch model gets 99.38% accuracy on the training set and 98.93% accuracy on the validation set. Then tested the model with CMATERdb 3.1.1 dataset and got 98.58% accuracy. Figure 6 shows the loss value and accuracy of the training set and the validation set and the Confusion Matrix.

For CMATERdb 3.1.1 dataset, after 30 epoch model gets 99.05% accuracy on the training set and 99.42% accuracy on the validation set. After the train tested the model with the BanglaLekha-Isolated dataset and got 92.65% accuracy. The test

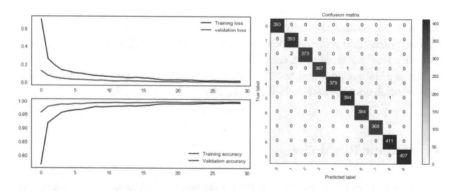

Fig. 5 ISI handwritten character database

Table 1 Result comparison in different dataset

Datasets	Tr. Loss	Val Loss	Tr. Acc.	Val Acc.	Test Acc.
ISI	0.0221	0.0090	99.35	99.74	99.58
BanglaLekha	0.0199	0.0510	99.38	98.93	98.58
CMATERdb	0.0339	0.0207	99.05	99.42	92.65
Mixed dataset	0.0256	0.0181	99.23	99.43	99.51

Table 2 Result comparison in different model

ISI numeral dataset		CMATERDB 3.1.1	
Work	Accuracy (%)	Work	Accuracy (%)
Bhatt and Chaudhuri	98.20	Haider et al. [11]	94
Wen and He [12]	96.91	Hassan et al. [13]	96.7
Nasir and Uddin [14]	96.80	Sarkhel el at. [15]	98.23
Akhnad et al. [16]	97.93	Basu et al. [17]	95.10
HybridHOGPPC8 [18]	99.03	HybridHOGPPC8 [18]	99.17
Proposed model	99.74	Proposed model	99.42

accuracy is too poor because CMATERdb 3.1.1 contains the noise-free image that fails to learn the noisy image of the BanglaLekha-Isolated dataset. Figure 7 shows the loss value and accuracy of the training set and the validation set and the Confusion Matrix.

For combined dataset, after 40 epoch models get 99.23% accuracy on the training set and 99.43% accuracy on validation set and 99.51% accuracy on the test set. Figure 8 shows the loss value and accuracy of the training set and the validation set and the Confusion Matrix.

4.3 Result Comparison

See Table 2.

5 Error Observations

From Fig. 9, for these twelve cases, it is clear that this model is doing great jobs in classifying digits. Some of those errors can also make by humans. Variance in handwriting model is getting confused with some digits, especially 5, 6 and 9, 1. Figs are sorted by error rate in ascending order.

6 Conclusion and Future Work

This paper presented a new CNN model that performs better classification accuracy in the different dataset for both train and validation set for lesser epochs and less computation time compared to the other CNN model. Also, the cross-validation from different distribution's data proposed model achieve a great result that makes it as a

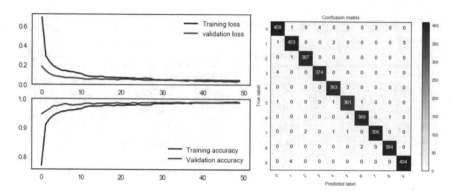

Fig. 6 BanglaLekha-Isolated evaluation

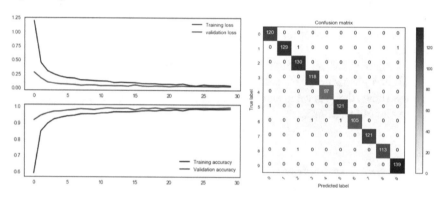

Fig. 7 CMATERdb evaluation

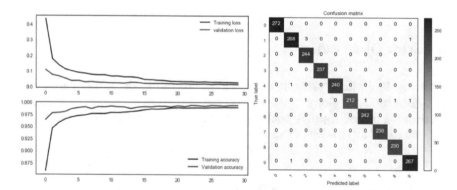

Fig. 8 Mixed datasets evaluation

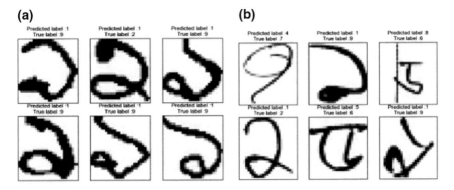

Fig. 9 **a** Validation set error, **b** Test set error

robust model that improve any other previous model. Also, this model got 99.55% accuracy on English MNIST handwritten dataset.

Sometimes proposed model confused to understand overwritten character and dataset contained some incorrect labeling images. Also, the model performed poorly if the train on noise-free data. In future work fixing dataset and overcoming the limitation of overwriting digit should fix.

References

1. Bhattacharya, U., Chaudhuri, B.: Handwritten numeral databases of indian scripts and multistage recognition of mixed numerals. IEEE Trans. Pattern Anal. Mach. Intell. **31**, 444–457 (2009). https://doi.org/10.1109/TPAMI.2008.88
2. Biswas, M., Islam, R., Gautam, K.S., Shopon, Md., Mohammed, N., Momen, S., Abedin, A.: BanglaLekha-Isolated: a multi-purpose comprehensive dataset of Handwritten Bangla Isolated characters. Data Brief. **12**, 103–107 (2017). https://doi.org/10.1016/j.dib.2017.03.035
3. Sarkar, R., Das, N., Basu, S., Kundu, M., Nasipuri, M., Basu, D.K.: Cmaterdb1: a database of unconstrained handwritten Bangla and Bangla-English mixed script document image. Int. J. Doc. Anal. Recogn. (IJDAR) **15**(1), 71–83 (2012)
4. LeCun, Y., Bottou, L., Bengio, Y., Haffner, P.: Gradient-based learning applied to document recognition. In: Proceedings of the IEEE, vol. 86(11), pp. 2278–2324, Nov 1998
5. Pal, U., Chaudhuri, B.: Automatic Recognition of Unconstrained Off-Line Bangla Handwritten Numerals **1948**, 371–378 (2000). https://doi.org/10.1007/3-540-40063-x_49
6. Alom, Md.Z., Sidike, P., Tarek, M.T., Asari, V.: Handwritten Bangla Digit Recognition using Deep Learning (2017)
7. Kingma, D.P., Ba, J.: Adam: A Method for Stochastic Optimization, Dec 2014. arXiv:1412.6980
8. Srivastava, N., Hinton, G., Krizhevsky, A., Sutskever, I., Salakhutdinov, R.: Dropout: A simple way to prevent neural networks from overfitting. J. Mach. Learn. Res. **15** 1929–1958 (2014)
9. Janocha, K., Czarnecki, W.M.: On Loss Functions for Deep Neural Networks in Classification (2017). arXiv:1702.05659
10. Schaul, T., Zhang, S., and LeCun, Y.: No More Pesky Learning Rates (2012). arXiv:1206.1106

11. Khan, H.A., Helal, A.A., Ahmed, K.I.: Handwritten Bangla digit recognition using sparse representation classifier. In: 2014 International Conference on Informatics, Electronics & Vision (ICIEV), pp. 1–6. IEEE (2014)

12. Wen, Y., He, L.: A classifier for Bangla handwritten numeral recognition. Expert Syst. Appl. **39**(1), 948–953 (2012)

13. Hassan, T., Khan, H.A.: Handwritten Bangla numeral recognition using local binary pattern. In: 2015 International Conference on Electrical Engineering and Information Communication Technology (ICEEICT), pp. 1–4. IEEE (2015)

14. Nasir, M.K., Uddin, M.S.: Handwritten Bangla numerals recognition for automated postal system. IOSR J. Comput. Eng. **8**(6), 43–48 (2013)

15. Sarkhel, R., Das, N., Saha, A.K., Nasipuri, M.: A multi-objective approach towards cost-effective isolated handwritten Bangla character and digit recognition. Pattern Recogn. **58**, 172–189 (2016)

16. Mahbubar Rahman, S.I.P.S. Md., Akhand, M.A.H., Rahman, M.M.H.: Bangla handwritten character recognition using convolutional neural network. I. J. Image Graph. Signal Process. (IJIGSP) **7**(3), 42–49 (2015)

17. Basu, S., Sarkar, R., Das, N., Kundu, M., Nasipuri, M., Basu, D.K.: Handwritten Bangla digit recognition using classifier combination through ds technique. In: Pattern Recognition and Machine Intelligence, pp. 236–241. Springer (2005)

18. Sharif, S.A.M., Nabeel, M., Mansoor, N., Momen, S.: A hybrid deep model with HOG features for Bangla handwritten numeral classification. In: 2016 9th International Conference on Electrical and Computer Engineering (ICECE), pp. 463–466 (2016)

Distinction Between Phases of Human Sleep Cycle Using Neural Networks Based on Bio-signals

Trishita, Simran Kaur Bhatia, Gaurav Kumar and Aleena Swetapadma

Abstract Sleep disorders can be monitored by analyzing the various stages of sleep. The stages of human sleep cycle can be broadly classified into three types Awake, rapid eye movement (REM) sleep and non-REM sleep stages. In this work a neural network based method is proposed to distinguish between Awake, REM sleep and non-REM sleep stages. Various types of bio-signals such as electro-occulogram (EOG), electromyogram (EMG), and electroencephalogram (EEG) are used as input to the neural network based method. Accuracy of the proposed neural network based method is found to be 100%. The results of the method are promising, hence can be used to monitor sleep disorders.

1 Introduction

Human sleep cycle consist of various stages during which the process such as consciousness, inhibition of sensory activity and voluntary muscles occurs repeatedly. Human sleep cycle generally has alternating phases of REM and Non-REM sleep stages. If a person does not have any sleep disorder, then quality of sleep is normal in general. To monitor the sleep disorders various methods has been suggested by researchers. The sleep waves have been monitored with various data mining techniques in [1]. In [2], a hierarchical classification for sleep stage classification using forehead EEG signals has been proposed. The method has various stages such as wake detection rule, feature extraction and feature selection, and classification with SVM.

Trishita · S. K. Bhatia · G. Kumar · A. Swetapadma (✉)
KIIT University, Bhubaneswar 751024, Odisha, India
e-mail: aleena.swetapadmafcs@kiit.ac.in

Trishita
e-mail: trishitasingh997@gmail.com

S. K. Bhatia
e-mail: simranbhatia8986@gmail.com

G. Kumar
e-mail: shakyagaurav108@gmail.com

© Springer Nature Singapore Pte Ltd. 2019
A. Abraham et al. (eds.), *Emerging Technologies in Data Mining and Information Security*, Advances in Intelligent Systems and Computing 755,
https://doi.org/10.1007/978-981-13-1951-8_12

Drowsiness is one of the effects which occur when a person is desired or inclined to sleep. In [3], a hybrid method using wavelet transform and fuzzy logic is proposed for driver drowsiness classification. In [4], single-channel EEG signals are used for sleep stage classification using metric learning approach. The learning method used is a k-nearest neighbor classification rule with Euclidean metric. In [5], sleep diagnosis has been carried out for obstructive sleep apnea patients based on cardio respiratory signals. In [6], single-lead ECG signals are used for sleep stage and apnea epoch classification. In [7], sleep apnea severity is classified. In [8], ECG signals are used for recognition of sleep apnea syndrome using support vector machines. After studying all the above methods it seem like there is still a need for performance enhancement in sleep disorder monitoring.

In this work a neural network based approach is proposed for distinction between awake, REM and Non-REM sleep stages. The remaining of the paper is organized in the following manner: general idea about human sleep cycle which is followed by NN classifier. Proposed method is followed by performance of the method which is finally followed by conclusion.

2 Human Sleep Cycle

According to American Academy of Sleep Medicine (AASM) manual the human sleep is divided into various stages such as awake stage, REM stage, non-REM stage 1, non-REM stage 2, non-REM stage 3. In REM sleep stage dreaming and rapid eye movements under closed eyelids occur which constitutes about 20–25% of sleep [4]. The Non-REM sleep of the sleep cycle is classified into other stages. In this work only a broad classification of sleep stages are carried out.

3 Neural Network

NN classifiers have been applied in many fields of engineering. Among the various types of NN back-propagation NN is one of the important neural networks. In back-propagation NN the weight update algorithm used is gradient-descent method. Back-propagation NN is feed-forward network with input layer, hidden layer and an output layer [9]. The output of the neuron is given by the following relationship in (1).

$$O = f(net) = f\left(\sum_{j=1}^{n} w_j x_j\right), \tag{1}$$

where the parameters represent the transfer or activation function f (net) and are obtained as the scalar product of input (x_j) and weight (w_j). In this work neural

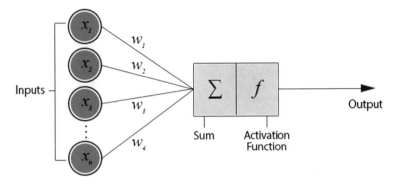

Fig. 1 Neural network structure

network is used as a classifier to classify the sleep stages. A neural network is shown in Fig. 1 with all its components.

4 Proposed Method

4.1 Data Acquisition

To carry out the proposed work, signals are obtained from physionet sleep dataset [10]. The data set contain EOG and EEG sampled at 100 Hz. The recordings also contain the EMG signals sampled at 1 Hz. The signals are then divided into two parts, 90% of the signals are used for training and 10% is used for testing. The proposed method is described in next sub-section.

4.2 Distinction of Sleep Stages Using Neural Network

In this work ANN is used as classifier to distinguish sleep stages by taking bio-signals as inputs. The targets of the network is designed as [1 0 0] for awake, [0 1 0] for REM and [0 0 1] for non-REM. After trying various networks the final neural network is obtained. The optimal neural network obtained is a two-layered network with five neurons in hidden layer and tan-sig transfer function. Figure 2 shows the final neural network obtained for the proposed method. Figure 3 shows the mean square error obtained for the optimal network obtained. The results of the method are discussed in detail in next section.

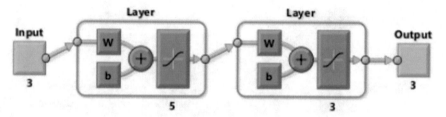

Fig. 2 Final neural network obtained

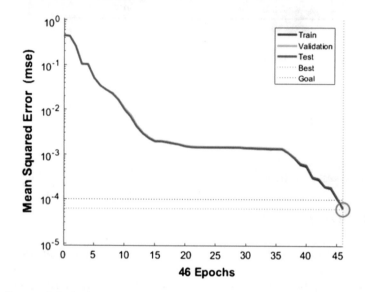

Fig. 3 Error goal obtained

5   Result

Performance of the proposed method is evaluated under different configuration of the artificial neural network such as error goal, number of neurons, number of hidden layers, etc. Performance measures used in this work is % accuracy. Results of the proposed NN-based method are discussed in the sub-section below.

5.1   Performance Varying Mean Square Error

Performance of the method is evaluated varying mean square error goal. The results are shown in Table 1 for various networks. The accuracy of the method is highest with the network with mean square error of 0.001 and 0.0001. Confusion matrix of both is shown in Fig. 4. From Fig. 4 it can be observed that there are some misclassified data

Table 1 Performance varying mse

Error goal	Accuracy (%)
0.1	91.7
0.01	99.9
0.001	100.0
0.0001	100.0

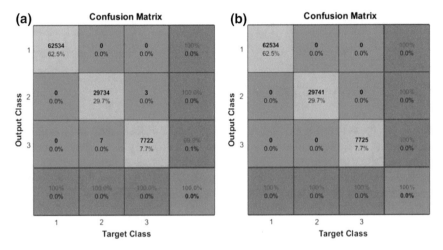

Fig. 4 Confusion matrix **a** with mse $=0.001$ **b** with mse $=0.0001$

with mse $=0.001$ but all data are classified accurately with mse $=0.0001$. Hence the mse of the optimal neural network is chosen to be 0.0001.

5.2 Performance Varying Number of Neurons

Performance of the method is evaluated considering varying number of neurons. The results for various neurons are shown in Table 2. From Table 2, it can be observed that the method is having highest accuracy with 5 and 10 neurons. But as the accuracy is achieved with lower neurons, it is considered as optimal. Hence five neurons are chosen as appropriate for the proposed method.

Table 2 Performance varying number of neurons

Neurons	Accuracy (%)
1	85.6
5	100.0
10	100.0

Table 3 Comparison with other methods

Different technique	Methods used	Accuracy (%)
Redmond et al. [5]	Quadratic discriminant classifier	87
Yilmaz et al. [6]	Beat-to-beat interval based classification	90
Eiseman et al. [7]	k-NN, SVM and NB	89
Khandoker et al. [8]	SVM	93
Proposed method	ANN	100

5.3 Comparison with Other Methods

Various methods have been proposed by researchers for sleep study and monitoring by researchers. Comparative study of some of the already developed schemes is given in Table 3. From Table 3, it can be observed that in comparison to other methods proposed method has highest accuracy, i.e., 100%. Hence proposed NN-based method for sleep stage classification can be used efficiently.

6 Conclusion

In this work a NN-based method is proposed for distinction of various sleep stages such as awake, REM, and non-REM sleep. The input signals used to the NN based classifier are EOG, EMG, and EEG signals. After checking with all the signals taken, it was found that accuracy of the NN-based method is 100%. The NN based method is very simple, robust and it has high accuracy. Hence the proposed method can be used to monitor sleep disorders.

References

1. Swetapadma, A., Swain, B.R.: A data mining approach for sleep wave and sleep stage classification. In: IEEE International Conference on Inventive Computation Technologies Coimbatore, pp. 1–6 (2016)
2. Huang, C.S., Lin, L., Ko, W., Liu, S.Y., Sua, T.P., Lin, C.T.: A hierarchical classification system for sleep stage scoring via forehead EEG signals. In: IEEE Symposium on Computational Intelligence, Cognitive Algorithms, Mind, and Brain, Singapore, 1–5 (2013)
3. Khushaba, R.N., Kodagoda, S., Lal, S., Dissanayake, G.: Driver drowsiness classification using fuzzy wavelet-packet-based feature-extraction algorithm. IEEE Trans. Biomed. Eng. **58**(1), 121–131 (2011)
4. Phan, H., Do, Q., Do, T.L., Vu, D.L.: Metric learning for automatic sleep stage classification. In: 35th Annual International Conference of the IEEE EMBS, pp. 5025–5028 (2013)

5. Redmond, S.J., Heneghan, C.: Cardio respiratory-Based Sleep Staging in Subjects With Obstructive Sleep Apnea. IEEE Trans. Biomed. Eng. **53**(3), 485–496 (2006)
6. Yilmaz, B., Asyali, M.H., Arikan, E., Yetkin, S., Özgen, F.: Sleep stage and obstructive apneaic epoch classification using single-lead ECG. Biomed. Eng. Online **9**, 39 (2010)
7. Eiseman, N.A., Westover, M.B., Mietus, J.E., Thomas, R.J., Bianchi, M.T.: Classification algorithms for predicting sleepiness and sleep apnea severity. J. Sleep Res. **21**, 101–112 (2010)
8. Khandoker, A.H., Palaniswami, M., Karmakar, C.K.: Support vector machines for automated recognition of obstructive sleep apnea syndrome from ECG recordings. IEEE Trans. Inf Technol. Biomed. **13**(1), 37–48 (2009)
9. Abraham, A.: Artificial neural networks. In: Sydenham, P.H. (ed.) Handbook of Measuring System Design. Wiley, New York (2005)
10. Goldberger, A.L., Amaral, L.A.N., Glass, L., Hausdorff, J.M., Ivanov, P.C., Mark, R.G., Mietus, J.E., Moody, G.B., Peng, C.K., Stanley, H.E.: PhysioBank, PhysioToolkit, and PhysioNet: components of a new research resource for complex physiologic signals. Circulation **101**(23), 215–220 (2000)

Genetic Algorithm Based Load Evaluation Approach for Salvation of Complexities in Allocation of Budget Assets

Neelima Chilagani and S. S. V. N. Sarma

Abstract In this innovation fast-growing environment the genetic algorithm is an iterative approach; it is concerning the intellectual based tryout and to get the fault mechanism to discover an inclusive best possible occurrences. Previously some of the cash based management models are developed by the researchers, many models are proposed and executed to solve the various complex problems, but still some challenging problems are waiting for the solutions. The Liquid Cash equal distribution is big challenge for many developing countries such as India, Srilanka, and China. These countries will plan their financial years in the form of year-wise budget models. But, here the realistic problem will starts when the allocated liquid cash assets are lesser or greater than the required cash asset amount of particular objective and the problem will be raised once it has implicated in practical executable mode. This paper described the full comparative analysis of various asset-cash balancing distribution of different state wise annual budget formation in table forms along with their department wise allotted asset value, objective and percentage of increment. Here described the year-wise improvement scenario with item based description along with the total expenders, revenue deficit, fiscal deficit, and primary deficit. It will delineate the genetic algorithm implications are implied to the major problem of asset equal distribution with the genetic-reproduction, genetic-crossover and genetic-mutation principles then to exchange the financial based services like as asset values and resources to each other using the traditional based mean difference methodology and recompose the cluster based inequality asset values into equality with concerns of the asset objectives and derived resources. This methodology will more helpful to the economist and finance profession for equally distributing the total liquid asset values into resource-based-and-liquid-based asset values.

N. Chilagani (✉)
Computer Science Department, Kakatiya University,
Warangal 506009, Telangana, India
e-mail: neelima.chilagani@gmail.com

S. S. V. N. Sarma
Computer Science & Engineering Department, Vaagdevi College
of Engineering, Warangal 506001, Telangana, India
e-mail: ssvn.sarma1@gmail.com

© Springer Nature Singapore Pte Ltd. 2019
A. Abraham et al. (eds.), *Emerging Technologies in Data Mining and Information Security*, Advances in Intelligent Systems and Computing 755,
https://doi.org/10.1007/978-981-13-1951-8_13

131

Keywords Search space (SS) · Traditional-Gradient-based process (TGbP)
Genetic algorithm (GA) · Local-active-maxima (LAM)
Global-active-maxima (GAM)

1 Introduction

A genetic algorithm (GA) consists the technical objectives and composed hybrid techniques to complex problems in medical, engineering, aeronautical, biological, and financial sectors, it is deriving the iterative approach with involvement of intellectual assessments and inaccuracy. Its main intension is to generate the global optimum. Nature-based corresponding occurrences are the process of progression over time, where many generations are formed, and each genetic based inhabitants turn into enhanced and superior personalized to its situation. Nature based corresponding occurrences are the process of progression over time, where many generations are formed, and each genetic-based inhabitants turn into enhanced and superior personalized to its situation.GA is the deriving the approach of bunch of search procedures with modulated on the dynamic-mechanics of expected assortment based progression. In this generalized evolutionary procedure, the genetic-reproduction techniques are used to find best outfit generation other techniques such as genetic-based genetic-crossover and genetic-mutation are to find finest solutions to financial, engineering and mathematical problems. GA is the most sophisticated active tools for the out-sized asymmetrical search space based problems where a global optimum is requisite. In general, in Traditional-gradient-based process of optimization has come across some technical problems when the search space is multi-modal as shown in the Fig. 1 as they tend to become stuck at local-active-maxima (LAM) rather than the global-active-maxima (GAM). GA is attention to suffer less from this problem of premature convergence [1].

Fig. 1 Search-space (SS) is multi-modal in traditional-gradient-based process (TGbP) of optimization [1]

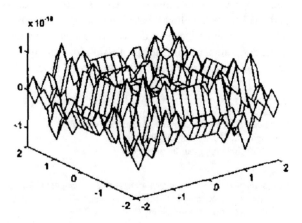

In economical and financial sectors the asset-based cash balance is a typical difficulty in fixed budget based financial transactions [2]. For this purpose, the Load Balancing Approach of GA has been proposed for Liquid Cash equal distribution in the year-wise budget-based financial issues. Previously some of the cash based management models are developed by the researchers, such as the Baumol and Tobin has proposed the main cash flow management models in 1950, later the Miller-Orr has proposed the Miller-Orr model in 1960, and their further developments since the 1980 and so on. Many models are proposed and executed to solve the various complex problems, but still some challenging problems are waiting for the solutions. In the development of economic-asset-investment circulation form the largest part of the implications are executed the identical hypothesis which is very closer to the originated hypothesis such as the essential approach of Miller-Orr implicational sequences which is partially contrary with the stochastic representations of the concern financial problem variations cited below along with their author names.

The development model by the researchers of Tapiero and Zuckerman [3], another researcher's Milbourne [4], and another team of researchers of Hinderer and Waldmann [5], author Baccarin [6], author Premachandra [7], author Volosov et al. [8] and another researcher of Baccarin [9] development variations. The Liquid Cash equal distribution is big challenge for many developing countries such as India, Sri Lanka and China. These countries will plan their financial years in the form of year-wise-budget models. The country economists and finance ministry will allocate the liquid cash assets based on purpose, type of usage, location, area wise population and their objectives. The finance ministry will get the successes when the allocated liquid cash assets are equal to required cash asset amount and also successful utilization of allocated cash assets. But, here the realistic problem will starts that most of time allocated liquid cash assets are lesser than or greater than the required cash asset amount of particular objective and real also problem will be raised once they started to execute. Managing, controlling the asset based liquid cash balance issues are difficult and important in business and financial based administration [2]. The use of evolutionary computational algorithms along with the genetic algorithms can reduce limitations when developing more complex models for reducing the various constrains such as WZSAV and SZEAV and making the computational implementation easier in according towards to firm-based business and financial transaction based environments such as banks, stock markets, budget preparations, and so on. But GA is having good computational mechanism for generating the best-fit asset value for making it balance for further execution.

The asset-cash balanced based supervision models are showing the alternatives between the asset-cash based resource supervision and the type of objective speculation selection, and the typically the extremely high range assets like as stoke exchange assert and investment based bonds concern with prosperity, possibility and liquidity and also it will try to slow down the sensible realistic applications when the computationally concern with active representational procedures. In the concern of liquid cash balancing, the designed architectures represent and derive the collections of liquid asset values from different resources in cases of shortage of liquid cash assets lack of resources scarcity concern with the various levels of cash-provisions, mentioned time periods and financial transaction expenditures.

Many researchers are proposed the various probability circulations such as the Brownian motion and Poisson distribution are signify in Wiener process. In the Poisson distribution is concern with relevant case studies by concentrating the credit–debit-pay rolls over the precise time periods, which is what, justifies its application [2]. The authors groups of Simutis et al. [10], the effective cash management based advanced algorithms to truthfully forecast the cash supply-and-demand which is consent to the banks to pro-actively administer the practice of currency all through their association. Recently, some authors attempted to optimize the cash by modeling and forecasting the demand. The more general problem of modeling cash flow has been first investigated by Miller and Orr in 1966 [11], concluding that the cash flow is mostly unpredictable. The cash balance is occurrences irregularly over the mentioned time [12] based on their implicated approaches [7]. Then it has sustained with the implications of Miller-Orr finale. The deficient in the demand of visibility is frequently measured the most important test-case in cash management-optimization approaches [7]. Based on the present and future cash balance demand, the liquid cash should be maintained and uploaded into the machines or hand-to-hand cash supply or e-transactions. The basis of stochastic modeling approaches are forecasting and withdrawals the support to maintain the sufficient liquid cash due to the future demand and which is hard to implement due to high unpredictability and non-stationary of demand [12].

2 Identified Drawbacks of Asset Distribution in Annual Budget Implementation

In India, some of finance profession are facing the technical hitches for equal asset allotment of purpose based liquid asset values in every year budget matters regarding their total asset value and objective based sufficient asset and insufficient assets and so on. Various states are willing to prepare the plan of year-wise budget-based on their allotted total assets into available resources and infrastructure based on the purpose as shown in the Tables 1, 2, 3 and 4. Some states are having insufficient funding resources and infrastructure for a particular purpose and event, in the same time the other or nearby state is having extra surplus recourses for a same event that could be lack in previous state. Here they have the environment to make it balance and convert insufficient allocation converted it into balance allocation. This methodology is having the idea to overcome this problem by using the genetic-based algorithm implications.

Table 1 Telangana state annual budget allocation sheet [14] of Data for 2016–17 (in Rs crore) [14]

Department	Revised budget of 2015–16	Revised rear 2016–17	% change from RE 2015–16 to BE 2016–17	Provision of budget asset values for 2016–17
Irrigation	Rupees of 8,966	Rupees of 24,132	169.1	66.2% o f the allocation to b e spent on creating capital assets related to irrigation. such as dams
Social welfare	Rupees of 10.007 crores	Rupees of 14,617 crores	46.1	Scheduled Caste (SC): Allocation of Rs 7,122 crore. which is an increase of 59.6% over' 2015–16 Scheduled Tribe (ST): Allocation of Rs 3.752 crore. which is an increase of 48.7% over 2015–16. Backward Class (BQ: Rs 2,538 crore, which is an increase of 32% over 2015–16
School education	Rupees of 8,266 crores	5,575 crores	3.7	Residential school hostels of SC. ST and BC minorities
Agriculture	Rupees of 6,312 crores	6,611 crores	4.7	Rupees of 4,250 crore to be spent on relieving debt of farmers in the form of Soil Health Cards
Rural development	Rupees of 6,736 crores	6,345 crores	−5.5	Rupees of 2.3 32 to be spent on MNREGA and for other health care problems government spent the rupees of 2,878 crores of funding
Medical and health	Rupees of 4,148 crores	5,967 crores	43.8	Rupees of 600 crores for hospital medical equipment for new buying or maintenance. Rupees of 316 crores for medical diagnostic purpose
Percentage of total expenditure	44.4%	50.8%	–	
Other departments	Rupees of 55.627 crores	Rupees of 64,169 crores		

3 Methodology Implications for Asset Equal Distribution

GA principles can helpful for balancing the uncertainties of budget asset values. This paper proposing the load balancing approach of genetic algorithm for make the balance and equal distribution of budget asset values in year-wise financial transitions, firstly we gather the total budget asset values from individual neighbor states as clusters, secondary to identify the strong zone excess-asset and week zone scarcity-asset values of a particular objective, here it generate alert signals based on the zones [13].

Table 2 Telangana Budget 2016–17—Key figures (in Rs crore) [14]

Items	2014-15 Actual	2015-16 Budgeted	2015-16 Revised	% change from BE 2015-16 to RE of 2015-16	2016-17 Budgeted	% change from RE 2015-16 to BE 2016-17
Total Expenditure	62,306	1,15,689	1,00,062	13.5%	1,30,416	30.3%
A. Borrowings	9,580	19,630	20,327	3.6%	25,580	25.8%
B. Receipts (except borrowings)	52,893	96,056	79,835	-16.9%	1,04,849	31.3%
Total Receipts (A+B)	62,473	1,15,686	1,00,162	-13.4%	1,30,429	30.2%
Revenue Deficit (-)/Surplus(+)	369	531	61		3,718	
As % of state GDP	0.07	-	0.01		0.55	
Fiscal Deficit (-)/Surplus(+)	-9,410	-16,969	-16,912		-23,467	
As % of state GDP	1.78	-	2.90		3.50	
Primary Deficit (-)/Surplus(+)	-4,184	-9,414	-9,749		-15,761	
As % of state GDP	0.79	-	1.67		2.35	

Table 3 Telangana state expense asset values in budget 2016–17 (in Rs crore) [14]

Item	2014–15 actuals	2015–16 budgeted	2015–16 revised	% change from BE 2015–16 to RE 2015–16 (%)	2016–17 budgeted	% change from RE 2015–16 to BE 2016–17 (%)
Capital expenditure	11,633	22,089	20,810	−5.8	33,209	59.6
Revenue expenditure	50,673	93,600	79,252	−15.3	97,206	22.7
Total expenditure	62,306	1,15,689	1,00,06,2	−13.5	1,30,416	30.3
A. Debt repayment	1,777	3,714	3,765	1.4	3,149	−16.4
B. Interest payments	5,227	7,555	7,163	−5.2	7,706	7.6
Debt servicing (A+B)	7,004	11,269	10,928	−3.0	10,856	−0.7

Table 4 Chhattisgarh state annual budget allocation sheet [15] of Data for 2016–17 (in Rs crore) [14]

Department	2015–16 budgeted	2016–17 budgeted	% change from BE 2015–16 to BE 2016–17 (%)	Budget provisions for 2016–17
General education	11,645	12,921	11.0	Rs 2,200 crore for Sarva Shiksha Abhiyaan and Rs 682 crore for Madhvamik Shiksha Abhivaan
Food and storage	7,976	4,570	−42.7	Allocation of Rs 3,324 crore for Mukhvamantri Khadyaan Sahayata Yojana for 60 lakh beneficiaries
Roads and bridges	4,517	5,625	24.5	Allocation of Rs 4,640 crore for construction of bridges, upgradation of roads, etc.
Pension	3,780	5,183	37.1	Allocation of Rs 5,183 crore has been made for pensions. In 2015–16 here was a 51% increase from 2014–15
Medicine along with public health care	2,724	3,335	22.4	Rs 388 crore for Primary Health Centers along with Rs 950 crore for National Health Mission
% of total expenditure	47.1%	45.2%		
Other departments	34,371	38,425	11.8	

And finally to frame transition trading rules based on the load balancing approach with help of genetic algorithm genetic-reproduction, genetic-crossover and genetic-mutation principles then to exchange the services financial asset data to each other. Genetic algorithm is applied to each objective based cluster and this centralized cluster job schedule the rules which considers load balancing to prevent the node-based asset value from getting strong zone excess-asset or become week zone scarcity-asset values. Using GAs, every initial population solution is encoded as a sequence of fixed length with help of finite initial population variables A1, A2 and B1, B2 for the two different type's objective based investments I1 and I2. Each objective based investments of initial population samples are treated as an individual cluster groups.

Methodology implementation basically starts with as arbitrarily generated population of size as SZEA, WZSA from cluster group A and cluster group B. With various levels of iterations, a new active cum perfect population of same or different size is generated from the current cum initial population sample by using three basic GA implication functionalities like as selection, crossover, mutation and reproduction with concern looping structure based on the individuals cluster groups of the population. Crossover: Single point—part of the first initial population sample asset value is copied and the remaining considered as in the same order as in the second first initial population sample asset value. In dual points, two basic initial population sample asset values of the first investment population are copied and the rest between is taken in the same order as in the second first investment population and in neutral point, no need to use crossover, offspring is accurate reproduction of initial population sample asset values. Mutation: Through normal-random procedure only some initial population asset values are chosen and exchanged, through random procedure only few recovering like as only some initial population asset values are randomly selected and interchanged only when they progress solution based on fitness, through systematic procedure only getting better initial population asset values are systematically selected and interchanged only when they progress solution based on fitness.

Through random improving procedure, the same as random, only getting better, but before this the normal-random mutation functionality has performed. Through the systematic improving procedure, the same procedure as systematic, only improving, but before this the normal-random mutation functionality has implemented and through neutral procedural, no need apply mutation functionality for further improvements.

3.1 State 1-Telangana Budget (2016–17) Analysis and Case Study

And also year-wise improvement scenario with item based description along with the total expenders, revenue deficit, fiscal deficit and primary deficit.

3.2 State 2-Chhattisgarh Budget (2016–17) Analysis and Case Study

See Tables 4, 5 and 6.

3.3 Methodology Implications on State Wise Budget with Genetic Case Studies

See Figs. 2, 3 and Table 7.

Table 5 Chhattisgarh Budget 2016–17—Key figures (in Rs crore) [15]

Items	2014-15 Actuals	2015-16 Budgeted	2015-16 Revised	% change from BE 2015 -16 to RE of 2015-16	2016-17 Budgeted	% change from RE 2015-16 to BE 2016-17
Total Expenditure	46,207	65,013	65,898	1.4%	70,059	6.3%
A. Borrowings (Public Debt)	5,103	6,258	6,248	-0.2%	7,524	20.4%
B. Receipts (except borrowings)	41,017	58,677	59,566	1.5%	62,448	4.8%
Total Receipts (A+B)	46,120	64,935	65,814	1.4%	69,972	6.3%
Revenue Deficit (-)/Surplus(+)	-1,564	4,227	6,948		5,037	
As % of state GDP	-0.70	1.85	1.57		1.79	
Fiscal Deficit (-)/Surplus(+)	-8,075	-6,836	-6,832		-8,111	
As % of state GDP	3.62	3.00	2.72		2.88	
Primary Deficit (-)/Surplus(+)	-6,412	-4,754	-4,750		-5,539	
As % of state GDP	2.87	2.07	1.89		1.97	

Table 6 Expenditure Chhattisgarh Budget 2016–17 (in Rs crore) [15]

Item	2014–15 actual	2015–16 budgeted	2015–16 revised	% change from BE 2015–16 to RE 2015–16 (%)	2016–17 budgeted	Change from RE 2015–16 to BE 2016–17 (%)
Capital expenditure	6,621	11,000	10,749	−2.3	13,004	21.0
Revalue expenditure	39,497	53,730	54,866	2.1	56,390	2.8
Loans and advances	90	283	283	0.0	665	135.0
Total expenditure of which	46,207	65,013	65,898	1.4	70,059	6.3
Interest payments	1,664	2,081	2,081	0.00	2,572	23.6

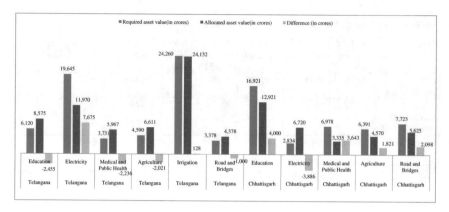

Fig. 2 Graphical representation of state and department wise asset allocated and required values and their lagging differences

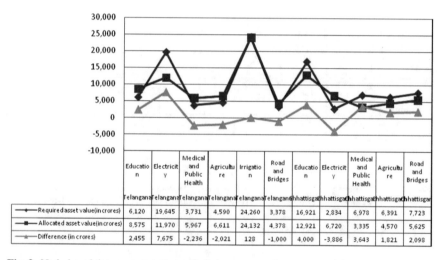

Fig. 3 Node-based data representation with their concern values

3.4 Genetic Cluster Based Objective Algorithm [13]

Statement-0: Begin
Statement-1: Cluster Group – A (Investment1)
Statement-2: Identify the strong zone excess-asset (SZEA) and week zone scarcity-
                asset (WZSA) values of a particular objective.
Statement-3: If Required-Asset (I1) < Available-Budget-Asset
Statement-4: Then   SZEA → Population sample (A1)
                f(A1) = Available budget Asset - Required asset
                Otherwise
Statement-5: WZSA → Population sample (A2)
                f (A2) = Available budget Asset - Required asset

Table 7 State and department wise asset allocated and required values and their differences

State	Department	Required asset value (in crores)	Allocated asset value (in crores)	Difference (in crores) SZEA, WZSA ->Population sample (A) f (A)= Available budget asset—required asset	Status
Telangana	Education	6,120	8,575	−2,455	WZSA
	Electricity	19,645	11,970	7,675	SZEA
	Medical and public health	3,731	5,967	−2,236	WZSA
	Agriculture	4,590	6,611	−2,021	WZSA
	Irrigation	24,260	24,132	128	SZEA
	Road and bridges	3,378	4,378	−1,000	WZSA
Chhattisgarh	Education	16,921	12,921	4,000	SZEA
	Electricity	2,834	6,720	−3,886	WZSA
	Medical and public health	6,978	3,335	3,643	SZEA
	Agriculture	6,391	4,570	1,821	SZEA
	Road and bridges	7,723	5,625	2,098	SZEA

Statement-6: Cluster Group – B (Investment2)
Statement-7: Identify the strong zone excess-asset (SZEA) and week zone scarcity-asset (WZSA) values of a particular objective
Statement-8: If Required-Asset (I2) < Available-Budget-Asset
Statement-9: Then SZEA → Population sample (B1)
            f(B1) = Available budget Asset - Required asset
            Otherwise
Statement-10: WZSA → Population sample (B2)
            f(B2) = Available budget Asset - Required asset
Statement-11: Genetic-Selection:
Statement-12: Choose the models based on SZEA (A1, B1) and WZSA (A2, B2) values of a particular objective (Investment 1and 2) into new population
Statement-13: Genetic Mutation, Accept and replace:
Statement-14: SZEA (A1, B1) and WZSA (A2, B2) asset values
Statement-15: f(SZEA(A1)) = f(WZSA(B2)) and f(SZEA(A2)) = f(WZSA(B21))
Statement-16: End

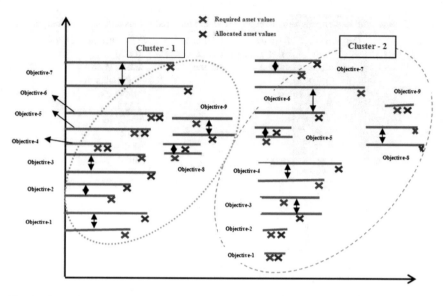

Fig. 4 Cluster-based objective concern budget asset values, required and allocated along with their differences and status

3.5 Genetic Cluster Based Objective Approach

This generated and accepted new outcome of best-fit can be implicated for further generations for getting advanced realistic outcomes. GA is produced the different optimal strategies, aimed at identifying a set of coefficient rules, it can able to reduce the wastage of asset values and to undertaking it into the availability at the same time to others whose are having same type of requirement.

The cluster based objective concern budget asset values, required and allocated along with their differences and status as shown in the Fig. 4. The red colored text indicating the information regarding required-asset-values and blue colored text indicating the information regarding allocated-asset-values, this implication can be implicated by using the mention genetic algorithmic [13] sequences (Fig. 5).

4 Conclusion and Future Scope

The Liquid Cash equal distribution is big challenge for many developing countries such as India, Sri Lanka, and China. These countries will plan their financial years in the form of year-wise-budget models. Managing, controlling the asset based liquid cash balance issues are difficult and important in business and financial based administration. But, here the realistic problem will starts that most of time allocated liquid cash assets are lesser or greater than the required cash asset amount of particular objective and real also problem will be raised once they started to execute.

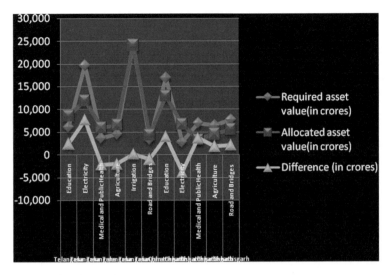

Fig. 5 State-wise Cluster-based objective concern budget asset values, required and allocated along with their differences and status

This paper described the full comparative analysis of various asset-cash balancing distribution of different state wise annual budget formation in table forms along with their department wise allotted asset value, objective and percentage of increment. Here described the year-wise improvement scenario with item based description along with the total expenders, revenue deficit, fiscal deficit and primary deficit. It will delineate the genetic algorithm implications are implied to the major problem of asset equal distribution with the genetic-reproduction, genetic-crossover and genetic-mutation principles then to exchange the financial based services like as asset values and resources to each other using the traditional based mean difference methodology and recompose the cluster-based inequality asset values into equality with concerns of the asset objectives and derived resources. This methodology will more helpful the economist and finance profession for equally distributing the total liquid asset values into resource-based- and liquid-based asset values.

References

1. Jackson, A.: Genetic Algorithms for Use in Financial Problems
2. da Costa Moraes, M.B., Nagano, M.S., Sobreiro, V.: Stochastic cash flow management models: a literature review since the 1980s. In: Decision Models in Engineering and management, Decision Engineering. Springer international Publications, Switetzerland (2015)
3. Tapiero, C.S., Zuckerman, D.: A note on the optimal control of a cash balance problem, Vol. 4, n. 4, pp. 345–352. JBF (1980)
4. Millbourne, R.: Optimal-Money-Holding Under Uncertainty, vol. 24, n. 3, pp. 685–698. IER (1983)

5. Hinderer, K., Waldmann, K.H.: Cash Management in Randomly Varying Environment. EJOR E **130**(3), 468–485 (2001)
6. Baccarin, S.: Optimal Impulse Control for Cash Management with Quadratic Holding-Penalty Costs, vol. 25, n. 1, pp. 19–32. DEF (2002)
7. Premachandra, M.: A diffusion approximation model for managing cash in firms: an alternative approach to the miller-orr model. Eur. J. Oper. Res. **157**(1), 218–226 (2004). Smooth and Nonsmooth Optimization
8. Volosov, K., Mitra, G., Spangolo, F., Carisma, C.L.: Treasury Management Model with Foreign Exchange Exposure, vol. 39, n. 1–2, pp. 179–207. CPA (2005)
9. Matinez, M., Garcia-Nieto, S., Sanchis, J., Blasco, X.: Genetic Based Algorithms Optimization Techniques for Normalized Normal Constraint Model Under Pareto Construction, vol. 40, n. 4, pp. 260–267. AES (2009)
10. Simutis, R., Dilijonas, D., Bastina, L., Friman, J., Drobinov, P.: Optimization of cash balance management for ATM based network, vol. 36, no. 1A, pp. 117–121. ITC, Kaunas, Technologija (2007)
11. Miller, M.H., Orr, R.: A method of the demand of money for firms. Quart. J. Econ. **80**, 413–435 (1966)
12. Armenise, R., Birtolo, C., Sangianantoni, E., Troiano, L.: Optimizing ATM cash management by GAs. IJCISIMA **4**, 598–608 (2012). ISSN 2150-7988
13. Neelima, Ch., Sarma, S.S.V.N.: Load balancing approach of genetic algorithm for balancing and equal distribution of budget asset values in finance transactions. In: 2017 Second IEEE International Conference on Electrical, Computer and Communication Technologies, pp. 1153–1158, Conducted at SVS Collgege of Engineering on 22–24 Feb 2017
14. http://www.prsindia.org/administrator/uploads/general/1458039598_Telangana%20Budget%20Analysis%202016-17.pdf
15. http://www.prsindia.org/administrator/uploads/general/1457610311_Chhattisgarh%20Budget%20Analysis%202016-17.pdf

A Theoretic Approach to Music Genre Recognition from Musical Features Using Single-Layer Feedforward Neural Network

Sourav Das and Anup Kumar Kolya

Abstract Musical genres are categorical classifications that are used to distinguish between different types of music. Each genre differs from other genres in certain musical features. In pre-computational intelligence era, music genre categorization has traditionally been performed manually, mostly due to the lack of modern human—computer interaction concept, and obviously for the lack of enough computational processing abilities of the computers. However, with the ever-increasing number of digital music and vast features, genre recognition using Neural Network is producing a wide range of results across a variety of experiments recently. By studying and extracting information on such features, with applying relevant Neural Network algorithm and technique, and also exploring some new recognition techniques on the same dataset which has been used in established research works, we hope to discover and gather new information about genre classification, and further understand future potential directions and prospects that could improve the art of computational musical genre recognition, decomposition of the clustered data corpus, and as a whole construction thereafter.

Keywords Music genre recognition · Artificial neural network · Single-Layer feedforward neural network · Unsupervised learning

1 Introduction

Genre classification is a popular topic for research, particularly in the fields of Music Information Retrieval (MIR) and Neural Network. Since Artificial Neural Network with a variety of learning methods have been introduced in this domain, there have been many research works done on methods for classifying genres according to the

S. Das (✉) · A. K. Kolya
Deptartment of CSE, RCC Institute of Information Technology, Kolkata 700015, India
e-mail: sourav.das17.91@gmail.com

A. K. Kolya
e-mail: anupkolya@gmail.com

© Springer Nature Singapore Pte Ltd. 2019
A. Abraham et al. (eds.), *Emerging Technologies in Data Mining and Information Security*, Advances in Intelligent Systems and Computing 755,
https://doi.org/10.1007/978-981-13-1951-8_14

diversity of their features. Human brain computes in a distinctly different way rather than the conventional computers. Keeping this concept in mind, the research work on Artificial Neural Network has been motivated. The main ideology of a Neural Network is to represent a linear or nonlinear and parallel computing architecture. Researchers have also suggested methods for music genre classification using multiple feature vectors and pattern recognition ensemble approach. Music segments are also decomposed according to time segments obtained from the beginning, middle and end parts of the original music signal (time-decomposition), alongside with distinguished musical features such as pitch, intensity, beat, drop and so on, which is which is unlikely to be the same for even two distinct music files. In the present scenario, one of the biggest bottlenecks in many Music Information Retrieval (MIR) tasks is the access to large amounts of music data and their features, in particular to audio features extracted from commercial music recordings [1]. However as of now, at least hypothetically these challenges can also be addressed by using multilayer neural network with more than one hidden layers within it, and with backpropagation feedback signaling. In this paper, we present a robust theoretical work to show how the music genre recognition can be done from distinct musical features using a single layer feedforward neural network combining with an unsupervised learning algorithm (Hebbian learning) and propose the paths for future enhancements.

2 Related Works

Related work contains a brief but orderly description of research works carried out on this domain so far. To understand the evolving trend of content-based genre recognition in Music Information Retrieval using Neural Network over the past years, we have carried out the literature survey thoroughly from the early 2000s to the most recent and relevant works. Each research work has made a positive impact in our way of approach towards the matter and objectifying new scope and possibilities.

Tzanetakis et al. [2] stated that manual categorization of music is commonly used to classify and describe music [2]. But the increment of digital music on Internet over the time has led to the inevitability of Music Information Retrieval. Through their research work, they explored the algorithms for automatic genre classification. They also proposed a set of features for representing the music surface and rhythmic structure of audio signals. The performance of the training dataset represented by them was measured and evaluated by training statistical pattern recognition classifiers using audio collections collected from compact disks, radio, and the web relevant at that time.

In another research work, 109 musical features were extracted from music playback and recordings, and furthermore, were classified according to their genres by McKay [3]. The features were based on the distinct characteristics of the music files that were used for the experiment. The classification experiment was carried out using different sets of features for different hierarchical levels, which were actually determined by using genetic algorithms.

Pattern recognition combined with vector-based approach has also been made for music genre classification problem [4]. The researchers labeled music genre classification by binary classifiers, and finally concluded the result by merging all the results gathered from each test case.

Each audio file was decomposed based on time segments such as beginning part, middle part, and the end part of each song. Besides using conventional Machine Learning Algorithms, authors have also implemented Support Vector Machines (SVM) and Multilayer Perceptron Neural Networks.

Deep Belief Network (DBN) can also be used to extract features from audio files. A Deep Belief Network is basically a Multilayer Feedforward Neural Network, with multiple (deep) layers of hidden layers between the input and output layer. In their work, Hamel et al. [5] trained the activation functions of this network with a nonlinear Support Vector Machine [5]. This training eventually led them to solve the music genre identification problem up to an extent with an accuracy of 84.3%.

Dieleman et al. [6] developed a Convolutional Neural Network with unsupervised learning method which was train to produce results such as artist recognition, genre recognition, etc. [6]. They experimented with the "Million Song Dataset" [7]. The audio dataset was fragmented and labeled, and the Multilayer Perceptrons were trained with those labels for each task. Henceforth, these researchers observed that their system gave an improved accuracy for both genre and artist recognition. They also stated that unsupervised method of learning ensured comparatively a more rapid convergence of produced accurate results.

Tacit knowledge earned by a neural network can also be used to extract the common features from music [8]. Rather than only audio-based music information retrieval, Humphrey et al. [8] made an approach to implementing the deep signal processing architecture. Through their work, the researchers observed that improved results about music information can be gathered through breaking a large network into smaller and simpler fragments or parts. They have also observed that applying the learning methods in flexible machines can really make a positive impact on musical feature extraction.

It is also observed that using Deep Neural Network for audio event classification can produce better results [9]. It is said to perform better than Support Vector Machine for the same task. The neural network was trained with the Restricted Boltzman Machine, which itself is a Genetic Algorithm's selection method. Boltzman Machine is used to produce a stable distribution of all available sample test population.

A drawback of applying neural network for Music Information Retrieval can be relatively larger training time of the network, with respect to the training process often gets stuck with local minima, i.e., considering the current best result as the universally best result in a widely distributed neural network architecture, when it is actually not. Sigtia et al. [10] experimented with the ways to improve the learning capacity of a neural network in reduced time with the help of Rectified Linear Units (ReLU) as the activation functions, and expansion-conjugation methods such as Hessian-Free Optimization [10].

Dai et al. [11] stated that, since musical genres do not consist the "multilingual" feature, hence they implemented the "nearest neighbor algorithm" to label the large musical database again afterwards classification [11]. They trained a multi deep neural network (DNN) using a large musical database, for so the DNN can learn quickly and transfer the musical files to the similar but smaller dataset, which is the target dataset. Finally, the authors evaluated the performance using a benchmark for many such popular music databases.

Jeong and Lee [12] explained the framework for a mundane audio feature learning system using the deep neural network, leading it to the implementation of music genre classification [12]. Also, they observed the conventional spectral feature learning framework, and modified it with cepstral modulation spectrum domain for better performance, which already has been applied previously in many successful speech or music feature learning experiments.

Choi et al. [13] introduced a convolutional neural network for music tagging by extracting the local musical features and summarizing those extracted features [13]. Furthermore, they compared the performance of the Convolutional recurrent neural network that they used, with the performance of a general convolutional neural network with respect to input parameters and training time per feature. They observed and presented that the CRNN they proposed, actually delivers a better performance due to the heterogenicity of this network in music feature extraction and summarization.

3   Motivation

The literature survey leads us to identify possible research areas which can be extended further. The aspects which can be carried out for further research are

1. The primitive motivation for proposed approach is to enhance the Natural Language Processing capabilities of machines and overall reflecting the growth for the betterment of human–computer interaction. It is already established that natural language processing can be related to music information retrieval [14]. However, it is yet unchallenged that how will neural network perform in enhancing the learning capacity of machines by classifying natural and distinct features of music files.
2. The neurons of each input unit from the input layer of a neural network accept bipolar (i.e., 0 or 1) values as the primary training input. Hence for, labeling different genres of music using binary values (for, e.g., 001, 010, 011, etc.) is needed to build the input training patterns for a neural network. It would be interesting to observe the learning and thereafter recognition performance of a neural network trained with such perquisite values.

4 Dataset

The widely popular dataset for music genre recognition is the GTZAN [15] dataset, which was originally developed by George Tzanetakis. It is a dataset of size 1.5 GB, consisting 100 music examples of 1o different musical genres consisting of 30 s sample for each music, such as Hip Hop, Rock, Metal, Jazz.

Another available and versatile dataset is the Million Song Dataset by Bertin-Mahieux et al. [7]. The Million Song Dataset contains 280 GB of data with 1,000,000 songs per file, 44,745 unique artists, etc.

The split validation of the dataset should be in a ratio of 0.66 and 0.34, i.e. 66% of the dataset should be implemented for training purpose, while the rest 34% of data should be applied for testing purpose after the training is completed.

5 Representation of the Neural Network

As we have mentioned in Sect. 1, there are several classes of Neural Network, such as Single Layer Feedforward Neural Network, Multilayer Feedforward Neural Network, Recurrent Network, etc. A single-layer network actually comprises two layers, one input, and one output layer. Whereas, multilayer network is made up of multiple layers, i.e., input, output, and one or more intermediary layers called hidden layers.

For our work, we have followed the simplistic approach of Single-Layer Feedforward Neural Network. In SLNN, only output layer can perform the computation, hence it is called the single layer. In addition, the synaptic links carrying the weights [16], connect every input neuron to the output neurons, but not the reverse. Hence it is called the feedforward network.

We show the diagram of a Single-Layer Feedforward Neural Network in Fig. 1.

5.1 Mathematical Representation

For output neuron y_1

$$y\_in1 = x_1 w_{11} + x_2 w_{21} + \cdots + x_n wn_1 = \sum_{i=1}^{n} x_i w_i \qquad (1)$$

In vector notation

$$y\_in1 = [x_1 \ x_2 \ \ldots \ x_n] \times \begin{bmatrix} w_{11} \\ w_{21} \\ w_{n1} \end{bmatrix} = X \times W_1$$

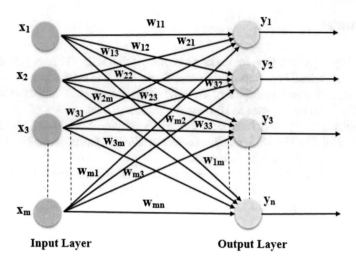

Fig. 1 Single layer feedforward neural network

where $X = [x_1 \ x_2 \ ... \ X_n]$ is the input vector of the input signal, and W_{*j} is the first column of the weight matrix as

$$W = \begin{bmatrix} w_{11} \ w_{12} \ w_{1m} \\ w_{21} \ w_{22} \ w_{2m} \\ w_{n1} \ w_{n2} \ w_{nm} \end{bmatrix}$$

In general, the net input y_{inj} to the output unit Y_j is given by

$$Y\_in_j = [x_1 x_2 \ ... x_n] \times \begin{bmatrix} w_{1j} \\ w_{2j} \\ w_{nj} \end{bmatrix} = X \times W_{*j} \tag{2}$$

If Y_in denotes the prime vector for all the net inputs to the array of input units,

$$\sum Y\_in = [y\_{in1} + y\_{in2} + \cdots + Y\_{inn}]$$

Then the next input to the entire array of output can be expressed concisely in matrix notation as

$$Y\_in = X \times W \tag{3}$$

The signals transmitted by the output units, depending on the nature of the Activation Function [17].

6 Training

A neural network can be trained or it can learn by several learning methods, such as Supervised Learning, Semi-Supervised Learning, Unsupervised Learning, Reinforced Learning, etc. The most commonly used learning methods are Supervised Learning and the Unsupervised Learning, which can be further categorized [18]. In Supervised Learning method, every input pattern that is used to train the neural network is associated with an output pattern. On the other hand, in Unsupervised Learning, the output is not presented to the network, rather the network learns by its own by discovering to structural features of input patterns. We are considering Hebbian Learning technique (Unsupervised Learning) featured with music genre labeling for the learning process of single-layer neural network.

Hebb Rule, or Hebbian Learning technique tautologize repeatedly in an input-output pattern, which can also be depicted as training vectors.

In Hebbian Learning, if two cells fire simultaneously, then the strength of the connection or weight between them should be increased. The increment in weight between two neurons is proportional to the frequency at which they operate together. Moreover, as the Hebb Rule accepts input/output patterns in bipolar form, the distinct music genres can be labeled as different binary values, and henceforth, we can proceed by creating a training set by using AND function.

First, we will consider the general training pattern of a neural network with Hebb's Rule, then we will proceed further with the case specific training for our approach.

Let us assume that, we have a pair of training vectors α and β, in correspondence with the input vector h.

According to Hebb's rule, let us first assign all the initial weight to 0, i.e.,

$$W_{ij} = 0, \text{ where } i = 1 \ldots m \text{ and } j = 1 \ldots n \tag{4}$$

For each input–output training epoch, the new input variable in the current input set will be

$$\alpha_i = h_i \ (i = 1 \ldots m)$$

With correspondence, the new output variable to the current output set will be

$$\beta_j = g_j \ (j = 1 \ldots n)$$

Henceforth, we have to adjust the weight using the following equation

$$W_{ij(new)} = w_{ij(old)} + h_i g_j, \ (i = 1 \ldots m \ \& \ j = 1 \ldots n) \tag{5}$$

From the above-mentioned equation, we can set the activation units for the binary output results of the neural network

Table 1 Training input table generated using Hebbian learning algorithm

h_0	h_1	h_2	Output
1	1	1	1
1	1	−1	−1
1	−1	1	−1
1	−1	−1	−1

$$g_j = 1, h > 0 \ (1)$$
$$= 0, h \leq 0 \ (-1)$$

Following is the training input table: (Table 1).
During training, all weights are initialized to 0.
So, $w_0 = w_1 = w_2 = 0$
The initial equation for Heb Rule

$$\Delta W_i = h_i \times t$$

Now, the revised equation

$$\Delta w_i \ (new) = w_i \ (old) + \Delta w_i \ (where, \ \Delta w_i = w_i \times t) \qquad (6)$$

In succession of this equation, we present the following learning iterations in a tabular form (Table 2).

With the above-generated model learning inputs, we also need to show the musical features in correspondence with a similar binary form, from which the neural network will learn and will be able to recognize such kind of features after a number of epochs. Hence, we explain the bipolar labeling of the different musical features as mentioned earlier. These features were initially proposed by Tzanetakis and Cook [2]. By following these features, Tao Li et al. made an accuracy table [19] of different Machine Learning methods tested on GTZAN dataset [15], which we have mentioned in Sect. 4. We take a reference to only four distinct musical and speech recognition features from their experiment trained by them using Support Vector

Table 2 Learning iterations with complete input vector set

No. of training patterns	h_0	h_1	h_2	t	Δw_0	Δw_1	Δw_2	w_0	w_1	w_2
								0	0	0
1	1	1	1	1	1	1	1	1	1	1
2	1	1	−1	−1	−1	−1	1	0	0	2
3	1	−1	1	−1	−1	1	−1	−1	1	1
4	1	−1	−1	−1	−1	1	1	−2	2	2

Table 3 Certain musical features in Single and Multi-class objective functions

Features	Methods	
	SVM1	SVM2
Beat	26.5 (3.30)	21.5 (2.71)
FFT	61.2 (6.74)	61.8 (3.39)
MFCC	58.4 (3.31)	58.1 (4.72)
Pitch	36.6 (2.95)	33.6 (3.23)

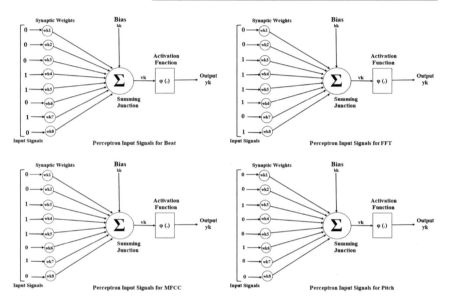

Fig. 2 Input patterns (or Input Signals) of perceptrons for different musical features

Machine. These features are Beat, Fast Fourier Transform (FFT) which decomposes a signal into a number of signals in a particular given time point, Mel-Frequency Cepstral Coefficient (MFCC) which is a collection of visibly specific speech recognition features, and finally, Pitch.

The table is represented (Table 3 and Fig. 2).

Here SVM1 suggests single class objective function, and SVM2 suggests multi-class objective function. As we are using a Single-Layer Feedforward Neural Network for our approach, we take each feature's distinct value in SVM1 and convert them into binary values. Henceforth we get the following Values (Table 4).

Now, we can represent these distinct binary values for each feature as the input patterns (or input signals) of our neural network, as shown in Fig. 3.

Further following these distinct input patterns for each musical feature, the whole feedforward neural network can be trained for n number of iterations as per the 66% of the total dataset. After the training epoch is completed, the rest of the dataset (34%) should be applied for recognizing the musical features, which should also be

Table 4 Conversion of
Single-class objective
function values into
corresponding binary
representation

Features	Method
	SVM1
Beat	00011010
FFT	00111101
MFCC	00111010
Pitch	00100100

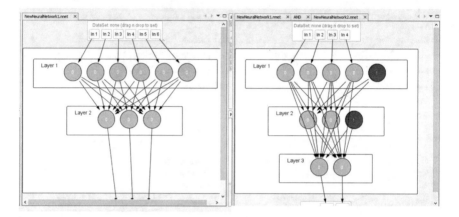

Fig. 3 Single Layer Neural Network versus Multilayer Neural Network in Neuroph Simulator

represented through the binary forms of the same musical features that are used to train the system (i.e., Beat, FFT, MFCC, Pitch).

7 Conclusion and Future Work

Through this paper, we represented a theoretical idea of how an approach of music genre recognition can be done from the distinct musical features, using one of the unsupervised learning techniques in the simplest neural network, i.e., Single-Layer Feedforward Neural Network. We proposed the basic structure and training (or learning method) of the neural network, and the different musical features as the inputs of the perceptrons used in that neural network, which can be implemented in this task. However, for our future work, we want to carry this idea as an extension and perform this similar task on a more complicated network, i.e., Multilayer Feedforward Network (Backpropagation Network) with real-time visual and performance simulation with the help of any neural network simulator, such as Java Neural Network Framework platform (Neuroph) or Python-based neural network simulator (NEST). We show a single layer versus multilayer neural network structure comparison done by us in Neuroph platform in Fig. 3. Alongside with that, as an extension of this work, we will also test and compare the performance of several Supervised and Unsupervised neural network learning techniques in this particular task.

References

1. Porter, A., Bogdanov, D., Kaye, R., Tsukanov, R., Serra, X.: Acousticbrainz: a community platform for gathering music information obtained from audio. In: International Society for Music Information Retrieval Conference, Oct 2015
2. Tzanetakis, G., Cook, P.: Musical genre classification of audio signals. IEEE Trans. Speech Audio Process. **10**(5), 293–302 (2002)
3. McKay, C.: Automatic Genre Classification of MIDI Recordings. Doctoral Dissertation, McGill University (2004)
4. Silla Jr., C.N., Koerich, A.L., Kaestner, C.A.: A machine learning approach to automatic music genre classification. J. Braz. Comput. Soc. **14**(3), 7–18 (2008)
5. Hamel, P., Eck, D.: August learning features from music audio with deep belief networks. In: ISMIR, pp. 339–344 (2010)
6. Dieleman, S., Brakel, P., Schrauwen, B.: Audio-based music classification with a pretrained convolutional network. In: 12th International Society for Music Information Retrieval Conference (ISMIR-2011), pp. 669–674. University of Miami (2011)
7. Bertin-Mahieux, T., Ellis, D.P., Whitman, B. Lamere, P.: The million song dataset. Ismir. **2**(9), p. 10 (2011)
8. Humphrey, E.J., Bello, J.P., LeCun, Y.: Moving beyond feature design: deep architectures and automatic feature learning in music informatics. In: ISMIR. pp. 403–408 Oct 2012
9. Kons, Z., Toledo-Ronen, O.: Audio event classification using deep neural networks. In: INTER-SPEECH. pp. 1482–1486 (2013)
10. Sigtia, S., Dixon, S.: Improved music feature learning with deep neural networks. In: Acoustics, Speech and Signal Processing (ICASSP), 2014 IEEE International Conference on, pp. 6959–6963. IEEE, May 2014
11. Dai, J., Liu, W., Ni, C., Dong, L., Yang, H.: Multilingual deep neural network for music genre classification. In: Sixteenth Annual Conference of the International Speech Communication Association (2015)
12. Jeong, I.Y., Lee, K.: Learning temporal features using a deep neural network and its application to music genre classification. In: ISMIR. pp. 434–440 (2016)
13. Choi, K., Fazekas, G., Sandler, M., Cho, K.: Convolutional recurrent neural networks for music classification. In: 2017 IEEE International Conference on Acoustics, Speech and Signal Processing (ICASSP), pp. 2392–2396. IEEE, Mar 2017
14. Amiri, N.: Natural Language Processing and Music Information Retrieval (2016)
15. Sturm, B.L.: The GTZAN dataset: its contents, its faults, their effects on evaluation, and its future use (2013). arXiv:1306.1461
16. Tavşanoğlu, V.: Neural Networks. Yıldız Technical University, Turkey
17. Karlik, B., Olgac, A.V.: Performance analysis of various activation functions in generalized MLP architectures of neural networks. Int. J. Artif. Intell. Expert Syst. **1**(4), 111–122 (2011)
18. Sathya, R., Abraham, A.: Comparison of supervised and unsupervised learning algorithms for pattern classification. Int. J. Adv. Res. Artif. Intell. **2**(2), 34–38 (2013)
19. Li, T., Ogihara, M. Li, Q.: A comparative study on content-based music genre classification. In: Proceedings of the 26th annual international ACM SIGIR conference on Research and development in informaion retrieval, pp. 282–289. ACM, July 2003

Analysis on Efficient Handwritten Document Recognition Technique Using Feature Extraction and Back Propagation Neural Network Approaches

Pramit Brata Chanda

Abstract Today, Handwritten recognition becomes a very much thrust area in the field of pattern recognition and image processing. Handwritten recognition methods are used in real-life fields such as banking checks, car plates number identification, recognition of ZIP code, mail sorting, reading of different commercial forms, etc. The work is presented in this paper a system of handwriting of English document recognition based on feature extraction of the character. Almost 400 handwritten numerals are collected from different datasets as for sample for the classification purposes. The main work is presented here are consisting of several steps like preprocessing, feature extraction and Multi-Layer Perceptron model of neural network for classifying handwritten digits separately. Basically, Offline Recognition of English handwriting using multilayer perceptron network or back propagation networks are described throughout the entire work. First, the English alphabets are used as features introducing features sets of different English handwritten documents. Then the neural network is used tos train the datasets. Different types of training methods of back propagation network are used for calculating performance of the entire system. The recognition methods are based on back propagation network for analyzing the classification performance of handwritten documents. Here the system achieves the accuracy more than 90% using this efficient back propagation neural network based classification and feature extraction methods using morphological operations based zones separation scheme of digits. Here, the performance performance parameter like sensitivity, specificity, recall, accuracy provides more than 90% of rates indicates that better classification of handwritten documents.

Keywords Handwritten character recognition · Recognition accuracy · Back propagation · Learning · Resilient · Recall · MSE · Feature extraction · Sensitivity

P. B. Chanda (✉)
Department of Computer Science and Engineering, Kalyani Government Engineering College, Kalyani, Nadia, West Bengal, India
e-mail: pramitcse@gmail.com

© Springer Nature Singapore Pte Ltd. 2019   157
A. Abraham et al. (eds.), *Emerging Technologies in Data Mining and Information Security*, Advances in Intelligent Systems and Computing 755,
https://doi.org/10.1007/978-981-13-1951-8_15

1 Introduction

Today a large amount of research has been going on different techniques that is being used for reducing the time of execution and providing higher recognition rate. In general, handwritten recognition is categorized into two types as offline and online recognition techniques. In case of offline recognition, the writing is generally captured by using a scanner and the scanned document is available as a gray image. But, for the cases of online system the two dimensional coordinates of successive points are represented as a function of time. The online methods are more superior to their offline parts in recognizing different document due to the information available is temporal compare with the previous. However, for the cases of offline systems, the neural networks have been successfully used to get higher accuracy level. Basically Offline recognition of Character is a technique of recognition of English alphabets from a set of character images and transfer it into its equivalent machine readable document format [1–3]. This work is a type of automation process where the interface between man and machine for several type applications are improved in different area [2, 4]. Recognition of Handwritten document is one of upcoming areas of research involved in pattern identification and artificial intelligence (AI) [4–6]. Handwritten Recognition consists of recognition of handwritten alphabets, symbol, or document Image. Different applications are used for many practical applications in different areas like ZIP code recognition [6, 7] mail sorting, commercial forms reading, reading of bank checks, and number recognition in car plates [8], Signature Verification [9, 10].

For working in this domain, primarily requires a standard benchmark database. Several standard databases, like as NIST, MNIST, CEDAR, and CENPARMI are available for recognition study. This task illustrates a new techniques for normalizing the size of the images, feature extraction and classification of handwritten documents using Artificial Neural Networks (ANNs). A MNIST and CEDAR database has been used in this task for the experimental purposes [11].

In this work the proposed technique are used for recognizing isolated characters using MLP with Back Propagation network having single hidden layer. This MLP network is trained by different training function for comparison of performance. The performance of the system is measured on the basis of different type of factors like MSE, Accuracy, sensitivity, and specificity factors. Here, I have taken a lot of training sample images for calculating the recognition accuracy.

2 Proposed Handwritten Recognition Methodologies

Here for designing a handwritten system for the feature extraction and training using different training function of neural networks are used to recognize the 26 letters of the alphabet and 10 numbers for different types of handwritten gray images [14, 12, 13]. Here a lot of training samples are taken from different datasets and each letters are represented as a matrix of boolean values which consisting of edge pixels. The character matrix is created according to edge pixels. It is made for different shapes of handwritten digits. The Offline Handwritten Recognition System is designed through different steps for simulating it into MATLAB. It consists of different steps like Preprocessing, Feature Extraction, Training and Testing through neural network and classification using confusion matrix for measurement of performance. These are consisting of different steps which are given (Fig. 1).

2.1 Preprocessing

The main objectives of preprocessing are to removing the variability among characters due to difference in size, and style of characters in different documents. Preprocessing operations are generally based on morphological-based approaches. For Preprocessing, I have taken a gray-scale image as input and then remove the noise part using median filter. After that Conversion of the gray-scale image to binary image using threshold value and then normalize the image to size of 40×40. Then thinning is used for Conversion of the normalized image to a single pixel image [9, 15].

2.2 Feature Extraction

After preprocessing, for capturing the most relevant characteristics of the handwritten to be recognized is applied through feature extraction. This techniques identifies 16 features by applying to the image and computing those values by moving direction in diagonally. The first stage is reading a preprocessed image. Each and Every handwritten character image of size 40×40 pixels is sub-divided into 16 equal zones for set of pixels in each character, each zone of size 10×10 pixels are considered for each and individual sub images of characters. The features of numerals are extracted from each zone pixels by moving along the diagonals of its respective 10×10 pixels. Each zone consisting of 19 diagonal lines and the foreground pixels are present along each diagonal line is summed to get a single sub-feature and thus 19 sub-features are obtained from the each zone. Finally, 16 features are extracted for each an individual image [16, 17].

2.3 Neural Network

There are two different stages where ANNs are used for processing of any kind of data, a training stage and a classification stage. These two stages are required for recognition of handwritten documents. In the classification stage, training data are

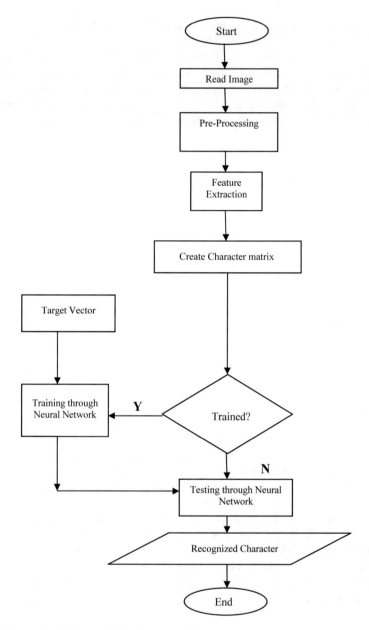

Fig. 1 Proposed technique in handwritten recognition system

passed as input vector into the back propagation network, resulting in a output vector produces by the ANN to be the most correct output results. However, to be successful in the stages of work, classification must be done by a training stage in which the ANN is given a set of inputs and the corresponding set of correct outputs. Training samples dataset are presented to the neural network during training and networks parameters are basically used to adjust towards minimizing the error value. Testing samples are used for measuring of network performance during time of training. A multilayer feed forward network is used for training the input data in which activation values are moves from input layer to output layer. Here the log-sigmoid transfer function used as activation functions. Neural network train to character matrices consisting of 1 for accurate edges are belonging and other for 0 for non-edge parameters [14, 18]. The hidden neurons values are approximately taken as per training requirement of the dataset given using guessing. The performance parameter varies after changing the values of each and every iterations the networks travels from different area.

2.4 Back Propagation Method

The Back propagation training algorithm is used for back propagating error for minimizing the system error and betterment of classification performance using multilayer feed forward network [19, 20]. There are several types of Back Propagation training algorithm are used in order to train the ANN and use log-sigmoid network. In case of batch mode, the values of weights are modified accordingly after all patterns are presented there, while in the incremental mode, the weights values are changed at each and every iteration after input pattern is presented [14, 21]. Back Propagation network requires some followed steps that are used here training as given below

Step1. Select the of the training pair data from the given training data set.
Step2. Calculate the output of the network as used training method.
Step3. Calculate the mean square error between the network output and the desired output value (the target vector from the training pair)
Step4. Set the weight vector value in a way for minimization of error.
Step5. Repeat steps 1–4 value until it belongs in the training datasets until the error value for given set of data comes to lower range of values.

3   Results and Analysis

Here, the performance analysis is shown where the relation between the iterations and the mse (mean squared error)/Performance factors are shown. It shows that if the number of iterations are increasing then the mse values are decreasing at a particular rate and after a certain limit it will be unchanged or remain constant. The no of hidden neurons values are changed on the basis of requirement of the problem and whatever situation demand according to dataset. If the no of iterations and values of hidden neurons are changed then the training rate, its factors affecting the performance is also changed. These results are collected using matlab. So, Resilient model and

Learning Vector model produce better performance in terms of less amount of error (MSE). Here the rate of recognition is altered in case of changing the hidden neurons values. So, Resilient model and Learning Vector model produce better Recognition Accuracy rather than comparing to Gradient Descent Training Algorithm (Tables 1, 2, 3, 4 and 5).

The confusion matrix consists of different parameters like Recall, Precision, Sensitivity, Specificity, Accuracy for measuring classification accuracy of handwritten system. Here the back propagation network based training methods are given better confusion parameters value than normal image segmentation scheme.

Table 1 Gradient descent training with momentum and adaptive learning based handwritten recognition

Hidden neurons values	Execution time	Maximum iteration	Mean Square error	Recognition rate (%)
10	0:00:03	280	0.5142	80
12	0:00:04	383	0.4595	85
13	0:00:03	178	0.5568	90
14	0:00:05	404	0.43544	80
16	0:00:06	464	0.3234	90
17	0:00:07	493	0.2068	85
18	0:00:08	555	0.2014	80
20	0:00:08	640	0.1937	80
22	0:00:09	689	0.17706	85

Table 2 Resilient back propagation neural network based handwritten recognition

Hidden neurons values	Execution time	Maximum iteration	Mean square error	Recognition rate (%)
10	0:00:03	183	0.3262	90
12	0:00:04	233	0.23905	90
13	0:00:04	254	0.2052	90
14	0:00:05	358	0.1737	90
16	0:00:05	421	0.1556	90
17	0:00:06	471	0.1237	100
18	0:00:07	544	0.1106	90
20	0:00:07	576	0.09806	90
22	0:00:08	665	0.08866	100

Table 3 Learning vector quantization back propagation neural network based handwritten recognition

Hidden neurons values	Execution time	Maximum iteration	Mean square error	Recognition rate (%)
10	0:00:05	111	0.05081	80
12	0:00:08	138	0.03051	85
13	0:00:09	154	0.02852	80
14	0:00:10	208	0.02678	80
16	0:00:11	221	0.02577	80
17	0:00:16	245	0.02325	80
18	0:00:16	288	0.02089	85
20	0:00:18	315	0.02007	85
22	0:00:18	380	0.01983	85

Table 4 Comparative results of different techniques

Different approach of handwritten recognition	Recognition accuracy	Execution time
Back Propagation Network	84.4	0.123
Gradient Descent Training Algorithm with momentum and adaptive learning	86.6	0.092
Resilient Back propagation Neural Network	90.33	0.087
Learning Vector Quantization Back propagation Neural Network	87.75	0.188
OCR using Image Segmentation and classification	78.12	0.234

Table 5 Calculation from confusion matrix

Techniques	Recall	Precision	Sensitivity	Specificity	Accuracy
Back Propagation	78.8	81.19	88.15	86.545	90.90
Gradient Descent	79.03	82.2	81.8	86.667	90.057
ResilientBack propagation	78.93	82.89	82.711	88.953	94.75
Learning Vector Quantization	80.24	84.65	82.356	88.365	92.6
OCRusingImage Segmentation and classification	77.07	80.48	78.861	85.802	88.95

3.1 Performance Characteristics

Here the graphs are given below which show that the mean square error value is decreasing after gradually increasing the no of iterations the neural network training. These figure given are used for assessing the performance of different techniques like as Gradient Descent Training Algorithm with momentum and adaptive Learning, Resilient Back propagation Neural Network other is Learning Vector Quantization Back propagation Neural Network Model for handwritten recognition (Figs. 2, 3, 4, 5, 6 and 7).

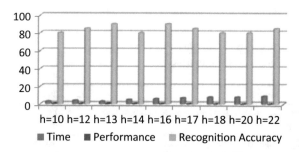

Fig. 2 Performance, time, recognition accuracy changes with hidden layer values for recognition system with gradient descent training algorithm with momentum and adaptive learning based recognition

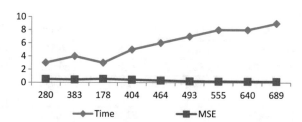

Fig. 3 Gradient descent training algorithm with momentum and adaptive lr back propagation algorithm based recognition (Time and performance vs. no of iterations in neural network)

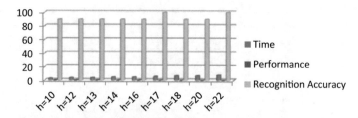

Fig. 4 Performance, time, recognition accuracy changes with hidden layer values for recognition system with resilient back propagation neural network based recognition

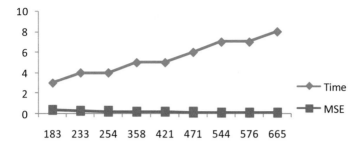

Fig. 5 Resilient back propagation neural network algorithm based recognition (Time and performance vs. no of iterations in neural network)

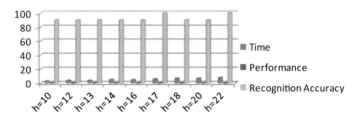

Fig. 6 Performance, time, recognition accuracy changes with hidden layer values for recognition system with learning vector quantization back propagation neural network based recognition

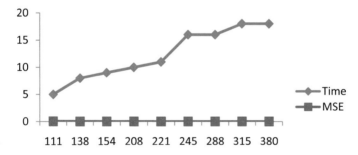

Fig. 7 Learning vector quantization back propagation neural network algorithm based recognition (Time and performance vs. no of iterations in neural network)

The performance plot shows here the time of execution of neural network and mean square error of the system. Here this two figure shows (i) the changes of performance, Recognition Accuracy with hidden layer values, (ii) the changes of performance, time with respect to number of iterations of neural network. These comparative results are used for analyzing the performance for Resilient, Gradient Descent, Learning

Vector Quantization training based back propagation neural network model based recognition system. Learning Vector Quantization Back propagation based system provides better recognition rate and performance rather than other Back propagation model.

4 Conclusion

In this work, the proposed technique based on feature extraction methods for betterment of the recognition performance of different types of handwritten digits, and neural network based system is used for classifying offline handwritten digits. The back propagation method used for these handwritten documents for higher recognition accuracy. The overall accuracy of the algorithm is more than 90% which is better than the previous approaches.

Here, experimental study shows that back propagation network using different training methods provides recognition accuracy of more than 90%. Here, handwritten set pattern is trained and test through neural network for classification using different types of training techniques. The training parameter is used as like number of hidden neurons, mean square error, training rate are compared with different training methods. The accuracy, Sensitivity, Specificity measures for different training methods varies the accuracy performance. These comparative results illustrate the recognition performance of each individual handwritten character with different back propagation methods for classifying the datasets accurately.

References

1. Ding, K., Liu, Z., Jin, L., Zhu, X.: A comparative study of GABOR feature and gradient feature for handwritten 17hinese character recognition. In: International Conference on Wavelet Analysis and Pattern Recognition, pp. 1182–1186. Beijing, China, 2–4 Nov 2007
2. Pranob, K., Charles, Harish, V., Swathi, M., Deepthi, CH.: A review on the various techniques used for optical character recognition. Int. J. Eng. Res. Appl. 2(1), 659–662, (2012)
3. Sharma, O P., Ghose, M. K., Shah, K.B.: An improved zone based hybrid feature extraction model for handwritten alphabets recognition using euler number. Int. J. Soft. Comput. Eng. 2(2), 504–58 (2012)
4. Pradeepa, J., Srinivasana, E., Himavathib, S.: Neural network based recognition system integrating feature extraction and classification for english handwritten. Int. J. Eng. 25(2), 99–106 (2012)
5. Cheng-Lin, L., Nakashima, K., Sako, H., Fujisawa, H.: Handwritten digit recognition: investigation of normalization and feature extraction techniques. Pattern Recognit. 37(2), 265–279 (2004)
6. Deshmukh, S., Ragha, L.: Analysis of directional features—stroke and contour for handwritten character recognition. IEEE International Advance Computing Conference, India, pp. 1114–1118, 6–7 Mar 2009
7. Srihari, S.N.: Recognition of handwritten and machine printed text for postal address interpretation. Patterns Recogn. Lett. 14, 291–302 (1993)

8. Dudarin, A., Kovacic, Z.: Alphanumerical Character Recognition Based on Morphological Analysis. IEEE (2010)
9. Suen, C.Y., Nadal, C. et al.: Computer recognition of unconstrained handwritten numerals. Proc. IEEE. **80**, 1162–1180 (1992)
10. Neves, R.F.P. et al.: A new technique to threshold the courtesy amount of brazilian bank checks. In: Proceedings of 15th IWSSIP. IEEE Press, June 2008
11. https://cedar.buffalo.edu/Databases/
12. Patil, V., Shimpi, S.: Handwritten english character recognition using neural network. Elixir. Comput. Sci. Eng. **41**, 5587–5591 (2011)
13. Chanda, P.B., Datta, S., Choudhury, J.P.: Analysis of character recognition using back propagation neural network algorithm. In: Proceedings of National Conference on Brain and Conciousness, Sept 2013
14. Chanda, P.B., Datta, S., Mukherjee, S., Goswami, S., Bisi, S.: Comparative analysis of offline character recognition using neural network approaches. In: ETCC 2014, LNEE. Springer 22–23 Mar 2014
15. Fujisawa, H.: Forty years of research in character and document recognition—an industrial perspective. Pattern Recogn. **41**(8), 2435–2446 (2008)
16. Sivanandam, S.N., Deepa, S.N.: Principles of Soft Computing
17. Kowaliw, T., Kharma, N., Jensen, C., Mognieh, H., Yao, J.: Using competitive co-evolution to evolve better pattern recognizers. Int. J. Comput. Intell. Appl. **5**(3), 305–320 (2005)
18. Singh, D., Dutta M., Singh, SH.: Neural network based handwritten hindi character recognition. ACM Int. J. Mach. Learn. Comput. **2**(4) (2012)
19. Amit, G., Kosta, Y. P., Gaurang, P., Chintan, G.: Initial classification through back propagation in a neural network following optimization through ga to evaluate the fitness of an algorithm. Int. J. Comput. Sci. Inf. Technol. (IJCSIT), **3**(1) (2011)
20. Pal, A., Singh, D.: Handwritten English character recognition using neural network. Department of Computer Science & Engineering, U.P. Technical University, Lucknow, India Int. J. Comput. Sci. Commun. **1**(2), pp. 141–144. (20100
21. Binti, N., Hamid, A.: The effect of adaptive parameters on the performance of back propagation. PhD Disseration, Faculty of Computer Science and Information Technology. University Tun Hussein Onn Malaysia, (2012)
22. Liu, C.-L., Jaeger, S., Nakagawa, M.: Online recognition of Chinese characters: the stateof-the-art. IEEE Trans. Pattern Anal. Mach. Intell. **26**(2), 198–213 (2004)
23. Dey, E.: Recognition Bangla and English Text from the Same Document, July 2009
24. Pal, U., Choudhuri, B.B.: Indian script character recognition:a survey. Pattern Recogn. **37**, 1887–1899 (2004)
25. Mori, S., Suen, C.Y., Kamamoto, K.: Historical review of OCR research and development. Proc. IEEE **80**(July), 1029–1058 (1992)
26. Singh, M.P., Dhaka, V.S.: Handwritten Character Recognition Using Modified Gradient Descent Technique of Neural Networks and Representation of Conjugate Descent for Training Patterns
27. Mathur, S., Aggarwal, V., Joshi, H., Ahlawat, A.: Offline handwriting recognition using genetic algorithm. 6th International Conference on Information Research and Applications, Varna, Bulgaria, June-July (2008)
28. Mamedov, F., Abu Hasna, J.F.: Character Recognition using Neural Network. Near East University, North Cyprus, Turkey via Mersin-10, KKTC
29. Devireddy, S.K., Rao, S.A.: Handwritten character recognition using back propagation network. JATIT (2005–2009)
30. Gonzalez, R.C., Woods, R.E.: Digital Image Processing, 2nd edn. Prentice-Hall, USA (2002)

31. Singh, R., Yadav, C.S., Verma, P., Yadav, V.: Optical character recognition (OCR) for printed devnagari script using artificial neural network. Int. J. Comput. Sci. Commun. **1**(1), 91–95 (2010)
32. Ganapathy, V., Liew, K.L.: Handwritten character recognition using multiscale neural network training technique. World Acad. Sci. Eng. Technol. **39** (2008)
33. Cheriet, M., Kharma, N., Liu, C.-L., Suen, C.Y.: Character Recognition Systems, a Guide for Students and Practitioners. Wiley, New Jersey (2007)

Microarray Gene Expression Analysis Using Fuzzy Logic (MGA-FL)

Daksh Khanna, Tanupriya Choudhury, A. Sai Sabitha and Nguyen Gia Nhu

Abstract Fuzzy logic is an arrangement to sort with registering in light for "degrees of truth" instead of the standard thing "genuine or false" (1 or 0) Boolean logic. A Deoxyribonucleic corrosive (DNA) microarray (in like manner normally known as Deoxyribonucleic Acid chip or bio-chip) is a collection of small DNA-spots annexed to a hard surface (Mukhopadhyay et al., Analysis of microarray data using multiobjective variable string length genetic fuzzy clustering, 2009) [1]. DNA microarrays are used to gage the explanation hierarchy of significant amounts of characteristics in the meantime or to geno-type distinctive areas of a genome. Each Deoxyribonucleic Acid spot consist of micro-moles (appx. 11 mol) of a particular Deoxyribonucleic Acid progression, justified as tests (or writers or oligos). It can be a small territory of a quality or other DNA segment which is used to hybridize a complementary DNA/complementary RNA (in like manner called antagonistic to identify Ribonucleic Acid) test (called centre) under high-stringency terms. Test object hybridization is normally recognized and acquired through distinguishing proof of "fluorophore-, silver-, or chemiluminescence"—named centres to choose relatively abundant nucleic destructive game plans for the goal. The principal nucleic destructive displays were expansive scale groups around 9 cm $\times$ 12 cm and the fundamental modernized picture based examination was circulated in 1981. Microarray information investigation includes a few unmistakable advances. Changing any of the means will change the result of the examination, so the MAQC Project was made to recognize an arrangement of standard procedures (Mukhopadhyay et al., Analysis

D. Khanna (✉) · A. S. Sabitha
Amity University, Noida, Uttar Pradesh, India
e-mail: dakshkhanna97@gmail.com

A. S. Sabitha
e-mail: saisabitha@gmail.com

T. Choudhury
University of Petroleum and Energy Studies, Dehradun, India
e-mail: tanupriya1986@gmail.com

N. G. Nhu
Duy Tan University, Da Nang, Vietnam
e-mail: nguyengianhu@duytan.edu.vn

© Springer Nature Singapore Pte Ltd. 2019   169
A. Abraham et al. (eds.), *Emerging Technologies in Data Mining and Information Security*, Advances in Intelligent Systems and Computing 755,
https://doi.org/10.1007/978-981-13-1951-8_16

of microarray data using multiobjective variable string length genetic fuzzy clustering, 2009) [1], (Li, International Seminar on Future BioMedical Information Engineering, 2008) [4]. Organizations exist that utilization the MicroArray/Sequencing
Quality Control (MAQC) conventions to play out a total examination. The human
genome contains roughly 21,000 qualities. At any given minute, each of our cells
has some mix of these qualities turned on, and others are killed (Pujari, Data Mining
Techniques, 2001) [2].

Keywords Bioinformatics · Microarray · Fuzzy logic · Association · Clustering

1 Introduction

Fuzzy logic helps in overcoming the problem of PC comprehension of normal language. Characteristic Language is not with no effort turned into the outright terms
of 1 and 0. It might see Fuzzy Algorithm in accordance to thinking truly helps and
paired or Boolean Algebra is essentially a unique instance of it. Fuzzy logic consist
of 1 as strict case of truth (or "the state of matters" ˆ "fact") and 0 and also embraces
the multiples states of truth in b/w so that, for ex., the outcome of a comparison b/w
two things could be not "tall" or "short" but ".48 of shortness. The centre rule for
micro-arrays is hybridization b/w 2 DNA-strands, the property of reciprocal nucleic
corrosive arrangements for explicitly matching with every one by framing H_2 bonds
b/w integral nucleotide base sets. A high no. of integral base matches in a nucleotide
arrangement implies more tightly non-covalent holding b/w the 2 strands [3]. In the
wake of washing off non-particular holding successions, just unequivocally matched
strands will remain hybridized. Fluorescently named object groupings that predica-
ment in a test succession create a flag that relies upon the hybridization terms, (for
example, Temp.), and cleaning after hybridization. Add up to quality from the flag,
with a spot (highlight), relies on the measure of sample test official to the tests exhibit
on that spot. Microarrays utilize relative quantisation in which the power of an element is contrasted with the force of a similar component under an alternate condition,
and the personality of the element is known by its position. Microarrays can be created utilizing an assortment of advances, incorporating printing with needle-pointed
pins into glass slides, photolithography utilizing pre-made veils, photolithography
utilizing dynamic micro-mirror gadgets, ink-stream printing, or electronic-chem. on
micro-electrode exhibits.

1.1 In Spotted Microarrays

The tests are 'oligonucleotides', complementary DNA or little parts of Polymerase
Chain Reaction items that compare to messenger RNAs. The tests are incorporated
before affidavit on the exhibit surface and are then "spotted" to glass. A typical

methodology uses a variety of fine sticks or needle restrained by an automated arm that is drowned into wells containing Deoxyribonucleic acid tests and afterward storing every test in assigned areas on the cluster surface. The subsequent "framework" of tests speaks to the nucleic corrosive profile of readied tests and is prepared to get corresponding cDNA or complementary Ribonucleic Acid 'targets' originated from exploratory or clinical specimens. These exhibits might be effortlessly modified for each examination, since scientists can pick the tests and printing areas on the clusters, blend the tests in their own lab (or working together office), and recognize the clusters [4]. They would then be able to produce their own marked specimens for hybridization, hybridize the examples to the exhibit, lastly filter the clusters with their own gear. This gives a moderately ease microarray that might be redone for each examination, and maintains a strategic distance from the expenses of obtaining regularly more costly business clusters that may speak to tremendous quantities of qualities that are not important for agent. Productions exist which show in-house spotted microarrays may not give a similar level of affectability contrasted with business oligonucleotide arrays, preferably attributable to the little clump size and lessened printing accuracies if contrasted with modern makes of oligo clusters.

1.2   In Oligonucleotide Microarrays

The tests are minimal arrangements intended for coordinating parts of the succession from re-knowned or anticipated perusing outlines. In spite of the fact that oligonucleotide tests are regularly utilized as a part of "spotted" microarrays, the expression "oligonucleotide cluster" frequently alludes to a particular method of assembling. Oligonucleotide exhibits are created by printing short oligonucleotide groupings intended to speak to a solitary quality or group of quality graft variations by incorporating this succession straightforwardly onto the cluster surface as opposed to storing in place arrangements. Arrangements might be huge or shorter contingent upon the coveted reason; Bigger tests seem to be particular to singular sample qualities, smaller tests might be seen in high thickness over the exhibit and are less expensive to make. One method adopted to create oligonucleotide exhibits incorporate photolithographic combination (Affymetrix) on a silica substrate where fragile-delicate covering specialists are utilized to "construct" an arrangement 1 nucleotide at any given moment over the whole cluster. Each pertinent test is specifically "unmasked" preceding washing the cluster in an answer of a solitary nucleotide, at that point a concealing response happens and the following arrangement of tests are unmasked in planning for an alternate nucleotide presentation [5]. After numerous reiterations, the groupings of each test turn out to be completely built (Fig 1).

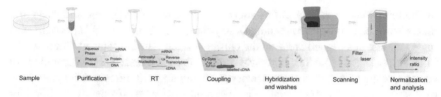

Fig. 1 The steps required in a microarray experiment

2 Literature Review

Micro array technology is kind of latest advanced technologies in molecular biology. It enables visualizing the expression thousands of genes (virtually the entire genome) simultaneously. A huge variation of various ways is suggested for analysing the GE data that includes "hierarchical clustering, self-organizing maps, and k-means approaches".

In [18], Zuoliang Chen et al. (2008) has proposed an affiliated order approach to be specific Classif^n with Fuzzy Association Rules (CFAR). Here the Fuzzy rationale procedure is utilized for apportioning the areas. The exploratory outcomes construed that CFAR creates preferable understand ability over the conventional methodologies.

In this paper Jukic et al. [13] (2011) has proposed qualified affiliation rules mining. This proceeded connection proceeds mining is an expansion of the affiliation rules information mining strategy, which finds already obscure connections under a few conditions. This strategy goes for enhancing the activity comes about. Different trials are completed to show how the qualified guidelines increment the adequacy when contrasting and standard affiliation rules.

In [19], Yuchun Tang et al. (2008) suggested another 'Fuzzy Granular Support Vector Machine Recursive Feature Elimination calculation' (FGSVM-RFE). This calculation kills the superfluous, repetitive, or uproarious qualities in various granules at various stages and chooses profoundly educational qualities with possibly extraordinary organic capacities.

In [4], the author Dongguang Li (2008) presented information about instructions to find helpful data in view of the DNA microarray articulation information gathered from mouse tests for the SRBCT explore. Numerous strategies including fluffy sets, neural systems, hereditary calculations, harsh sets, wavelets, and their hybridizations, are recommended to give inexact arrangements.

In [17], Qilian Liang et al. (2000) has proposed an algo. for constructing interval Type-II fuzzy algorithm systems. The model of lower and upper member func^n is familiarized and defined in this paper.

In the article [12], Muzeyyen Bulut Ozek et al. (2007) has presented the latest Type-II Fuzzy Logic Toolkit in MATLAB prog. lang. The primary goal is to support user to recognize & implement Type-II fuzzy logic systems with an ease.

In [14], Nilesh N. Karnik et al. (1999) has proposed a Type-II fuzzy logic system that can handle rules suspicions. Type-II Fuzzy algorithm is applied to time-erratic channel equalization and it is proved that the Type-II fuzzy logic system provides enhanced performance than the existing Type-I Fuzzy and nearest neighbour classifier.

3 Problem Formulation and Methodology

It suggests a strategy for choosing genes which depends on the idea of fuzzy pattern. Quickly, given an arrangement of micro arrays which are very much grouped, for every group we can enhance a fuzzy pattern (FP) from Using FP for Gene Selection and Data Red^n on Micro array Data 1089 'Fuzzy Microarray Descriptor' (FMD) related to every possible microarrays [6]. The Fuzzy Microarray Descriptor is an intelligible portrayal for every quality as far as one from the accompanying phonetic marks: HIGH, MEDIUM, and LOW. Along these lines, the fuzzy pattern is a model of the Fuzzy Microarray Descriptor's having a place with a similar class where the enrolment rule of every quality to the fluffy example of the class is recurrence based. Clearly, this reality can be of intrigue, if the arrangement of starting perceptions is marked with a similar sort of tumour. The example's nature of fluffiness is given by the way that the chose marks originate from the phonetic names characterized amid the change into FMD of an underlying perception. In addition, if a particular mark of one component is exceptionally basic in every one of the cases having a place with a similar class, this element is chosen to be incorporated into the example.

The SRBCT Dataset is downloaded from Markov Blanket-Embedded Genetic Algorithm for Gene Selection website consist of 2308 Genes with their equivalent SRBCT values (0.0448, 0.72) [7]. Micro array gene data contains noisy and inconsistent data. Preprocessing is the method of removing the inconsistent data and to derive necessary information. Table 1 show the microarray gene data downloaded consists of empty spots. The preprocessing phase is used for extracting needed information from the data that consists of two main processes (Fig 2).

In the preprocessing step the vacant spots are replaced with null values using is empty method. The empty spots are replaced by unique elements in dataset using unique method. Then the Ø values in the dataset are swapped by the maximum exclusive samples by following max method [8]. Table 2 shows the preprocessed data. After replacing all the null values in the microarray gene data, the preprocessed gene expression values are given as i/p for the further process, the fuzzification (Table 3).

	Dataset	#Genes	#Instances	#Class [10]
Table 1 Micro array dataset and it's information	SRBCT	2308	83	4

```
'Procedure VariousFuzzyPatterns (i/p: ListFP; o/p: ListDFP)
{
0. begin
1. initialize_VFP: FP ← ∅
2. for every fuzzy pattern(FPi) ∈ ListFP do
3. Initialize: VFPi ← ∅
4. for every FPj ∈ ListFP and FPᵢ <>FPⱼ do
5. for every gen g ∈ GetGenes(FPᵢ) do
6. if (g ∈ GetGenes(FPⱼ)) &
                        (GetLabel(FPᵢ,ₘ) <> GetLabel(FPⱼ,ₘ)) then
            7. AddMember(VFPi, Member(FPᵢ,ₘ))
8. AddtoListof_VFP: Add(ListVFP, VFPi)
9. end.'
}
```

Fig. 2 Pseudo-code for fuzzy algorithm

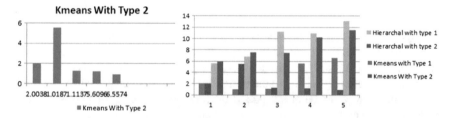

Fig. 3 Total no. of genes and their clusters

Table 2 Microarray gene data with empty spots and preprocessed microarray gene data

TP rate	FP rate	Precision	Recall	F-Measure	ROC area	Class
0.931	0.019	0.961	0.931	0.947	0.956	1
1	0	1	1	1	1	2
1	0.092	0.75	1	0.857	0.954	3
0.76	0.017	0.95	0.76	0.844	0.871	4
0.904	0.032	0.918	0.904	0.904	0.936	
0.966	0.259	0.667	0.966	0.789	0.853	1
0.273	0	1	0.273	0.429	0.636	2
0.611	0	1	0.611	0.759	0.806	3
0.92	0.069	0.852	0.92	0.885	0.926	4
0.783	0.111	0.839	0.783	0.763	0.836	

3.1   Fuzzy Association Based Gene Clustering

The numeric quantitative estimations of quality information are changed over into fluffy terms utilizing fuzzy logic. After fuzzification, the Fuzzy esteems are given as contribution for the following stage, the finding of quality affiliation. In the second stage, to locate the Fuzzy affiliation design lpq ljk the relationship between the etymological terms lpq and ljk are discovered. The microarray quality information

Table 3 Gene data with fuzzy values and gene data with fuzzy terms

TP rate	FP rate	Precision	Recall	F-Measure	ROC area	Class
0.931	0.019	0.964	0.931	0.947	0.956	1
1	0	1	1	1	1	2
1	0.092	0.75	1	0.857	0.954	3
0.76	0.017	0.95	0.76	0.844	0.871	4
1	0.111	0.829	1	0.906	0.983	1
0.818	0	1	0.818	0.9	1	2
0.889	0	1	0.889	0.941	0.988	3
0.84	0.034	0.913	0.84	0.875	0.987	4
0.904	0.049	0.914	0.904	0.904	0.988	

with three states contains nine conceivable relationships as indicated by the quality articulation states [9]. The quality information contains Fuzzy association designs like (Tables 4 and 5).

The gene positions corresponding to a particular association are grouped as shown in Table 5. For example 51, 69, 3, 5, indicate the gene position numbers in which the associations are present.

The uncertainty associated with the fuzzy association patterns can be modelled by using confidence measure. The confidence measure is defined as the probability of the pattern Pr(lpq ljk) [11]. The weight of evidence measure W(lpq ↔ ljk) is calculated to handle the uncertainties (Table 6).

Table 4 Association patterns and its gene numbers

Rule No.	Associations	Gene numbers
1	L → A	21, 23, 45, 776, 42, 567, 122, 12
2	L → H	1, 2, 3, 4, 5, 6, 7, 8, 9, 10, 11, 12, 13, 14
3	A → H	2308, 1200, 1234, 1423, 1567, 2200, 1231, 1001
4	A → L	17, 18, 19, 20, 21, 22, 23, 24, 25, 26, 27, 28
5	H → L	284, 285, 286, 287, 288, 289, 290, 291
6	H → A	755, 556, 575, 578, 575, 576
7	L → L	0
8	A → A	0
9	H → H	0

Table 5 Total number of genes for association pattern

Association pattern	Total number of genes
L → A	60
L → H	2000
H → L	12

Table 6 Calculated weights for five clusters

Associations	Cluster-1	Cluster-2	Cluster-3	Cluster-4	Cluster-5
L → A	0.9244	2.0651	1.5113	1.3343	1.2766
L → H	0.1601	0.4445	0.3157	0.2018	0.7770
A → H	1.2572	0.7772	1.0058	1.1137	0.2212
A → L	0.8278	0.1681	0.4507	0.8376	0.1432
H → L	0.2404	0.3483	0.3066	0.2091	0.1669
H → A	1.2256	2.5980	1.9409	1.6590	0.4744
L → L	5.5859	6.3768	6.0424	5.5453	9.4915
A → A	1.2196	1.3639	1.2969	1.2515	1.7763
H → H	1.5323	1.2921	1.4433	1.2193	1.1851

Table 7 Accuracy values for kmeans and type-1 fuzzy and precision values of hierarchial and type 1 fuzzy

Gene No.	Hierarchal without fuzzy	Hierarchal + type 1 fuzzy
Gene1	1.5113	2.0038
Gene2	0.3157	1.0187
Gene3	1.0058	1.1137
Gene4	0.4507	5.6096
Gene5	1.9409	6.5574

Table 8 Accuracy values for kmeans and type 1 fuzzy

Gene No.	Kmeans without fuzzy	Kmeans + type 1 fuzzy
Gene1	1.5113	2.0038
Gene2	0.3157	1.0187
Gene3	1.0058	1.1137
Gene4	0.4507	5.6096
Gene5	1.9409	6.5574

After the calculation of weight, a pair of gene data accumulated from a set of N' genes from previously unseen gene expression data are collected. To predict the accuracy the FP previously extracted in each class is reviewed to check for the matching pattern with new expression profile [12]. The weight of evidence W'(lpq ljk) supporting the assignment of new class is defined as:

$$W\prime(lpq \leftrightarrow ljk) = W(lpq \leftrightarrow ljk) * \mu Fpq \qquad (1)$$

Table 9 Outcomes for type-(I & II) fuzzy

Type 1		Type 2	
Kmeans	Hierarchal	Kmeans	Hierarchal
2.0038	5.60	2.0038	5.90
1.0187	6.79	5.5453	7.59
1.1137	11.19	1.2515	7.53
5.6096	10.93	1.2193	10.27
6.5574	13.07	0.9425	11.51

3.2 Clustering and Verification Phase

The fuzzy association patterns are discovered and the accuracy is calculated in the second phase. After the pattern discovery the clustering is done in third phase. 2 kinds of clustering algos are in the picture. One is "k-means and other is hierarchical clustering." Besides clusterification the fuzzy logic technique is processed for the clusterification outcomes. An evaluation is done with the outcomes from Kmeans w/o fuzzy method and Kmeans with fuzzy logic [13]. Evaluation is done for hierarchical clustering equivalent to Kmeans. The contrast is done for both Type-I and Type-II fuzzy. After comparison, the outcomes are noted down and justified for both types of clusters. After the authentication of results, it is observed that there is an increase in accuracy after using fuzzy logic Algorithm [14]. The accuracy values for Kmeans clustering w/o using fuzzy method and with Type-I fuzzy method are depicted in Table 9.

4 Conclusion and Future Directions

Fuzzy Algorithm is a part of Boolean logic that has been enhanced to handle the concept of partial truth, where the value of truth might vary b/w fully true and totally false. We have proposed a methodology consisting of a framework (MGA-FL) for enhancing the suggested idea. The primary idea of our proposed work is to apply fuzzy method to microarray gene data to find fuzzy association patterns. The association uncertainties are handled by calculating weight and the accuracy of gene clusters are predicted. The fuzzy associations are proposed for both Types of fuzzy (Type 1 and 2). In this study the microarray gene dataset is implemented and the clustering is done using fuzzy logic associations [15]. The various snapshots obtained in the Sect. 4 prove that the proposed idea is found to have more accuracy than existing clustering algos. Kmeans and Hierarchical clustering. We have also compared the proposed work with existing Type 1 fuzzy and it is found that the accuracy results of the suggested Type-II fuzzy algo. is increased when compared to

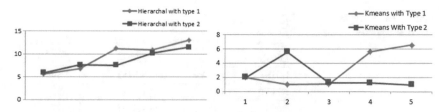

Fig. 4 Accuracy outcomes for Kmeans and hierarchical clusterification

Type-I fuzzy approach. The proposed approach clusters microarray gene data based on fuzzy association rules. The same can be implemented for classifying microarray gene data and compared with famous algorithms like K-NN and SVM (Table 8).

5   Implementation Results and Discussion

The experimental results and comparative study of the Fuzzy techniques and two algorithms are presented in this section [15]. The membership functions are described in fuzzy logic toolkit. The membership function for Type 2 fuzzy is depicted in Fig. 4. The membership func^n for Type 2 fuzzy is represented by trimf function type.

The number of clusters and the number of genes linked with some individual cluster is represented in the shape of bar graph in Fig. 3. From the bar graph it is observed that max. nos. of gene lie beneath the cluster-2 (1.0187) for fuzzy associations A → H, H → A, H → H, A → A.

The bar diagram for the different association patterns and total number of genes is represented in Fig. 7. Based on the analysis of results from the Fig, it is inferred that the association patterns like A → H, H → A, A → A, and H → H occurred frequently in most of the clusters [16]. The comparison of clustering algorithm accuracies for Type-I and Type-II fuzzy is done and it is shown in the Table 7.

The result for Type-I fuzzy and Type-II fuzzy are analysed and compared and discussed in the form of line chart. The accuracy values for Kmeans and hierarchical clustering are plotted in line chart and it is shown in Fig. 8 and Fig. 9. The scrutiny of the clustering outcome implies that the accuracy values are incremented by opting fuzzy algorithm techniques when equated to the vintage clustering algos. From the applied algos., outcomes and the precision can be further incremented by applying the suggested Type-II fuzzy method (MGA-FL) compared to the usual Type-I fuzzy methodology.

The technique proposed in the research work using Type 2 fuzzy for finding association patterns to cluster microarray data is found to have highest accuracy than Type 1 fuzzy and other traditional clustering algorithms (Fig 5).

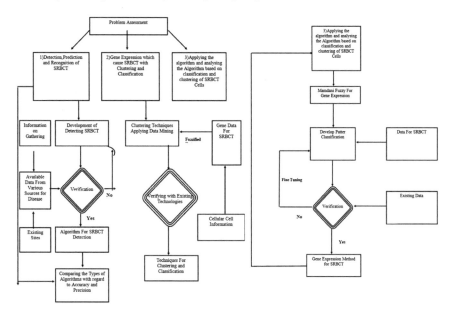

Fig. 5 Flow chart of the research outline

References

1. Mukhopadhyay, A., Bandyopadhyay, S., Maulik, U.: Analysis of microarray data using multi-objective variable string length genetic fuzzy clustering. IEEE (2009). ISBN 978-1-4244-2959
2. Pujari, A.K.: Data Mining Techniques. Universities Press (India) Limited (2001). ISBN-81-7371-3804
3. Larose, D.T.: Introduction to data mining. In: Discovering Knowledge in Data: An Introduction to Data Mining (2005). ISBN 0-471-66657-2
4. Li, D.: DNA microarray expression analysis and data mining for blood cancer. International Seminar on Future BioMedical Information Engineering (2008)
5. Huerta, E.B., Duval, B., Hao, J-K.: Fuzzy logic for elimination of redundant information of microarray data. **6**(2), (2008)
6. Hellman, M.: Fuzzy logic introduction. Info. Ctl. **12**, 94–102 (1968)
7. Castro, J.R. Castillo, O., Martinez, L.G.: Interval type-2 fuzzy logic toolbox. Eng. Letters, **15**(1), EL_15_1_14 (2007)
8. Tari, L., Baral, C., Kim, S.: Fuzzy c-means clustering with prior biological knowledge. J. Biomed. Inform. (2008)
9. Luscombe, N.M., Greenbaum, D., Gerstein, M.: Bio informatics: a proposed definition and overview of the field. IMIA yearbook of medical informatics: digital libraries and medicine. Int. J. Comp. Sci. Eng. Appl. (IJCSEA) **2**(2): 83–99 (2012)
10. Zhu, Z., Ong, Y.S., Dash, M.: Markov blanket-embedded genetic algorithm for gene selection. Pattern Recogn. **49**(11), 3236–3248 (2007)
11. Pourhashem, M.M., Kelarestaghi, M., Pedram, M.M.: Missing value estimation in microarray data using fuzzy clustering and semantic similarity. Global J. Comp. Sci. Technol. **10**, 18 Oct 2010
12. Ozek, M.B., Akpolat, Z.H.: A Software Tool: Type-2 Fuzzy Logic Toolbox. Wiley Periodicals Inc. (2007)

13. Jukic, N., Nestorov, S., Velasco, M., Eddington, J.: Uncovering actionable knowledge in corporate data with qualified association rules. Int. J. Bus. Intell. Rese. **2**(2), 1–21, Apr-June (2011)
14. Karnik, N.N., Mendel J.M., Liang, Q.: Type-2 fuzzy logic systems. IEEE Trans. Fuzzy Syst. **7**(6), Dec 1999
15. Munoz, P.M., Francisco, J., Velo, M.: Fuzzy CN2: an algorithm for extracting Fuzzy classification rule lists. WCCI 2010 IEEE World Congress on Computational Intelligence, CCIB, Barcelona, Spain, 18–23 July (2010)
16. Woolf, P.J., Wang, Y.: A Fuzzy logic approach to analyzing gene expression data. Physiol. Genomics **3**, 9–15 (2000)
17. Liang, Q., Mendel, J.M.: Interval type-2 fuzzy logic systems: theory and design. IEEE Trans. Fuzzy Syst. **8**(5) (2000)
18. Chen, Z., Chen, G.: Building an associative classifier based on Fuzzy association rules. Int. J. Comput. Intell. Syst. **1**(3), 262–273 (2008)
19. Tang, Y., Zhang, Y-Q., Huang, Z., Hu, X., Zhao, Y.: Recursive Fuzzy granulation for gene subsets extraction and cancer classification IEEE Trans. Inf. Technol. Biomed. **12**(6) Nov 2008

Stratification of String Instruments Using Chroma-Based Features

Arijit Ghosal, Suchibrota Dutta and Debanjan Banerjee

Abstract Identification of instrument type from acoustic signal is a challenging issue. It is also an interesting and popular research area having several promising applications in music industry. Researchers have already been able to classify instruments into several broad categories like String, Woodwind, Percussion, and Keyboard etc. using acoustic features like Mel-Frequency Cepstral Coefficients (MFCCs), Zero Crossing Rate (ZCR) etc. MFCC has been found to be excessively used. In this work an alternative acoustic feature of MFCC has been proposed. Chroma is an octave independent estimation of strength of all possible notes in Western 12 note scale at different points of time. Sound envelope originated by a note reflects the signature of an instrument and this can be used to stratify String instruments into various categories. The proposed work relies on Chroma-based low-dimensional feature vector to categorize String instruments. For classification purpose, simple and popular classifiers like Neural Network, k-NN, Naïve Bayes' have been exercised.

Keywords Instrument classification · Chroma · Note · Neural network
Classification

A. Ghosal (✉)
Department of Information Technology, St. Thomas' College
of Engineering & Technology, Kolkata 700023, West Bengal, India
e-mail: ghosal.arijit@yahoo.com

S. Dutta (✉)
Department of Information Technology and Mathematics,
Royal Thimphu College, Thimphu, Bhutan
e-mail: suchibrota@gmail.com

D. Banerjee
Department of Management Information Systems,
Sarva Siksha Mission, Kolkata 700042, West Bengal, India
e-mail: debanjanbanerjee2009@gmail.com

© Springer Nature Singapore Pte Ltd. 2019   181
A. Abraham et al. (eds.), *Emerging Technologies in Data Mining and Information
Security*, Advances in Intelligent Systems and Computing 755,
https://doi.org/10.1007/978-981-13-1951-8_17

1 Introduction

India is known as an inheritor as one of the most ancient and grown music systems in the world. From the history of India it is learnt that so many different types of String instruments were being used from the past. This work deals with the difficulty of musical instrument categorization especially for String instruments from monophonic audio sources. A good audio classification system serves as foundation of numerous applications e.g. audio indexing, audio retrieval, music and instrument classification, etc. Music data analysis is a very popular research area in recent years. Advancement in the field of signal processing and data mining techniques also caused development of research work in the domain of audio retrieval, detection of musical instruments and their classification.

The oldest recognized system of categorizing instruments was Chinese at the time of fourth-century BC [1]. That system classified instruments according to the materials by which they are made of. Since then, musicologists are constantly in search of proposing a good instrument classification system. Lots of schemes have been proposed by different researchers in connection with musical instrument classification but unfortunately none of them is recognized throughout the world to meet the needs of all applications. This has made the research work of classification of musical instrument hard to propose a single classification method to convince all the users or applications. Hence, classification of musical instrument has become not only a subject for musicologists, but also a significant research field. Most of the recent research works in the field of audio signal processing were related to speech recognition, but very little work has been performed for classifying instruments. Researchers have already categorized instruments into broad categories like String, Woodwind, Percussion, Keyboard, etc., in their past works using different acoustic features. But they have done less work so far to sub-classify broad categories of instruments. This work aims to propose a simple classification scheme to sub-classify String instruments. Introduction is followed by description of previous work done related to musical instrument classification in Sect. 2. Section 3 deals with approach of the proposed scheme. Experimental result with comparative analysis has been illustrated in Sect. 4.

2 Related Works

Researchers have worked with various acoustic features to propose a good feature set for instrument classification. Agostini et al. [2] have worked with spectral features for musical instrument timbres classification. They have considered pitch, centroid bandwidth, skewness, and zero crossing rate (ZCR) based features in their work. Zhu et al. [3] have also worked with spectral features. They have generated spectrum of instruments. They have considered pop, jazz and rock instrumentals only. A hierarchical scheme for classification of musical instruments has been proposed by Essid et al. [4]. For classification they have used Support Vector Machine (SVM). In their work temporal features like autocorrelation coefficients and ZCR, cepstral features like Mel-Frequency Cepstral Coefficients (MFCC), spectral features like MPEG-7 audio features, wavelet features, perceptual features like loudness and sharpness have been used.

Sinith and Rajeev [5] have worked with South Indian classical music. They have used Hidden Markov Model in their work. Deng [6] has classified instruments into different broad categories like string, brass, woodwind, and percussion using perception-based features, MPEG-7 audio features and MFCC. In the work of Gunasekaran and Revathy [7] fusion of multiple classifiers has been used. Temporal, spectral, perceptual, harmonic and statistical features have been used in their work. Senan et al. [8] have worked with Malay Musical Instruments. They have used 37 features which comprises of MFCC and perception-based features. Liu and Xie [9] have explored SVM based approach for classification of Chinese and western musical instruments. They have classified instruments into three broad categories—wind, percussion and string. North Indian musical instruments like Flute, Sitar, Dholak, Bhapang, and Mandar have been classified by Kumari et al. [10] using spectral features in combination with MFCC. Barbedo and Tzanetakis [11, 12] have worked with polyphonic instrumental signals.

Ghosal et al. [13] have used hierarchical approach for classification of instruments. They have used wavelet-based and MFCC-based features to classify instruments into broad categories like String, Woodwind, Percussion and Keyboard categories. Grindlay and Ellis [14] have worked with polyphonic music. Chroma Toolbox for Matlab has been used by Müller and Ewert [15]. Chandwadkar and Sutaone [16] have studied the significance of features and classifiers on accuracy for the purpose of identification of musical instruments. They have used spectral features along with MFCC-based features. Temporal, spectral and MFCC based features have been used by Nadgir and Joshi [17]. Gaikwad et al. [18] have classified Indian classical instruments by applying Principal Component Analysis (PCA) based cepstrum features as well as spectral features. They have considered amplitude and spectral range along with MFCC as spectral features. Arora and Behera [19] have worked with Probabilistic Latent Component Analysis (PLCA) for instrument identification. Indian instrumental music has been classified by Dandawate et al. [20] by considering Bhairav, Bhairavi, Todi and Yaman Raga. They have considered temporal features to achieve their goal. Abeber and Weib [21] have worked with polyphonic type classical music for recognition of instruments. They have classified instruments into brass, woodwinds, strings, piano, and vocal categories. Kumar et al. [22] have followed musical onset detection approach. They have considered Carnatic percussion instruments like ghatam, mridangam, kanjira, thavil and morsing in their work. Bhalke et al. [23] have also used MFCC-based features. They have built their system by using Counter Propagation Neural Network (CPNN). Ghisingh and Mittal [24] have classified musical instruments by using methods applied for speech signal processing. They have designed their feature vector by using MFCC, Spectral Centroid (SC), ZCR and signal energy. They have calculated these acoustic features by applying methods used in different speech signal processing.

3   Proposed Methodology

From the past work it is quite clear that Mel-Frequency Cepstral Coefficients (MFCC) has been used several times by many researchers for instrument classification. MFCC is an excellent feature for categorization of instruments as it reflects the spectral power distribution of an instrument very well. But overuse of MFCC has created a scope of research in the field of instrument classification to propose a new feature for classification of instruments. In this work the aim is focused to sub-classify Indian String instruments into different categories using a suitable acoustic feature which will work as an alternative of MFCC.

Sounds are produced in String instruments using the vibration generated from string or chord. String instruments can be played in several ways. Based on how a String instrument is played, String instruments can be divided into several categories—*Plucking, Bowing,* and *Striking.* Plucking is a method of playing on instrument using a thumb, finger, or quills (now plastic plectra) to pluck strings. Veena, Sitar, Guitar are example of this type of String instruments. Bowing is a method of playing a String instrument with the help of a bow. Violin, cello are bowing type String instruments. Striking type String instruments are played by striking the string. Hammered dulcimer, Santoor are examples of striking String instruments. In the following sub-sections feature extraction and classification technique adopted has been described respectively. Proposed approach is split into two modules—extraction of features and then classification of input audio signal based upon the extracted features. Figure 1 depicts the layout of the suggested scheme.

3.1   Feature Extraction

It is known that a good feature set has the characteristics that it will help to serve the classification job with high accuracy while maintaining its dimension as small as possible. It is observed that the *Plucking, Bowing* and *Striking* types of String instruments are played with different postures and their notes are also different.

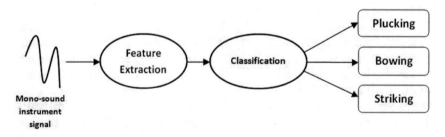

Fig. 1 Layout of proposed scheme

Presence of different number of strings in those different types of String instruments is also one of the reasons of their differences in their notes.

In music, it is known that a note is the representation of pitch and duration of a sound. Note is nothing but a small bit of sound which is similar concept of syllable while speaking a language or it can be said as single musical event. Notes generated from different String instruments will obviously vary because pitch of different String instruments is not same as well as musical events are also not same for them. Different types of String instruments may generate distinctive tones, but all of them have one thing in common—ability to produce different note. When notes are played in orderly fashion or in concordance with the others, music is created. A note is in truth just sound waves, based on some sort of vibration, of a specific wavelength. The amount of those vibrating waves that hit eardrum of a person each second is called as the "frequency". So, notes are considered as the building blocks of music specifically instruments. This observation has motivated to propose a feature set which can reflect characteristics of note of different String instruments well to achieve the goal with high accuracy keeping feature vector length small.

3.1.1   Chroma-Based Features

Chroma features are the representation of music audio where the entire spectrum of the input is projected into 12 distinct semitones or Chroma of the musical octave. Chroma-based features are also denoted as pitch class profiles. It is a powerful tool for analyzing music especially instrumental music. It is fact that Chroma features have the characteristics that they can capture the harmonic and melodic nature of music. From hearing the sounds generated by Plucking, Bowing and Striking types of String instruments also it can be felt that these sounds vary harmonically and melodically. So Chroma features are the best features to represent this. In music, it is known that the notes are just one octave away from each other, are recognized as analogous. Presuming the equal tempered scale, anyone can consider twelve Chroma values represented by the set

$$\{C, C\#, D, D\#, E, F, F\#, G, G\#, A, A\#, B\}$$

It denotes the 12 pitch spelling attributes used in Western 12 note scale notation. Specifying the Chroma values, the set of Chroma values can be identified as a set of integers $\{1, 2, …, 12\}$, where 1 denotes to Chroma C, 2 to C#, and so on. A pitch class can be described as the collection of all pitches which share same Chroma. Chroma denotes the angle of pitch rotation as it passes through the helix. Two octave-related pitches will have the same angle in Chroma circle which is depicted in Fig. 2. It is such a relation that is not possible to be captured by a linear pitch scale or even by using Mel. This is the advantage of using Chroma-based features over Mel-based features like MFCC.

To analyze western tonal music this angle has been quantized into 12 positions or pitch classes. Intensity of each of these 12 individual pitch classes is denoted by

Fig. 2 Chroma circle

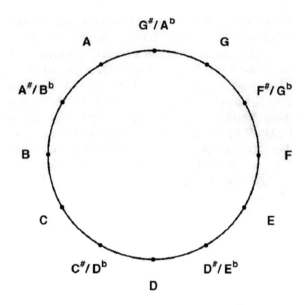

Chroma features. Chroma features indicate perceptual differences of pitches within an octave.

Instrumental signals are not stationary in general. To capture Chroma features in a better way audio data is broken into 50% overlapping frames. Frames are overlapped to avoid missing of any boundary characteristics of a certain frame. For all the frames Chroma features are computed and then their mean has been considered. So, Chroma-based features produce 12 distinct values.

Plots for Chroma values with magnitude for Plucking, Bowing and Striking type String instruments are depicted in Fig. 3, Fig. 4 and Fig. 5, respectively.

These plots indicate that Chroma features vary noticeably for the three different types of String instruments. The chromagram toolbox [15] has been used for extraction of chroma features.

3.1.2 Skewness-Based Feature

The third-order moment of the spectral distribution is denoted by spectral skewness. It is a statistical feature. The asymmetry present in a certain normal distribution regarding its mean position can be measured by skewness. The three types of String instruments—Plucking, Bowing, and Striking have some amount of asymmetry in their normal distribution. The skewness s is represented by the below Eq. (1).

$$k = \frac{E(a - \mu)^3}{\sigma^3},$$

(1)

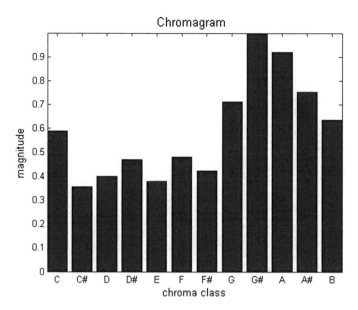

Fig. 3 Plot of Chroma values against magnitude for plucking type string instrument

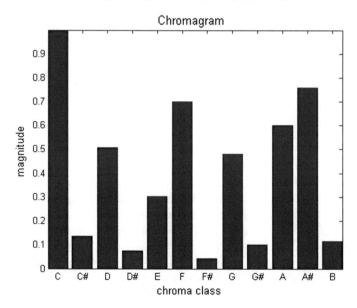

Fig. 4 Plot of Chroma values against magnitude for Bowing type string instrument

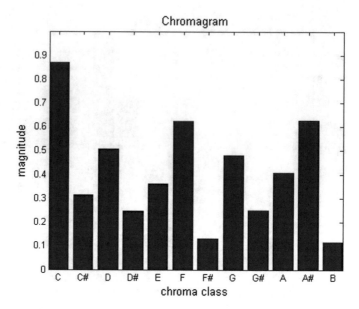

Fig. 5 Plot of Chroma values against magnitude for striking type String instrument

where μ represents the mean of sampled data a, σ is the standard deviation of a, and the expected value of the quantity x is denoted by $E(x)$.

3.2 Classification

The features extracted from the input audio data contains 13 feature values. These values contain 12 Chroma-based features and 1 skewness based feature. This 13-dimensional feature vector has been fed to the classifiers for classifying String instruments. For classification work, it is expected to split the entire dataset used in the system into two parts, where each part represents either testing or training data. A total of 300 audio files has been used where each of the three types of String instruments have 100 audio files. This dataset is split into testing and training parts each with 150 files where each category of String instruments has 50 files. k-NN has been used using 'City Block' as distance metric and 'Nearest Neighbour' rule for tie-break. Multi Layer Perceptron (MLP) has been used as Neural Network having 13 neurons in the input layer representing 13 feature values, 3 neurons in the output layer corresponding to 3 types of String instruments and 7 neurons in the only hidden layer present. Naïve Bayes' classifier has also been used here.

4 Experimental Results

A custom dataset has been used here comprised of 300 mono audio files having equal distribution for all three categories. All of these audio files are of duration 90 s having sampling frequency 22,050 Hz and 16-bit per sample. These audio files are recordings collected from various sources like Internet, live performances, audio CD etc. Some audio files in the dataset are noisy also. 50% of the dataset has been has used for training purpose and the rest 50% has been chosen for testing purpose. Again the experiment is performed reversing the training and testing data set. Average of them is tabulated in Table 1.

4.1 Comparative Analysis

The performance of the proposed scheme has been compared with some other works done so far. Aiming to propose an acoustic feature alternative to MFCC, this work has been compared with those recent works where MFCC has been used. The custom dataset used in this work has also been used for implementation of systems proposed by Kumar et al. [22], Bhalke et al. [23] and Ghisingh and Mittal [24].

The comparative analysis has been tabularized in Table 2 from where it is observed that the proposed feature set not only classifies String instruments with good accuracy, but it is also capable of categorizing Carnatic percussion instruments (ghatam, kanjira, mridangam, morsing and thavil) and broad classification of instruments (String, percussion) also.

Table 1 Classification accuracy for classification of String instruments

Classification scheme	Accuracy (in %) of classification for proposed work		
	Plucking	Bowing	Striking
Naïve Bayes	94.2	93.3	94.1
Neural network (MLP)	92.2	91.1	92.8
k-NN	91.4	90.7	91.3

Table 2 Accuracy (in %) of comparative analysis of proposed work with other works

Name of the method	Plucking	Bowing	Striking
Kumar et al. [22]	91.6	90.2	91.1
Bhalke et al. [23]	90.7	89.8	90.4
Ghisingh and Mittal [24]	92.8	90.6	92.2

5 Conclusion

In this work a new feature set has been propounded to classify String instruments into several categories based on how those instruments are played. The feature set is designed based on Chroma-based features which reflects the characteristics of notes produced by diverse type of String instruments very well. The proposed feature set performs better than MFCC-based features whereas MFCC is very popular and at the same time overused too. Experimental results exhibit that the proposed feature set works better than other existing works done. Computationally this feature set is simple as well as it is of low dimension. The proposed feature set will be tested to classify other types of instruments in future. To highlight the classification strength of the proposed feature set, simple and popular classification scheme like Neural Network, k-NN and Naïve Bayes' has been adopted.

References

1. Wikipedia contributors, Musical Instrument Classification. http://en.wikipedia.org/w/index.p hp?title=Musical_instrument_classification&oldid=325554329
2. Agostini, G., Longari, M., Pollastri, E.: Musical instrument timbres classification with spectral features. EURASIP J. Appl. Signal Process. **2003**, 5–14 (2003)
3. Zhu, J., Xue, X., Lu, H.: Musical genre classification by instrumental features. ICMC (2004)
4. Essid, S., Richard, G., David, B.: Hierarchical classification of musical instruments on solo recordings. In: Proceedings of the IEEE International Conference on Acoustics, Speech and Signal Processing (ICASSP 2006), vol. 5 (2006)
5. Sinith, M.S., Rajeev, K.: Pattern recognition in South Indian classical music using a hybrid of HMM and DTW. In: IEEE International Conference on Computational Intelligence and Multimedia Applications, vol. 2 (2007)
6. Deng, J.D., Simmermacher, C., Cranefield, S.: A study on feature analysis for musical instrument classification. IEEE Trans. Syst. Man Cybern. Part B **38**(2), 429–438 (2008)
7. Gunasekaran, S., Revathy, K.: Recognition of Indian musical instruments with multi-classifier fusion. In: IEEE International Conference on Computer and Electrical Engineering (ICCEE 2008), pp. 847–851 (2008)
8. Senan, N. et al.: Feature extraction for traditional malay musical instruments classification system. In: IEEE International Conference on Soft Computing and Pattern Recognition (SOCPAR, 2009) pp. 454–459 (2009)
9. Liu, J., Xie, L.: SVM-based automatic classification of musical instruments. In: International Conference on Intelligent Computation Technology and Automation (ICICTA, 2010), vol. 3, pp. 669–673 (2010)
10. Kumari, M., Kumar, P., Solanki, S.S.: In: Classification of North Indian Musical Instruments using Spectral Features, GESJ: Computer Science and Telecommunication, vol. 6, pp. 11–24 (2010)
11. Barbedo, J.G.A., Tzanetakis, G.: Instrument identification in polyphonic music signals based on individual partials. In: IEEE International Conference on Acoustics Speech and Signal Processing (ICASSP, 2010), pp. 401–404 (2010)
12. Barbedo, J.G.A., Tzanetakis, G.: Musical instrument classification using individual partials. IEEE Trans. Audio Speech Lang. Process. **19**(1), 111–122 (2011)

13. Ghosal, A., Chakraborty, R., Dhara, B.C., Saha, S.K.: Automatic identification of instrument type in music signal using wavelet and MFCC. In: Venugopal, K.R., Patnaik, L.M. (eds.) Computer Networks and Intelligent Computing. Communications in Computer and Information Science, vol. 157, pp. 560–565. Springer, Berlin, Heidelberg (2011)

14. Grindlay, G., Ellis, D.P.W.: Transcribing multi-instrument polyphonic music with hierarchical Eigen instruments. IEEE J. Sel. Top. Signal Process. **5**(6), 1159–1169 (2011)

15. Müller, M., Ewert, S.: Chroma toolbox: MATLAB implementations for extracting variants of chroma-based audio features. In: Proceedings of the 12th International Conference on Music Information Retrieval (ISMIR). hal-00727791, version 2–22 (2011)

16. Chandwadkar, D.M., Sutaone, M.S.: Role of features and classifiers on accuracy of identification of musical instruments. In: IEEE 2nd National Conference on Computational Intelligence and Signal Processing (CISP, 2012), pp. 66–70 (2012)

17. Joshi, M., Nadgir, S.: Extraction of feature vectors for analysis of musical instruments. In: IEEE International Conference on Advances in Electronics, Computers and Communications (ICAECC, 2014), pp. 1–6 (2014)

18. Gaikwad, S., Chitre, A.V., Dandawate, Y.H.: Classification of Indian classical instruments using spectral and principal component analysis based cepstrum features. In: IEEE International Conference on Electronic Systems, Signal Processing and Computing Technologies (ICESC, 2014), pp. 276–279 (2014)

19. Arora, V., Behera, L.: Instrument identification using PLCA over stretched manifolds. In: IEEE Twentieth National Conference on Communications (NCC), pp. 1–5 (2014)

20. Dandawate, Y.H., Kumari, P., Bidkar, A.: Indian instrumental music: Raga analysis and classification. In: IEEE 1st International Conference on Next Generation Computing Technologies (NGCT, 2015), pp. 725–729 (2015)

21. Abeber, J., Weib, C.: Automatic recognition of instrument families in polyphonic recordings of classical music. In: 16th International Society for Music Information Retrieval Conference (2015)

22. Kumar, P.A.M., Sebastian, J., Murthy, H.A.: Musical onset detection on carnatic percussion instruments. In: IEEE Twenty First National Conference on Communications (NCC, 2015), pp. 1–6 (2015)

23. Bhalke, D.G., Rao, C.B.R., Bormane, D.S.: Automatic musical instrument classification using fractional Fourier transform based-MFCC features and counter propagation neural network. J. Intell. Inf. Syst. **46**(3), 425–446 (2016)

24. Ghisingh, S., Mittal, V.K.: Classifying musical instruments using speech signal processing methods. In: IEEE Annual India Conference (INDICON, 2016), pp. 1–6 (2016)

Liver Disorder Prediction Due to Excessive Alcohol Consumption Using SLAVE

Sahil Saxena, Vikas Deep and Purushottam Sharma

Abstract The disturbances caused in the liver are recognized as liver disorder which results in illness. The liver plays a vital role and exclusive functions in the body. Liver injury/rupture can lead to massive damage to the body. Liver disease is a widely used term that covers all the probable issues that cause the liver to fail in order to perform its selected functions. Overconsumption of alcohol can lead to liver disorder which ultimately leads to many other problems. SLAVE is an associate degree inductive learning calculation that employs concepts in lightweight of down like principle hypothesis. This hypothesis can be used to perceive an authentic device for enhancing the comprehension of the knowledge, which cannot be inherited if just the purpose of reading by the user is known. Besides, SLAVE utilizes associate reiterative approach for learning. The research tends to propose an associate adjustment of the underlying reiterative approach used as a region of SLAVE.

Keywords Fuzzy logic · Genetic algorithms · Learning systems · SLAVE

1 Introduction

The liver is the biggest organ of the body. It weighs around three avoirdupois unit (1.36 kg). It is dark reddish in color and is partitioned into four flaps of unequal size and form. The liver lies on the right of the abdomen hole. Blood is sent to the

---

S. Saxena · V. Deep (✉) · P. Sharma
Department of Information Technology, Amity University Noida,
Noida 201313, Uttar Pradesh, India
e-mail: vikasdeep8@gmail.com

S. Saxena
e-mail: sahil.saxena98@gmail.com

P. Sharma
e-mail: psharma5@amity.edu

© Springer Nature Singapore Pte Ltd. 2019
A. Abraham et al. (eds.), *Emerging Technologies in Data Mining and Information Security*, Advances in Intelligent Systems and Computing 755,
https://doi.org/10.1007/978-981-13-1951-8_18

liver through 2 substantial vessels referred to as the venous blood vessel also called the entrance vein. The heptic route conveys oxygen-rich blood from the arteries (a noteworthy vessel within the heart). The entree vein convey blood containing processed nourishment from the tiny gastrointestinal tract. These veins subdivide within the liver quite once, ending in very little vessels. Each slim prompts a lobe [1]. Liver tissue is created out of thousands of lobules, each lobe is comprised of viscous cells that the essential metabolic cells of the liver. This paper portrays excessive utilization of liquor that causes an intense or constant aggravation of the liver and will even cause other organs to misbehave, liquor initiated liver ill remains a stimulating issue. The paper depicts the biopsy taken at the purpose once a person is influenced to liver issue like antacid phosphatase, alanine transaminase, and aspartate transaminase.

The research is inclined towards the introduction of the reformulation of slave learning formula which includes some very important modifications to original slave learning. The foremost very important modification consists of introducing associate degree alternate technique for scheming the positive and negative example of fuzzy rule [2]. This rule selection decides the foremost effective rule selection for slave employment example in each step. During this iteration, the selection of best individual rule is carried with no knowledge concerning antecedent learned.

2 Risk Level Prediction for Liver

The risk level for liver disorder or malfunction can be predicted using data mining tools like keel by using different algorithms for this paper we will use Bupa.

3 Liver Disease Dataset Description

The first five variables square measure all blood tests that square measure thought to be sensitive to liver disorders which may be generated due to alcohol consumption in high capacity. Every line within the dataset constitutes the record of one male individual [3] (Figs. 1, 2, 3, 4, 5, 6 and 7).

Liver Disorder Dataset [4]:

```
DataSet View   Attribute Info   Charts 2D
Content
@relation bupa
@attribute mcv integer [65, 103]
@attribute alkphos integer [23, 138]
@attribute sgpt integer [4, 155]
@attribute sgot integer [5, 82]
@attribute gammagt integer [5, 297]
@attribute drinks real [0.0, 20.0]
@attribute selector {1, 2}
@inputs mcv, alkphos, sgpt, sgot, gammagt, drinks
@outputs selector
@data
85, 92, 45, 27, 31, 0.0, 1
85, 64, 59, 32, 23, 0.0, 2
86, 54, 33, 16, 54, 0.0, 2
91, 78, 34, 24, 36, 0.0, 2
87, 70, 12, 28, 10, 0.0, 2
98, 55, 13, 17, 17, 0.0, 2
88, 62, 20, 17, 9, 0.5, 1
88, 67, 21, 11, 11, 0.5, 1
92, 54, 22, 20, 7, 0.5, 1
```

Fig. 1 Liver disorder dataset

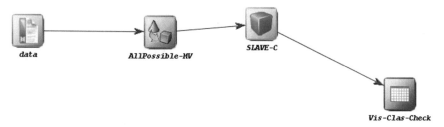

Fig. 2 KEEL experiment design

4 Experiment Design Format with KEEL

Bottom is a product instrument created to assemble and utilize distinctive data mining models [5]. We might want to comment this is the primary programming apparatus of this sort containing a free code Java library of Evolutionary Learning Algorithms [6, 7]. The principle highlights of KEEL are:

- It contains pre-handling calculations: change, discretization, occurrence determinations, and highlight choices.
- It additionally contains a Knowledge Extraction Algorithms Library, regulated and unsupervised, commenting the fuse of different developmental learning calculations.
- It has a measurable investigation library to break down calculations.
- It contains a user-friendly interface, oriented to the analysis of algorithms.

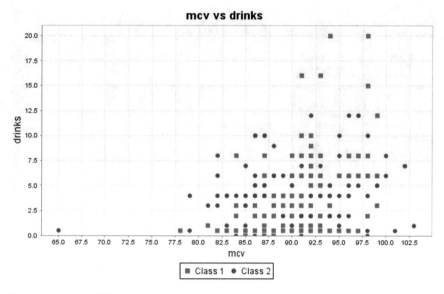

Fig. 3 mcv versus drinks

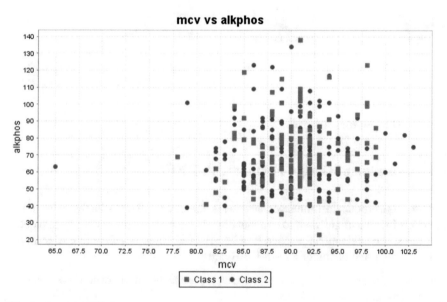

Fig. 4 mcv versus alkphos

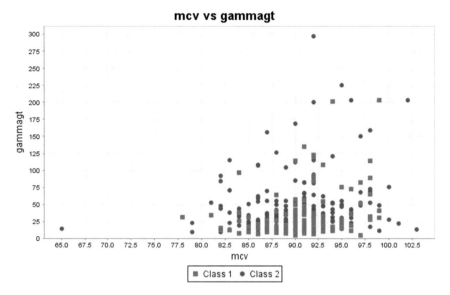

Fig. 5 mcv versus gammagt

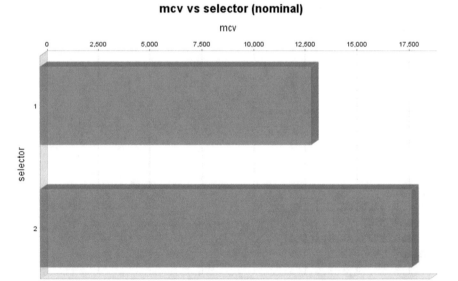

Fig. 6 mcv versus selector

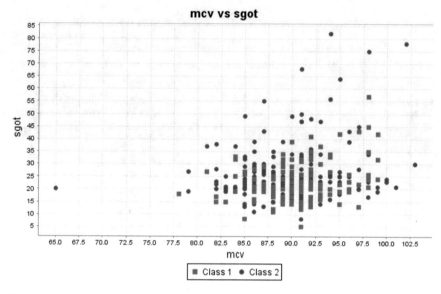

Fig. 7 mcv versus sgot

Fuzzy Set domain:

```
{\tiny }

Variable: X0
========================
Name: L0
[65.0,65.0,65.0,74.5]
Name: L1
[65.0,74.5,74.5,84.0]
Name: L2
[74.5,84.0,84.0,93.5]
Name: L3
[84.0,93.5,93.5,103.0]
Name: L4
[93.5,103.0,103.0,103.0]

Variable: X1
========================
Name: L0
[23.0,23.0,23.0,51.75]
Name: L1
[23.0,51.75,51.75,80.5]
Name: L2
```

```
[51.75,80.5,80.5,109.25]
Name: L3
[80.5,109.25,109.25,138.0]
Name: L4
[109.25,138.0,138.0,138.0]

Variable: X2
=========================
Name: L0
[4.0,4.0,4.0,41.75]
Name: L1
[4.0,41.75,41.75,79.5]
Name: L2
[41.75,79.5,79.5,117.25]
Name: L3
[79.5,117.25,117.25,155.0]
Name: L4
[117.25,155.0,155.0,155.0]

Variable: X3
=========================
Name: L0
[5.0,5.0,5.0,24.25]
Name: L1
[5.0,24.25,24.25,43.5]
Name: L2
[24.25,43.5,43.5,62.75]
Name: L3
[43.5,62.75,62.75,82.0]
Name: L4
[62.75,82.0,82.0,82.0]

Variable: X4
=========================
Name: L0
[5.0,5.0,5.0,78.0]
Name: L1
[5.0,78.0,78.0,151.0]
Name: L2
[78.0,151.0,151.0,224.0]
Name: L3
[151.0,224.0,224.0,297.0]
Name: L4
[224.0,297.0,297.0,297.0]
```

```
Variable: X5
========================
Name: L0
[0.0,0.0,0.0,5.0]
Name: L1
[0.0,5.0,5.0,10.0]
Name: L2
[5.0,10.0,10.0,15.0]
Name: L3
[10.0,15.0,15.0,20.0]
Name: L4
[15.0,20.0,20.0,20.0]

Variable: Class
========================
Name: 1
[0.0,0.0,0.0,0.0]
Name: 2
[1.0,1.0,1.0,1.0]
```

Ruleset1

```
Number of rules = 6
IF X0 = {L3} X1 = {L2 L3 L4} X3 = {L0 L1}
X4 = {L0 L2} X5 = {L0 L2 L3 L4}
THEN Class IS 1 W 0.584614828325317

IF X5 = {L4} THEN Class IS 1 W 1.0

IF X0 = {L0 L1 L4} X1 = {L2 L3 L4}
X2 = {L3}   X4 = {L0 L1 L4}
THEN Class IS 1   W 1.0

IF X1 = {L0 L2 L3 L4} X2 = {L1 L2 L3 L4} X3 = {L0}
X4 = {L0 L2 L4} X5 = {L2 L3 L4}
THEN Class IS 1 W 0.5765281790538619

IF X3 = {L1 L2 L3}
THEN Class IS 2   W 0.5845253918939388

IF X4 = {L1 L2 L3 L4}
THEN Class IS 2   W 0.12340756634756407
```

Ruleset2
Number of rules = 6

```
IF X0 = {L0 L1 L3} X2 = {L1 L2 L3 L4} X3 = {L0 L1}
X4 = {L0 L3} X5 = {L0 L2 L3 L4}
THEN Class IS 1   W 0.5648164506054825

IF X1 = {L0 L1 L2 L4} X3 = {L0 L2 L3 L4}
X4 = {L0 L1 L4} X5 = {L3}
THEN Class IS 1   W 0.9125780553077609

IF X0 = {L0 L1 L4} X1 = {L2 L3 L4} X2 = {L3 L4}
X4 = {L0 L1} THEN Class IS 1   W 1.0

IF X0 = {L0 L3 L4} X1 = {L0 L1 L2 L4} X2 = {L2 L3 L4}
X3 = {L0 L1} X4 = {L0 L1 L4} X5 = { L2 L3 L4 }
THEN Class IS 1   W 0.7740899582345862

IF X0 = {L0 L2} X1 = {L0 L3} X2 = {L1 L4}
X4 = {L1 L3 L4} X5 = { L2 L3 L4 }
THEN Class IS 1   W 0.35474929749411277

IF X0 = {L0 L2 L3 L4} X1 = {L1 L2 L3 L4}
X3 = {L1 L2 L3 L4} X5 = {L0 L1 L2}
THEN Class IS 2   W 0.5983891215619033
```

5 Classification with Respect to Class

The graphs mentioned below are representing the two respective attributes information taken from Bupa.

6 Conclusion

In this study, we have got explored KEEL, that is a software program bundle device for the analysis of biological manner studying approaches applied to facts processing problems. This device allows the researchers to attention on the evaluation of their new studying models compared with the existing ones. Moreover, the device permits researchers with little or no understanding of biological process computation ways to use evolutionary studying algorithms to their paintings. A complete reformulation of the training algorithmic program SLAVE has been done in the research that features some important modifications to the initial SLAVE learning algorithmic program. The foremost necessary modification consists of proposing an alternate technique for computing the positive and negative examples for a fuzzy rule. The rule choice of SLAVE decides the most effective rule for the coaching examples in every step within the repetitious approach of SLAVE, and the choice of the most effective individual rule is distributed with no information concerning the foundations learned. The experiments show the 12 rules in two ruleset and the accuracy is more better.

References

1. Quinlan, J.R.: C4.5: Programs for Machine Learning. Morgan Kauffman Publishers, San Mateo, California (1993)
2. Frawley, W.J., Piatetsky-Shapiro, G.: Knowledge Discovery in Databases: An Overview. The AAAI/MIT Press, Menlo Park, CA (1996)
3. McDermott, J., Forsyth, R.S.: Diagnosing a disorder in a classification benchmark. Pattern Recogn. Lett. **73** (2016)
4. Lichman, M.: UCI machine learning repository. University of California, School of Information and Computer Science, Irvine, CA (2013). http://archive.ics.uci.edu/ml
5. Triguero, I., Gonzlez, S., Moyano, J.M., Garca, S., Alcal-Fdez, J., Luengo, J., Fernndez, A., del Jesus, M.J., Snchez, L., Herrera, F.: KEEL-3.0: an open source software for multi stage analysis in data mining. Int. J. Comput. Intell. Syst. **10**, 1238–1249 (2017)
6. Alcal Fdez, J., Snchez, L., Garca, S., del Jesus, M.J., Ventura, S., Garrell, J.M., Otero, J., Romero, C., Bacardit, J., Rivas, V.M., Fernndez, J.C., Herrera, F.: KEEL: a software tool to assess evolutionary algorithms to data mining problems. Soft Comput. **13**(3), 307–318 (2009)
7. Alcal-Fdez, J., Fernandez, A., Luengo, J., Derrac, J., Garca, S., Snchez, L., Herrera. F.: KEEL—data mining software tool: data set repository, integration of algorithms and experimental analysis framework. J. Mult. Valued Logic Soft Comput. **17**(2–3), 255–287 (2011)

Analysis on Multi-objective Optimization Problem Techniques

Aditi Jaiswal

Abstract In past few years, Web-based application and services are growing rapidly and this growing demands needs different Quality of Services (QoS) requirements for efficient use of such web-based services. The purpose behind utilizing these application resources could be tarnished if the fundamental communication network does not fulfill the QoS requirements. However, different applications have distinct QoS necessities as each application have different priorities. The main concern is to come across such solution which will optimize the network not in the terms of minimum number of hops but in terms of Qos parameters of network, relies upon application running over that network. This issue comes under Multi-objective Optimization Problem (MOOP) and Genetic Algorithm (GA) is one of the techniques which can possibly control numerous parameters all together, and hence GA is applied to solve MOOP, which can enhance the QoS. This paper surveys the various MOOP techniques and then gives the best solution among them.

Keywords Multi-objective optimization problem · Evolutionary algorithms Quality of service (qos) · Genetic algorithm (ga)

1 Introduction

Computer network is an interrelated set of computers with the capacity to send and receive data [1]. Nowadays, wireless networks are the heart of modern method for communication. Qos on routing is an important concern that has a major impact on the networks related issues.

An ideal routing algorithm should seek to discover the best possible path for packet transmission in a precise time to assure the QoS. The routing protocols which we are using today use a simple metric, i.e., hop count and shortest path algorithm so

A. Jaiswal (✉)
Computer Science and Engineering Department, Maulana Azad National Institute
of Technology, Bhopal 462003, Madhya Pradesh, India
e-mail: aditijaiswal1393@gmail.com

© Springer Nature Singapore Pte Ltd. 2019
A. Abraham et al. (eds.), *Emerging Technologies in Data Mining and Information Security*, Advances in Intelligent Systems and Computing 755,
https://doi.org/10.1007/978-981-13-1951-8_19

as to find out the best possible routes [1]. In QoS routing, we can determine routes by requirements based on aspect of how data flows in the network, such as bandwidth, packet delay, packet loss, and cost. There are two important objectives that are needed to be accomplished by the QoS routing algorithm. The primary goal is to discover a path that satisfies the QoS necessities corresponding to the application running over the network [2]. The other goal is to optimize the overall network resource consumption.

Quality of services varies from application to application. Some applications have high requirement of bandwidth, or require a more scale of reliability or while some expect a small amount of response time and so on. For example, an application like an E-mail claim for high reliability and can bear longer delay than the real-time applications such as videoconferencing which cannot tolerate more delay otherwise we face the problem of buffering. So main aim is to find a routing strategy which is not only economic, decisive, or correct but assure the QoS measure as according to applications concern also.

There are several multi-objective optimization methods, some of them are purely mathematical, and others are based on Particle Swarm Optimization or Ant Colony Optimization. Evolutionary algorithms are considers as one among the foremost flourishing strategies for solving the multi-objective optimization problems. The main advantage for using these is that, they are population based, in this manner; it can discover more than one interesting solution in a single run. MOEAs advance a population of candidate solutions for look out at gathering of best solutions in a solitary run. This group of optimal solution is called Pareto-optimal front while the solutions generated are called nondominated solutions.

Evolutionary Algorithms are known as a good means for explaining NP hard and NP-complete problems because of the potential of Multi-objective Optimization (MOO) [3] as Evolutionary approach has been used earlier in many multi-objective optimization problems such as optimization of distributed network topology [4], bicriteria transportation problems [5], and many more. Thus, the heuristics based on EAs set as the best decision for discovery of the optimal routes according to QoS procedures because they have an immense potential to handle multiple constraint problem and discrete variable.

In this paper, we have talked about the various methods for solving multi-objective optimization problem. The QoS consider here acts as a parameter such as bandwidth, reliability, and response time [6] and also cost and delay [7]. For any application, only some QoS measures are of major importance while the requirement for other QoS measures is less rigorous. This survey paper is organized as follows: in the next section, we define the fundamental concept about multi-objective optimization. Then, we are going to illustrate few multi-objective optimization techniques. Next, we describe about the objective which can be considered. Furthermore, we give one of the solutions of MOO using evolution technique and finally, we made conclusion and few proposals for future work.

2 Fundamental Concept

2.1 Definition of MOO Problem

Multi-objective optimization means when we have various objectives and all have to be optimized at a single run. This type of situation occurs in various fields like in, economics and engineering, when one of the finest decisions are required to be taken in the existence of trade-offs involving different contradictory objectives [2]. There are three main points of MOO problems. The first one is that, the objective are contradictory. It means that there will be no single solution present till now which has the capacity to simultaneously optimize each objective. The second one is that optimizing a result w.r.t one objective can effect in undesirable results w.r.t other objectives, i.e., no other objective functions can be enhanced in value without debasing some of the other objective values. The last one is that, we want to find, not only one solution , but a number of solutions, which correspond to the best possible trade-offs among all the objectives that we plan to optimize.

A Multi-objective Optimization Problem (MOOP) [8] can be defined as follows: Given an m-dimensional vector of decision variables $y = y_1, y_2, \ldots, y_m$, in solution space Y. We have to find a particular vector y_k, that minimizes/maximizes a given set of P objective functions $f(y_k) = f_1(y_k), \ldots, f_p(y_k)$, where f_i is objective i. A function $f : Y \rightarrow X$ is used to evaluate the quality of a specific solution by assigning it to an objective vector $x_1, x_2, \ldots, x_k$ in the objective space X. The mapping takes place between an m-dimensional solution vector and a p-dimensional objective vector.

2.2 Pareto Optimality

The solution of MOOPs is contained in a set of trade-off solutions called Pareto-optimal set which will hold the entire nondominated solutions [6]. This set signifies one of the best possible trade-offs among the various objectives present. A feasible solution $y_k \in Y$ can be termed as Pareto-optimal iff, there is no additional solution $y \in Y$ such that y dominates y_k [8]. Thus, we cannot enhance a pareto-optimal solution w.r.t any objective without deteriorate at least one of the other objective. Pareto-optimal set is the set of all possible nondominated solutions, and for a specified pareto-optimal set, the equivalent objective function values in the objective space are called the pareto front. For many issues, the quantity of pareto-optimal solutions is large, and also be incomputable. The ultimate goal of MOO algorithms is to spot solutions within the pareto-optimal set. However, it is impractical to distinguishing the complete pareto-optimal set for several multi-objective issues due to its size [7].

3 Multi-objective Optimization Solution Techniques

In this section, we tend to discuss about MOO solution techniques.

3.1 Scalarization-Based MOO Function Formulation

3.1.1 Weighted Global Criterion Methodology

One of the trendy general scalarization techniques for multi-objective optimization is that the global criterion methodology within which all objective functions is combined to create one function. Though, a global criterion is also a mathematical function with no correlation to preferences [9]. One of the most common utility functions is expressed in its simplest type as the weighted exponential sum:

$$U = \sum_{i=1}^{k} w_i [F_i(x)]^p, \; F_i(x) > 0 \; \forall i \tag{1}$$

$$U = \sum_{i=1}^{k} [w_i F_i(x)]^p, \; F_i(x) > 0 \; \forall i \tag{2}$$

The most common extensions of (1) and (2) are

$$U = \sum_{i=1}^{k} w_i [F_i(x) - F_i^\circ]]^p]^{\frac{1}{p}} \tag{3}$$

$$U = \sum_{i=1}^{k} w_i^p [F_i(x) - F_i^\circ]]^p]^{\frac{1}{p}} \tag{4}$$

Here, w is a vector of weights typically set by the decision maker such that $\sum_{i=1}^{k} w_i = 1$ and $w > 0$, U is called an achievement function. The solutions to these types of approaches depend on the value of p. Generally, p is proportional to the amount of significance placed on minimizing the function with the largest difference between $F_i(x)$ and F_i° (Koski and Silvennoinen 1987).

3.1.2 Weighted Sum Approach

The most general approach to multi-objective optimization is the weighted sum [10].

$$U = \sum_{i=1}^{k} w_i F_i(x) \qquad (5)$$

This is a form of (1) or (2) with p = 1. If every single weight is positive, then the least of (5) is Pareto optimal, i.e., minimizing (5) is sufficient for Pareto optimality. In any case, the plan does not give a necessary condition for Pareto optimality.

3.2 Meta-Heuristics

In this section, we will discuss the following well-understood meta-heuristics to resolve MOO problems:(1) Evolutionary algorithms (2) Ant Colony Optimization method (3) Particle Swarm Optimization method.

3.2.1 Evolutionary Algorithms

EAs measure a collection of contemporary met heuristics, used with success in several applications with great complexness. Meta-heuristics direct different heuristic algorithms while exploring through the feasible solution space for distinguishing an optimal solution. The underlying thought is to rapidly converge toward an optimal solution [9]. Numerous NP-complete issues have been comprehended using this technique. A typical kind of EA is the genetic algorithm (GA). NSGA is one of the MOEA algorithms used to solve MOO problems.

Nondominated Sorting Genetic Algorithm (NSGA) NSGA is well-known non-domination based genetic algorithm for multi-objective optimization. It is a awfully effective and successful algorithm, however has been typically criticized for its computational complexity, insufficiency or absence of elitism and for selecting the optimal parameter value for sharing parameter. An altered variant, NSGA-II was produced, which has an improved sorting algorithm, fuses elitism, and no sharing parameter should be picked from the earlier. Srinivas and Deb [11] depict a two-step NSGA: fast nondominated sorting and crowding distance calculation. Fast nondominated sorting highlights elitism to save the best individuals till the present generation while crowding distance calculation promotes resolution diversity. Here, the first step is to initialize the population, which depends on the range of problem and constraints present in the problem. After, the population initialization, it is arranged in light of the nondomination into every front. The primary front being completely nondominant set within the present population and the second front being dominated by the individuals within the first front solely and it goes on like this. Every individual within each front are assigned rank (fitness) values or based on front in which they fit in to. The fitness values are provided to the individuals, and first front individuals are given a fitness value of 1, then 2 and so on. The other parameter which is equally important, called crowding distance, which is calculated for every individual. Now, the calculation of crowding distance is done by observing how close an individual is to its neighbors, which would be the sum of distances to the nearby neighbors [12]. We set the crowding distance of best solutions in every objective to infinity. Large average crowding distance will lead to higher diversity in the population. We use the binary tournament selection for selection of Parents from the population, which is based on the rank and crowding distance.

3.2.2 Ant Colony Optimization (ACO)

Swarm intelligence is a generally new way to deal with critical thinking that takes motivation the social behaviors of insects and of different creatures. In particular, from ants, various strategies and techniques are introduced, among which the most considered is the general purpose optimization technique called as ant colony optimization [13]. ACO takes motivation from the scavenging conduct of some ant species. To mark some favorable path, these ants store pheromone on the ground that ought to be followed by different individuals of the colony through indirect communication. The optimal path in the graph support behavior of ants seeks a path between their colony and supply of food. To begin with, scouting ants start looking for sources of food. Once the ants come back to the colony, they leave pheromones, as markers on the way prompting to the food, which can provide different ants directions to seek out the food [13, 14]. Additional pheromones on a path mean a much better path with higher chance of finding food. This eventually prompts finding a superior (or best) path. To utilize this technique, an optimization problem ought to be changed over to a problem of finding a best path on a weighted graph. ACO is beneficial for dynamic applications like finding routes under progressively changing network topologies [14].

3.2.3 Particle Swarm Optimization (PSO)

Particle Swarm Optimization (PSO) is a population-based technique developed by Dr. Eberhart and Dr. Kennedy in 1995, galvanized by social activities of bird flocking or fish schooling. PSO imparts several similarities like that of Genetic Algorithms (GA). The framework is instated with a population of irregular solution and looks for optima by changing generations. In any case, PSO does not comprise of evolution operators such as crossover and mutation, like GA. When varieties of operations performed, a bunch of variables have their values balanced nearer to the member whose value is closest to the target at any given moment. Imagine a flock of birds circling over a region, where they will smell a hidden supply of food. The one who is nearest to the food tweets the loudest and hence the remaining birds swing around him. If any of the other surrounding birds comes nearer to the target than the first, it twitters louder and the others come over to him. This algorithm monitors three global variables:

Table 1 Summarizes above methods

Techniques	Pros	Cons
Weighted sum	Computationally proficient in generating a robust nondominated solution	Contingent on weight coefficients, concave trade-off curve might not be determined
Evolutionary algorithms (EA)	Gives heuristic however near optimal solutions	Computationally costly fitness functions that generally create local optima
Nondominated sorting genetic algorithm (NSGA)	Preserves solution diversity supported the crowding distance comparison process; High potency by safeguarding optimal solutions in view of elitism	Poor performance under the failure of creating crowd solutions under different objectives
Ant colony optimization (ACO)	Helpful for dynamic applications	Helpful for dynamic applications
Particle swarm optimization (PSO)	High-quality solution with less time	May be troublesome to spot the best optimal parameters sometimes

1. Target esteem or condition.
2. Global best (gBest) value indicating which particle's data is currently closest to the target.
3. Stopping value showing when the algorithm should stop if the Target is not discovered [15] (Table 1).

4 Objectives to Be Considered for Networking Problems

There are mainly four objectives for any networking problem, for which we want to find the optimal value:

Cost The aggregate cost function can be expressed as the summation of cost of connecting link along the path from the source to the destination [6].

Delay The delay of a network specifies how much time a bit of data takes to move over the network from source to destination [6].

Response Time Response time is the time required for a packet to move over the network path from source to destination [7].

Reliability The path which will be generated must be reliable. Therefore, we take primary path and backup path simultaneously. So that if chosen path for routing fails, then backup path should be present for communication. By this, we can avoid situation of connection failure [7].

5 Required Steps for Solving MOO Problem Using Evolution

In communication network, first, initialization of routing path is done then apply essential genetic operators on the population. The operators which we included here are selection, crossover, and mutation. The subsequent subsections explain these operators.

5.1 Initialization

First, we will encode the routing path which is done with the help of string of positive integers that correspond to the IDs of nodes. There are two types of encoding techniques discussed here.

5.1.1 Random-Based Encoding

The chromosome is basically an inventory of nodes on the constructed path, $S \rightarrow N_1 \rightarrow N_{k-1} \rightarrow N_k \rightarrow D$. The initial population consists of a particular variety of chromosomes. A random path is explored from source node S to destination node D by arbitrarily selecting a node N from the inventory of n nodes. After that, an additional node N_k is haphazardly selected from the inventory of nodes. This procedure is rehashed until the point that destination D is reached. But we want that the path formed must be free from loop, so we do not include that nodes which are used previously. In this manner, duplicate entry of same node is avoided [7].

5.1.2 Priority-Based Encoding

Some problem may arise when an arbitrary grouping of edges does not correspond to a path. To defeat such challenges, an indirect approach is adopted by training some information to create a path. Strategy starts from the source node and ending at the destination node n. To seek out a path from source node to destination node, the nodes which are associated to source node has to be determined first. At each and every step, sometimes we have to consider many nodes. Every node is allotted a priority with random means and then finally includes those nodes which have the maximum priority in the path [7]. Gene in a chromosome is described by two elements: locus and allele. In this encoding scheme, the node ID signifies the position of gene and node priority is symbolized by its value for creating a path between candidates. From this encoding proposal, unique path can be determined.

5.2   Tournament Selection

Selection plays a vital role in enhancing the average quality of the population by passing the prime quality chromosomes to successive generation [16]. The individual which is having most reduced front number is chosen, if the two individuals are from completely different fronts. And If the individual is from the identical front then the individual with the maximum crowding distance is chosen. In every iteration, the N existing individual parents produce N new individual offspring. Then, both parents and offspring try to win from each other for inclusion in the next iteration [6].

5.3   Crossover and Mutation

The first genetic operation performed to the chromosomes is crossover. The thought behind crossover is to make a data trade, between two chromosomes. After doing this, the algorithm will search new paths and ideally have the capacity to discover better paths in the process. Two crossover schemes that are projected here are

(1) Node-Based Crossover (NBX) [16].
(2) Partially Mapped Crossover (PMX) [16].

5.3.1   Node-Based Crossover

The cross over scheme is an adjustment of the one-point cross. Here, we arbitrarily select a node or locus from every pair and then we match the node ID of the locus with the genes in the further chromosomes. If we find that there is a match, then we perform the crossover operation else two fresh paths are selected for crossover till the mating pool is vacant [7].

5.3.2   Partially Mapped Crossover

Here, in this method, we randomly select two crossover points and then their positions are exchanged, which ensures that all positions that will be discovered precisely once in every offspring [7]. PMX is done as follows:

(1) First, two chromosomes are associated.
(2) Two crossing sites are chosen consistently at indiscriminately along the strings, characterizing an identical section.
(3) Now we use the identical section to cross through position-by-position exchange action.
(4) Finally, alleles are moved to their fresh positions within the offspring.

In this method, the source node and the destination nodes are permanent, it cannot be changed. Every partial route is exchanged and bring together and in this way, two new routes are created. In NBX, when crossover is performed, there is a chance of routes having loops. So as to stay away from this, repair function is used as a countermeasure. So, repair function discovers and removes loops in a routing path without rising computation cost. It can fix loops in a chromosome by searching recurring nodes in the chromosome. The nodes in between the recurring nodes are then disposed of. But in PMX loop creation can be avoided. Here, there will be no recurrence of nodes present. The recurrence of nodes will be avoided by mapping function. That's why PMX discover numerous new paths without rising computational complexity. Here, there is no need of repair function.

Mutation is another genetic operator used to preserve genetic diversity from one generation of a population of genetic algorithm chromosomes to the next. Mutation alters one or more gene values in a chromosome from its initial state. In mutation, the solution may change entirely from the previous solution [7].

6 Conclusion

In the above segments we show the different fields of multi-objective optimization. In general, it requires more effort to compute multi-objective optimization than single-objective optimization. If we involve multiple objectives in any problem, then we get large set of Pareto-optimal solutions instead of a single globally optimal solution, we cannot consider that these Pareto-optimal solutions are superior than the others on the Pareto front without having any additional information. Therefore, MOO algorithms are used for finding several Pareto-optimal solutions as possible. The selection of a specific method depends on the type of information that is provided in the problem, the user's preferences, the solution requirements, and the availability of software. Based on all this information, we would like to generate further new interesting algorithms. For example, we would like to develop a new crossover operator specifically designed for multi-objective optimization.

References

1. Rouskas, G.N., Baldine, I.: Multicast routing with end-to-end delay and delay variation constraints. IEEE J. Sel. Areas Commun. **15**(3), 346–356 (1997)
2. Craveirinha, J., Giro-Silva, R., Clmaco, J.: A meta-model for multiobjective routing in MPLS networks. Cent. Eur. J. Oper. Res. **16**(1), 79–105 (2008)
3. Deb, K.: Multi-objective Optimization Using Evolutionary Algorithms. Wiley, New York, NY (2001)
4. Pierre, S., Legault, G.: A genetic algorithm for designing distributed computer network topologies. IEEE Trans. Syst. Man Cybern. Part B (Cybernetics) **28.2**, 249–258 (1998)
5. Gen, M., Li, Y.-Z.: Spanning tree-based genetic algorithm for bicriteria transportation problem. Comput. Ind. Eng. **35**(3), 531–534 (1998)

6. Kumar, D., et al.: Routing path determination using QoS metrics and priority based evolutionary optimization. In: 2011 IEEE 13th International Conference on High Performance Computing and Communications (HPCC). IEEE (2011)
7. Chitra, C., Subbaraj, P.: Multiobjective optimization solution for shortest path routing problem. Int. J. Comput. Inf. Eng. **4**(2), 77–85 (2010)
8. Yu, X., Gen, M.: Introduction to Evolutionary Algorithms. Springer Science & Business Media (2010)
9. van Veldhuizen, D.A., Lamont, G.B.: Multiobjective evolutionary algorithms: analyzing the state-of-the-art. Evol. Comput. **8**(2), 125–147 (2000)
10. Coello, C.A.: An updated survey of GA-based multiobjective optimization techniques. ACM Comput. Surv. (CSUR) **32**(2), 109–143 (2000)
11. Srinivas, N., Deb, K.: Muiltiobjective optimization using nondominated sorting in genetic algorithms. Evol. Comput. **2**(3), 221–248 (1994)
12. Deb, K., et al.: A fast and elitist multiobjective genetic algorithm: NSGA-II. IEEE Trans. Evol. Comput. **6.2**, 182–197 (2002)
13. Blum, C.: Ant colony optimization: introduction and recent trends. Phys. Life Rev. **2**(4), 353–373 (2005)
14. Dorigo, M., Stutzle, T.: Ant Colony Optimization. MIT Press, Cambridge, MA (2004)
15. Lee, K.Y., Park, J.-B.: Application of particle swarm optimization to economic dispatch problem: advantages and disadvantages. In: Power Systems Conference and Exposition, 2006. PSCE'06. 2006 IEEE PES. IEEE (2006)
16. Pangilinan, J.M.A., Janssens, G.: Evolutionary Algorithms for the Multi-objective Shortest Path Problem (2007)
17. Mishra, K.K., Kumar, A., Misra, A.K.: A variant of NSGA-II for solving priority based optimization problems. In: IEEE International Conference on Intelligent Computing and Intelligent Systems, 2009. ICIS 2009, vol. 1. IEEE (2009)
18. Fleming, P.J., Pashkevich, A.P.: Computer aided control system design using a multiobjective optimization approach. Control **85**, 174–179 (1985)

Part II
Cloud Computing

Proposed Methodology to Strengthen the Performance of Adaptive Cloud Using Efficient Resource Provisioning

Lata J. Gadhavi and Madhuri D. Bhavsar

Abstract The delivery of services as a computing and management of the resources like CPU, memory, software, information, and devices for end users are the key responsibilities of cloud computing. To enable the services as per the demand of end users and provisioning the resources to its hosted applications are defined as an approach in this paper. The dynamic and complexity of cloud environment create some challenges in managing the resources to fulfill the need of fluctuating resources. For the commercial and scientific applications or jobs, resource management has to be managed as per their current requirement. Improving the runtime performance of adaptive cloud for cloud-based services using efficient resource provisioning strategy is the key terminology in this paper. Adaptive cloud is to be built to analyze process, classify, and manage the data for cloud-based services. To make the cloud more intelligent and to adapt the dynamic data analysis, it should be trained to accept the runtime need of end users.

1 Introduction

A recent year shows the distinctiveness of cloud computing such as auto-scaling of resources, services provisioning as per the need and the utility-based services. The cloud computing paradigms are widely adopted by various industries and academicians [1]. The dynamic and complexity of cloud environment create some challenges in managing the resources to fulfill the needs of fluctuating resources. For the commercial and scientific applications or jobs, resource management has to be managed as per their current requirement. To fulfill the requirement of semantic data processing is more complicated. Several scientific jobs require data processing of its database and to produce the needed result. The pool of resource can be provisioned and re-

L. J. Gadhavi (✉) · M. D. Bhavsar
Institute of Technology, Nirma University, Ahmedabad 382481, Gujarat, India
e-mail: 12extphde92@nirmauni.ac.in

M. D. Bhavsar
e-mail: madhuri.bhavsar@nirmauni.ac.in

© Springer Nature Singapore Pte Ltd. 2019
A. Abraham et al. (eds.), *Emerging Technologies in Data Mining and Information Security*, Advances in Intelligent Systems and Computing 755,
https://doi.org/10.1007/978-981-13-1951-8_20

provisioned dynamically to the hosted applications as per requirement of the specific application and available resources. Here cloud computing covers two criteria which are dynamic resource management and adaptive cloud.

(a) Adaptive Cloud:

Adaptive resource management module performs the adjustment operation, add or release resources for application through dynamic cloud resource management interface model. To host the several types of application on cloud, adaptive resource management criteria should be considered first. Adaptive resource management consist the efficient, scalable, dynamic, and autonomous operations. Adaptive cloud is to be built more intelligently to process data and execute the service instance generated by data processing with resource management [2, 3].

(b) Efficient Resource Provisioning and Management

In the cloud computing, resource management is described to manage the resources as per the need of current and future periods for their hosted applications [4]. To enhance the performance of adaptive cloud, efficient resource provisioning will enable the provisioning of virtualized resources dynamically as per the need of end users. On-demand provisioning of resources can measure the performance based on through-put and response time for the execution of service instance. To make best use of the limited resources and to enable the scalable nature of cloud computing, adaptive cloud has to be built to provision resources dynamically and on-demand. As per the requirement and workload of application, resource allocation has to be done efficiently. More resources should be allocated, when workload becomes high. When workload is low then resource allocation strategy should keep away from allocating unnecessary resources. Static model for allocation of resources remains unchanged if once the resource is allocated to the application in the starting phase. This strategy may waste the resource if resource utilization is low and put the service-based application in the waiting mode if resource requirement is high. If allocation of resources configured adaptively, it maximizes the utilization of resources by considering configuration of the system performance and workloads. If the resource management is done manually, it will be time overwhelming and complex. Thus automatic and adaptive scenario is desirable to provision the resources dynamically and on demand. Section 2 presents the literature survey about the domain. Section 3 describes proposed work and case study for the efficient resource provisioning in cloud computing scenario. Section 4 describes about the configuration of private cloud environment. Section 5 concludes the whole work and shows the future enhancement.

2 Literature Survey

To make the cloud more intelligent and to satisfy the dynamic resource management need, it enables the on-demand provisioning of services to its hosted. In following papers, experiments are carried out with some limitations. Wei et al. in [5] presented

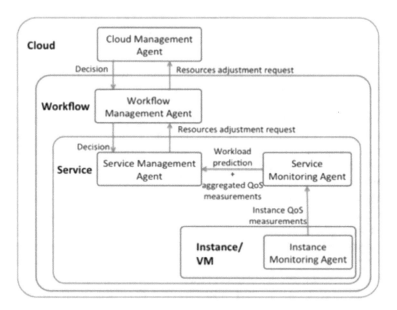

Fig. 1 Agent-based service workflow management framework [5]

an approach which exploits the management of virtualized resources for service workflows. An adaptive algorithm is applied based on reactive and predictive policy. The work does not consider the load balancing, optimization of the performance and not working on future behavior. Figure 1 shows architecture of an agent-based service workflow management framework.

Following methodology is adopted by authors in this paper.

It shows that traditional method for resource management was not able to fulfill the requirements when job becomes more intricate and complex. In this paper authors have investigated the issue regarding dynamic management of resources for service workflows in a cloud environment. An adaptive resource management algorithm is used to make resource management decisions based on reactive and predictive results. Methodology applied in this paper works on present and past statics of the service, which is not for future behavior and not suitable for performance optimization.

Iqbal et al. in [6] described a methodology to scale-up and down the resources dynamically for web applications based on predictive results. It is working on reactive and predictive policy, not on future behavior. IT shows that when restricted accesses are defined, the system allocates the resources required by the application dynamically to determine the identified restricted access and retain response time requirements. It will perform the resource management for applications as per the need of end users. In this paper software configuration management and future behavior policy are not addressed.

Zhang et al. in [7] explored an approach which focuses on integration of resource consumption and on-demand allocation of resources. To make best use of the resource utilization of an application and to allocate the resources, authors have proposed an approach to manage resources by the consumption of resources in optimized way and loop of resource for allocation. Proposed methodology is time-consuming because it considers factor of performance. If the performance is considered to be unsatisfactory, it will first try to optimize the system configuration before allocating additional resources.

Feng et al. in [8] presented an approach of adaptive resource provisioning in cloud computing environment for revenue maximization. Resource allocation problem solved by Queuing theory based mathematical formulation are presented. This methodology works for pricing model but not considered data processing based on predictive or future behavior.

Wang et al. in [9] proposed heuristics iterative optimization methods to achieve the minimum completion time of workflow tasks and economic cost of cloud resources. Authors have not focused on reliability, and fault tolerance.

Nagavaram et al. in [2] presented dynamic resource allocation strategy in the cloud computing. Proposed approach is monitoring the complete workflow implementation, and the system repeatedly select to add or release resources. This work only considers homogeneous resources which do not work on heterogeneous resources.

3 Proposed Model

To improve the runtime performance of adaptive cloud for service-based application, there are two aspects of technical issues to be addressed. The first one is the balancing of large amount of data on existing resources. The second is dynamic resource provisioning which can adjust the amount of resources dynamically in order to adapt the time-varying workload. Cloud computing has several characteristics such as dynamic scaling, on-demand provisioning, elasticity, security, efficiency, and cost-effective utility that is widely accepted by industrial and academic users. With the ever-increasing cloud computing acceptance, users are creating and generating more complex and complicated commercial processes and scientific jobs for hosting in the cloud. Such complicated processes and scientific jobs consist of many interdependent sub-processes or tasks with semantic data, and are represented in the form of service instances. With the advent of the parallel computing and service technologies, data processing applications become more and more complex. To fulfill the requirement of semantic data processing is more complicated than large data processing. Several scientific jobs require informative data processing of its database and to produce the needed result. Service instance is considered as an application which is analyzed, processed, classified, and trained on adaptive cloud. It requires sophisticated controls of the service instance to enable data to be processed on a large pool of distributed resources to generate informative and useful data. To host the several types of service instances on cloud, adaptive resource management criteria should be considered first. Adaptive resource management consists the efficient, scalable, dynamic, and autonomous operations for resources allocation. To represent the ability of system, weather forecast application can be considered to produce the large amount of semantic (meaningful) data from satellite images to analyze the area covered by sun rays on earth for setup the solar plants.

Adaptive cloud can represent the result of informative data processing periodically and execute service instance on allocated resource to maximize the performance. To make the cloud more intelligent to adapt the dynamic data analysis, it should be trained to accept the runtime need.

Adaptive methodology is to be built based on following parts:

a. Monitor and store—These phases collect data from various sources and store in the data repository. To convert the large data into meaningful and semantic data filtering process should be handled. That task to be executed in the data processing phase.
b. Analyze and Process—These phases analyze and process the collected data based on some predefined policy or guide to determine if a change is necessary. It generates the dataset as per the predefined policy. This dataset is to be trained in the following phase to create the service instance.
c. Training—This section trains the generated dataset as per the requirement of service instance creation. Created trained data set is to be considered as a service instance and it will send to execute on cloud resources.
d. Configure and execute—This section puts service instance on allocated resources as per the requirement of the application. It executes the service instances on allocated resources which are to be managed in next phase.
e. Management—This phase manages the allocated resources adaptively and as per the need of the service instance. It will provision the services as a part of resources to the specific application.

Adaptive cloud for data processing and adaptive resource management scenario are depicted in the following Fig. 2.

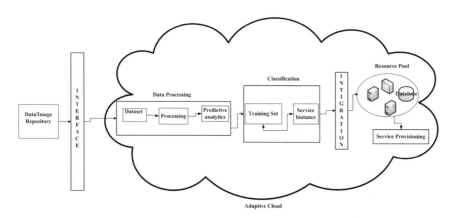

Fig. 2 Adaptive cloud

Data/image repository is the storage of collected data which is to be processed on adaptive cloud via interface of connecting database and cloud. After the collection of data, it is to be analyzed and processed based on some predefined policy or guide to determine if a change is necessary. It generates the dataset as per the predefined policy. After the data processing phase, generated dataset is to be trained in classification phase to create the service instance as per the requirement of application. Cloud is to be built intelligent to take the self decision for resource provisioning services which is considered as an adaptive cloud. The architecture of adaptive cloud for service instance and resource management represents the adaptive resource management strategy to manage resources dynamically for specific application. Resource management phase allocate resources adaptively as per the need of the service instance. It will provision the services as a part of resources to the specific application. Figure 2 represents the overall environment of the adaptive cloud for data processing.

3.1 Scope

In order to meet earlier mentioned research objectives the scope of research work is twofold:

Adaptive cloud:
It is proposed to go as per following action plan:

- Study of available tools, gaps, and applications requirements.
- To build the cloud using open source middleware.
- Design architecture of adaptive cloud for data processing and dynamic resource management based on future behavior to enhance the performance of adaptive cloud.
- Design an algorithm for data processing and dynamic resource management on adaptive cloud to measure, analyze, and strengthen the performance of adaptive cloud for deployed application.
- To implement an automated resource and service provisioning methodology.

When task will run on cloud, sometimes it has to wait when resources are not available. So there is a need of adaptive cloud which can provide resources based on predictive behavior. To alleviate the above-mentioned problems, it is necessary to design an adaptive cloud that integrates various aspects of factors as well as parameters to provide optimal configuration. In future to build up the adaptive cloud, predictive model has to be trained on the target platform.

Case Study:
Solar system deployment model—The solar system application analyzes satellite images and applies the semantic data processing. For the solar energy plant setup, satellite images will be taken to understand the specific location to setup the solar plant efficiently. Data manager analyze the satellite images and transform (convert)

it into common format. Data extraction tool extract specific application's data and load it into the database or data repository.

- Prerequisites for deployment of application

    - Geographical site location
    - Space available for mounting the solar panels
    - Average sunlight available without shadow

- Attributes to be considered for solar energy system

    - Temperature
    - Sunlight
    - Shadow light
    - Availability of average sunlight per year

To run the number of tasks with variant input size in predictive basis, training is required by influential machine learning technique for each cluster of datasets. It uses the SVM (support Vector Machine) regression model. Support Vector Machines (SVMs) is used to obtain the predicted information of a testing data set. It applies the clustering technique along with Euclidean distance metric to group datasets with similar input size. Then task find the Vm which is match with the pattern for required utilization of CPU, disk usage, memory usage and number of CPU cores.

- Formulation of problem for data processing

    Assume given a set of data to make tasks,

    $$D_{in} = \{ d_i, d_{i+1} \ldots d_n\} \tag{1}$$

    For list of N tasks,

    $$Tn = \{D_{in1}, D_{in2}, \ldots D_{in10}\} \tag{2}$$

- Task Management

For n tasks, allocate Vm instances to specified task which is match with the highest probability of execution time of task. For 50 numbers of input data, it is clustered in 10 different tasks with similar pattern. It creates five tasks $T1$, T_2, T_3, T_4, and T_5 with different input size. These five tasks run through SVM and learn the supervised learning methodology to train the other predictive tasks. Following table shows the properties to train the predictive tasks. In initial stage tasks T1 to T5 run on every Vm and show the calculated execution time for every task. It measures the probability value (P_{stat}) for execution of task on Vm which is matching with the required parameters. P_{stat} value will be compared with every Vm for individual task and it selects Vm which have highest value of P_{stat} compared to others.

$$P_{stat} = \max(T_{exec})_{vm} / \sum_{vm=1 \text{ to } n} (T_{exec}) \tag{3}$$

Based on formula (1) and (2) P_{stat} is measured and Vm is allocated to the tasks on reactive strategy. As shown in the formula (3). This reactive strategy can be improved through proactive and predictive behavior. Task is managed as described in the algorithm 1.

Algorithm 1 :Task Management
Input: Data Repository D_i; where i $\in$ 1…..N Output: Task Management

Step 1. Extract each D_i from Data Repository
 	
Step 2. For each D_i calculate temp, sunlight, shadow light, visibility
 	
Step 3. Create dataset for task T_i, i $\in$ 1….N
        For each dataset create task,
        $T_{i=}D_i$
        Identify T_{max}, i $\in$ 1 ….N
 	
Step 4. Identify make span m for each task
 	
    T_m ,m is size of task
 	
Step 5. Task distribution

4   Configuration

In this section Setup of private Cloud Environment with Eucalyptus (IaaS) is described as shown in Fig. 3. Private cloud setup is established with configuration of virtual environment as per our previous work [10]. Configuration steps are performed by installing the CEntOS 6 in the frontend node. In that Eucalyptus 3.3 is installed for Frontend. Various required components CC, CLC, SC and Walrus are configured. Same steps are configured for the node controller.

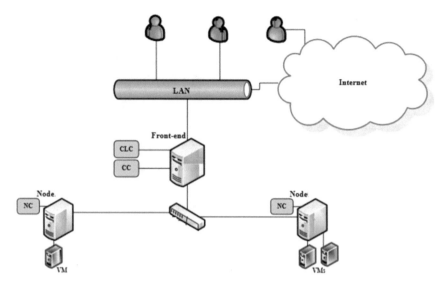

Fig. 3 Test bed of private cloud using 3 physical machines

5 Conclusions and Future Work

To build the adaptive cloud for data processing and adaptive resource management, four sub-domains such as data processing, adaptive cloud computing, resource management, and service provisioning are elaborated in this paper. Proposed methodology to improve the performance of cloud using efficient resource provisioning is described. Configuration of eucalyptus-based test bed is defined to prepare the cloud environment. In future, we create the resource pool in cloud environment, Identify specific resource to execute service instances.

References

1. Jiang, Y., Perng, C.-S., Li, T., Chang, R.: Self adaptive cloud capacity planning. In: Ninth International Conference on Service Computing, pp. 73–80. IEEE (2012)
2. Nagavaram, A., et al.: A cloud-based dynamic workflow for mass spectrometry data analysis. In: 2011 International Conference on Cloud Computing and Service Computing, eScience, 2011, pp. 219–226. IEEE (2011)
3. Buyya, R., Pandey, S., Vecchiola, C.: Cloudbus toolkit for market-oriented cloud computing. In: Proceedings of CloudCom Conference (2009)
4. Huo, J., Shang, S., Zhang, Z.: ACRUM: an adaptive cloud resource utilization model, pp. 34–46. IEEE (2013)
5. Wei, Y., Blake, M.B., Saleh, I.: Adaptive resource management for service workflows in cloud environments. In: 2013 IEEE 27th International Symposium on Parallel & Distributed Processing Workshops and Ph.D. Forum, 2013, pp. 2147–2156. IEEE (2013)

6. Iqbal, W., Matthew, N., Carrera, D., Janecek, P.: Adaptive resource provisioning for read intensive multi-tier applications in the cloud. Future Gen. Comput. Syst. **27**(6), 871–894 (2011). Springer

7. Zhang, Y., Huang, G., Liu, X., Mei, H.: Integrating resource consumption and allocation for infrastructure resources on-demand. In: 2010 IEEE 3rd International Conference on Cloud Computing, IEEE 2010, pp. 75–82 (2010)

8. Feng, G., Garg, S., Buyya, R., Li, W.: Revenue maximization using adaptive resource provisioning in cloud computing environment. In: 2012 ACM/IEEE 13th International Conference on Grid Computing, 2012 IEEE, pp. 192–200 (2012)

9. Wang, L., Duan, R., Li, X., Lu, S., Hung, T., Calheiros, R.N., Buyya, R.: An iterative optimization framework for adaptive workflow management in computational clouds. In: 11th IEEE International Symposium on Parallel and Distributed Processing With Applications (ISPA), Melbourne, Australia, 16–18 July (2013)

10. Gadhavi, L.J., Bhavsar, M.D.: Efficient and dynamic resource provisioning strategy for data processing using cloud computing. Int. Rev. Comput. Softw. (IRECOS) **11**(8), 2016 (2016)

A Cloud-Based Vertical Data Distribution Approach for a Secure Data Access on Mobile Devices

Jens Kohler and Thomas Specht

Abstract Vertical database partitioning and the distribution of the partitions to different clouds improves the level of security and privacy, as the respective partitions from one cloud are worthless without the others. *SeDiCo* (a framework for a secure and distributed cloud data store) is an implementation of such an approach. However, the data partitioning and distribution approach demands a powerful client with sufficient hardware capabilities to efficiently join the partitioned data. Mobile devices have limited hardware resources in terms of memory, processing power, and storage capacity. The aim of this paper is to adopt the *SeDiCo* approach to mobile devices and to prove its feasibility. For this purpose, an Android application that implements the vertical partitioning approach is conceptualized and developed. Furthermore, the implementation is evaluated with the TPC-W benchmark. As a result, mobile devices should be enabled to use a secure and distributed cloud storage to foster a broader dissemination of Cloud Computing for mobile devices. Furthermore, this approach fosters bring-your-own-device (BYOD) strategies for enterprises, as the level of data security and privacy is improved.

Keywords Cloud computing security · Data security and privacy · Mobile cloud platform

1 Introduction

Steadily increasing network bandwidths, pay-per-use models, uniform access, and dynamic scalability of resources according to the current demand are promising features of Cloud Computing. However, using resources from an external cloud provider requires an enormous amount of trust in the provider. Hence, data security

J. Kohler (✉) · T. Specht
Institute for Enterprise Computing, Paul-Wittsack-Str. 10, 68163 Mannheim, Germany
e-mail: j.kohler@hs-mannheim.de

T. Specht
e-mail: t.specht@hs-mannheim.de

© Springer Nature Singapore Pte Ltd. 2019
A. Abraham et al. (eds.), *Emerging Technologies in Data Mining and Information Security*, Advances in Intelligent Systems and Computing 755,
https://doi.org/10.1007/978-981-13-1951-8_21

and privacy-related challenges are still the main focus of Cloud Computing. Encryption, as an approach to address these challenges, is often used nowadays and in the center of recent research work, e.g., [1–3]. Apart from that, alternative approaches use data distribution in order to disseminate data chunks to different cloud providers. Here, *SeDiCo* (a framework for a secure and distributed cloud data store) uses a vertical database partitioning mechanism that logically separates database data into disjoint subsets (so-called partitions) and distributes these partitions across different clouds. Thus, every cloud provider only receives a small logically independent chunk of the entire dataset which cannot be exploited without the others. The framework user, i.e., the client, is the only authorized entity that is able to reconstruct the entire dataset. Figure 1 illustrates the vertical partitioning and distribution approach with a motivating example.

On the one hand, this improves the level of security and privacy despite the use of external cloud capabilities. On the other hand, the approach suffers from severe performance challenges, i.e., response time, when data are accessed. More detailed analyses considering the performance of the *SeDiCo* framework can be found in [4, 5]. This paper focuses on the adaption of the approach to mobile devices.

With respect to the uniform access to the clouds, the question of how to realize a security and privacy-aware access (to partitioned and distributed data) with manifold and heterogeneous devices comes up. Closely related to this is another current challenge with particular importance for enterprises: bring-your-own-device (BYOD). The goal here is to integrate a plethora of different devices and software (i.e., apps) into the IT architecture of an enterprise. This is why we aim at adopting the *SeDiCo* approach to mobile devices and to show that it is a feasible approach also for mobile devices with limited hardware capabilities. Therefore, the paper outlines the general vertical partitioning and distribution approach for mobile devices and implements it on an Android operating system. It further evaluates the performance for the required join of the partitioned data, which we call response time.

As already motivated with Fig. 1, a source database is vertically partitioned and these partitions are distributed across several clouds. The current implementation [6] supports the following databases: MySQL, Oracle, MariaDB, and PostgreSQL. Moreover, as cloud infrastructures Amazon EC2 and Eucalyptus are supported. Above that, all systems can be used interchangeably and it is even possible to mix them and use them simultaneously, e.g., for the usage of different database systems for the partitions.

However, this approach requires comparatively powerful hardware (e.g., a laptop or a server) to realize the join of the vertically partitioned and distributed data in a reasonable time frame. Recent studies show that the usage of solid state disks (SSDs) instead of traditional hard disk drives (HDDs) and the application of sophisticated join algorithms (i.e., hash or sorted-merge join [7]) contribute to an improved response time for all database operations, i.e., *create*, *read*, *update*, and *delete* (CRUD). Figure 2 illustrates a *read* response time measurement based on a current standard laptop with an Intel Core i7 2.4 GHz processor, 8 GB RAM, and 500 GB HDD to give an impression about the response time of the basic *SeDiCo* implementation with-

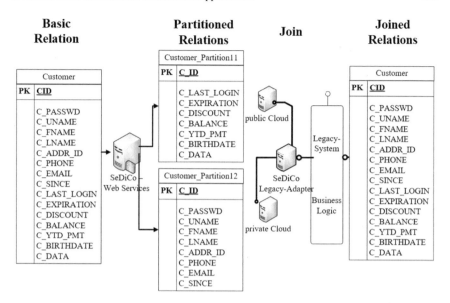

Fig. 1 Motivating example

Fig. 2 Initial *SeDiCo*
performance

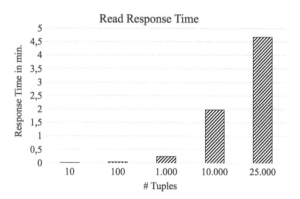

out any optimizations. These values can also be found in [8] with a more detailed
technical analysis and conclusion.

To sum up the central aspect of this paper, we pick up the current *SeDiCo* implementation and transfer it to a mobile architecture. Thus, we demonstrate the technological feasibility and we determine and investigate the limits of a mobile approach.

This closes the introductory part of this paper and the contributions can be summarized as follows:

- the adoption of a vertical partitioning and distribution approach to mobile devices;
- the development of an architecture for a secure and privacy-aware data-centric
  usage with BYOD policies;

- a demonstration of a secure and privacy-aware data-centric usage with BYOD policies;
- a performance evaluation of the vertical partitioning and distribution approach on mobile devices; and
- a demonstration of the technological feasibility of the vertical partitioning and distribution approach for mobile devices.

The remainder of this work is structured as follows. First, the approach is outlined in more detail in the next section, which is followed by a concrete implementation. This implementation is then evaluated in a subsequent section. After this evaluation, a closer look is taken at related work in the field of mobile databases, sophisticated synchronization mechanisms, and approaches to increase the level of security and privacy for mobile databases. Finally, the last section gives an outlook about future work challenges which have to be addressed in the context of the further development of the approach.

2 Approach

The main aspect of the vertical partitioning and distribution approach that has also be considered for mobile devices is that the mobile client is the only location where data are entirely reconstructed. Regarding the current (nonmobile) implementation, the reconstruction logic, i.e., the logic to join the data has to be adopted: not all Java libraries and especially Hibernate are not applicable to mobile devices. Furthermore, the usage of Hibernate (for the database abstraction) and jclouds (for the cloud inter-face abstraction) is a challenging aspect, as there are no analogous implementations for mobile devices. Here, new approaches have to be investigated. This results in a redesign of the original *SeDiCo* framework architecture which is illustrated in Fig. 3.

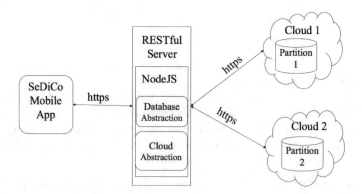

Fig. 3 Mobile architecture

This redesign (Fig. 3) transfers the two crucial components: the *Database* and the *Cloud Abstraction* to a so-called REST*ful* Server between the mobile clients and the vertically partitioned and distributed databases (*Partition*1 and *Partition*2).

Yet, another challenging aspect with this architecture arises: both abstraction components require complete information about the partitioning and distribution schema of the dataset. Unfortunately, this contradicts the key principle of the original *SeDiCo* approach. Thus, a way that keeps this sensitive information on the mobile client (and only there) must be introduced.

Therefore, REST [9] as a stateless communication protocol is considered feasible. It offers four basic methods: GET, PUT, POST, and DELETE, which are mapped to the previously mentioned database CRUD operations. Table 1 shows this mapping in greater detail.

3   Implementation

The implementation now maps the CRUD operations that contain sensitive information about the database partitioning schema, the locations in the clouds, the concrete values for the partitions, and the database metadata to the REST methods. Furthermore, this information is encoded in a way, such that the complete information about the entire dataset is never stored at the REST*ful* Server. Figure 4 illustrates the entire approach with a *read* method (i.e., a *select* statement), issued against a vertically partitioned and distributed database.

Figure 4 illustrates the adoption of the vertical partitioning and distribution approach, such that it is applicable for mobile devices. First of all, an original query

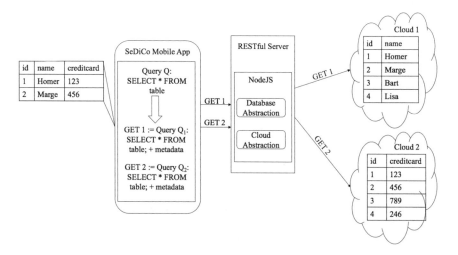

Fig. 4 Mobile vertical partitioning and distribution approach

Table 1 Mapping between REST and CRUD methods

REST methods	CRUD operations
GET	Read
PUT	Create, update, delete
POST	Create, update, delete
DELETE	Delete

Q is decomposed into two so-called *reconstruction queries* Q_1 and Q_2. Then, these *reconstruction queries* are enriched with metadata, containing information about the location of the partition in the cloud, the type of the database system, the database user and password, etc. Hence, the respective *reconstruction query* and the additional metadata are encoded as the following REST method: in case of a read operation, this results in a GET request (see Table 1). This request is then passed to the REST*ful* Server where the *Database* and *Cloud Abstraction* components read the respective query and its metadata to forward the request to the correct cloud and database system. After that, the query is issued against its partition and its result set is collected.

Finally, the transfer of the result set to the mobile app is analogous to the above-depicted approach, except that the result set is encoded in the corresponding GET response. Furthermore, it has to be noted that not the Database Abstraction component in the REST*ful* Server joins the result sets coming from the cloud partitions, as this would contradict the vertical partitioning and distribution approach. Instead, the REST*ful* Server simply forwards the respective responses (including the corresponding result sets) to the mobile app. Thus, the join of the result sets is performed on the mobile device and the vertical partitioning and distribution principle is maintained.

Lastly, it has to be stated that the entire traffic between the mobile app, the REST*ful* Server and the corresponding clouds is encrypted with the HTTPS protocol, which implements an SSL/TLS encryption.

This finishes the outline of the implementation with its technologies and the following section evaluates this implementation.

4 Evaluation

For this evaluation, we used the TPC-W benchmark [10], which is an industry standard benchmark for relational databases. We evaluated the implementation outlined in the previous section with respect to its response time. Therefore, the architecture depicted in Fig. 4 was implemented and instantiated with the CUSTOMER table from the TPC-W benchmark.

The original source table contains 288 K rows and 16 attributes which results in a dataset size of $\sim$147 MB. We partitioned this table evenly with respect to the number of attributes evenly. This results in a size of $\sim$56 MB and 9 attributes for *partition*1 (CUSTOMER_p1) and $\sim$113 MB and 8 attributes for *partition*2 (CUSTOMER_p2), respectively. The bigger size of the partitions in MB and the additional attribute

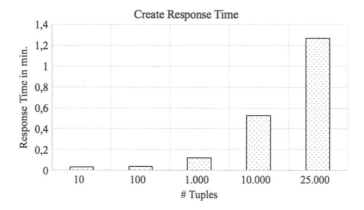

Fig. 5 Create response time

result in the replication of the primary key. The deviation between the size of the two partitions comes from the different data types of the attributes. We conducted the measurements with a Motorola LG G4 (6-core 64-Bit Qualcomm Snapdragon 808 1.8 GHz processor and 3 GB RAM). We set the maximum of available RAM for an Android application to 64 MB. Currently, it is impossible to allocate more RAM for a single Android application. This mobile device was connected via Wi-Fi (IEEE 802.11ac) with the REST*ful* Server which was operated on an Intel i7-2600 (4-core) processor with 3.4 GHz and 8 GB RAM. In order to avoid further network deviations or unpredictable workloads of a cloud environment, the database partitions were also operated on this REST*ful* Server. Every measurement was performed three times, and the following figures illustrate the average of these measurements (Fig. 5).

5 Conclusion

First of all, the figures show that not all 288 K rows of the CUSTOMER table could be considered on a mobile device because of the limited RAM. The measurement further showed that the maximum number of tuples for the mobile device was ∼50 K. However, we stopped the measurement at 25 K rows because of the impractical response times depicted in Fig. 6. Yet, the evaluation showed that the vertical partitioning and distribution approach is viable to a limit of ∼1 K tuples per table. With larger tables, e.g., ∼10 K tuples, the response time increases up to ∼50 min (Fig. 6) which is unbearable in practical usage scenarios.

Furthermore, there is a significant difference in the response time between the *create* and the *read*, *update*, and *delete* operations. The reason for this is that the join for the data is an extremely expensive operation on the mobile device and this is the dominating factor in the response time. The *create* operation does not require a join; therefore, creating a tuple is significantly faster than the other operations (*read*,

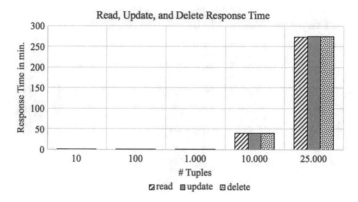

Fig. 6 Read, update, and delete response time

update, and *delete*). In our implementation, obviously, the *read* operation requires a join and so do *update* and *delete* operations. Before a tuple is updated on the mobile device, it first has to be read (at least once until it is in the database of the mobile device). Then, this tuple is updated and propagated to the respective cloud partitions again. The same holds for the *delete* operation. A tuple that is deleted must exist on the mobile device and this again requires a read operation before, in order to check if the tuple eventually exists. Thus, the *read* operation is included in the *update* and *delete* operations, respectively. Figure 6 shows that the response times for the update and delete operations are slightly slower than the response time for the read operation. Thus, subtracting the *update* and *delete* from the *read* response time results in the fact that updating and deleting a tuple without reading it before is in the same order of magnitude as creating one (Fig. 5).

Last but not least, it has to be mentioned that in our setup, the vertical partitions are operated in the REST*ful* Server and not in a dedicated cloud infrastructure (Fig. 4). We decided to use this to avoid unpredictable side effects from a network and due to possibly unknown workloads in a cloud infrastructure that would distort the measurements. However, additional measurements in a cloud infrastructure show that the network overhead increases the response time averagely by factor ∼2.

To sum all up, the results of this work show that the vertical partitioning and distribution approach is also applicable for mobile devices with smaller datasets, i.e., less than ∼1 K tuples per table. Absolutely, the response time is increased with this approach, but it can be stated that the increased response time is bearable as long as the dataset remains below ∼1 K tuples, which is the case for most enterprise databases as [11] in their study with 12 companies and ∼74 K tables per company show. Thus, this work proved promising results for further BYOD scenarios with possibly more advanced and sophisticated compliance rules that have to be mapped to the proposed vertical partitioning and distribution approach.

6 Related Work

Despite increasing network bandwidths for mobile devices and the growing hardware capabilities (e.g., storage capacity), mobile devices still suffer from comparatively small hardware dimensions. As network bandwidths grow, the amount of data grows accordingly. Therefore, using Cloud Computing for a ubiquitous access to all kinds of data (e.g., music, videos, but also personal and sensitive data) is still a viable approach. However, this requires effective data synchronization mechanisms and this is a broad field of research. With respect to this, the authors of [12] propose a database synchronization protocol for databases on mobile devices based on the calculation of hash values. These hash values enable an effective analysis for updated, outdated, or data that is still up-to-date and therefore do not have to be transferred to the mobile device. Thus, the amount of data that has to be synchronized between a cloud and a mobile device is minimized. The work also outlines strategies to handle synchronization conflicts. Calculating hash values for data is a common approach in mobile databases as other works, e.g., [13, 14] or [15] show. These promising approaches are considered viable and build an excellent foundation for the further development of the framework presented in this paper.

Other works, e.g., [16] shift the synchronization overhead to the programming layer with specially designed cloud data types. Hence, programmers are able to use these cloud types without thinking about the data synchronization. Above that, there are approaches to move as much computational complexity from the mobile device to a cloud infrastructure and to just use the mobile device as a front-end. Here, virtualization, a basic technology for Cloud Computing offers promising features. [17] describe an approach that transfers mobile applications to a cloud infrastructure. Therefore, the mobile app is transferred to a virtualized app in the cloud which is operated in a virtual machine similar to the mobile operating system.

7 Outlook & Future Work

All in all, it can be summarized that the limiting factor is the join on the mobile device due to limited hardware resources. In order to overcome this limiting factor, the following two aspects are of particular interest:

- The join can be realized in the REST*ful* Server if additional security measures, like placing this server in an internal network or securing the access to it via a virtual private network (VPN) are taken into account.
- As [11] showed, the average size of a relation is 1.500 rows. Thus, if the dataset is small enough, the vertical partitioning and distribution approach can be applied as is.

For future work challenges, we aim at testing the approach against a real-world BYOD scenario with a larger number of users and a real-world dataset that is viable for the approach with respect to its size.

Other future work tasks deal with optimization techniques for the outlined mobile approach. Yet a concrete optimization challenge is to reduce the number of tuples that have to be joined on the mobile device. Basically, not always all tuples that are processed by the mobile device have to be transferred to it, e.g., if some tuples are already (joined) on the mobile device. In such cases, these tuples (or their ids) could also be encoded as metadata and added to the respective CRUD operation. Then, the framework would not consider these tuples which would reduce the overall number of tuples that would be joined.

Above that, future work challenges consider the adoption of the approach to NoSQL databases. Recent studies [4, 18] show that the vertical partitioning and distribution approach is also feasible for such new kinds of databases (currently with some restrictions that are elaborated in more detail in the respective papers). Due to the weaker consistency criteria of such databases, the overhead is significantly smaller, and therefore, we assume that with NoSQL databases, larger amounts of data can be processed on a mobile device.

References

1. Ferretti, L., Colajanni, M., Marchetti, M.: Supporting security and consistency for cloud database. In: Proceedings of the 4th International Conference on Cyberspace Safety and Security, pp. 179–193 (2012)
2. Juels, A., Oprea, A.: New approaches to security and availability for cloud data. Commun. ACM 56(2), 64–64 (2013)
3. Neves, B.A., Correia, M.P., Bruno, Q., Fernando, A., Paulo, S.: DepSky: dependable and secure storage in a cloud-of-clouds. ACM Trans. Storage (TOS) 9, 31–46. ACM (2013)
4. Kohler, J., Specht, T., Simov, K.: An approach for a security and privacy-aware cloud-based storage of data in the semantic web. In: Proceedings of The First IEEE International Conference on Computer Communication and the Internet (ICCCI 2016), Wuhan, China. IEEE Computer Society (2016)
5. Kohler, J., Simov, K., Fiech, A., Specht, T.: On the performance of query rewriting in vertically distributed cloud databases. In: Proceedings of The International Conference Advanced Computing for Innovation ACOMIN 2015, Sofia, Bulgaria (2015)
6. Kohler, J.: GitHub Page of SeDiCo Implementation (2016)
7. Balkesen, C., Alonso, G., Teubner, J., Ozsu, M.T.: Multi-core, main-memory joins: sort vs. hash revisited. Proc. VLDB Endow. 7(1), 85–96 (2014)
8. Kohler, J., Specht, T.: Vertical query-join benchmark in a cloud database environment. In: Proceedings of The 2nd IEEE World Conference on Complex Systems, Agadir, Morocco (2014)
9. Fielding, R.T.: Architectural styles and the design of network-based software architectures. Ph.D. thesis, University of California (2000)
10. TPC: TPC Benchmark W (Web Commerce) Specification Version 2.0r (2003)
11. Krueger, J., Kim, C., Grund, M., Satish, N., Schwalb, D., Chhugani, J., Plattner, H., Dubey, P., Zeier, A.: Fast updates on read-optimized databases using multi-core CPUs. Proc. VLDB Endow. 5(1), 61–72 (2011)
12. Domingos, J., Simões, N., Pereira, P., Silva, C., Marcelino, L.: Database synchronization model for mobile devices. In: Iberian Conference on Information Systems and Technologies, CISTI (2014)

13. Alhaj, T.A., Taha, M.M., Alim, F.M.: Synchronization wireless algorithm based on message digest (SWAMD) for mobile device database. In: Proceedings 2013 International Conference on Computer, Electrical and Electronics Engineering: Research Makes a Difference, ICCEEE 2013, pp. 259–262 (2013)

14. Balakumar, V., Sakthidevi, I.: An efficient database synchronization algorithm for mobile devices based on secured message digest. In: 2012 International Conference on Computing, Electronics and Electrical Technologies, ICCEET 2012, pp. 937–942 (2012)

15. Choi, M.-Y.C.M.-Y., Cho, E.-A.C.E.-A., Park, D.-H.P.D.-H., Moon, C.-J.M.C.-J., Baik, D.-K.B.D.-K.: A database synchronization algorithm for mobile devices. IEEE Trans. Consum. Electron. **56**(2), 392–398 (2010)

16. Burckhardt, S., Fähndrich, M., Leijen, D., Wood, B.P.: Cloud types for eventual consistency. In: Lecture Notes in Computer Science (Including Subseries Lecture Notes in Artificial Intelligence and Lecture Notes in Bioinformatics), vol. 7313, pp. 283–307. LNCS (2012)

17. Chun, B., Ihm, S., Maniatis, P., Naik, M., Patti, A.: Clonecloud: elastic execution between mobile device and cloud. In: Proceedings of the Sixth Conference on Computer Systems, pp. 301–314 (2011)

18. Kohler, J., Lorenz, R., Gumbel, M., Specht, T., Simov, K.: A security by distribution approach to manage big data in a federation of untrustworthy clouds. In: Sharvari Tamane, N.D., Solanki, V.K. (eds.) Privacy and Security Policies in Big Data. IGI Global (2017)

A Priority-Based Process Scheduling Algorithm in Cloud Computing

Misbahul Haque, Rakibul Islam, Md. Rubayeth Kabir,
Fernaz Narin Nur and Nazmun Nessa Moon

Abstract Nowadays, cloud computing is in demand as it provides progressive pliable resource allocation, for unfailing and guaranteed services in the pay-as-you-use scheme, to cloud service users. So, there is a dispensation that all resources are made available to requesting users in an efficient manner to satisfy their needs. Process scheduling has become the key issue in cloud computing. In this paper, we have presented a priority-based process scheduling (PRIPSA) algorithm, which is developed with the block-based queue in cloud computing. It concentrates on the preemptive part as well as it calculates the energy consumption and reducing starvation of process for scheduling the process in the cloud. We provide a priority-based algorithm which considered preempt able task scheduling with block-based queue using burst time and lead time. This job is being performed by the dynamic voltage and frequency scaling (DVFS) controller in our algorithm. The load management, energy consumption, reducing the starvation problem of the processes, and maximizing the revenue are the key motives of our consideration.

Keywords Block-based queue · Cloud computing · Energy efficiency
Preemptive · Priority · Process scheduling

M. Haque (✉) · R. Islam · Md. Rubayeth Kabir · F. Narin Nur · N. Nessa Moon
Department of Computer Science and Engineering,
Daffodil International University, Dhaka, Bangladesh
e-mail: misbahul3434@diu.edu.bd

R. Islam
e-mail: rakib3406@diu.edu.bd

Md. Rubayeth Kabir
e-mail: kabir2990@diu.edu.bd

F. Narin Nur
e-mail: narin@daffodilvarsity.edu.bd

N. Nessa Moon
e-mail: moon@daffodilvarsity.edu.bd

© Springer Nature Singapore Pte Ltd. 2019
A. Abraham et al. (eds.), *Emerging Technologies in Data Mining and Information
Security*, Advances in Intelligent Systems and Computing 755,
https://doi.org/10.1007/978-981-13-1951-8_22

239

1 Introduction

Cloud computing is known as an Internet-based approach, which consists of many interrelated computing resources in a complex manner. These resources are shared to computers or other devices on demand. Resources in the cloud are abidingly accumulated on the server and provisionally cached for the clients. Utility-based computing makes the software more attractive and allocates the resources on the pay-per-use basis [1, 2]. All sorts of data in the cloud are stored online. The data is accumulated in a virtualized pool of storage. Generally, a third-party vendor drives the virtualized storage pool. Infrastructure as a service (IaaS) provides vital services such as networks, storage, processing, and computational resources over the Internet to the clients [3]. Cloud computing has different types of features such as shared infrastructure, self-service-based usage mode, dynamic provisioning, self-managed platform, network access, manager metering, rapid elasticity, commercialization virtualization, self-managed platform, resource pooling, and multi-tenancy [3].

Process scheduling problems is a combinational optimization problem, which searches an optimal resembling from a finite data set in such a way that the data set of enforceable solutions is ordinarily finite than continuous. In scheduling-based solution, scheduler decides which process is to be run next. To make an efficient scheduling, the scheduler is only activated at some starting point that decides which process to run when [4].

In the literature, we have found a good number of process scheduling algorithms. Two task-based scheduling algorithms are proposed in [1], where authors discussed the two process scheduling algorithms which are priority-based and earliest deadline first process scheduling algorithm. The processes are assigned on the basis of their priority and the process having higher priority get scheduled first. In this work, the authors introduced a waiting queue. In the queue, the processes are scheduled according to their priority. A scheduling plan for the tasks is proposed in [5], which concentrates on the preemptive solution and it calculates the energy expenditure on cloud computing servers for scheduling the processes. The computing servers are allocated to the process based on the best fit as per their energy requirements. This paper focused on the priority of the jobs and also reduced the power cost. In these algorithms, the processes with low priority face starvation problem, which degrades the reliability of the system. A cluster-based method is proposed in [6], which delineates the notion of clustering the processes based on their burst time. Traditional process scheduling methods such as first come first serve, shortest job first, EASY, and combination of backfill and improved backfill using balance spiral method create fragmentation in the system. In this paper, jobs are allocated to different queue depending on their burst time but they did not consider the lead time of the jobs.

A large number of researchers worked on job scheduling algorithm. However, the objective of most of these methods is to diminish the overall completion without regarded the minimization of the overall cost of the service. One of the major problems of these algorithms is the starvation problem. So a proper scheduling method is needed to schedule the resources in any cloud properly. In this paper, we have proposed a *priory-based process scheduling* algorithm that reduces the starvation problem as well as it also reduces the waiting time of the processes in the queue. It also improves the cumulative throughput of the system. The main contributions of this paper are as follows:

- We have developed a priority-based process scheduling algorithm.
- The proposed algorithm reduces the starvation problem by considering the burst time and lead time of the processes.
- The proposed algorithm also improves the overall network performance.
- Finally, we have implemented it in visual studio result show that excellent efficiency and throughput.

The remaining part of the paper is incorporated as follows. We have described related works in Sect. 2. In Sects. 3 and 4, the main components and algorithm are discussed with numerical example. Finally, Sects. 5 and 6 contain the simulation results and Conclusion.

2 Related Work

In recent, cloud computing has taken the intentness of computer scientists and information technologists. A lot of researchers worked on process scheduling algorithm. Process scheduling algorithm can be centralized or distributed. Distributed process scheduling algorithm is very much harder than centralized process scheduling algorithm. The main purpose of process scheduling algorithm is to acquire ideal performance in computing and better throughput from the system.

In cloud computing environment, Ergu et al. [7] presented a model for process-oriented resource allocation. The pairwise correspondence matrix approach is ranked the resource allocation process and user preferences and the available resources are provided by the analytical hierarchy process. According to the rank of processes computing resources are allocated to identify the inconsistent elements inaugurated bias matrix and different types of processes are assigned to improve the consistency ratio when weights are conflicting. To validate the proposed method, we introduced two illustrative examples.

Li et al. [8] propose an accommodated resource allocation preemptive algorithm for the cloud in which depending on the updating of the actual process execution algorithms adjust the resource allocation adaptability. For static resource allocation, static process scheduling includes adaptive list scheduling and adaptive min-min scheduling algorithms which are used for process scheduling, is generated offline. The online adaptive method is used to reevaluate the static resource allocation. The process scheduler again calculates the finishing time of the submitted processes, not the processes which are already assigned to the cloud in each reevaluation process.

Liu et al. [9], implemented an optimized scheduling strategy to reduce power consumption satisfying the task response time during scheduling. Choose the minimum number of most proficient servers for processing is a greedy approach. The processes are miscellaneous in nature so that they create various energy consumption levels and have different process response times. The optimum task is based on lowest energy consumption and lowest execution time of a process on a specific machine.

For the high priority process, solutions are provided by priority preemptive process scheduling algorithm. For this reason, sometimes lower priority task cannot get scheduled and the starvation problem occurred. To the best of our knowledge, the starvation problem is not discussed in the state-of-the-art works. Existing process scheduling algorithm gives ideal throughput and cost-effective but they don't think reliability and availability. To overcome these problems we need a proper algorithm which will reduce the starvation problem as well as increase the reliability and availability.

3   Protocol Operation

We have proposed a priority-based process scheduling algorithm named as PRIPSA. The operation of the proposed algorithm is based on the priority of the jobs for selecting the VM (virtual machine) for executing the processes. Our paper concentrates on the preemptive operation as well as it calculates the energy consumption. It also reduces the starvation of processes for scheduling in the computing servers.

3.1   System Components

In the proposed solution, a central controller collects the requests from different client machines and schedules them as per their requirements. After submitting processes by clients, processes are sent to the VMs manager. VMs manager priorities the processes using the proposed algorithm. Then finally, it sends to the server for execution. Figure 1 shows the system model for the proposed mechanism. A number of components that have been used in our proposed algorithm are given below.

Virtual Machine Manager (VMM): For the virtualized datacenter, VMM is a management solution to manage and configure virtualization host, networks, and storage resources to deploy and create virtual machines and services to the private clouds.

Virtual Machine Server (VMs): Through emulation and virtualization different operating systems perform as full computing platforms on their own. In different environments for saving money software, developers use VM servers for testing software without getting the actual hardware for those particular environments. VM server restrains damages to the hardware, which appear with potentially buggy software. A VM server hosts multiple virtual machines at once so that multiple procedures can be done simultaneously.

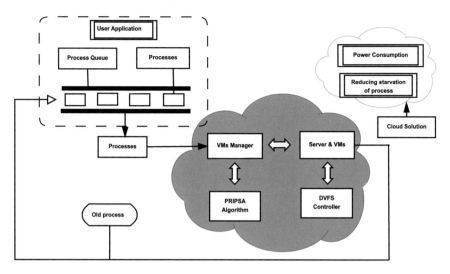

Fig. 1 System model

Dynamic Voltage and Frequency Scaling (DVFS): DVFS is a very feasible technique for reducing power consumption. Generally, it does not fit perfectly for individual regulation. It uses just two voltages, high and low, for voltage scaling.

3.2 Proposed Priority-Based Scheduling Algorithm

The proposed method amplifies the scheduler by grouping different lead and burst time-based processes into the queue. It mitigates the starvation problem using burst and lead time-based allocation system of the processes to choose the appropriate processes from betwixt the remaining processes and does not remission the performance of the system. In this paper, we proposed a processed scheduling algorithm for real-time services with process relocation. Our method fixes the priority for the processes and technically migrate the processes when it misses the deadline in order to diminish the response time and therefore attain better performance.

The proposed algorithm provides a better solution for process abortion when it misses its deadline, and most probably processes with the higher priority are taken for the execution. Our algorithm works at scheduling point and the priority of the processes are given according to the burst time and lead time of each process. The proposed approach provides importance to all the processes and a different number of clients in cloud computing and each client wants more requirements and expectation. It grants different importance according to their burst time and lead time. Figure 2 shows the proposed approach. In the process pool, processes are not in organized form. It organizes it according to the priority for execution. Here, the different types

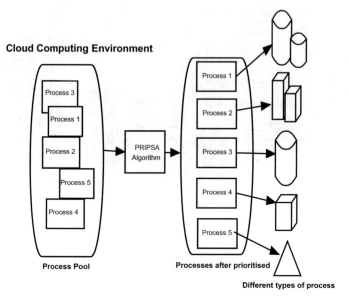

Fig. 2 The proposed scheduling approach

of processes are symbolized by different types of symbol. Processes are submitted to the VM Manager by the clients. VMs manager finds the best suitable process to be executed among the available.

In the priority-based process scheduling algorithm, there is a time machine (TM), where the time period is measured in λ seconds and after this λ seconds, TM auto reset. The algorithm rearranges the processes after every second according to their Lead time and burst time which are waiting in the process pool. The processes are scheduled using the following equation that contains a constant α and its value is fixed. This mechanism gives the most priority to the lead time rather than the burst time. That's why processes are gets scheduled before they die. In this way, the proposed mechanism improves the throughput and reduces the starvation problem. Each burst time (B) and lead time (T) is multiplied by α and $1 - \alpha$. Here, α is greater than 0.5 and less than 1. In the equation, α is multiplied by burst time (B), and $(1 - \alpha)$ is multiplied by lead time (T), so that the lead time gets the greater priority. Here, lower value gets the highest priority. The equation is given below:

$$P = \alpha \times B + (1 - \alpha) \times T \tag{1}$$

These processes are sent to the blocks with new priority. Every block has fixed length. Each block can store the maximum n number of processes in a particular time. Then, these processes are executed according to the block. The working procedure of the proposed algorithm is shown in Fig. 2. The detailed working procedure of the proposed algorithm is discussed below:

a. Let $P = P_1, P_2, P_3 \ldots P_n$ is a set of process.
b. For each process create a priority value as per Eq. 1 for step (a), priority value depends on the variables B and T. Here, lowest value gets the highest priority.
c. After fixed the priority, the process is moved into the multilevel block and each block hold max n processes.
d. Then, process executes according to the block.

4 Numerical Example

In this section, we provide a numerical example for priority-based process scheduling algorithm. There are total 10 jobs submitted to VM and split it into the two blocks where each block contained five processes and each process are executed according to their priority which is created by the proposed priority-based scheduling algorithm as shown in Table 1.

When processes arrive, they are submitted to VM and the proposed algorithm creates priority value for each process. After calculating the priority value according to Table 1, the submitted processes are divided into block based on their priority value. In the first block, the processes $(P_2, P_4, P_1, P_5, P_3)$ are executed one after one. Then in the second block, $(P_7, P_{10}, P_9, P_6, P_8)$ are executed one after one.

5 Performance Evaluation

Our proposed mechanism is simulated to explore the system performance to achieve the less waiting time, high throughput, and also reduce the starvation problem in the system. Here, we compared our priority-based scheduling mechanism with the preemptive algorithm [7]. The Preemptive algorithm gives the schedule to higher

Table 1 Priority value calculation for processes

Processes	Arrival time	Burst time	Lead time	Priority
P_1	0	5	15	8
P_2	1	4	10	5.8
P_3	2	6	20	10.2
P_4	3	4	12	6.4
P_5	4	5	17	8.6
P_6	5	7	25	12.4
P_7	6	3	10	5.1
P_8	7	8	45	19.1
P_9	8	6	25	11.7
P_{10}	9	5	22	10.1

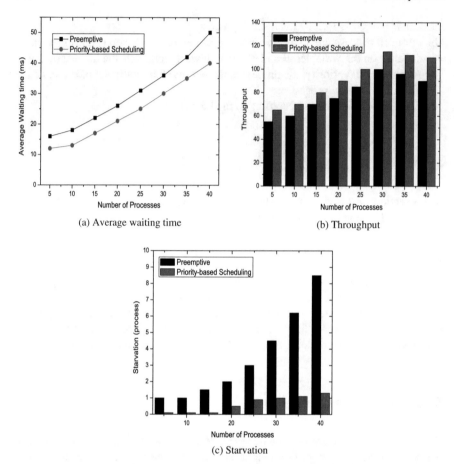

(a) Average waiting time

(b) Throughput

(c) Starvation

Fig. 3 Impacts of number of processes on performances of the studied mechanisms

priority-based processes but lower priority-based processes did not get scheduled. We have taken average of 10 simulation results for each graph points.

Average waiting time is measured as the time the processes wait in a queue to be executed. Figure 3a shows the average waiting time of preemptive and priority-based scheduling algorithm. Here, the average waiting time of preemptive algorithm is exponentially increasing rather than the priority-based scheduling algorithm. So, the priority-based scheduling algorithm shows the better performance considering waiting time.

Throughput counts the number of processes completed within the fixed time unit. In Fig. 3b, we can see that our proposed algorithm gives better throughput compared to the preemptive algorithm, as in our proposed work one after one process is completing and gradually throughput is increasing.

Starvation is a kind of problem that some processes are never allocated though resources are available for allocation. In the graph Fig. 3c, starvation for preemptive algorithm is increasing linearly where for priority-based scheduling algorithm, increasing rate is very low. Priority-based scheduling algorithm specially focused on starvation. That's why get excellent output for starvation. In the preemptive algorithm, the major problem is the starvation problem. Our proposed algorithm gives the high priority to the processes based on their lead time then burst time. In this way, it is able to solve the starvation problem.

6 Conclusion

This paper presents an efficient priority-based process scheduling algorithm in cloud computing. Priority-based scheduling algorithm focused on reducing the starvation problem. In the existing method, processes are scheduled preemptively or non-preemptively. Only highest priority-based processes are selected for execution first. The lowest priority-based process didn't get the scheduled for execution until the highest priority-based processes didn't complete. Sometimes, some processes have reached the deadline and discarded. In order to minimize these challenges, this paper proposed a priority-based processed scheduling algorithm which gives the priority to every task according to their burst time and lead time. The performances of the proposed method show that this method significantly outperforms the preemptive and non-preemptive process scheduling algorithm.

References

1. Gupta, G., Kumawat, V.K., Laxmi, P.R., Singh, D., Jain, V., Singh, R.: A simulation of priority based earliest deadline first scheduling for cloud computing system. In: 2014 First International Conference on Networks & Soft Computing (ICNSC), pp. 35–39 (2014)
2. Dhinesh Babu, L.D., Krishna, P.V.: Honey bee behavior inspired load balancing of tasks in cloud computing environments. Appl. Soft Comput. 2292–2303 (2013). Elsevier
3. Shenai, S., et al.: Survey on scheduling issues in cloud computing. Procedia Eng. **38**, 2881–2888 (2012). Elsevier
4. Casati, F., Shan, M.-C.: Definition, execution, analysis, and optimization of composite e-services. IEEE Data Eng. Bull. **24**(1), 29–34 (2001)
5. Patel, S., Bhoi, U.: Priority based job scheduling techniques in cloud computing: a systematic review. Int. J. Sci. Technol. Res. **2**(11), 147–152 (2013)
6. Karthick, A.V., Ramaraj, E., Subramanian, R.G.: An efficient multi queue job scheduling for cloud computing. In: 2014 World Congress on Computing and Communication Technologies (WCCCT), pp. 164–166 (2014)
7. Ergu, D., Kou, G., Peng, Y., Shi, Y., Shi, Y.: The analytic hierarchy process: task scheduling and resource allocation in cloud computing environment. J. Supercomput. (11), 1–14 (2013). Springer

8. Li, J., Qiu, M., Niu, J.-W., Chen, Y., Ming, Z.: Adaptive resource allocation for preemptable jobs in cloud systems. In: 2010 10th International Conference on Intelligent Systems Design and Applications (ISDA), pp. 31–36 (2010)
9. Liu, N., Dong, Z., Rojas-Cessa, R.: Task scheduling and server provisioning for energy-efficient cloud-computing data centers. In: 2013 IEEE 33rd International Conference on Distributed Computing Systems Workshops (ICDCSW), pp. 226–231 (2013)

Efficient Model of Cloud Trustworthiness for Selecting Services Using Fuzzy Logic

Rashi Srivastava and A. K. Daniel

Abstract The area of organizational management has rapid development in information technology and information science and playing a vital role over cloud computing. The cloud computing has emerged as a new paradigm for delivering and hosting services. Various types of services are provided by different organizations to the user to enhance the information system through cloud services. The infrastructure (software/hardware), processing, operation, and cost of the cloud service are effective. The paper proposed a framework to provide an analysis of cloud service to the service provider with a set of parameters as agility, finance, usability, security, system performance using fuzzy logic. The proposed model represent the performance of different cloud service provider with respect to the different parameter for providing services according to the set of requirement to the user as a quality of service (QoS).

Keywords Trustworthiness measurement · Cloud service · Fuzzy logic · QoS
Agility · Usability

1 Introduction

Cloud computing [1] is a new computing paradigm for delivery and hosting services on the internet. Cloud computing is increasing interest in research focused towards academic and industry. The cloud computing means a type of Internet-based computing services such as the server, storage, and application are delivered to an organization. Cloud services are categorized into three group as Software as a Service (SaaS), Infrastructure as a Service (IaaS), Platform as a Service (PaaS). PaaS and IaaS provider is further known as infrastructure provider/cloud are often part of the same

R. Srivastava (✉) · A. K. Daniel
Department of Computer Science and Engineering, Madan Mohan
Malaviya University of Technology, Gorakhpur 273010, India
e-mail: rashi03.sri@gmail.com

A. K. Daniel
e-mail: danielak@rediffmail.com

© Springer Nature Singapore Pte Ltd. 2019
A. Abraham et al. (eds.), *Emerging Technologies in Data Mining and Information
Security*, Advances in Intelligent Systems and Computing 755,
https://doi.org/10.1007/978-981-13-1951-8_23

249

Fig. 1 General cloud
computing parameters

organization [2]. IaaS provide computing infrastructure, including virtual machine, computer hardware, and database as a service. Including software framework development and operating system support PaaS provide platform layered resource. SaaS implies to providing on-demand applications over the internet. There are different types of Cloud as follows. Public clouds provide services of their resources to the general public; Including no shifting of risk to infrastructure provider and no initial capital investment; it also offers several key benefits to service providers. Second is Private Clouds which is mainly designed for exclusive use by a single organization and it also knows as internal clouds. It may be built and managed by external providers. Third is Hybrid Clouds which is the combination of public and private cloud models that tries to address the drawback of each approach. In comparing with public and private clouds Hybrid clouds offer more flexibility and last Virtual private cloud an alternative solution to addressing the drawback of both public and private clouds is called virtual private cloud (VPC). It is an essential platform that runs on top of public clouds.

Cloud provides a genuine environment for an organization and customer to bring out the transaction with cloud providers. The major problem is that if any user wants to use resources for less duration they may pay for full duration that is overload to the user. To measure the trustworthiness of cloud service different researcher assumes the different set of parameter. The cloud service trustworthiness is the degree of confidence of cloud service to meet the set of requirement [3].

The role of different parameter as mentioned in Fig. 1, performed by cloud service provider is discussed below:

Agility: Agility is a intellectual activity performed by service provider to provide services and draw conclusion to the user in more quick and easy way.
Usability: When the user request service from Cloud service provider, the system must provide efficient output, i.e., Hit ratio (successful rate) should be maximum. For example if normal user want to access the information from bank of other customer depend upon validity of the customer (user) bank will decide whether it should be provided or not but if income tax department or government of India need to access the bank will supply the data and hence the level of security stands.
Finance: System and operating cost is called as finance cost.

Security: Encrypted data stored in CSP and whenever user demand it will be authenticated by the system and data provider must ensure that it should be given or not.
System performance: The system performance should be maximum output i.e. Hit ratio must be maximum.

The proposed model assumes the different set of the parameter as agility, finance, usability, security, system performance to compute the cloud service trustworthiness. As the technology changes user requirement changes so the trustworthiness value of a cloud service may be changed. For the selection of cloud services consumer has to consider trustworthiness as services of cloud with better trustworthiness can provide more reliance and confidence. The paper proposed fuzzy logic based cloud service trustworthiness selection model. The paper is ordered as follows: Sect. 2 related work. Section 3 proposed work. Section 4 analysis and validation using fuzzy logic. Section 5 simulation result and Sect. 6 Conclusions.

2 Related Works

Definition of trustworthiness of cloud service varies with varying organizational structures and application. Trust evaluation is one of the growing and challenging issues in many fields such as E-commerce, MANET, software services and much more. The significant research has been carried out for deciding the trustworthiness concept, parameter, definition, and method for enhancing the cloud service trustworthiness. Evaluating the cloud service made by the group or individual trustworthiness value depends on the selected set of requirement. Following are some of the proposed model:

Pandey and Daniel [3] has proposed fuzzy logic based cloud service trustworthiness model (CSTM) framework for assessing the cloud service trustworthiness. CSTM Model specifically focused on finance parameter, as a factor in selection of cloud service and thereby improved the quality of cloud service.

In [4] Sarvesh and Daniel have proposed we propose a QoCS and Cost Based Cloud Service Selection Framework based on fuzzy logic. With multidimensional perceptive it help to analyze any cloud service. By using five set of parameter as Finance, Security, Reliability, Maintainability, and Usability they specifically analyzed the trustworthiness of cloud service.

In [5] Marsh formalized trust as a computational concept in computer science, author argues about trust from the point of vision of artificial agent. With the increase in the trust value Trustworthiness increases. With the range $[-1, 1]$ the trust value expressed as a real number.

In [6] Zhang et al. have proposed a fuzzy comprehensive evaluation model to evaluate the trustworthiness of a software service. They have considered parameter such as availability, reliability, safety, security, and maintainability for the evaluation of trustworthiness.

In [7] Ouzzani and Bouguettaya proposed a query and optimization model for providing complex query capabilities to web services. This model selects and combines virtual operation and concrete operation based on the quality of web services (QoWS), relevance, rating matching degrees and feasibility and provided query optimization model.

In [8] Ian Lumb et al. proposed a taxonomy and survey on various existing cloud computing services invented by the various project such as Google, Amazon, force.com. They also used taxonomy and result to identify area requiring further research.

In [9] Zhou et al. has focused on various cloud computing system provider and concerned on an issue of security and privacy and find these are not adequate and added more aspect (availability, confidentiality, data integrity, control, audit).

Cloud Armor [10] a platform for credibility-based trust management of cloud services, delivers trust as a service (TaaS). Cloud Armor relies on decentralized architecture having components as trust data provisioning, trust, and credibility, assessment, and Trust-based cloud service recommendation.

In [11] Hu et al. have proposed trustworthiness fusion model for service cloud platform based on D-S evidence theory in which they described the meaning of trustworthiness and computation method for trustworthiness fusion.

In [12] Rong and Jian-Xun have proposed Trustworthiness fusion of web service based on D-S Evidence theory in which they discussed the definition and computation method of trustworthiness.

In [13] Supriya et al. have proposed comparison of cloud service providers based on direct and recommended trust rating so the user can choose the best plan for them.

In [14] Supriya et al. has proposed model for Trust Management based on Fuzzy Logic which specifically focused on Agility, Finance, Performance and help the customer for choosing the appropriate CSP as per their demand. Below is the table of Trust Value of CSPs.

In [15] Supriya et al. have proposed the hierarchical trust model to compare the trust values of various cloud service providers for varying levels of trustworthiness based on Infrastructure as a Service.

3 Proposed Model

The CSTM framework model [3] was designed and used only financial parameter to provide analyses of cloud service by single service provider. The proposed model consists of five parameters such as agility, finance, security, usability, system performance as a set of the parameter in order to provide analysis of cloud services provided by the different service provider and give services according to the requirement of users. Proposed framework consider following parameter as agility, finance, usability, security and system performance. The goal of this model is by computing and evaluation of trustworthiness we provide a framework of cloud service. For selection of appropriate cloud service the proposed model helps the requestor of cloud service in order to reach the goal for which it requested, every cloud service generally have some primary characteristic as define in the framework special parameter are enhanced by only single cloud service. These special parameter are managed by using ontology alignment approach [16]. For all cloud service some parameter are common such as history, risk, feedback, recommendation, and time. In cloud service the trustworthiness value will dynamically change and become more precise when requestor start giving review or feedback when they start using the services depend upon collection of feedback provided by requestor. By using set of multi-criteria analyses such as trust vector aggregation, minimum polyhedron, support vector clustering, etc., the trustworthiness value is provided by agent of cloud service. Fuzzy comprehensive evaluation model [17, 6] is used to evaluate the cloud service trustworthiness. The parameters P define as P1 for agility, P2 for Finance, P3 for Usability, P4 for security and P5 for system performance.

The weight of each parameter is determined according to the level of importance of cloud service having different demand toward trustworthiness parameter. The high quality of security is needed in military, aerospace and space system. According to level of importance weight of each parameter are required to determined by expert.

Let Wi be weight of Pi and the weight set is

$$W = \{W1, W2, W3, W4, W5\}, 0 \le Wi \le 1, \qquad \Sigma Wi = 1$$

Expert evaluation method, different parameters have sub-parameters which are evaluated by quantity and some by quality. Each evaluation method is divided into five levels. $L = \{VH, H, M, L, VL\}$, where VH (very high), H (high), M (medium), L (low), and VL (very low).

Let us assume these levels as, $L = \{L1, L2, L3, L4, L5\}$.

According to the collected data and information the membership degree to the five comments set of each parameter Pij are ($di j1\ di j2\ di j3\ di j4\ di j5$).

The evaluation result of Ki factors can be represented by Ki × 5 order fuzzy matrix Fi, where i = 1, 2, 3, 4, 5.

$$F(n) = \begin{bmatrix} di11 & di12 & di13 & di14 & di15 \\ di21 & di22 & di23 & di24 & di25 \\ di31 & di32 & di33 & di34 & di35 \\ \vdots & \vdots & \vdots & \vdots & \vdots \\ diki1 & diki2 & diki3 & diki4 & diki5 \end{bmatrix}$$

Here, Fi represents the membership degree of all the sub-parameters of a parameter Pi of a cloud service in specified trustworthiness levels. Fi is a single factor of fuzzy comprehensive evaluation matrix Pi, and dijm is the membership degree of a sub-parameter pij as grade m. As the weight set Wi is determinate, fuzzy comprehensive evaluation matrix for the Pi can be evaluated using the Min-Max composition as follows:

Let X to Y be fuzzy relation of Wi and Y to Z is fuzzy relation of Fi, the composition of Wi and Fi is WioFi which is a fuzzy relation from X to Z and is shows as:

$$Bi = Wi \bigcirc Fi \leftrightarrow dwi \bigcirc Fi\,(x, z)$$

$$Bi = \vee y\{Wi(x, y) \wedge fi\,(y, z)\}$$

Min-Max composition with $\wedge = $ Max, $\vee = $ Min, $\bigcirc = $ operator. Applying different fuzzy operators according to the situation and operation results of cloud service. Bi is a fuzzy comprehensive evaluation matrix.

B represents the membership degree of the cloud service trustworthiness of different parameters as agility, finance, usability, security, system performance.

$$Bi = \begin{bmatrix} Wi1 & Wi2 & Wi3 & \ldots & Wi5 \end{bmatrix} \bigcirc \begin{bmatrix} d11 & d12 & d13 & d14 & d15 \\ d21 & d22 & d23 & d24 & d25 \\ d31 & d32 & d33 & d34 & d35 \\ \vdots & \vdots & \vdots & \vdots & \vdots \\ diki1 & diki2 & diki3 & diki4 & diki5 \end{bmatrix}$$

Now, on applying composition of Min-Max the final result of membership degree of Parameter Xi:

$$Bi = \begin{bmatrix} Bi1 & Bi2 & Bi3 & Bi4 & Bi5 \end{bmatrix}$$

In specified trustworthiness levels, Bi describes the membership degree of the parameter Pi (agility, finance, usability, security, system performance) of a cloud service.

4 Validations and Analysis Using Fuzzy Logic

The Poisson distribution concept is used for Data Set representation for computation followed by using Fuzzy Logic.

Consider the weight set for every service provider as follows:

$$\text{weight set } W = \{0.6,\ 0.8,\ 0.5,\ 0.75,\ 0.5\}$$

The comment set $L = \{L1,\ L2,\ L3,\ L4,\ L5\}$, $L1 = VH$, $L2 = H$, $L3 = M$, $L4 = L$, $L5 = VL$

For cloud service provider 1
The weight set of the service provider 1 are

Table 1 Trust value of CSPs

CSP	Agility	Financial	Performance	Trust
CSP A	0.45	0.856	0.347	Poor
CSP B	0.451	0.5	0.348	Good
CSP C	0.151	0.5	0.346	Very good
CSP D	0.45	0.857	0.8	Excellent
CSP E	0.451	0.5	0.15	Very poor

$$W = \{0.6,\ 0.8,\ 0.5,\ 0.75,\ 0.5\}$$

The comment set $L = \{L1,\ L2,\ L3,\ L4,\ L5\}$; $L1 = VH$, $L2 = H$, $L3 = M$, $L4 = L$, $L5 = VL$.

The membership degree is provided by expert
$B_{11} = \{0.4,\ 0.6,\ 0.9,\ 0.6,\ 0.2\}$, $B_{12} = \{0.4,\ 0.5,\ 0.6,\ 0.8,\ 0.7\}$, $B_{13} = \{0.95,\ 0.88,\ 0.72,\ 0.4,\ 0.2\}$, $B_{14} = \{0.74,\ 0.89,\ 0.7,\ 0.5,\ 0.3\}$ $B_{15} = \{0.3,\ 0.4,\ 0.7,\ 0.6,\ 0.45\}$

$$B = W \bigcirc F$$

Evaluating the value of B using MIN-MAX composition, to find membership value of a given cloud service as follows:

$$B = \begin{bmatrix} 0.6\ 0.8\ 0.5\ 0.75\ 0.5 \end{bmatrix} \bigcirc \begin{bmatrix} 0.4 & 0.6 & 0.9 & 0.6 & 0.2 \\ 0.4 & 0.5 & 0.6 & 0.8 & 0.7 \\ 0.95 & 0.88 & 0.72 & 0.4 & 0.2 \\ 0.74 & 0.89 & 0.71 & 0.5 & 0.3 \\ 0.3 & 0.4 & 0.7 & 0.6 & 0.45 \end{bmatrix}$$

$$B1 = \begin{bmatrix} 0.74\ 0.75\ 0.71\ 0.80\ 0.70 \end{bmatrix}$$

The B1 represents the membership degree of the overall cloud service in specified trustworthiness levels (Table 1).

For cloud service provider 2
$B_{21} = \{0.46,\ 0.6,\ 0.9,\ 0.8,\ 0.4\}$, $B_{22} = \{0.4,\ 0.65,\ 0.85,\ 0.67,\ 0.55\}$, $B_{23} = \{0.8,\ 0.9,\ 0.75,\ 0.4,\ 0.2\}$, $B_{24} = \{0.5,\ 0.68,\ 0.9,\ 0.68,\ 0.61\}$, $B_{25} = \{0.54,\ 0.62,\ 0.91,\ 0.70,\ 0.4\}$
$B2 = \begin{bmatrix} 0.5\ 0.6\ 0.8\ 0.75\ 0.72 \end{bmatrix}$

For cloud service provider 3
$B_{31} = \{0.75,\ 0.83,\ 0.7,\ 0.5,\ 0.2\}$, $B_{32} = \{0.79,\ 0.88,\ 0.6,\ 0.4,\ 0.3\}$, $B_{33} = \{0.82,\ 0.92,\ 0.68,\ 0.60,\ 0.35\}$, $B_{34} = \{0.3,\ 0.4,\ 0.6,\ 0.8,\ 0.7\}$, $B_{35} = \{0.98,\ 0.90,\ 0.6,\ 0.3,\ 0.1\}$

$$B3 = \begin{bmatrix} 0.79 & 0.8 & 0.6 & 0.75 & 0.7 \end{bmatrix}$$

For cloud service provider 4

B_{41} = {0.4, 0.5, 0.6, 0.9, 0.7}, B_{42} = {0.5, 0.66, 0.7, 0.65, 0.4}, B_{43} = {0.71, 0.78, 0.6, 0.3, 0.2}, B_{44} = {0.72, 0.8, 0.7, 0.5, 0.3}, B_{45} = {0.3, 0.4, 0.6, 0.9, 0.77}

$$B4 = \begin{bmatrix} 0.72 & 0.75 & 0.7 & 0.65 & 0.6 \end{bmatrix}$$

For cloud service provider 5

B_{51} = {0.5, 0.7, 0.8, 0.75, 0.4}, B_{52} = {0.78, 0.88, 0.7, 0.4, 0.25}, B_{53} = {0.7, 0.77, 0.8, 0.5, 0.3}, B_{54} = {0.95, 0.87, 0.7, 0.3, 0.1}, B_{55} = {0.3, 0.4, 0.6, 0.9, 0.7}

$$B5 = \begin{bmatrix} 0.78 & 0.8 & 0.7 & 0.6 & 0.5 \end{bmatrix}$$

For cloud service provider 6

B_{61} = {0.6, 0.8, 0.85, 0.7, 0.4}, B_{62} = {0.78, 0.85, 0.67, 0.55, 0.3}, B_{63} = {0.5, 0.75, 0.87, 0.7, 0.42}, B_{64} = {0.6, 0.82, 0.92, 0.68, 0.4}, B_{65} = {0.4, 0.8, 0.9, 0.75, 0.5}

$$B6 = \begin{bmatrix} 0.78 & 0.8 & 0.75 & 0.68 & 0.5 \end{bmatrix}$$

For cloud service provider 7

B_{71} = {0.96, 0.85, 0.65, 0.4, 0.2}, B_{72} = {0.97, 0.89, 0.7, 0.3, 0.1}, B_{73} = {0.5, 0.65, 0.85, 0.77, 0.6}, B_{74} = {0.4, 0.75, 0.84, 0.7, 0.3}, B_{75} = {0.75, 0.87, 0.65, 0.54, 0.2}

$$B7 = \begin{bmatrix} 0.8 & 0.8 & 0.75 & 0.7 & 0.5 \end{bmatrix}$$

5   Simulation Result

We have considered a set of CSP randomly and uniformly assigned Parameter Agility, Finance, Usability, Security, System Performance.The simulation are carried out using the C language and MATLAB and Table 2 Present the performance of different cloud service provider with respect to the different parameter agility, Finance, Usability, security, and system performance. The Cloud service provider CSP1 has security the highest trustworthiness value (0.80) whereas in cloud service provider CSP2 has usability parameter the highest trustworthiness value (0.80) and so on. The proposed framework helps the user for selecting the appropriate cloud service. Figures 2, 3, 4, 5 and 6 show the trustworthiness evaluation of different parameters varies on different cloud service provider. The model response is shown in Fig. 8 with respect to various parameter of different service provider. The Reference model [14] "estimating trust value for cloud service providers using fuzzy logic model" is display in Fig. 9 using Table 1 and Fig. 7 defines overall system parameters using Table 2.

Table 2 Trustworthiness value of different parameters

	CSP1	CSP2	CSP3	CSP4	CSP5	CSP6	CSP7
Agility	0.74	0.50	0.79	0.72	0.78	0.78	0.80
Finance	0.75	0.60	0.80	0.75	0.80	0.80	0.80
Usability	0.71	0.80	0.60	0.70	0.70	0.75	0.75
Security	0.80	0.75	0.75	0.65	0.60	0.68	0.70
System performance	0.70	0.72	0.70	0.60	0.50	0.50	0.50

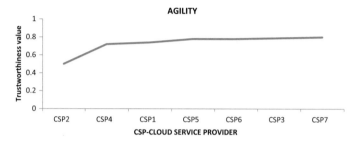

Fig. 2 Trustworthiness of agility parameter

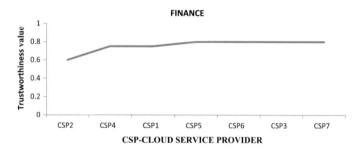

Fig. 3 Trustworthiness of finance parameter

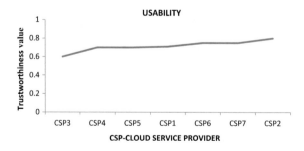

Fig. 4 Trustworthiness of usability parameter

Fig. 5 Trustworthiness of security parameter

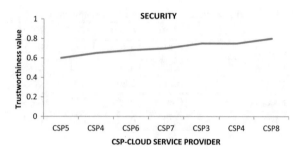

Fig. 6 Trustworthiness of system performance

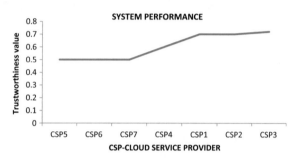

Fig. 7 Overall system parameters

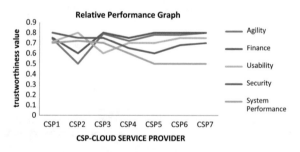

Fig. 8 Trustworthiness of cloud service provider with its parameters

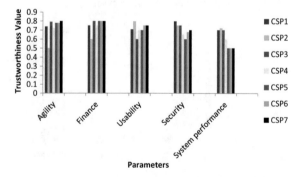

Fig. 9 Trustworthiness of cloud service provider [14]

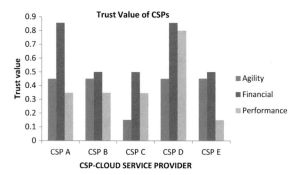

The proposed model is analyzed with model [14] and the result show that system performance of the proposed model is improved model and is efficient and providing high QoS.

6 Conclusion

In this paper the proposed framework is providing efficient cloud service to the service provider with a set of parameters as agility, finance, usability, security, system performance using fuzzy logic. The proposed framework helps the user for selecting the appropriate cloud service. The performance of the proposed model is providing QoS to the provider.

References

1. Patidar, S., Rane, D., Jain, P.: A survey paper on cloud computing. In: 2012 Second International Conference on Advanced Computing & Communication Technologies, Rohtak, Haryana, 2012, pp. 394–398
2. Zhang, Q., Cheng, L., Boutaba, R.: Cloud computing: state-of-the-art and research challenges. J. Internet Serv. Appl. **1**(1), 7–8 (2010)
3. Pandey, S., Daniel, A.K.: Fuzzy logic based cloud service trustworthiness model. In: 2016 IEEE International Conference on Engineering and Technology (ICETECH), Coimbatore, 2016, pp. 73–78
4. Pandey, S., Daniel, A.K.: QoCS and cost based cloud service selection framework. Int. J. Eng. Trends Technol. (IJETT) **48**(3), 167–172 (2017)
5. Marsh, S.P.: Formalising trust as a computational concept (1994)
6. Zhang, Y., Zhang, Y., Hai, M.: An evaluation model of software trustworthiness based on fuzzy comprehensive evaluation method. In: 2012 International Conference on Industrial Control and Electronics Engineering, Xi'an, 2012, pp. 616–619
7. Ouzzani, M., Bouguettaya, A.: Efficient access to web services. IEEE Internet Comput. **8**(2), 34–44 (2004).https://doi.org/10.1109/mic.2004.1273484
8. Rimal, B. P., Choi, E., Lumb, I.: A taxonomy and survey of cloud computing systems. In: 2009 INC, IMS and IDC, NCM'09 Fifth International Joint Conference on IEEE, 2009

9. Zhou, M., Zhang, R., Xie, W., Qian, W., Zhou, A.: Security and privacy in cloud computing: a survey. In: 2010 Sixth International Conference on Semantics, Knowledge and Grids, Beijing, 2010, pp. 105–112

10. Noor, T.H., Sheng, Q.Z., Zeadally, S., Yu, J.: Trust mangement of service in cloud environment. ACM Comput. Surv. **46**(1), 1–30 (2013)

11. Hu, R., Liu, J., Liu, X.F.: A trustworthiness fusion model for service cloud platform based on D-S evidence theory. In: 2011 11th IEEE/ACM International Symposium Cluster Cloud Grid Computing, pp. 566–571 (2011)

12. Rong, H., Jian-Xun, L.: Trustworthiness fusion of web service based on D-S evidence theory. In: Proceedings of 6th International Conference on Semantic Knowledge Grid, SKG 2010, pp. 343–346 (2010)

13. Supriya, M., Sangeeta, K., Patra, G.K.: Comparison of cloud service providers based on direct and recommended trust rating. In: 2013 IEEE International Conference Electronics, Computing and Communication Technology, CONECCT 2013, pp. 1–6

14. Supriya, M., Sangeeta, K., Patra, G.K.: Estimating trust value for cloud service providers using fuzzy logic. Int. J. Comput. Appl. **48**(19) (2012)

15. Supriya, M., Sangeeta, K., Patra, G.K.: Estimation of trust values for varying levels of trustworthiness based on infrastructure as a service. In: Proceeding of the 2014 International Conference on Interdisciplinary Advances in Applied Computing. ACM

16. Jiang, Y., Wang, X., Zheng, H.-T.: A semantic similarity measure based on information distance for ontology alignment. Inf. Sci. (Ny) **278**, 76–87 (2014)

17. Wang, X., Sun, F.-Y., Ge, Q.-Y.: Application of fuzzy comprehensive evaluation method in performance knowledge management. J. Inf. **4**, 8–10 (2007)

CS-PSO based Intrusion Detection System in Cloud Environment

Partha Ghosh, Arnab Karmakar, Joy Sharma and Santanu Phadikar

Abstract Cloud Computing is a provider of different types of services to the user in the Internet to make it cost-effective, time-effective and contributes an effort-less environment. For all these reasons it attracts huge number of users and makes it vulnerable to security threats. Cloud deals with huge number of data and to prohibit intrusion with huge dataset is quite hectic, so to overcome these security issues Intrusion Detection System (IDS) are deployed. To predict all the intrusions instantly and accurately an appropriate training is required for the IDS. Existence of trivial feature set in the training data enhances the memory space and training time. In this paper the authors have implemented a novel CS-PSO-based IDS to classify attacks rapidly and easily. Here NSL-KDD dataset have been chosen to demonstrate the proficiency of the IDS.

Keywords Cloud computing · Intrusion detection system (IDS)
Host intrusion detection system (HIDS) · Network intrusion detection system
(NIDS) · Cuckoo search (CS) · Particle swarm optimization (PSO)
NSL-KDD dataset

P. Ghosh · A. Karmakar (✉) · J. Sharma
Netaji Subhash Engineering College, Kolkata 700152, India
e-mail: arnabkarmakar007@gmail.com

P. Ghosh
e-mail: partha1812@gmail.com

J. Sharma
e-mail: sharmajoy96@gmail.com

S. Phadikar
Maulana Abul Kalam Azad University of Technology, Kolkata 700064, India
e-mail: sphadikar@yahoo.com

© Springer Nature Singapore Pte Ltd. 2019   261
A. Abraham et al. (eds.), *Emerging Technologies in Data Mining and Information
Security*, Advances in Intelligent Systems and Computing 755,
https://doi.org/10.1007/978-981-13-1951-8_24

1 Introduction

Cloud Computing provides different kinds of applications and services to the clients or the users over the Internet. The services are provided from different servers or Cloud which are remotely located away from all clients. Cloud Computing enables a user to store data in the Cloud use different kinds of software without installing in the user's system which eases the user from not going through the complexity of installing as well as maintaining the applications. Nowadays due to these facilities there is an increasing demand for Cloud which leads to take security measures of it, as more the demand more is the threat to its security [1]. The organizations or the Cloud Service Provider (CSP) needs to strengthen the safety of the data stored in the Cloud and that leads to the introduction of IDS [2]. An illegitimate entry to the system by anyone is known as intrusion. To provide a secure environment a good IDS is necessary which will act as a security guard and protect any unknown element to enter. Nowadays most of the CSPs are using IDS to secure Cloud data. IDS analyses the network and checks for any malicious activity. If malicious activity occurs, it alerts the network administrator at once. It sometimes even block IP address of the user from accessing the network who is trying to intrude. IDS can be differentiated into two types, one of them is HIDS and the other one is NIDS. HIDS monitors the intrusions that has happened in any host machine in the local system. While analyzing, they use the logging and other information of the host machine to check whether there is any deviation from the normal behavior [3]. A NIDS audits traffic on a network looking for dubious activity, which could be an attack or illegitimate activity. Changes in the server core components can also be detected by an NIDS server [4]. NIDS informs the system administrator that an attack happens or had happened and it takes suitable measures to prevent same kind of attack on the system that is why it is also termed as passive IDS [5]. To classify all the attacks from dataset a proficient IDS is required. Eliminating all the irrelevant data from training dataset is a better way to enhance the classification accuracy and yield a proficient IDS. Here the authors have proposed a novel feature selection method CS-PSO for producing a proficient IDS. To demonstrate the performance of the proposed model the authors have used NSL-KDD benchmark dataset, which has four components [6]. Among the four segments of NSL-KDD dataset the authors have chosen KDDTrain+ for training purpose and KDDTest+ for testing purpose. The training and testing dataset contains 1,25,973 and 22,544 records respectively. In this paper the authors' target is to reduce the dataset by deducting the redundant features without compromising the accuracy of it while yielding a proficient IDS.

2 Related Work

Cloud Centers have become more vulnerable because it is situated at different parts in the world. In 2013, Hashizume et al. [7] have listed a number of threats that is possible in the Cloud. They have also tried to provide recommendation to avert the risks and vulnerabilities. The security problem is best taken care by the IDS, which is demonstrated by Kene and Theng [8]. In their paper they have briefed different types of IDS and the number of attacks that affects the integrity and confidentiality of Cloud. Several approaches of data mining are used for designing IDS. A collaborated work of multi-threaded HIDS and NIDS in a model proposed by Ghosh et al. [9], where Information Gain (IG) helps in selecting relevant features and a hybrid KNN_NN classifier is used to detect attacks. Suguna and Thanushkodi [10] proposed a model where they have combined Genetic Algorithm (GA) and K-Nearest Neighbor (KNN) to prepare good classification algorithm to get a better solution. In the same literature they have implemented feature selection technique by combining Rough Set Theory and Bee Colony Optimization (BCO). Datti and Verma [11] proposed a model where they have used Linear Discriminant Analysis (LDA) method to extract the features and Back Propagation method to classify the IDS. In that paper they have used NSL-KDD dataset for experimental purpose and yielded better outcome. Ghosh et al. [12] proposed a model on penalty-reward-based instance selection method, where they have removed the noisy points and then applied Reverse Nearest Neighbor Reduction (RNNR) method to reduce the dataset as well as decrease the training time. Their model reduced the NSL-KDD dataset almost by one fourth of its own size. To reduce dataset, different feature selection methods are also shown indicating its usefulness in various sectors. Pereira et al. [13] proposed a meta-heuristic approach of CS algorithm to figure out the robustness of that algorithm to extract features as an optimisation problem. In the same purpose they have compared their binary version of CS algorithm with three different nature inspired optimisation techniques. Seal et al. [14] used PSO technique for selecting features in their work on thermal face recognition. They have accurately made use to search the best features which are noiseless and relevant. Ghosh et al. [15] presented a model where they have applied Rough Set Theory and Information Gain method on NSL-KDD dataset to remove irrelevant features and a hybrid KNN_NN classifier to classify different attacks. In the proposed method, the authors have tried to combine the PSO and CS technique, applied on NSL-KDD dataset to prepare more accurate feature selecting algorithm, thus resulting a proficient IDS.

3 Proposed Model

Cloud Computing provide us a means to access applications as utilities, over the Internet. Due to large amount of storage, backup and restore facilities Cloud is attracting a huge number of mass and companies. And due to this increasing demand of its utilities it is becoming vulnerable to security threats. Firewalls are not always perfect in protecting the Cloud from the intruders, that is why IDS is introduced. It provides a more proficient way of protecting the Cloud from intrusions. Datasets are used to train these IDSs before deploying it in the Cloud. These datasets are usually large in size which consists of different kinds of attacks accompanying with many irrelevant data as well. In order to reduce the training time as well as the memory allocation of the dataset, different Feature Selection methods are introduced. These methods usu-

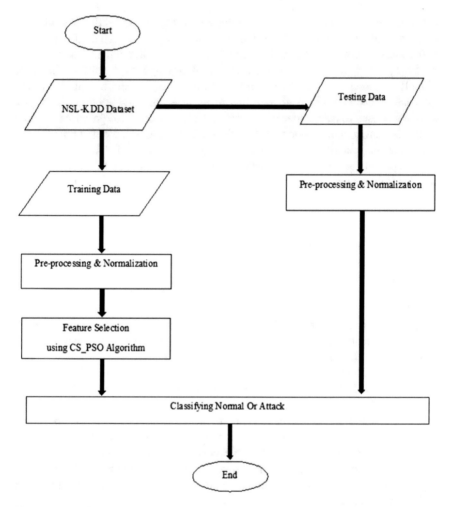

Fig. 1 Flow of the proposed model

ally removes all the irrelevant data present in the dataset without compromising the accuracy. In this paper authors have combined two meta-heuristic feature selection methods to prepare the IDS.

NSL-KDD dataset is applied on the proposed model. Initially the dataset are preprocessed and normalized. After preprocess and normalization, the proposed CS-PSO algorithm reduced the dataset by removing all the irrelevant features. This reduced dataset can be used to classify different kinds of attacks. Figure 1 depicts the progress of proposed model.

In the proposed CS-PSO algorithm, for experiment purpose authors have taken N data points from the dataset to perform the experiment in the first round. Then the value of objective function for each data point is calculated followed by the calculation of fitness value of these data points. The pbest and gbest value of each data points are allocated to keep a check on the individual best value and best value produced by the group. Initializing λ and setting the limit of iteration authors keep on searching nest positions using the Levy flight technique, given in Eqs. (1) and (2).

$$x_{pq}^{t+1} = x_{pq}^{t} + s_{pq} \times \alpha \times Levy(\lambda) \tag{1}$$

$$Levy(\lambda) = \left| \frac{\Gamma(1+\lambda) \times \sin\left(\frac{\pi \times \lambda}{2}\right)}{\Gamma\left(\frac{1+\lambda}{2}\right) \times \lambda \times s^{\frac{(\lambda-1)}{2}}} \right|^{\frac{1}{\lambda}}, \tag{2}$$

where λ is a constant that lies between $(1 \leq \lambda \leq 3)$ and α is a random value between $(-1 \leq \alpha \leq 1)$. Here, $s > 0$ is the step size which should be associated to the scales of the problems of interest. If s is too large then the position obtained will be too far whereas if s is too small then the position generated will be too close to be significant and the search will not be efficient. The formula is given in Eq. (3).

$$s_{pq} = x_{pq}^{t} - x_{fq}^{t} \tag{3}$$

where f is a randomly generated index and it has to be different from p. Following this step if any position value to be violating its limit then that value is clamped maximum or least limit. Again the objective function for each point are calculated and the fitness of those points are evaluated. After this step authors have calculated the probability of the egg ($prob_p$) to be discovered by the host bird, which is done by Eq. (4).

$$prob_p = \left(\frac{0.45 \times Fit_p}{pbest_p} \right) + \left(\frac{0.45 \times Fit_p}{gbest_p} \right) + 0.1, \tag{4}$$

where Fit_p is the fitness value of that particular egg. In order to check where the egg will be identified by the host bird, generated a random value p_a. If the probability of the survival of the egg is less than the random variable p_a then the egg will be discovered by the host bird. In this situation a new nest position is searched by the formulas given in Eqs. (5) and (6).

$$V_{pq}^{(t+1)} = V_{pq}^{t} + c_1 r_1 \left(x_{pbest_{pq}} - x_{pq} \right) + c_2 r_2 \left(x_{gbest_{pq}} - x_{pq} \right) \tag{5}$$

$$N\_N = x_{pq}^{t} + V_{pq}^{(t+1)} \tag{6}$$

This new nest positions (N_N) of the respective eggs are added to the set of data points that already in the dataset. Then again calculate the objective value of these data points along with the new nest positions following which the fitness values are calculated. After evaluating the fitness value of these data points, sort the fitness values in descending order and change the order of the positions accordingly. From these sorted list of positions, take only the first N number of points and proceed to the next generation. After a certain number of iterations the feature subset that tends to possess highest fitness value will contain the most relevant features.

In this experiment, to get a desired solution authors have used the algorithm CS-PSO which is prepared by combining the CS and PSO methods. In original PSO algorithm, the position of the swarms are updated in each iteration. Here the new positions are added in the existing positions which increases the exploration phase. The concept of local best and global best have been incorporated in the CS-PSO algorithm. So that solution proceeds towards the global best value without compromising with its own local best value.

In our experiment, the CS-PSO algorithm demonstrates its effectiveness by solving the trade compromisation between exploitation and exploration. In the proposed model, authors have used the effectiveness of PSO in finding the best value in a local area and is a very good example of exploitation concept. To take care of the exploration process the CS method does its job with the help of Levy Flight technique. It jumps into random positions to get better global optimized value. The combination of these two nature inspired methods have brought some fruitful results with high accuracy and thus producing an efficient IDS in Cloud Environment.

Algorithm:
Input: NSL-KDD Training Dataset.
Output: Best Feature Subset.

Begin
Take N data points randomly.
For each data point find the fitness value.
For each data point find the pbest and gbest.
Set the value of λ, c_1, c_2 and set the total no. of iteration as generation.
While(iteration <= generation)
     do

               For each data points search nest position using levy flight using equation (1)
               For each points find the fitness value and update pbest and gbest
               For each data point:
                    Find the probability of survival of egg($prob_p$) in the
                    host nest using equation(4)
                    Generate random number,say pr_p
                    If $pr_p > prob_p$
                       Calculate new nest position using equation number(6)
                       Add $newNest$ to the set of existing data points.
                    End if
               For each new build nests find the fitness value and pbest.
               Sort the data points according to the fitness value in descending order.

Take first N data points in consideration for the next iteration.
Iteration += 1
Endwhile

The nest containing best fitness value will be selected as Best Feature Subset.
End

4 Experimental Results

To classify all the attacks properly and proficiently, IDS is required for securing the Cloud data. For getting a better classification accuracy, a proper training of IDS is significant. The presence of irrelevant features in training data set increases training time, deteriorate classification accuracy as well as increases memory representation (Fig. 2).

Designing of CS-PSO contains two phases. The proposed CS-PSO algorithm is used for feature selection in the first phase and in second phase different classifiers are required for determining the classification accuracy of the proposed IDS model. NSL-KDD dataset contains 41 features but after applying CS-PSO method, without compromising the classification accuracy, successfully authors can reduce a number of irrelevant attributes. To verify the performance of the IDS model, three classifier Logistic Regression (LR), AdaBoost (AB) and Random Forest (RF) are selected for classification purpose. Confusion matrix for LR classifiers are depicted as follows (Tables 1 and 2).

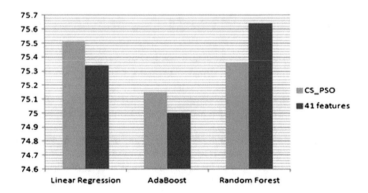

Fig. 2 Classification accuracy using different classifiers on separate datasets

Table 1 Confusion matrix for testing data using LR classifier (trained by KDDTrain+ using 41 features)		Attack	Normal
	Attack	7910	4923
	Normal	635	9076

Table 2 Confusion matrix for testing data using LR Classifier (trained by reduced training set)

	Attack	Normal
Attack	7592	5241
Normal	279	9432

5 Conclusion

Using network, Cloud is getting very much efficient with increase in technology which requires IDS for the security of user's data. IDS helps in preventing users activities over the Cloud. In the proposed algorithm, authors have designed an IDS, exploring the datasets mentioned in the paper. In this technique, CS and PSO algorithm plays a vital role in forming the complete approach. Authors have developed feature selection through CS-PSO algorithm over NSL-KDD dataset. Feature Selection from a high dimensional dataset is a better approach with respect to training time and memory storage which the CS-PSO algorithm shows by boosting the proficiency of the IDS.

References

1. Moorthy, M.S., Rajeswari, M.: Virtual host based intrusion detection system for cloud. Int. J. Eng. Technol. **5**(6), 5023–5029 (2013)
2. Ghosh, P., Shakti, S., Phadikar, S.: A cloud intrusion detection system using novel PRFCM clustering and KNN based dempster-shafer rule. Int. J. Cloud Appl. Comput. **6**(4), 18–35 (2016). IGI Global
3. Vijayarani, S., Sylviaa, M.: Intrusion detection system- a study. Int. J. Secur. Priv. Trust Manag. **4**(1), 31–44 (2015)
4. Prabhu, G.N., Jain, K., Lawande, N., Kumar, N., Zutshi, Y., Singh, R., Chinchole, J.: Network intrusion detection system. Int. J. Eng. Res. Appl. **4**(4), 69–72 (2014)
5. Kumar, B.S., Sekhara, T.C., Raju, P., Ratnakar, M., Baba, S.D., Sudhakar, N.: Intrusion detection system- types and prevention. Int. J. Comput. Sci. Inf. Technol. **4**(1), 77–82 (2013)
6. Tavallaee, M., Bagheri, E., Lu, W., Ghorbani, A.: A detailed analysis of the KDD CUP 99 data set. In: IEEE Symposium on Computational Intelligence in Security and Defense Applications (CISDA 2009), pp. 1–6 (2009)
7. Hashizume, K., Rosado, D.G., Fernández-Medina, E., Fernandez, E.B.: An analysis of security issues for cloud computing. J. Internet Serv. Appl. **4**(1), 1–13 (2013)
8. Kene, S.G., Theng, D.P.: A review on intrusion detection techniques for cloud computing and security challenges. In: IEEE 2nd International Conference on Electronics and Communication Systems (ICECS 2015), pp. 227–232 (2015)
9. Ghosh, P., Mandal, A.K., Kumar, R.: An efficient cloud network intrusion detection system. In: Information Systems Design and Intelligent Applications, Advances in Intelligent Systems and Computing, vol. 339, pp. 91–100. Springer, New Delhi (2015)
10. Suguna, N., Thanushkodi, K.: An improved k-nearest neighbor classification using genetic algorithm. Int. J. Comput. Sci. **7**(4), 18–21 (2010)
11. Datti, R., Verma, B.: Feature reduction for intrusion detection using linear discriminant analysis. Int. J. Comput. Sci. Eng. **2**(4), 1072–1078 (2010)

12. Ghosh, P., Saha, A., Phadikar, S.: Penalty-reward based instance selection method in cloud environment using the concept of nearest neighbor. Procedia Comput. Sci. **89**, 82–89 (2016). Science Direct, Elsevier

13. Pereira, L.A.M., Rodrigues, D., Almeida, T.N.S., Ramos, C.C.O., Souza, A.N., Yang, X., Papa, J.P.: A binary cuckoo search and its application for feature selection. In: Cuckoo Search and Firefly Algorithm, Studies in Computational Intelligence, vol. 512, pp. 141–154. Springer, Cham (2014)

14. Seal, A., Ganguly, S., Bhattacharjee, D., Nasipuri, M., Gonzalo-Martin, C.: Feature selection using particle swarm optimization for thermal face recognition. Applied Computation and Security Systems, Advances in Intelligent Systems and Computing, vol. 304, pp. 25–35, Springer, New Delhi (2015)

15. Ghosh, P., Debnath, C., Metia, D., Dutta, R.: An efficient hybrid multilevel intrusion detection system in cloud environment. IOSR J. Comput. Eng. **16**(4), 16–26 (2014)

Matrix-Based Data Security in Cloud Computing Using Advanced Cramer–Shoup Cryptosystem

Y. Mohamed Sirajudeen and R. Anitha

Abstract Cloud computing allows users to store data in a distributed cloud storage. So the data are stored in different geographical locations which will predominantly increase the vulnerability of the data. The natural way to safeguard the data from the intruders is to encrypt the data before storing in the cloud. This paper proposes an Advanced Cramer–Shoup Cryptosystem to resolve the security threats in cloud environment. The proposed encryption method withstands well against the adaptive chosen cipher text attack (CCA2). While considering high velocity, encryption method has to work as fast as possible to improve the efficiency. The proposed ACS Cryptosystem supports batch encryption and decryption to increase the efficiency and reduces the computation overhead. The experimental analysis for the proposed method is carried out and observed that the proposed encryption technique improves the security of the data and results are compared with existing security algorithms.

Keywords Cloud security · Advanced cramer–shoup cryptosystem
Cloud encryption · Data security

1 Introduction

Cloud computing is a technology which has an ability to lend hardware components, infrastructures to the customers via online. According to the National Institute of Standards and Technology (NIST) cloud computing has been defined as, model for enabling ubiquitous, on-demand access to a shared pool of configurable computing resources which can be rapidly provisioned and released with minimal management efforts [1]. The reason behind the rapid growth of cloud computing technology in

Y. Mohamed Sirajudeen (✉) · R. Anitha
DST-SERB Cloud Research Lab, Sri Venkateswara College of Engineering,
Sriperumbudur 602117, Tamil Nadu, India
e-mail: ducksirajsmilz@gmail.com

R. Anitha
e-mail: ranitha@svce.ac.in

© Springer Nature Singapore Pte Ltd. 2019
A. Abraham et al. (eds.), *Emerging Technologies in Data Mining and Information Security*, Advances in Intelligent Systems and Computing 755,
https://doi.org/10.1007/978-981-13-1951-8_25

recent times is reduced upfront license cost and reduced hardware cost. It also offers many benefits such as on-demand access to the resources, pay as per you go, easier maintenance and service availability anywhere, anytime, Cloud offers three major services to the cloud user, (i) Infrastructure as a service (ii) Software as a service (iii) Platform as a Service. In which, user sensitive data has to be uploaded safely to the cloud. User sensitive data should be kept clean without inside as well as outside data breaches. User sensitive data has to be downloaded safely from the cloud. The above all three states need a proper security mechanism to avoid data breaches. Data is claimed to be secure if it fulfills three conditions (i) Confidentiality (ii) Integrity (iii) Availability. Transferring the data and storing it in cloud storage makes the data vulnerable, because any unauthorized person can access, modify data. So the data has to be kept secure. The evolvements and arise in the security threats has turned the researchers to focus on the cloud security.

Cloud environment is made of different local systems connected together and includes the members from multiple environments, so providing security is much complicated. The best way to give protection to data is encryption. After encrypting the plain text, the data can be stored as cipher text in the cloud. This technique is so-called cryptography. The chosen cryptographic mechanism in the cloud by the Cloud Service Provider (CSP) should be made as powerful as possible to avoid data breaches.

Ronald Cramer and Victor Shoup came up with the cryptosystem called Cramer–Shoup Cryptosystem [2] which holds well against Adaptive Chosen Cipher Text attack (CCA2) based on the Elgamal and Diffie-Hellman algorithm. Since cloud has a large number of users, Cramer–Shoup Cryptosystem is adopted to increase the security against CCA2 attack. The strength and complexity of the key generations are less. So the proposed work has modified the Cramer–Shoup Cryptosystem to work with matrixes and so the key strength is increased. Modified Cramer–Shoup Cryptosystem also increases the performance of multi message encryption or batch encryption. Even though the time complexity increases, the Advance Cramer–Shoup Cryptosystem holds good for a sensitive data to be stored in Cloud environment.

The rest of the paper is organized as follows: Sect. 2 summarizes the related works. Section 3 describes the system architecture model and discusses about the design of the system model in detail. Section 4 describes matrix-based Advanced Cramer–Shoup cryptosystem, key generation, encryption, decryption of proposed model. The performance evaluation based on the prototype implementation is given in Sects. 5 and 6 concludes the paper.

2   Related Works

Cryptography is a term which is been in practice for more than thousand years. It is a technique used to represent a secret text in a cipher form. It is classified into two major techniques depending upon its working mechanism (i) Symmetric encryption (ii) Asymmetric encryption. Time taken to encrypt the plain text using symmetric key encryption is very less compare to the asymmetric encryption model. But it uses single key to encrypt and decrypt the input data where as it creates a security threat in case of key leakage. After 1970, researchers started to focus on the asymmetric key encryption (i.e.) public-key cryptography. The idea of Public-key cryptography is to work with a pair of keys to encrypt the plain text and decrypt the cipher text. Any person is allowed to encrypt message using the public-key of the receiver, but such cipher text can only be decrypted with the receiver's private key. It must be computationally easy for a user to generate a public and private key pair. In most of the asymmetric encryption algorithm keys are generated in a random manner. The strength of a public-key cryptography relies on the degree of difficulty for a properly generated private key to be determined from its corresponding public key. Using Public-key Encryption in cloud environment will increase the security. This section, discuss about the previous work carried out in cloud security and also the drawbacks of existing Cramer–Shoup Cryptosystem is elaborated.

2.1 Traditional Asymmetric Key Algorithms

In 1976, Diffie et al. [3] proposed an asymmetric key encryption, which is one of the earliest techniques to implement the public-key exchange between two parties. It is also called as Diffie-Hellman key exchange algorithm which is based on the exponentiation in a finite field (i.e.) chosen prime number should be finite or it should be able to computable in a polynomial time. If the algorithm chooses a large prime number, it will lead to a discrete logarithm problem. It takes $O\ (e^{\ \log n \log \log n})$ time to compute a discrete logarithm problem. Even a high end system cannot work efficiently. Later in 1977, Rivest et al. [4] built an asymmetric key encryption, so-called RSA algorithm. It is one of the most popular public-key algorithm and widely used for secure transmission of data. In RSA, the prime number should be kept as a secret. It is based on the difficulty of factoring the large products. Three major components of the RSA algorithm are exponentiation, inversion and modular operation. Time complexity of the key generation is $O\ (N^2)$ and encryption, decryption takes $O\ (N^3)$. In 1985, two algorithms were proposed, Elgamal and Elliptic curve cryptography. Elgamal algorithm was proposed Taher Elgamal. It is based on the difficulty of computing logarithms over finite fields [5]. Due to the randomization in the ciphering operation, the cipher text for a given message 'm' is not repeated in the Elgamal Algorithm, (i.e.) even if message is encrypted twice, the cipher text will not be repeated. Elliptic curve cryptosystem [6] is based on algebraic structure of elliptic curve over the finite fields. The advantage of this algorithm is it requires smaller keys compare to the previous security mechanisms. But the disadvantage of ECC is, it generate a larger cipher text compare to the RSA algorithm. Later in 1999, Paillier et al. proposed a probabilistic-based asymmetric-based cryptosystem [7]. This algorithm has the property of additive homomorphic encryption which helps to work on top of the cipher text.

2.2 Matrix-Based Security Mechanism

Matrix is a defined as a rectangular array of numbers for in which operations like addition, subtraction, multiplication, and transpose can be performed. Over the evolvement of information security, matrix has placed a major role. In this section, the recent works related to the matrix and security algorithms is discussed.

Mamvong et al. [8] came up with a new idea of combining the Elgamal cryptosystem with a matrix encryption to improve the security in the distributed storage systems. This scheme combines the advantages of speed implementation over typical public-key cryptosystems, as well as the advantage of secure key distribution over typical secret key cryptosystems. It provides a viable option to maintaining communication privacy even in insecure cloud environments. Anitha et al. [9] proposed a model which involves the attributes of metadata for key generation and distribution techniques in which the Elgamal cryptosystem is used for the generator of a cyclic group formed by the elliptic curve, finite state machines and key matrices obtained from the fibonacci sequences. Here, Q matrixes are used to the secure key. The encryption is done by the finite state machine and the Elgamal algorithm. Chowdhury et al. [10] presented an efficient method for hiding sensitive data in an image and sends to the receiver in a safe manner. This mechanism does not need any key for embedding and extracting data. Also it allows hiding four bits in a block of size 5×5 with minimal distortion. Anitha et al. [11, 12] proposed a practical and efficient security system where the cipher text is generated from the attributes of metadata. Jasvinder Kaur et al. e proposed a new idea to hide the data for the secret communication by using matrix matching technique [13, 14]. This algorithm changes the bits of the pixel value according to the matrix that is developed. Triple M Method (i.e.) Matrix Matching Method makes the cryptanalyst harder to break the code because the stress of this method is not on same specific bits.

2.3 Existing Cramer–Shoup Cryptosystem

Cramer et al. [2] came up with a new public-key cryptography which is provably secure against the adaptive chosen cipher text attack (i.e.) the intruder sends number of cipher text to be decrypted then uses the cipher result to select the subsequent cipher text. The previously well-known algorithms like Elgamal, Diffie-Hellman does not secure against the adaptive chosen cipher text attack. Moreover, the public-key cryptosystem also uses the universal hash function for encryption.

Consider a group G of prime order q, where q is large and presume the original messages or the encoded messages are to be stored as the element of G. The key generation algorithm chose the random element g_1, $g_2 \in G$ and random elements x_1, x_2, y_1, y_2, $z \in Z_q$ are chosen. Next step, is to compute the group element. $c = g_1^{x_1} \cdot g_2^{x_2}$, $d = g_1^{y_1} \cdot g_2^{y_2}$, $h = g_1^z$. Then one-way hash function (SHA-1) is chosen. The public keys is (g_1, g_2, c, d, h, H) and the private key is x_1, x_2, y_1, y_2, $z \in Z_q$.

The encryption algorithm encrypts the given message $M \in G$. It chooses the value $r \in Z_q$. $u_1 = g_1^r$, $u_2 = g_2^r$, $e = h_r m$, $\alpha = H(u_1, u_2, e,)$, $v = c_r d_r \alpha$. Where, H denotes the hash function. And the cipher text of the message is (u_1, u_2, e, v) transfer through Internet to the receiver. The decryption algorithm decrypts the cipher text $(u_1, u_2, e,$ and $v)$. Then it computes $\alpha = H(u_1, u_2, e,)$ and checks, $u_1 x_1 + y_1 \alpha + u_2 x_1 + y_2 \alpha = v$. It the conditions in satisfied then message can be decrypted by $m = e/u_{1z}$.

The problem with the existing Cramer–Shoup cryptosystem is, it does not allow batch encryption or decryption. As far as the system is conferred about the cloud environment, the velocity of the data will be high. Since the Cramer–Shoup does not support the batch encryption and decryption, the time taken for the security system to encrypt each file will be large. So in the proposed work, existing Cramer–Shoup Cryptosystem has been modified to work with the batch encryption using matrix model. In next section, the modifications on the existing CSC is elaborated.

3   System Model

The architecture diagram of the proposed model is shown in Fig. 1. The Key generation module generates the public and private keys using a random matrix value generation based on the size of the message or plain text. The encryption module encrypts the plain text with the keys generated. This encryption module supports the batch encryption or multi-message encryption. The encrypted text, so-called cipher text is stored in the cloud storage. The final module in the architecture diagram is decryption module, which allows the users to decrypt the cipher text. The private key values are transferred through a secure channel.

Cramer–Shoup cryptosystem works well with the cyclic group of values. It also supports homomorphic multiplicative operations on the cipher text. So the proposed works intends to perform the Cramer–Shoup Cryptosystem on a matrix rather than integer. Since the algorithm works with more number of key values compared to the previous technique, the strength of the key will be increased.

The proposed Matrix-based Advanced Cramer–Shoup cryptosystem involves three functionalities (i) Key Generation (ii) Encryption (iii) Decryption.

3.1   Key Generation

Consider a group G of prime order q, where q is large and presume that the original messages or the encoded messages are to be stored as the element of G. Normally one-way hash function is used for encryption.

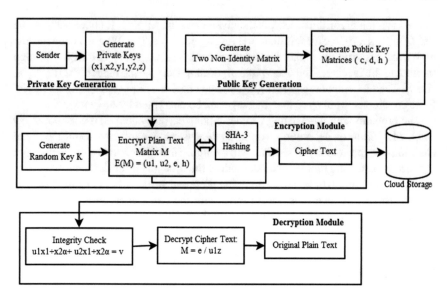

Fig. 1 Architecture diagram of proposed advanced Cramer–Shoup Cryptosystem. Public and private keys are generated in the first phase. Later the encryption module explains about how the encryption takes place and cipher text gets stores in cloud storage. The final phase has integrity check to ensure the credibility of the cipher text before decryption

Pseudo code for key generation:

(a) Generate private keys: x_1, x_2, y_1, y_2 and $z \in Z_q$
(b) Generate public keys:
Create two matrixes, g_1 and g_2, where $g_1 * g_2 = g_2 * g_1$

$$
g_1 = \begin{bmatrix} a_{1,1} \ a_{1,2} \dots a_{m,n} \\ a_{2,1} \ a_{2,2} \dots a_{m,n} \\ a_{3,1} \ a_{3,2} \dots a_{m,n} \end{bmatrix} \quad g_2 = \begin{bmatrix} b_{1,1} \ b_{1,2} \dots b_{m,n} \\ b_{2,1} \ b_{2,2} \dots b_{m,n} \\ b_{3,1} \ b_{3,2} \dots b_{m,n} \end{bmatrix}
$$

$$
\text{Compute} \begin{bmatrix} c_{1,1} \ c_{1,2} \dots c_{m,n} \\ c_{2,1} \ c_{2,2} \dots c_{m,n} \\ c_{3,1} \ c_{3,2} \dots c_{m,n} \end{bmatrix} = \begin{bmatrix} a_{1,1} \ a_{1,2} \dots a_{m,n} \\ a_{2,1} \ a_{2,2} \dots a_{m,n} \\ a_{3,1} \ a_{3,2} \dots a_{m,n} \end{bmatrix}^{x1} \begin{bmatrix} b_{1,1} \ b_{1,2} \dots b_{m,n} \\ b_{2,1} \ b_{2,2} \dots b_{m,n} \\ b_{3,1} \ b_{3,2} \dots b_{m,n} \end{bmatrix}^{x2}
$$

$$
\text{Compute} \begin{bmatrix} d_{1,1} \ d_{1,2} \dots d_{m,n} \\ d_{2,1} \ d_{2,2} \dots d_{m,n} \\ d_{3,1} \ d_{3,2} \dots d_{m,n} \end{bmatrix} = \begin{bmatrix} a_{1,1} \ a_{1,2} \dots a_{m,n} \\ a_{2,1} \ a_{2,2} \dots a_{m,n} \\ a_{3,1} \ a_{3,2} \dots a_{m,n} \end{bmatrix}^{y1} \begin{bmatrix} b_{1,1} \ b_{1,2} \dots b_{m,n} \\ b_{2,1} \ b_{2,2} \dots b_{m,n} \\ b_{3,1} \ b_{3,2} \dots b_{m,n} \end{bmatrix}^{y2}
$$

$$\text{Compute} \begin{bmatrix} h_{1,1} \ h_{1,2} \ \ldots h_{m,n} \\ h_{2,1} \ h_{2,2} \ \ldots h_{m,n} \\ h_{3,1} \ h_{3,2} \ \ldots h_{m,n} \end{bmatrix} = \begin{bmatrix} a_{1,1} \ a_{1,2} \ \ldots a_{m,n} \\ a_{2,1} \ a_{2,2} \ \ldots a_{m,n} \\ a_{3,1} \ a_{3,2} \ \ldots a_{m,n} \end{bmatrix}^z$$

(c) Public keys:

$$\begin{bmatrix} c_{1,1} \ c_{1,2} \ \ldots c_{m,n} \\ c_{2,1} \ c_{2,2} \ \ldots c_{m,n} \\ c_{3,1} \ c_{3,2} \ \ldots c_{m,n} \end{bmatrix}, \begin{bmatrix} d_{1,1} \ d_{1,2} \ \ldots d_{m,n} \\ d_{2,1} \ d_{2,2} \ \ldots d_{m,n} \\ d_{3,1} \ d_{3,2} \ \ldots d_{m,n} \end{bmatrix}, \begin{bmatrix} h_{1,1} \ h_{1,2} \ \ldots h_{m,n} \\ h_{2,1} \ h_{2,2} \ \ldots h_{m,n} \\ h_{3,1} \ h_{3,2} \ \ldots h_{m,n} \end{bmatrix}$$

(d) The generated public keys are published.

3.2 Encryption

Encryption is a process of converting data into form in which it prevents the unauthorized access. The input message data which needs to be encrypted is already been preprocessed and converted into M [m*n] matrix. The proposed Advanced Cramer–Shoup encryption technique has an efficient way to encrypt the data. The encryption mechanism has given below.

Pseudo code for encryption:

(a) Generate a Random number k.
(b) Input should be in a Matrix format

$$\text{Message m} = \begin{bmatrix} m_{1,1} \ m_{1,2} \ \ldots m_{m,n} \\ m_{2,1} \ m_{2,2} \ \ldots m_{m,n} \\ m_{3,1} \ m_{3,2} \ \ldots m_{m,n} \end{bmatrix}$$

(c) $E(M) = (u_1, u_2, e, v)$; Encryption of a message with u_1, u_2, e, v.

$$\begin{bmatrix} u1_{1,1} \ u1_{1,2} \ \ldots u1_{m,n} \\ u1_{2,1} \ u1_{2,2} \ \ldots u1_{m,n} \\ u1_{3,1} \ u1_{3,2} \ \ldots u1_{m,n} \end{bmatrix} = \begin{bmatrix} a_{1,1} \ a_{1,2} \ \ldots a_{m,n} \\ a_{2,1} \ a_{2,2} \ \ldots a_{m,n} \\ a_{3,1} \ a_{3,2} \ \ldots a_{m,n} \end{bmatrix}^k$$

$$\begin{bmatrix} u2_{1,1} \ u2_{1,2} \ \ldots u2_{m,n} \\ u2_{2,1} \ u2_{2,2} \ \ldots u2_{m,n} \\ u2_{3,1} \ u2_{3,2} \ \ldots u2_{m,n} \end{bmatrix} = \begin{bmatrix} b_{1,1} \ b_{1,2} \ \ldots b_{m,n} \\ b_{2,1} \ b_{2,2} \ \ldots b_{m,n} \\ b_{3,1} \ b_{3,2} \ \ldots b_{m,n} \end{bmatrix}^k$$

$$\begin{bmatrix} e_{1,1} \ e_{1,2} \ \ldots e_{m,n} \\ e_{2,1} \ e_{2,2} \ \ldots e_{m,n} \\ e_{3,1} \ e_{3,2} \ \ldots e_{m,n} \end{bmatrix} = \begin{bmatrix} h_{1,1} \ h_{1,2} \ \ldots h_{m,n} \\ h_{2,1} \ h_{2,2} \ \ldots h_{m,n} \\ h_{3,1} \ h_{3,2} \ \ldots h_{m,n} \end{bmatrix}^k \begin{bmatrix} m_{1,1} \ m_{1,2} \ \ldots m_{m,n} \\ m_{2,1} \ m_{2,2} \ \ldots m_{m,n} \\ m_{3,1} \ m_{3,2} \ \ldots m_{m,n} \end{bmatrix}$$

$$\begin{bmatrix} v_{1,1} & v_{1,2} & \dots & v_{m,n} \\ v_{2,1} & v_{2,2} & \dots & v_{m,n} \\ v_{3,1} & v_{3,2} & \dots & v_{m,n} \end{bmatrix} = \begin{bmatrix} c_{1,1} & c_{1,2} & \dots & c_{m,n} \\ c_{2,1} & c_{2,2} & \dots & c_{m,n} \\ c_{3,1} & c_{3,2} & \dots & c_{m,n} \end{bmatrix}^{k} \begin{bmatrix} d_{1,1} & d_{1,2} & \dots & d_{m,n} \\ d_{2,1} & d_{2,2} & \dots & d_{m,n} \\ d_{3,1} & d_{3,2} & \dots & d_{m,n} \end{bmatrix}^{k\alpha}$$

where $\alpha = H(u_1, u_2, e)$: SHA-3 as one-way hash function.

SHA-3 algorithm is a 512-bit hash function which resembles the earlier SHA-1 algorithm with significant changes in the digests bits. It was designed by the National Security Agency (NSA) to be part of the Digital Signature Algorithm. It is used to find whether the data is modified or tampered by comparing the hash value of the downloaded file with the previously published hash result. The inputs are fed to the system as 1024 bits and the system produces the 512 bits as output. After all the 1024 bit-block have been processed, as a final result 512 bit message digest is produced, which is a function of all the bits of your plain text.

3.3 Decryption

Decryption mechanism of the Advanced Cramer–Shoup Cryptosystem is illustrated in this section.

Pseudo code for decryption:

(a) Check for owner's information
(b) Computes $\alpha = H(u_1, u_2, e,)$ and checks,

$$\begin{bmatrix} u1_{1,1} & u1_{1,2} & \dots & u1_{m,n} \\ u1_{2,1} & u1_{2,2} & \dots & u1_{m,n} \\ u1_{3,1} & u1_{3,2} & \dots & u1_{m,n} \end{bmatrix}^{x_1+y_1\alpha} + \begin{bmatrix} u2_{1,1} & u2_{1,2} & \dots & u2_{m,n} \\ u2_{2,1} & u2_{2,2} & \dots & u2_{m,n} \\ u2_{3,1} & u2_{3,2} & \dots & u2_{m,n} \end{bmatrix}^{x_2+y_2\alpha} = \begin{bmatrix} v_{1,1} & v_{1,2} & \dots & v_{m,n} \\ v_{2,1} & v_{2,2} & \dots & v_{m,n} \\ v_{3,1} & v_{3,2} & \dots & v_{m,n} \end{bmatrix}$$

(c) If step (b) equals then decrypts the message,

$$\begin{bmatrix} m_{1,1} & m_{1,2} & \dots & m_{m,n} \\ m_{2,1} & m_{2,2} & \dots & m_{m,n} \\ m_{3,1} & m_{3,2} & \dots & m_{m,n} \end{bmatrix} = \frac{\begin{bmatrix} e_{1,1} & e_{1,2} & \dots & e_{m,n} \\ e_{2,1} & e_{2,2} & \dots & e_{m,n} \\ e_{3,1} & e_{3,2} & \dots & e_{m,n} \end{bmatrix}}{\begin{bmatrix} u1_{1,1} & u1_{1,2} & \dots & u1_{m,n} \\ u1_{2,1} & u1_{2,2} & \dots & u1_{m,n} \\ u1_{3,1} & u1_{3,2} & \dots & u1_{m,n} \end{bmatrix}} z$$

Since division is not possible in matrix $A * A^{-1}$ can be used. (i.e.)

$$\left[\begin{bmatrix} u1_{1,1} & u1_{1,2} & \dots u1_{m,n} \\ u1_{2,1} & u1_{2,2} & \dots u1_{m,n} \\ u1_{3,1} & u1_{3,2} & \dots u1_{m,n} \end{bmatrix}^{z}\right]^{-1} = \frac{1}{\det(u1)} adj \begin{bmatrix} u1_{1,1} & u1_{1,2} & \dots u1_{m,n} \\ u1_{2,1} & u1_{2,2} & \dots u1_{m,n} \\ u1_{3,1} & u1_{3,2} & \dots u1_{m,n} \end{bmatrix}^{z}$$

$$\begin{bmatrix} m_{1,1} & m_{1,2} & \dots m_{m,n} \\ m_{2,1} & m_{2,2} & \dots m_{m,n} \\ m_{3,1} & m_{3,2} & \dots m_{m,n} \end{bmatrix} = \begin{bmatrix} e_{1,1} & e_{1,2} & \dots e_{m,n} \\ e_{2,1} & e_{2,2} & e_{m,n} \\ e_{3,1} & e_{3,2} & e_{m,n} \end{bmatrix} * \left[\begin{bmatrix} u1_{1,1} & u1_{1,2} & \dots u1_{m,n} \\ u1_{2,1} & u1_{2,2} & \dots u1_{m,n} \\ u1_{3,1} & u1_{3,2} & \dots u1_{m,n} \end{bmatrix}^{z}\right]^{-1}$$

The message is transferred securely. The above figure represents the architecture of the proposed matrix-based Cramer–Shoup Cryptosystem. It has three major sections. User level models deals with the key generation. (i.e.) five random values and two non-identity matrix g_1, g_2 are created to generate the public-key values (c, d, f) and that are sent to the receiver through a secure transmission channel. Encryption module is used to encrypt the message to be transmitted. In the last module the decryption process takes place after the validation process. In the next section, the example of the proposed algorithm is discussed.

4 Working Model of Advanced Cramer–Shoup Cryptosystem

The working model of the Advanced Cramer–Shoup Cryptosystem explains about how the matrix values are encrypted as well as decrypted in the cloud environment.

Key Generation. Consider the cyclic group $G = \{0, 1, 2, 3, 4, 5, 6\}$ and the message transferred to be a Matrix M, the same size of g_1.

$$\text{Let Matrix: } g_1 = \begin{bmatrix} 2 & 3 \\ 5 & 7 \end{bmatrix} g_2 = \begin{bmatrix} 7 & 5 \\ 9 & 8 \end{bmatrix}$$

Randomly chosen values are, $x_1 = 2$, $x_2 = 1$, $y_1 = 1$, $y_2 = 1$, $z = 2$ all belong to set Z_q. Compute the following matrix elements:

$$\text{Matrix } c = \begin{bmatrix} 376 & 311 \\ 891 & 737 \end{bmatrix} \text{Matrix } d = \begin{bmatrix} 41 & 34 \\ 98 & 81 \end{bmatrix} \text{Matrix } h = \begin{bmatrix} 19 & 27 \\ 45 & 64 \end{bmatrix}$$

$$\text{Public Keys are, } c = \begin{bmatrix} 376 & 311 \\ 891 & 737 \end{bmatrix} d = \begin{bmatrix} 41 & 34 \\ 98 & 81 \end{bmatrix} h = \begin{bmatrix} 19 & 27 \\ 45 & 64 \end{bmatrix}$$

Private keys are, $x_1 = 2$, $x_2 = 1$, $y_1 = 1$, $y_2 = 1$, $z = 2$.

Encryption. Choose any random value for 'k' from Z_q.

Let us assume $k=2$, $u_1 = \begin{bmatrix} 19 & 27 \\ 45 & 65 \end{bmatrix}$; $u_2 = \begin{bmatrix} 19 & 27 \\ 45 & 65 \end{bmatrix}$. Input matrix $M = \begin{bmatrix} 1 & 2 \\ 3 & 4 \end{bmatrix}$.

Then find the value of $e = \begin{bmatrix} 8299 & 12116 \\ 19668 & 28714 \end{bmatrix}$.

Values of u_1, u_2 and e are stored in a separate text file for hashing. The result of the hashing will produce a 32 bit cipher text, which will be transferred to the cloud storage.

$$\text{Matrix } v = \begin{bmatrix} 1941334613 & 86526799 \\ 189355311 & 65649694 \end{bmatrix}$$

Bob sends the cipher text as,

$$\left\{ \begin{array}{c} u_1 = \begin{bmatrix} 19 & 27 \\ 45 & 65 \end{bmatrix}, u_2 = \begin{bmatrix} 19 & 27 \\ 45 & 65 \end{bmatrix}, e = \begin{bmatrix} 8299 & 12116 \\ 19668 & 28714 \end{bmatrix} \text{ and} \\ v = \begin{bmatrix} 1941334613 & 86526799 \\ 189355311 & 65649694 \end{bmatrix} \end{array} \right\}$$

Decryption. If $v = u_1^{x+y_1\alpha} + u_{21}^{x+y_2\alpha}$ then decrypt the message received. This helps in identifying the authenticated user or receiver.

$$\text{Matrix } v = \begin{bmatrix} 1941334613 & 86526799 \\ 189355311 & 65649694 \end{bmatrix} \text{ and verify}$$

$$\text{Matrix } u_1^{x_1+y_1\alpha} + u_2^{x_1+y_2\alpha} = \begin{bmatrix} 1941334613 & 86526799 \\ 189355311 & 65649694 \end{bmatrix}$$

Both are equal so decrypt. Decryption can be done by $M = e/u_1^z$. Matrix multiplication is not possible. So $e * (u_1^z)^{-1}$ will give the resultant matrix.

$$\text{The decrypted cipher text is } M = \begin{bmatrix} 1 & 2 \\ 3 & 4 \end{bmatrix}$$

5 Experimental Results

The proposed model is analyzed by executing set of experiments in the open source eucalyptus tool. The experiments are carried out in a cloud setup using eucalyptus tool which contains cloud controller and walrus as storage controller on a five-node cluster. Each node has two 3.06 GHz Intel (R) Core TM Processors, i-7 2600, CPU @ 3.40 GHZ, 4 GB of memory and 512 GB hard disks, and running eucalyptus in

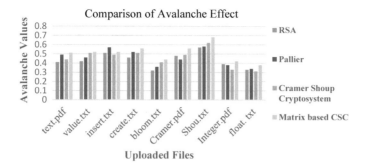

Fig. 2 Comparison of Avalanche effect

it. The files are uploaded to the cloud using Matrix-Based Advanced Cramer–Shoup cryptosystem and downloaded verifying the owner's authenticity. Since, proposed algorithm deals with more number of random generation, figuring out the level of improvement in the avalanche effect is important. Avalanche effect is by changing only one bit in a matrix, leads to a large change in the existing key, hence it is hard to perform an analysis of cipher text, when trying to come up with an attack. Higher the avalanche effect, higher the strength of the cipher key (Fig. 2).

The avalanche effect is calculated by the formula

$$Avalanche\ Effect = \frac{Number\ of\ values\ changed\ in\ the\ cipher\ key}{Total\ numbers\ of\ values\ in\ the\ cipher\ key}$$

To identify the key strength improvement in the proposed matrix-based Cramer–Shoup Cryptosystem, the comparison chart is created. The above graph represents the difference in the avalanche effect between the existing CSC the proposed Matrix-based Advanced Cramer–Shoup Cryptosystem.

6 Conclusion and Future Work

This paper presents how matrix-based Cramer–Shoup Cryptosystem can be used to provide security to the data stored at cloud storage servers. The matrix-based Cramer–Shoup is a promising technique, which can be applied in any remote storage systems. It carries the advantages of Cramer–Shoup Cryptosystem (i.e.) it holds strong against the adaptive chosen cipher text attack. Moreover, the key strength is increased in the proposed cryptosystem. Since the security system deals with the matrix operation, it can be used in batch encryption, which will highly reduce the time of encryption and decryption. In future work, we will try to reduce the time taken for encryption and decryption in a 16 * 16 Matrix system. And, converting the Advanced Cramer–Shoup cryptosystem to work with the additive and multiplicative Homomorphic encryption is our prime focus.

Acknowledgements The work of this paper is financially supported by Science and Engineering Research Board (SERB), Department of Science and Technology, Government of India. The Grant number is ECR/2016/000546.

References

1. Mell, P., Grance, T.: The NIST definition of cloud computing. National Institute of Standards and Technology, NIST Special publication, pp. 800–145 (2011)
2. Cramer, R., Shoup, V.: A practical public key cryptosystem provably secure against adaptive chosen cipher text attack. Int. Cryptol. Conf. **1462**, 13–25 (1998)
3. Diffie, W., Hellman, E.: New direction in cryptography. IEEE Trans. Inf. Theory **22**(6), 644–654 (1976)
4. Rivest, R.L., Shamir, A., Adleman, L.: A method for obtaining digital signatures and public-key cryptosystems. Commun. ACM **21**(2), 120–126 (1976)
5. Elgamal, T.: A public key cryptosystem and a signature scheme based on discrete logarithms. In: Advances in Cryptology-CRYPTO '84, pp. 10–18 (1985)
6. Miller, V.S.: Use of elliptic curves in cryptography. In: Advances in Cryptology CRYPTO '85, pp. 418–426 (1986)
7. Paillier, P.: Public key cryptosystems based on composite degree residuosity classes. In: EURO-CRYPT, pp. 223–238. Springer (1999)
8. Mamvong, N.J., Aboiyar, T., Gbaden, T.: A hybrid cryptosystem using elgamal algorithm and matrix encryption. African J. Comput. ICT **8**(3), 43–50 (2015)
9. Anitha, R., Mukherjee, S.: Metadata driven Efficient CRE based cipher key generation and distribution in cloud security. Int. J. Secur. Appl. **8**(3), 377–392 (2014)
10. Chowdhury, N.: An efficient method of stenography using matrix approach. Int. J. Intell. Syst. Appl. **1**, 32–38 (2012)
11. Anitha, R., Pratheepan, P., Yogesh, P., Mukherjee, S.: Data storage security in cloud using meta data. In: International Conference on Machine Learning and Computer Science, pp. 26–30 (2014)
12. Libert, B., Yung, M.: Adaptively secure non-interactive threshold cryptosystem. Theor. Comput. Sci. **478**, 76–100 (2013)
13. Kaur, J., Duhan, M., Kumar, A., Yadav, R.K.: Matrix matching method for secret communication using image steganography. Int. J. Eng. **3**, 45–49 (2013)
14. Zhang, S., Li, X., Wang, B.: Study on the protection method of data privacy based on cloud storage. Int. J. Inf. Comput. Sci. **1**(2), 46–51 (2012)

Makespan Efficient Task Scheduling in Cloud Computing

Y. Home Prasanna Raju and Nagaraju Devarakonda

Abstract Cloud computing is an emerging technology in modern era of online processing of customizable resources gathered commonly for several remote server accesses through on-demand access. Cloud Service Provider (CSP) renders cloud computing infrastructure in pay per use scheme in various formats. Thus, CSP provides a major role in optimization of Task Scheduling (TS) in trade off with cost afford by the end user. In proposed scheme, to create efficient utilization of resources and balanced cost of rendering service to end user, Modified Fuzzy Clustering Means algorithm (MFCM) along with Modified Ant Colony Optimization (MACO) technique is used thereby minimizing the cost of using a cloud computing structure and with reduced makespan along with load balancing capability. Proposed strategy provides better results than existing strategies of various modifications on ACO alone that concentrates on optimizing lineup of Virtual Machine (VM).

Keywords Cloud service provider · Modified ant colony optimization · Modified fuzzy clustering means · Task scheduling · Virtual machine

1 Introduction

Cloud computing is fundamentally an internet dependent service offers a cloud storage, which will be accessible only under through subscription or pay-per-use scheme. Such a typical cloud environment contains several fundamental components like Data center, Virtual Machine, Cloudlet, Broker etc. Among these components, data center is the one of the hot research area picks that include various domains like comput-

Y. Home Prasanna Raju
Department of CSE, Acharya Nagarjuna University, Guntur 522510, Andhra Pradesh, India
e-mail: yhprasannaraju@gmail.com

N. Devarakonda (✉)
Department of IT, Lakireddy Bali Reddy College of Engineering,
Vijayawada 521230, Andhra Pradesh, India
e-mail: dnagaraj_dnr@yahoo.co.in

© Springer Nature Singapore Pte Ltd. 2019
A. Abraham et al. (eds.), *Emerging Technologies in Data Mining and Information Security*, Advances in Intelligent Systems and Computing 755,
https://doi.org/10.1007/978-981-13-1951-8_26

ing, networking, management, and so on. Recent technologies and developments are required to keep the speed of the data capacity growth and to deal with the underlying system complexity, power consumption, etc. A data center is a reservoir of all facilities like bandwidth, RAM, storage etc., which are all the resources required for storage and processing of cloud information. Because of dynamic sharing of elements like I/O devices, memory and processor, the design of data centers are very critical in system architecture. Ismaeel et al. [1] explained the energy minimization strategies in cloud data centers by forecasting the low energy consumption VMs. Later how reduced latency for localized traffic is benefited in cloud computing is well demonstrated by Yang [2]. After that Levy and Hallstrom [3] came up with a new approach, low power wireless sensors which can be used in data center infrastructure for real-time monitoring and management. Parameters included in monitoring are power, temperature, humidity, airflow, etc. For better resource sharing in cloud computing environments, cloud data centers can be integrated with cloud federation techniques [4]. Next [5] Babukartik, R. G., and P. Dhavachelvan showed how makespan can be effectively reduced by introducing the cuckoo approach with the combination of ACO. In [6] Sharma Suruchi, and PratyayKuila proposed a new heuristic algorithm for cloud environment to reduce the period of scheduling. An improved ACO algorithm can be used for reducing the transportation cost in construction lay out [7]. A multiobjective optimal scheduling algorithm was proposed for better results of both performance and cost in cloud task scheduling [8]. Li et al. proposed LBACO algorithm [9] for load balancing the system by reducing the cloud task's makespan. Most of the task scheduling algorithms work on cloud task resource requirements which include CPU memory, execution time, and execution cost. But there is one more parameter that can be considered is network bandwidth. Razaque et al. [10] worked with network bandwidth parameter for better results of task scheduling. In cloud computing platform, one of the major challenges faced by CSP is Task Scheduling (TS). General issue in TS is Non-deterministic Polynomial (NP)-hard optimization problems like traveling salesman problem, combinatorial problems like integer programming and addressing problem. These problems occur due to the allocation of hundreds and thousands of Virtual Machines (VM) to cloud resources that causes delay in the performance of TS. To solve these problems, for clouds, Ant Colony Optimization technique is more suitable. In this paper, modified FCM (MFCM) clustering algorithm by modifying FCM algorithm [11, 12] and modified ACO (MACO) optimization algorithm by modifying ACO algorithm [13, 14] are proposed and implemented as an effective approach that focus to reduce makespan and to remain in balanced load.

2   Basic Entities in CloudSim

CloudSim is the simulator to simulate the cloud computing environment. The basic entities of CloudSim are given below [9].

a. Data Center and Cloudlets
b. Data Center Broker
c. CIS (Cloud Information Service)

Data center and Cloudlets. Data centers are the resource providers for brokers. Virtual machines (VMs) and cloudlets are used to perform certain tasks with specified allocation of essential parameters. Cloudlet (a set of instructions that was demanded by the user which has to be performed by the cloud computing machine) recognizes a task in Cloudsim which is helpful in performing and submission of tasks. Data centers have N number of task; each and every task is clustered into multiple set of tasks. Each cluster is carried over to corresponding VM by broker.

Data center Broker. Depending on user's requirements, services are generated between users and service providers by the Data center broker. Brokers perform service to transport tasks across clouds which are created from user developed programming algorithms. Brokers are allocated in TS as serving medium between Data center and other parts.

CIS. It maps user requests to suitable cloud providers. It maintains the latest details of data center characteristics such as processing elements, RAM, bandwidth etc.

Data centers, brokers, VM and cloudlets have been created for performing the required task by using Cloudsim toolkit to satisfy customer's demand. Cloudsim is a framework for simulation of cloud infrastructure, where data centers are resource providers, brokers help in creation and destruction of virtual machines, cloudlets for performing and submission of task and to provide information about the RAM size, bandwidth and number of processor allocations.

3   Fuzzy C Means Algorithm (FCM)

Bezdek et al. [11], Yaikhom [12] clustering algorithm play an important role in grouping similar task into set of tasks. Aim of cluster analysis done in this work is for dividing the N number of tasks into C number of clusters in such a way that tasks within the cluster are high similar to each other and tasks in other clusters are not similar to one another. Clustering means organize tasks into groups based on similarity criteria. FCM is a clustering technique which allows number of similar tasks from a whole bunch of tasks will be placed under one cluster. In FCM where each job belong to more than one cluster, where degree of membership for each task is given by a probability distribution over the clusters. It is useful when required to process a number of clusters which are pre-determined. It does not calculate absolute membership of data points instead it calculate degree of membership that a data point will belong to that cluster. It is fact because it does not calculate absolute membership. The main idea in fuzzy clustering technique is to partition the data into a collection of clusters non-uniquely. In FCM algorithm, membership values are given to data points for each of the clusters moreover, clusters are allowed to grow into their natural shapes. The clustering of cloudlets can be done by minimizing the following objective function presented in Eq. (1).

$$J = \sum_{i=1}^{N} \sum_{j=1}^{C} \mu_{ij} \|x_i - c_j\|^2 \tag{1}$$

where

J   Ojective function of FCM clustering algorithm
C   Total number of clusters
x_i Length of ith Cloudlet
c_j Length of Centroid
N   Total number of cloudlets in the cloud environment and C depicts the total number of clusters.
μ_{ij} ith data point (cloudlet) membership value with respect to jth cluster and it is calculated as below

$$\mu_{ij} = \frac{1}{\sum_{k=0}^{C} \left(\frac{\|x_i - c_j\|}{\|x_i - c_k\|} \right)^{\left(\frac{2}{m-1} \right)}} \tag{2}$$

m   the Fuzziness Co-efficient

In this paper, the number of cluster defined based on the number of different kinds of cloudlet presented in the following Eq. (3).

$$C = CL_R + 1 \tag{3}$$

From the above Eq. 3 where, C indicates number of clusters and CL_R represents number of different kinds of cloudlets and additional one for reserve cluster of cloudlet. Once the number of cluster has been defined, next thing is to select number of centroids randomly for every C and it will update using the following Eq. (4) at the end of iteration. At initial stage of clustering, membership values between each data points with respect to each and every cluster can be defined randomly in between 0 to 1 bounded to constraints like for each data point summation of all membership should be equal to one. After a successive iteration the membership values can be updated using the membership update function presented in the Eq. (5).

$$c_j = \frac{\sum_{i=1}^{N} \mu_{ij} . x_i}{\sum_{i=1}^{N} \mu_{ij}} \tag{4}$$

$$\mu_{ij} = \frac{1}{\sum_{k=1}^{C} \left(\frac{\|x_i - c_j\|}{\|x_i - c_k\|} \right)} \tag{5}$$

where,

c_j	Updating of jth centroid
x_i	represents the ith data point
$x_i - c_j$	Euclidean distance between the ith datapoint with respect to jth centriod
$x_i - c_k$	Euclidean distance between the ith datapoint with respect to all other centroids except jth centriod

With the aid of Eqs. (4) and (5), calculated centroids and membership values are updated after an iteration. This process is repeated until minimum value of J stated in Eq. (1) is achieved.

4   Ant Colony Optimization

The basic ACO optimization is based on the statement an ant possess ability to find an optimal way from its living place to the source of food. While moving to its nest after getting food they will put down a chemical fluid called pheromone throughout the path. When an ant finds an existing trail of pheromone on its way, ant's decision will be based on a higher probability of following the trail. By that, the trail will get even more pheromone by the additional lay out of one more ant. Probability for an ant choosing a particular way depends on the pheromone amount in that way. If more ants chose a particular way its pheromone level increases and it attracts even more ants. As a positive feedback mechanism this results in ants to find an optimal way. Each ant will have a Tabu or tour table which consists of the history of tour made by the ant in order of first visited place by the ant to the last visited place by it [9, 13, 14].

In ACO scheme ants are initially placed at all VMs randomly and then they are initialized with a level of pheromone according to MIPS, band width and number of processors according to their initial VM. Then ants are allowed to move from one VM to another randomly by a process called selection of next VM. In selection process, Ant will choose next VM which is not visited or not in its Tabu or tour table history. After selecting the VM, a mark is made in the corresponding Tabu or tour table stating that the selected VM is visited. Probability of an ant for choosing next VM can be given as following.

$$P_{ij}(VM) = \frac{[\tau_{ij}(t)]^\alpha [\eta_{ij}]^\beta}{\sum_{allowed} [\tau_{ij}(t)]^\alpha [\eta_{ij}]^\beta}, \tag{6}$$

where

$P_{ij}(VM)$	Probability to choose another VM by an ant
$\tau_{ij}(t)$	Pheromone amount deposited between VMs at indices i and j
η_{ij}	Visibility for heuristic algorithm found by its calculation which expresses execution time and processing speed of virtual machine

α Relative Pheromone Deposit
β Visibility of other pheromones

Parameters α, β provides description about relative pheromone deposit and visibility of other pheromones respectively.

$$\tau_{ij} = (1 - \rho)\tau_{ij} + \sum_k \Delta\tau_{ij}. \tag{7}$$

5 Methodology

In a cloud computing environment, proper task scheduler policy should be implemented in order to regulate its scheduling technique for fluctuating situation occurs in a structure. In domain of cloud computing, VM is a major part that is based on number of processors that has to be accommodated in TS. Moreover, VM is a committed part to boot up an operating system with use of which, TS will incorporate VM, data centers, cloudlets and brokers. Time span is the total time elapses from beginning to end of processing a cloudlet in a VM. Processes reach the demand of user query; also increase performance of computing platform. In this paper, majorly two points are highlighted. Initially, for reducing makespan, MACO algorithm is utilized (used for Cloudlet scheduler policy) with the intention of ranking up every processor according to their efficiency and cost of usage. i.e., Modified Ant Colony Optimization (MACO) is utilized to attain a fine tuning procedure.

Secondary aim is clustering every task into clusters according to different VM ability and cost, for that MFCM algorithm is utilized. In order to reduce evaluation cost, parameters like makespan is used. The basic workflow of Cloudsim architecture with proposed scheme is illustrated in Fig. 1. In this kind of TS, every task are equally distributed to multiple processors in Cloudsim and each task are clustered into set of tasks, for clustering purpose MFCM algorithm is utilized. With the help of MFCM, system can easily avoid data collision, incorrect data alignment and data missing or redundancy and over utilization of expensive resources. MACO is utilized to attain a fine tuning procedure.

5.1 Pseudo Code for Modified FCM Algorithm

Modified FCM algorithm pseudo code is given below.

Fig. 1 Working procedure
of proposed methodology

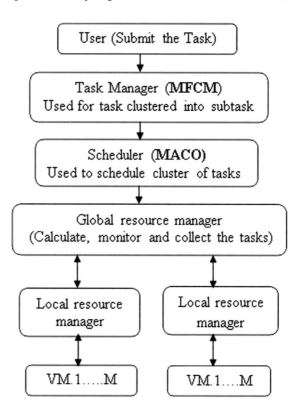

Algorithm MFCM

{

 Input. Takes the cluster1 and cluster2 centers from the output of FCM running on given cloudlets.

 Output. Creates the cloudlet clusters of type heavy load and less loads by checking the load balance.

 Begin

 1. Store the cluster1 center value into c1 and cluster2 center value into c2.

 2. Sort the initial cloudlets according to their lengths and store them into 'sorted' list.

 3. Normalize the cluster center values to the size of cloudlet lengths for making cluster heads 'cluster h1', 'cluster h2'.

 4. Call the function createCloudletLists(c1,c2,sorted);
  // returns costlier and cheap cloudlet cluster lists 'cloudlet_list1', 'cloudlet_list2' under cluster heads 'cluster h1' and 'cluster h2'.

 5. Call function checkLoad(cloudlet_list1, cloudlet_list2);
  // returns less load and heavy load clusters. Heavy load (>75% load) clusters are responsible for booting reserved virtual machines.

 End

}   // end algorithm

```
function createCloudletLists(c1,c2,sorted)
        {
            t1:=c1,t2:=c2;
            s1:=c1,s2:=c2;
            for(int i=0;i<sorted.size();i++) {
            t1+:=sorted.get(i);
            // gives ith cloudlet length
            t2+:=sorted.get(i);
            x:=t1;
            if(t2<x) {
            x:=t2;
            s2+:=sorted.get(i);
            t1:=s1;
            t2:=s2;
            cloudlet_list2.add(i);
            // adds ith cloudlet to list
            } else {
            x:=t1;
            s1+:=sorted.get(i);
            t1:=s1;
            t2:=s2;
            cloudlet_list1.add(i);
            }    // end if
            }    // end for
           return  cloudlet_list1, cloudlet_list2
        } // end function


function checkLoad(cloudlet_list1, cloudlet_list2)
       {
         A:= cloudlet_list1, B:= cloudlet_list2;
         sum:=0,sumA:=0,sumB:=0,sumC:=0,sumD:=0;
        for(int=0;k<Total_cloudletList.size();k++)
         {
        sum=Total_cloudletList.get(k)
             .getCloudletLength();
         } // end for

        for(int i=0;i<A.size();i++){
        sumA +=A.get(i).getCloudletLength();
        } // end for

         for(int i=0;i<B.size();i++){
         sumB +=B.get(i).getCloudletLength();
         } // end for

         if(((sumA/sum)*100)>75){
         Listx.addAll(B);
```

```
    getPar();      // get initial parameters of
                   // FCM algorithm
    runFCM(A);     // redo FCM algorithm
} else if(((sumB/sum)*100)>75) {
Listx.addAll(A);
getPar();      // get initial parameters of
               // FCM algorithm
runFCM(B);     // redo FCM algorithm
} // end if
Return cloudlet_list1, cloudlet_list2,
       Listx;

}// end function
```

General FCM algorithm does not maintain the balance load in cluster length and it leads to imbalance of workload in cloud computing. To overcome this MFCM ensures the load balance by checking the length of two clusters and booting of third reserve virtual machine with additional cluster in case of heavy load on cloud computing framework.

5.2   Modified ACO Algorithm

TS process will occur as a fundamental process in all Cloudsim architectures; In a TS process, sometimes makespan increasing possibilities are presently happening may be due to chances of load imbalance, large size of task, or task complexity, during that time makespan will get so long. For reducing the makespan purpose ACO algorithm can be employed. Proposed scheme will perform ACO optimization for all processors in VMs based on VM's capability of every individual processor and store it up as an optimized list by modified ACO optimization. Main responsibility of MACO algorithm is improving performance efficiency of TS by reducing makespan. But, ACO scheme possess the inability of assigning task to individual processor instead it will assign to a whole VM which leads only one processor to work and others be idle. To overcome this pitfall and for efficient utilization of resources MACO is proposed in this work. Initially all ants are presented on processors randomly with initial pheromones. Now ants move to processors randomly with the attraction of pheromones laid in their path and register the visited processors in their tour or visited history. After visiting the processor, ants will place pheromone in the processor according to its processing speed, amount of traffic to reach and all. Similarly all ants will tend to move in similar way. Modification that made MACO better than ACO is it takes distance between processors instead of VM and keeps an optimal order of individual processors instead of sorting VM. By this proposed scheme developer can feed more number of tasks to high efficiently processing VM and less number to slow processing VM. Thus developer need not to worry about utilization of resources by framework and left alone of some best working processors in their virtual machine. The pseudo-code for the proposed scheme is shown below:

```
Algorithm MACO
    {
        Input.    Collect cloudlet list cloudlet_list1,
                  cloudlet_list2, Listx from the MFCM
                  algorithm.
        Output.   Scheduling of cloudlets by taking the
                  distance between processors.

        Begin.

        1. Initialize all the input parameters as in
           ACO algorithm.
        2. Call the function comp_dist();
           //to initialize the tour distance
        3. for(try=0; try<max_tries;try++){
           While(no_tours<max && AntBestTourLength >
           optimal){
           for each ant {
                Next VM is chosen by the intensity of
                Pheromone trail
                       } // end for loop
           Update_pheromoneTrail();
           } // end while loop
           AntTour←Store ant tour;
           } // end for
        4. Call the function ant_bestTour(AntTour);
           // returns best available processors in
           // order
        5. Submit the cloudlets from the lists cloud-
           let_list1, cloudlet_list2, Listx to VMs in
           the order returned by the step5. If Listx is
           not empty then an additional VM is booted
           for executing cloudlet tasks.

        End.
    } // end algorithm
```

```
function comp_dist()
  {
    // Compute Distance between Processors by terms
    // of processing capability. Listx is a list of
    // cloudlets taken from MFCM algorithm.

    for  i=0 to Pelist.size() do
         for j=0 to VmPelist.size() do
             dist[i][j]=Listx.get(i).getCloudletLen
             gth()/VmPelist.get(j).getMips();
         end for
    end for

  }// end function
```

```
function ant_bestTour(AntTour)
    {
        k ← index of AntTour
        for i←AntTour.length()-1
          for j←AntTour.length()
            if(AntTour [i]> AntTour [j])
                Tmp=k[i];
                k[i]=k[j];
                k[j]=Tmp;
            end if
          end for
        end for
        return k;    } // end function
```

Cloudlet Scheduler Policy will get the results of Best Tour given out by the MACO algorithm and it schedules the cloudlet based on it, in the meantime it removes previously processed cloudlets from the execution list of VM in order to make room for new cloudlets. So, that MACO can assign more cloudlets to a VM for efficient utilization. Execution time of a cloudlet is calculated as below.

Execution time T_E: If workload of jth task which is hosted on ith resource with processing capacity P_i then execution time of jth task is given in Eq. (8).

$$T_E = \frac{W_j}{P_i}, \tag{8}$$

where

T_E Execution Time
W_j Workload of jth cloudlet
P_i Processing time of ith processor

It is assumed that execution time is proportional to execution cost which is used for the calculation of makespan parameter.

6 Experimental Result

Proposed strategy can be put it into field by a java platform IDE namely Java NetBeans 8.2 version running on an 32 bit operating system, with support of 8 GB RAM along with all other basic requirements. In MACO, initially values of α and β are taken as 1 and 2 respectively. And we assumed that one cloudlet is created for one task. Initially with two VMs (VM1, VM2) and one VM (VM3) for reserve purpose are created. After user submits the cloudlets (i.e., tasks), MFCM creates two clusters with total cloudlet length of both being equal and creates one more cluster of cloudlet in case of overloading in system with a length sufficient to balance the load at that time. The initial two clusters are assigned to VM1 and VM2. And at last VM3 is used if the system is overloaded. As a part of this strategy, MACO reduces makespan greatly. In MACO, 25 ants are created with an assumption that they are spread out randomly throughout all processors with a minimal criteria of each processor contains at least one ant. MACO is mainly focused to diminish makespan in TS by effectively allocating cluster cloudlets to processors with in the VMs, which results in proposed algorithm to give greater results than any other traditional algorithm like basic ACO. In upcoming chart (Fig. 2) and table (Table 1) it is depicted the makespan (total execution time) of proposed methodology utilizing the low cost VMs1 and 2 created in datacenter 1 with keeping an expensive VM3 created in datacenter 3 as reserve backup which, will be utilized only when emergency heavy load criterion occurs. And it has been identified that VM3 has not been utilized so that its execution time is zero.

Fig. 2 Makespan of each Virtual Machine versus number of tasks

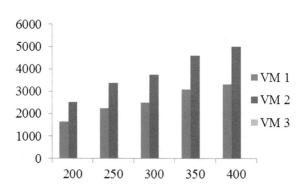

Table 1 Execution time of each VM at various number of Tasks

No. of Tasks/VM	VM1	VM2	VM3
200	1640.21	2540.1	0
250	2246.88	3380.1	0
300	2500.21	3750.1	0
350	3093.54	4610.1	0
400	3333.54	5000.1	0

Table 2 Makespan of various methodologies

Methodology	Makespan for 100 tasks ($\times 10^3$ s)	Makespan for 200 tasks ($\times 10^3$ s)	Makespan for 300 tasks ($\times 10^3$ s)
FCFS [8]	48	76	82
ACO [8]	27	47	73
Min-Min [8]	26	46	70
PBACO [8]	24	42	68
MACO-MFCM	1.46	2.540	3.75

Table 3 MIPS utilization % of each VM for various set of tasks

No. of Tasks\VM	VM1	VM2	VM3
200	40.83	63.25	0
250	49.5	74.25	0
300	40.83	63.25	0
350	51.66	77.25	0
400	59.66	84.5	0

Fig. 3 MIPS utilization versus number of tasks

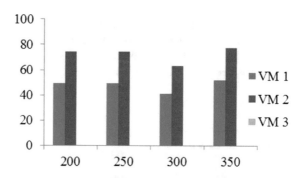

Table 2 lists the makespan comparison results of proposed methodology with various methodologies. Table 3 and Fig. 3 list MIPS utilization percentage for various set of tasks. From the results, it is understood that moderate efficient and affordable machine to work for longer time to compensate loss of work done by an expensive machine. So, the makespan also distributed evenly among all working VMs and load also balanced efficiently. Scheduling is done in a way such that moderate machine will share a major load as it can process without any traffic. Thus, it gives the ability to avoid higher expensive virtual machine in order to operate system at lower cost. Above results show that combination of MACO and MFCM algorithm helps to reduce traffic and improve efficiency of utilization along with inexpensive cost of utilization. Hence, proposed methodology can able to satisfy its desire in an efficient manner.

7 Conclusion

Many ways have been emerged for TS in CloudSim, the proposed methodology is applicable for CSP to provide makespan effective cloud infrastructure to people. Implementation of such approach provides for significant growth to cloud computing environment. Thus a fusion of MACO and MFCM makes remarkable changes in makespan as well as efficient real time utilization of a CloudSim model. Proposed scheme will be more helpful in serving the society in needs of a faster, cheaper and reliable cloud computing environment which makes computations as well as storage simple and effective for anyone to handle with a less complex manner. In future, this remarkable growth can be further expanded by improving it to more number of cluster formations in shorter span and reducing more utility to a one or more dedicated machines in a standard environment.

References

1. Ismaeel, S., Miri, A., Al-Khazraji, A.: Energy-consumption clustering in cloud data centre. In: 3rd MEC International Conference on Big Data and Smart City (ICBDSC). IEEE (2016)
2. Yang, Qi.: Design of optical data vortex cluster network for large data center network. In: 2016 10th International Symposium on Communication Systems, Networks and Digital Signal Processing (CSNDSP). IEEE (2016)
3. Levy, M., Hallstrom, J.O.: A new approach to data center infrastructure monitoring and management (DCIMM). In: 7th Annual Computing and Communication Workshop and Conference (CCWC). IEEE (2017)
4. Murudi, V., Kumar, K.M., Kumar, D.S.: Multi data center cloud cluster federation-major challenges & emerging solutions. In: 2016 IEEE International Conference on Cloud Computing in Emerging Markets (CCEM). IEEE (2016)
5. Babukartik, R.G., Dhavachelvan, P.: Hybrid algorithm using the advantage of ACO and cuckoo search for job scheduling. International Journal of Information Technology Convergence and Services 2(4), 25 (2012)
6. Sharma, S., Pratyay K.: Design of dependable task scheduling algorithm in cloud environment. In: Proceedings of the Third International Symposium on Women in Computing and Informatics, pp. 516–521. ACM (2015)
7. Calis, G., Yuksel, O.: An improved ant colony optimization algorithm for construction site layout problems. J. Build. Constr. Plan. Res. 3(4), 221 (2015)
8. Zuo, L., et al.: A multi-objective optimization scheduling method based on the ant colony algorithm in cloud computing. *IEEE Access* 3, 2687–2699 (2015)
9. Li, K., Xu, G., Zhao, G., Dong, Y., Wang, D.: Cloud task scheduling based on load balancing ant colony optimization. In: Sixth Annual Chinagrid Conference (ChinaGrid), pp. 3–9. IEEE (2011)
10. Razaque, A., Vennapusa, N.R., Soni, N., Janapati, G.S.: Task scheduling in Cloud computing. In: Long Island Systems, Applications and Technology Conference (LISAT), pp. 1–5. IEEE (2016)
11. Bezdek, J.C., Ehrlich, Robert, Full, William: FCM: the fuzzy c-Means clustering algorithm. Comput. Geosci. 10(2–3), 191–203 (1984)

12. Yaikhom, G.: Implementing the Fuzzy c-Means Algorithm. Public domain
13. Dorigo, M., Blum, C.: Ant colony optimization theory: a survey. Theor. Comput. Sci. **344**(2–3), 243–278 (2005). https://doi.org/10.1016/j.tcs.2005.05.020
14. Dorigo, M., Birattari, M., Stutzel, T.: Ant colony optimization. IEEE Comput. Intell. Mag. 28–39 (2006). https://doi.org/10.1109/MCI.2006.329691

Improved Lattice-Based Encryption with LP Solver for Secured Outsourced Data in Cloud Computing

Vemuri Sudarsan Rao and N. Satyanarayana

Abstract Cloud computing has created an intense impact over different applications with limited computational resources. Since, it works on pay-per-use manner; a massive amount of computational power is utilized. Anyhow, Security is the most vital thing for the outsourced data in cloud systems. Thus, secure outsourcing mechanisms are in great need to not only protect sensitive information by enabling computations with encrypted data, but also protect customers from malicious behaviors by validating the computation result. Such a mechanism of general secure computation outsourcing was recently shown to be feasible in theory, but to design mechanisms that are practically efficient remains a very challenging problem. In this paper, we focus on providing security to the outsourced data via improved Lattice-Based Encryption (LBE). Identifying hard computational problems which are amenable for cryptographic use is a very important task. With the help of Linear Programming (LP) solver, the cloud data are encrypted using LBE model which provides strong security proofs for the outsourced data. Initially, we will discover the highly significant sensitive attributes and stored in the lattice structure. LP solver is used as the cost objective function that minimizes the computational cost for every larger computational data used for outsourcing to the cloud server. Time is the research metric used for validating the proposed model via effectiveness, efficiency and outsourcing key generation. It is evident from the analysis that our proposed data outsourcing model ensures lessened overhead with lessened time taken for computing larger number of users.

Keywords Cloud computing · Data outsourcing · Data computation
Lattice structure · Linear programming and time

V. S. Rao (✉)
Department of CSE, Khammam Institute of Technology
and Sciences, Khammam 507002, Telangana, India
e-mail: sudharshan.cse@gmail.com

N. Satyanarayana
Nagole Institute of Technology and Sciences, Nogole,
Ranga Reddy, Hyderabad, Telangana, India
e-mail: nsn2008@gmail.com

© Springer Nature Singapore Pte Ltd. 2019
A. Abraham et al. (eds.), *Emerging Technologies in Data Mining and Information
Security*, Advances in Intelligent Systems and Computing 755,
https://doi.org/10.1007/978-981-13-1951-8_27

299

1 Introduction

The recent developments made in business environments impressed tremendous amount of web users. Computing has embedded into every fabric of our world [1]. This kind of exponential growth changes the global information exchanges and inter-action. Data outsourcing facilitates greater resilience towards cloud deployed on vari-ant environments. By doing so, the business world achieves lessened computational overheads from server and client side systems. It also eliminates the capital expenses of the computing resources. The recent innovation, cloud technologies supports on-demand network access for the outsourced data without any restriction on devices. Since the works are outsourced, the clients enjoy unlimited computing over hardware and software environment with lessened overhead [2]. In spite of these benefits, data outsourcing deprived over the public cloud imposes several security issues towards promising computation model. Generally, the outsourced data may contain sensitive information such as transaction records, research data, health records, etc. Thus, data outsourcing enforces a better security model.

However, security has become major concern on the data outsourcing paradigm [3]. When the data owner outsources their data into the cloud, inevitably, the owner lost their dominance on the outsourced data. The cloud server will take complete responsibilities of the outsourced data like input processing, computational results, etc. Thus, the end-user's privacy is totally relied on the cloud. Furthermore, the cloud could prone the outsourced data and return fake results to the end-users. Some of the outsourced data plays vital role in business environments. Henceforth, computation outsourcing [4] is a key component of the cloud computing. It enables the resource constrained end-users to outsource their computational tasks to the cloud servers. Then the tasks are processed in the cloud servers and solutions are returned to the end-users. The technical and economic advantages make computation outsourcing a promising application for cloud computing [5].

The contributions made in the paper are

- We observed the current security scenario of the data outsourcing system.
- To achieve cost efficiency of the cloud model, LP solver is utilized for achieving optimal decision making systems.
- To maintain the confidentiality of the data, it is encrypted using Lattice-Based Encryption system.
- To preserve the computational resources, the efficiency, effectiveness of the query vectors are analyzed.
- To generate the outsourcing key within stipulated period of time, we have analyzed the data lattices to the number of users which assures a better and outsourcing model.

The paper is arranged as follows: Sect. 2 describes the prior works; Sect. 3 describes the proposed model; Sect. 4 describes experimental analysis of the pro-posed work and concludes in Sect. 5.

2 Related Work

This section reveals the prior security mechanism suggested for the outsourced data in the cloud environment. Authorization and authentication are the two security parameters [6]. Initially, a credential-based cloud management system was introduced. Relied upon the credentials, the data is classified and a security framework is developed. It also assisted to find out the infrastructure maintenance with reference to credentials based on the context. These two parameters also find the appropriate users, services classification and services recovery. MiLAMob [7] is a mobile software application that provides IaaS cloud applications. It provides services to the data consumers with reduced HTTP traffic. Fermicloud [8] is another application which satisfies the authentication and authorization based on public key infrastructure model. It is basically developed from the Open Nebula that provided better user interfaces and user management systems. This also extended the security services by creating trust among the users.

The author in [9] suggested collaborative mechanisms that satisfy the access control requirements like centralized facilities, agility, homogeneity, and outsourcing trust. It included Authentication as a Service (AaaS) which solved the issues in multi-tenant architecture. Then, the trust factor [10] is analyzed with the cryptographic Role-Based Access Control (RBAC) which helps to securely outsource the data. Role inheritance is also similar concept that ensures the trustworthship of the cloud ecosystems. The author in [11] discussed about the user-centric model for enhancing authentication services. Their model eliminated the apportioning of usernames and passwords over the cloud services.

Specifically, it executed on the OAuth2 a protocol that runs the services on behalf of end-users. In [12], the input and output feasibility of the privacy maintenance of the results are analyzed. It depicted that the correctness of the cloud results that plays vital role in the secure computation.

The author in [13] depicted the secured protocols that involved algebraic computation of the outsourcing model. It does not deal with the huge complexity with variant modular exponentiation. The generation of the public key is expensive. In [14], secured outsourcing model is developed using matrix multiplication model. The message passing mechanism involves higher computational overhead. The author in [15] studied about the computational power of the asymmetric model. The other problem is security asymmetry because no party alone knows all problem input, leading to difficulty in result validation. In [16, 17], they presented additive group of users with enhanced cryptographic techniques. Since, the cloud server is not trusty system, constraint matrix was evolved for the preserving the computational resources. In [18], permutation matrix is studied to provide feasible solution for the real time cloud environment. The author in [19] presented the privacy protection model for the outsourced data. It presented higher computational overhead from the client side. Index privacy is studied by the author in [20] that doesn't yield better performance in terms of supporting larger number of users.

3 Proposed Methodology

This section states the proposed methodology designed for secure data outsourcing in the cloud environment. Since security is the major concern of the cloud environment, we aim to provide confidentiality of the outsourced data. Generally, the outsourced data may contain sensitive information such as proprietary data, research data etc. In order to track the sensitive information, the cloud server incessantly investigates the encrypted output. Most of the security threats are processed by the malicious behavior of the Cloud Server (CS). The design goals of or study are:

(a) Correctness: The decrypted data should be correctly retrieved by the cloud clients.
(b) Traceability: In case of cloud sharing, the data should be easily traceable within stipulated period of time.
(c) Integrity: The suggested mechanism should retrieve the accurate data without any modification.
(d) Efficiency: Despite of n number of data, the mechanism should efficiently deal with the outsourced data.

3.1 Improved Lattice-Based Encryption (ILBE)

With the help of Linear Programming (LP) solver, a secure and practical outsourcing scheme for the problem input/output is analyzed. Linear Programming is an optimization model that helps to clear out the different decision variables. It generally works on the constraints. Relied upon the given constraints, the objective function is designed with linear equation and inequalities. Initially, the linear objective function is defined as:

$$minimize\ C^T X\ subject\ to\ Px = Q, \quad X \geq 0 \tag{1}$$

where, x is the decision variable vectors (n * 1);
  P is the m *n matrix.
  C and Q are the n * 1 vectors.
   The above Eq. (1) is further generalized to enhance the security of the multi-user systems. To improve the security strength of the LP at outsourcing, the feasible region should be hided from the encryption region X. Based on the decision variables, i.e., registered sensitive input, then the secret key is mapped. Consider A as the random n * n non-singular matrix and d as n * 1 vector where X is affined as $y = A^{-1}(x + d)$. Then, the decision variable y,

$$Minimize\quad C^T Ay - C^T d \tag{2}$$

$$\text{Subject to } PAy = b + Ad$$
$$QAy \geq Bd \tag{3}$$

The proposed Lattice based Encryption via LP solver is explained as follows:

(a) *System Initialization* (1^λ):
Security parameter λ is taken with the attributes picked from finite field $F = Z_q$ where q, prime chosen θ (2^λ), then the output is given as, public params p_{params} and master secret key *msk*.

(b) *TrapGen* (1^λ):

- It generates the one-way function for enhancing the security of the sensitive attributes. A random matrix n * m where $A_0 \in Z_q^{n*m}$ with m-vector $T_{A_0} \in Z_q^{n*m}$ such that $T_{A_0} \in \wedge_q^\tau (A_0)$ and $\|T_{A_0}\| \leq m.w(\sqrt{\log m})$
- Choosing a matrix $Q \in Z_q^{n*m}$
- Every registered attributes $i \in U_i$, then the matrix $A_i \in Z_q^{n*m}$
- Then, the w-vector $S_T = (s_1, s_2, s_n \ldots) \in Z_p^w$
- Output: Public parameter $p_{params} = (A_0, B, \{A_i\}_i \in U, s)$ and Master private key T_{A0}

(c) *KeyGen* (p_{params}, *msk*, (*M*, ρ)):

- It intends to provide outsourcing key Out $_k$.
- Using the public parameters and master private key, the outsourcing key is
- generated with an access policy (M, ρ) where

$$M \in Z_q^{l*n}, \rho = [w] - \rightarrow Z_p$$

- Along with the linear solver, the vectors are given as $y_1, y_2,\ldots y_w \in Z_q^n$. It is further given in matrix form such as:

$$y_j^i = \begin{matrix} s_1 & y_{11} & y_{1n} \\ s_2 & y_{21} & y_{2n} \\ s_w & y_{w1} & y_{wn} \end{matrix}^T$$

Where $i = 1 \ldots l$ and $j = 0, 1, 2 \ldots n$

- Assume $Ay_i = (\lambda_1^{(i)}, \lambda_2^{(i)} \ldots .\lambda_l^{(i)})^T$
- Finally, with the msk, the outsourcing key is generated for their registered attributes, as $sk_{\rho(i)} = (u1_{\rho(1)}^{(i)}, u2_{\rho(21)}^{(i)}, \ldots \ldots un_{\rho(n)}^{(i)}$ is passed over the secure channel of the intended user.

(d) Encryption (p_{params}, $sk_{\rho(i)}$, data):

- The ciphertext c(t) is created with the public $_{params}$, outsourcing key and the data.
- An arbitrary j—vector is chosen for $x \in Z_q^n$

Fig. 1 System architecture

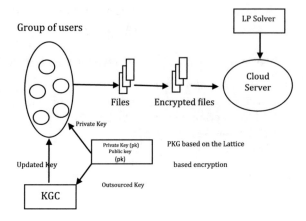

- Developing a n-dimensional vector for the Encry (e) which expands w-vector, that again sets Encry(e) $_i$ = s$_i$, i ≤ 1, Encry(e) $_i$ = 0, 1 < i ≤ n
- To avoid channel errors, low—gaussian noise is deployed for creating cipher-text c(t). It is computed as follows:

$$c_0 = x^T + X_0 + \left[\frac{q}{2}\right] data \in Z_q$$
$$c = x^T A_0 + x \in Z_q^{2m}$$
$$c_i = x^T (A_i + B)x_i^T \in Z_q^{2m}$$

Finally, the ciphertext c(t) = (S, C$_0$, C′, C$_i$)

(e) *Decrypt (c(t))*:

- It takes public $_{params}$, master secret key *msk*, and ciphertext c (t).
- Initially, it decrypts the set of shared attributes S in rows M.
- It computes the constants w$_i$ ∈ Z$_q$
- Finally, the decrypt [c(t)] = $C_0 - \sum_{j=1}^{l} \sum_{i \in l} w_i \, (c, c_i) e_{\rho(i)}^{(j)}$, decrypt [c(t)] < $\frac{q}{4}$,

Output true else false (Figs. 1 and 2).

4 Experimental Results and Analysis

This section depicts the experimental analysis of our proposed outsourcing model. The practical possibility of the proposed outsourcing model with experiments is discussed without real cloud deployment. The source LP problem φ and the encrypted φ (c (t) are solved. The test benchmark analyzed the LP problems ranges from 50 to 3000 cloud users. Time is the motivational metric used for validating our proposed algorithm. The test cases are analyzed in time metrics for solving the LP$_{original}$ prob-

Fig. 2 Proposed workflow

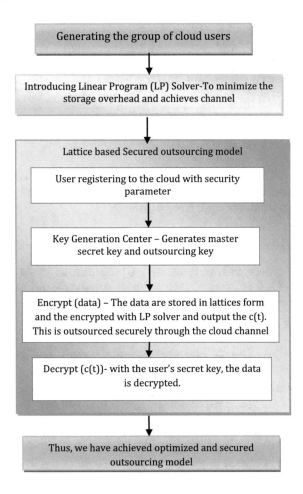

lem, LP $_{client}$, and LP $_{cloud\ server}$. Since, each cloud users possess different outsourcing key, thus, the time taken for generating keys are investigated. The parameters setting is given in Table 1. The performance analysis is carried out in the multi-user application of the public cloud storage. It deploys in tree-based key agreement structure.

Table 1 Parameter setting

r	0.1	0.2	0.3	0.4	0.5	0.6	0.7	0.8	0.9	0.95
Setup	8.4									
Encrypt	2	4	5	7	8	9	10	10	11	11
Decrypt	4	6	9	12	14	15	16	18	20	20

(a) *Effectiveness*:

The analysis of effectiveness depicts the vector commitment over the registered attributes. Consider that l-query features $l = |F_q|$, which maps the elements in the w-vector are equal to ρ, is given as

$$s_i[j_k] = \rho_{[j_k]} \quad 1 \leq k \leq l$$

Since, each query vector $q_i[j_k]$ is transmuted to the $c(t)_l$ and other components are set to 0 which indicates that our proposed mechanism does not involve false negative into the *TrapGen* and serves as an indicator to the Cloud Server.

(b) *Efficiency*:

From the perspective of data user, the response of the searched query is depicted with its LP solver for every data vector. It is efficiently computed by the cloud server via lattices of n-dimensional vectors. Aside from these computations, the encryption of the vector requires additive group of systems. The given data is converted into matrix form with its LP objectives. This lattice matrix conversion is highly reduced the computational cost of the multi-user systems.

(c) *Generation speed of outsourced key*:

Table 2 Performance results

Set up	C (T)	Effectiveness (data vector into lattices (s))	Efficiency (query search time (s))	Outsourced key generation (original to the cloud server) (s)
0.1	4	0.164	24.3	0.923
0.2	6	0.230	43.6	0.945
0.3	9	0.350	22.6	0.825
0.4	12	0.620	21.6	0.369
0.5	14	1.023	25.1	0.756
0.6	15	3.089	27.02	0.459
0.7	16	4.780	27.23	0.365
0.8	18	4.589	28.69	0.289
0.9	20	5.089	28.36	0.897

The generation speed of the outsourced key depicts the preservation of computing resources between the cloud clients and cloud server using the proposed mechanism. It is the time analysis of LP solver towards the cloud clients. In some cases, the trapdoor generation cannot be yielded by the cloud server. Based upon the search request, the queries may be altered which incurs different linear solver.

It is evident from the Table 2, that the proposed outsourcing model yields better performance in terms of effectiveness, efficiency and outsourced key generation. It means that the end-user is confident and secured about their data resided at the cloud. Since the each data vector into lattice form, the collision of servers can be avoided.

5 Conclusion

Cloud technology is the recent innovation introduced by the Information and Communication Communities (ICC). Its objectives are to achieve optimal computing resources with less overhead. In spite of the benefits, clients outsources their data to the cloud server with the assistance of sensitive attributes registration. Thus, security of the outsourced data is the major concern of our research study. Linear Computational Programming (LCP) is the recent optimization study widely adopted for secured data outsourcing model. Prior works studied about the Linear Programming (LP) solver under categorization of the cloud data storage. Since, a practical possibility of secured data outsourcing is not suggested. In this paper, we have designed an efficient and secured data outsourcing model in the cloud system. We have adopted Linear Program (LP) solver with the lattice-based encryption systems. Our intention is to achieve optimal data outsourcing model in term of time metrics. The use of LP solver is to design an objective function that capable to handle multi-cloud users in the similitude time. Initially, cloud data are stored in the lattice which additives data vector irrespective of the number of users. The performance analysis is carried out in the multi-user application of the public cloud storage. Time taken by the query vector processing, response to the searched query vector and generation of the outsourced key are examined which states that the proposed mechanism yielded less computational time on the public cloud environment.

References

1. Wang, C., et al.: Secure and practical outsourcing of linear programming in cloud computing. In: IEEE INFOCOM (2011)
2. Cloud Security Alliance: Security guidance for critical areas of focus in cloud computing. (2009). http://www.cloudsecurityalliance.org
3. Gentry, C.: Computing arbitrary functions of encrypted data. Commun. ACM **53**(3), 97–105 (2010)
4. Sun Microsystems, Inc.: Building customer trust in cloud computing with transparent security (2009). https://www.sun.com/offers/details/suntransparency.xml

5. Atallah, M.J., Pantazopoulos, K.N., Rice, J.R., Spafford, E.H.: Secure outsourcing of scientific computations. Adv. Comput. **54**, 216–272 (2001)
6. Hohenberger, S., Lysyanskaya, A.: How to securely outsource cryptographic computations. In: Proceedings of the TCC, pp. 264–282 (2005)
7. Atallah, M.J., Li, J.: Secure outsourcing of sequence comparisons. Int. J. Inf. Sec. **4**(4), 277–287 (2005)
8. Benjamin, D., Atallah, M.J.: Private and cheating-free outsourcing of algebraic computations. In: Proceedings of the 6th Conference on Privacy, Security, and Trust (PST), pp. 240–245 (2008)
9. Gennaro, R., Gentry, C., Parno, B.: Non-interactive verifiable computing: outsourcing computation to untrusted workers. In: Proceedings of the CRYPTO'10, Aug. 2010
10. Atallah, M., Frikken, K.: Securely outsourcing linear algebra computations. In: Proceedings of the ASIACCS, pp. 48–59 (2010)
11. Yao, A.C.-C.: Protocols for secure computations (extended abstract). In: Proceedings of the FOCS'82, pp. 160–164 (1982)
12. Gentry, C.: Fully homomorphic encryption using ideal lattices. In: Proceeding of the STOC, pp. 169–178 (2009)
13. Luenberger, D., Ye, Y.: Linear and Nonlinear Programming, 3rd edn. Springer (2008)
14. Liu, X., Deng, R.H., Ding, W., Lu, R., Qin, B.: Privacy-preserving outsourced calculation on floating point numbers. IEEE Trans. Inf. Forensics Secur. **11**(11), 2513–2527 (2016)
15. Even, S., Goldreich, O., Lempel, A.: A randomized protocol for signing contracts. Commun. ACM **28**(6), 637–647 (1985)
16. Liu, X., Deng, R.H., Choo, K.-K.R., Weng, J.: An efficient privacypreserving outsourced calculation toolkit with multiple keys. IEEE Trans. Inf. Forensics Secur. **11**(11), 2401–2414 (2016)
17. Blanton, M., Atallah, M.J., Frikken, K.B., Malluhi, Q.: Secure and efficient outsourcing of sequence comparisons. In: Computer Security, pp. 505–522. Springer (2012)
18. Yu, J., Ren, K., Wang, C.: Enabling cloud storage auditing with verifiable outsourcing of key updates. IEEE Trans. Inf. Forensics Secur. **11**(6), 1362–1375 (2016)
19. Hu, S., Wang, Q., Wang, J., Qin, Z., Ren, K.: Securing SIFT: Privacy-preserving outsourcing computation of feature extractions over encrypted image data. IEEE Trans. Image Process. **25**(7), 3411–3425 (2016)
20. Goldwasser, S., Kalai, Y.T., Rothblum, G.N.: Delegating computation: interactive proofs for muggles. In: Proceedings of the STOC, pp. 113–122 (2008)

A Dynamic Resource Allocation Strategy to Minimize the Operational Cost in Cloud

Chinnaiah Valliyammai and Rengarajan Mythreyi

Abstract Cloud computing has gained momentum in the recent times, due to the features it provides, like rapid elasticity and on-demand service. It involves the interaction between the user and a Resource Broker. The Resource Broker accepts the user jobs along with the requirements, and provides the results and the status of the job back to the user. The user jobs can be data intensive or computational intensive. The resource is allocated according to the type of the user job. The proposed Particle Swarm Optimization technique with migration optimizes the allocation process using computation and network based parameters. Migration efficiently eliminates the problems of over-utilization of resources. The clustering of virtual machines has also been explored in two dimensions namely resource clustering and idle clustering to increase the utilization of resources.

Keywords Live migration · Particle swarm optimization · Idle clustering

1 Introduction

Cloud computing is a model for enabling ubiquitous, convenient, on-demand network access to a shared pool of configurable computing resources (e.g., networks, servers, storage, applications, and services) which can be dynamically provisioned and released. It is based on Virtualization, which separates the computational resources from the physical hardware. In simple terms, cloud computing means accessing the stored data and programs over the Internet instead of the user's computer's hard

C. Valliyammai (✉)
Department of Computer Technology, MIT, Anna University Chennai, Chennai 600044, India
e-mail: cva@annauniv.edu

R. Mythreyi
Goldman Sachs Group, Inc., Mumbai, India
e-mail: mythreyirengarajan@gmail.com

© Springer Nature Singapore Pte Ltd. 2019
A. Abraham et al. (eds.), *Emerging Technologies in Data Mining and Information Security*, Advances in Intelligent Systems and Computing 755,
https://doi.org/10.1007/978-981-13-1951-8_28

drive. It relies on sharing of a pool of physical or virtual resources, rather than local deploying of personal hardware and software. The advantages of cloud computing are numerous, a few being low costs, re-provisioning of resources and remote accessibility. Resource allocation strategies are required to manage cloud service allocation properly. It integrates cloud providers for utilizing and allocating scarce resources and it should avoid the following problems such as resource contention, resource scarcity, resource fragmentation, over-provisioning, and under-provisioning.

Live migration is the movement of a Virtual Machine (VM) from one physical host to another while continuously powered-up and it allows an administrator to take a virtual machine offline for maintenance or upgrading without subjecting the system's users to downtime. Particle swarm optimization (PSO) is a computational method which optimizes a problem through iterations to improve a candidate solution with respect to a given measure of quality. Cluster computing provides high computation through parallel programming which uses many processors simultaneously to solve many or a single problem with fault tolerance ability.

2   Related Work

Hybrid cloud computing paradigm has been advocated as a solution for SaaS providers to handle dynamic user requests. The providers can extend their local services into public cloud for request processing with local servers and public cloud capacity. Optimization of operational costs becomes an important measure. Existing approaches require a prior knowledge of the user demands and the VM prices, or an accurate prediction. The dynamics of user requests were not considered. The theoretical model of Lyapunov Optimization framework is tailored according to the real world challenges. An Online Dynamic Provisioning Algorithm was proposed with no prior information of renting prices [1]. A Cloud Resource Allocating Algorithm via Fitness-enabled Auction was proposed for cloud resource allocation [2].

Network optimization and management in cloud data centers were achieved by re-mapping VMs to the substrate servers. Live migration of Virtual machines was used to implement the re-mapping through moving VMs from the initial servers to the target servers. Considering VM migration planning as an optimization problem with computation and bandwidth constraints, a formulation was proposed where multiple VMs were migrated simultaneously when resource constraints were satisfied [3]. A migration progress management system, called Pacer was proposed with better tradeoff between performance and migration time [4]. The challenges of dynamic migration of VMs were studied and a triple objective optimization model was proposed by considering the parameters such as energy consumption, communication between VMs and migration cost [5].

A hybrid Particle Swarm Optimization provides better performance in terms of execution ratio and average schedule length [6]. An improved particle algorithm considering the characteristics of complex networks was proposed for optimized resource load balancing in the cloud environment [7]. This mechanism considers the characteristics of networks which are complex in nature into consideration to establish a corresponding resource-task allocation model. This approach only considers the case in which the sub-tasks are standalone to each other when executing in parallel. Analysis were not been done in cases where the population sizes are different.

A new asking/bidding strategy, the dynamic asking/bidding strategy had been introduced into the bargaining model, so as to improve the auction efficiency in cloud environment [8]. Server consolidation was done to power off the least utilized nodes when the workload concentration was fewer in certain nodes to reduce the carbon emission and also to increase the utilization of cloud resources even in idle periods by forming virtual clusters [9]. A resource allocation scheme was proposed without communication overhead between the users and the servers depend on the arrival of the job and cutting off the scheduling overhead from the job's critical path [10]. The idle virtual machines are suspended to reduce the energy consumption and the active virtual machines are mapped into different energy clusters using fuzzy C-means algorithm [11]. This paper mainly focuses the Particle Swarm Optimization with migration for resource allocation by considering computational and network parameters and also idle resource clustering which increases the resource utilization.

3 Proposed Resource Allocation Strategy

The users communicate with the Resource Broker to submit a particular job and to get the result of the submitted job. A job may be data intensive or computation intensive depending on its requirements. The proposed resource allocation strategy is shown in Fig. 1.

The users submit the jobs to the Resource Broker with requirements of the job for execution. The Resource Broker has a Match Mapper, which does the task of mapping a particular job to VM's according to its requirements. The Resource Monitor performs the overall monitoring of the available resources. Once a user submits a job, the Resource Broker queues the job. The Resource Mapper then maps the job to a host, in the Resource Allocation process. If the VM resources are sufficient, the Resource Allocator does the job of allocating a VM to the job. If the resources are under-provisioned, then the VM is migrated to another suitable resource using the Migration Manager. This scenario takes place when the user scales up the VM resources. The Resource Mapper, chooses the first job from the queue of jobs, and tries to map it to a host. If the resource availability is split across nodes, the Resource

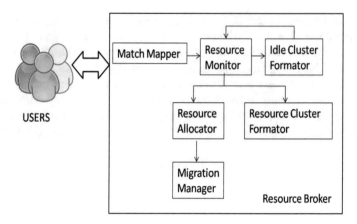

Fig. 1 The proposed resource allocation strategy

Cluster Formator is used to cluster multiple resources and allocate a job to it. If all the resources are busy, then their energy consumption parameter (CPU frequency) is considered to form a cluster using the Idle Cluster Formator. The response about the success of the submitted job and the user required result is sent back to the user.

3.1 Virtual Machine Migration Using Particle Swarm Optimization

The PSO algorithm works by having a population (swarm) of candidate solutions (particles). These particles are moved around in the search space according to a few parameters. The movement of the particles is guided by their own best known position in the search space and the swarm's best known position. The movements of the swarm are modified when better positions are discovered. The process is repeated until the best solution is discovered. The complexity of the process is O(MX) where M is the number of initial particles in the search space and X is the number of iterations. The pseudo-code for the Virtual Machine Migration using Particle Swarm Optimization is illustrated in Algorithm 1.

Algorithm 1: Virtual Machine Migration using Particle Swarm Optimization

---

Input: VM's and their Physical Nodes (PN's) with their requirements
Output: Virtual Machine Migration to appropriate PN's
Initialize particles
Repeat
For each particle i=1 to n do
    If job requirement<=VM Capacity & CPU Freq_req<=unused CPU Freq
           Calculate the cost for the particle
  End if
    If cost(x_i) < cost ($lBest_i$) then
      $lBest_i = x_i$
    End if
    If cost($lBest_i$) < cost($gBest_i$) then
      $gBest_i = lBest_i$
    End if
End for
For each particle i=1 to n do
    Find the new velocity and particle position
End for {Until Maximum iterations}

3.2 Resource Clustering

A Resource Cluster is formed to group multiple resources (VM's) and allocate a job to it for efficient management of resources, if the resource availability is split across nodes. The job requirements such as memory and processing speed are considered. It finds the status of the VM's including the CPU Frequency and Memory allocated and results the least number of VM's that can efficiently run the job as a cluster. The complexity of the algorithm is O(N) where N is the number of Virtual Machines. The pseudo code for resource clustering is illustrated in Algorithm 2.

Algorithm 2: Resource Clustering

Input: Set of all the VM's without a job allocation, Job requirement
Output: A cluster for the job
For each VM, i=1 to n
   Find the status of the VM
     Sum=0
     If (CPU Frequency = 0)
         Cluster = x_i
     End if
  End for
  For each VM in cluster j=1 to m
    Arrange the VM's in the descending order
  End for
  For each VM in cluster j=1 to m
    Sum += mem_j
  ResCluster = x_j
     If (sum <= jobMem)
        Continue
     End if
  End for
 Return ResCluster

The energy consumption parameter, CPU frequency is monitored to cluster the Idle VM's for forming the Idle Cluster. The resource clustering finds the status of the VM and group the idle VM's so that the cluster accept the jobs to run. The complexity of the algorithm is O(N) where N is the number of VMs.

4 Experimental Setup and Results

The cloud environment is set up using Open Nebula [12]. There is a frontend and two hosts which are connected to the frontend using a cluster. The Open Nebula dashboard in the frontend machine is used to allocate jobs to virtual machines in various hosts. The Resource Allocation process is randomized to any host in the cluster. The processing speed and memory requirements of the host can be directly monitored, while latency and bandwidth are monitored indirectly from the packets sent and received, and the bytes sent and received respectively. The ganglia monitored data is available in.rrd files which are converted into.txt files using the rrd tool. The data in a required time interval (5 min) is extracted from a file into a hash map, as a key-value pair. The four key values such as memory, processor, bandwidth and latency are obtained for all the three hosts and these form the input of the migration using the Particle Swarm Optimization Algorithm 1.

The migration of VM's, to the most appropriate host in terms of it's requirements, is proposed for better optimization results. PSO method is used to predict the efficient

Fig. 2 Response time for
resource allocation

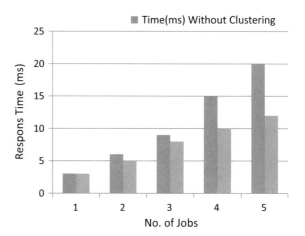

host for the VM. It works iteratively for finding the local and the global best at each step. The number of iterations is fixed according to the number of hosts used in the process. The various velocities with the corresponding fitness values of the particle at that location are obtained. The user is required to provide information on, whether the user job is processor intensive or memory intensive. Accordingly, two different fitness functions are derived as following,

Data intensive:

$$\text{Fitness Value} = (\text{Bandwidth}/8 * \text{Latency}/8)/(\text{Memory}/2 * \text{Processor}/4) \quad (1)$$

Computation intensive:

$$\text{Fitness Value} = (\text{Bandwidth}/8 * \text{Latency}/8)/(\text{Memory}/4 * \text{Processor}/2) \quad (2)$$

The resulting fitness value is obtained in terms of time (seconds) and hence the host with the least fitness value is chosen as the global best solution and the VM is migrated to that host. Depends on the CPU usage, the VM's are sorted in descending order of their memory and then form a resource cluster according to the user requirement. If the CPU is not used at all by the VM, the VM is idle and all such VM's are clustered together to form the Idle Cluster. The clustering of VM's has a positive effect in the process of resource allocation, as the jobs need not wait for idle hosts, but it begins execution in one of the idle clusters. The response time of the resource allocation with and without the clustering approach is shown in Fig. 2.

The experimental results show that the clustering approach reduces the response time of resource allocation, hence increasing the number of jobs done in unit time and the efficiency of job completion. Various user jobs, both data and computation intensive, have been submitted to the proposed system, and the resource allocation has been done quite efficiently which is shown by the number of jobs successfully completed over a period of time in Fig. 3.

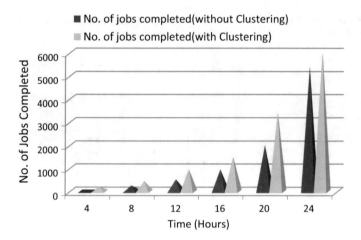

Fig. 3 No. of jobs completed in 24 h

5   Conclusion and Future Work

Cloud computing has enormous usage in today's world, which can be put to better use, if the resources utilized are optimized. A migration strategy based Particle Swarm Optimization has been proposed for the efficient resource utilization. The user may submit computation-intensive jobs or data-intensive jobs. The network parameters like bandwidth and latency have been used in the computation of the fitness function along with the host parameters like memory and processor. The proposed system increases the resource utilization. Clustering has been applied to the idle resources to form resource clusters and idle clusters to minimize the under-utilization of resources. The system can be tested for scalability with varying loads. More network parameters like round trip time and delay can be included to increase the efficiency of resource the allocation strategy.

References

1. Li, S., Zhou, Y., Jieo, L., Yan, X., Wang, X., Lyu, M.R.: Towards operational cost minimization in hybrid clouds for dynamic resource provisioning with delay-aware optimization. IEEE Trans. Serv. Comput. **8**(3), 398–409 (2015)
2. Kang, Z., Wang, H.: A novel approach to allocate cloud resource with different performance traits. In: IEEE International Conference on Services Computing, pp. 128–135 (2013)
3. Liu, J., Su, L., Jin, Y., Li, Y., Jin, D., Zeng, L.: Optimal VM migration planning for data centers. IEEE Global Communications Conference, pp. 2332–2337 (2014)
4. Zheng, J., Ng, T.E., Sripanidkulchai, K., Liu, Z.: Pacer: a progress management system for live virtual machine migration in cloud computing. IEEE Trans. Netw. Serv. Manage. **10**(4), 369–382 (2013)

5. Tao, F., Li, C., Liao, T.W., Laili, Y.: BGMBLA: a new algorithm for dynamic migration of virtual machines in cloud computing. IEEE Trans. Serv. Comput. **9**(6), 910–925 (2016)
6. Sridhar, M., Babu, G.R.M.: Hybrid Particle Swarm Optimization scheduling for cloud computing. In: IEEE Conference on Advance Computing Conference, pp. 1196–1200 (2015)
7. Pan, K., Chen, J.: Load balancing in cloud computing environment based on an improved particle swarm optimization. In: IEEE Conference Publications on Software Engineering and Service Science, pp. 595– 598 (2015)
8. Wang, H., Kang, Z., Wang, L.: Performance-aware cloud resource allocation via fitness-enabled auction. IEEE Trans. Parallel Distrib. Syst. **27**(4), 1160–1173 (2016)
9. Selvi, S.T., Valliyammai, C., Sindhu, G.P., Basha, S.S.: Dynamic resource management in cloud. In: IEEE Sixth International Conference on Advanced Computing, pp. 287–291 (2014)
10. Nahir, A., Orda, A., Raz, D.: Resource allocation and management in cloud computing. In: IEEE International Symposium on Integrated Network Management, pp. 1078–1084 (2015)
11. Valliyammai, C., Uma, S., Surya, P.: Efficient energy consumption in green cloud. In: IEEE International Conference on Recent Trends in Information Technology, pp. 1–4 (2014)
12. OpenNebula. http://archives.opennebula.org (2016)

Review Paper on Cloudlet Allocation Policy

Rachna Anuragi and Manish Pandey

Abstract Cloud computing is the service that enables us to use computing resources such as processing entities, storage, and applications as on-demand over the web. It begins to influence many areas, e.g., government, finance, telecommunications, and education. Cloudlet scheduling is a major issue which is greatly influencing the performance of cloud computing environment. The user requests are given to datacenter broker and data center broker allotted user requests to suitable VM with the assistance of cloudlet allocation policy. So, cloudlet allocation policy must be sufficient to execute user request on VM as early as possible because several users wait to execute their request for accessing cloud services. The main aim is to use the resources effectively and get maximum profit. This paper demonstrates review of an existing cloudlet allocation policy that assists in the allocation of cloudlets on the suitable virtual machines (VMs). It utilizes all offered resources effectively and upgrades the QoS. Cloudlet allocation policy uses CloudSim Toolkit-3.0.3 for their implementation by only changing the desired classes.

Keywords Cloudlet · Quality of service (QoS) · Cloud service provider

1 Introduction

Cloud computing is a pool of resources where cloud user having the flexibility to access services anytime from anywhere. It provides services to cloud user consistent with their demand on pay-per-use basis [1, 2]. Collection of user task is known as cloudlet. The Cloudlets are scheduled on the Virtual Machine (VM) at PaaS layer. Cloud User demanded cloud services on a timely basis, usually in hours or minutes. So the cloudlet algorithm has to be sufficient to use the VM effectively. Allocation of

R. Anuragi (✉) · M. Pandey
Maulana Azad National Institute of Technology, Bhopal 462003, Madhya Pradesh, India
e-mail: rachna49anuragi@gmail.com

M. Pandey
e-mail: contactmanishpandey@yahoo.co.in

© Springer Nature Singapore Pte Ltd. 2019
A. Abraham et al. (eds.), *Emerging Technologies in Data Mining and Information Security*, Advances in Intelligent Systems and Computing 755,
https://doi.org/10.1007/978-981-13-1951-8_29

cloudlets on suitable VMs effectively may be a major drawback in cloud computing. So, need of efficient cloudlet allocation. Cloudlet allocation algorithm which allocate the cloudlet to VM in an optimal way.

Cloudlet allocation policy includes scheduling policy which called as cloudlet scheduling policy that schedules the cloudlets on the VMs. Different type of cloudlet allocation policy available in cloud computing. Datacenter Broker is an important module in cloudlet allocation policy which is responsible for processed user request to resources or VMs and also improving the Quality of Service (QoS) of the complete system [3, 4]. Cloudlet allocation policy is selected on the premise of different parameter like cloudlet size, VM size, cloudlet priority, cloudlet arrival time. Essential work of cloudlet allocation policy is to upgrade the performance of overall system as reducing the completion time alongside make span of VM.

2   Scheduling

Cloudlet scheduling is that the methodology of assigning the cloudlets to the VM for execution of cloudlet. Datacenter Broker policy decides that scheduling policy used for execution of cloudlets [3]. The QoS measurements for cloudlet scheduling are as per the following: Response time, TurnAround time, Execution time, Throughput, Resource Utilization and makespan.

2.1   Types of Cloudlet Scheduling

Cloudlet Scheduling is classified in numerous ways that primarily based on the idea within which problem is solved. Different types of cloudlet scheduling are as follows [5].

2.1.1   Static Scheduling

In static scheduling, all information about cloudlets and VMs are known before the starting of execution.

2.1.2   Dynamic Scheduling

In dynamic scheduling, all information about cloudlets and VMs are not known before the starting of execution. This scheduling depends on all the information about cloudlet and VM provided earlier and additionally the present standing of the system.

3 Literature Review

3.1 Min–Min [6]

It starts with all unallocated cloudlets. It complete its execution in two steps. In the first step, the Minimum completion time for all cloudlet on all VMs is computed. In the second step, choose the cloudlet with the smallest completion time among all cloudlets and allocate that cloudlet on the VM on which it takes less time to process and availability status of that VM is shared to all the remaining cloudlets [3]. A main problem with the Min-min algorithmic is to assign the smaller jobs on the VM with relatively extreme computational power.

3.2 Max–Min [6]

It is same as Min–Min, however a definite difference present within the second step. This Max–Min, computes the minimum execution time for each cloudlet on all VMs. Then choose the cloudlet that has maximum value of the execution time among all cloudlets having minimum execution time on any VM and allocate that cloudlet on the virtual machine on that it takes the minimum time and availability status of that VM is shared to all the remaining cloudlets. Within the Max-min algorithm, the cloudlet having low MI value may wait for large cloudlet to be executed.

3.3 Round Robin Allocation (RRA) Policy

During this policy, All cloudlets are sent to the Datacenter broker and then broker distributes the cloudlets on the most readily accessible available VMs [7]. For example, Assume there are four cloudlets (CL_1, CL_2, CL_3, CL_4) and two VMs (VM_1, VM_2) want to assign that four cloudlet to two VM. Table 1 [8] highlights the allocation style. In step with this policy, cloudlet allocated to VM_1, CL_2 allocated to VM_2 and CL_3 allocated to VM_1, because VM_1 get earlier available than VM_2. All cloudlets which are remaining left, allocated on the available VM within the same manner as above. RRA is not efficient when large size cloudlet allocated to smaller size VM, it takes more time to execute the cloudlet and Because of this expanding the waiting time and therefore the response time of the cloudlets. Apart from that it may also happen when large MIPS VMs get the smaller cloudlets and hence its resource utilization become useless and at the same moment decreasing the overall performance.

Table 1 Shows the RRA policy style

Cloudlet	Virtual machine
CL_1	VM_1
CL_2	VM_2
CL_3	VM_1
CL_4	VM_2

3.4 Rasa

This algorithm joins the benefits of Min–min as well as Max-min. The Min-Min algorithm is used when the quantity of accessible VM is odd [9]. Otherwise, Max-Min algorithm is employed. The complete procedure can be split into several stages wherever in each stage two cloudlets are assigned to suitable VMs by one of the two policies, instead. The rule is, if the Min-Min algorithm is applied to assign the primarily accessible cloudlet in the present stage then next cloudlet of that stage will be assigned on the VM by utilizing Max-Min algorithm [9].

3.5 Conductance Algorithm

This algorithm maintains the processing power of every VM as a pipe. It computes the Conductance (processing capacity) according to Eq. (1) for every VM. The conductance of VM is that the proportion of its MIPS value to the adding MIPS value of every VMs within the system [10].

$$Conductance_i = \text{MIPS}_j / \sum_{i=1}^{n} \text{MIPS}_j \qquad (1)$$

After computation of conductance for all VM, find the strip length for all VMs by using Eq. (2). It decides the amount of cloudlets the VM can process.

$$Striplength_j = Conductance_j \times (Length\, of\, cloudlet\, list) \qquad (2)$$

The main drawback of conductance algorithm is that the low MIPS VMs occasionally get free too rapidly thus wasting its resources and the high MIPS VMs occasionally get overloaded when the length of greatest cloudlets are very large.

3.6 The Opportunistic Load Balancing (OLB)

OLB algorithm is applied to keep every VM busy without remembering that what amount of cloudlets already offered within the VM. The Opportunistic Load Balancing algorithm [7] utilizes all VM maximum amount as attainable by assigning cloudlets on the VM. OLB chooses a cloudlet from the cloudlet list randomly and allocated it to the succeeding VM which seems to be available, without considering the cloudlets execution on that VM either efficient or not.

3.7 Modified Shortest Job First

In this algorithm, maintain the queue of cloudlets in an increasing order on the premise of cloudlet size and calculate the mean of the cloudlet size for all cloudlets [11]. It examines cloudlet size for all cloudlets, if cloudlet size is a smaller amount than the mean of cloudlet size and variety of cloudlets in VM1 less than the variety of cloudlets in VM2, then the cloudlet is going to be allotted to VM1, else to VM2. And at last, calculate the average response time and makespan of every VM. MSJF contains many parameters which are as follows:

Mean of cloudlet size ($Mean.clsize$) are often calculable by using Eq. (3). This algorithm has enhanced the completion time and makespan of the VMs similarly the host(tos). This algorithm has enhanced the completion time and makespan of the VMs similarly to the host(s). Wherever VMP is the processing power of VM. Completion Time (CT) is that the total of the ExT of all the earlier cloudlet and ExT_m of the current cloudlet assigned to the similar VM as indicated in the Eq. (5).

Makespan is that the completion time of all cloudlets. [10] it is defined in Eq. (6). Where CT_n is the completion time of last the cloudlet. Response Time (RT) It is calculated by using Eq. (7). CT: completion time of cloudlet, SB: Submission time of cloudlet. Then calculate the average of response time for each VM using Eq. (8).

$$Mean.clsize = \sum_{i=1}^{n} CL_1/n \tag{3}$$

$$ExT = CL/VMP \tag{4}$$

$$CT = ExT_m + \sum_{i=1}^{m-1} ExT \tag{5}$$

$$Makespan = CT_n \tag{6}$$

$$RT = \sum_{i=1}^{n} CT_i + SB_i \tag{7}$$

$$Avg.RT = RT/N \tag{8}$$

3.8 New Cloudlet Allocation Policy [10]

In this cloudlet allocation policy, all the VMs are maintained in dropping order on the idea of their MIPS (Million Instruction per Second) and keep in vm_list [12]. Then, compute the maximum load capacity (MLC) of every VM which is calculated by Eq. (9) [10]. The cloudlet size of each cloudlet (cl_size) is estimated by using Eq. (10). Then, VM which has the highest capacity will be assigned first.

$$MLC(\%) = (MIPS_j/total\ MIPS\ of\ all\ VM) \times 100 \qquad (9)$$

$$clsize_i = (MI_i/Total\ MI\ of\ all\ cloudlet) \times 100 \qquad (10)$$

the And after assignment of highest capacity VM, the remaining load capacity (RLC) of that VM will be computed and compare that RLC value with the MLC of the other VMs within the vm_list. If the RLC value is greater than the MLC value of other alternative VMs present within the vm_list, then that VM could resume with any cloudlet till RLC value becomes lesser than the MLC value of the other VMs. This procedure can continue till all the cloudlets in cloudlet_list can get allotted into the VMs [10]. This algorithm has enhanced the completion time and makespan of the VMs similarly to the host(s).

$$RLC_i(\%) = MLC_i - clsize_i \qquad (11)$$

3.9 Range Wise Busy-Checking 2-Way Balanced

The RB2B algorithm allocates cloudlet to an acceptable VM on the premise of cloudlets size and cloudlets are distributed equally to VM for reducing the load on particular VM. This algorithm optimizing the resource utilization likewise as a finish time. It completes its process in three stages. The stages are as follows: (a) VM categorization stage, (b) Two round busy-checking stage and (c) Cloudlet-still not-allocated (CSNA) stage. There is a condition that balance the distribution of cloudlets among the VM as equally as possible, known as two-way balancing conditions.

3.9.1 Phases of RB2B [8]

VM Categorization Phase

In this phase, ADCB is a module by which RB2B algorithm is implemented. Initially, ADCB describes the cloudlet acceptableness length for each VM which is described in Table 2[10]. All VM are arranged in upward order on the premise of their MIPS. And then ADCB selects an acceptable VM for cloudlet per the cloudlet length.

Table 2 Shows the VM acceptability range

VM	Lower limit	Upper limit
VM_0	C_{min}	$C_{min} + f_0 x$
$VM_1 \ldots$	$C_{min} + f_0 x + 1 \ldots$	$C_{min} + f_0 x + f_1 x \ldots$
$VM_m \ldots$	$C_{min} + f_0 x + f_1 x + \cdots + f_{m-1} + 1 \ldots$	$C_{min} + f_0 x + f_1 x + \cdots + f_m x \ldots$
VM_{n-1}	$C_{min} + f_0 x + f_1 x + \cdots + f_{n-2} x + 1$	$C_{min} + f_0 x + f_1 x + \cdots + f_{n-1} x = C_{max}$

Selected VM is called as target VM at this stage. wherever C_{min} and C_{max} are the smallest and largest cloudlet length (In MI). f_i denote the MIPS of VM_i.

Two Round Busy-Checking Stage

In this stage, ADCB tests whether that targeted VM is free or not. If that targeted VM is free then ADCB will test the condition of Balance threshold. If the condition has been fulfilled by the VM then that VM get allocated by cloudlet the Else,

ADCB finds for the other alternative VMs by the VM then that VM get allocated by cloudlet. Else, ADCB finds for the other alternative VMs by the following two steps: (i) If the targeted VM does not contain maximum MIPS, then ADCB examines whether the succeeding VM with higher MIPS is available and whether or not it meets Balance threshold condition. If this VM fulfill Balance threshold condition, hen the cloudlet is allotted to it. (ii) In this step, testing the lowest MIPS VM till the cloudlet is allotted to an acceptable VM.

Cloudlet-Still-not-Allocated Stage (CSNA)

It is the last stage of RB2B. Once finishing the two stages if arriving cloudlet allotted to a VM that fulfilling the balance factor condition and so ADCB can test the next cloudlet. If the coming cloudlet is still not assigned then the ADCB can take a look at the earliest finish time of VM for that cloudlet, and follows the two-way balance conditions.

4 CloudSim

CloudSim is a simulation toolkit developed by the GRIDS laboratory of University of Melbourne that implement basic features of cloud [13]. It permits user to test and analyze their new approaches for generation and allocation of VM to cloudlets for execution. Cloudsim 3.0.3 consists various classes for implementation of VM schedulers, Datacenters, Cloudlet scheduler, Datacenter Broker and VM allocation policies. CloudSim 3.0.3 toolkit has some important module which are as follows.

4.1   Cloud Information Service (CIS)

CIS allocates user request to suitable cloud provider. It contains all information about entities like cloudlets, virtual machine and so on.

4.2   Cloudlet

It is a collection of user tasks. A cloudlet is dispatch from the user for processing to the DC. Each cloudlet is measured in MI (Million Instruction). Each cloudlet has various fields like cloudlet ID, cloudlet length, time of arrival etc.

4.3   Virtual Machine (VM)

Virtual machine run inside the host by using various resources of hosts. Cloudlets schedule within virtual machine by using scheduling policy which is determined by datacenter broker. Process power of virtual machine is measured in MIPS (Million Instruction Per Second).

4.4   Datacenter Broker (DCB)

Datacenter broker act as a mediator between cloud provider and cloud user. It take care of user requirements. Scheduling algorithms and cloudlet allocation policies are executed by Datacenter Broker method.

4.5   VM Scheduler

VM scheduler is a class in cloudsim. It is executed by a Host component. It schedules processing element of the host to virtual machines.

5 Conclusion

In this review paper, we have surveyed several cloudlet allocation policies in cloud computing. The main aim of cloudlet allocation policy is that the reduction of completion time similarly as makespan of VM. The upgradation of QoS rely on the choice of cloudlet scheduling policy. From the literature survey, cloudlet allocation policy has covered several factors like execution time, cost, response time, flow time, throughput and resource utilization. however still enhancement desired in some factors e.g. makespan, time and space complexity and execution cost. So, Using Dynamic Programming to develop effective cloudlet allocation policy that allotted user requests to the accessible virtual machine in an optimal way that may optimize several QoS factor e.g. completion time, execution cost, makespan and so on.

References

1. Foster, I., et al.: Cloud computing and grid computing 360-degree compared in grid computing environments workshop. In: GCE'08. IEEE (2008)
2. Ahmed, M., et al.: An advanced survey on cloud computing and state-of-the-art research issues. Int. J. Comput. Sci. Issues (IJCSI) (2012)
3. Sindhu, S.: Task scheduling in cloud computing. Int. J. Adv. Res. Comput. Eng. Technol. **46**, 3019–3023 (2016)
4. Lei, X., Zhe, X., Shao Wu, M., Xiong Yan, T.: Cloud computing and services platform construction of telecom operator. In: Broadband Network & Multimedia Technology, IC-BNMT'09. 2nd IEEE International Conference on Digital Object Identifier, pp. 864–867 (2009)
5. Geetha, V., et al.: Performance comparison of cloudlet scheduling policies. In: International Conference on Emerging Trends in Engineering, Technology and Science (CENTERS). IEEE (2016)
6. Etminani, K., Naghibzadeh, M.: A min-min max-min selective algorithm for grid task scheduling. In The Third IEEE/IFIP International Conference on Internet, Uzbekistan (2007)
7. Banerjee, S., et al.: An approach toward amelioration of a new cloudlet allocation strategy using cloudsim. Arab. J. Sci. Eng. 1–24 (2017)
8. Banerjee, S., et al.: Development and analysis of a new cloudlet allocation strategy for QoS improvement in the cloud. Arab. J. Sci. Eng. **40**(5), 1409–1425 (2015)
9. Parsa, S., Reza E.-M.: RASA: a new task scheduling algorithm in a grid environment. World Appl. Sci. J. (Special issue of Computer & IT) **7**, 152–160 (2009)
10. Roy, S., et al.: Development and analysis of a three-phase cloudlet allocation algorithm. J. King Saud Univ.-Comput. Inf. Sci. (2016)
11. Al Warafi, M.A., et al.: An improved SJF scheduling algorithm in cloud computing environment. In 2016 International Conference on Electrical, Electronics, Communication, Computer and Optimization Techniques (ICEECCOT). IEEE (2016)
12. Chatterjee, T., et al.: Design and implementation of an improved datacenter broker policy to improve the QoS of a cloud. In: Proceedings of the Fifth International Conference on Innovations in Bio-Inspired Computing and Applications IBICA 2014. Springer, Cham (2014)
13. Wickremasinghe, B., Calheiros, R.N., Buyya, R.: Cloud analyst: a cloudsim-based visual modeler for analyzing cloud computing environments and applications. In: 2010 24th IEEE International Conference on Advanced Information Networking and Applications (AINA), pp. 446–452. IEEE (2010)

Part III
Computational Mathematics

Modelling and Analysis of Bio-convective Nano-fluid Flow Past a Continuous Moving Vertical Cylinder

Debasish Dey

Abstract Analysis of two-dimensional boundary layer flow of nano-fluid past a continuous moving cylinder has been carried out under the influence of bio-convection. The fluid flow consists of nano-sized particles and gyrotactic micro-organisms. Micro-organism is induced to stabilize the movement of nano-sized particles. Mathematical modelling of the fluid flow has been done by using conservation principles of mass, momentum, energy, species concentration, and micro-organism concentration respectively. Similarity transformations have been used to convert partial differential equations into ordinary differential equations. Equations are solved numerically using MATLAB built in bv4pc solver technique. Results are discussed graphically and numerically for various flow parameters involved in the equation.

Keywords Brownian diffusion · Gyrotactic micro-organism · Bio-convection
Nano-fluid · Peclet number

1 Introduction

One of the major advances in the studies of fluid flow is the mechanics of nano-fluid flow. Nano-fluid is a medium for heat transfer as it increases the conductivity of fluid flow. It is a mixture of nano-sized particles in a base fluid and often termed as super coolants due to its high heat assimilation ability. Choi [6] is the pioneer of nano-fluid flow, he developed it by adding nano-sized particles in a base fluid. Application of nano-fluid is seen in power generation, various chemical processing, nano-technology etc. Recently various research works [3, 11, 15, 17, 21, 25] have characterize the flow behaviour of nano-fluid using various geometries.

Bio-convection is the convection of fluid motion due to density gradient caused by swimming of micro-organisms. Micro-organisms may be used to enhance biodegradable polymeric nanomaterials and improve various medical characteristics like

D. Dey (✉)
Department of Mathematics, Dibrugarh University, Dibrugarh 786004, India
e-mail: debasish41092@gmail.com

© Springer Nature Singapore Pte Ltd. 2019
A. Abraham et al. (eds.), *Emerging Technologies in Data Mining and Information Security*, Advances in Intelligent Systems and Computing 755,
https://doi.org/10.1007/978-981-13-1951-8_30

331

bioavailability, biocompatibility, encapsulation, DNA embedding in gene therapy, protein deliverability etc. Examples of bio-nano-polymers are polylactic acid, chitosan, gelatin, poly caprolactone and poly alkyl cyano-acrylate are seen in [25]. In bio-convection process, the motion of micro-organisms induces a convective motion in the fluid. The oxytactic micro-organisms lead to fluidic convection which transports cell and oxygen. Problem of mixed bio-convective flow of nano-fluid over a solid sphere in a porous medium has been studied in [24] using Keller Box method. Free convective boundary layer flow of nano-fluid past a horizontal flat surface containing gyrotactic micro-organisms has been analysed in [4]. Behaviour of bio-convective flow of nano-fluid through horizontal channel using homotopy analysis method (HAM) has been investigated in [26]. Analysis of nano-fluid flow containing micro-organism has been carried out by other researchers in [1, 2, 7–10, 12, 13, 16, 19, 20, 22, 23].

The objective of the present study is to find out the effects Brownian diffusion and theromo-phoresis with Peclet number on bio-convective nano-fluid flow past a vertical cylinder consisting micro-organisms. The governing equations of motion are solved numerically using MATLAB built in solver bv4pc technique. This work is relevant to the application of bio-nano-polymers used in manufacturing processes [5] and in various food processing industries. Poor mechanical and barrier properties of bio-polymer leads to the inclusion of nano-particles to form composites [18]. Bio-nano-particle may be used as a promising alternative food packaging applications instead of conventional plastics [18].

2 Mathematical Formulations

A steady two-dimensional boundary layer flow governed by nano-fluid past a vertical cylinder (moving continuously) of radius a has been investigated. A cylindrical polar co-ordinate system (r, θ, z) has been used to study the problem. The motion is taken along the z-axis. Let $(u, 0, w)$ be velocity of the governing fluid motion. Let T, C and n be temperature, concentration of nano-particles and concentration of micro-organism respectively. Temperature, nano-particle concentration and density of micro-organisms at the free stream are denoted by T_∞, C_∞ and n_∞ respectively. The geometry of the problem is given by Fig. 1. To study the governing fluid motion, we have used the following assumptions [14]:

- Due to axial symmetry, the system is independent of θ, i.e., $\frac{\partial}{\partial \theta}() = 0$.
- There is a continuous motion of cylinder with uniform velocity U.
- At the stretching surface of the cylinder, temperature of fluid, nano-particle concentration and motile micro-organism density are assumed constant and greater than temperature, nano-particle concentration and density of motile micro-organism at free stream respectively.
- It is assumed that the nano-particles don't have any effect on motion of micro-organisms.

Fig. 1 Geometry of the problem

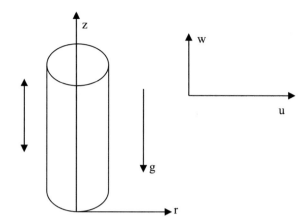

- All the fluidic properties are assumed constant except the variation of density in the buoyancy term (Boussinesq Approximation).
- Fluid and nano-particles are in thermal equilibrium.
- Relative velocities between motile micro-organisms, nano-particles and fluid are assumed to be zero.

Using the above assumptions, the governing equations of fluid motion are given as follows:

$$\frac{\partial u}{\partial r} + \frac{\partial w}{\partial z} + \frac{u}{r} = 0, \tag{1}$$

$$u\frac{\partial u}{\partial r} + w\frac{\partial u}{\partial z} = -\frac{1}{\rho}\frac{\partial p}{\partial r} + v\left(\frac{\partial^2 u}{\partial r^2} + \frac{1}{r}\frac{\partial u}{\partial r} - \frac{u}{r^2}\right)$$

$$+ g\beta(1 - C_\infty)(T - T_\infty) - g\frac{(\rho_p - \rho)}{\rho}(C - C_\infty)$$

$$- g\gamma\frac{(\rho_m - \rho)}{\rho}(n - n_\infty), \tag{2}$$

$$u\frac{\partial T}{\partial r} + w\frac{\partial T}{\partial z} = \alpha\left(\frac{\partial^2 T}{\partial r^2} + \frac{1}{r}\frac{\partial T}{\partial r}\right)$$

$$+ \tau\left[D_B\left(\frac{\partial T}{\partial r}\frac{\partial C}{\partial r}\right) + \frac{D_T}{T_\infty}\left(\frac{\partial T}{\partial r}\right)^2\right]$$

$$+ Q(T - T_\infty), \tag{3}$$

$$u\frac{\partial C}{\partial r} + w\frac{\partial C}{\partial z} = D_B\left(\frac{\partial^2 C}{\partial r^2} + \frac{1}{r}\frac{\partial C}{\partial r}\right) + \frac{D_T}{T_\infty}\left[\frac{\partial^2 T}{\partial r^2} + \frac{1}{r}\frac{\partial T}{\partial r}\right], \tag{4}$$

$$u\frac{\partial n}{\partial r} + w\frac{\partial n}{\partial z} + \frac{bW_c}{C_\infty}\left[\frac{\partial}{\partial r}\left(n\frac{\partial C}{\partial r}\right)\right] = D_n\left(\frac{\partial^2 n}{\partial r^2} + \frac{1}{r}\frac{\partial n}{\partial r}\right), \tag{5}$$

where, ρ, ρ_1 and ρ_2 are densities of fluid, nano-particles and micro-organisms respectively, v viscosity of fluid, g acceleration due to gravity, β co-efficient of volume expansion, γ average volume of micro-organisms, β^* co-efficient of volume transfer due to mass transfer, α thermal diffusivity, τ ratio of specific heat capacities of nano-particles and fluid respectively, D_B diffusivity of nano-particles, D_T thermophoretic diffusion co-efficient, W_c maximum cell swimming speed, D_n micro-organism diffusion co-efficient.

The boundary conditions of the problem for solving the Eqs. (2–5) along with the continuity Eq. (1) are

$$\left.\begin{array}{l} u = 0, w = U, T = T_w, C = C_w, n = n_w \; at \; r = a \\ w \to 0, T \to T_\infty, C \to C_\infty, n \to n_\infty, \; at \; r \to \infty \end{array}\right\}, \quad (6)$$

3 Method of Solution

To solve the above equations subject to the boundary conditions given in (6), similarity transformation and the similarity variables are given as follows:

$$u = -\frac{1}{2}U\frac{f(\eta)}{\sqrt{\eta}}, w = Uf', \theta = \frac{T - T_\infty}{T_w - T_\infty}, \phi = \frac{C - C_\infty}{C_w - C_\infty}, \eta = \frac{r^2}{a^2}, \chi = \frac{n - n_\infty}{n_w - n_\infty}, \quad (7)$$

The set of ordinary differential equations using (7) are

$$-3Rff'\eta + 3Rf^2 + 3\frac{G}{Re^2}(\theta - Nr\phi - Rb\chi)\eta\sqrt{\eta} + 10f = 24f''\eta^2, \quad (8)$$

$$4\eta\theta'' + 4\theta' + 4RPrNb\theta'\phi'\eta + 4RPrNt\theta'^2\eta + Q_HRPr\theta + RPrf\theta' = 0, \quad (9)$$

$$4\eta\phi'' + 4\phi' + 4\frac{Nt}{Nb}(\theta' + \theta''\eta) + f\theta'RSc = 0, \quad (10)$$

$$-RSnf\chi + 2Pe(2\eta\chi'\phi' + \chi\phi' + 2\chi\phi''\eta) = 4\chi' + 4\chi''\eta, \quad (11)$$

Corresponding boundary conditions for solving the Eqs. (8–11) are

$$f = 0, f' = \theta = \phi = \chi = 1 \; at \; \eta = 1 \; \& \; f' = \theta = \phi = n \to 0 \; at \; \eta \to \infty, \quad (12)$$

where,

$R = \frac{Ua}{v}$ Reynolds number,

$\frac{G}{R^2} = \frac{4ag\beta}{U^2}(T_w - T_\infty)(1 - C_\infty)$ Richardson number,

$Pr = \frac{v}{\alpha}$ Prandtl number,

$Nr = \frac{(\rho_1 - \rho)(C_w - C_\infty)}{\rho\beta(T_w - T_\infty)(1 - C_\infty)}$ Buoyancy ratio,

$Pe = \frac{bW_c(C_w - C_\infty)}{D_n}$ Peclet number,

$Sc = \frac{\nu}{D_B}$ Schmidt number,

$Sn = \frac{\nu}{D_n}$ Schmidt number for micro-organisms,

$Rb = \frac{\gamma(\rho_2-\rho)(n_w-n_\infty)}{\rho\beta(T_w-T_\infty)(1-C_\infty)}$ Rayleigh number(bio-convection),

$Nb = \frac{\tau D_B(C_w-C_\infty)}{Ua}$ Brownian diffusion parameter,

$Nt = \frac{\tau D_T(T_w-T_\infty)}{T_\infty Ua}$ thermo-diffusion parameter,

$Q_H = \frac{Q}{Ua}$ external heat agent

4 Results and Discussions

The steady two-dimensional boundary layer flow of nano-fluid past a continuously moving vertical cylinder has been studied. The fluid motion is influenced by the presence of bio-convection with heat and mass transfer. The objective of the present study is to find out the effects of Peclet number, Brownian diffusion and thermal diffusion on governing fluid flow. Equations (7–10) are solved numerically using MATLAB built in bvp4c solver technique subject to the boundary conditions (11). Results are analysed graphically and numerically (in tabular form) with a fixed set of parameters R = 0.2, = 0.5, Pr = 7, $\frac{Gr}{R^2}$ = 10, Q_H = 2, Sc = 0.7 and Sn = 0.5. We have taken positive values of Q_H (=2) for external heat source. From the application point of view, rate of heat, mass and motile organism transfer are very important. Their dimensionless forms: Nusselt number or rate of heat transfer (12), Sherwood Number or rate of mass transfer (13) and transfer rate of motile organisms (14) at the surface are given as

$$NU = \left(-\frac{\partial\theta}{\partial\eta}\right)_{\eta=0}, \tag{12}$$

$$SH = \left(-\frac{\partial\phi}{\partial\eta}\right)_{\eta=0}, \tag{13}$$

$$MT = \left(-\frac{\partial\chi}{\partial\eta}\right)_{\eta=0}, \tag{14}$$

Figures 2 represent the velocity profile against the displacement variable for various values of Peclet number and bio-convective Rayleigh number. Peclet number characterizes the cell swimming speed and the higher values of Peclet number enhance the speed. Growth in cell swimming speed of micro-organisms accelerates the fluid motion. There is a growth in 33.33% approximately during the enhancement of Peclet number by 15 times. Also, it is revealed that fluid is accelerated in the neighbourhood of the surface then speed slows down and gradually it tends to zero in the free stream region. Rayleigh number (Rb) characterizes the effects of bio-convection against the convection due to Buoyancy force and its increasing values retard the fluid motion. It may be interpreted that the inertia force of the fluid

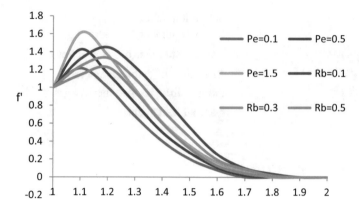

Fig. 2 Velocity profile against displacement variable for $R = 0.2$, $\frac{Gr}{R^2} = 10$, $Nr = 0.5$, $Pr = 7$, $Nb = 0.2$, $Nt = 0.1$, $Q_H = 2$, $Sc = 0.7$, $Sn = 0.5$

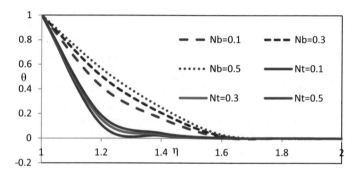

Fig. 3 Temperature field against displacement variable for $R = 0.2$, $\frac{Gr}{R^2} = 10$, $Nr = 0.5$, $Pr = 7$, $Rb = 0.3$, $Pe = 0.5$, $Q_H = 2$, $Sc = 0.7$, $Sn = 0.5$

motion overshadowed by bio-convection. There is a reduction in speed by 15.48% approximately during the enhancement of Rb by 400% approximately.

Figure 3 represent the variation of temperature against the displacement variable for various values of Brownian diffusion (Nb) and thermal diffusion (Nt). Nb characterizes the random movement of nano-particles and its higher value increases the random motion. These random motions generate the heat into the system. As a result, there is an increment in temperature by 34.61% with the rise in Brownian diffusion parameter by 400% approximately. A growth in temperature in temperature by 68.10% is observed during 400% growth in thermal diffusion. Physically it may be interpreted that during thermal diffusion, particles are moved away from hotter region to cooler region. This movement of particles produces heat and temperature rises. Also, it is noticed that temperature decreases as we move gradually from the stretching surface.

Figure 4 shows that behaviour of concentration of nano-particles against the displacement variable. It is seen that concentration of the nano-particles is higher in the

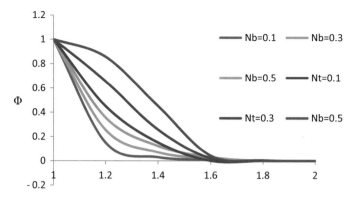

Fig. 4 Concentration field of nano-particles against displacement variable for R = 0.2, $\frac{Gr}{R^2}$ = 10, Nr = 0.5, Pr = 7, Rb = 0.3, Pe = 0.5, Q_H = 2, Sc = 0.7, Sn = 0.5

neighbourhood of the surface and then it gradually decreases as we proceed towards the free stream region. As we have seen that, during the enhancement of Nt, more and more particles will move from heated region to cooled one. Thus there is a significant enhancement in concentration of the nano-particles by 136.02% approximately with the increase in thermo-diffusion parameter (Nt) by 400% approximately. An similar phenomenon (Growth in concentration by 89.92% approximately) is observed during 400% growth of Brownian diffusion parameter(Nb).

Effects of Peclet number and Brownian diffusion on motile density profile are shown by Fig. 5. It is observed that motile density follows a decline trend with displacement variable. Peclet number is inversely proportional to the motile diffusivity, so the increasing values of Peclet number reduces the density of micro-organism. There is a reduction in 27.78% approximately during the growth of Peclet number from 0.1 to 1. A significant drop in motile density is experienced during the rise in Brownian diffusion.

Table 1 represents the effects of various fluid flow parameters on skin friction (ST), rate of heat transfer (NU), rate of mass transfer (SH) and transfer rate of gyrotactic micro-organism (MT) respectively. The fluid flow on the stretching surface of the cylinder forms viscous drag or skin friction. There is an increment in shearing stress at the surface during the enhancement of bio-convective Rayleigh number, Brownian diffusion and thermal diffusion but an opposite trend is noticed during the growth of Peclet number. This encourages in selecting the fluidic properties in such a manner that the viscous drag at the surface is lesser, so that there is less damage at the surface. Nusselt number describes the rate of heat transfer at the surface and it is reduced by the enhancement of Brownian diffusion and thermo-diffusion. There is an enhancement of Sherwood number with the increase of Brownian diffusion. This motivates us to choose the fluidic properties in such a way to get the required nature of rate of heat transfer and rate of mass transfer. Increment in Peclet number enhances the transfer rate of gyrotactic micro-organisms.

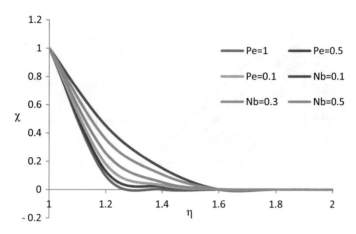

Fig. 5 Density of motile micro-organisms against displacement variable for R $= 0.2$, $\frac{Gr}{R^2} = 10$, $Nr = 0.5$, Pr $= 7$, $Rb = 0.3$, $Nt = 0.2$, $Q_H = 2$, Sc $= 0.7$, Sn $= 0.5$

Table 1 Numerical representation of Shearing stress or viscous drag at the surface (St), Nusselt number or rate of heat transfer (Nu), Sherwood number or mass transfer rate (Sh), transfer rate of density of motile micro-organisms (Mt) for R $= 0.2$, $Nr = 0.5$, Pr $= 7$, $\frac{Gr}{R^2} = 10$, $Q_H = 2$, Sc $= 0.7$, Sn $= 0.5$

Cases	Nb	Nt	Pe	Rb	St	Nu	Sh	Mt
I	0.1	0.1	0.1	0.1	0.8811	0.1866	0.1039	1.6459
II	0.2	0.1	0.1	0.1	1.2201	0.1513	0.1473	———
III	0.1	0.2	0.1	0.1	1.3302	0.1439	———	———
IV	0.1	0.1	0.5	0.1	———	———	———	1.7861
V	0.1	0.1	0.1	0.3	1.0202	———	———	———

5　Conclusions

A numerical investigation on bio-convective flow with nano-particles through a vertical cylinder moving continuously has been carried out. From this study, the following conclusions are prepared:

- It is possible to stabilize the fluid motion by reducing its speed with Bio-convective Rayleigh number.
- Micro-organisms swimming speed helps to accelerate the fluid motion, so this encourages to reduce the resistance of viscosity
- Brownian diffusion and thermal diffusion enhance temperature of fluid motion through the generation of heat.
- heat transfer rate is controlled by Brownian and thermal diffusion parameters.
- There is a reduction in motile micro-organisms density during the growth of Peclet number.

References

1. Akbar, N.S., Khan, Z.H.: Magnetic field analysis in a suspension of gyrotactic microorganisms and nanoparticles over a stretching surface. J. Magn. Magn. Mater. **410**, 72–80 (2016)
2. Alsaedi, A., Khan, M.I., Farooq, M., Gull, N., Hayat, T.: Magnetohydrodynamic (MHD) stratified bioconvective flow of nanofluid due to gyrotactic microorganisms. Adv. Powder Technol. **28**, 288–298 (2017)
3. Ashorynejad, H.R., Sheikholeslami, M., Pop, I., Ganji, D.D.: Nanofluid flow and heat transfer due to a stretching cylinder in the presence of magnetic field. Heat Mass Transfer. **49**, 427–436 (2013)
4. Aziz, A., Khan, W.A., Pop, I.: Free convection boundary layer flow past a horizontal flat plate embedded in porous medium filled by nanofluid containing gyrotactic microorganisms. Int. J. Therm. Sci. **56**, 48–57 (2012)
5. Basir, M.F.M., Uddin, M.J., Ismail, A.I.M., Beg, a.: nanofluid slip flow over a stretching cylinder with schmidt and péclet number effects. AIP Adv. http://dx.doi.org/10.1063/1.4951675
6. Choi, S.U.S., Eastman, J.A.: enhancing thermal conductivity of fluids with nanoparticle. In: ASME International Mechanical Engineering Congress & Exposition, San Francisco, 12–17 Nov 1995. https://www.osti.gov/scitech/servlets/purl/196525
7. Khan, W.A., Uddin, M.J., Ismail, A.I.M.: Free convection of non-newtonian nanofluids in porous media with gyrotactic microorganisms. Transp. Porous Med. **97**, 241–252 (2013)
8. Kuznetsov, A.V.: The onset of nanofluid bioconvection in a suspension containing both nanoparticles and gyrotactic microorganisms. Int. Commun. Heat Mass Transfer **37**, 1421–1425 (2010)
9. Kuznetsov, A.V.: Nanofluid biothermal convection: simultaneous effects of gyrotactic and oxytactic micro-organisms. Fluid Dyn. Res. **43**, 055505 (2011). https://doi.org/10.1088/0169-5983/43/5/055505
10. Mahdy, A.: Gyrotactic microorganisms mixed convection nanofluid flow along an isothermal vertical wedge in porous media. Int. Sch. Sci. Res. Innov. **11**(4), 840–850 (2017)
11. Makinde, O.D., Aziz, A.: Boundary layer flow of a nanofluid past a stretching sheet with a convective boundary condition. Int. J. Therm. Sci. **50**, 1326–1332 (2012)
12. Makinde, O.D., Animasaun, I.L.: Bioconvection In MHD nanofluid flow with nonlinear thermal radiation and quartic autocatalysis chemical reaction past an upper surface of a paraboloid of revolution. Int. J. Therm. Sci. **109**, 159–171 (2016)
13. Martínez, M.S.H., Smyth, W.D.: Trapping of gyrotactic organisms in an unstable shear layer. Cont. Shelf Res. **36**, 8–18 (2012)
14. Mehryan, S.A.M., Kashkooli, F.M., Soltani, M., Raahemifar, K.: Fluid flow and heat transfer analysis of a nanofluid containing motile gyrotactic micro-organisms passing a nonlinear stretching vertical sheet in the presence of a non-uniform magnetic field; numerical approach. PLoS One (2016). https://doi.org/10.1371/journal.pone.0157598
15. Murthy, P.V.S.N., Reddy, C.R., Chamkha, A.J., Rashad, A.M.: Magnetic effect on thermally stratified nanofluid saturated non-darcy porous medium under convective boundary condition. Int. Commun. Heat Mass Transfer **47**, 41–48 (2013)
16. Mutuku, W.N., Makinde, O.D.: Hydromagnetic bioconvection of nanofluid over a permeable vertical plate due to gyrotactic microorganisms. Comput. Fluid **95**, 88–97 (2014)
17. Mustafa, M., Hayat, T., Alsaedi, A.: Unsteady boundary layer flow of nanofluid past an impulsively stretching sheet. J. Mech. **29**, 423–432 (2013)
18. Othman, S.H.: Bio-nanocomposite materials for food packaging applications: types of biopolymer and nano-sized filler. Agric. Agric. Sci. Proced. **2**, 296–303 (2014)
19. Raees, A., Raees-ul-Haq, M., Xu, H., Sun, Q.: Three-dimensional stagnation flow of a nanofluid containing both nanoparticles and microorganisms on a moving surface with anisotropic slip. Appl. Math. Modell. **40**, 4136–4150 (2016)
20. Raju, C.S.K., Sandeep, N.: Heat and mass transfer In MHD non-newtonian bioconvection flow over a rotating cone/plate with cross diffusion. J. Mol. Liq. **215**, 115–126 (2016)
21. Rashidi, M.M., Abelman, S., Mehr, N.F.: Entropy generation in steady MHD flow due to a rotating disk in a nanofluid. Int. J. Heat Mass Transfer **62**, 515–525 (2013)

22. Siddiqa, S., Sulaiman, M., Hossain, M.A., Islam, S., Gorla, R.S.R.: Gyrotactic bioconvection flow of a nanofluid past a vertical wavy surface. Int. J. Therm. Sci. **108**, 244–250 (2016)
23. Tausif, M.S., Das, K., Kundu, P.K.: Multiple slip effects on bioconvection of nanofluid flow containing gyrotactic microorganisms and nanoparticles. J. Mol. Liq. **220**, 518–526 (2016)
24. Tham, L., Nazar, R., Pop, I.: Mixed convection flow over a solid sphere embedded in a porous medium filled by a nanofluid containing gyrotactic microorganisms. Int. J. Heat Mass Transfer **62**, 647–660 (2013)
25. Turkyilmazoglu, M., Pop, I.: Heat and mass transfer of unsteady natural convection flow of some nanofluids past a vertical infinite flat plate with radiation effect. Int. J. Heat Mass Transfer **59**, 167–171 (2013)
26. Xu, H., Pop, I.: Fully developed mixed convection flow in a horizontal channel filled by a nanofluid containing both nanoparticles and gyrotactic microorganisms. Eur. J. Mech. B/Fluids **46**, 37–45 (2014)

Unsteady MHD Flow of Viscoelastic Fluid Through a Porous Medium in a Vertical Porous Channel

Sonam Dorjee and Utpal Jyoti Das

Abstract In this paper, an analytical solution of an unsteady MHD free convective flow of viscoelastic fluid (Walter's liquid model) through a porous medium in a vertical porous channel in presence of heat sink parameter and thermal radiation is presented. Hall current and Soret effect are taken into account. The equations of motion, energy, and concentrations are solved. The effects of various parameters involved in the primary and secondary velocity are shown graphically while the skin friction, Nusselt, and Sherwood number obtained in terms of amplitude and phase angles are discussed through the tabular form.

Keywords Viscoelastic fluid · Radiation · Heat sink · Soret effect

1 Introduction

The problem of magnetohydrodynamic flow with heat and mass transfer under the influence of chemical reaction through porous media arises in engineering, chemical technology, and industries. Applications related to convective flows in porous media can be found in Nield and Bejan [1]. Attia and Kotb [2] studied MHD flow between two porous parallel plates with constant suction and injection at the plates. Mangles et al. [3] have studied the Soret and hall effects on heat and mass transfer in MHD free convective flow through a porous medium in a vertical channel. The role of thermal radiation effect on MHD flow and heat and transfer problem has an

S. Dorjee
Rajiv Gandhi University, Itanagar 791112, Arunachal Pradesh, India
e-mail: sdojee8@gmail.com

U. J. Das (✉)
Gauhati University, Guwahati 781014, Assam, India
e-mail: utpaljyotidas@yahoo.co.in

© Springer Nature Singapore Pte Ltd. 2019
A. Abraham et al. (eds.), *Emerging Technologies in Data Mining and Information Security*, Advances in Intelligent Systems and Computing 755,
https://doi.org/10.1007/978-981-13-1951-8_31

Fig. 1 Schematic
presentation of the physical
problem

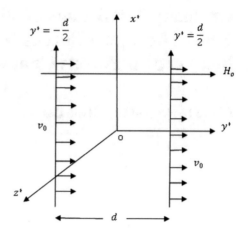

industrial importance to design many advanced energy conversion systems at a higher temperature. In view of important application of radiative heat, [4, 5] have studied the effect of radiation on fluid flow problem in different physical situations. Makinde [6] studied the free convection flow with thermal radiation and mass transfer past a moving vertical plate. Ibrahim et al. [7] have studied the effect of the chemical reaction and radiation absorption on the unsteady MHD free convection flow past a semi-infinite vertical permeable moving plate with heat source and suction. Reddy et al. [8] have studied radiation and chemical reaction effects on MHD flow along a moving vertical porous plate. The study of the flow properties of viscoelastic fluid with simultaneous heat and mass transfer through a porous medium has attracted the attention of researchers from different areas due to its importance associated with many branches of science and technology. Various flow problems have been studied by several investigators [9–13] which include the analysis of viscoelastic fluid flow through a porous medium.

This paper extends the study by Manglesh [3]. The objective of the current work is to study the influence of chemical reaction parameter, thermal radiation parameter, and viscoelastic parameter in an unsteady MHD free convection flow of a viscoelastic fluid (characterized by Walter's model B') in a vertical porous channel through a porous medium in presence of Soret effect and Hall effect (Fig. 1).

2 Formulation of Problem

We consider the flow of an unsteady MHD free convective flow of an incompressible, electrically conducting viscoelastic fluid through a porous medium bounded between two infinite vertical nonconducting porous plates distance d apart taking heat source parameter, hall current, thermal radiation, chemical reaction, and Soret effect into account. The x^*-axis is taken vertically upward along the centerline of the channel, and y^*-axis is perpendicular to the plate of the channel. The injection and suction velocities at both the plates are assumed to be a constant v_0. The temperature and concentration at one of the plates is oscillating. It is assumed that there is first-order chemical reaction between the diffusing species and the fluid. A magnetic field of uniform strength H_0 is applied along the y^*-axis and induced magnetic field is assumed to be negligible in comparison with the applied transverse magnetic field. Since the plates are infinite in extent, all the physical quantities except the pressure are functions of y^* and t^*. The schematic presentation of the flow problem is shown in Fig 1. If u^* and w^* are the components of velocity along x^* and z^* direction, respectively; then by usual Boussinesq approximation, the flow of fluid characterized by Walter's model B' [14] is governed by the following equations:

$$\frac{\partial u^*}{\partial t^*} + v_0 \frac{\partial u^*}{\partial y^*} = -\frac{1}{\rho}\frac{\partial P^*}{\partial x^*} + \vartheta \frac{\partial^2 u^*}{\partial y^{*2}} - \frac{\sigma \mu_e H_0^2}{\rho(1+m^2)}(u^* + mw^*) + g\beta T^* + g\beta^* C^*$$
$$-\frac{\vartheta u^*}{K_p^*} - \frac{K_0}{\rho}\left(v_0 \frac{\partial^3 u^*}{\partial y^{*3}} + \frac{\partial^3 u^*}{\partial y^{*2}\partial t^*}\right) \tag{1}$$

$$\frac{\partial w^*}{\partial t^*} + v_0 \frac{\partial w^*}{\partial y^*} = -\frac{1}{\rho}\frac{\partial P^*}{\partial y^*} + \vartheta \frac{\partial w^*}{\partial y^*} - \frac{\sigma \mu_e H_0^2}{\rho(1+m^2)}(mu^* - w^*) - \frac{\vartheta w^*}{K_p^*}$$
$$-\frac{K_0}{\rho}\left(v_0 \frac{\partial^3 w^*}{\partial y^{*3}} + \frac{\partial^3 w^*}{\partial y^{*2}\partial t^*}\right) \tag{2}$$

$$\frac{\partial T^*}{\partial t^*} + v_0 \frac{\partial T^*}{\partial y^*} = \frac{k}{\rho C_p}\frac{\partial^2 T^*}{\partial y^{*2}} + Q_0 T^* - \frac{1}{\rho C_p}\frac{\partial q_r^*}{\partial y^*} \tag{3}$$

$$\frac{\partial C^*}{\partial t^*} + v_0 \frac{\partial C^*}{\partial y^*} = D_m \frac{\partial^2 C^*}{\partial y^{*2}} + Kr^* C^* + \frac{D_m K_T}{T_m}\frac{\partial^2 T^*}{\partial y^{*2}} \tag{4}$$

The corresponding boundary conditions are

$$u^* = w^* = T^* = C^* = 0 \qquad\qquad \text{at}\quad y^* = -\frac{d}{2}$$
$$u^* = w^* = 0, T^* = T_o\cos\omega^* t^*, C^* = C_o\cos\omega^* t^* \quad \text{at}\quad y^* = \frac{d}{2} \tag{5}$$

where $P^*, \rho, \vartheta, \mu_e, \sigma, m, T^*, C^*, K_p^*, K_0, g, \kappa, C_p, Q_0, K_r^*, D_m, T_m, K_T, \beta, \beta^*$ are pressure, density, kinematic viscosity, magnetic permeability, electrical conductivity, Hall parameter, temperature, concentration, dimensional porous permeability parameter, dimensional viscoelastic parameter, acceleration due to gravity, thermal conductivity, specific heat at constant pressure, dimensional heat source parameter, dimensional chemical reaction parameter, coefficient of mass diffusivity, mean temperature of fluid, coefficient of thermal diffusion, and coefficient of heat and mass expansions, respectively.

Using Cogley et al. [4] for thin optically fluid, the radiative flux q_r^* is given by

$$\frac{\partial q_r^*}{\partial y^*} = 4\alpha_1^2 T^*$$

(6)

where α_1 is the mean radiation absorption coefficient.

The following nondimensional quantities are introduced:

$$u = \frac{u^*}{v_0}, w = \frac{w^*}{v_0}, x = \frac{x^*}{d}, u = \frac{y^*}{d}, \theta = \frac{T^*}{T_0}, C = \frac{C^*}{C_0}, t = \frac{t^*\vartheta}{d^2}, \omega = \frac{\omega^* d^2}{\vartheta},$$

$$K_p = \frac{K_p^*}{d^2}, P = \frac{P^* d}{v_0 \mu}, \lambda = \frac{v_0 d}{\vartheta}, G_r = \frac{g\beta T_0 d^2}{v_0 \vartheta}, G_m = \frac{g\beta^* C_0 d^2}{v_0 \vartheta}, M = \frac{\sigma \mu_e^2 H_0 d^2}{\mu},$$

$$P_r = \frac{\mu C_p}{k}, S = \frac{Q_0 d^2}{\vartheta}, S_c = \frac{\vartheta}{D_m}, S_r = \frac{D_m K_T}{T_m} \frac{T_0}{C_0 \vartheta}, N = \frac{4\alpha_1^2 d^2}{\mu C_p},$$

$$K_r = \frac{K_r^* d}{v_0}, K_1 = \frac{K_0}{\rho d^2}$$

(7)

Using (6) and (7), Eqs. (1)–(4) reduce to the following nondimensional form:

$$\frac{\partial u}{\partial t} + \lambda \frac{\partial u}{\partial y} = -\frac{\partial P}{\partial x} + \frac{\partial^2 u}{\partial y^2} - \frac{M}{(1+m^2)}(u+mw)$$

$$+ G_r\theta + G_m C - \frac{u}{K_p} - K_1\left[\lambda \frac{\partial^3 u}{\partial y^3} + \frac{\partial^3 u}{\partial y^2 \partial t}\right]$$

(8)

$$\frac{\partial w}{\partial t} + \lambda \frac{\partial w}{\partial y} = -\frac{\partial P}{\partial y} + \frac{\partial^2 w}{\partial y^2} - \frac{M}{(1+m^2)}(mu-w)$$

$$- \frac{w}{K_p} - K_1\left[\lambda \frac{\partial^3 w}{\partial y^3} + \frac{\partial^3 w}{\partial y^2 \partial t}\right]$$

(9)

$$\frac{\partial \theta}{\partial t} + \lambda \frac{\partial \theta}{\partial y} = \frac{1}{P_r}\frac{\partial^2 \theta}{\partial y^2} + S\theta + N\theta$$

(10)

$$\frac{\partial C}{\partial t} + \lambda \frac{\partial C}{\partial y} = \frac{1}{S_c}\frac{\partial^2 C}{\partial y^2} + S_r\frac{\partial^2 \theta}{\partial y^2} - \lambda K_r C$$

(11)

The corresponding boundary conditions are

$$u = w = \theta = C = 0 \qquad \text{at } y = -\frac{1}{2}$$

$$u = w = 0, \quad \theta = C = \cos\omega t \quad \text{at } y = \frac{1}{2}$$

(12)

where $G_r, G_m, M, P_r, N, S_r, S_c, K_1$ are the heat Grashof number, mass Grashof number, Hartmann number, Prandtl number, Radiation parameter, Soret number, Schmidt number, and viscoelastic parameter, respectively.

For the oscillatory channel flow, we assume

$$-\frac{\partial p}{\partial x} = A\cos\omega t \quad \text{and} \quad -\frac{\partial p}{\partial y} = 0 \tag{13}$$

Using (13) and introducing $F = u + iw$, Eqs. (8) and (9) can be combined into single equation of the form

$$\frac{\partial F}{\partial t} + \lambda\frac{\partial F}{\partial y} = A\cos\omega t + \frac{\partial^2 F}{\partial y^2} - \frac{M}{(1+m^2)}(1-im)F + G_r\theta + G_mC - \frac{F}{K_p} - K_1\lambda\frac{\partial^3 F}{\partial y^3}$$

$$- K_1\frac{\partial^3 F}{\partial y^2\partial t} \tag{14}$$

The corresponding boundary conditions are

$$\left.\begin{array}{l} F = 0, \quad \theta = C = 0 \quad \text{at} \quad y = -\frac{1}{2} \\ F = 0, \quad \theta = C = \cos\omega t \quad \text{at} \quad y = \frac{1}{2} \end{array}\right\} \tag{15}$$

3 Solution of the Problem

In order to solve (10), (11) and (14), we assume the solution of the following form:

$$F(y, t) = F_0(y)e^{i\omega t}$$
$$\theta(y, t) = \theta_0(y)e^{i\omega t}$$
$$C(y, t) = C_0(y)e^{i\omega t}$$
$$-\frac{\partial p}{\partial x} = Ae^{i\omega t} \tag{16}$$

The transformed boundary conditions become

$$F_0 = 0, \theta_0 = C_0 = 0 \quad \text{at} \quad y = -\frac{1}{2}$$

$$F_0 = 0, \theta_0 = C_0 = 1 \quad \text{at} \quad y = \frac{1}{2} \tag{17}$$

Substituting (16) into Eqs. (10), (11) and (14), then comparing the harmonic and nonharmonic terms, we obtain the following equations:

$$i\omega F_0 + \lambda \frac{\partial F_0}{\partial y} = A + \frac{\partial^2 F_0}{\partial y^2} - \frac{M}{1+m^2}(1-im) + G_r\theta_0 + G_m C_0 - \frac{F_0}{K_p} - K_1\lambda \frac{\partial^3 F_0}{\partial y^3}$$
$$- i\omega K_1 \frac{\partial^2 F_0}{\partial y^2} \qquad (18)$$

$$\frac{\partial^2 \theta_0}{\partial y^2} - \lambda P_r \frac{\partial \theta_0}{\partial y} + P_r(S + N - i\omega)\theta = 0 \qquad (19)$$

$$\frac{\partial^2 C_0}{\partial y} + S_r S_c \frac{\partial^2 \theta_0}{\partial y^2} - \lambda S_c \frac{\partial C_0}{\partial y} + S_c(\lambda Kr + i\omega)C_0 = 0 \qquad (20)$$

Solving (19) and (20) under the boundary conditions (17), we obtain

$$\theta = \{A_0 e^{ry} + B_0 e^{sy}\} e^{i\omega t} \qquad (21)$$

$$C = \{A_1 e^{r_1 y} + A_2 e^{s_1 y} + A_3 e^{ry} + A_4 e^{sy}\} e^{i\omega t} \qquad (22)$$

To solve Eq. (18) we note that $|K_1| << 1$ for viscoelastic fluid with small shear rate and so we can assume

$$F_0 = F_{00} + K_1 F_{01} \qquad (23)$$

Using (23) and equating the coefficient of zeroth order and first order of K_1, and then solving under the appropriate boundary conditions, we get

$$F(y, t) = \{A_{23} e^{r_3 y} + A_{24} e^{s_3 y} + A_{25} e^{r_2 y} + A_{26} e^{s_2 y} + A_{28} e^{ry} + A_{29} e^{sy}$$
$$+ A_{30} e^{r_1 y} + A_{31} e^{s_1 y} + B_1\} e^{i\omega t} \qquad (24)$$

Skin friction coefficient τ at left plate of the channel in terms of amplitude and phase angle is

$$\tau = \left[\frac{\partial F_0}{\partial y} - K_1 \frac{\partial^2 F_0}{\partial y \partial t}\right]_{y=-\frac{1}{2}} e^{i\omega t} = |D| \cos(\omega t + \alpha) \qquad (25)$$

where $|D| = \sqrt{D_r^2 + D_i^2}$ and $\alpha = \tan^{-1}\left(\frac{D_i}{D_r}\right)$,

$$D_r + i D_i = \{A_{31} e^{r_3 y} + A_{32} e^{s_3 y} + A_{33} e^{r_2 y} + A_{34} e^{s_2 y} + A_{35} e^{ry} + A_{36} e^{sy}$$
$$+ A_{37} e^{r_1 y} + A_{38} e^{s_1 y}\}$$

Heat transfer in terms of Nusselt number (Nu) at left plate of the channel in terms of amplitude and phase angle is

$$Nu = \left(\frac{\partial \theta}{\partial y}\right)_{y=-\frac{1}{2}} = \left(\frac{\partial \theta_0}{\partial y}\right)_{y=-\frac{1}{2}} e^{i\omega t} = |H| \cos(\omega t + \beta) \qquad (26)$$

where $|H| = \sqrt{H_r^2 + H_i^2}$ and $\beta = tan^{-1}\left(\frac{H_i}{H_r}\right)$, $H_r + iH_i = \{rA_0e^{ry} + sB_0e^{sy}\}$

Mass transfer coefficient in terms of Sherwood number (Sh) at left plate of the channel in terms of amplitude and phase angle is

$$Sh = \left(\frac{\partial C}{\partial y}\right)_{y=-\frac{1}{2}} = \left(\frac{\partial C_0}{\partial y}\right)_{y=-\frac{1}{2}} e^{i\omega t} = |G|cos(\omega t + \gamma) \qquad (27)$$

where $|G| = \sqrt{G_r^2 + G_i^2}$ and $\gamma = tan^{-1}\left(\frac{G_i}{G_r}\right)$,

$G_r + iG_i = \{r_1 A_1 e^{r_1 y} + s_1 A_2 e^{s_2 y} + rA_3 e^{ry} + sA_4 e^{sy}\}$

The constants are not presented here for the sake of brevity.

4   Result and Discussion

Selected computation has been carried out for the problem by assigning the following numerical values: $P_r = 3$, $m = 0.5$, $M = 5$, $S = 0.1$, $G_r = 5$, $G_C = 4$, $S_r = 1$, $S_c = 0.60$, $K_p = 0.5$, $K_r = 0.5$, $N = 1$ and $K_1 = 0.1$, $\lambda = 0.5$ unless otherwise stated.

Figures 2, 3, 4, 5, 6, 7, 8 and 9 present the variations of primary velocity (u) and secondary velocity (w) for various values of Magnetic field parameter M, Porous permeability parameter K_p, Soret number S_r, Radiation parameter N, Chemical reaction parameter K_r and Heat sink parameter S, respectively. From the figures,

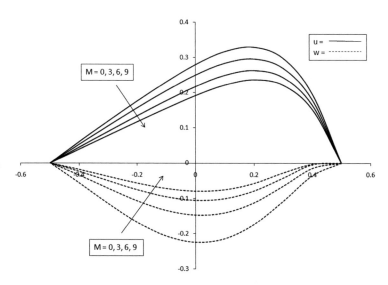

Fig. 2 Effects of M on u and w

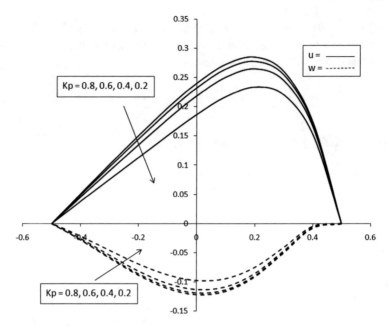

Fig. 3 Effects of K_p on u and w

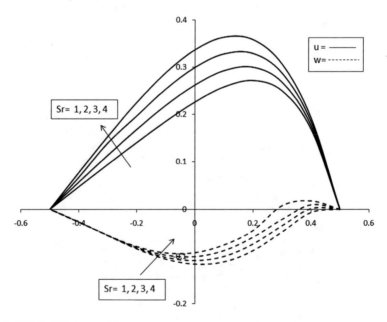

Fig. 4 Effects of S_r on u and w

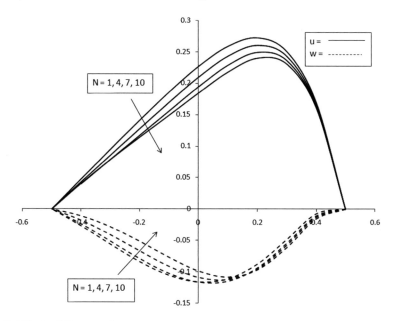

Fig. 5 Effects of N on u and w

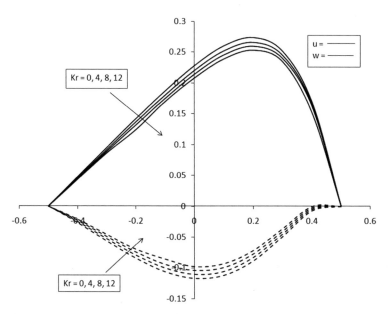

Fig. 6 Effects of K_r on u and w

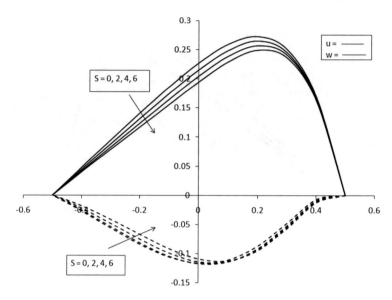

Fig. 7 Effects of S on u and w

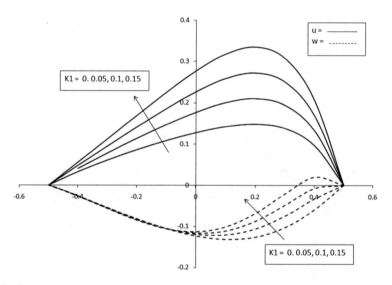

Fig. 8 Effects of K_1 on u and w

it is seen that both the velocity distributions are nearly parabolic in nature with its maximum value attained toward the right side of the channel. From Fig. 2, it is seen that an increase in M from the nonconducting, i.e., purely hydrodynamic case $M = 0$ through 3, 6, 9 there is a clear decrease in primary velocity while secondary velocity increases. This suggests that the effect of buoyancy force parameter can be

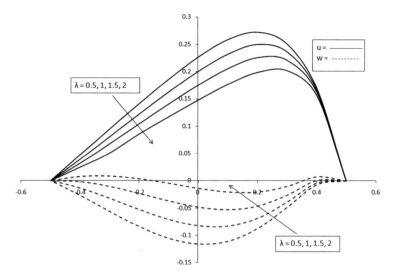

Fig. 9 Effects of λ on u and w

controlled by increasing the magnetic field parameter. In Fig. 3, it is observed that primary velocity rises with an increase in K_p whereas the reverse effect is seen for secondary velocity.

Figure 4 shows that the Soret number S_r increases both the primary and secondary velocity. The Soret number S_r defines the effect of the temperature gradients including significant mass diffusion. Figure 5 shows that an increase in radiation parameter $N = 1$ through 3, 6, 9 decreases the primary velocity because an increasing N correspond to an increased predominance of conduction over radiation, and correspondingly decreases the buoyancy force and in turn the primary velocity whereas secondary velocity increases in the left half side of the channel. But, in the right half side of the channel, secondary velocity increases nearly up to 0.1, and then, reverse effect is observed because of the presence of hot plate in this half.

From Fig. 6, it is observed that an increase in chemical reaction parameter K_r from nonchemical reaction $K_r = 0$ through 4, 8, 12 decreases the primary velocity whereas secondary velocity increases. From Fig. 7, it is observed that an increase in heat sink parameter S from 0 (i.e., no heat source/sink is present) through 2, 4, 6, the primary velocity across the channel decreases, this is due to the fact that buoyancy forces decrease as heat is absorbed which retards the primary velocity whereas secondary velocity increases in the left half side of the channel. Again, it is observed that due to the effect of heat sink parameter the secondary velocity increases nearly up to 0.1 and then decreases in the right half of the channel where the hot plate is present.

Figure 8 shows that an increase in viscoelastic parameter K_1 from Newtonian fluid $K_1 = 0$ through 0.05, 0.10, 0.15 increases both the primary and secondary velocities. It is observed from Fig. 9 that an increase in suction velocity decreases the primary

Table 1 Values of skin friction, phase angle, and amplitude at $\omega = 10$, $P_r = 3$, $\omega t = 0$

m	S	S_r	S_c	K_p	K_r	N	λ	K_1	$|D|$	α	τ
0.5	0.1	1	0.60	0.5	0.5	1	0.5	0.1	1.0874	−1.2528	0.3399
1	0.1	1	0.60	0.5	0.5	1	0.5	0.1	1.1513	−1.0264	0.5963
0.5	0.15	1	0.60	0.5	0.5	1	0.5	0.1	1.0871	−1.2532	0.3395
0.5	0.1	2	0.60	0.5	0.5	1	0.5	0.1	1.5269	−1.0990	0.7104
0.5	0.1	1	0.75	0.5	0.5	1	0.5	0.1	0.8834	−1.3250	0.2149
0.5	0.1	1	0.60	1	0.5	1	0.5	0.1	1.1760	−1.1448	0.4868
0.5	0.1	1	0.60	0.5	1	1	0.5	0.1	1.0815	−1.2297	0.3618
0.5	0.1	1	0.60	0.5	0.5	2	0.5	0.1	1.0804	−1.2569	0.3308
0.5	0.1	1	0.60	0.5	0.5	1	1	0.1	0.5197	−0.9553	0.3001
0.5	0.1	1	0.60	0.5	0.5	1	0.5	0.2	2.7782	−1.3327	0.6551

Table 2 Values of $|H|$, β, and Nu ($\omega t = 0$)

S	P_r	λ	ω	N	$|H|$	β	Nu
0.1	3.0	0.5	5	1	0.6429	−0.7815	0.6375
0.15	3.0	0.5	5	1	0.6463	−0.7843	0.6409
0.1	3.5	0.5	5	1	0.5973	−0.8945	0.5924
0.1	3.0	1	5	1	0.4813	−0.8497	0.4773
0.1	3.0	0.5	10	1	0.4986	−1.2750	0.0945
0.1	3.0	0.5	5	2	0.5463	−1.3264	0.5418

velocity indicating the usual fact that suction stabilizes the boundary layer growth whereas secondary velocity increases.

Table 1 represents the numerical values of skin friction coefficient in terms of amplitude and phase angle. It is observed from the table that an increase in Hall parameter, heat sink parameter, Soret number, porous permeability parameter, and viscoelastic parameter leads to an increase in the amplitude of the skin friction coefficient whereas an increase in Schmidt number, chemical reaction parameter, radiation parameter, and suction parameter leads to decrease in amplitude of the skin friction. The value of phase angle increases with an increase in the value of hall parameter, Soret number, chemical reaction parameter, and suction parameter, while phase angle decreases with an increase in heat sink parameter, Schmidt number, porous permeability parameter, radiation parameter, and viscoelastic parameter. Also, it is observed that an increase in hall parameter, Soret number, porous permeability parameter, and viscoelastic parameter leads to an increase in skin friction coefficient, while an increase in heat sink parameter, Schmidt number, chemical reaction parameter, radiation parameter, and suction parameter leads to decrease in the value of skin friction coefficient.

Table 2 shows the numerical values of Nusselt number in terms of amplitude and phase angle. It is observed that an increase in Prandtl number, suction parameter,

Table 3 Values of $|G|$, γ, and Sh ($\omega t = 0$)

S_r	S_c	S	λ	K_r	$|G|$	γ	Sh
1	0.60	0.10	0.5	0.5	1.3508	−0.8863	0.8256
2	0.60	0.10	0.5	0.5	1.9051	−0.8707	1.2275
1	0.75	0.10	0.5	0.5	1.3261	−1.0616	0.6465
1	0.60	0.15	0.5	0.5	1.3070	−0.8867	0.8259
1	0.60	0.10	1	0.5	1.0633	−0.8994	0.6616
1	0.60	0.10	0.5	1	1.2814	−0.8808	0.8156

frequency parameter of oscillation, and radiation parameter leads to an increase in both the amplitude and value of Nusselt number, while an increase in heat sink parameter leads to decrease in the value of amplitude and Nusselt number. The value of the phase angle β_1 decreases with an increase in the value of heat sink parameter, Prandtl number, suction parameter, frequency parameter, and radiation parameter.

Table 3 shows the numerical values of the Sherwood number in terms of amplitude and phase angle γ. It is noticed from the table that both the values of amplitude and phase angle decreases with an increase in the value of Schmidt number, heat sink parameter, suction parameter, and chemical reaction parameter, while an increase in Soret number leads to an increase in all the values of amplitude, frequency, and value of the Sherwood number. Also, it is observed that an increase in heat sink parameter leads to an increase in the value of Sherwood number, while an increase in Schmidt number, suction parameter, and chemical reaction parameter leads to decrease in the value of Sherwood number. The obtained behaviors and results are compared with that of Manglesh [3] for Newtonian fluid in absence of chemical reaction parameter and thermal radiation parameter and are found to be in good agreement.

5 Conclusion

The conclusions of the study are as follows:

1. Primary velocity accelerated with an increase in radiation parameter, heat sink parameter, viscoelastic parameter, Soret effect, and porous permeability parameter, while an increase in chemical reaction parameter and magnetic field parameter (Hartmann number) decelerated the primary velocity. Secondary velocity accelerated with an increase in chemical reaction parameter, viscoelastic parameter, and magnetic field parameter, while an increase in porous permeability parameter and Soret effect leads to a deceleration in the secondary velocity.
2. Skin friction increases with an increase in hall current, Soret effect, porous permeability parameter, and viscoelastic parameter, while an increase in heat sink parameter, chemical reaction parameter, radiation parameter, Schmidt number, and suction parameter leads to decrease in the value of skin friction.

3. Nusselt number increases with an increase in Prandtl number, radiation parameter, and suction parameter, while an increase in heat sink parameter decreases the Nusselt number.
4. Sherwood number increases with an increase in heat sink parameter and Soret number, whereas it decreases with an increase in chemical reaction parameter, Schmidt number, and porous permeability parameter.

References

1. Nield, D., Bejan, A.: Convection in Porous Media. Springer, New York (1999)
2. Attia, H.A., Kotb, N.A.: MHD flow between two parallel plates with heat transfer. Acta Mech. **117**, 215–220 (1966)
3. Manglesh, A., Gorla, M.-G., Chand, K.: Soret and hall effect on heat and mass transfer in MHD free convective flow through a porous medium in a vertical porous channel, proc. Natl. Acad. Sci., India, Sect. A phys. Sci. **84**, 63–69 (2014)
4. Cogley, A.C., Vincenti, W.-G., Gilles, S.E.: Differential approximation for radiation in a non-gray gas near equilibrium. AIAA J. **6**, 551–563 (1968)
5. Hossain, M.A., Takhar, H.S.: Radiation effect on mixed convection along a vertical plate with uniform surface temperature. Heat Mass Transfer. **31**, 243–248 (1996)
6. Makinde, O.D.: Free convection flow with thermal radiation and mass transfer past a moving vertical plate. Int. Commun. Heat Mass Transfer. **32**, 1411–1419 (2005)
7. Ibrahim, F.S., Elaiw, A.-M., Bakr, A.A.: Effect of the chemical reaction and radiation absorption on the unsteady MHD free convection flow past a semi-infinite vertical permeable moving plate with heat source and suction. Commun. Nonlinear Sci. Numer. Simul. **13**, 1056–1066 (2008)
8. Reddy, G.V.R., Reddy, N.-B., Gorla, R.S.R.: Radiation and chemical reaction effects on MHD flow along a moving vertical porous plate. Int. J. Appl. Mech. Eng. **21**, 157–168 (2016)
9. Ariel, P.D.: The flow of a viscoelatic fluid past a porous plate. Acta Mech. **107**, 199–204 (1994)
10. Cortell, R.: Toward an understanding of the motion and mass transfer with chemically reactive species for two classes of viscoelastic fluid over a porous stretching sheet. Chem. Engng. Process. Intensif. **46**, 982–989 (2007)
11. Mohiuddin, S.G., Prasad, V.R., Varma, S.V.K., Beg, O.A.: Numerical study of unsteady free convective heat and mass transfer in a walter-B visco-elastic flow along a vertical cone. Int. J. Appl. Math. Mech. **6**, 88–114 (2010)
12. Choudhury, R., Das, U.J.: Heat transfer to MHD oscillatory visco-elastic flow in a channel filled with porous medium. Phy. Res. Int. Article ID 879537 (2012)
13. Das, U.J.: Heat and mass transfer of MHD visco-elastic oscillatory flow through an inclined channel with heat generation/ absorption in presence of chemical reaction. Lat. Am. Appl. Res. **47**, 47–52 (2017)
14. Walters, K.: Non-Newtonian effects in some general elastico-viscous liquids. In: Proceedings of the I.U.T.A.M. Symposium, Hafia, Isreal pp. 507–515 (1962)

Heat Equation-Based ECG Signal Denoising in The Presence of White, Colored, and Muscle Artifact Noises

Prateep Upadhyay, S. K. Upadhyay and K. K. Shukla

Abstract In this paper, we have derived a novel solution of heat equation which comes out in the form of wavelet transformation and we have applied this solution to the signals of the MIT-BIH normal sinus rhythm database from PhysioBank in the presence of white Gaussian noise, colored noises, and muscle artifact (MA) noise respectively. It was found that the proposed method outperforms the recently reported method by Hamed Danandeh Hesar et al. in their specified SNR range of noises.

Keywords Heat Equation · Wavelets · Multiresolution analysis
ECG signals · Denoising

1 Introduction

The measure of electrical activity of the heart is known as Electrocardiogram (ECG). The ECG signals are obtained by positioning the surface electrodes on the subject's chest at the standardized locations [1]. While recording, the ECG signals get corrupted by several kinds of noises and artifacts, for example, powerline interference, electromyographic (EMG) noise due to motion artifacts, muscle artifacts and muscle

P. Upadhyay (✉)
DST-CIMS Banaras Hindu University, Varanasi 221005, Uttar Pradesh, India
e-mail: prateepster@gmail.com

S. K. Upadhyay
Department of Mathematical Sciences, IIT (B.H.U.), & DST-CIMS Banaras
Hindu University, Varanasi 221005, Uttar Pradesh, India
e-mail: sk_upadhyay2001@yahoo.com

K. K. Shukla
Department of Computer Science and Engineering, IIT (B.H.U.) Varanasi,
Varanasi 221005, Uttar Pradesh, India
e-mail: kkshukla.cse@iitbhu.ac.in

© Springer Nature Singapore Pte Ltd. 2019
A. Abraham et al. (eds.), *Emerging Technologies in Data Mining and Information
Security*, Advances in Intelligent Systems and Computing 755,
https://doi.org/10.1007/978-981-13-1951-8_32

contraction, baseline wanders, instrumentation noise, etc. [2]. The unwanted components developed from the non-cardiac reasons that act as a source of error in the signal analysis are known as noise [3, 4]. ECG signals help in the investigation of a few types of abnormal heart function which also include arrhythmias, conduction disturbances, and heart morphology. It also helps to assess the performance of pacemakers. ECG signals give useful information not only for the purpose of describing the functioning of heart but also for other body systems, for example, circulation or nervous systems [5].

ECG signals denoising has always been an interesting area for researchers. Many model and non-model methods are proposed for denoising of ECG signals. Major proposed non-model-based methods are principal component analysis [6], independent component analysis [7, 8], neural network [9], wavelet denoising [10–12], ensemble averaging [13], and adaptive filtering [14, 15]. The major proposed model-based methods for ECG signals denoising include a three-state nonlinear dynamical model in Cartesian space for ECG synthesis by McSharry et al. [16]. For denoising and feature extraction from ECG signals, an ECG dynamical model (EDM) proposed by McSharry et al. associated with Bayesian filtering approach is used by many researchers, e.g., [17], and Sameni et al. simplified the EDM given by McSharry et al. to a two-state polar EDM so as to implement an extended Kalman filter (EKF) based algorithm (known as EKF2) for denoising ECG signals [17]. This model was further extended by Sayadi et al. In this extended model, the characteristic parameters of polar EDM were added to state model and they were considered as autoregressive (AR) states [18, 19]. Sayadi et al. [20] used a Gaussian wave-based state space model to represent some of the segments of ECG wave, e.g., P, QRS, and T, by making some modifications in polar EDM.

Some researches have studied ECG signals denoising by applying particle filters (PF). Detailed information about particle filters can be found in [21]. Lee et al. [22] used a particle filter-based framework associated with a modified polar EDM for the extraction of the atrial signal from ECG signals, for the detection of atrial flutter and atrial tachycardia. Lin et al. [23] proposed marginalized particle filter (MPF) for ECG denoising to overcome the computational complexity of particle filters. The triangular model proposed by Schon et al. [24] was implemented by Lin et al. The triangular model needed some modification of the original two-state polar EDM with some limitation on the choices of linear states. In this method, the PF is supposed to approximate few linear AR states which give rise to increased computational complexity. Hamed Danandeh Hesar et al. proposed a model-based Bayesian filtering framework known as marginalized particle extended Kalman filter (MP EKF) algorithm for ECG signals denoising. Authors claim this algorithm does not suffer from the shortcoming of extended Kalman filter (EKF) [25]. Omkar Singh et al. studied about the ECG signals denoising in the presence of baseline wander correction and powerline interference using empirical wavelet transform [26].

One of the most promising techniques for the denoising of ECG signals is by using wavelet transforms. Several studies have been performed to show the use of different wavelets with different thresholding techniques. Using proper thresholding techniques, the different decomposition of ECG signals can be analyzed [27]. Maniewski et al. [28] studied the spectro-temporal methods of high-resolution ECG analysis based on three different methods, namely, fast Fourier transform, autoregressive estimation, and wavelet transform. The sensitivity of these methods was evaluated. Authors found that sensitivity obtained from wavelet transform was lower than that of the fast Fourier transform and autoregressive methods. Kania et al. [29] investigated the reduction of noise in the case of multichannel high-resolution ECG signals using wavelet denoising methods. The authors used fast wavelet transform. Janusek et al. [30] studied the denoising of T-wave alternans (TWA) analysis in high-resolution ECG maps using wavelets. Wissam jenkal et al. studied about ECG signal denoising based on adaptive dual threshold filter (ADTF) and discrete wavelet transform (DWT) [31].

2 Technical Background

2.1 Wavelets

Definition: From [32], a wavelet is a function $\Psi \in L^2(\mathbb{R})$ which satisfies the condition

$$C_\psi = \int\limits_{-\infty}^{\infty} \frac{|\hat{\Psi}(\omega)|^2}{|\omega|} d\omega < \infty$$

where $\hat{\Psi}(\omega)$ is the Fourier transform of $\Psi(t)$.

2.2 Wavelet Transformation

From [32], based on the idea of wavelets as a family of functions constructed from translation and dilation of a single function Ψ, called the mother wavelet (or affine coherent states), we define wavelets by

$$\Psi_{a,b}(t) = \frac{1}{\sqrt{|a|}} \Psi\left(\frac{t-b}{a}\right), a, b \in \mathbb{R}, a \neq 0 \tag{1}$$

where a is called a scaling parameter which measures he degree of compression or scale, and b is a translation parameter, which determines the time location of wavelets.

2.3   Denoising Using Wavelets

Denoising means restoring useful signal from the additive noise corrupted observations [33].
The simplest statistical model of denoising is thus given by

$$X_t = g_t + \pi_t, t = 1 \ldots n$$

where g is an unknown function, $(X_t)_{1 \leq t \leq n}$ are observed variables, and $(\pi_t)_{1 \leq t \leq n}$ is a centered Gaussian white noise with an unknown variance σ^2. We then have to reconstruct the signal $g(t)_{1 \leq t \leq n}$ or to estimate the function g solely on the basis of observations.

2.4   Algorithm of Denoising

The basic algorithms of denoising consist of three steps [33]:

  i Decomposition,
  ii Selection or thresholding of the coefficients, and
 iii Reconstruction.

Using the discrete transform, the signal is first decomposed over an orthogonal wavelet basis. In the next step, through thresholding, a part of coefficients is selected while the coefficients of approximation of a suitably chosen level are kept intact. In the last stage, the signal is reconstructed using thresholded coefficients by applying inverse discrete transform to them. This finally obtained signal is called denoised signal.

3   Proposed Method

We propose a novel solution to heat equation by changing the boundary conditions. Our proposed solution turns out to be in the form of wavelet transformation. We apply this solution to the different signals of [34] as reported in [25] with symlet 8 wavelet. The results obtained outperform the recently reported results of [25] and the reported results in [17].
For a clear understanding of our proposed method, the following aspects are required.

3.1 Proposed Solution of Heat Equation

From [32], the one-dimensional heat equation with no sources or sinks involved can be written as

$$\frac{\partial u}{\partial t} = k \frac{\partial^2 u}{\partial x^2}, \; where - \infty < x < \infty \tag{2}$$

where k is an arbitrary constant in Eq. (2). We assume that the boundary conditions are given by

$u(x, 0) = |a|^{\frac{-1}{2}} f(\frac{x}{a})$ and $u(x, t) \to 0$ as $x \to \pm\infty$, $a \in \mathbb{R}$, $a \neq 0$

Applying Fourier transform on both sides of Eq. (2), we get

$$\mathscr{F}\left(\frac{\partial u}{\partial t}\right) = k\mathscr{F}\left(\frac{\partial^2 u}{\partial x^2}\right) \tag{3}$$

From the definition of Fourier transform, we have

$\mathscr{F}(u) = \hat{u}(\omega, t) = \int\limits_{-\infty}^{\infty} u(x, t)e^{-i\omega x}dx$

We can write

$$\frac{\partial}{\partial t} \int\limits_{-\infty}^{\infty} u(x, t)e^{-i\omega x}dx = \frac{\partial \hat{u}(\omega, t)}{\partial t} \tag{4}$$

Now, we have

$$\mathscr{F}\left(\frac{\partial u}{\partial x}\right) = \int\limits_{-\infty}^{\infty} \frac{\partial u(x, t)}{\partial x}e^{-i\omega x}dx \tag{5}$$

Applying integration by parts, we get

$$\int\limits_{-\infty}^{\infty} \frac{\partial u(x, t)}{\partial x}e^{-i\omega x}dx = i\omega\hat{u}(\omega, t) \tag{6}$$

From Eq. (3), we have
$\mathscr{F}\left(\frac{\partial^2 u}{\partial x^2}\right) = -\omega^2 \mathscr{F}(u)$

$$\mathscr{F}\left(\frac{\partial^2 u}{\partial x^2}\right) = -\omega^2 \hat{u}(\omega, t) \tag{7}$$

Substituting the values of Eqs. (4) and (7) in Eq. (3), we get

$$\frac{\partial \hat{u}(\omega,t)}{\partial t} + k\omega^2 \hat{u}(\omega, t) = 0$$

On solving, we get $\hat{u}(\omega, t) = ce^{-k\omega^2 t}$

where c is a constant w.r.t the partial differential variable t, which means c can be a function of ω. Thus, we can write

$$\hat{u}(\omega, t) = c(\omega)e^{-k\omega^2 t} \tag{8}$$

at $t = 0$

$$\hat{u}(\omega, 0) = c(\omega)$$

In other words, $c(\omega)$ is the Fourier transform of the function of $u(x, 0)$

Thus, now we can write

$$c(\omega) = \int\limits_{-\infty}^{\infty} u(x, 0)e^{-i\omega x} dx$$

$$= \int\limits_{-\infty}^{\infty} \overline{|a|^{-\frac{1}{2}} f\left(\frac{x}{a}\right)} e^{-i\omega x} dx$$

$$c(\omega) = |a|^{-\frac{1}{2}} \int\limits_{-\infty}^{\infty} \overline{f\left(\frac{x}{a}\right)} e^{-i\omega x} dx \tag{9}$$

On solving, we get

$$c(\omega) = |a|^{\frac{1}{2}} \overline{\hat{f}(a\omega)} \tag{10}$$

substituting the value of $c(\omega)$ from (10) in (8) we get

$$\hat{u}(\omega, t) = |a|^{\frac{1}{2}} \overline{\hat{f}(a\omega)} e^{-k\omega^2 t} \tag{11}$$

Now let us assume

$$\hat{g}(\omega) = e^{-k\omega^2 t} \tag{12}$$

From Eqs. (11) and (12), we get

$$\hat{u}(\omega, t) = |a|^{\frac{1}{2}} \overline{\hat{f}(a\omega)} \hat{g}(\omega) \tag{13}$$

Taking inverse Fourier transform on both sides of Eq. (13), we get

$$\mathscr{F}^{-1}(\hat{u}(\omega, t)) = |a|^{\frac{1}{2}} \mathscr{F}^{-1}(\overline{\hat{f}(a\omega)\hat{g}(\omega)}) \tag{14}$$

R.H.S. of Eq. (14)

$$|a|^{\frac{1}{2}} \mathscr{F}^{-1}(\overline{\hat{f}(a\omega)\hat{g}(\omega)}) = |a|^{\frac{1}{2}} \int\limits_{-\infty}^{\infty} e^{i\omega x} \overline{\hat{f}(a\omega)}\hat{g}(\omega) d\omega$$

$$= |a|^{\frac{1}{2}} \int\limits_{-\infty}^{\infty} e^{i\omega x} (\int\limits_{-\infty}^{\infty} e^{ai\omega t} \overline{f(t)} dt) \hat{g}(\omega) d\omega \tag{15}$$

by taking $at = u$ Eq. (15) becomes

$$= |a|^{\frac{-1}{2}} \int\limits_{-\infty}^{\infty} g(x+u) \overline{f\left(\frac{u}{a}\right)} du \tag{16}$$

If we put $x + u = v$, Eq. (16) becomes

$$|a|^{\frac{-1}{2}} \int\limits_{-\infty}^{\infty} g(v) \overline{f\left(\frac{v-x}{a}\right)} dv = \mathscr{W}_f[g](a, x) \tag{17}$$

Thus, solution of Eq. (2) can be written in form of continuous wavelet transform of function w.r.t. wavelet of g.

$$\mathscr{F}^{-1}(\hat{u}(\omega, t)) = u(x, t) \tag{18}$$

From (17) and (18), we get

$$u(x, t) = \mathscr{W}_f[g](a, x) \tag{19}$$

The flowchart of the proposed method is given in Fig. 1. An example of the application of the proposed method is depicted in Fig. 2.

Fig. 1 Flowchart of the
proposed method

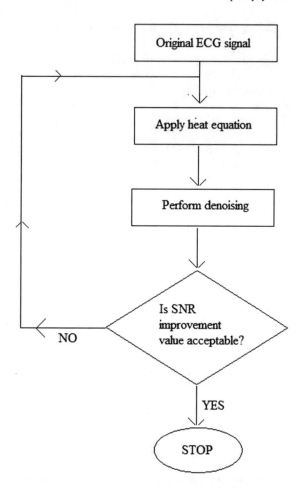

4 Results and Discussion

The benchmark datasets used for our experiment is MIT-BIH normal sinus rhythm
database from PhysioBank [34]. We performed our experiments under similar conditions as reported in [25], i.e., we have chosen two hundred signal segments from
different subjects of the database [34]. Further, similar to [25], we have selected
approximately 20 segments from each record. Each segment is approximately 30s in
signal length. These signals contain normal ECG cycles with no noticeable arrhythmias. We compare the proposed method with the recently reported method in [25]
and with some other methods reported in [17] with four different types of noises,
namely, Gaussian white noise, pink noise, brown noise, and MA noise. The input
SNR values for these noises are chosen to be in the range 10B to −5 dB. The chosen
SNR values for our purpose are −5, −4, −3, −1, 0, 1, 2, 4, 6, and 8 dBs, respec-

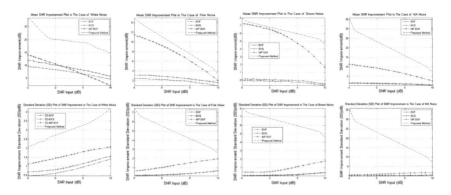

Fig. 2 Mean SNR improvement and SNR improvement SD plot in the case of different noises

tively. Gaussian white noise, pink noise, and brown noise are generated from the spectral density

$$S(f)\alpha\frac{1}{f^\beta}$$

where $S(f)$ and f are the noise spectral density function and frequency in Hz, respectively. Parameter β takes the values 0, 1, and 2 for Gaussian white noise, pink noise, and brown noise, respectively [25]. The MA noises were introduced from MIT-BIH noise stress test database [35]. For quantitative comparison, we have used SNR improved (SNR$_{imp}$) measure [18].

$$SNR_{imp} = 10log\left(\frac{\Sigma_i|x_n(i) - x_0(i)|^2}{\Sigma_i|x_d(i) - x_0(i)|^2}\right)$$

where x_0 is the original signal, x_n is the noisy signal, and x_d is the denoised signal, respectively.

Figure 1 depicts the flowchart of the proposed method. Figure 2 represents the mean SNR improvement and SNR improvement standard deviation plots of different noises in different SNR environments. Figure 3 represents the comparison performances of different denoising methods as reported in [17, 25] to that of the proposed method applied on record number 19140 of [34], with white Gaussian noise of SNR value 2 dB. Figure 4 represents the comparison of denoising methods as reported in [17, 25] to that of the proposed method applied on record number 19090 of [34] in the presence of MA noise in SNR 6 dB environment. Tables 1, 2, 3 and 4 give the performance evaluation of EKF, extended Kalman smoother (EKS) [17], MP EKF, and proposed method in terms of "multi scale entropy-based weighted distortion measure" or commonly known as MSEWPRD [36] in the presence of Gaussian white noise, pink noise, brown noise, and real muscle artifact noise, respectively, for different SNR values.

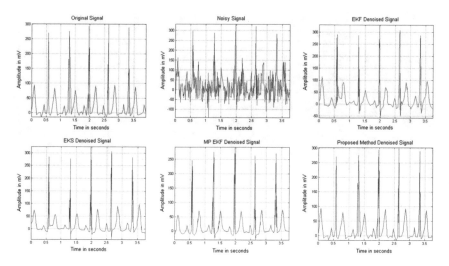

Fig. 3 Denoising results from different methods in the case of white Gaussian noise of SNR 2dB for the record 19140

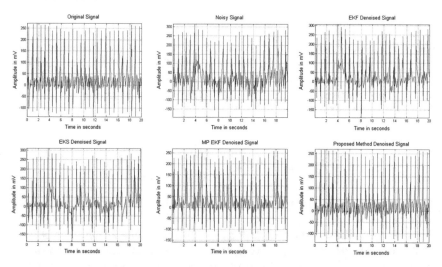

Fig. 4 Denoising results from different methods in the case of MA noise of SNR 6dB for the record 19090

It can be clearly seen from Tables 1, 2, 3 and 4 that the proposed method outperforms the methods reported in [25] and methods reported in [17] for Gaussian white noise, pink noise, brown noise, and real muscle artifact noise, respectively, for SNR values 0, −3, and −5 dB.

From Figs. 2, 3 and 4, it can again be very easily seen that the proposed method outperforms the results reported in [17, 25].

Table 1 Performance comparison of proposed method, MP EKF, EKS, and EKF in the presence of white Gaussian noise in terms of MSEWPRD (mean ± SD)(mV)

Method name	0dB	−1dB	−3dB	−5dB
Proposed method	1.134 ± 0.121	1.205 ± 0.143	1.339 ± 0.142	1.424 ± 0.15
MP EKF	1.284 ± 0.225	1.416 ± 0.224	1.434 ± 0.231	1.552 ± 0.242
EKS	1.358 ± 0.2	1.458 ± 0.196	1.678 ± 0.237	1.923 ± 0.288
EKF	1.677 ± 0.183	1.824 ± 0.2	2.158 ± 0.242	2.663 ± 0.297

Table 2 Performance comparison of proposed method, MP EKF, EKS, and EKF in the presence of pink noise in terms of MSEWPRD (mean ± SD)(mV)

Method name	0dB	−1dB	−3dB	−5dB
Proposed method	1.024 ± 0.11	1.014 ± 0.115	1.312 ± 0.171	1.503 ± 0.181
MP EKF	1.322 ± 0.204	1.387 ± 0.202	1.55 ± 0.209	1.814 ± 0.238
EKS	2.192 ± 0.288	2.415 ± 0.325	2.939 ± 0.414	3.585 ± 0.528
EKF	2.395 ± 0.31	2.65 ± 0.351	3.251 ± 0.451	4 ± 0.612

Table 3 Performance comparison of proposed method, MP EKF, EKS, and EKF in the presence of brown noise in terms of MSEWPRD (mean ± SD)(mV)

Method name	0dB	−1dB	−3dB	−5dB
Proposed method	1.225 ± 0.101	1.301 ± 0.105	1.451 ± 0.103	1.57 ± 0.102
MP EKF	1.335 ± 0.204	1.408 ± 0.205	1.585 ± 0.219	1.81 ± 0.251
EKS	2.231 ± 0.299	2.46 ± 0.337	3.001 ± 0.431	3.672 ± 0.552
EKF	2.399 ± 0.315	2.655 ± 0.356	3.261 ± 0.457	4.013 ± 0.584

Table 4 Performance comparison of proposed method, MP EKF, EKS, and EKF in the presence of muscle artifact noise in terms of MSEWPRD (mean ± SD)(mV)

Method name	0dB	−1dB	−3dB	−5dB
Proposed method	1.31 ± 0.118	1.431 ± 0.103	1.512 ± 0.114	1.665 ± 0.117
MP EKF	1.58 ± 0.199	1.552 ± 0.2	1.747 ± 0.223	1.987 ± 0.255
EKS	2.933 ± 0.455	3.247 ± 0.514	3.992 ± 0.656	4.92 ± 0.836
EKF	3.057 ± 0.473	3.393 ± 0.534	4.021 ± 0.643	5.179 ± 0.866

5   Conclusion

In this paper, we have applied the novel solution of heat equation to denoise ECG signals from MIT-BIH normal sinus rhythm database from PhysioBank [34] in the presence of white Gaussian noise, colored noises, viz., pink noise, brown noise, and muscle artifact noises. The SNR levels of noises are chosen similar to the SNR levels of [25] that is in the range −5, −4, −3, −1, 0, 1, 2, 4, 6, and 8 dBs, respectively. It was found the proposed method outperforms and gives much promising results than that of the recently reported method in [25] and methods reported in [17].

References

1. Kaergaard, K., Jensen, S. H., Puthusserypady, S.: A comprehensive performance analysis of EEMD-BLMS and DWT-NN hybrid algorithms for ECG denoising. Biomed. Signal Process. Control **25**, 178–187 (2016)
2. Wang, J., Ye, Y., Gao, Y., Qian, S., Gao, X.: Fractional compound integral with application to ECG signal denoising. Circuits Syst. Signal Process. **34**, 1915–1930 (2015)
3. Wang, Z., Wan, F., Wong, C. M., Zhang, L.: Adaptive Fourier decomposition based ECG denoising. Comput. Biol. Med. **7**, 195–205 (2016)
4. Pal, S., Mitra, M.: Empirical mode decomposition based ECG enhancement and QRS detection. Comput. Biol. Med. **42**(1), 83–92 (2012)
5. Gacek, P., Adam, W.: ECG Signal Processing, Classification and Interpretation a Comprehensive Framework of Computational Intelligence. Springer (2012)
6. Moody, G.B., Mark, R.G.: QRS morphology representation and noise estimation using the Karhunen-Loeve transform. Proc. Comput. Cardiol. 269–272 (1989)
7. Barros, A.K., Mansour, A., Ohnishi, N.: Removing artifacts from electrocardiographic signals using independent components analysis. Neurocomputing **22**(1), 173–186 (1998)
8. He, T., Clifford, G., Tarassenko, L.: Application of independent component analysis in removing artefacts from the electrocardiogram. Neural Comput. Appl. **15**(2) 105–116 (2006)
9. Clifford, G., Tarassenko, L., Townsend, N.: One-pass training of optimal architecture auto-associative neural network for detecting ectopic beats. Electron. Lett. **37**(18), 1126–1127 (2001)
10. Kestler, H., Haschka, M., Kratz, W., Schwenker, F., Palm, G., Hombach, V., Hoher, M.: Denoising of high-resolution ECG signals by combining the discrete wavelet transform with the Wiener filter. In: Proceedings, Computers in Cardiology, pp. 233–236 (1998)
11. Popescu, M., Cristea, P., Bezerianos, A.: High resolution ECG filtering using adaptive Bayesian wavelet shrinkage. In: Proceedings, Computers in Cardiology, pp. 401–404 (1998)
12. Agante, P.M., Sa, J.P.M.D.: ECG noise filtering using wavelets with soft thresholding methods. In: Proceedings, Computers in Cardiology, pp. 535–538 (1999)
13. Lander, P., Berbari, E.J.: Time frequency plane Wiener filtering of the high-resolution ECG: development and application. IEEE Trans. Biomed. Eng. **44**(4), 256–265 (1997)
14. Thakor, N.V., Zhu, Y. S.: Applications of adaptive filtering to ECG analysis: noise cancellation and arrhythmia detection. IEEE Trans. Biomed. Eng. **38**(8), 785–794 (1991)
15. Laguna, P., Jane, R., Meste, O., Poon, P. W., Caminal, P., Rix, H., Thakor, N.V.: Adaptive filter for event-related bioelectric signals using an impulse correlated reference input: comparison with signal averaging techniques. IEEE Trans. Biomed. Eng. **39**(10), 1032–1044 (1992)
16. McSharry, P.E., Clifford, G.D., Tarassenko, L., Smith, L.A.: A dynamical model for generating synthetic electrocardiogram signals. IEEE Trans. Biomed. Eng. **50**(3), 289–294 (2003)
17. Sameni, R., Shamsollahi, M.B., Jutten, C., Clifford, G.D.: A nonlinear Bayesian filtering framework for ECG denoising. IEEE Trans. Biomed. Eng. **54**(12), 2172–2185 (2007)
18. Sayadi, O., Shamsollahi, M.B.: ECG denoising and compression using a modified extended Kalman filter structure. IEEE Trans. Biomed. Eng. **55**(9), 2240–2248 (2008)
19. Sayadi, O., Shamsollahi, M.B.: A model-based Bayesian framework for ECG beat segmentation. Physiol. Meas. **30**(3), 335–352 (2009)
20. Sayadi, O., Shamsollahi, M.B., Clifford, G.D.: Robust detection of premature ventricular contractions using a wave-based bayesian framework. IEEE Trans. Biomed. Eng. **57**(2), 353–362 (2010)

21. Arulampalam, M.S., Maskell, S., Gordon, N., Clapp, T.: A tutorial on particle filters for online nonlinear/non-Gaussian Bayesian tracking. IEEE Trans. Signal Process. **50**(2), 174–188 (2002)
22. Lee, J., McManus, D.D., Bourrell, P., Sörnmo, L., Chon, K.H.: Atrial flutter and atrial tachycardia detection using Bayesian approach with high resolution time frequency spectrum from ECG recordings. Biomed. Signal Process. Control **8**(6), 992–999 (2013)
23. Lin, C., Bugallo, M., Mailhes, C., Tourneret, J.Y.: ECG denoising using a dynamical model and a marginalized particle filter. In: Proceedings, 45th Asilomar Conference on Signals, Systems and Computers (ASILOMAR), pp. 1679–1683 (2011)
24. Schon, T., Gustafsson, F., Nordlund, P.J.: Marginalized particle filters for mixed linear/nonlinear state-space models. IEEE Trans. Signal Process. **53**(7), 2279–2289 (2005)
25. Hesar, H.D., Mohebbi, M.: ECG denoising using marginalized particle extended kalman filter with an automatic particle weighting strategy. IEEE J. Biomed. Health Inform. **21**(3) (2017)
26. Singh, O., Sunkaria, R.K.: ECG signal denoising via empirical wavelet transform. Australas. Phys. Eng. Sci. Med. **40**(1), 219–229 (2017)
27. Banerjee, S., Gupta, R., Mitra, M.: Delineation of ECG characteristic features using multiresolution wavelet analysis method. Measurement **45**(3), 474–487 (2012)
28. Maniewski, R., Lewandowski, P., Nowinska, M., Mroczka, T.: Time-frequency methods for high-resolution ECG analysis. In: 18th Annual International Conference IEEE on Proceedings, Engineering in Medicine and Biology Society. Bridging Disciplines for Biomedicine (1996)
29. Kania, M., Fereniec, M., Maniewski, R.: Wavelet denoising for multi-lead high resolution ECG signals. Meas. Sci. Rev. **7**(4) (2007)
30. Janusek, D., Kania, M., Zaczek, R., Fernandez, H.Z., Zbieć, A., Opolski1, G., Maniewski, R.: Application of wavelet based denoising for T-wave alternans analysis in high resolution ECG maps. Meas. Sci. Rev. **11**(6) (2011)
31. Jenkal, W., Latif, R., Toumanari, A., Dliou, A., B'charri, O.E., Maoulainine, F.M.R.: An efficient algorithm of ECG signal denoising using the adaptive dual threshold filter and the discrete wavelet transform. Biocybern. Biomed. Eng. **36**, 499–508 (2016)
32. Debnath, L.: Wavelet Transforms and Their applications, pp. 63–371. Birkhäuser, Boston (2002)
33. Misiti, M., Misiti, Y., Oppenheim, G., Poggi, J.M.: Wavelets and Their Applications, pp. 197–206. ISTE
34. The MIT-BIH Normal Sinus Rhythm Database, PhysioNet. http://www.physionet.org/physiobank/data-base/nstdb/
35. The MIT-BIH Noise Stress Test Database, PhysioNet. http://www.physionet.org/physiobank/data-base/nstdb/
36. Manikandan, M.S., Dandapat, S.: Multiscale entropy-based weighted distortion measure for ECG coding. IEEE Signal Process. Lett. **15**, 829–832 (2008)

Second-Order Fluid Through Porous Medium in a Rotating Channel with Hall Current

Hridi Ranjan Deb

Abstract Thermal and mass diffusion of time-dependent hydromagnetic second-order fluid through porous medium has been considered. The porous medium is formed between two vertical parallel plates. The buoyancy force generates the free convection. In this investigation the effect of external heat agency is also considered. The governing equations of the flow field are solved using regular perturbation technique. The expressions for velocity, temperature concentration and skin friction are obtained analytically. The variation of skin friction with the combination of different flow parameters computed using MATLAB software is represented graphically.

Keywords Second-order fluid · Heat and mass transfer · MHD · Porous media
Hall current

1 Introduction

The study of flow in a rotating porous media is motivated by its practical applications in various natural phenomena such as in the study of inland waters, the physical and biological aspects of the ocean. The electrically conducting incompressible visco-elastic fluids in a rotating channel in the presence of applied magnetic field has gained considerable interest in different branches of science and engineering, e.g., the solar cycle in the solar physics dealing with sun spot problems. A list of the key references in the vast literature concerning this field is given in [5, 8, 11, 12, 14, 18]. Recent additions considering flow of non-Newtonian fluids with heat and mass transfer and along with MHD are given in [1, 3, 6, 13, 15, 16].

When a solid material carries an electric current and placed in a magnetic field normal to the direction of current, then a transverse electric field is developed. This phenomenon is known as Hall effect. The magnetic field exerts a force on the moving charged particles that constitute the electric current and hence electric field is

H. R. Deb (✉)
Silchar Collegiate School, Tarapur, Silchar 788003, Assam, India
e-mail: hrd2910@gmail.com

© Springer Nature Singapore Pte Ltd. 2019   369
A. Abraham et al. (eds.), *Emerging Technologies in Data Mining and Information Security*, Advances in Intelligent Systems and Computing 755,
https://doi.org/10.1007/978-981-13-1951-8_33

developed. The effect of Hall current has been studied under different conditions by several authors [7, 9, 17]. The objective of this work is to carry out the effect of visco-elasticity on buoyancy driven force flow through a rotating system under the influence of Hall current, thermal and mass diffusion characterized by second-order fluid [4, 10].

2 Mathematical Formulation

We consider the time-dependent flow of second-order electrically conducting fluid flow with external heat agent. The porous medium is formed between two vertical parallel non-conducting plates. The two plates are kept at $z=0$ and at $z=L$. A uniform transverse magnetic field B_0 is applied perpendicular to the channel. Both the plates and fluid rotate with the same angular velocity Ω in its undisturbed state taking Hall current into account.

The equations governing the fluid flow through porous medium between two parallel plates channel in presence of Hall current are as follows:

$$\frac{\partial u'}{\partial t'} - 2\Omega v' = -\frac{1}{\rho}\frac{\partial p'}{\partial x} + v_1\frac{\partial^2 u'}{\partial z^2} + v_2\frac{\partial^3 u'}{\partial t'\partial z^2} - \left(\frac{\sigma B_0^2}{\rho(1+m^2)} + \frac{v_1}{k}\right)u' + g\beta(T - T_0)$$
$$+ g\beta^*(C - C_0) \tag{1}$$

$$\frac{\partial v'}{\partial t'} + 2\Omega u' = v_1\frac{\partial^2 v'}{\partial z^2} + v_2\frac{\partial^3 v'}{\partial t'\partial z^2} - \left(\frac{\sigma B_0^2}{\rho(1+m^2)} + \frac{v_1}{k}\right)v') \tag{2}$$

$$-\frac{1}{\rho}\frac{\partial p'}{\partial z} = 0 \tag{3}$$

$$\frac{\partial T'}{\partial t'} = \frac{k}{\rho C_p}\frac{\partial^2 T'}{\partial z^2} - \frac{Q_0}{\rho C_p}(T' - T_0) \tag{4}$$

$$\frac{\partial C'}{\partial t'} = D\frac{\partial^2 C'}{\partial z^2} - K_c(C' - C_0), \tag{5}$$

where ρ is the fluid density of the fluid, p is the pressure including centrifugal force, σ is the electrical conductivity, t' is the time, B_0 is the strength of the applied magnetic field, C_p is specific heat at constant pressure, T' is the temperature, C' is the concentration, k is the thermal conductivity, m is the hall parameter, β is the coefficient of volume expansion due to temperature, β^* is the coefficient of volume expansion due to concentration, D is the chemical diffusivity, v_1 is the kinematic viscosity, v_2 is the visco-elasticity, Q_0 is the dimensional heat generation/absorption coefficient, u and v are the velocity components in x and y-direction respectively.

The flow of fluid is oscillatory Hartmann convective so that we consider the pressure P' as

$$P' = 2R_1 x\cos(\omega t) + J(y) + H(z) \tag{6}$$

It is noticed from Eqs. (1)–(3) that pressure p' is constant along z-axis i.e., $\frac{\partial p}{\partial z} = H'(z) = 0$. The absence of pressure gradient term $\frac{\partial p'}{\partial z} = J'(y)$ in Eq. (2) implies that there is a net cross flow in y-direction.

The slip boundary condition for fluid velocity are (Beavers et al. [2]):

$$\mu\frac{du'}{dz} = -\beta_1 u' \text{ and } \mu\frac{dv'}{dz} = -\beta_1 v' \text{ at } z = 0 \tag{7}$$

$$\mu\frac{du'}{dz} = -\beta_1 u' \text{ and } \mu\frac{dv'}{dz} = -\beta_1 v' \text{ at } z = L \tag{8}$$

Here, the coefficient of dynamic viscosity and coefficient of sliding friction μ and β_1 respectively.

Boundary conditions relevant to the geometry of the problem are

$$T' = T_0, C = C_0 \text{ at } z = 0 \tag{9}$$

$$T' = T_0 + (T_w - T_0)\cos\omega t,$$
$$C' = C_0 + (C_w - C_0)\cos\omega t \text{ at } z = L, \tag{10}$$

where $T_0 < T' < T_w$, $C_0 < C' < C_w$, T_0, C_0, T_w, C_w are temperature and concentration at the plate $z = 0$ and $z = L$.

We introduce the following non-dimensional quantities:

$$\varsigma = \frac{x}{L}, \eta = \frac{z}{L}, u = \frac{u'L}{v_1}, v = \frac{v'L}{v_1}, t = \frac{t'v_1}{L^2}, p = \frac{L^2 p'}{\rho v_1^2}, T = \frac{T' - T_0}{T_w - T_0}, C = \frac{C' - C_0}{C_w - C_0},$$

$$Sc = \frac{v_1}{D} (Schmidt \ number)$$

$$K^2 = \frac{\Omega L^2}{v_1} (Rotation \ parameter)$$

$$K_1 = \frac{K'}{L^2} (Permeability \ parameter)$$

$$M = \sqrt{\frac{\sigma B_0^2 L^2}{\rho v_1}} (magnetic \ parameter)$$

$$Pr = \frac{v_1 \rho C_p}{K_1} (Pr \, andtl \ number)$$

$$Gr = \frac{g\beta(T_w - T_0)L^3}{v_1^2} (Grashof \ number \ for \ heat \ transfer)$$

$$Gm = \frac{g\beta^*(C_w - C_0)L^3}{v_1^2} (Grashof \ number \ for \ mass \ transfer)$$

$$\phi = \frac{Q_0 L^2}{\rho v_1 C_p} (Heat \ generation/absorption \ coefficient)$$

$$D_0 = \frac{v_2}{L^2} (visco - elastic \ parameter) \tag{11}$$

The non-dimensional governing equations are

$$\frac{\partial u}{\partial t} - 2K^2 v = -\frac{\partial p}{\partial \varsigma} + \frac{\partial^2 u}{\partial \eta^2} + D_0 \frac{\partial^3 u}{\partial t \partial \eta^2} - (\frac{M^2}{1 + m^2} + \frac{1}{K_1})u$$
$$+ GrT + GmC \qquad (12)$$

$$\frac{\partial v}{\partial t} + 2K^2 u = \frac{\partial^2 v}{\partial \eta^2} + D_0 \frac{\partial^3 v}{\partial t \partial \eta^2} - (\frac{M}{(1 + m^2)} + \frac{1}{K_1})v \qquad (13)$$

$$\frac{\partial T}{\partial t} = \frac{1}{Pr} \frac{\partial^2 T}{\partial \eta^2} - \phi T \qquad (14)$$

$$\frac{\partial C}{\partial t} = \frac{1}{Sc} \frac{\partial^2 C}{\partial \eta^2} - K_c C \qquad (15)$$

Boundary conditions (7)–(10), in dimensionless form are:

$$\left. \begin{array}{l} u = -\alpha \frac{\partial u}{\partial \eta} \quad and \quad v = -\alpha \frac{\partial v}{\partial \eta} \quad at \ \eta = 0 \\ u = \alpha \frac{\partial u}{\partial \eta} \quad and \quad v = \alpha \frac{\partial v}{\partial \eta} \quad at \ \eta = 1 \end{array} \right\} \qquad (16)$$

$$\left. \begin{array}{l} T = 0, C = 0 \qquad\qquad at \ \eta = 0 \\ T = \cos \omega t, \ C = \cos \omega t \ at \ \eta = 1 \end{array} \right\} \qquad (17)$$

where $\alpha = \frac{\mu}{\beta_1 L}$ is the slip parameter and $\omega = \frac{\omega' L^{2\prime}}{v_1}$ is the frequency parameter. Equations (12) and (13), in compact form, become

$$\frac{\partial F}{\partial t} + 2iK^2 F = -\frac{\partial p}{\partial \varsigma} + \frac{\partial^2 F}{\partial \eta^2} + D_0 \frac{\partial^3 F}{\partial t \partial \eta^2} - (\frac{M^2}{1 + m^2} + \frac{1}{K_1})F + GrT + GrC \quad (18)$$

where F = u + iv.

Boundary conditions (16), in compact form, are

$$F + \alpha \frac{\partial F}{\partial \eta} = 0 \quad at \ \eta = 0 \qquad (19)$$

$$F - \alpha \frac{\partial F}{\partial \eta} = 0 \quad at \ \eta = 1 \qquad (20)$$

The flow of fluid past a plate is induced due to due to applied and oscillatory pressure gradient and also plates are maintained at different temperature so by heating of fluid. Therefore, pressure gradient $\frac{\partial p}{\partial \varsigma} = 0$, fluid velocity F(η, t) and fluid temperature T(η, t) are assumed, in non-dimensional form,

$$\frac{\partial p}{\partial \varsigma} = R(e^{i\omega t} + e^{-i\omega t}) \qquad (21)$$

$$F(\eta, t) = F_1(\eta)e^{i\omega t} + F_2(\eta)e^{-i\omega t} \qquad (22)$$

$$\theta(\eta, t) = T_1(\eta)e^{i\omega t} + T_2(\eta)e^{-i\omega t} \qquad (23)$$

$$C(z, t) = C_1(\eta)e^{i\omega t} + C_2(\eta)e^{-i\omega t}, \qquad (24)$$

where $R_1 < 0$ for favourable pressure.

Equations (14), (15) and (18) with the use of (21)–(24) reduce to

$$\frac{d^2 T_1}{d\eta^2} - Pr(\phi + i\omega)T_1 = 0 \tag{25}$$

$$\frac{d^2 T_2}{d\eta^2} - Pr(\phi + i\omega)T_2 = 0 \tag{26}$$

$$\frac{d^2 C_1}{d\eta^2} - Sc(k_c + i\omega)C_1 = 0 \tag{27}$$

$$\frac{d^2 C_2}{d\eta^2} - Sc(k_c + i\omega)C_2 = 0 \tag{28}$$

$$(1 + D_0 i\omega)\frac{d^2 F_1}{d\eta^2} - (\frac{M}{1 + m^2} + \frac{1}{K_1} + i(2K^2 + \omega))F_1 = R - GrT_1 - GmC_1 \tag{29}$$

$$(1 - D_0 i\omega)\frac{d^2 F_2}{d\eta^2} - (\frac{M}{1 + m^2} + \frac{1}{K_1} + i(2K^2 - \omega))F_2 = R - GrT_2 - GmC_2 \tag{30}$$

Boundary conditions (17), (19) and (20) becomes

$$\text{At } \eta = 0 : F_1 + \alpha \frac{\partial F_1}{\partial \eta} = 0 \quad and \quad F_2 + \alpha \frac{\partial F_2}{\partial \eta} = 0, \; T_1 = T_2 = 0,$$

$$C_1 = C_2 = 0 \tag{31}$$

$$\text{At } \eta = 1 : F_1 - \alpha \frac{\partial F_1}{\partial \eta} = 0 \quad and \quad F_2 - \alpha \frac{\partial F_2}{\partial \eta} = 0, \; T_1 = T_2 = 1/2,$$

$$C_1 = C_2 = 1/2 \tag{32}$$

3 Results and Discussions

The resultant skin friction at the plate $\eta = 1$ in the direction of primary and secondary velocities are respectively given by

$$St = \frac{\partial^2 F}{\partial \eta^2} + D_0 \frac{\partial^3 F}{\partial t \partial \eta^2}$$

Nusselt number and Sherwood number at the plate $\eta = 1$ are

$$\frac{\partial T}{\partial Z} = A_1(A_2 - A_3)e^{i\omega t} + B_1(B_2 - B_3)e^{-i\omega t}$$

$$\frac{\partial C}{\partial Z} = C_3(C_4 - C_5)e^{i\omega t} + D_1(D_2 - D_3)e^{-i\omega t}$$

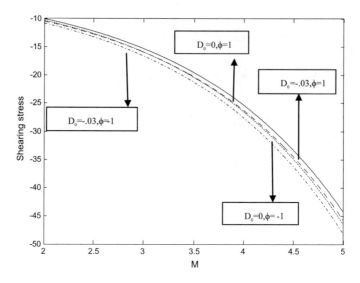

Fig. 1 Variation of shearing stress (C_f) against M for Gr = 5, Gm = 3, Pr = 0.71, m = 1, k = 0.5, R = −1, t = 1, α = 1, ω = 1, Sc = 0.3, Kc = 0.5

In this study, the visco-elastic effect is exhibited through the non-Newtonian parameter D_0 and its zero value gives the result for Newtonian fluid. The values of the parameter α = 1, K = 0.5, Kc = 0.5, Sc = 0.5 Pr = 0.71, R = −1, ωt = $\pi/2$, α = 1. Heat generation coefficient ϕ = −1, heat absorption coefficient ϕ = 1 are kept fixed throughout the discussions. The figures are plotted using MATLAB software.

Figures 1–8, gives the pattern of viscous drag formed during the motion of Newtonian and non-Newtonian fluids.

Figures 1 and 2, represent the resultant viscous drag against magnetic parameter(M) and Hall Parameter(m) at the plate η = 1. It is evident from the Figs. 1 and 2 for both heat generation and absorption fluids resultant shearing stress follow an diminishing trend with the rise of magnetic parameter(M) and Hall parameter(m). But it is also observed from Fig. 1 that shearing stress follow an diminishing trend for ϕ < 0 as compared to ϕ > 0 and also for the increase of absolute value of visco-elastic parameter. But reverse trend is observed is observed in Fig. 2.

Figures 3 and 4 depict that profile of resultant shearing stress improved along with the amplified values of Grashof number for mass transfer(Gm) and thermal Grashof number(Gr). Also the profile of shearing stress follow an rising trend ϕ < 0 as compared to ϕ > 0.

It is also observed from the expression of T and C that the temperature field and concentration field are independent of visco-elastic parameter so the rate of heat transfer and rate of mass transfer are not affected by the visco-elastic parameters.

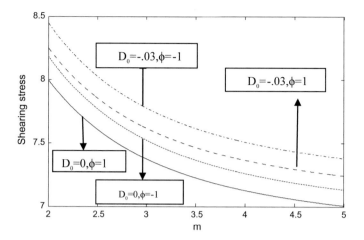

Fig. 2 Variation of shearing stress (C_f) against m for M = 5, Gr = 5, Gm = 3, Pr = 0.71, m = 1, k = 0.5, R = −1, t = 1, ϕ = 1, ω = 1, Sc = 0.3, Kc = 0.5

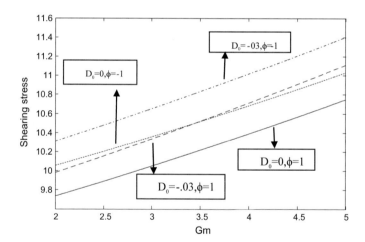

Fig. 3 Variation of shearing stress (C_f) against Gm for M = 5, Gr = 5, Pr = 0.71, m = 1, k = 0.5, R = −1, t = 1, ϕ = 1, ω = 1, Sc = 0.3, Kc = 0.5

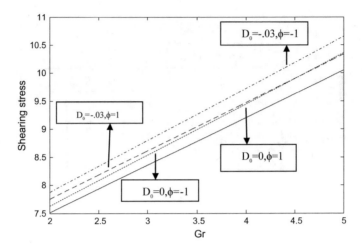

Fig. 4 Variation of shearing stress (C_f) against Gr for $M = 5$, $Gm = 3$, $Pr = 0.71$, $m = 1$, $k = 0.5$, $R = -1$, $t = 1$, $\alpha = 1$

4 Conclusions

The following conclusions are made from the study of hydromagnetic convective flow through porous medium in a rotating parallel plate channel with hall effects are as follows:

(i) The shearing stress formed at the plate $\eta = 1$, is subdued for the increasing values of magnetic parameter (M) and hall parameter (m).

(ii) The shearing stress shows a rising trend for Grashof number for heat and mass transfer(Gr, Gm).

(iii) The shearing stress is significantly affected by visco-elastic parameter.

(iv) The rate of heat and mass transfer are not affected by visco-elastic parameter.

Acknowledgements I gratefully acknowledge to Dr. R. Choudhury, Professor, Department of Mathematics, Gauhati University for her valuable help during the preparation of the paper.

References

1. Ariel, P.D.: The flow of a viscoelastic fluid past a porous plate. Acta Mech. **107**, 199–204 (1994)
2. Beavers, G.S., Joseph, D.D.: Boundary conditions at a naturally permeable wall. J. Fluid Mech. **30**(1), 197–207 (1967)
3. Choudhary, R., Das, U.J.: Heat transfer to MHD oscillatory viscoelastic flow in a channel filled with porous medium. Phys. Res. Int. (2012). https://doi.org/10.1155/2012/879537
4. Coleman, B.D., Noll, W.: An approximation theorem for functional, with applications in continuum mechanics. Arch. Ration. Mech. Anal. **6**, 355–370 (1960)

5. Hayat, T., Asghar, S., Siddiqui, A.M.: Periodic unsteady flows of a non-Newtonian fluid. Acta Mech. **131**, 169–173 (1998)
6. Hayat, T., Hutter, K., Nadeem, S., Asghar, K.: Unsteady hydromagnetic rotating flow of a conducting second grade fluid. Z. Angew. Math. Mach. **55**, 626–641 (2004)
7. Hossain, M.A., Rashid, R.I.M.I.: Hall effect on hydromagnetic free convection flow along a porous flat plate with mass transfer. J. Phys. Soc. Jpn. **56**, 97–104 (1987)
8. Jana, R.N., Datta, N.: Couette flow and heat transfer in a rotating system. Acta Mech. **26**, 301–306 (1977)
9. Kumar, R., Chand, K.: Effect of slip conditions and Hall current on unsteady MHD flow of a viscoelastic fluid past an infinite vertical porous plate through porous medium. Int. J. Eng. Sci. Tech. **3**, 3124–3133 (2011)
10. Markovitz, H., Coleman, B.D.: Incompressible second order fluids. Adv. Appl. Mech. **8**, 69–101 (1964)
11. Pop, I., Gorla, R.S.R.: Second order boundary layer solution for a continuous moving surface in a non-Newtonian fluid. Int. J. Eng. Sci. **28**, 313–322 (1990)
12. Puri, P., Kulshreshtha, P.K.: Rotating flow of a non-Newtonian fluids. Appl. Anal. **4**, 131–140 (1974)
13. Seth, G.S., Nandkeolyar, R., Ansari, MdS: Unsteady MHD convective flow within a parallel plate rotating channel with thermal source/sink in a porous medium under slip boundary conditions. Int. J. Eng. Sci. Technol. **2**(11), 1–16 (2010)
14. Singh, K.D.: Exact solution of MHD mixed convection periodic flow in a rotating vertical channel with heat radiation. Int. J. Appl. Mech. Eng. (IJAME) **18**, 853–869 (2013)
15. Singh, K.D.: Viscoelastic mixed convection MHD oscillatory flow through a porous medium filled in a vertical channel. Int. J. Phy. Math. Sci. **3**, 194–205 (2012)
16. Singh, A.K., Singh, N.P.: MHD flow of a dusty viscoelastic liquid through a porous medium between two inclined parallel plates. In: Proceedings of the National Academy of Sciences, India, vol. 66A, pp. 143–150 (1966)
17. Sulochana, P: Hall current effects on unsteady MHD convective flow of heat generating/absorbing fluid through porous medium in a rotating parallel plate channel. Int. J. Adv. Res. Ideas Innov. Tech. **2**(6), 1–11 (2016)
18. Vidyanidhuand, V., Nigam, S.D.: Secondary flow in a rotating channel. J. Math. Phys. Sci. **1**, 85(1967)

Adept-Disseminated Arithmetic-Based Discrete Cosine Transform

K. B. Sowmya and Jose Alex Mathew

Abstract Disseminated arithmetic (DA) based construction of DCT for less circuit cost and less power consumption is presented here. Using disseminated arithmetic-less number of additions is used to the Discrete Cosine Transform by exploiting the time property of the DCT. The planned One-D DCT architecture is implemented on the Xilinx FPGA. The document describes the design of two-dimensional discrete cosine transform (DCT) which is widely used in image and video compression algorithms. The intention of this paper is to design a totally parallel distributed arithmetic (DA) architecture for two-dimensional DCT. DCT requires great amount of statistical computations including addition and multiplication. Multipliers are finally avoided in the projected design as an alternative DA-Based ROM and ROM accumulators are used, thereby rich-throughput DCT designs have been taken to fit the requirements of instantaneous applications. Disseminated arithmetic is a method of adaptation at bit stream for SOP or vector dot product to partition the multiplications. The speed is increased in the wished-for design with the fully corresponding approach. In this work, existing DA architecture for two-dimensional DCT and the proposed area efficient fully parallel DA architecture for two-dimensional DCT are realized. The modeling and synthesizing is performed using Xilinx ISE.

Keywords FPGA (Field programmable gate arrays) · Two-dimensional discrete cosine transform (two-dimensional DCT) · Disseminated arithmetic (DA)

K. B. Sowmya (✉)
R V College of Engineering, Bengaluru 560059, Karnataka, India
e-mail: kb.sowmya@gmail.com

J. A. Mathew
Srinivas Institute of Technology, Mangaluru 574219, Karnataka, India
e-mail: aymanamkuzhy@gmail.com

© Springer Nature Singapore Pte Ltd. 2019
A. Abraham et al. (eds.), *Emerging Technologies in Data Mining and Information Security*, Advances in Intelligent Systems and Computing 755,
https://doi.org/10.1007/978-981-13-1951-8_34

1 Introduction

Data compression maps from a high dimensional space to a small dimensional space. Video and Image Multimedia interactions need high volume of data transmission.

The main aspire of the compression of image is to characterize a picture with least amount of bits with an acceptable quality of picture.

Conventional method of implementing DCT requires a large amount of adders and multipliers for direct implementation. Multipliers consume more power and hence Disseminated arithmetic (DA) is used to develop DCT without multiplier and adders.

Discrete Cosine Transform transforms the data in time province into frequency province. By transforming the data into frequency province, the spatial idleness in the time domain is reduced. Transformed data's energy is mainly strong in low frequency province. In the block based DCT coding the transformation is not useful to the complete picture, but it is applied over fixed blocks each of size usually 8×8 or 16×16.

Discrete Cosine Transform needs great amount of numerical calculations which includes product calculation and accumulations. Deduction in the count of multipliers using butterfly architecture is obtained which results in an uneven architecture with extended design instance duration. For competent method, an associated architecture is described in with Disseminated Arithmetic [1, 2].

2 Literature Review

The discrete cosine transform (DCT) was pioneered by Ahmed and his colleague in 1974 [1]. The discrete cosine transform (DCT) is a family of the Discrete Fourier Transform. Similar to Discrete Fourier Transform, Discrete Cosine Transform gives in order about the signal in the frequency province. The DCT coefficient of a real signal is real-valued [1]. This is the key motivation for the fame of these transformation techniques. DCT expresses a sequence of finitely many data points in terms of a sum of cosine functions oscillating at different frequencies. In image firmness for compression the DCT transform is widely worn [1–4].

3 Disseminated Arithmetic-Based One-Dimensional DCT and Error Reduction Technique

For a two-dimensional data $X(i, j)$, $0 \leq i \leq 7$ and $0 \leq j \leq 7$, 8×8 2-D DCT [3] is given by Eq. 1.

$$F(u, v) = \frac{2}{8} C(u)C(v) \sum_{i=0}^{7} \sum_{j=0}^{7} X(i, j)$$

$$\times \cos\left(\frac{(2i + 1)u\pi}{16}\right) \cos\left(\frac{(2j + 1)v\pi}{16}\right), \tag{1}$$

where $0 \le u \le 7$ and $0 \le v \le 7$ and
c(u), c(v) = 1/2 for u, v = 0,
c(u), c(v) = 1 otherwise.

Implementation computation is reduced by decomposing in two 8×1 one-dimensional DCT given by Eq. 2,

$$F(u) = \frac{1}{2} C(u) \sum_{i=0}^{7} X(i) \cos\left(\frac{(2i + 1)u\pi}{16}\right), \tag{2}$$

where $0 \le u \le 7$
c(u) = 1/2 for u = 0,
c(u) = 1 otherwise.

The one-dimensional DCT is cut down as by applying i = 0 to 7

F(0) = [X(0) + X(1) + X(2) + X(3) + X(4) + X(5) + X(6) + X(7)]P

F(1) = [X(0) − X(7)]A + [X(1) − X(6)]B + [X(2) − X(5)]C + [X(3) − X(4)]D

F(2) = [X(0) − X(3) − X(4) + X(7)]M + [X(1) − X(2) − X(5) + X(6)]N

F(3) = [X(0) − X(7)]B + [X(1) − X(6)](−D) + [X(2) − X(5)](−A) + [X(3) − X(4)](−C)

F(4) = [X(0) − X(1) − X(2) + X(3) + X(4) − X(5) − X(6) + X(7)]P

F(5) = [X(0) − X(7)]C + [X(1) − X(6)](−A) + [X(2) − X(5)]D + [X(3) − X(4)]B

F(6) = [X(0) − X(3) − X(4) + X(7)]N + [X(1) − X(2) − X(5) + X(6)](−M)

F(7) = [X(0) − X(7)]D + [X(1) − X(6)](−C) + [X(2) − X(5)]B + [X(3) − X(4)](−A), (3)

where,

$$M = 1/2 \cos(\pi/8),$$
$$N = 1/2\cos(3\pi/8),$$
$$P = 1/2\cos(\pi/4),$$
$$A = 1/2\cos(\pi/16),$$
$$B = 1/2\cos(3\pi/16),$$
$$C = 1/2\cos(5\pi/16),$$
$$D = 1/2\cos(7\pi/16); \tag{4}$$

Disseminated arithmetic is used to work out the Eqs. 3 and 4 where cosine terms are expressed in Disseminated Arithmetic form. Implementation is realized by using shift register, Look Up Table and adder components [5, 6]. Shift register output data are indicated by reduced count of bits and hence reducing adder bit-width resulting in less hardware cost. For Discrete Cosine Transform computation picture data is

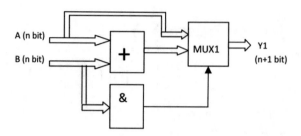

Fig. 1 Architecture for calculation of eight point discrete cosine transform in pipeline manner for computation of F(0) and F(4) [5, 6]

Fig. 2 Architecture for computation of discrete cosine transform in pipeline manner for computation of remaining solution [5, 6]

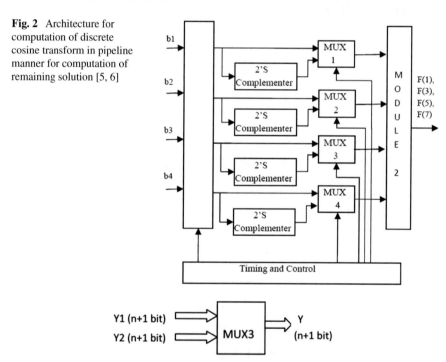

Fig. 3 VLSI architecture for computation of 8 point DCT in pipeline manner for computation of F(2) and F(6)

represented in signed 2's complement appearance. Bit-width of shift register output data is determined by number of times shift operation is performed. So different bit-width intermediate data are present which are to be added (Figs. 1, 2 and 3).

Table 1 Result comparison for the 1-D DCT architecture, proposed 1-D DCT architecture, 2-D DCT architecture

	1-D DCT architecture	Proposed 1-D DCT architecture	2-D DCT architecture
# of 4 input LUTs	1268	696	2522
# of slices	694	370	1701
# of slice flip flops	0	97	1025
# of IOB flip flops	88	0	–
Min. period (ns)	32.6	16.29 (Freq. 61.38 MHz)	45.173
Power (W)	13.1	2.06	0.751

4 Implementation Results and Comparisons

Verilog code is written in Xilinx for the implementation of one-dimensional Discrete Cosine Transform architecture [2] in conventional and proposed architecture. Code is synthesized in Xilinx virtex 5 FPGA device [7]. Table 1 shows the device utilization summary and power consumption for the one-dimensional DCT architecture, proposed one-dimensional DCT architecture, two-dimensional DCT architecture in FPGA implementation.

5 Conclusion

Disseminated arithmetic (DA) based construction of Discrete Cosine Transform for less circuit cost and less power consumption [8, 9] is developed where area utilization is improved with improvement in architecture where power consumption in one-dimensional Discrete cosine transform [2] is 13.1 W and in Disseminated arithmetic (DA) based DCT IS 2.06 W and in two-dimensional Discrete Cosine Transform power consumption is 0.751 W.

References

1. Shams, A., Chidanandan, A., Pan, W., Bayoumi, M.: A low power high throughput DCT architecture. IEEE Trans. Signal Process. **54**(3) (2006)
2. Peng, C., Cao, X., Yu, D., Zhang, X.: A 250 MHz optimized distributed architecture of 2D 8 × 8 DCT. In: 7th International Conference on ASIC, pp. 189–192, Oct 2007
3. An, S., Wang, C.: Recursive algorithm, architectures and FPGA implementation of the two-dimensional discrete cosine transform. IET Image Process **2**(6), 286–294 (2008)
4. Makkaoui, L., Lecuire, V., Moureaux, J.-M.: Fast zonal DCT-based image compression for wireless camera sensor networks. In: 2nd International Conference on Image Processing Theory Tools and Applications (IPTA), pp. 126–129 (2010)

5. Vinetha Kasturi, V., Syamala, Y.: VLSI architecture for DCT based on distributed arithmetic. Int. J. Eng. Res. Technol. (IJERT) **2**(5)

6. Kassem, A., Hamad, M., Haidamous, E.: Image compression on FPGA using DCT. In: International Conference on Advances in Computational Tools for Engineering Applications, 2009, ACTEA '09, pp. 320–323, 15–17 July 2009

7. Lin, Y.K., Li, D.W., Lin, C.C., Kuo, T.Y., Wu, S.J., Tai, W.C., Chang, W.C., Chang, T.S.: A 242 mW 10 mm^2 1080 pH. 264/AVC high-profile encoder chip. In: IEEE International Solid-State Circuits Conference (ISSCC 2008), pp. 314–316. https://doi.org/10.1109/isscc.2008.4523183. ISBN: 978-1-4244-2010-0

8. Indumathi, S., Sailaja. M.: Optimization of ECAT through DADCT. IOSR **3**(1), 39–50 (2012)

9. Wang, Y., Ostermann, J., Zhang, Y.: Video Processing and Communications, 1st edn. Prentice-Hall, Englewood Cliffs, NJ (2002)

10. Chen, Y.-H., Chang, T.-Y., Li, C.-Y.: High throughput DA based DCT with high accuracy error-compensated adder tree. IEEE Trans. Very Large Scale Integr. (VLSI) Syst. (99), 1–5 (2010)

A Survey Road Map on Different Algorithms Proposed on Protein Structure Prediction

Kunal Kabi, Bhabani Shankar Prasad Mishra and Satya Ranjan Dash

Abstract The protein is not just a simple word. It is a backbone of every living organism from small to big. Protein monomer units consist of structures which are constructed by some definite process which is called protein folding. Protein folding process gives brief idea about the sequence's structural occurrence from primary to quaternary state. In this process one important aspect is protein structure prediction which helps predicting the structural formation from a previous structural state. In bioinformatics field most researchers concentrate on protein structure prediction for better drug discovery. In this paper, we studied the protein secondary structure prediction using different algorithms like simple Artificial Neural Network (ANN), Machine Learning (ML) with multiple ANN (ML-ANN), and Deep Neural Network (DNN) in Restricted Boltzman Machine(RBM) on one dataset called Protein Data Bank (PDB). We compare the accuracy result of these three techniques with the same dataset.

Keywords Protein structure prediction · Machine learning · Artificial neural network · Deep neural network

K. Kabi (✉) · B. S. P. Mishra
School of Computer Engineering, KIIT, Deemed to be University,
Bhubaneswar 751024, Odisha, India
e-mail: kunal018kabi@gmail.com

B. S. P. Mishra
e-mail: bsmishrafcs@kiit.ac.in

S. R. Dash
School of Computer Applications, KIIT, Deemed to be University,
Bhubaneswar 751024, Odisha, India
e-mail: sdashfca@kiit.ac.in

© Springer Nature Singapore Pte Ltd. 2019
A. Abraham et al. (eds.), *Emerging Technologies in Data Mining and Information Security*, Advances in Intelligent Systems and Computing 755,
https://doi.org/10.1007/978-981-13-1951-8_35

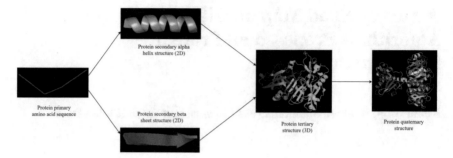

Fig. 1 Protein folding process

1 Introduction

Protein is the main ingredient of all biological organisms in spite of different shape or size whether it is a plant or animal. It contains 20 different amino acids which have some definite structure. These structures represent definite function for different biological aspects, for instance in case of human color and texture of skin, tendon (muscle to bone interface), hair growth, etc. The amino acids basically contain two important units' amino and carboxyl connected to a main carbon atom. A side chain attached to the amino acid for linking to another amino acid.

These amino acid sequences are then folded into a definite structure having some definite characteristics. Different folded structures have different characteristics. Protein is folded into four different structures.

(i) Primary structure—It is a linear sequence of different amino acids in the polypeptide chain. These amino acids are connected through peptide bonds.

(ii) Secondary structure—It is the secondary or two-dimensional stage after first folding process where three new shapes are generated such as alpha helix, beta sheet, and coil. The peptide bonds are twisted here.

(iii) Tertiary structure—Here the three different shapes alpha helix, beta sheets and coil are combined and formed three-dimensional structure. It refers to a compact globular structure which iss held by hydrogen bonding.

(iv) Quaternary structure—Two or more tertiary structures are joined to form the final structure of protein. This structure gives definite characteristics of the quaternary structure.

An example of Protein folding process is given in Fig. 1.

Protein folding problem [1] defines how the amino acid linear primary structure is folded into a quaternary structure. It has three parts

i. Folding Code
ii. Structure Prediction
iii. Pathway Prediction

Most concentration during the folding process goes on structure prediction and pathway prediction. From a given primary sequence what will be the secondary (alpha helix, beta sheets or coils) or tertiary (combination of alpha helices, beta sheets, and coils) is called structure prediction. By which way it is folded is called pathway prediction. Here we take consideration of protein secondary structure prediction [2] because it confirms the structure of alpha helices, beta sheets, or coils.

We studied three different but most common algorithms used for protein related computation simple Artificial Neural Network (ANN), Machine Learning with ANN (ML-ANN) in sliding window approach and Deep Neural Network (DNN) in Restricted Boltzman Machine (RBM).

2 Literature Survey

2.1 Machine Learning

Machine Learning approach is often indicated as in silico method. It has intelligence to vitally study and enhance from past knowledge without certainly programmed. Machine learning emphasizes on up gradation of computer programs. The process of studying begins with examining the data from direct past knowledge and decides better performance result for future. The main goal is to learn computer without any human interface.

2.2 ANN

Artificial Neural Network (ANN) is taken from neurons present in human brain. It is a part of deep learning mechanism which is used to solve entangled problems like pattern recognition or signal processing. ANN contains large amount of processors in a parallel manner and sequenced as tiers. First tier collects raw input data like human brain neurons and process it and transfers to the successor tier. Each tier holds some definite information which includes previous stored observations programmed or designed for itself.

2.3 Deep Neural Network

Deep neural network (DNN) is the near upgraded approach for ANN. It is the mixture of Deep Learning and ANN. Its complexity is far more than ANN. It avails more sophisticated mathematical models to process dataset in complex manner.

3   Related Work

In May 2017 Hasic et al. [4] proposed a hybrid approach for predicting protein secondary structure by combining machine learning (ML) and multiple ANN for predicting secondary structures like *alpha helices* or *beta sheets* or *coils* with a sliding window method [5]. In protein folding this ML mechanism is used to train the computer from previously available protein structure and test and compare against the new dataset. Mostly the accuracy based on simple ANNs [3] give maximum 60%. The author Hasic et al's new approach gives up to 66–70% accuracy.

In August 2017 Harrison et al. [6] proposed a deep neural network (DNN) for predicting only secondary alpha helices from given amino acid primary sequence. The DNN is a deep learning with neural network approach which gives better performance result than simple ANN. Restricted Boltzmann Machines (RBM) [7] is used for DNN. It gives up to 77.3% accuracy.

3.1   Algorithm for Hybrid Method Based on Machine Learning and Multiple Artificial Neural Networks

```
Function PREDICTPROTEINSECONDARYSTRUCTURE
Require:
s - Size of the ensemble;
w - Sliding window size;
data[] - Datasets used for training;
annP[] - Network parameters;
iSeq - Amino acid chain;


Step 1: k - k-fold cross-validation parameter;
Step 2: for i < s; i=i + 1 do
          : bIn = binarizeInputs(data[n], w);
          : bOut = binarizeOutputs(data[n]);
          : dataset = combine(bIn, bOut);
          : [t1,t2,val] = crossValidation(dataset, k);
          : ann = constructANN([t1,t2,val],annP[i]);
          : result[i] = ann.predict(iSeq);
          : end for
Step 3: secondaryStructure = cMethod(results);
Step 4: return secondaryStructure
```

3.2 Algorithm for Deep Neural Network (DNN)

```
Algorithm 1: modified RBM Learning
Step 1: process LEARN
Step 2: given vectors V, H, Ve , He
        : given matrix W
Step 3: for m < len(H) do
           : for n < len(V ) do
           : P = Anm · Vn · Hm
           : dot = Vn · Hm
           : f1 = exp(P)
           : f2 = exp(P)
           : dam = -(f1 - f2)/(f1 + f2)
           : if dot > 0 then
           : dam = -dam
           : end if
           : Anm -= dot + dam
           : end for
           : end for
Step 4: return A
           : end process
Algorithm 2: RBM reconstruct
Step 1: process RECONSTRUCT
Step 2: given vectors V, H, Ve, He
        : given matrix W
Step 3: for m < len(H) do
           : for n < len(V ) do
           : k += Vn ·Anm
           : end for
Step 4: Hm = logistic(k)
Step 5: end for
Step 6: end process
```

Table 1 Accuracy result

Sequence length	Techniques	Accuracy (%)
320	Simple ANN [5]	60.1
320	ML-ANN [4]	66.5
320	DNN [6]	77.3

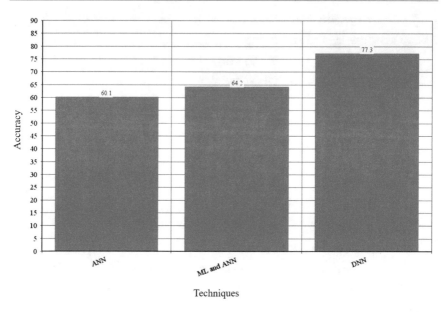

Fig. 2 Accuracy comparison

The logistic function for the given algorithm is

$$f(k) = \frac{1}{1 + \exp(-k)} \tag{1}$$

3.3 Result Analysis

Here we take PDB (Protein Data Bank) [8] dataset from RCSB web page having sequence length of 320 having PDB id 4UWW.

3.4 Evaluation

We analyze the PDB dataset applying in simple ANN, ML-ANN, and DNN. In Table 1 the accuracy result and in Fig. 2 the accuracy comparison is presented.

4 Conclusion and Future Work

In this paper we studied three algorithms such as simple ANN, ML-ANN, and DNN and compare the accuracy of these algorithms according to their folding structure prediction. But still we do not achieve more than 80% accuracy while predicting the secondary structure. Further if we use any other hybrid technique we can achieve better performance accuracy.

References

1. Chan, H., Dill, K.: The protein folding problem. Phys. Today **46**, 24–32 (1993)
2. Rost, B., Sander, C., Schneider, R.: Redefining the goals of protein secondary structure prediction. J. Mol. Biol. **235**, 13–26 (1994). [PubMed]
3. Chandonia, J., Karplus, M.: Neural networks for secondary structure and structural class predictions. Protein Sci. **1995**(4), 275–285 (1994)
4. Hasic, H., Buza, E., Akagic, A.: A hybrid method for prediction of protein secondary structure based on multiple artificial neural networks. In: 2017 40th International Convention on Information and Communication Technology, Electronics and Microelectronics (MIPRO), pp. 1195–1200. IEEE (2017)
5. Chen, K., Kurgan, L., Ruan, J.: Optimization of the sliding window size for protein structure prediction. In: 2006 IEEE Symposium on Computational Intelligence and Bioinformatics and Computational Biology, 2006. CIBCB'06. IEEE (2006)
6. Harrison, R., McDermott, M., Umoja, C.: Recognizing protein secondary structures with neural networks. In: 2017 28th International Workshop on Database and Expert Systems Applications (DEXA). IEEE (2017)
7. Salakhutdinov, R., Mnih, A., Hinton, G.: Restricted boltzmann machines for collaborative filtering. In: Proceedings of the 24th International Conference on Machine Learning, ICML'07, pp. 791–798. New York, NY, USA (2007)
8. RCSB Protein Databank www.rcsb.org

Further Reading

9. Tchoumatchenko, I., Vissotsky, F., Ganascia, J.-G.: How to make explicit a neural network trained to predict proteins secondary structure'. ACASA, LAFORIA-CNRS, Université Paris VI, 4 Place Jussieu, 75 252 Paris, CEDEX 05, France (1993)
10. Heffernan, R., Paliwal, K., Lyons, J., Dehzangi, A., Sharma, A., Wang, J., Sattar, A., Yang, Y., Zhoub, Y.: Improving prediction of secondary structure, local backbone angles, and solvent accessible surface area of proteins by iterative deep learning. Sci. Rep. **4** (2015)
11. Kneller, D.G., Cohen, F.E., Langridge, R.: Improvements in protein secondary structure prediction by an enhanced neural network. J. Mol. Biol. **214**(1), 171–182 (1990)

Graph Theoretic Scenario in Period Doubling and Limit Cycle Circumstances in Two-Dimensional Maps

Tarini Kumar Dutta, Debasmriti Bhattacherjee and Debasish Bhattacharjee

Abstract In this paper two-dimensional discrete dynamical systems have been considered. The period doubling bifurcation points of period 2^n corresponding periodic points of the dynamical system $x_{k+1} = 1 - ax_k^2 + y_k$, $y_{k+1} = \beta x_k$ where a is a parameter and β is constant are calculated for three different values of β, i.e., $\beta = 0.2$, $\beta = 0.02$ and $\beta = 0.01$. It has been seen that the relative position of the x coordinate of the Henon map follows a Mathematical model, which can be used to discuss some graph theoretic properties, up to some values of n and this value of n increases as value of β decreases. For the dynamical system $x_{n+1} = ax_n(1 - x_n) - bx_n y_n$, $y_{n+1} = -cy_n + dx_n y_n$ a limit cycle has been considered for a particular value of a, b, c, d and graph theoretical scenario has been put forward where degree of every points have been calculated by using a suitable computer program.

Keywords Period-Doubling bifurcation · Periodic points · Horizontal visibility graph · Limit cycle

2010 AMS Classification 37 G 15 · 37 G 35 · 37 C 45 · 05 C 07

1 Introduction

In recent years many authors have studied symbolic dynamics with the help of complex network which has made a combination of two fields, i.e., dynamical system and complex network theory and has become very much attractive in the literature. Lacasa et al. [12] presented a simple and fast computational method that convert

T. K. Dutta · D. Bhattacherjee · D. Bhattacharjee (✉)
Gauhati University, Guwahati 781014, India
e-mail: debabh2@gmail.com

T. K. Dutta
e-mail: tkdutta2001@yahoo.co.in

D. Bhattacherjee
e-mail: bdebasmriti@yahoo.com

© Springer Nature Singapore Pte Ltd. 2019   393
A. Abraham et al. (eds.), *Emerging Technologies in Data Mining and Information Security*, Advances in Intelligent Systems and Computing 755,
https://doi.org/10.1007/978-981-13-1951-8_36

time series into graph. The author has introduced horizontal visibility graph which maps a time series by geometric criteria to a directed network [13]. This horizontal visibility algorithm now a days has become very much attractive as it is a very good connection between time series, nonlinear dynamics and graph theory [14].

Dutta et al. showed in their paper [7] a Mathematical modeling exists in period doubling scenario of the dynamical system $x_{n+1} = ax_n(1 - x_n)$, where "a" is the control parameter. To obtain the Mathematical modeling they have taken the period doubling bifurcation points of period 2^n at period doubling bifurcation parameter. They have arranged the sets so obtained in ascending order. They have taken the smallest periodic point as 0 and assuming that 0 iterates 1, 1 iterates 2, 2 iterates 3 and so on. They have written the elements of the set in the notation $x_{a,b}$ or (a, b) which will denote that the position of a (which is obtained from 0 after number of iterations) is b. Also it is clear that position of 0 will be 0. After that they have taken up to 2^{10} periodic sets and have done the same transformation and observed that those sets (which obtained after transformation) follow a Mathematical model, i.e., they used Mathematical induction to obtain the set V_{n-1} from V_n. They have given the definition in the following way.

Let $V_1 = \{(0, 0), (1, 1)\}$. Let V_{n-1} be the set containing 2^{n-1} elements then we consider the following collections:

$$V'_{n,1} = \left\{ (2k, i) | (k, i) \in V_{n-1}, i = 0, 1, 2, \ldots, 2^{n-2} - 1 \right\}$$
$$V'_{n,2} = \left\{ (2k + 4, i) | (k, i) \in V_{n-1}, i = 2^{n-2}, \ldots, 2^{n-1} - 2, and, n > 2 \right\}$$
$$V'_{n,3} = \left\{ (2k + 4 - 2^n, 2^{n-1} - 1) | (k, 2^{n-1} - 1) \in V_{n-1} \right\}$$
$$V'_{n,4} = \left\{ (2k + 1, 2^{n-1} + i) | (k, i) \in V_{n-1}, i = 0, 1, 2, \ldots, 2^{n-1} - 1 \right\}$$

Then $V_n = \bigcup_{i=1}^{4} V'_{n,i}$

After obtaining the set V_n a horizontal visibility graph $G_n(V_n, E_n)$ is formed whose vertices sets are elements of V_n and the edge set E_n is defined in the following way:

$E_n = \{((n_1, i), (n_2, j)) | N((n_1, i), (n_2, j)) = 0\}$, where $N((n_1, i), (n_2, j))$ represents the number of elements of the form (k, k_1) such that $k_1 > i$ or j and $n_1 < k < n_2$.

In this paper we have considered two discrete dynamical systems. They are

$$x_{n+1} = 1 - ax_n^2 + y_n, \, y_{n+1} = \beta x_n \tag{1}$$

where a is a control parameter and β is a constant [3, 10] and

$$x_{n+1} = ax_n(1 - x_n) - bx_n y_n, \, y_{n+1} = -cy_n + dx_n y_n \tag{2}$$

where a, b, c, d are constants [9].

Since the system (1) exhibits period doubling bifurcation [1, 2, 4, 6, 9, 10] so the periodic points are calculated for three different values of β and the graph theoretical scenario has been highlighted.

The system (2) has a limit cycle for the values of $a = 2.63, b = 2, c = \frac{-2a-d+ad}{a}, d = a/(a-1) + 0.01$ [8]. The degree of the points of the limit cycle has been calculated by a suitable computer program.

In this paper we shall use the notation (a, b) instead of $x_{a,b}$.

Main Results:

2 Calculation of Periodic Points of the Henon Map $f(a, x, y) = (a, 1 - ax^2 + y, \beta x)$

As the graph theoretic scenario [5] is to be considered on the periodic points of Henon map at the period doubling bifurcation points for different values of β, first of all the periodic points have been extracted at the bifurcation points of Henon map with a suitable mathematica program.

(a) **For $\beta = 0.2$**

The bifurcation diagram of Henon map for $\beta = 0.2$ is shown below (Fig. 1)

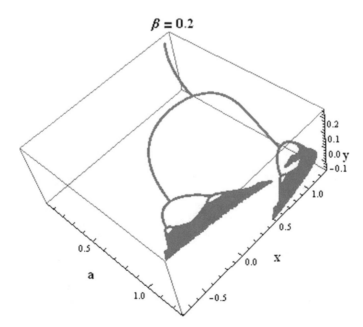

Fig. 1 Bifurcation diagram of the Henon map when $\beta = 0.2$

Table 1 Bifurcation points of period 2^n of the Henon map and one of the periodic points for $\beta = 0.2$

SL No	Period	Bifurcation points	One of the periodic points
1	2	1.0	$(1.12111, -0.0642221)$
2	4	1.115	$(-0.551766, 0.235847)$
3	8	1.1419	$(-0.596246, 0.234821)$
4	16	1.147526	$(1.16569, -0.0300364)$
5	32	1.148721	$(-0.607286, 0.235337)$
6	64	1.1489768	$(1.14496, 0.0245148)$
7	128	1.14903155	$(1.14488, 0.0245507)$
8	256	1.14904326	$(1.14534, 0.0243182)$
9	512	1.1490458	$(-0.481504, 0.228972)$

For $\beta = 0.2$ the bifurcation parameter and one of the periodic points is shown in the following Table 1:

(b) **For $\beta = 0.02$**

The bifurcation diagram of Henon map for $\beta = 0.02$ is shown below (Fig. 2)

For $\beta = 0.02$ the bifurcation parameter and one of the periodic points is shown in the following Table 2

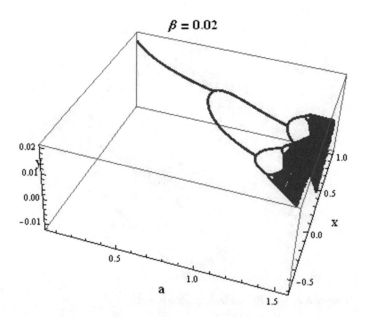

Fig. 2 Bifurcation diagram of the Henon map when $\beta = 0.02$

Table 2 Bifurcation points of period 2^n of the Henon map and one of the periodic points for $\beta = 0.02$

SL No	Period	Bifurcation points	One of the periodic points
1	2	1.2204951	$(0.9809500458170805, -0.0035599442269702215)$
2	4	1.337	$(0.0553248, 0.0167455)$
3	8	1.3645	$(-0.0302028, 0.0173215)$
4	16	1.3701999999	$(0.003762320881684966, 0.016993291545897873)$
5	32	1.37145	$(0.875674, -0.00648069)$
6	64	1.3717397	$(0.874729, -0.00650136)$
7	128	1.371802	$(0.880059, -0.00637952)$
8	256	1.14904326	$(1.14534, 0.0243182)$

(c) **For $\beta = 0.01$**

The bifurcation diagram of Henon map for $\beta = 0.01$ is shown below (Fig. 3)

For $\beta = 0.01$ the bifurcation parameter and one of the periodic points is shown in the following Table 3

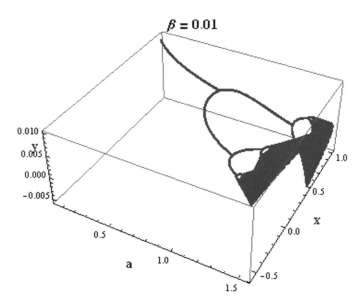

Fig. 3 Bifurcation diagram of the Henon map when $\beta = 0.01$

Table 3 Bifurcation points of period 2^n of the Henon map and one of the periodic points for $\beta = 0.01$

SL No	Period	Bifurcation points	One of the periodic points
1	2	1.2351	$(0.9733013186261127, -0.0017174678862886285417)$
2	4	1.353	$(-0.362297, 0.0100365)$
3	8	1.3792	$(-0.0271255, 0.00861597)$
4	16	1.384799	$(-0.344557700997621, 0.00985826011459505)$
5	32	1.386042	$(0.875576, -0.00311112)$
6	64	1.38629997	$(0.8719662512886216, -0.003152566685130255)$
7	128	1.3863577	$(0.870136, -0.00317345)$
8	256	1.3863695	$(0.8754592890324209, -0.003112094621676397)$

3 Modeling of the Periodic Points of the Henon Map on the Basis of Its x-Coordinate

Let W_n be the periodic points of period 2^n at nth period doubling bifurcation parameter of the Henon map as given in Tables 1 or 2 or 3 depending on the values of β. The elements of W_n are arranged in ascending order [7] depending on the x coordinate of the periodic points. Let (x_0, y_0) be the first element of W_n if it is assumed that x_0 is the smallest of all the x coordinates of all the periodic points of the set W_n. Let (x_n, y_n) be the nth periodic points which is iterated by (x_0, y_0) after n number of iterations. Let us now consider x_0 as 0, x_1 as 1, x_2 as 2 and so on. Therefore, W_n is transformed according as. Let a new set U_n be formed by eliminating the y_i from each element $i, y_i, \ i = 0, 1, 2, ..., 2^n - 1$ of W_n keeping the arrangement of the elements same as W_n. Again another set X_n is formed by writing the elements of U_n in the form $\{(a, b) | a \in \{0, 1, 2, ..., 2^n - 1\}, \ b$ represents the corresponding position of $a\}$. We consider the position of the first element of U_n to be 0, position of the second element of U_n to be 1 and so on. So it can be said that $b \in \{0, 1, 2, ..., 2^n - 1\}$.

As a result of above notations, the set X_n for Henon map for different values of n becomes

(a) **For $\beta = 0.2$**

$$X_1 = \{(0, 0), (1, 1)\}$$

$$X_2 = \{(0, 0), (2, 1), (1, 2), (3, 3)\}$$

$$X_3 = \{(0, 0), (4, 1), (6, 2), (2, 3), (1, 4), (5, 5), (3, 6), (7, 7)\}$$

$$X_4 = \{(0, 0), (8, 1), (12, 2), (4, 3), (6, 4), (14, 5), (10, 6), (2, 7), (1, 8), (9, 9), (13, 10), (5, 11),$$
$$(3, 12), (11, 13), (7, 14), (15, 15)\}$$

$$X_5 = \{(0, 0), (16, 1), (24, 2), (8, 3), (12, 4), (28, 5), (20, 6), (4, 7), (6, 8), (22, 9), (30, 10), (14, 11), (10, 12)$$
$$(26, 13), (18, 14), (2, 15), (1, 16), (17, 17), (25, 18), (9, 19), (13, 20), (29, 21), (21, 22), (5, 23), (3, 24),$$
$$(19, 25), (27, 26), (11, 27), (7, 28), (23, 29), (31, 30), (15, 31)\}$$

$$X_6 = \{(0, 0), (32, 1), (48, 2), (16, 3), (24, 4), (56, 5) \ldots \ldots \ldots, (63, 62), (47, 63), (15, 64)\}$$

(b) **For $\beta = 0.02$**

$$X_1 = \{(0, 0), (1, 1)\}$$

$$X_2 = \{(0, 0), (2, 1), (1, 2), (3, 3)\}$$

$$X_3 = \{(0, 0), (4, 1), (6, 2), (2, 3), (1, 4), (5, 5), (3, 6), (7, 7)\}$$

$$X_4 = \{(0, 0), (8, 1), (12, 2), (4, 3), (6, 4), (14, 5), (10, 6), (2, 7), (1, 8), (9, 9), (13, 10), (5, 11),$$
$$(3, 12), (11, 13), (7, 14), (15, 15)\}$$

$$X_5 = \{(0, 0), (16, 1), (24, 2), (8, 3), (12, 4), (28, 5), (20, 6), (4, 7), (6, 8), (22, 9), (30, 10), (14, 11),$$
$$(10, 12), (26, 13), (18, 14), (2, 15), (1, 16), (17, 17), (25, 18), (9, 19), (13, 20), (29, 21), (21, 22), (5, 23),$$
$$(3, 24), (19, 25), (27, 26), (11, 27), (7, 28), (23, 29), (15, 30), (31, 31)\}$$

$$X_6 = \{(0, 0), (32, 1), (48, 2), (16, 3), (24, 4), (56, 5), (40, 6), (8, 7), (12, 8), (44, 9), (60, 10),$$
$$(28, 11), (20, 12), (52, 13), (36, 14), (4, 15), (6, 16), (38, 17), (54, 18), (22, 19), (30, 20),$$
$$(62, 21), (46, 22), (14, 23), (10, 24), (42, 25), (58, 26), (26, 27), (18, 28), (50, 29),$$
$$(34, 30), (2, 31),$$
$$(1, 32), (33, 33), (49, 34), (17, 35), (25, 36), (57, 37), (41, 38), (9, 39), (13, 40), (45, 41),$$
$$(61, 42), (29, 43), (21, 44), (53, 45), (37, 46), (5, 47), (3, 48), (35, 49), (51, 50), (19, 51), (27, 52),$$
$$(59, 53), (43, 54), (11, 55), (7, 56), (39, 57), (55, 58), (23, 59), (15, 60), (47, 61), (31, 62), (63, 63)\}$$

$X_7 = \{(0, 0), (64, 1), (96, 2), (32, 3), (30, 4), (48, 5), (112, 6), (80, 7), (16, 8), (24, 9), (88, 10),$
$\quad (120, 11), (56, 12), (40, 13)(104, 14), (72, 15), (8, 16), (12, 17), (76, 18), (108, 19), (44, 20), (60, 21),$
$\quad (124, 22), (92, 23), (28, 24), (20, 25), (84, 26), \ldots\ldots\ldots, (95, 126), (127, 127), (63, 128)\}$

(b) **For $\beta = 0.01$**

$$X_1 = \{(0, 0), (1, 1)\}$$

$$X_2 = \{(0, 0), (2, 1), (1, 2), (3, 3)\}$$

$$X_3 = \{(0, 0), (4, 1), (6, 2), (2, 3), (1, 4), (5, 5), (3, 6), (7, 7)\}$$

$X_4 = \{(0, 0), (8, 1), (12, 2), (4, 3), (6, 4), (14, 5), (10, 6), (2, 7), (1, 8), (9, 9), (13, 10), (5, 11),$
$\quad (3, 12), (11, 13), (7, 14), (15, 15)\}$

$X_5 = \{(0, 0), (16, 1), (24, 2), (8, 3), (12, 4), (28, 5), (20, 6), (4, 7), (6, 8), (22, 9), (30, 10), (14, 11),$
$\quad (10, 12), (26, 13), (18, 14), (2, 15), (1, 16), (17, 17), (25, 18), (9, 19), (13, 20), (29, 21), (21, 22), (5, 23),$
$\quad (3, 24), (19, 25), (27, 26), (11, 27), (7, 28), (23, 29), (15, 30), (31, 31)\}$

$X_6 = \{(0, 0), (32, 1), (48, 2), (16, 3), (24, 4), (56, 5), (40, 6), (8, 7), (12, 8), (44, 9), (60, 10),$
$\quad (28, 11), (20, 12), (52, 13), (36, 14), (4, 15), (6, 16), (38, 17), (54, 18), (22, 19), (30, 20),$
$\quad (62, 21), (46, 22), (14, 23), (10, 24), (42, 25), (58, 26), (26, 27), (18, 28), (50, 29),$
$\quad (34, 30), (2, 31),$
$\quad (1, 32), (33, 33), (49, 34), (17, 35), (25, 36), (57, 37), (41, 38), (9, 39), (13, 40), (45, 41),$
$\quad (61, 42), (29, 43), (21, 44),$
$\quad (53, 45), (37, 46), (5, 47), (3, 48), (35, 49), (51, 50), (19, 51), (27, 52),$
$\quad (59, 53), (43, 54), (11, 55), (7, 56), (39, 57), (55, 58), (23, 59), (15, 60), (47, 61), (31, 62), (63, 63)\}$

$$X_7 = \{(0, 0), (64, 1), (96, 2), (32, 3), (48, 4), (112, 5), (80, 6),$$
$$(16, 7), (24, 8), (88, 9), (120, 10),$$
$$\cdots,$$
$$(47, 123), (31, 124), (95, 125), (63, 426), (127, 127)\}$$

$$X_8 = \{(0, 0), (128, 1), (192, 2), (64, 3), (96, 4), (224, 5), (160, 6),$$
$$\ldots\ldots\ldots (95, 251), (63, 252)(191, 253), (127, 254), (255, 255)\}$$

$$X_9 = \{(0, 0), (256, 1), (384, 2), (128, 3), (192, 4), (448, 5), (320, 6), (64, 7), (96, 8), (352, 9),$$
$$\ldots\ldots (319, 505), (447, 506), (191, 507), (127, 508), (383, 509), (255, 510), (511, 511)\}$$

$$X_{10} = \{(0, 0), (512, 1), (768, 2), (256, 3), (384, 4), (896, 5), (640, 6), (128, 7),$$
$$(192, 8), (704, 9) \ldots\ldots (191, 1015), (127, 1016), (639, 1017), (895, 1018), (383, 1019),$$
$$(255, 1020), (767, 1021), (511, 1022), (1023, 1023)\}$$

4 Comparison Of the Set X_n with the Set V_n [7]

When $\beta = 0.2$ the set X_n is same as V_n up to $= 4$. When $\beta = 0.02$ the set X_n is same as V_n up to $n = 6$.

When $\beta = 0.01$ the set X_n is same as V_n for at least $n = 9$.

5 Conclusion

We observe that the set X_n becomes equal to V_n for higher values of n as we decrease the value of β. This is because for $f(x_n, y_n) = (x_{n+1}, y_{n+1})$, as $\beta \to 0$, the effect of y-coordinate in x-coordinate becomes negligible and so the x-coordinate behaves like unimodal logistic map. So it can be said that the set X_n follows the following Mathematical model [7] for higher values of n as β becomes less.

Let $X_1 = \{(0, 0), (1, 1)\}$. Let V_{n-1} be the set containing 2^{n-1} elements then we consider the following collections:

$$X'_{n,1} = \left\{(2k, i) | (k, i) \in X_{n-1}, i = 0, 1, 2, \ldots, 2^{n-2} - 1\right\}$$
$$X'_{n,2} = \left\{(2k + 4, i) | (k, i) \in X_{n-1}, i = 2^{n-2}, \ldots, 2^{n-1} - 2, and, n > 2\right\}$$
$$X'_{n,3} = \left\{(2k + 4 - 2^n, 2^{n-1} - 1) | (k, 2^{n-1} - 1) \in X_{n-1}\right\}$$
$$X'_{n,4} = \left\{(2k + 1, 2^{n-1} + i) | (k, i) \in X_{n-1}, i = 0, 1, 2, \ldots, 2^{n-1} - 1\right\}$$

Then $X_n = \bigcup_{i=1}^{4} X'_{n,i}$

If the horizontal visibility graph [7] $G_n(X_n, E_n)$ is formed whose vertices set is X_n and the edge set is defined as $E_n = \{((n_1, i), (n_2, j))|N((n_1, i), (n_2, j)) = 0\}$ where $N((n_1, i), (n_2, j))$ represents the number of elements of the form (k, k_1) belonging to X_n such that $k_1 > i$ or j and $n_1 < k < n_2$ then the graph $G_n(V_n, E_n)$ (as defined in Chap. 3, Sect. 3.2) and $X_n(V_n, E_n)$ will have the same Graph theoretical properties up to some values of n depending on the value of β.

6   Limit Cycle and Its Graph Theoretical Scenario for the Dynamical System Governed by $x_{n+1} = ax_n(1 - x_n) - bx_n y_n$, $y_{n+1} = -cy_n + dx_n y_n$ are Constants.

The authors [8] considered the dynamical system $x_{n+1} = ax_n(1 - x_n) - bx_n y_n$, $y_{n+1} = -cy_n + dx_n y_n$ where a, b, c, d are constants for the values of $a = 2.63, b = 2, c = \frac{-2a-d+ad}{a}, d = a/(a - 1) + 0.01$ and obtained a limit cycle.

The diagram of limit cycle for the said dynamical system with the above values of the parameters is given bellow (Fig. 4):

To obtain a part of the time series scenario of the above limit cycle, let the 4,00,001 iterated point of $(0.5, 0.1)$ be (x_1, y_1) and similarity 4,00,002 number iterated point be (x_2, y_2) and so on. Then $(4,00,001, x_1, y_1)$ is reconsidered as $(1, x_1, y_1)$ and so on. Then $(1, x_1, y_1)$ to $(300, x_300, y_300)$ makes following diagram which looks like spiral.

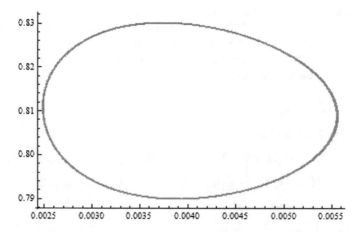

Fig. 4 Represents the limit cycle which is obtained after dropping 4,00,000 iterated points when $(0.5, 0.1)$ is iterated keeping the next 10,000 points

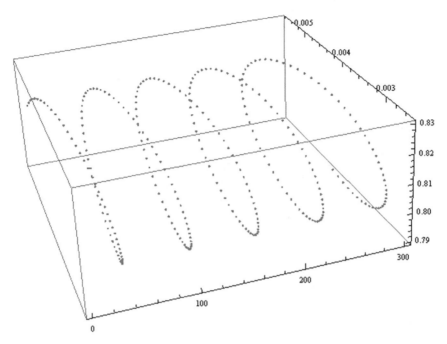

Fig. 5 Represents the limit cycle which is obtained after dropping 4,00,000 iterated points when (0.5, 0.1) is iterated keeping the next 10,000 points

A graph theoretical scenario of the limit cycle obtained in Fig. 5:

Let $(1, x, y)$ denote the point (x, y) which has been considered first to form the limit cycle as in Fig. 5 and (n, a, b) denote the point (a, b) which has been obtained from (x, y) after nth iteration.

A graph has been constructed with the vertices sets as $V =$ $\{(1, x_1, y_1), (2, x_2, y_2) \ldots (300, x_300, y_300)\}$ which has formed the spiral in Fig. 5 and the edge set is defined as two vertices (p, x_1, y_1) and (q, x_2, y_2) are adjacent if and only if for all (n, a, b) such as $(p < n < q)$ then $(b < y_1)$ as well as $(b < y_2)$.

With this condition a computer program in the software MATHEMATICA is developed to get the degree of each points of the said limit cycle.

```
f[{x_,y_}]:={a x (1-x)-b x y, -c y+d x y}
n[a_,{b_,c_}]:={a,b,c}
out[{x1_,x2_,x3_},{y1_,y2_,y3_}]:={{x1,x2,x3},{y1,y2,y3}}
a=2.63;
b=2;
c=(-2 a-d+a d)/a;
d=a/(a-1)+0.01;
iteration=9000;(*This is the number of points to be considered *)
noofpoints=300;(*This is the number of points which will be considered as ver-
tices and whose edgesets will be found and hence the degree; noofpoints should be
less than iteration value as the points will be taken from there*)
t= Nest[f,{0.5,0.1},400000];
ListPlot[NestList[f,t,10000]]
i=0;v=Table[i=i+1;n[i,z],{z,NestList[f,t,iteration]}];
fr=iteration-noofpoints;
v1=Drop[v,-fr]
ListPointPlot3D[v1]
edge=Permutations[v1,{2}];
degree=Table[sd[k]=Select[edge,#[[1,1]]==k&];
e={};
For[i=1,i<=Length[sd[k]],i++,
   If[sd[k][[i,1,1]]<sd[k][[i,2,1]],mark1=sd[k][[i,1,1]];mark2=sd[k][[i,2,1]],mark2
=sd[k][[i,1,1]];mark1=sd[k][[i,2,1]]]For[count=0;j=mark1+1,j<mark2,j++,If[Selec
t[v1,#[[1]]==j&][[1,3]]>sd[k][[i,1,3]]||Select[v1,#[[1]]==j&][[1,3]]>sd[k][[i,2,3]],
count=count+1;Break]];If[count==0,e=Union[e,{sd[k][[i]]}]]]
];e
,{k,1,Length[edge]}];
edgeset=Flatten[Drop[Union[degree],1],1]
Table[{i,Length[Select[edgeset,(#[[1,1]]==i)&]]},{i,1,Length[v1]}]
```

References

1. Alligood, K.T., Sauer, T.D., Yorke, J.A.: Chaos: An Introduction to Dynamical Systems. Springer, New York (1997)
2. Baker, G.L., Gollub, J.P.: Chaotic Dynamics: An Introduction. Cambridge University Press (1996)
3. Benedicks, M., Carleson, L.: The dynamics of the Hénon map. Ann. Math. **133**(1), 73–169 (1991)
4. Devaney, R.L.: An Introduction to Chaotic Dynamical Systems. Addison-Wesley (1989)
5. Diestel, R.: Graph Theory, 3rd edn. Graduate Texts in Mathematics 173. Springer, Berlin, Heidelberg (2005)

6. Dutta, T.K., Bhattacharjee, D.: Dynamical behaviour of two dimensional non-linear map. Int. J. Mod. Eng. Res. **2**(6), 4302–4306

7. Dutta, T.K., Bhattacherjee D., Bhattacharjee D.: Modelling of one dimensional unimodal maps to its corresponding network, Int. J. Statistika Mathematika **12**(1), 55–58 (2014)

8. Dutta, T.K., Jain A.K.: Neimark-Sacker bifurcation on a discrete-chaotic map. Int. J. Adv. Sci. Tech. Res. (IJASTR) **5**(3), 492–510 (2013). ISSN 2249-9954

9. Elsadany, A.E.A., Metwally, H.A.E., Elabbasy, E.M., Agiza, H.N.: Chaos and bifurcation of a nonlinear discrete prey-predator system. Comput. Ecol. Softw. **2**(3), 169–180 (2012)

10. Henon, M.: A two dimensional mapping with a strange attractor. Commun. Math. Phys. **50**, 69–77 (1976)

11. Lacasa, L., Luque, B., Ballesteros, F., Luque, J., Nuno, C.J.: From time series to complex networks: the visibility graph. PNAS **105**(13), 4972–4975 (2008)

12. Lacasa, L., Nuñez, A., Roldan, E., Parrondo, J.M.R., Luque, B.; Time series irreversibility: A visibility graph approach. arXiv:1108.1691V1

13. Lacasa, L., Toral, R.: Description of stochastic and chaotic series using visibility graph. Phys. Rev. E **82**, 036120 (2010)

14. Nuñez, A.M., Lacasa, L., Gomez, J.P., Luque, B.: Visibility algorithms: a short review. In: Zhang, Y. (ed.) New Frontiers in Graph Theory (2012). ISBN: 978-953-51-0115-4

Part IV
Computational Modeling

Enjoy and Learn with Educational Game: Likhte Likhte Shikhi Apps for Child Education

Md. Walid Bin Khalid Aunkon, Md. Hasanuzzaman Dipu, Nazmun Nessa Moon, Mohd. Saifuzzaman and Fernaz Narin Nur

Abstract Nowadays, children (under age 2–6 years) become so much affected to smartphone. It is impossible to take back the device when they play a game on that device. So, we realized if this affection is possible to convert to an educational game, it can be better. We decide to build an educational game that can help to teach them how to write and read a letter. A child can easily learn the technique how to write a letter properly by this game. After completing the writing part this app will play the actual pronunciation of that letter. The Graphical User Interface of this game is very attractive as well as user friendly. A child can learn both Bangla and English letter writing and reading by this game. This game is completely offline and dynamic game app. The game size is smaller than other games. And it can run smoothly on any of Android Device such as Smartphone, Tab and Smart TV.

Keywords Bangla letter · English letter · Interactive learning environments
Learning apps · Learning motivational game

Md. W. B. K. Aunkon (✉) · Md. H. Dipu · N. N. Moon (✉) · Mohd. Saifuzzaman · F. N. Nur
Department of Computer Science and Engineering, Daffodil International University, Dhaka,
Bangladesh
e-mail: walid15-1669@diu.edu.bd

N. N. Moon
e-mail: moon@daffodilvarsity.edu.bd

Md. H. Dipu
e-mail: hdipu2646@diu.edu.bd

Mohd. Saifuzzaman
e-mail: saifuzzaman.cse@diu.edu.bd

F. N. Nur
e-mail: narin@daffodilvarsity.edu.bd

© Springer Nature Singapore Pte Ltd. 2019
A. Abraham et al. (eds.), *Emerging Technologies in Data Mining and Information
Security*, Advances in Intelligent Systems and Computing 755,
https://doi.org/10.1007/978-981-13-1951-8_37

1 Introduction

Nowadays, everybody knows, Smartphones and tablets are now the most popular devices which is used for gaming among children in the age range of 2–17 according to a report released 2015 by the NPD Group [1]. The report, titled "Kids and Gaming 2015". The decline is most prominent among children ages 2–5 it said CNET [2]. Digital Trends [2] said Gaming among kid's ages 2–5 has increased the most and 91 percent of kids play video games.

Likhte Likhte Shikhi Pro (লিখতে লিখতে শিখি) is an Educative Application. It is developed for the children of 2–5 years who has not started school yet or may be studies in nursery class. But anyone can use this App. This App will help children to learn writing Bangla and English alphabets as well as reading by using smart devices.

Sometimes learning new stuffs could consume a decent amount of time and little bit boring for kids. That's why this App is developed as interesting as possible so that the kids pay their attention to the lessons.

We use Unity 3d Game Engine to develop this game application and for playing this game application on smartphone we use Android Software Development Kit (SDK). We use JavaScript and C# languages for developed the game application.

The application is developed specially for Bangla and English language. It is very useful for child learning. Here player will be learning how to write and read a letter or a number. Any illiterate people also can learn how to write and read by this game application.

The main challenge is to get the children concentration on the game. It is very challenging to move child concentration from usual game to educational game or apps. We will try to build this game so much attractive to the children so that they can easily involve to this game.

To develop this type of educational app is so much difficult because we must take care about lot of things like children mind reading, take their concentration, their entertainments etc. Maintaining all the things are difficult.

2 Experimental Method

This research work has been divided into two phases.

2.1 Research/Paper Work

Paper work is done actually based on ideas, secondary information from various journals, published book, and newspapers along with internet. The study is qualitative and unique in nature.

2.2 Developed Work

To make a complete study, here we developed our apps using Unity 3D game engine and for playing this game on smartphone, we use android software development kit (SDK). JavaScript and C# languages are used to develop the game application.

3 System Architecture

To demonstrate the feasibility and effectiveness of this system, we need to make a schematic architecture by which we can show the entire process at a glance (Fig. 1).

4 Methodology

To demonstrate the feasibility and effectiveness of this system, we used simple flow of methodology which can easily be understood by the following Fig. 2.

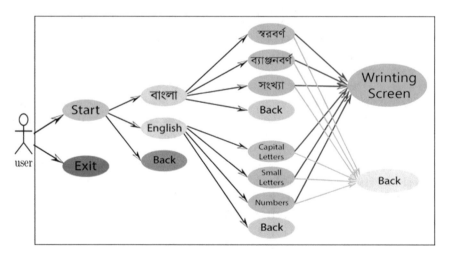

Fig. 1 System architecture of Likhte Likhte Shikhi apps

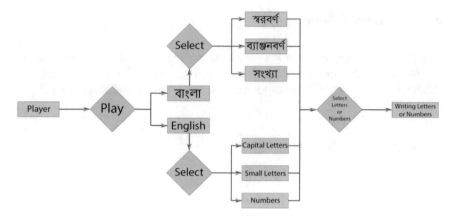

Fig. 2 Flow of methodology of Likhte Likhte Shikhi apps

5 Literature Review

In general, this research cover the literature review from different sources which is targeted various aspect of learning based game and its impact. Gaming has grown in popularity and become a defining characteristics of young learners.

Sadera et al. [3] featured on his research that Game based learning can able to improve learning through increased motivation and interaction. Tobias et al. [4] ensured that motivational game can assure specific knowledge and skills to our young learners. Papastergiou [5] also proposed a computer game for learning concept for Greek high school for investigating the learning effectiveness and motivational appeals of the students and the result was tremendous. Erhel and Jamet [6] highlighted on his research that learning is deeper with game based learning and its feedback is beneficial. Hwang et al. [7, 8] developed a learning-based problem solving activities for student and it also showed not also significant but also an improved the game-based learning approach. Sung et al. [9] proposed a game-based learning approach to a natural science course and it improved the learners learning achievement, skills and attitude of learning. Garris et al. [10] proposed a research and practice model in his research and showed the effectiveness of motivational game towards learning. Hwang et al. [11], Burguillo [12] and Charsky and Ressler [13] found out from their research that games are not only promotes the learning motivation, but also it can be able to improve the learning achievement. Qian et al. described in his review paper, the effectiveness and use of game based learning in 21st century and showed a tremendous result.

From above description, it can be said that game is not only an entertainment fact today, nowadays it is an educational instrument for effective learning and we actually develop an apps which can be able to make a huge contribution on the educational sector to improve the student's self-efficacy and skill.

6 Tests and Results

At first, we made Interaction design to evaluate how we will arrange our entire apps, what will we need to do further research and development as shown in Fig. 3.

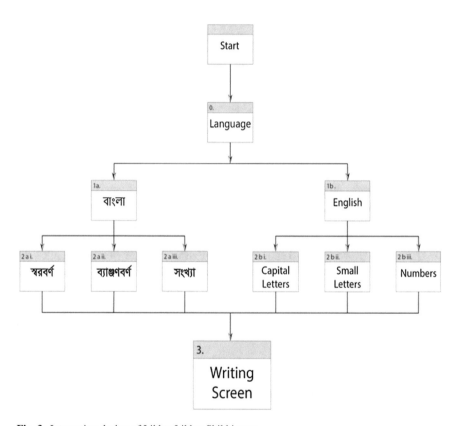

Fig. 3 Interaction design of Likhte Likhte Shikhi apps

Fig. 4 The initial starting GUI of the Likhte Likhte Shikhi

6.1 Graphical User Interfaces

User can interact with the Front-End GUI by the back-end work. We will discuss about the initial front-end design in this part. The following Fig. 4 shows the initial starting layout of the Likhte Likhte Shikhi Game application.

The following Fig. 5 shows Select language GUI and Select Languistic Alphabet to Write. The following Fig. 6 shows the GUI that will be shown when a child will select a type alphabet he/she want to write. The following Fig. 7 left side shows the Final GUI that will be shown when a child will select a letter or number from Bangla or English Alphabet or Number boards. The Child will learn to read and write by this Screen. The following Fig. 7 right side shows the Final GUI that will be shown when a child complete writing a letter or numbers.

6.2 User Review from Play Store

After developing our desired apps, we are intended to upload it to playstore for users view and see the result and comment was tremendous as shown in Fig. 8.

Fig. 5 Select language GUI and Select Languistic Alphabet to Write

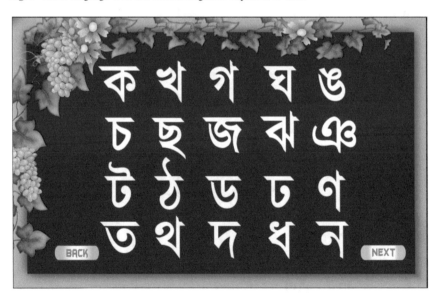

Fig. 6 Bangla Alphabet board GUI Layout for Select Letter

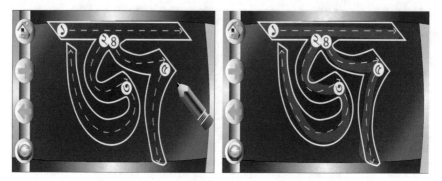

Fig. 7 Letter Writing Graphical User Interface

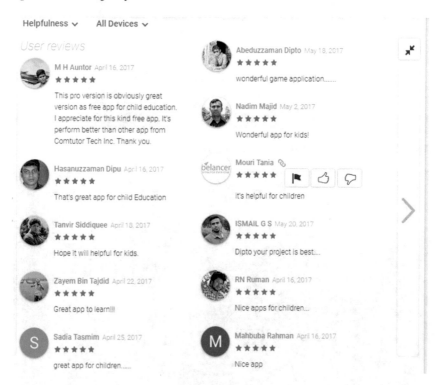

Fig. 8 User review from playstore

7 Conclusion and Future Scope

The smartphone is an effective portable device and this game application can be install and used on any smartphone supporting Android operation system. By this game application, a child can learn a letter and how to write it properly. It is a foundation way to learn read and write to a child. It will be helping to learn how to read and write as well as reduce the pressure of parent or guardian about using smartphone by their children.

The Limitation of this game application is actually this game application will work only on Android Smartphone or Tab, Minimum Android API level 16 "Jelly Bean 4.1".

The future developments of this game application are

- Arabic Letters can be included for reading and writing.
- These apps will be converted as Computer game.

Acknowledgements Author acknowledges the support of the faculty members of Daffodil International University to develop this proposed system and also acknowledge those who previously worked on this concept.

References

1. NeoGAF BELIEVE, CNET. http://www.neogaf.com/forum/showthread.php?t=111606
2. CNET—Kids pick mobile devices over PCs, consoles for gaming. https://www.cnet.com/new s/kids-now-pick-mobile-devices-over-pcs-consoles-for-gaming-npd-group/
3. Sadera, W.A., Li, Q., Song, L., Liu, L.: Digital Game-Based Learning. Comput. Sch. Interdiscip. J. Pract. Theory Appl. Res. **31**(1–2) (2014). https://doi.org/10.1080/07380569.2014.879801
4. Tabias, S., Fletcher, J.D., Wind, A.P.: Game-based learning. In: Spector, J., Merrill, M., Elen, J., Bishop, M. (eds.) Handbook of Research on Educational Communications and Technology, pp. 485–503. Springer, New York, NY (2013). https://doi.org/10.1007/978-1-4614-3185-5_38
5. Papastergiou, M.: Digital game-based learning in high school computer science education: impact on educational effectiveness and student motivation. Comput. Educ. **52**(1), 1–12 (2009). https://doi.org/10.1016/j.compedu.2008.06.004 (Elsevier, ScienceDirect)
6. Erhel, S., Jamet, E.: Digital game-based learning: impact of instructions and feedback on motivation and learning effectiveness. Comput. Educ. **67**, 156–167 (2013). https://doi.org/10. 1016/j.compedu.2013.02.019 (Elsevier, ScienceDirect)
7. Hwang, G.J., Wu, P.H., Chen, C.C.: An online game approach for improving students' learning performance in web-based problem-solving activities. Comput. Educ. **59**(4), 1246–1256 (2012). https://doi.org/10.1016/j.compedu.2012.05.009 (Elsevier, ScienceDirect)
8. Hwang, G.J., Yang, L.H., Wang, S.Y.: A concept map-embedded educational computer game for improving students' learning performance in natural science courses. Comput. Educ. **69**, 121–130 (2013). https://doi.org/10.1016/j.compedu.2013.07.008 (Elsevier, ScienceDirect)
9. Sung, H.Y., Hwang, G.J.: A collaborative game-based learning approach to improving students' learning performance in science courses. Comput. Educ. **63**, 45–51 (2013). https://doi.org/10. 1016/j.compedu.2012.11.019 (Elsevier, ScienceDirect)
10. Garris, R., Ahlers, R., Driskell, J.E.: Games, motivation, and learning: a research and practice model. SAGE J. (2002)
11. Hwang, G.J., Sung, H.Y., Hung, C.M., Huang, I., Tsai, C.C.: Development of a personalized educational computer game based on students' learning styles. Educ. Technol. Res. Dev. **60**(4), 623–638 (2012). https://doi.org/10.1007/s11423-012-9241-x (Springerlink)
12. Burguillo, J.C.: Using game theory and competition-based learning to stimulate student motivation and performance. Comput. Educ. **55**(2), 566–575 (2010)
13. Charsky, D., Ressler, W.: Games are made for fun: Lessons on the effects of concept maps in the classroom use of computer games. Comput. Educ. **56**, 604–615 (2011)

Technique for Data-Driven Mining in Physiological Sensor Data by Using Eclat Algorithm

Shraddha Kalbhor and S. V. Kedar

Abstract In this paper, we discuss the technique developed for mining the rules automatically in time series data which represent the physiological parameters in the clinical situation. Basically, the technique which mined the prototypical pattern for physiological time series data is known as data-driven technique. The patterns which are generated by the data-driven technique are explained in the format of text that seizes the tendency in each physical parameter and its connection with the data. In the following work, the sensor information in the multiparameter intelligent monitoring in intensive care (MIMIC) online database was utilized for appraisal, in which the mined transitory rules that were identified with different clinical conditions are expressed as a printed representation. Moreover, the proof that removed the tenet set for a specific medical condition was particular from different conditions. Moreover, for generating the patterns of rule, the system uses the FP-growth algorithm. But use of FP-growth algorithm may slow down the system. To solve this problem, the proposed system uses Eclat algorithm for generating the patterns of rule. Additionally, in this system, authorization of user is done; data can be shared with the authorized users only. Rule generation and pattern generation are performed at doctor side, and it can be accessed by authorized users and another doctor. If the system can refer the user to another doctor, then the only user can visit other doctors. At the doctor side, rules are encrypted by ECC algorithm. Doctor can prevent the disease, diagnose the disease, and also give suggestions to the patient. It gives better performance than previous algorithms. Evaluation of results shows the time and memory comparison of the proposed system with the existing system.

S. Kalbhor (✉)S. V. Kedar
Department of Computer Engineering, Rajarshi Shahu College of Engineering,
Savitribai Phule Pune University, Pune 411048, Maharashtra, India
e-mail: shraddhakalbhor@yahoo.com

S. V. Kedar
e-mail: seema_kedar@yahoo.com

© Springer Nature Singapore Pte Ltd. 2019   419
A. Abraham et al. (eds.), *Emerging Technologies in Data Mining and Information Security*, Advances in Intelligent Systems and Computing 755,
https://doi.org/10.1007/978-981-13-1951-8_38

1 Introduction

In hospital for monitoring and interpreted the patients data, there is a need for appropriate clinical datasets which indicate the data of patients properly. Sensors are used for collecting the data. Currently, the rate of collecting the physiological data is faster than the examining and modeling the data (Chen et al.). The parameters are further used for analyzing the different clinical conditions for early diagnosis the diseases. Some of the examples of parameters are monitoring the records of pulse rate, metabolism rate, and glucose state, which are the vital functions in medical stages. The measurements of physical credits are linear information in time string.

Nowadays, the frequently rise of fitness evidence in clinical information enhances the influence on fitness, for obtaining the knowledge there is a need to apply the data mining technique on the hospital data. For obtaining patient-specific data, different data mining techniques are used by the different decision support systems (Banaee et al.).

Cao et al. represent the predictive modeling technique on the basis of extracted trends and feature from time series data of heart rate and blood pressure. Author Rutledge et al. introduced a Bayesian network for modeling the intensive care unit for deriving a descriptive model of physiological states of the patients. Buchman and Riordan et al. introduced the use of examining the heart rate measurements for predicting and diagnose various clinical applications [1, 2]. Furthermore, some of the work has been done in clinical setting associated with vital signs for operating room monitoring system. Agarwal et al. studied on context-aware framework for examining physiological data, which collect data used for surgical procedure for detecting the significant changes and events [3, 4].

The existing system, for rule mining, association rule mining is used. In that, FP-growth algorithm is used for mining the data. But it does not give exact rule set at the output. It requires too much time to execute the dataset. Pattern abstraction is a method where the system generates a time series of each parameter design. Time series generate a series of parameters at an equal interval of time. Series generation means that to discrete the series in a segmentation (St = s1, s2, . . . , Sn).

After that, with different ideas time series separation, a sliding window method is used by almost all famous computations. In a sliding window concept, the dimension of aperture, i.e., w, is shown with extension along two regular windows. Clustering technique is used to form the cluster of subsequence series. The benefit of using the clustering technique is that the pattern generated by the technique is issued in a data-driven path that does not include any domain information for customizing the prototypical pattern. K-means clustering algorithm is applied to form clusters on each segment. The connection and the denotation for the sets of item that are used showing predecessor and the resultant in a written form are put by the natural language text. The proposed system uses Eclat algorithm for rule mining. In association rule mining, FP-growth algorithm is used, but in the existing system, it does not provide a satisfactory result and also takes more time to execute. Eclat is time efficient algorithm and also saves memory. Additionally, authentication of each user is done, i.e., only authorized user can access the data. Security is provided by using the ECC algorithm.

2 Literature Survey

In [1], the authors represent an ABAC approach mining algorithm. The proposed formula rehearses over ordered pair for users authorized connection by using selected ordered pair as kernel that develops applicant laws, for attempting to make statement for every applicants law to overcome extra ordered pair that substitutes the connection of authorized users conjuncts to assign expression along requirements.

This paper [2] formally defines the difficulties that clients face while developing disposable ingress command legislation and gives a ritual mechanism that handles everything effectively. The research has begun including speculative education that included different experts. The main aim was that to make a checklist and find difficulties with regard to the administration of access command legislation deposit and observe the handling of the difficulties by the experts.

In this article [5], the authors propose a role mining algorithm and presents the idea of weighted structural complexity measure and that mines RBAC frameworks with low structural complexity. Also, focus on the issue that has not been sufficiently tackled by existing role mining techniques is the way for finding roles with semantic meanings. At the point when the main data is user permission relation, they also proposed a technique for finding roles whose semantic meaning depends on formal concept lattices?

The paper [6] examines rule excavating along turbulent data as input also recommends separating issue with two different stages: no is discard also, applicant job creation. The system introduces a methodology with extreme position having decreased grid resolution that recognizes sound also investigation ally demonstrates its effectiveness of recognizing sound at authorized command information. Client and consent impute are in future utilized for enhancing perfection.

In paper [7], the author presents a structure for assessing role mining algorithms. The algorithm is classified into two steps: first one is output a sequence of prioritized roles while the second is complete RBAC states.

In [8], the authors proposed a novel algorithm for role mining. The proposed algorithm is used for optimizing numerous plans of standard benchmark that are built on plan proportions, benchmarks that are formed from understandability with due esteem to clients job related to imputed information also the amalgamation of the grade where dimensions and illustration are counted.

In paper [9], the authors present a parameterized RBAC (PRBAC) structure, wherein customers at the side of assent own credits implied within the variable jobs and may be utilized in role definitions, upgrading the scalability of RBAC by using parameterization. This paper offers algorithms for mining.

3 System Overview

In this section, the proposed system is discussed.

3.1 Proposed System Overview

Techniques used to implement this system are as follows:

- Prototypical Pattern Abstraction:
  Basically, this technique is developed for supporting a club of representative design taken from the unprocessed sequential information, which occurred in time series. For this task, two phases are proposed:
  (1) Discretization and
  (2) Clustering.
- Generating Rules:
  For generating a meaningful set of rules, association rule generation technique is used. These rule mining techniques are satisfactory methods for discovering logical connection among the multivariable patterns.
- Temporal Rule Set Similarity:
  It proposes the similarity function which calculates proportion among numerous laws taken away among R1 and R2 (R1 and R2 are law or rule records). These secular rules are extracted that show the behavior individually of important terms in a current situation.
  This similar nature of the function can assess the distinction of temporal rule sets.
- Temporal Rule Representation:
  The rules are represented by generating a textual representation for the systems end user.
- Data Encryption:
  Rules are encrypted for the security purpose, ECC algorithm is used for encryption. Steps of ECC algorithm are discussed in the algorithm section.
- User Authentication:
  User cannot access the data directly from the clinical portal. Initially, the user can register to the system for accessing the data. After successful login, the user can access the data, as the data is in encrypted format; initially, rules are decrypted by using the symmetric and asymmetric technique (Fig. 1).

3.2 Algorithm Used

3.2.1 Algorithm 1: ECLAT Algorithm for Rule Generation

1. For each item get tidlist.
2. Tidlist of a is similar list of transactions containing {a}.

Fig. 1 System architecture

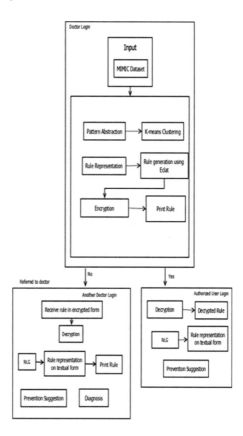

3. Intersect tidlist of a with the tidlists of all other items, resulting in tidlists of {a, b}, {a, c}, {a, d}, ... = {a} − conditional database (if {a} is removed).
4. Repeat from step 1 on {a} − conditional database.
5. Repeat for all other items.

3.2.2   Algorithm 2: ECC Algorithm

1. Key Generation:
   ECC algorithm used to generate the public and private keys.
   Public key is generated using this formula,
   $Q = d * P$
   where d = random number selected in the range of (1 to (n − 1)),
   P: point on the curve,
   Q: public key, and
   d: private key.

2. Encryption:

Let M be the message that we are sending.
Random select k from $[1 - (n - 1)]$.
Two cipher texts are generated; they are C1 and C2
C1 = k * P
C2 = M + k * Q
where C1 and C2 will be sent.

3. Decryption:

We get back the message M that is sending to us,
M = C2 − d C1
M can be represented as C2 − d C1 put (C2 = M + k Q and C1 = k * P)
C2 − d C1 = (M + k Q) − d (k P) put Q = d p
= M + k d P − d k P (canceling out k d P)
= M (Original Message).

The algorithm performs encryption, decryption, and reliability. By using this algorithm, key authority generates the public keys. These keys are used by users and senders of the system. In this system, the doctor encodes the data and stores in the database and gives access to the authenticated users. At last, the user gets to store the information at the storage on one condition if the client fulfills the access approach of the encrypted data.

3.3 Mathematical Model

System S = {Input, Process, Output}
   Input:D = is a set of input MIMIC dataset. The mimic dataset contains the three parameters having HR, RR, BP (Heart Rate, Respiration Rate, and Blood Pressure).

Process:

(1) Transaction Dataset:
   T = { t1, t2, …, tn}
   where
   T is each transaction in dataset.
(2) To find Segmentation:
   E= {S1, S2, …, Sn}
   where
   E is the segmentation.
(3) Finding rules:
   R = r1, r2, …, rn
   where
   R is the rules of set generated by FP-growth. (4) Rule Generation:

- Encryption:
  C1 = K p,
  C2 = M + kQ,
  C1 and c2 = cipher text,
  K= random number between $\{1 - (n - 1)\}$,
  P = point on curve,
  M = original message, and
  Q = public key.

- Decryption:
  M = c2 − d c1 and
  M = original message.
  Output:
  Textual-based rule set is given as an output
  O = O1, O2, …, On

4 Results and Discussion

4.1 Experimental Setup

The system is developed by using Java framework on the window platform. The development tool Netbeans 8.0.2 is used as a development tool.

4.2 Dataset Used

In this system, we use MIMIC dataset. The parameters or attributes of the dataset are pulsation, essential hypertension, and inspiration and expiration time, oxygen saturation, gained through intensive care unit monitors [4]. The whole database counts numerous recordings with different lengths of computation (from 1 to 77 h), which are obtained from 90 different patients that differ in age and gender.

4.3 Evaluation of Result

In the below section, we discussed the experimental result of the proposed system. The Table 1 shows the time comparison between the proposed system and the existing system. The following table shows that the time required for the existing system by using FP-growth algorithm is more than the time required for the proposed system by using Eclat classifier.

Table 1 Time efficiency comparison

Number of nodes	Existing system
Existing system using FP-growth	40,000,000 ms
Proposed system using Eclat	27,000,000 ms

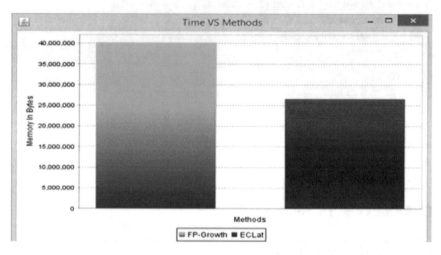

Fig. 2 Time efficiency comparison

Figure 2 shows the time comparison graph of the proposed system with the existing system. The time required for implementing the existing system by using FP-growth algorithm is more than the time required for the proposed system by using Eclat classifier.

The performance of the system measure on the time and memory parameter is discussed above. The proposed system used FP-growth algorithm which generates unique rules which are classified using the SVM algorithm. This method overcomes the data redundancy and increases the accuracy of the result as well as enhances the system performance.

5 Conclusion

The paper proceeds toward unmanned rule mining and depicting physiological sensor data by taking into consideration its singularity of rules in medical conditions. The principle part of extracting an information steer system is displaying transient examples content in streams of physiological information to create subject tenet set in a medical setting. Nine clinical conditions are taken into consideration in the initiated system, for example, cardiac infarction, septic infection, inhaling disappointment, alongside (heart rate, circulatory strain, and breath rate) these three sensor information. To improve the rule generation technique, the proposed system uses Eclat algorithm instead of FP-growth algorithm that already exists in the present method. Experimental results show the comparison between the existing and proposed system by using FP-growth and Eclat algorithm. ECC algorithm is used for providing the security to the rules. Also, user authentication process is performed and the doctor can diagnose the disease of the patient. From the results, it is concluded that the system with Eclat algorithm takes less time and low memory than the existing system. In the future, the system can be developed to generate more accurate rules. Means accuracy of the rule can be improved by which accurate results may be obtained.

References

1. Banaee, H., Loutfi, A.: Data-Driven rule mining and representation of temporal patterns in physiological sensor data. IEEE J. Biomed. Health Inf. **19**(5) (2015)
2. Hu, Y., Wan, X.: PPSGen: learning-based presentation slides generation for academic papers. In: IEEE Transactions on Knowledge and Data Engineering, vol. 27, no. 4, Apr 2015
3. Conrad, S., Schluter, T.: About the analysis of time series with temporal association rule mining. IEEE Symp. Comput. Intell. Data Min. pp. 325–332 (2011)
4. Dudek, D.: In: Measures for comparing association rule sets. Artificial Intelligence and Soft Computing, pp. 315–322. Springer, New York, NY, USA (2010)
5. Conrad, S., Schluter, T.: About the analysis of time series with temporal association rule mining. IEEE Symp. Comput. Intell. Data Min. 325–332 (2011)
6. Marlin, B.M., Kale, D.C., Khemani, R.G., Wetzel, R.C.: Unsupervised pattern discovery in electronic health care data using probabilistic clustering models. In: Proceedings of 2nd ACM SIGHIT, pp. 389–398 (2012)
7. Loutfi, A., Banaee, H., Ahmed, M.U.: A framework for automatic text generation of trends in physiological time series data. IEEE Int. Conf. Syst. Man Cybern. 3876–3881 (2013)
8. Zolhavarieh, S., Aghabozorgi, S., Wah, Y.: The a review of subsequence time series clustering. Sci. World J. **2014**, Article ID 312–521 (2014)
9. Umarani, V., Punithavalli, M.: Sampling based association rules mining-a recent overview. (IJCSE) Int. J. Comput. Sci. Eng. **02**(02), 314–318 (2010)
10. Wahyuningsih, Y., Muflikhah, L.: Fuzzy rule generation for diagnosis of coronary heart disease risk using subtractive clustering mMthod. Softw. Eng. Appl. **6**, 372–378 (2013). 25

A New Homogeneous Droplet Transportation Algorithm and Its Simulator to Boost Route Performance in Digital Microfluidic Biochips

Rupam Bhattacharya, Pranab Roy and Hafizur Rahaman

Abstract In the last few years, bishops-based on digital microfluidics give us proficient option in the area of clinical diagnostics. Cost of this alternative is relatively low which is also portable as well as disposable. A group of DMFBs can manipulate a nano-liter volume of discrete liquid globule using a series of electrodes arranged in a form of 2D array where actuation of liquid globule is based on the principal of electrowetting on dielectric. An important issue in design automation of DMFBs is parallel, time-synchronized routing of liquid globule called a droplet from its source to its destination inside the 2D array. The necessity of droplet routing is to schedule the movement of a set of droplets by minimizing utilization of resources while maintaining best possible latest arrival time. A liquid globule may be categorized either as heterogeneous or homogeneous. In case of homogeneous type droplet, at the time of routing, the main aim is to share the electrodes between liquid globule route ways so that we can minimize the number of electrodes to be used, that is utilization of resources will be high. In other words, we can say our aim is to allow a maximum number of contaminations at routing paths. In heterogeneous liquid globule routing, our aim is to minimize the number of collisions between more than one globule at the time of routing. In this paper, we suggest an algorithm for routing of homogeneous liquid globule and a complete simulator for that routing which shows details of routing, performance of routing within a given layout of DMFBs. We have used test benches given in benchmark suite I and III to test the performance of our algorithm.

R. Bhattacharya (✉)
Institute of Engineering & Management, Salt Lake, Kolkata 700091, India
e-mail: th_rup@yahoo.co.in

P. Roy
School of VLSI Technology, Indian Institute of Engineering Science
and Technology Shibpur, Howrah 711103, India
e-mail: ronmarine14@yahoo.co.in

H. Rahaman
Information Technology, Indian Institute of Engineering Science
and Technology Shibpur, Howrah 711103, India
e-mail: rahaman_h@yahoo.co.in

© Springer Nature Singapore Pte Ltd. 2019
A. Abraham et al. (eds.), *Emerging Technologies in Data Mining and Information Security*, Advances in Intelligent Systems and Computing 755,
https://doi.org/10.1007/978-981-13-1951-8_39

Keywords Digital microfluidics · Transportation · Bioassays · Algorithm
Routing region

The steady and huge development of micro fabrication methodology in the last few years helps the development of a novel set of appliance known as microfluidic biochips. Microfluidics implies machines, structure and rules to work with a nano-liter amount of liquid globule. It is an automated and highly parallel gadget primarily used for biological investigations. Digital microfluidic biochips are accomplished to manage the unit volume of liquid globule which can be controlled and manipulated separately on a planar substrate build with a set of electrode in a form of a 2D array. DMFBs operates on nano liter sized liquid globule which is similar to distinct valued functions so termed as digital microfluidic biochips. Management of liquid globule can be done using many actuation techniques, for example electrowetting, dielectrophoresis, thermocapillary transmission and surface acoustic signal transport. Electrowetting on Dielectrics (EWOD) has been considered to be the most efficient mechanism employed for DMFB-based applications [2]. When a liquid globule is placed on an electrode, a wetting power can generate upon use of an electric using those electrodes. That power performing in the Tri stage contact line can be changed by placing varied electrical potential to the liquid globule using the electrode. So any planned series of electric potential at successive electrodes can build a force disparity of those wetting power. The DMFBs is structured by a set of basic units called cells. A biochip is composed of two glass walls, which are parallel to each other. A sequenced array of independently controlled electrodes is represented by the ground glass wall. The top glass wall is covered by a continuing ground electrode. The stuffing medium, for example, silicon oil with a liquid globule to test is packed within the two glass walls. It is likely to adjust the interfacial tension of the liquid globule by applying time-varied voltage rates to switch on and off the electrodes of the DMFB. This can complete movement of liquid globules inside the total 2D array of electrode and operation of basic microfluidic processes for different bioassays. Method of transportation of a specific liquid globule from its source to destination discussed above permits to finds globule paths inside a given layout with dynamism. In routing of DMFBs, same cell may be shared by different droplet in a time synchronize way. This dynamic reconfigurability of liquid globule transmission and required mixing gives important advantages in contrast to long-established design automation methodology used in VLSI. The transmission path for a liquid globule at a given time can be controlled by microcontroller present outside of the chip which may be programmed to separately access any location represented by a cell in the given 2D array. A DMFB works under software control where protocols execute in a same way of conventional bench top techniques. The difference is that, DMFBs use better-quality level of automated system and smaller sample size with high level of parallelism (Fig. 1).

An important step in synthesis for DMBs is liquid globule transportation known as routing. A liquid globule needs to be transmitted or routed between modules, between the reservoir present on chip and modules and between modules and waste sinks.

Fig. 1 A representative drawing of DMFB [1]

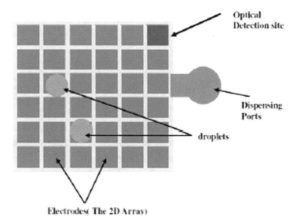

Optical Detection site

Dispensing Ports

droplets

Electrodes(The 2D Array)

Since DMFBs are dynamically reconfigurable, liquid globule or droplet movement paths may be considered as virtual route, which is a significant difference with routing used in traditional VLSI design. A liquid globule under test can be classified as homogeneous or heterogeneous. The homogeneous type implies liquid globule of the same kind or acquired from the same source. On the other hand heterogeneous droplets are of different kinds got from different sources of dissimilar environment. For homogeneous type liquid globule routing, a vital focus is on the utilization of cell that is electrode optimization along with minimizing the latest arrival time. In case of heterogeneous type liquid globule routing, the main aim is to avoid or reduce contamination between different globules while optimizing route time and resource utilization.

In this work we have proposed homogeneous type liquid globule routing algorithm. We have also developed a simulator of above algorithm which is an application based on JAVA applet. Our simulator graphically shows the initial position of the entire set of liquid globules which are predefined for the selected sub problem of a given test bench and then it will show routing paths along with different details of routing. It can also display individual routing path with required details.

In this work we have proposed homogeneous type liquid globule routing algorithm. We have also developed a simulator of above algorithm which is an application based on JAVA applet. Our simulator graphically shows the initial position of the entire set of liquid globules which are predefined for the selected sub-problem of a given test bench and then it will show routing paths along with different details of routing. It can also display individual routing path with required details.

The remaining of the paper is arranged as follows: Sect. 2 briefly depicts recent works which are related to liquid globule routing in DMFBs. In Sect. 3 we devise the problem of parallel routing in DMFBs. At part IV we describe the inspiration of our work. In Sect. 5 we describe our proposed method of homogeneous routing. Segment VI shows a descriptive illustration of homogeneous globule routing for a given arrangement of the initial positions of source–target pairs. In Sect. 7 we have given and demonstrate the computed outcome and performance of our proposed algorithm. At division VIII, we provide details about our routing simulator.

1 Related Works

Transportation of liquid globules is one of the most important procedures in digital microfluidic biochip which raises a significant design automation matter to obtain optimized results in parallel bioassay in tandem. In the last few years a significant number of methods have been developed for optimized liquid globule transportation in DMFBs. Su et al. [3] investigated dynamically reconfigurable microfluidic array. This proposed technique starts with an initial placement. Then at the phase of module placement, at different time length, it will get a configuration of a set of 2D placement. After that, optimized transportation paths are selected to finalize liquid globule routing. In [4], a graph coloring approach was used.

Here direct addressing scheme was used where, depends on the liquid globule time, an acyclic graph was generated and coloring was done based on parallel liquid globule routing. In [5], cross referencing based routing in DMFBs was proposed. In this technique, liquid globule transportation problem has been cracked by a model developed using graph clique. ILP that is Integer linear programming (ILP) was used in [6] to the modeled direct referencing mode of liquid globule transmission. In [7], a method based on network flow was introduced for routing of liquid globules in DMFBs. That algorithm was built on non-intersecting bounding box method. In [8], a liquid globule transportation methodology of cross-contamination avoidance was introduced. This method tries to reduce the number of used cell for transportation by aiming transmission paths of liquid globules which are disjoint to each other. Soukup's routing method for transmission path assignment of more that one liquid globule was discussed in [9]. An algorithm based on A* search technique was proposed in [10]. It used a graph representation method to represent the position of the source–destination pair at different time. After that, optimum path from source to destination has been found using an A* based algorithm. In [11], stalling and detour was taken into account in routing scheme. In this method, selection of liquid globules was done on the basis of the Manhattan paths between source and destination. A partition oriented technique was developed in [12] for pin constrained design. A broadcast depended addressing of routing was developed in [13]. This approach was mainly for pin constrained digital microfluidic biochips.

2 Simultaneous Routing Problem in DMFBs

Digital microfluidic bishop can be represented by a collection of electrode arranged in a form of 2D array with x number of rows and y number of columns. A liquid globule under test is placed on the top of a hydrophobic plane over an electrode. Every liquid globule is transmitted to a predetermined destination, where they get mixed with probes which are pre placed, fluorescently labeled library. The smallest virtual rectangle encapsulates a source–target combination of a liquid globule is termed as *routing region*. Routing finds the paths for the liquid globule under test through the rooms represented by the electrodes. Routing of liquid globules must be taken place in parallel while satisfying certain essential restrictions. To achieve effective routing, algorithms for parallel transmission of liquid globule need to be formulated which should give efficient routing by satisfying time and utilization of the cell.

The liquid globules with one source and target under test routed through pins of several reservoirs on chip or modules are called 2-pin net. There is another type of net consists with two sources, a mixer and a destination called 3-pin net. For storage, splitting, mixing its modules is placed on the chips known as *Hard Blockages* or *Obstacle*.

The following constraints are applied to routing. To achieve effective liquid globule transportation, there must always have at least one cell space between two liquid globules, so that, there will not have any accidental collision between two or more globules. Cells which constitute *Hard Blockages* are reserved, that is, those cells cannot be used for droplet transportation. In case of homogeneous type liquid globules, an important objective is to use same electrodes for routing in the time multiplexed manner. This will improve resource utilization. We should take care so that one cell should not be shared by more than three liquid globules. This is because, after transportation, there will be residual of the liquid which will be large enough to overlap more than one cell if same cell has been used for the transportation of more than three liquid globules. While routing is on, the algorithm must ensure that at the same time, no two globule passes through the same cell. The sharing of a cell should be taking place in a time sliced manner.

Fluidic constraint entails a liquid globule at position (m, n) at time stamp p, all surrounding cell $(m, n - 1)$, $(m + 1, n + 1)$, $(m + 1, n - 1)$, $(m - 1, n - 1)$, $(m - 1, n + 1)$, $(m + 1, n)$, $(m - 1, n)$ and $(m, n + 1)$, are restricted for any other globule to enter at the time stamps p and $p + 1$.

3 Inspiration

Concurrent routing of droplet is very important to boost the performance of the DMFBs. This is because parallel routing maximizes the resource utilization while optimizing the latest arrival time. It is important to build an efficient algorithm for homogeneous droplet routine which should minimize cell utilization hence resource utilization will be maximized. The algorithm should also satisfy all the constraints discussed in Sect. 3.

The majority of the previous effort depicts in Sect. 2, concurrent routing has been approached for those droplets which are free from collision and blockages. In this work, we devise an algorithm for homogeneous droplet routing, which allow complete parallelism at the same time ensure maximum contamination by satisfying all the Constraints.

4   Homogeneous Droplet Transportation Algorithm

INPUT: Pool by P net, pool of hardened blockages
OUTPUT: transportation path, contaminations, Cell used, LAT, AAT

STEP 1: Let there are n droplets.
STEP 2: Arrange the droplets in decreasing Manhattan distance.
STEP 3: Priority of a net (source–target pair) will be same as its Manhattan distance.
The higher priority value indicates higher priority.
STEP 4: For the droplet with highest priority, out of two Manhattan paths, it will
select that which is moving through the maximum number of routing zones of other
droplets.

For i=2 to n do
    For j=1 to n do
        If i<> j AND Prioriry (Di) < Priority (Dj) **Then**
            Droplet Di(Si, Ti) evaluates and determine the routing path in following way
            Init_pos=Si
        WhileInin_pos!=Ti
          a.   Starts with two Manhattan paths, whenever there is a contamination with droplet Dj,
              It will check the next cells towards TJ for more contaminations.
          b.   From that contamination point it will evaluates all the paths towards Ti and deviates
              through the contaminated path.
          c.   Count number of contaminations for each path of Di.
          d.   The maximum allowed contaminations in a cell will be 3.
          e.   The collision should be avoided by stalling of low priority liquid globule
          f.   While evaluating the routing path of Di,detour is not allowed. Routing will be inside
              routing zone of Di.
          g.   **If** hard blockages found then
          i.   Detour from the blocked cell to avoid hard blockages.
          ii.   Backward movement is not allowed.
          iii.   From blockage, evaluates two detoured paths.
          iv.   Select that detoured path where the number of cells to be traversed is less.
          v.   If two detoured path have a same number cell, then select that path which
              moves through the maximum number of routing zone.
        End if
          h.   Finally select that path for droplet Di (Si, Ti) where number of contaminations is
              highest and avoid all hard blockages.
      End while
      End if
    End for
End for

5   An Elestrative Example of Heterogeneous Droplet Routing Algorithm

In this example, we have taken a 2D grid with 11 rows and 11 columns where we
have five 2-pin nets represented by respective source–target of five liquid globules.
Here, V1, V2, V3, V4, and V5 are representing the source and R1, R2, R3, and R5
are targets respectively (Fig. 2).

Rooms marked by gray color are represented as hard blockages. All five droplets
run simultaneously on the way to respective targets. From each source to target, we

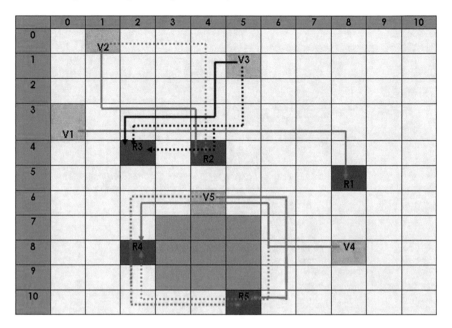

Fig. 2 Evaluation of paths

have considered two possible Manhattan paths. A virtual rectangle defined by (3, 0) to (5, 8) from the routing region of liquid globule (V1, R1). For droplet (V1, R1), the solid line represents the selected path. That path has been selected because that path is moving through a number of routing regions of other droplets, so the chance of contaminations is relatively high compared to another Manhattan path. Similarly, for droplet (V2, R2) we have considered tow Manhattan paths. At cell (3, 1) one Manhattan paths find a contamination. Now, from that point, it will check the next cell towards destination for more contaminations. Since there is contamination in cell (3, 2), that path has changed direction from the cell (3, 1) towards (3, 2). Change of direction of the path is allowed, but the detour is no allowed. The same way we have considered second Manhattan path. Again the solid line represents the selected path out of two possible Manhattan paths. That path has been selected because there are more contaminations comparable to other path.

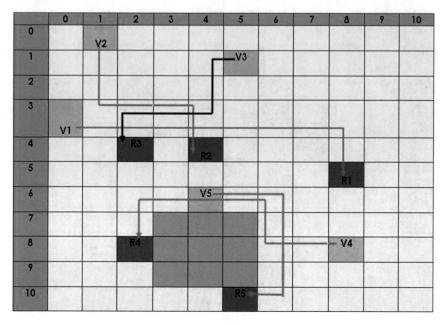

Fig. 3 Selected path of each droplet

For droplet (V4, R4) since there are hard blockages, droplet detoured from cell (8, 6) and examine both possible paths to reach target. Considering the routing region of droplet (V5, R5) it will select the path represented by the solid line. For droplet (V5, R5), the algorithm selects the solid line path because a lesser number of cells need to traverse in that direction to reach the target compare to other path.

Figure 3 shows the selected paths from each source to respective destinations.

6 Experimental Results

Here we have used test benches like Invitro-1, Invitro-2, protein1, and protein2 which are under benchmark suite II. Apart from that we have used benchmark suite III to test our algorithm discussed in Sect. 5. At Table 1 we depict the detail result of our algorithm on test bench Invitro-2 and at Table 2, we have depicted a summarize report on all above-mentioned test benches. Our main focus is to maximize cell utilization that is contaminations while optimizing biggest arrival time.

Table 1 Detailed result of homogeneous routing of Invitro-2

Sub problem	Droplet type		Homogeneous routing			
	2 Pin	3 Pin	LAT	Avg. arrival time	Cell used	Contamination
1	2	0	13	12.5	13	1
2	1	1	11	11	15	6
3	4	1	15	11	40	18
4	3	0	12	7	14	0
5	3	1	8	8	23	8
6	2	0	12	9.5	17	0
7	1	0	8	8	3	0
8	5	1	14	8	36	4
9	0	2	15	17	31	8
10	1	0	4	4	3	0
11	1	0	12	12	11	0
12	2	0	3	2	3	0
13	2	0	9	7	12	0
14	1	0	12	12	11	0
15	1	0	5	5	4	0
Total	29	6	168	134	236	45

Table 2 Summarized result of Homogeneous routing of benchmark suite II and III

Test bench				Homogeneous routing result			
Bench suit	No. of sub prob	2-pin droplet	3-pin droplet	Max route time	Avg. route time	Contamination	Cell used
INVITRO1	11	20	6	19	129.34	75	229
INVITRO2	15	29	6	15	134	45	236
PROTEIN1	64	170	8	24	717.07	205	1570
PROTEIN2	76	153	8	28	542.63	130	945
TEST12_12_1	12	12	0	15	14.75	74	59
TEST12_I2_2	12	12	8	19	11.42	89	57
TEST12_12_3	12	0	12	21	11.17	101	61
TEST16_16_1	16	16	0	30	15.75	141	101
TEST16_16_2	16	16	8	25	14.4	140	106
TEST16_16_3	16	16	0	34	15.44	162	107
TEST16_16_4	16	16	0	31	14.73	151	114
TEST24_24_1	24	24	8	35	23.25	331	219
TEST24_24_2	24	24	8	43	20.17	329	211

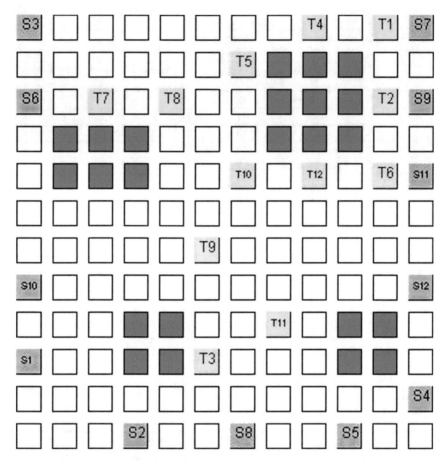

Fig. 4 Preliminary placement of Test12-12-1

7 Our Simulator

We have designed a simulator of our homogeneous liquid globules algorithm. This simulator is built using JAVA Applet which gives liquid globule transportation details of our algorithms. Figure 4 shows the primary placement of the Test12-12-1. In Fig. 5 shows the homogeneous globule transmission path and other information of Test12-12-1.

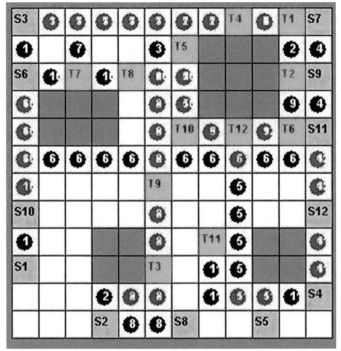

CELL USED: 59

TOTAL CONTAMINATION: 74

LAT : 15   **AAT :** 14.75

Fig. 5 Red bubbles specifies one or more contamination

References

1. Ren, H., Paik, P., Fair, R.B., Pollack, M., Pamula, V., Srinivasan, V.: Electrowetting-based on-chip sample processing for integrated microfluidics. In: Proceedings of the IEDM Technical Digest, pp. 32.5.1–32.5.4 (2003)
2. Su, F., Zeng, J.: Computer-aided design and test for digital microfluidics. In: IEEE Design & Test of Computers, pp. 60–70, Jan–Feb 2007
3. Su, F., Chakrabarty, K., Hwang, W.: Droplet routing in the synthesis of digital microfluidic biochips. In: Proceedings of Design Automation and Test in Europe (2006)
4. Griffith, E.J., Goldberg, M.K., Akella, S.: Performance characterization of a reconfigurable planar-array digital microfluidic system. IEEE Trans. Comput.-Aided Des. Integr. Circuits Syst. **25**(10), 340–352
5. Chakrabarty, K., Xu, T.: A cross-referencing-based droplet manipulation method for high-throughput and pin constrained digital microfluidic arrays. In: Proceedings of DesignAutomation and Test in Europe, pp. 552–557, Apr 2007
6. Yuh, P.H., Sapatnekar, S., Yang, C.-L., Chang, Y.-W.: A progressive-ILP based routing algorithm for cross referencing biochips. In: Proceedings of Design Automation Conference, pp. 284–289, June 2008

 7. Chang, Y.W., Yang, C.L., Yuh, P.H.: Bioroute: a network flow based routing algorithm for digital microfluidic biochips. In: Proceedings of IEEE/ACM International Conference on Computer Aided Design, pp. 752–757 (2007)
 8. Chakrabarty, K., Zhao, Y.: Cross-contamination avoidance for droplet routing in digital microfluidic biochips: In: Proceedings of the DATE (2009)
 9. Rahaman, H., Dasgupta, P., Roy, P.: A novel droplet routing algorithm for digital microfluidic biochips. In: GLSVLSI, Providence, Rhode Island, USA (2010)
10. Boahringer, K.F.: Modeling and controlling parallel tasks in droplet based microfluidic systems. IEEE Trans. Comput. Aided Des. Integr. Circuits Syst. **25**(2), 334–344 (2006)
11. Soukap, J.: Fast maze router. In: Proceedings of the 15th ACM/IEEE Design Automation Conference, pp. 100–102
12. Chakrabarty, K., Xu, T.: Droplet-trace-based array partitioning and a pin assignment algorithm for the automated design of digital microfluidic biochips. In: Proceedings of IEEE/ACM International Conference on Hardware/Software Codesign and System Synthesis, pp. 112–117 (2006)
13. Chakrabarty, K., Xu, T.: Broadcast electrode addressing for pin-constrained multi-functional digital addressing for pin-constrained multi-functional digital microfluidic biochips. In: Design Automation Conference, Anaheim, California USA, pp. 173–158, June 2008

A New Combined Routing Technique in Digital Microfluidic Biochip

Rupam Bhattacharya, Pranab Roy and Hafizur Rahaman

Abstract In last one decade, Digital Microfluidic Biochips (DMFBs) are providing a well-organized platform in field of biochemical study. Particularly in the area of clinical diagnostic related applications, DMFBs provides low priced, movable and disposable tools. In a large section of DMFBs, actuation of droplet is accomplished by the method of electro wetting-on-dielectric, where nano-liter volume liquid or droplet can be controlled and manipulated on two-dimensional array of electrode. A most important design automation concern in DMFBs is the parallel transportation of droplet in a time multiplexed way inside a 2D array of electrode. The requirement of droplet routing is to organize the transportation of droplets in parallel by minimizing resource usage by satisfying maximum allowed time for routing. A droplet can be either homogeneous or heterogeneous. For routing of homogeneous type droplets, main aim is to share same electrodes between several route paths of different droplets for minimizing cell usage. In other words, our aim is to maximizing the cross contaminations. For heterogeneous type droplets our main aim is to eliminate or minimize the cross contamination. In most of the previous works, algorithm has been proposed either for homogenous or for heterogeneous droplet routing. In this work, we proposed an algorithm for combined routing. Our algorithm as input takes a sub-problem consists with a set of homogeneous droplets and a set of heterogeneous droplets. Then our algorithm applies homogeneous routing for homogeneous droplets where it will maximize contaminations and for heterogeneous droplets it will minimize contaminations by applying heterogeneous routing. We have applied

R. Bhattacharya (✉)
Institute of Engineering & Management, Kolkata, India
e-mail: th_rup@yahoo.co.in

P. Roy
School of VLSI Technology, Indian Institute of Engineering Science and Technology, Shibpur, Howrah, India
e-mail: ronmarine14@yahoo.co.in

H. Rahaman
Information Technology, Indian Institute of Engineering Science and Technology, Howrah, India
e-mail: rahaman_h@yahoo.co.in

© Springer Nature Singapore Pte Ltd. 2019
A. Abraham et al. (eds.), *Emerging Technologies in Data Mining and Information Security*, Advances in Intelligent Systems and Computing 755,
https://doi.org/10.1007/978-981-13-1951-8_40

our algorithm on a test12_12_2 present in bench mark suite I. In test12_12_2 sub-problem, we have 12 droplets. We have assumed six droplets are homogeneous and remaining is heterogeneous.

Keywords Digital microfluidics · Routing · Droplet · Algorithm · LAT

1 Introduction

The significant growth in the field of micro-fabrication from last decade direct to the appearance of a new set of tools known as microfluidic biochips. Tools based on microfluidics deals with nano-liter or micro-liter volume of liquid or droplet. Biochips are highly parallel computerized devices mainly applied for biochemical analysis and test. The development of initial class of microfluidic based biochip was based on continuous liquid flow in fixed micro channels. Using this technology, integration and scaling was difficult. Fixed channel based systems results in limited reconfigurability and capability of fault tolerance was poor. A new technology has been developed where a unit volume of liquid can be controlled and manipulated on a grid of two-dimensional array of electrode. This technology known as digital microfluidics since tools based on this technology use nano-liter volume of independently controllable liquid droplet as operational unit. Handling of liquid droplet can be done using several technologies for example dielectrophoresis, electrowetting, and surface acoustic wave transport. Out of these technology, EWOD, i.e., Electrowetting on Dielectrics considered as best technology used for DMFB-related operation and application [1]. When a nano liter volume droplet is placed on an electrode a wetting power will generate on the tri phase contact point may be varied by introducing varied electrical potential to the droplet using the electrode. So sequences of programmed electrical voltage at successive electrode create a potency disparity of this wetting power. This will force a droplet to move from current electrode to next successive electrode. The two-dimensional micro-fluidic array is built with a set of electrode which represent a set of basic cells. The set is built with two parallel glass shields. The bottom one is individually controllable array of electrode. The top glass plate is covered with a continuous ground electrode. The droplet along with silicon oil as filler medium is squeeze in between two parallel glass plates. It is then possible to organize the interfacial pressure of the droplet using time changing voltage to switch off/on of the electrodes positioned on the DMFBs. This is the way to achieve routing of droplet throughout the entire two-dimension array and set of basic operations on droplet for bioassays. The technique we have discussed above for droplet routing allows us to determine routing paths at run time for a predefined layout. Same cell can be shared in DMFBs routing in a time synchronized way for basic operation of DMFBs on droplets like transportation, mixing, splitting, merging, etc. One can determine droplet routing path using a microcontroller placed outside of the chip. That microcontroller can be programmed so that individual access on any location on two dimensional array of electrode can be achieved. A digital biochip works under the control of software, where methods of work are almost same as conventional bench top methods. The only difference is that, in DMFBs, level of automation and parallelism is extremely high and the requirement of sample is very small in size. At geometry level synthesis one of the most vital step in DMFB is routing or transportation of droplet between several components of biochips like modules, reservoir, waste sinks, and photo detectors. Since DMFBs are dynamically reconfigurable, routing path of a droplet can be considered as virtual route path which is an important difference with routing technique applied on conventional design used in VLSI design. A droplet in DMFBs terminology can be termed as heterogeneous or homogeneous. Droplets from the same source are termed as homogeneous droplet where as droplets from different source as considered as heterogeneous type droplet. In case of homogeneous droplet routing, objective is to share same set of cells or electrodes between route paths of different homogeneous droplets. This will increase number of cross contaminations and number of cell or electrode requirement will be reduced. For heterogeneous droplet routing, one should try to minimize the cross contaminations because heterogeneous droplets are originating from different sources. In both the cases, we have to optimize latest arrival time of the droplet. In earlier works, methods has been designed either for homogenous or for heterogeneous droplet routing. In our work, we have developed an algorithm for combined routing. Our algorithm as input takes a sub-problem consists with a set of homogeneous droplets and a set of heterogeneous droplets. Then our algorithm applies homogeneous routing for homogeneous droplets where it will maximize contaminations and for heterogeneous droplets it will minimize contaminations by applying heterogeneous routing.

The remaining of the paper is arranged as follows: Sect. 2 depicts current works which are related to droplet routing in DMFBs. At Sect. 3 we have discussed the problem of parallel routing in DMFBs. At Sect. 4 we describe the motivation of this job. In Sect. 5 we describe our proposed method of combined routing. Section 6 shows a descriptive illustration of combined routing for a given arrangement of initial positions of source–target pairs. In Sect. 7 we have given and demonstrate the computed outcome for droplet routing.

2 Prior Works

Parallel transportation or routing of droplets is an important step in DMFBs. Routing must satisfy those constraints we have discussed previously. In most of the previous work, droplets are considered as either homogeneous or heterogeneous. In some cases, same methods have been used for both type droplets. In recent years a number of techniques have been designed for optimization of droplet routing in DMFBs. In [2], direct addressing mode of routing was developed using integer linear programming (ILP). Routing technique based of cross referencing technique was proposed in [3]. In this approach, graph clique has been used to identify the routing path. Graph coloring method has been devised in [4]. It was a direct addressing method. A acyclic graph was formed by considering droplet routing time. After that, depend on the parallel droplet transportation, coloring of the graph was done. Reconfigurable, dynamic, microfluidic array was investigated in [5]. The designed method starts by considering an initial arrangement. At module placement phase at different time length, it will generate a set of two dimensional placement. Then routing paths are selected to complete droplet routing. A* search based method was developed in [6]. Source–target position of droplet at different time was represented using graph. Final routing path of a droplet from source to target then determined using A* based method. In [7], assignment of routing paths of multiple droplets using Soukup's routing technique was discussed. An algorithm to avoid cross contamination in droplet routing was developed in [8]. This algorithm attempts to minimize cross contamination in droplet transportation by planning routing path of those droplets which are not overlapping to each other. In [9], an algorithm was developed using partition based approach for the design of pin constrained routing. In [10], broadcast-oriented method of droplet transportation was design. This method was primarily developed for pin constrained DMFBs. Network flow based model was proposed in [11] for droplet routing. The proposed method was based on non-overlapping bounding box method. Algorithm proposed in [12] where choice of droplet was depend on Manhattan distance between source and target of the droplet. Detour and stalling was considered in this technique.

3 Parallel Routing Problem in DMFBs

DMFBs can be treated as a collection of electrode-organized in a form of 2D array with u number of rows and v number of columns. A liquid droplet under test is kept on the surface of a hydrophobic plane over an electrode. Each droplet is routed to a destination which is predefined, where they get mixed with probes which are also pre placed, fluorescently labeled library. Some important conditions need to be satisfied while parallel routing of droplet is taking place. To achieve the goal of efficient routing, algorithm or method for concurrent transportation of droplet need to be device that must ensure efficient routing, at the same time follow their source utilization and time constraint.

A liquid droplet consists with one source and one destination or target can be termed as 2-pin net. There are another category of droplet with two sources a mixer and a target is termed as 3-pin net. Modules consists with cells on chip are reserved for operations like splitting, storage, mixing, etc., are termed as hard blockages. Those cells cannot be used for routing purpose.

There are certain constraints that must be satisfied at the time droplet routing. For efficient droplet transmission, between two droplets, there should be at least one cell space. Otherwise there is a chance of accidental mixing of two droplets which is not at all desirable particularly for heterogeneous type droplet. While sharing the cells for routing, one cell should not be shared by more than four droplets, because in that case left over of each droplet will form a larger size droplet which may overlap one cell and leads to malfunction.

So, at the time of droplet routing, algorithm must ensure that at equal time, no two droplets routed through the same electrode. Electrode sharing need to take place in time multiplexed manner.

Fluidic constraint demands a liquid droplet at position (s, t) at time stamp p, all adjacent electrode (s, t − 1), (s + 1, t + 1), (s + 1, t1), (s − 1, t − 1), (s − 1, t + 1), (s + 1, t), (s − 1, t) and (s, t + 1), are restricted for any other droplet to come inside at time stamps p and p + 1.

4 Motivation

Parallel routing is very important to uplift the performance of the DMFBs. Parallel routing must take place by satisfying fluidic constraint discussed in previous section.

Almost all previous work discussed in Sect. 2, either they have designed routing algorithm for homogeneous droplet or for heterogeneous droplets. In some cases same procedure has been applied on both homogeneous and heterogeneous type droplets.

The algorithm we have developed in this work, we have considered some of the droplets are homogeneous and remaining are heterogeneous in a given sub-problem. Our algorithm routed homogeneous droplets is a way so than contamination increases so resource sharing will maximize. For heterogeneous droplets, same algorithm follows the route procedure so that contamination will be minimized. In both cases, our algorithm satisfies the fluidic constraint and latest arrival time.

5 Algorithm for Heterogeneous Droplet Routing

INPUT-1: M set of homogeneous droplets and P set of heterogeneous droplet.
INPUT-2: H set of hard blockages.
OUTPUT: Routing paths, homogeneous contamination, heterogeneous contamination, LAT, AAT.
Step-1: Let there are total N droplets in a given sub-problem. Out of that, there are M number of homogeneous and P number of heterogeneous droplets.
Step-2: For i = 1 to N do

   a. For droplet Oi, whose source is Si and destination is Ti, consider two available Manhattan paths PT1 and PT2 from source to target.
   b. Let at path PT1, there are CPT1_1 number of homogeneous contamination and CPT1_2 number of heterogeneous contamination. Similarly at path PT2 there are CPT2_1 number of homogeneous contamination and CPT2_2 number of heterogeneous contamination.
   c. CPT1 = CPT1_1 + CPT1_2 and CPT2 = CPT2_1 + CPT2_2.
   d. If Oi is a homogeneous droplet then

    i.   To initiate routing, Select PT1 if CPT1_1 > CPT2_1 otherwise select CPT2.

   ii.   Let C_position = Si

  iii.   While C_position ! = Ti do

       1. Traverse one cell at a time stamp through selected path.

       2. If there is a contamination at the next cell with another droplet D and if D is homogeneous then allow the contamination.

       3. From that contamination, check for more homogeneous contamination towards Ti. If total homogeneous contaminations are more than selected PTi then deviate from PTi to newpath.

       4. If D is heterogeneous droplet then From the cell just before the contamination consider two paths NH1 and NH2 to avoid that contamination. Between NH1, NH2 and initially selected PTi, select that path where number of homogeneous contamination is maximum.

       5. If there is a hard blockage at the next cell, then consider two detoured paths DH1 and DH2 from the current cell. Select DH1 if number of cell to be traversed in DH1 is less that DH2.

          Otherwise select DH2.

          End while.

    End if.

e.   If Oi is a heterogeneous droplet then

      i.   To initiate routing, Select PT1 if CPT1 < CPT2 otherwise select CPT2.

     ii.   Let C_position = Si

    iii.   While C_position ! = Ti do

         1. If there is a contamination at the next cell with another droplet D.

         2. From the cell just before the contamination consider two paths NH1 and NH2 to avoid that contamination. Between NH1, NH2 or initially selected PTi, select that path where contamination is minimum.

         3. If there is a hard blockage at the next cell then follow d.iii.5 of step

         End while.

    End if.

6   An Elestrative Example of Heterogeneous Droplet Routing Algorithm

Here we are considering 11×11 grid chip where we have 11 rows and 11 columns. In this arrangement we have 6 droplets where U1, U2, U3, U4, U5 and U6 are source and V1, V2, V3, V4, V5, and V6 are respective targets. We are assuming that droplet D1(U1, V1), D2(U2, V2) and D3(U3, V3) are homogeneous droplets and D4(U4, V4), D5(U5, V5) and D6(S6, T6) are heterogeneous in nature. Cells marked by gray color are a hard obstacle (Fig. 1).

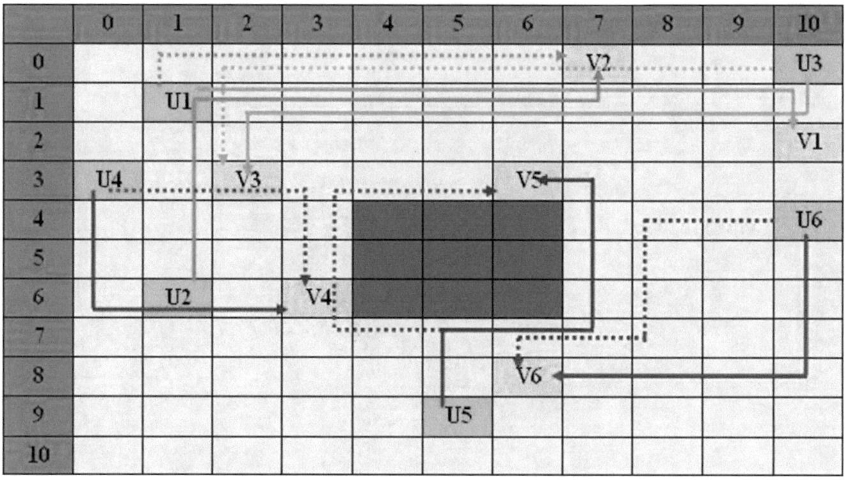

Fig. 1 Route paths of each droplet

Since droplet D2 is homogeneous, intention is to route it through a homogeneously contaminated path. Initially it will start traversing rows first because in that Manhattan paths there are more contamination than another Manhattan path. At cell (1, 1) it will encounter first contamination with D1. Since D1 is homogeneous with D2, so this is a homogeneous contamination. So we allow this contamination and check next cell for more contamination towards it target V2. In this new direction there are more homogeneous contaminations than initially selected path shown in dotted blue line. It will now deviate to new path marked with solid blue line from initially selected path. In this way, according to our algorithm, D1, D2, D3 will be routed.

On the other hand, D4 is heterogeneous in nature. So it will select solid green line path because in this path number if contamination is less compare to another Manhattan path shown in dotted green line. Contamination at cell (6, 1) is inevitable because apart from encountering hard blockages, detour is not allowed.

Droplet D5 encounter hard blockage at cell (7, 5). So according to our algorithm, D5 will consider two detoured path to avoid the hard blockages. Solid line path is the selected one as number of cell to be traversed in this path is less compare to other detoured path marked by dotted line (Fig. 2).

This diagram shows the final paths for routing. In this example we have 16 homogeneous contaminations, that is contaminations between D1, D2 and D3. We have 1 contamination between D4 and D2 at cell (6, 1). This is a homogeneous-heterogeneous contamination because S4 is heterogeneous droplet and S2 is homogeneous droplet. There is no contamination between two heterogeneous droplets.

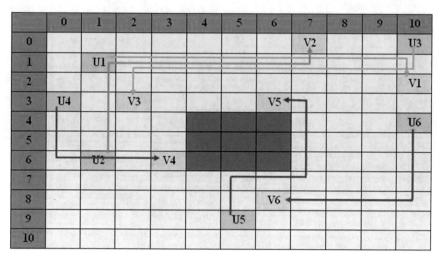

Fig. 2 Final route path

7 Experimental Results

We have used our algorithm depicts at section (V) on Test_12_2_1 of benchmark suite III. Here we assume that droplets D0, D2, D4, D6, D8, D10, D12 are heterogeneous and D1, D3, D5, D7, D9, D11 are homogeneous in nature. Following diagram shows the final route path determined by our algorithm (Fig. 3).

Here a heterogeneous droplet Di(Ui, Vi) marked with black color and its route path is marked by green solid line. Homogeneous droplet Di(Ui, Vi) marked with red color and route paths are marked by maroon solid line.

We have observed following detail of contaminations.

Contamination between Heterogeneous droplets	Contamination between Homogeneous droplets	Contamination between Heterogeneous and Homogeneous droplets	Cell used	LAT	AAT
3	46	22	79	21(D3)	13.166

Here LAT implies latest arrival time or maximum arrival time and AAT stands for average arrival time.

In this experiment we have got following detail about droplets and its routing.

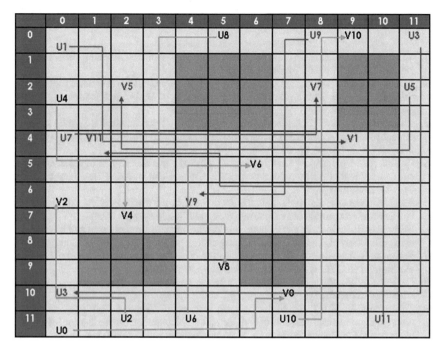

Fig. 3 Final route path for Test12_12_2

Droplet	Manhattan distance	Arrival time	Stall	Detour
D0	8	8	0	0
D1	13	13	0	0
D2	7	7	0	0
D3	21	21	0	0
D4	7	7	0	0
D5	9	16	2	2
D6	8	8	0	0
D7	10	10	0	0
D8	9	13	0	2
D9	10	10	4	0
D10	13	18	7	0
D11	16	16	0	0
Total			13	4

References

1. Su, F., Zeng, J.: Computer-aided design and test for digital microfluidics. In: IEEE Design & Test of Computers, pp. 60–70, Jan–Feb 2007
2. Yuh, P.-H., Sapatnekar, S., Yang, C.-L., Chang, Y.-W.: A progressive-ILP based routing algorithm for Cross referencing biochips. In: Proceedings of Design Automation Conference, pp. 284–289, June 2008
3. Chakrabarty, K., Xu, T.: A cross-referencing-based droplet manipulation method for high-throughput and pin-constrained digital microfluidic arrays. In: Proceedings of Design Automation and Test in Europe, pp. 552–557, Apr 2007
4. Griffith, E.J., Goldberg, M.K., Akella, S.: Performance characterization of a reconfigurable planar-array digital microfluidic system. IEEE Trans. Comput.-Aided Des. Integr. Circuits Syst. 25(10), 340–352
5. Su, F., Chakrabarty, K., Hwang, W.: Droplet routing in the synthesis of digital microfluidic biochips. In: Proceedings of Design Automation and Test in Europe (2006)
6. Boahringer, K.F.: Modeling and controlling parallel tasks in Droplet based microfluidic systems. IEEE Trans. Comput. Aided Des. Integr. Circuits Syst. 25(2), 334–344 (2006)
7. Rahaman, H., Dasgupta, P., Roy, P.: A novel droplet routing algorithm for digital microfluidic biochips. In: GLSVLSI, Providence, Rhode Island, USA (2010)
8. Chakrabarty, K., Zhao, Y.: Cross-contamination avoidance for droplet routing in digital microfluidic biochips. In: Proceedings of the DATE (2009)
9. Chakrabarty, K., Xu, T.: Droplet-trace-based array partitioning and a pin assignment algorithm for the automated design of digital microfluidic biochips. In: Proceedings of IEEE/ACM International Conference on Hardware/Software Codesign and System Synthesis, pp. 112–117 (2006)
10. Chakrabarty, K., Xu, T.: Broadcast electrode addressing for pin-constrained multi-functional digital addressing for pin-constrained multi-functional digital microfluidic biochips. In: Design Automation Conference, Anaheim, California USA, pp. 173–178, June 2008
11. Chang, Y.W., Yang, C.L., Yuh, P.H.: Bioroute: a network flow based routing algorithm for digital microfluidic biochips. In: Proceedings of IEEE/ACM International Conference on Computer Aided Design, pp. 752–757 (2007)
12. Soukap, J.: Fast maze router. In: Proceedings of the 15th ACM/IEEE Design Automation Conference, pp. 100–102

Optimized Multi-agent Personalized Search Engine

Disha Verma, Barjesh Kochar and Y. S. Shishodia

Abstract With the advent of personalized search engines a myriad of approaches came into practice. With social media emergence the personalization was extended to different level. The main reason for this preference of personalized engine over traditional search was need of accurate and precise results. Due to paucity of time and patience users didn't want to surf several pages to find the result that suits them most. Personalized search engines could solve this problem effectively by understanding user through profiles and histories and thus diminishing uncertainty and ambiguity. But since several layers of personalization were added to basic search, the response time and resource requirement (for profile storage) increased manifold. So it is time to focus on optimizing the layered architectures of personalization. The paper presents a layout of the multi-agent based personalized search engine that works on histories and profiles. Further to store the huge amount of data, distributed database is used at its core, so high availability, scaling, and geographic distribution are built in and easy to use. Initially results are retrieved using traditional search engine, after applying layer of personalization the results are provided to user. MongoDB is used to store profiles in flexible form thus improving the performance of the engine. Further Weighted Sum model is used to rank the pages in personalization layer.

Keywords Personalized search engine (PSE) · Weighted sum model (WSM)
Information retrieval · User profiling · Web personalization · MongoDB
Distributed database

D. Verma (✉)
VIPS, Guru Gobind Singh Indraprastha University, New Delhi 110034, India
e-mail: disha.verma.in@gmail.com

D. Verma · Y. S. Shishodia
Jagannath University, Jaipur 303901, India
e-mail: pvc@jagannathuniversity.org

B. Kochar
BPIT, Rohini, New Delhi 110085, India
e-mail: bjkochar@gmail.com

© Springer Nature Singapore Pte Ltd. 2019
A. Abraham et al. (eds.), *Emerging Technologies in Data Mining and Information Security*, Advances in Intelligent Systems and Computing 755,
https://doi.org/10.1007/978-981-13-1951-8_41

1   Introduction

Today, the World Wide Web provides us with a huge ever-growing source of information and has become a crucial part of our everyday lives. As an outcome of the speedy growth and dynamic content of the web, the traditional web search engines are becoming deficient.

Personalized search engines are replacing the traditional ones by catering the personalization on the basis of various parameters like user history, user profiles, etc.

With increased social media usage the users can be studied best using their social media accounts which further helps to produce fewer but personalized results. Thus next evident progression is integration of social media with search engines.

In this paper, we propose a multi-layered search personalization approach. The architecture of the proposed model differentiates it from previous research [1] as it does not defy the vertical search engine instead it uses the results fetched by it and applies an additional layer of personalization. Personalization is done through several parameters like history, interest, profile, social media account, etc. Also since layered architecture may increase response time thus backend used is unlike conventional databases. The concept of distributed database MongoDB is used as a core. The first part of the paper represents previous work in the field followed by theoretical explanation of the various layers used in the proposed architecture. Several parts of architecture like levels of personalization and distributed database technology Mongo DB is explained briefly. Following the conceptual description implementation is discussed comprehensively. The last section of paper focuses on Results and analysis. Future scope of the paper is also discussed in brief. On results concept introduced for the architecture (Evolution of Ontology in Multi-Agent Systems). Problem definition section explains the current problem and proposed solutions for the area. Following the problem definition section is proposed architecture section which discusses the proposed architecture in detail. After which the analysis and evaluation section compares the traditional search engine with the proposed architecture. The last section explains the future scope and concludes the research.

2   Literature

The Owing to need for accurate results personalization came into existence. Personalization is not new, and the need for results according to the user preferences has led to many researches working on it. A lot of work has been done in last decade in field of personalization. Personalization has been proposed through various ways.

A number of research groups have discovered and explored personalization and have broadly divided it into two categories: Explicit profiling and Implicit Personalization based Heuristics. Also Profiling could be done using several methods. Gauch et al. [1] explored user profiles from browsing history, Speretta and Gauch [2] created profiles using search history, and Chirita et al. [3] used profiles that users specified explicitly. Leung and Lee [4] on other hand proposed studying the logs creating profiles based on it. Captain Nemo project [5] implemented a functional search engine with personalized hierarchical search which extracts and displays search results according to retrieval models (personalization) and arrangement styles. In the WebNaut project [6] a multi-agent based search engine is proposed that consists of a set of interconnected agents and uses a meta-genetic algorithm for learning of the user's interests and personalizing search results. Contradicting experiment on small sample size showed the level of domain knowledge seems to have an effect on users search behavior, but not its effectiveness [7].

Another area explored was algorithm or model best suited for Ranking and personalization. Cho and Qiu [8] used Random Surfer Model for ranking pages. They also discussed extending the normal ranking model to Topic-Sensitive PageRank scheme (TSPR). This model was based on Topic Preference Vector. A Session-based personalized search algorithm was proposed by Daoud et al. which used correlation as background [9]. Shen et al. [10] proposed implicit profiling using decision-theory they also used TF-IDF weighting model for calculating information based on click-throughs. Author of Excalibur project proposed a personalized search engine that extracts users preference implicitly and re rank them by using the Naive Bayesian classifier and the resemblance measure [11].

Besides the several algorithms and models, Ontology-based personalization also helped to produce contextual results and improving strategic adaptation based on the knowledge obtainable from users' actions [12–15]. Understanding the Perspective or category can only be done if related words are and subcategories are already explored for the domain thus Open Directory Project is used in several ontology based personalization [13, 16, 17]. Radovanovic and Iva-novic [18] proposed a meta-search engine, called CatS that utilizes text classification techniques to improve the view of search results and displays a tree of topics derived from the dmoz which is an Open Directory topic hierarchy that can be traversed by user.

Another dimension explored in the area is multi-source personalization. A multisource profiling and multi-application personalization approach that leverages diverse usage data collected from multiple service domains such as mobile and web [19].

Some personalization focuses of exploring the new area called semantic web instead of WWW. Under the assumption of a shared model of semantic concepts, one can represent the content metadata (categories) as well as the semantics of consumption acts (purchase events, viewing or browsing sessions) with the same terms in the user profile (interests) [20]. According to the overlay approach, the profile created is composed of a set of {concept, value} where in concept refers to user's interest and value refers to degree of interests [21]. Understanding user and creating profiles not helps search engine to cater precise results but it also opens a lot of revenue streams for companies based on information model [22]. Some authors also created a relation between long long-term search activity history activity and short-term search session behavior [16]. The short-term context, which is the is in regards with information that emerges from the current user's information need in a single session. The other context is long-term which refers to the user interests that have been inferred from profile explicitly created by him or his past sessions [12]. Kanteev et al. [23] proposed a multi-agent content understanding system which is based on the semantics of pages. It generates the semantic descriptors as well as uses the knowledge about problem in the specific domain that is stored in the form of ontology. Very few researches focused of storage and optimization. When personalization or ranking is implemented, it may result in increase of response time which is imperative to be dealt with. [24] proposed the double-byte inverted index and virtual memory drives technology to ensure the system response time is not hampered [25]. Another concern is need for a database that is able to store huge amount of data produced by profiling. Normal relational database is not able to handle the concurrent queries and profiling a This leads to research in database technology one of which is NoSQL database. Compared to relational database, MongoDB (technology of NoSQL) supports schema-free storage (unlike relational), has great query performance with huge amount of data and provides easy horizontal scalability. It is more fitting for data storage in personalized search engines [25]. Cassandra another example of NoSQL database which is now deployed as the backend storage system for multiple services within Facebook [26]. Serge et al. [27] proposed an algorithm OPIC to decrease CPU utilization while calculating page visits.

It was essential to study all the above concepts for our study since our architecture has various layers which directly or indirectly implement the discussed concepts.

3 Background

This section of paper lays down the theoretical background of the architecture implemented.

3.1 Concept of Personalization

A traditional search engine returns the same results for the same query irrespective of user interest and choice. Unlike traditional system, personalized search engine cater to user needs.

Personalized search engine intends to customize search based on an individual user's interest, needs, requirements or his search history/pattern having an effect on the user's relevance assessment. Thus basic categorization of personalization could be done on the basis of degree of personalization required. The two major categories on which personalization is based are Explicit & Implicit profiling. Explicit profiling refers to creation of a profile by taking interest as input from users.

Explicit profiling is performed on server side where in every user interests are stored as separate entity. Implicit profiling is client side personalization which studies pattern of user search like his history, duration he spends on a page etc. After storing this data in cache and cookies an implicit profile can be created and thus results can be customized for specific user.

3.2 Multi-Criteria Decision-Making (MCDM)

Multi-criteria decision-making (MCDM) refers to traversing, prioritizing, selecting the alternatives from among a finite set of substitutes or options in terms of the multiple criteria. Weights play a significant role in MCDM models which provide the degree of importance of criteria under consideration. Several different methods have been developed to compare these criteria's in account. Mostly weights are inferred through judgments and adhoc approaches which make them vague and inaccurate in nature. Thus weights cannot be exactly evaluated or calculated with numerical values, leading "true" weights almost nonexistent. Even if the precise weights are possible, it is very difficult and time consuming making it impractical for use [28]. Here the rank ordering weighting methods plays an important role in providing an approximation of "true" weights when rank ordering information is known.

In our research,

We consider a problem with m page as alternatives $p_1, p_2, p_3, \ldots, p_m$, and personalization criteria $C_1, C_2, C_3, \ldots, C_n$.. For criteria, we have $w = [w_1, w_2, \ldots, w_n]$

Such that $w_1 + w_2 + \cdots + w_n = 1$ where w_j represents the weight of criterion

$$C_j, w_j \geq 0 (j = 1, 2, 3, \ldots, n)$$

Let x_{ij} denotes the performance value of each page alternative p_i in terms of criteria C_j. Where $(i = 1, 2, 3, \ldots, m; j = 1, 2, 3, \ldots, n)$

The decision matrix

$$D = (x_{ij})_{m \times n}$$

represents the evaluation score x_{ij} of each page p_i with respect of each criterion C_j.

All criterions are then normalized using following formula

$$C_{ij} = \frac{x_{ij} - \min_i x_{ij}}{\max_i x_{ij} - \min_i x_{ij}}$$

where x_{ij} is the score of ith page with respect to jth criterion before normalization.

After applying normalization higher C_{ij} is more preferred over lower C_{ij}

Let the normalized decision matrix be

$$C = (c_{ij})_{m \times n}$$

The method used in our research for re-ranking Simple Additive Weighting (SAW) which is also known as weighted linear combination or scoring method.

The three major steps in simple additive weighting is Scaling of scores so that they are comparable, applying criteria weights and sum the values along rows and rank the pages according to the final score of each page.

Final score of each alternative is calculated as:

$$S_i = \sum_{j=1}^{n} c_{ij} w_j$$

where S_i is score for ith page, and c_{ij} is the normalized score of ith page with respect to jth criterion and w_j is the weight of criteria j.

The final scores are used to re-rank the pages. Higher the value of S_i for a page higher its rank (1 being the highest rank). For calculating weights it was imperative to understanding importance of each criterion. So considering if we have n prioritized criteria (priority taken from user), each criteria has a has a rank. This rank is inversely proportional to the weight (r=1 denoting highest weight). After this understanding various methods were compared and ranks were converted to numerical weights. Further the rank sum method was used to calculate the weights of the criterions. The weight can be calculated as

$$n - rj + 1,$$

where n is total number of criterions and rj is the straight ranks assigned on the basis of importance (Tables 1 and 2).

Table 1 Weights and normalized weights of criterions

Criterion	Straight rank	Weight	Normalized weight	
A	4	2	0.133	
B	2	4	0.267	
C	5	1	0.067	
D	1	5	0.333	
E	3	3	0.200	
			15	1.00

The Weights can also be represented using a matrix as represented in table

Table 2 Pages scores corresponding to various criterions

	A	B	C	D	E
Page	0.133	0.267	0.067	0.333	0.200
P1	5	4	2	4	3
P2	6	3	1	3	1
P3	5	3	3	3	1
P4	4	3	4	3	2

Fig. 1 Normal weights and normalized calculated weight

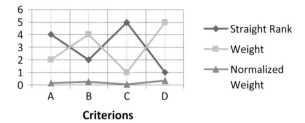

Fig. 2 Final scores corresponding to pages

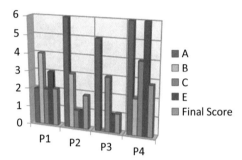

Figure 1 represents the normal weight and normalized weights corresponding to the various criterions. After deciding weight for the entire criterions Fig. 2 represents the final score using weighted sum model discussed previously.

4 Proposed Work

The proposed work deals with search engine that incorporates the feature of fetching results from a search engine, and further personalizing it using implicit and explicit profiling The following modules have been used in this personalized search engine:

4.1 Search Engine Module

This layers/module will fetch the results from search engine, in our paper traditional search engine is used as search engine. API's were used to fetch the contents from search engine and social media. The focus of this layer is to fetch the results from search engine in a format where in we can filter and perform other operations on it (Fig. 3).

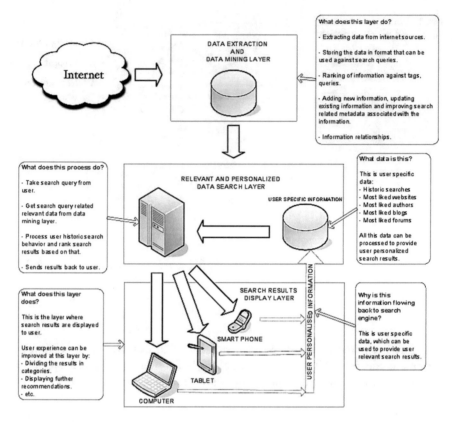

Fig. 3 High-level view of proposed architecture [29, 30]

4.2 Personalization Module

This module takes input from the previous layer and personalizes it through various parameters like history etc. This layer is based on two types of personalization. The first one being Profile based, in which user is required to input his/her interests explicitly also prioritize his result on the electronic form provided to him. The second type of personalization is heuristic based. User's browsing history plays an important role in depicting his interests. Thus the architecture proposed tracks user browsing history and re-rank the page accordingly. Another type of personalization implemented in our architecture is social media based. Our architecture fetches the user interests from his/her Facebook page and re-rank his results accordingly. The prerequisite of this type of personalization is Facebook Log-in. User will be required to log-in through his Facebook id and password. This type of personalization lays down a prototype that personalization can be done using any kind of social media profile. This is one of the features that differentiate this architecture from the previous ones.

4.3   Storage and Processing Module

As a backend to the second layer this layer is responsible in producing re-ranked results in optimized time. The process of fetching the result from search engine and re-ranking may increase response time. Thus this layer introduces the concept of distributed databases.

Relational databases have been underpinning search applications from long. But researchers now are increasingly considering alternatives to traditional relational infrastructure. There exist several motivations behind it. In few cases the motivation is technical that is necessity to handle new, multi-structured data types or scale beyond the current capacity contingencies of legacy systems. Another important motivation being agility or speed required to deal with mammoth amount of data generated by use of social media.

4.4   Presentation Module

The final module focuses on presenting the re-ranked result compatible to all the devices like desktops and hand held.

5   Dataset Characteristics and Experiment

In this section we will discuss the working prototype build for experiment and result analysis.

5.1   Working Model

The detailed architecture of the model is represented in Fig. 4.

The following steps are being performed to find refined results:

(a)   Search a query on the GUI provided
(b)   First layer will fetch results from traditional search engine corresponding to the query. Category agent categorizes the result into various categories and provides suitable score. Various weights and scores are assigned to all the parameters like history, interest social media etc. using algorithm discussed above.
(c)   The results are re-ranked using distributed databases.
(d)   The results are displayed on user's screen.

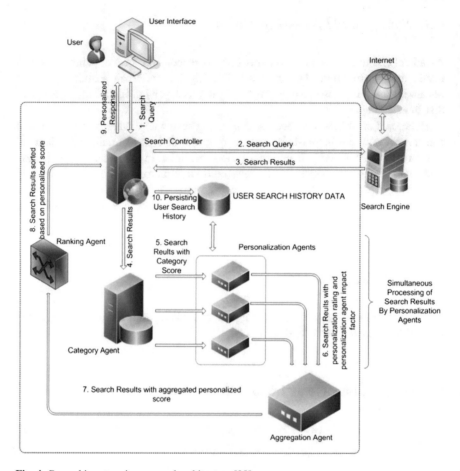

Fig. 4 Re-ranking steps in proposed architecture [30]

5.2 Data Maintained by Engine

Our engine is storing large amount of data few of the crucial tables are

(a) User interest table which is created on the basis of profile build on user interest.
(b) Hits table keep track of the user history and log
(c) Social Media Data table is responsible for fetching data from user's social media profile and storing it in semi structured form

Fig. 5 Home page of the engine

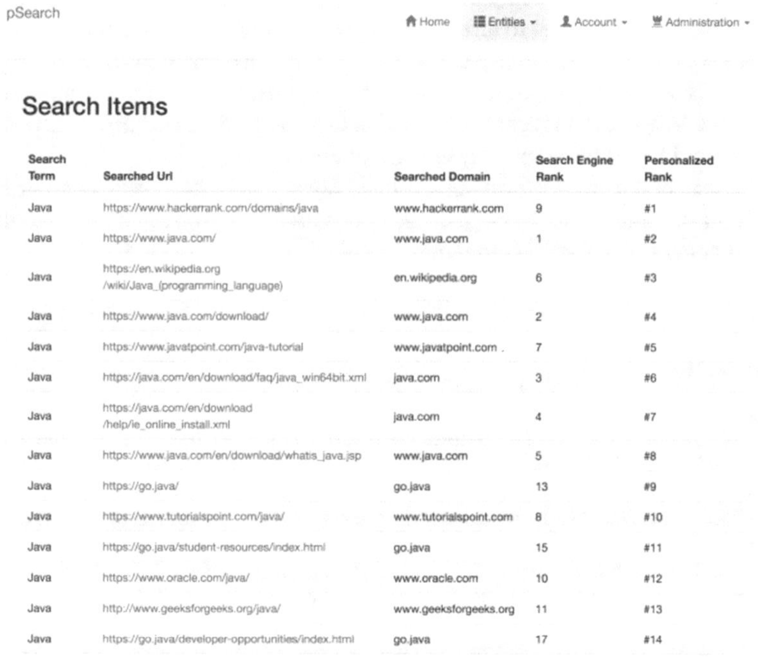

Search Items

Search Term	Searched Url	Searched Domain	Search Engine Rank	Personalized Rank
Java	https://www.hackerrank.com/domains/java	www.hackerrank.com	9	#1
Java	https://www.java.com/	www.java.com	1	#2
Java	https://en.wikipedia.org /wiki/Java_(programming_language)	en.wikipedia.org	6	#3
Java	https://www.java.com/download/	www.java.com	2	#4
Java	https://www.javatpoint.com/java-tutorial	www.javatpoint.com .	7	#5
Java	https://java.com/en/download/faq/java_win64bit.xml	java.com	3	#6
Java	https://java.com/en/download /help/ie_online_install.xml	java.com	4	#7
Java	https://www.java.com/en/download/whatis_java.jsp	www.java.com	5	#8
Java	https://go.java/	go.java	13	#9
Java	https://www.tutorialspoint.com/java/	www.tutorialspoint.com	8	#10
Java	https://go.java/student-resources/index.html	go.java	15	#11
Java	https://www.oracle.com/java/	www.oracle.com	10	#12
Java	http://www.geeksforgeeks.org/java/	www.geeksforgeeks.org	11	#13
Java	https://go.java/developer-opportunities/index.html	go.java	17	#14

Fig. 6 Re-ranked versus old rank results

5.3 Re-ranked Results

The last step of the engine is to re-rank the result on the basis of proposed methodology. We performed the test by fetching first ten results and then re-ranking them accordingly. The engine's home page is represented using Fig. 5. The output produced is shown in Fig. 6. Through an experiment we have shown how traditional search engine result can further be refined on the basis of user interests and history. Importantly, we also showed that how by changing backend technology the response time. Also since we have written generic code this engine can be customized to any vertical search engine provided an API is available for it. Also more them one social media personalization can be performed with few changes. The mongo DB used as background help to attain horizontal scalability.

6 Conclusion

Personalization could be done in different ways and different levels. The aim is to attain maximum precision in understanding user's interests and preferences. The search engine proposed and implemented in our paper focuses on working on layered architecture. It enables user to take advantage of traditional search engine as well as provide user the power to control the personalization on basis of his interest, history, social media account. The algorithm used is weighted sum with prioritized weight calculation. The backend technology used is distributed database MongoDB which helps in providing the re-ranked results in timely manner.

The focus of our work is providing refined fewer results in an efficient optimized manner. In our future studies we will apply this architecture with increased data set before making it live for users to work on.

References

1. Gauch, S., Chaffee, J.: A ontology based personalized search and browsing. Web Intell. Agent Syst. **1**(3), 219–234 (2003)
2. Speretta, M., Gauch, S.: Personalized search based on user search histories. In: Proceedings of the 2005 IEEE/WIC/ACM International Conference on Web Intelligence, pp. 622–628. IEEE, 19 Sept 2005
3. Chirita, P.A., Nejidi, W.: Using ODP metadata to personalize search. In: Proceedings of 28th Annual International ACM SIGIR Conference on Research and Development in Information Retrieval. ACM, Aug 2005
4. Leung, K.W.T., Lee, D.L.: Deriving concept-based user profiles from search engine logs. IEEE Trans. Knowl. Data Eng. **22**(7), 969–982 (2010)
5. Souldatos, S., Dalamagas, T., Sellis, T.: Captain Nemo: a metasearch engine with personalized hierarchical search space. Informatica-Ljubljana **30**, 173–182 (2006)
6. Zacharis, N., Panayiotopoulos, T.: SpiderServer: the meta-search engine of WebNaut. In: Proceedings of the 2nd Hellenic Conference on Artificial Intelligence, pp. 475–486 (2002)
7. Zhang, X., Anghelescu, H.G.B., Yuan, X.: Domain knowledge, search behavior, and search effectiveness of engineering and science students. Inf. Res. **10**(2), 217 (2005)
8. Qiu, F., Cho, J.: Automatic identification of user interest for personalized search. In: Proceedings of the 15th International Conference on World Wide Web, pp. 727–736. ACM, May 2006

9. Daoud, M., Tamine-Lechani, L., Boughanem, M.: Learning user interests for a session-based personalized search. In: Proceedings of the Second International Symposium on Information Interaction in Context (IIiX '08), pp. 57–64. ACM (2008). http://dx.doi.org/10.1145/1414694. 1414708

10. Shen, X., Tan, B., Zhai, C.: Implicit user modeling for personalized search. In: Proceedings of the 14th ACM International Conference on Information and Knowledge Management, pp. 824–831. ACM, Oct 2005

11. Yuen, L., Chang, M., Lai, Y.K., Poon, C.K.: Excalibur: a personalized meta-search engine. In: Proceedings 28th Annual International Computer Software and Applications Conference (COMPSAC'04), vol. 2, pp. 49–50 (2004)

12. Tamine-Lechani, L., Boughanem, M., Zemirli, N.: Personalized document ranking: Exploiting evidence from multiple user interests for profiling and retrieval. JDIM **6**(5), 354–365 (2008)

13. Xiang, B., Jiang, D., Pei, J., Sun, X., Chen, E., Li, H.: Context-aware ranking in web search. In: Proceedings of the 33rd International ACM SIGIR Conference on Research and Development in Information Retrieval, Geneva, Switzerland, 19–23 July 2010. https://doi.org/10.1145/183 5449.1835525

14. Prates, J.C., Siqueira, S.S.M., Using educational resources to improve the efficiency of web searches for additional learning material. In: 2011 11th IEEE International Conference on Advanced Learning Technologies (ICALT), pp. 563, 567, 6–8 July 2011

15. Joung, Y., Zarki, M.E., Jain, R.: A user model for personalization services. In: 4th International Conference on Digital Information Management, ICDIM 2009, pp. 247–252 (2009)

16. Bennett, P.N., White, R.W., Chu, W., Dumais, S.T., Bailey, P., Borisyuk, F., Cui, X.: Modeling the impact of short-and long-term behavior on search personalization. In: Proceedings of the 35th International ACM SIGIR Conference on Research and Development in Information Retrieval, pp. 185–194. ACM, Aug 2012

17. Middleton, S., Roure, D.D., Shadbolt, N.: Capturing knowledge of user preferences: ontologies in recommender systems. In: 1st International Conference on Knowledge Capture, pp. 100–107. ACM Press (2001)

18. Radovanovic, M., Ivanovic, M.: CatS: a classification-powered meta-search engine. In: Advances in Web Intelligence and Data Mining, vol. 23, pp. 191–200. Springer (2006)

19. Aghasaryan, A., et al.: Personalized application enablement by web session analysis and multisource user profiling. Bell Labs Tech. J. **15**(1), 67–76 (2010)

20. Shah, A., Jain, S., Chheda, R., Mashru, A.: Model for re-ranking agent on hybrid search engine for e-learning. In: 2012 IEEE Fourth International Conference on Technology for Education (T4E), pp. 247, 248, 18–20 July 2012

21. Brusilovsky, P., Millan, E.: User models for adaptive hypermedia and adaptive educational systems. In: Brusilovsky, P., Kobsa, A., Nejdl, W. (eds.) The Adaptive Web: Methods and Strategies of Web Personalization, pp. 136–154. Springer, Berlin, Heidelberg, New York (2007)

22. Beccue, M., Shey, D.: Service personalization: subscriber data management, subscriber profiling, policy control, real-time charging, location, and presence, ABI Research, 2Q (2009)

23. Kanteev, M., Minakov, I., Rzevski, G., Skobelev, P., Volman, S.: Multi-agent Meta-Search Engine Based on Domain Ontology. Springer, vol. 4476, pp. 269–274, July 2007. https://doi. org/10.1007/978-3-540-72839-9_22

24. Suping, X., Mu, L., Ziqi, Z.: Research of index technology for topic search engine. In: 2011 International Symposium on IT in Medicine and Education (ITME), vol. 1, pp. 93, 95, 9–11 Dec 2011

25. Gu, Y., Shen, S., Zheng, G.: Application of NoSQL database in web crawling. Int. J. Digit. Content Technol. Appl. **5**(6), 261–266 (2011)

26. Lakshman, A., Malik P, Cassandra: a decentralized structured storage system. ACM SIGOPS Oper. Syst. Rev. **44**(2) (2010). https://doi.org/10.1145/1773912.1773922

27. Serge, A., Mihai, P., Gregory, C.: Adaptive on-line page importance computation. In: Proceedings of the 12th International Conference on World Wide Web (WWW '03), pp. 280–290. ACM (2003). http://dx.doi.org/10.1145/775152.775192
28. Roszkowska, E.: Rank ordering criteria weighting methods—a comparative overview. Optimum. Stud. Ekono. **5**, 65 (2013)
29. Verma, D., Kochar, B.: Multi agent architecture for search engine (IJACSA). Int. J. Adv. Comput. Sci. Appl. **7**(3), 224–220 (2016)
30. Verma, D., Minocha, K., Kochar, B.: A multi-agent based personalized search engine with topical crawling capabilities. IUP J. Comput. Sci. **8**(3), 20–33 (2014)

Reversible Code Converters Based on Application Specific Four Variable Reversible Gates

Sanjoy Banerjee, Abhijit Kumar Pal, Mahamuda Sultana, Diganta Sengupta and Abhijit Das

Abstract   The rising research in reversible logic, estimating it to be latent alternative for CMOS, has paved the way for several proposals for reversible logic synthesis. Application oriented reversible circuit designs are witnessed in almost every aspect of digital communication. It is this very interest that the present study proposes binary to gray code converters and vice versa as two independent four variable reversible gates. The converters have been conceptualized as four variable reversible gates having potential to realize efficient parity generator/checker circuits exhibiting better peer comparison results. Hence the work in this paper may find acceptance in reversible cryptography as well as XOR intensive operations in image processors. Also by virtue of definition, reversibility supports lossless communication as information loss is arrested in subsequent stages of information transfer in a reversible function.

1   Introduction

Computations on digital hardware result in enormous amount of heat dissipation [1] owing to the irreversibility of computation in classical electronics [2]. Bennett [2] argued that if the operations are made reversible, then the amount of heat dissipated

S. Banerjee · A. K. Pal
Future Institute of Engineering & Management, Kolkata 700150, India
e-mail: banerjeesanjoy2003@gmail.com

A. K. Pal
e-mail: abhijitpalece10@gmail.com

M. Sultana
Techno India College of Technology, Kolkata 700156, India
e-mail: sg.mahamuda@gmail.com

D. Sengupta (✉)
Techno India – Batanagar, Kolkata 700141, India
e-mail: sg.diganta@ieee.org

A. Das
RCC Institute of Information Technology, Kolkata 700015, India
e-mail: ayideep@yahoo.co.in

© Springer Nature Singapore Pte Ltd. 2019
A. Abraham et al. (eds.), *Emerging Technologies in Data Mining and Information Security*, Advances in Intelligent Systems and Computing 755,
https://doi.org/10.1007/978-981-13-1951-8_42

can theoretically be equated to zero, something since long had been in the concept of Turing Machine in-principal.

With increase in miniaturization and chip density, CMOS has started facing physical threshold limitations. Another point of concern is the clocking speed in CMOS which has also reached near maximum (GHz). Quantum Dot Cellular Automata [3] has promised operating speeds in the order of THz and lower power dissipation due to coulombic repulsion, although in the nascent stage. Hence, powered with the vision of theoretically zero power dissipation and higher operating frequencies, Reversible Logic and Quantum Dot Cellular automata promise a viable alternative to CMOS in not so distant future. Several four variable application specific reversible gates have been provided in the last decade. A detailed survey on application specific existing four variable reversible gates can be found in [4]. We propose two novel four variable reversible gates which are basically Binary to Gray code converter and vice versa. The 4-bit Gray codes for a Binary bit stream are unique for all the $2^4 = 16$ combinations and obey a one-to-one mapping between the Binary and the Gray equivalent of a number. Hence, inherently the Binary–Gray combination is reversible according to Definition 1. This concept motivated us to design the four variable reversible gates BG_4 for Binary to Gray Code conversion and GB_4 for the reverse conversion.

Definition 1 A Boolean Function $f(n)$: $B^n \rightarrow B$ over an input vector $I_v = \{x_0, x_1, x_2 \ldots x_{n-1}\}$ is reversible if the output vector O_v is a permutation over all the elements of I_v. Hence, a reversible function is inherently *Bijective—One-One* and *Onto*. Also reversible gates promote FO1; fan out of 1; and do not support feedback.

The fundamental Reversible Gate Library comprises of the Toffoli [5], Fredkin [6], Feynman [7] and the Peres Gates [8] respectively. We have used Toffoli (Definition 2) Gates to design the proposed gates using the Unidirectional Algorithm of [9].

Definition 2 A Multiple Control Toffoli (*MCT*) is denoted as $TOFx(C, T)$; $C \cap T = 0$; where $C =$ Control Line Set controlling a single-target line T and x is the composite line count for C and T.

Hence, $TOF\{x_1\}$ is basically a *No-Control-Single-Target* gate having x_1 as the target line. Hence, signal passing through this gate is always inverted. $TOF2\{x_1, x_2\}$ is a *Single-Control-Single-Target* gate in which x_1 is the control line and x_2 is the target line which gets inverted if x_1 is high. In $TOF3\{x_1, x_2, x_3\}$ x_1 and x_2 are the control lines and x_3 is the target line which gets inverted when both x_1 and x_2 are high. The Boolean representation of the input–output pair for Toffoli Gates are $\{P, Q, R\} = \{P = A, Q = B, R = (A.B) \oplus C\}$ where $\{A, B, C\}$ and $\{P, Q, R\}$ are the respective $I_v - O_v$ vectors.

2 Related Work

To the best of our knowledge, till date nineteen application specific four variable reversible gates have been proposed. The vocabulary comprises of RI [10], R2 [10], TSG [11], HNG [12], SCG [13], RPS [14], HNFG [15, 16], MKG [15], IG [17], FAG [18], ALG [19], MTSG [20], DFG [18], DKG [21], MRG [21], CSMT [22], BVMF [23], s2c2 [24] and TCG [25] Gates. The targeted applications for some of these gates are Adders, Parity Generators and Checkers, Cryptographic Applications, etc. For example, a single HNG Gate/TSG Gate is capable of performing as a full adder. A single SCG Gate performs as a full adder and seconds as a full subtractor depending upon the value of a certain control line. The SCG gate can also perform as a single bit comparator and Match Logic independently. Comparators can also be realized using BVMF Gate. With certain control line combinations, BVMF gate can also be used as a Demultiplexer or a 2 to 4 Decoder. Reversible 4:2 compressors have been designed using TSG, MTSG and FAG Gates. Reversible sequential circuits have been designed using DFG Gates. Primitive reversible ALUs have been proposed using ALG, DKG, MRG, and TSG Gates. Reversible n-bit multipliers have been proposed using the MKG Gate as well as DFG Gates. RPS has been used to design reversible decimal adders. Some of the gates have been used to realize code conversions. CSMT [22] Gate is of particular interest as the authors have proposed Binary to Gray code conversion using the CSMT gate. Their proposal varies with our proposal in the design and the outputs. The truth table (Table 1) in page 3 of [22] claims to support Binary to Gray code conversion as per Fig. 5 in Page 4 of [22], but close observation reveals that subsequent outputs differ with a Hamming Distance of 2; whereas subsequent Gray Codes differ by a Hamming Distance of 1. Cascade of certain reversible gates have been used to design Parity Generator/Checker circuits. The work in this paper introduces two four variable reversible gates capable of converting Binary code to Gray code and vice versa. The proposed gates optimize parity generation and checking operation in peer comparison.

3 Proposed Four Variable Reversible Gates

In digital applications, Gray code is one of the most popular codes and is also referred to as Reflected Binary Code (RBC). This is a type of binary number system where two successive bit streams differ by a single bit; Hamming Distance (Definition 3) equals 1.

Definition 3 The Hamming Distance ($\delta(A, B)$) between two bit sequences A and B is the number of bit inversions in corresponding bit positions in the sequences.

The following subsections detail the proposed codes. The proposed gates have been synthesized using the tool *RC Viewer+* [26].

Table 1 Truth table for four-bit Binary to Gray code conversion

Input sequence				Output sequence				Hamming distance
A	B	C	D	P	Q	R	S	
0	0	0	0	0	0	0	0	0
0	0	0	1	0	0	0	1	0
0	0	1	0	0	0	1	1	1
0	0	1	1	0	0	1	0	1
0	1	0	0	0	1	1	0	1
0	1	0	1	0	1	1	1	1
0	1	1	0	0	1	0	1	2
0	1	1	1	0	1	0	0	2
1	0	0	0	1	1	0	0	1
1	0	0	1	1	1	0	1	1
1	0	1	0	1	1	1	1	2
1	0	1	1	1	1	1	0	2
1	1	0	0	1	0	1	0	2
1	1	0	1	1	0	1	1	2
1	1	1	0	1	0	0	1	3
1	1	1	1	1	0	0	0	3
Complexity								24

Fig. 1 Block diagram for BG$_4$ Gate

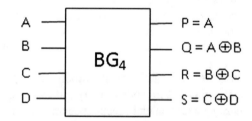

3.1 Binary to Gray Code Converter Gate (BG$_4$)

As mentioned earlier, the motivation of the work was the truth table for Binary to Gray code conversion which is inherently reversible; the gate has been designed based on the concept that the truth table itself reflects reversibility as shown in Table 1. It can be observed that the consecutive outputs in Table 1 conform to Gray Code having a Hamming Distance of 1 in contrast to the proposal in Table 1 of [22]. We claim that our proposal does support Binary to Gray code conversion using the single proposed BG$_4$ Gate.

The Complexity for the proposed gate is 24 according to Definition 4. Figure 1 presents the block diagram for the BG$_4$ Gate. The Toffoli synthesis has been done using the *Basic Unidirectional Algorithm* of [9] and is provided in Fig. 2.

Fig. 2 Toffoli Netlist for
BG_4 Gate

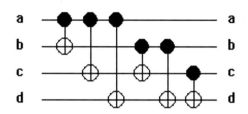

Table 2 Boolean function realization using BG_4

Input vector				Output vector				Boolean logic realization
A	B	C	D	P	Q	R	S	
A	0	0	0	A	A	0	0	Fan Out of 2 (FO2): Signal Duplication
0	A	0	0	0	A	A	0	Fan Out of 2 (FO2): Signal duplication
0	0	A	0	0	0	A	A	Fan Out of 2 (FO2): Signal Duplication
0	0	0	A	0	0	0	A	Through Signal (Delay)
A	1	1	1	A	$\overline{A}$	0	0	Negation and through Signal
1	A	1	1	1	$\overline{A}$	$\overline{A}$	0	Negation and Vcc Transmit
1	1	A	1	1	0	$\overline{A}$	$\overline{A}$	Negation and Vcc Transmit
1	1	1	A	1	0	0	$\overline{A}$	Negation and through Signal

Definition 4 The Complexity 'C_f' of a given m-variable reversible function f(X) is given by $\sum_{i=0}^{m-1} \delta(A_i, B_i)$; where A_i and B_i are the corresponding 2^m input–output patterns of f(X).

The Quantum Cost for the Toffoli Netlist in Fig. 2 has been calculated according to *Definition 5* and found to be 6. The Quantum Costs for *TOF1*, *TOF2*, *TOF3* and *TOF4* are 1, 1, 5 and 13 respectively.

Definition 5 For a given reversible function $f(X)$, The Quantum Cost (QC) is defined as $\sum_{i=1}^{n} TOFx_i$; where $TOFx_i$ is the cost of the ith Toffoli Gate.

Certain constant input values to the BG_4 gate produce certain Boolean function realizations as listed in Table 2 for input vector $\{A, B, C, D\}$. Since by definition, reversible gates possess a maximum fan out of 1, hence higher fan outs can be obtained using BG_4 as shown in Table 2 with certain input combinations.

3.2 Gray to Binary Code Converter Gate (GB₄)

The truth table for Gray to Binary code conversion is presented in Table 3. BG_4 being reversible in nature confirms the reversibility of GB_4. The complexity of the

Table 3 Truth table for four-bit Gray to Binary code conversion

Input sequence				Output sequence				Hamming distance
A	B	C	D	P	Q	R	S	
0	0	0	0	0	0	0	0	0
0	0	0	1	0	0	0	1	0
0	0	1	0	0	0	1	1	1
0	0	1	1	0	0	1	0	1
0	1	0	0	0	1	1	1	1
0	1	0	1	0	1	1	0	1
0	1	1	0	0	1	0	0	2
0	1	1	1	0	1	0	1	2
1	0	0	0	1	1	1	1	1
1	0	0	1	1	1	1	0	1
1	0	1	0	1	1	0	0	2
1	0	1	1	1	1	0	1	2
1	1	0	0	1	0	0	0	2
1	1	0	1	1	0	0	1	2
1	1	1	0	1	0	1	1	3
1	1	1	1	1	0	1	0	3
Complexity								24

Fig. 3 Block diagram for GB_4 Gate

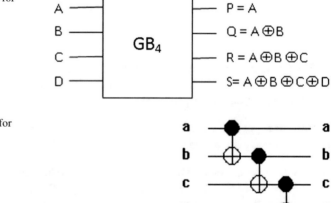

Fig. 4 Toffoli Netlist for GB_4 Gate

proposed GB_4 gate is also 24 calculated according to Definition 4. Figures 3 and 4 present the block diagram and the Toffoli Netlist respectively for the GB_4 Gate.

As in the case for BG_4, for certain input vectors, GB_4 also realizes a considerable amount of Boolean logic functions. The logic realizations for $I_v = \{A, B, C, D\}$ are presented in Table 4. Table 5 provides the relevant metrics for the proposed gates.

Table 4 Boolean function realization using GB_4

Input vector				Output vector				Boolean logic realization
A	B	C	D	P	Q	R	S	
0	0	0	A	0	0	0	A	Signal Duplication (Delay)
0	0	A	0	0	0	A	A	Fan Out of 2 (FO2): Signal Duplication
0	A	0	A	0	A	A	0	Fan Out of 2 (FO2): Signal Duplication
A	0	0	0	A	A	A	A	Fan Out of 4 (FO4): Signal Duplication
1	1	1	A	1	0	1	$\overline{A}$	Negation, Vcc Transmit
1	A	1	A	1	$\overline{A}$	A	$\overline{A}$	Negation, Vcc Transmit
A	0	1	1	A	$\overline{A}$	A	$\overline{A}$	Negation and Delay
A	B	0	0	A	$A \oplus B$	$A \oplus B$	$A \oplus B$	Fan Out of 3 (FO3): XOR
A	B	1	1	A	$A \oplus B$	$\overline{A \oplus B}$	$A \oplus B$	Fan Out of 3 (FO3): XOR; XNOR
X	X	A	B	X	0	A	$A \oplus B$	Delay; XOR; TOF2Gate (Definition 2)

X = Don't care

Table 5 Quantum metrics for the proposed Gates

Gate	No. of inputs/outputs	Gate count	Quantum cost	No. of two-qbit Gates
BG_4	4	6	6	6
GB_4	4	3	3	3

Fig. 5 Even/odd parity generator circuit

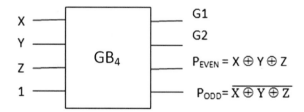

4 Parity Generator/Checker

Parity is used for error detection during transmission of bit streams. A parity bit is an extra bit added to the original message bit stream in order to make the bit count for '1's in the composite bit stream (message + parity) either even or odd depending upon application. If the no of 1's of the string is even then the added parity bit is known as even parity, else odd parity. An error is detected if the receiving parity does not match with the transmitted parity. Parity Generator generates the parity bit Parity Checker detects error at the receiving end. The proposed GB_4 Gate doubles both as a Parity Generator as well as a Parity Checker. Figures 5 and 6 present the block diagrams for three bit parity generator and checker designs with respective sets of input vectors.

Fig. 6 Even/odd parity checker circuit

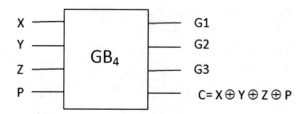

Table 6 Truth Table for even/odd parity generation

3 bit message			Parity bit	
X	Y	Z	Even	Odd
0	0	0	0	1
0	0	1	1	0
0	1	0	1	0
0	1	1	0	1
1	0	0	1	0
1	0	1	0	1
1	1	0	0	1
1	1	1	1	0

Fig. 7 Comparative analysis of literary proposals for Binary to Gray code converter

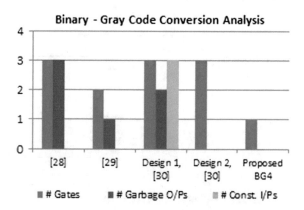

The garbage output count for the designs shown in Figs. 5 and 6 are two and three respectively. The input vector for Parity Generator is $I_v = \{X, Y, Z, 1\}$ where X, Y and Z are the three bits in the message and '1' is an Ancilla input. The output vector $O_v = \{G_1, G_2, P_{EVEN}, P_{ODD}\}$ comprises two garbage outputs G_1 and G_2 and the valid Parity bits according to the application whether Even or Odd parity is desirable. Tables 6 and 7 provide the truth table for Even/Odd Parity Generator and Even Parity Checker respectively.

'P' in the input vector $I_v = \{X, Y, Z, P\}$ in Fig. 8 is basically the Parity bit P_{EVEN}/P_{ODD} generated in Fig. 7 which is getting reflected in Table 7.

Table 7 Truth table for parity error checking—even parity checker

X	Y	Z	P	Parity check
0	0	0	0	0
0	0	0	1	1
0	0	1	0	1
0	0	1	1	0
0	1	0	0	1
0	1	0	1	0
0	1	1	0	0
0	1	1	1	1
1	0	0	0	1
1	0	0	1	0
1	0	1	0	0
1	0	1	1	1
1	1	0	0	0
1	1	0	1	1
1	1	1	0	1
1	1	1	1	0

Table 8 Comparative analysis for Binary to Gray code conversion and vice versa

Network	No. of gates	No. of garbage outputs	No. of constant inputs
Binary to Gray Code Converter [27]	3	3	0
Binary to Gray Code Converter [28]	2	1	0
Binary to Gray Code Converter [29], Design 1	3	2	3
Binary to Gray code converter [29], Design 2	3	0	0
Proposed BG$_4$	1	0	0
Gray to Binary Code Converter [27]	5	3	2
Proposed GB$_4$	1	0	0

5 Comparative Analysis

The proposed gates have been compared for both the domains; as Code Converter and as Parity Generator/Checker application. Both the applications have been compared with existing counterparts in literature and found to fare better. Table 8 provides comparison with respect to designs proposed by Saravanan et al. [27].

Table 9 Comparative analysis for parity generator/checker application

Application	Gate	No. of Gates	No. of inputs	No. of constant inputs	No. of garbage outputs	Parity type
Parity generator	Feynman [30]	2	3	0	2	Single
	Feynman [31]	3	3	1	3	Single
	Toffoli [31]	3	3	4	6	Single
	New Gate [31]	3	3	4	6	Single
	Peres [31]	3	3	4	6	Single
	Proposed BG_4	1	4	1	2	Both
Parity checker	Feynman [30]	3	4	0	3	Both
	Feynman [31]	3	4	0	3	Single
	Toffoli [31]	3	4	3	6	Single
	New Gate [31]	3	4	3	6	Single
	Peres [31]	3	4	3	6	Single
	Proposed GB_4	1	4	0	3	Both

Table 9 provides the comparative analysis for parity generator/checker circuits. Among all the literary proposals our gate excels in the fact that a single gate is sufficient to generate both the even and odd parities. Further comparison with [30] exhibits that our proposed gate excels over the design in [30] using Feynman Gates.

Mustafa and Beigh [32] have proposed parity generator and checker circuits in QCA focusing on XOR operators. Our design are also XOR specific as can be seen in Figs. 7 and 8. But the proposal in [32] is not reversible and hence we do not analyze that proposal in our comparisons. Figures 7 and 8 provide the graphical interpretation of the data in Table 8 for Binary to Gray and Gray to Binary Code Converters respectively. Figures 9 and 10 reflect the data in Table 9 for Parity Generator and Parity Checker respectively.

6 Conclusion

We have proposed two four variable reversible gates in this present communication. The primary objectives of the gates were to realize Binary to Gray code conversion and vice versa. The proposed gates double as efficient Parity Generator and Checker

Fig. 8 Comparative analysis of literary proposals for Gray to Binary converter

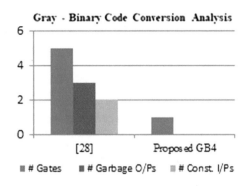

Fig. 9 Comparative analysis of literary proposals for parity generator circuits

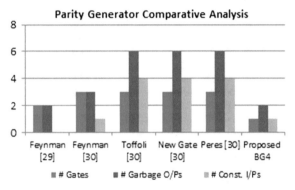

Fig. 10 Comparative analysis of literary proposals for parity checker circuits

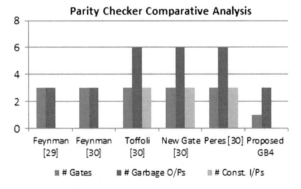

circuits when compared to literary counterparts. Since the proposed gates aid in code conversion and also exhibit XOR operations, hence we believe the gates can form efficient foundation for XOR intensive cryptographic and image processing environments. Code conversion capability also places the proposed gates as lossless information transmission agents.

References

1. Landauer, R.: Irreversibility and heat generation in the computing process. IBM J. Res. Dev. **5**(3), 183–191 (1961)
2. Bennett, C.H.: Logical reversibility of computation. IBM J. Res. Develop. **17**, 525–532 (1973)
3. Lent, C., Tougaw, P., Porod, W., Bernstein, G.: Quantum cellular automata. Nanotechnology **4**(1), 49–57 (1993)
4. Sultana, M., Prasad, M., Roy, P., Sarkar, S., Das, S., Chaudhuri, A.: Comprehensive quantum analysis of existing four variable reversible gates. In: 2017 Devices for Integrated Circuit (DevIC), pp. 116–120. IEEE, Kolkata (2017)
5. Toffoli, T.: Reversible Computing. Tech Memo MIT/LCS/TM-151. MIT Lab for Computer Science (1980)
6. Fredkin, E., Toffoli, T.: Conservative logic. Int. J. Theor. Phys. **21**, 219–253 (1982)
7. Feynman, R.: Quantum mechanical computers. Opt. News **11**, 11–20 (1985)
8. Peres, A.: Reversible logic and quantum computers. Phys. Rev. A **32**(6), 3266–3276 (1985)
9. Maslov, D., Dueck, G., Miller, D.: Synthesis of Fredkin–Toffoli reversible networks. IEEE Trans. Very Large Scale Integr. VLSI Syst. **13**(6), 765–769 (2005)
10. Vasudevan, D., Lala, P., Di, J., Parkerson, J.: Reversible-logic design with online testability. IEEE Trans. Instrum. Meas. **55**(2), 406–414 (2006)
11. Thapliyal, H., Srinivas, M.: Novel reversible 'TSG' Gate and its application for designing components of primitive reversible/quantum ALU. In: Fifth International Conference on Information, Communications and Signal Processing (2005)
12. Maity, G., Maity, S.: Implementation of HNG using MZI. In: Third International Conference on Computing Communication & Networking Technologies (ICCCNT), pp. 1–6 (2012)
13. Sengupta, D., Sultana, M., Chaudhuri, A.: Realization of a novel reversible SCG Gate and its application for designing parallel adder/subtractor and match logic. Int. J. Comput. Appl. **31**(9), 30–35 (2011)
14. James, R., Jacob, K., Sasi, S.: Design of compact reversible decimal adder using RPS gates. In: World Congress on Information and Communication Technologies (WICT), pp. 344–349 (2012)
15. Haghparast, M., Navi, K.: A Novel reversible full adder circuit for nanotechnology based systems. J. Appl. Sci. **7**(24), 3995–4000 (2007)
16. Haghparast, M., Navi, K.: A novel reversible BCD adder for nanotechnology based systems. Am. J. Appl. Sci. **5**(3), 282–288 (2008)
17. Islam, M., Rahman, M., Begum, Z.: Fault tolerant reversible logic synthesis: carry look-ahead and carry-skip adders. In: International Conference on Advances in Computational Tools for Engineering Applications, ACTEA '09, pp. 396–401 (2009)
18. Rashmi, S., Umarani, T., Shreedhar, H.: Optimized reversible montgomery multiplier. Int. J. Comput. Sci. Inf. Technol. **2**(2), 701–706 (2011)
19. Arun, M., Saravanan, S.: Reversible Arithmetic Logic Gate (ALG) for quantum computation. Int. J. Intell. Eng. Syst. **6**(3), 1–9 (2013)
20. Biswas, A., Hasan, M., Chowdhury, A., Babu, H.: Efficient approaches for designing reversible Binary Coded Decimal adders. Microelectron. J. **39**(12), 1693–1703 (2008) (Elsevier)
21. Biswas, P., Gupta, N., Patidar, N.: Basic reversible logic gates and it's QCA implementation. Int. J. Eng. Res. Appl. **4**(6), 12–16 (2014)
22. Shukla, V., Singh, O., Mishra, G., Tiwari, R.: Application of CSMT gate for efficient reversible realization of binary to gray code converter circuit. In: 2015 IEEE UP Section Conference on Electrical Computer and Electronics (UPCON), pp. 1–6 (2015)
23. Bhagyalakshmi, H., Venkatesha, M.: Design of a multifunction BVMF reversible logic gate and its applications. Int. J. Comput. Appl. **32**(3), 0975–8887 (2011)
24. Sultana, M., Chaudhuri, A., Sengupta, D., Chaudhuri, A.: Logic design and quantum mapping of a novel four variable reversible s2c2 gate. In: Nature, S., (ed.): CSI 2017—52nd Annual Convention of Computer Society of India, Kolkata (2018) (in Press)

25. Chaudhuri, A., Sultana, M., Sengupta, D., Chaudhuri, A.: A novel reversible two's complement gate (TCG) and its quantum mapping. In: 2017 Devices for Integrated Circuit (DevIC), pp. 252–256. IEEE, Kolkata (2017)
26. Arabzadeh, M., Saeedi, M.: RCViewer+: A viewer/analyzer for reversible and quantum circuits (2008–2013, version 2.5)
27. Saravanan, M., Manic, K.S.: Energy efficient code converters using reversible logic Gates. In: Proceedings of 2013 International Conference on Green High Performance Computing, India, Mar 2013
28. Kamani, K., Koneti, S., Bollampalli, U., Shankara, S.: Energy efficient reversible logic design for code converters. Int. J. Res. Appl. **1**(3), 132–136 (2014)
29. Haghparast, M., Hajizadeh, M., Bashiri, R.: On the synthesis of different nanometric reversible converters. Middle-East J. Sci. Res. **7**(5), 715–720 (2011)
30. Das, J., De, D.: Quantum-dot cellular automata based reversible low power parity generator and parity checker design for nanocommunication. Front. Inf. Technol. Electron. Eng. **3**, 224–236 (2016–17)
31. Gayathri, S., Ananthalakshmi, A.: Design and implementation of efficient reversible even parity checker and generator. In: International Conference on Science Engineering and Management Research (ICSEMR), pp. 1–4 (2014)
32. Mustafa, M., Beigh, M.: Design and implementation of quantum cellular automata based novel parity generator and checker circuits with minimum complexity and cell count. Indian J. Pure Appl. Phys. **51**, 60–66 (2013)

Regression-Based AGRO Forecasting Model

B. V. Balaji Prabhu and M. Dakshayini

Abstract Prediction plays an important role everywhere particularly in business, technology, and many others. It helps all types of organizations to improve profits and reduce the loss by taking timely decisions. Agriculture is also like an organization where farmers suffer with the loss most of the time in this business. This could be mainly because, there is no system to ensure the synchronization between the demand and supply for various food commodities required by the society. Science, enormous amount of data from different authorized sources like government websites revealing the demand and supply of various food commodities for forgoing period, this paper proposes a novel regression based AGRO forecasting model. This could help the farmers to make timely decisions and work towards fulfilling the actual needs of the society and avoiding putting themselves into the loss by growing unnecessary crops. Proposed model has been implemented using MapReduce parallel programming approach with Hadoop Distributed File System. This processes time series data with Regression model for predicting the demand, supply and price for the agricultural commodities in distributed environment. Resulting forecasted values are in the range of real values.

Keywords Prediction · Decision · Agriculture · Demand–supply · Forecast
Parallel programming model · Hadoop distributed file system · Regression
MapReduce · Time series

B. V. Balaji Prabhu (✉) · M. Dakshayini
Department of ISE, BMS College of Engineering, Bangalore 560019, Karnataka, India
e-mail: balajitiptur@gmail.com

M. Dakshayini
e-mail: dakshayini.ise@bmsce.ac.in

© Springer Nature Singapore Pte Ltd. 2019   479
A. Abraham et al. (eds.), *Emerging Technologies in Data Mining and Information Security*, Advances in Intelligent Systems and Computing 755,
https://doi.org/10.1007/978-981-13-1951-8_43

1   Introduction

Today one of the main reasons where the farmers incur loses is due to the variation in demand–supply of crop production in the market. In spite of reliable agriculture relevant data available from government [1–3], farmers in the developing countries like India currently facing the lot of problems. One of the main problems is Demand—Supply management Problem [4]. There is no synchronization in production and demand due to which either farmer fail to get good market prices to their produce or consumer suffers high prices due to less production. Hence there is a need for a decision support system where farmers can make timely decisions in crop production. To synchronize the demand–supply there is a need to forecast the same for the commodities being produced [5]. Based on the forecasted values, gap between demand and supply for a commodity can be identified earlier so that farmer community can be provided with an educated decision support mechanism while choosing crops to grow.

Time Series data play an vital role in predictions, the analysis of Time Series data can be performed with many machine learning algorithms such as regression, exponential smoothing, moving average, classification, clustering and model-based recommendation [6]. A forecasting model can be built with the time series data to predict the future, where the model will learn through the historic data and can predict the future. The available time series models will work efficiently with the small data. But the amount of data generated by modern digital world has reached the size up to petabytes and zetabytes, called as big data [7]. The ability to analyze and process this big data is a tedious job for any organization to obtain the useful insights to make better decisions to improve the business in time. Better idea is to split this big data into small chunks and store it on the distributed environment such as Hadoop Distributed File System (HDFS) [8], so that analytics can be performed in parallel on all the data chunks to get the insights [9]. For this purpose, the big data to be processed requires a high performance analytical system which runs on the distributed environment such as MapReduce [10].

In this work a parallel linear regression model has been implemented using a MapReduce framework which can efficiently handle very large datasets to effectively forecast the demand, supply, and price of the agricultural commodities in the market. Experimental results shows that there is a huge gap between demand and supply for the agricultural commodities which is the reason being the price variation in the market and the loss for the farmers in their crop production.

2 Related Works

2.1 Big Data and Predictive Models

Big data is a huge repository of structured or unstructured data which has been widely incorporated across industries in recent years. Big data can be defined as a huge considerable volume of data which is exponentially increasing day by day. Current predictive models failed to cope up with this quick updation of data, hence they use subset of existing data for prediction. Due to which organizations are failing to take up the right decision in the right time to enhance the business productivity. Parallel predictive models can make use of big data to leverage all the information available with large data rather than its subset [10]. Which provides an opportunity to create unprecedented business advantage over decision making and gain useful insights to make better decisions at right time without overhead [11].

2.2 Hadoop and MapReduce

Hadoop is an open source framework for distributed data storage where the data is stored on different clusters. A parallel programing model can be applied on all data clusters to process data effectively. MapReduce Programming model is a parallel processing engine used to process the distributed datasets stored on Hadoop clusters [9]. The data stored on Hadoop cluster can be processed using MapReduce in two phases such as Map phase and Reduce phase. The Map phase reads the data from different data clusters stored and produce an intermediate results in the form of <key, value> pair. The Map phase will write the output into a context object from where a reducer can read the intermediate results to process further. The reducer reads data from context object in the form of <key, value> pair and process input data to produce the output in same format.

2.3 Predictive Modeling and Techniques

Predictive modeling is the process of building a conceptual model to express the output variable as a function of explanatory variables to predict the outcome. In general predictive modeling can be categorized into three main techniques, they are traditional techniques, data adaptive techniques and model dependent techniques [12]. The traditional approach, such as linear regression and logistic regression models estimate parameters for linear predictors for predictions. The data adaptive techniques identify the most substantial factors in the given data which contributes to the prediction more effectively. Model-dependent techniques use the data predictor functions with the analytical methods to generate the prediction. Time series analysis is a data adaptive approach used to forecast the values using historic data. Time series data is a collection of values identified over a specific time interval for forecasting.

This survey shows that, there is no substantial work being done in predicting demand, supply and price of the agricultural commodities, due to which farmers fail to recognize the crop demands from which they can retrieve maximum profit. This paper proposes a parallel linear regression model using a MapReduce framework which can efficiently handle very large datasets to effectively forecast the demand, supply and price of the agricultural commodities. This model would help the farmers in crop selection based on the forecasted values to make crop production gainful.

3   Linear Regression Models

Liner Regression is the simplest and most commonly used machine learning technique being used in building predictive models. This method is mainly used to predict the value of a dependent variable given a set of independent variables. The basic idea behind the regression is that the output variable can be expressed as a linear combination of a set of input variables.

The linear regression model establishes relationship between independent and dependent variables by fitting a best line in such a way that the differences between the distances of data points from the curve or line is minimized. The regression line is represented by a linear Eq. 1.

$$Y = I + S * X + e, \tag{1}$$

where,

Y   is the output variable or dependent variable
X   is the input variable or independent variable
I   is an Intercept of the line
S   is the slope of the line
e   is the random error.

Based on the number of independent variables used to predict the value of dependent variable, regression models can be categorized into two types

1. Simple Linear Regression Model
2. Multiple Linear Regression Model

3.1   Simple Linear Regression Model

In simple linear regression model an output variable depends on the single input variable. A simple linear regression equation can be written as given in the Eq. 2.

$$Y_i = I + S_i * X_i + e_i, \tag{2}$$

where

Y_i is a ith dependent variable
X_i is the ith independent variables
I   is an Intercept of the line
S_i is the slope of the line
e_i is the error in ith value of X.

In this linear regression model, an output variable Y depends on each observations of predictor variable X.

3.2 Multiple Linear Regression Model

In multiple linear regression model the predictor variable depends on more than one independent variable. A multiple linear regression equation can be written as given in the Eq. 3.

$$Y_i = I + S_1 * X_{i,1} + S_2 * X_{i,2} + \cdots + S_n * X_{i,n} + e_i, \tag{3}$$

where

Y_i is a ith dependent variable
X_i is the ith independent variables
I   is an Intercept of the line
S_i is the slope of the line
e_i is the error in ith value of X

Slope of the regression line must be calculated using the Eq. 4

$$\text{Slope (S)} = \left(N \sum X * Y - \left(\sum X \right) \left(\sum X \right) \right) \Big/ \left(N \sum X^2 - \left(\sum X \right)^2 \right) \tag{4}$$

Intercept of the regression line is calculated using the Eq. 5

$$\text{Intercept (I)} = \left(\sum X - s \left(\sum Y \right) \right) \Big/ N \tag{5}$$

Substituting the value of intercept I and a slope S in the Eqs. 2 or 3 with the set of X values gives the prediction for the dependent variable Y.

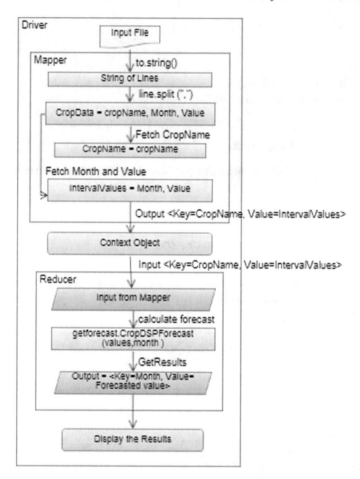

Fig. 1 MapReduce forecasting functionalities

4 Regression-Based AGRO Forecasting Model Using MapReduce

In this paper a linear regression model has been implemented using MapReduce approach with HDFS model to forecast the demand, supply and price of various agricultural commodities.

The supply, demand, and price data required for this analysis has been gathered from the authorized sources like Ministry-of-Agriculture-Agmarknet Directorate of marketing and Inspection Ministry-of-Agriculture and Farmers-Welfare Government of India [1], National Horticulture Board (NHB) India [2] and from [13] from the state Karnataka.

As the data collected will not be in the required form and contains noise values and missing values, these data must be pre-processed to remove the noise and build the missing values before applying forecasting algorithm for better performance. During this process, incomplete information is eliminated and all NA (Not Applicable) values are aggregated to the average value. As the data sets collected contain multiple attributes, only the required attributes are separated and stored as a time series data file (input file).

This input file is partitioned and stored in clusters of HDFS to be given as an input to the mapper phase. The mapper phase will process this input data stored on HDFS clusters and produce an intermediate results in the form of <key, value> pair.

Once the mapper finishes its job, it creates the output context object, which would be the input for the reducer. Reducer reads the intermediate results from the context object and process this <key, value> pair to forecast the values. The reducer will use CropDSPForecast (Value, Month) method to calculate the predictions for the specified month where Value is a set of values from reducer phase and month is specified by user.

The mapper and reducer functionalities are combined using Driver module by specifying the jar files which contains the implementation of mapper and reducer as shown in Fig. 1. The paths of input and output files are specified through the arguments to driver program to fetch the data and to produce the output. The <key, value> pair formats are also specified in the driver program to get the results. After all the parameters are set, the MapReduce job has been run to get the forecast.

Algorithms
Mapper function
Input: Partitioned Data sets
Output : <Key, Value> pair
Begin
      line = value.toString()
      CropData = line.split("\\t")
      CropName=(CropData[0]);
      IntervalValues = (CropData[1]+ "-" + CropData[2]);
      context.write(CropName, IntervalValues)
End
Output: <Key=CropName, value=IntervalValues> pair

Reducer function
Input: Intermediate Result from Mapper phase(<Key, Value> pair)
Output : Forecasted results(<Key, Value> pair)
Begin
      month = targetMonth;
      Result = cropDSP(values, month)
      Forecast = getforecast.cropDSPForecaste( )
      context.write(Forecast)
End
Output: <Key=month, value=forecasted value> pair

CropDSPForecast function

[Nomenclature:

X = Independent Variable (Months)

Y = Dependent variable (Demand/Supply/Price)

N = Initial Number of Months

TM = Target Month to forecast the values

S = Slope of the Line

I = Intercept of the Line

F^{tm} = Forecasted Value for the Target Month

FV^{tm} = Forecasted Values till Target Month

Y_f = Future Y values Till Target Month

T^P = Temporary Projected Values

]

    Input: month, values

    Output: forecasted value for the month specified

    Begin

        Calculate the Slope.

        Slope$= (N \sum X * Y - (\sum X)(\sum Y))/(N \sum X^2 - (\sum X)^2)$

        Calculate the Intercept.

        Intercept $= (\sum X - S (\sum Y))/N$

        Calculate the Forecast for the Target Month.

        $F^{tm} = I + (S*TM)$

        Calculate the Cumulative Values for the Target Month.

        $FV^{tm} = \sum Y_f + \sum Y$

        Return the Forecasted Values for the Target Month.

            $\sum Y_f = 0$, $T^P = 0$, $FV^{tm} = 0$

            for (i = N+1; i<=TM; i++)

                TP = I + (S*i)

                Y_f += TP

            End for

    End

    Output: Return Forecasted values

5   Results and Discussion

In this work, demand–supply and price values for the crop Tomato has been considered and taken for the foregoing period 2008–2016 for the state Karnataka, which has been analyzed and presented Fig. 2. The demand, supply, and price values have been forecasted using the proposed regression model for the years 2017 and 2018.

    Figures 2 and 3 shows the varied values of demand and supply for the crop Tomato for the foregoing period 2008–2016, and also the forecasted values using regression based AGRO forecasting model for the crop tomato for the years 2017 and 2018.

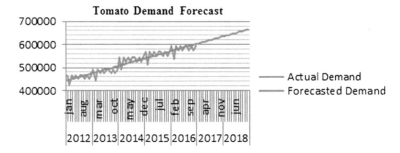

Fig. 2 Demand variation for the crop tomato for the foregoing period 2008–2016 and predicted demand for the years 2017 and 2018

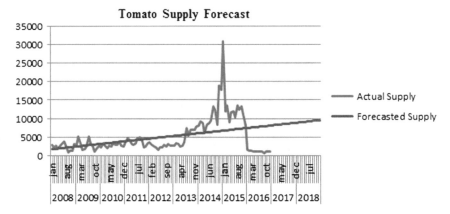

Fig. 3 Supply variation for the crop tomato for the foregoing period 2008–2016 and predicted supply for the years 2017 and 2018

In Fig. 4 the demand versus supply values for the crop tomato for the foregoing period 2008–2016 has been plotted to show the mismatch between demand and supply of the food crops in the market. This graph clearly indicates that there is a varied gap between demand and supply for the food commodity considered hence, has resulted in the price variation shown in Fig. 5 for the foregoing period 2008–2016. Figure 5 also shows the predicted price values for the years 2017 and 2018. From Fig. 5 it can be observed that fall in price when there is a huge supply and rise in price for the commodity when there is less supply.

Figures 6 and 7 gives the demand and supply forecast for the crop tomato till 2020. From the Figures it can be observed that there is a huge gap between the demand and supply for the crop tomato and is growing higher from year to year but the supply for these crops is not matching with the demand.

Table 1 compares the actual and forecasted values of demand and supply values for the crop Tomato for the years 2016 and 2017. Forecasted values are in par with the actual values.

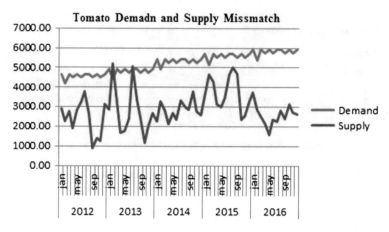

Fig. 4 Demand versus supply plot for the crop tomato for the foregoing period 2008–2016

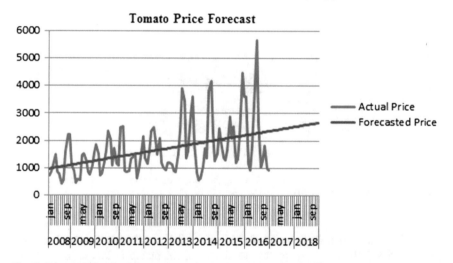

Fig. 5 Price variation for the crop tomato for the foregoing period 2008–2016 and predicted market price for the years 2017 and 2018

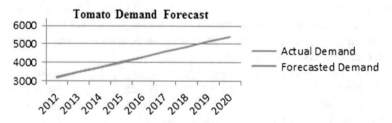

Fig. 6 Demand forecasting for the crop tomato up to 2020

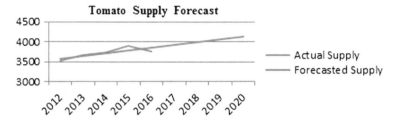

Fig. 7 Supply forecasting for the crop tomato up to 2020

Table 1 Comparison of actual versus forecasted values	Year	Demand (Lakh tons)		Supply (Lakh tons)	
		Actual	Forecasted	Actual	Forecasted
	2016	4289	4298	3755	3851
	2017	4611	4574	3814	3919

6 Conclusion

In this paper, Regression-based AGRO forecasting model is proposed using Map-Reduce approach with HDFS and Linear Regression algorithm to forecast the demand, supply and price for the agricultural commodities. From the forecasted results obtained, it has been observed that the gap between the demand for various agricultural crops from the customers and the supply of the same crops from the farmers is huge, which has been a reason for unexpected price variation in the market due to which both consumers and farmers suffer with the loss. A Regression based AGRO forecasting model developed in this work has addressed this problem by helping the system to guide the farmers in selecting the appropriate crops to grow satisfying the actual needs of the society (demand) and hence minimizing the loss for both farmers and consumers thus achieving equilibrium in demand and the supply of the crops that could effectively deal the current situation.

References

1. "agmarknet." http://agmarknet.gov.in/PriceTrends/Default.aspx
2. "horticulture." http://nhb.gov.in/OnlineClient/MonthwiseAnnualPriceandArrivalReport.aspx
3. "hopcoms." http://www.hopcoms.kar.nic.in/(S(qbaji2453cfnhs55umapkyzk))/default.aspx
4. Kumar, P., Joshi, P.K., Birthal, P.S.: Demand projections for foodgrains in India. Agric. Econ. Res. **22**, 237–243 (2009)
5. Balaji Prabhu, B.V., Dakshayini, M.: A novel cloud based data analytics framework for effective crop management. Int. J. Control Theory Appl. **9**(22), 257–264 (2016)
6. Arputhamary, B.: PAFHWKM: an enhanced parallel approach to forecast time series data using holt-winters and K-Means algorithm
7. Chen, M., Mao, S., Liu, Y.: Big Data: A Survey, pp. 171–209 (2014)

8. Shvachko, K.: The Hadoop distributed file system. In: IEEE 26th Symposium on Mass Storage Systems and Technologies, pp. 1–10 (2010)
9. Saldhi, A., Yadav, D., Saksena, D., Goel, A., Saldhi, A., Indu, S.: Big data analysis using Hadoop cluster. In: 2014 IEEE International Conference on Computational Intelligence and Computing Research, 2014. http://ieeexplore.ieee.org/document/7238418/
10. Elagib, S.B., Najeeb, A.R., Hashim, A.H., Olanrewaju, R.F.: Big data analysis solutions using MapReduce framework. In: Proceedings of 5th International Conference on Computer and Communication Engineering Emerging Technologies via Comp-Unication Convergence, ICCCE 2014, pp. 127–130 (2015)
11. İşlek, İ., Öğüdücü, Ş.G.: A retail demand forecasting model based on data mining techniques. In: 2015 IEEE 24th International Symposium on Industrial Electronics (ISIE), pp. 55–60 (2015)
12. Kaur, N., Kaur, A.: Predictive modelling approach to data mining for forecasting electricity consumption. In: 2016 6th International Conference, Cloud System and Big Data Engineering, pp. 331–336 (2016)
13. Planning, P., Cell, M.: Supply and demand for selected fruits and vegetables in Karnataka. Int. J. Farm Sci. 5(2), 184–197 (2012)

CRICRATE: A Cricket Match Conduction and Player Evaluation Framework

**Md. Ashraf Uddin, Mahmudul Hasan, Sajal Halder, Sajeeb Ahamed
and Uzzal Kumar Acharjee**

Abstract Cricket has appeared as one of the most favorite outdoor games in the present world. The cricket players represent a country and create economic, political, and diplomatic relations among nations. The cricket board of a country requires selecting the fittest players for the upcoming team among some good players. We propose an architecture called Cricket Match Conduction and Player Evaluation Framework by developing some algorithms to predict the score of the players as well as the algorithm to evaluate the man of the match in one day or test cricket match. We implemented the framework by Weka and web technology.

Keywords Player evaluation · Cricket · Local match · Score prediction
Man of the match

1 Introduction

Cricket is nowadays a favorite game which is played all around the world. Almost 105 countries [4] play cricket all around the world. So, it becomes essential to handle cricket data, and measurement of player performance is equally important as well as conducting a match. Although cricket is a favorite game, still the cricket board of the

Md. A. Uddin (✉)
Internet Commerce Security Lab, Federation University, Ballarat, Australia
e-mail: mdashrafuddin@students.federation.edu.au

M. Hasan · S. Halder · S. Ahamed · U. K. Acharjee
Department of Computer Science and Engineering, Jagannath University,
Dhaka, Bangladesh
e-mail: mahmudulcsejnu@gmail.com
S. Halder
e-mail: sajal@cse.jnu.ac.bd

S. Ahamed
e-mail: sajeebl@cse.jnu.ac.bd

U. K. Acharjee
e-mail: uzzal@cse.jnu.ac.bd

© Springer Nature Singapore Pte Ltd. 2019
A. Abraham et al. (eds.), *Emerging Technologies in Data Mining and Information Security*, Advances in Intelligent Systems and Computing 755,
https://doi.org/10.1007/978-981-13-1951-8_44

countries that take part in an international and national cricket match do not select their team based on the statistical analysis. We design a system where a cricket board can conduct a local game by registering on our network. The controller of the cricket board can create a new match, the team for that match, and select the best squad to play the match.

Our web-based development of the framework allows us to conduct local cricket match. There are three types of user namely, Unregistered, Registered, and Admin. The unregistered users can only view the contents of the website; the registered users can post and reply the comment on the blog named CricBlog and ask any query in the Frequently Ask Questions (FAQ) section. The admin can manipulate the function of the system like edit, delete the record or any operations which need higher permission. There is a live scoreboard for an ongoing match, and one can see that. Files of the players are stored in an organized database. The players are evaluated based on the stored information. Every player will obtain a ball by ball rating points concerning the assessed value. After finishing the match, we can find the most valuable batsman and the most valuable bowler and also the man of the match. Before starting a match, there is an option which can predict the score based on the previous records of the players.

Our contributions include player evaluation, ball by ball player rating, and selection of the man of the match by this rating and score prediction which are not provided by the most of the popular cricket website such as ESPNcricinfo, and CricBuzz. We review related works in Sect. 2. Section 3 describes our framework and algorithms. We show the performance analysis in Sect. 4 before concluding the paper.

2   Related Work

There are 10 full members, 39 associate members, and 56 affiliate members of ICC [5]. Cricket is played between two teams of eleven players. Players score runs (points) by running between two sets of three small, wooden posts called wickets. Each of the wickets is at one end of a rectangle of flattened grass called the pitch. The pitch is a much longer oval of grass called the cricket ground. When international or national cricket match holds, the run rate, balling rate, and total score update are normally maintained by a controller called scoreboard so that people get an update of the running cricket match from any corner of the world. With the popularity of this game, there is a demand for blogs, news portal, and answering site of any question of cricket to enrich the cricket knowledge of people. A number of websites have been developed in the last decade for this purpose. We explored several sites where those issues are handled individually or all together.

ESPNCricinfo [1] is one of the oldest cricket websites having some useful features such as news, columns, videos, fantasy sports game, and blog having a live scoreboard. The database of the site holds the statistics of player's performance, cricket boards, and details of future tournaments. CricBuzz [8] is an app and sports news website for cricket. It contains live streaming of cricket scores, ball by ball commentary, upcoming schedules, and player rankings. CricWaves is another cricket website that has already stored cricket data for the past 4 years. Sify Sports, Cricket Nirvana [9], Yahoo Cricket [7] and Sports, and NDTV are some other favorite cricket websites having the standard features of telecasting live score and cricket news. None of this framework is suitable for conducting local cricket match, and these structures does not predict the potential player for the future cricket match. In [11], the authors estimate the average score of batting team against balling team. Dynamic programming did the estimation. Raj and Padma [10] applied the mining approach of association rule and predicted the toss, batting first or second, and match result. Swartz et al. [13] estimate ball by ball outcome based on the Chain Monte Carlo and Bayesian Latent variable model. In [12], the authors proposed a prediction model of the score and next match-winning by using linear regression and naive Bayes classifiers.

Although there are some famous and popular cricket websites which provide score update of the international match, very few apps or website can accommodate a cricket board by conducting local cricket match and selection of the automatic man of the match. Normally, the man of the match is determined by the analog system. The selection of the man of the match is determined by commentators' votes or a selected panel which is formed by ICC. Therefore, we need to develop some algorithms to make the process automatic. To perform score prediction of the players and selection of the squad as per the prediction, we also need to develop some algorithms.

3   The CRICRATE Framework

In this section, first, we describe the feature of our proposed cricket framework. Next, we present our designed algorithm to evaluate player's score and man of the match. The feature of the framework is illustrated in Fig. 1. Our cricket match conducting framework called CRICRATE consists of three major components, namely, General Component, Match Component, and Special Component.

General Component: This component contains four module called CricBlog, Cricfaq, Fixture, and News update. The CricBlog is a blog where a registered user can publish his article on cricket game. The system lets the user publish article after checking the context relevancy. Usually, it extracts cricket-related keywords from the article. If the article contains the significant amount of keywords, the article is published. Cricfaq is a frequently asked question module where the user can show the ratings of their asked questions and put their answers. Fixture maintains the schedule for a local and international cricket match. News update publishes news about cricket match and players. Upcoming events are available on the News update module.

Match Component: The significant modules of this component are score prediction, man of the match, best batsman of the match, and best bowler of the match. In score prediction, we predict the score of the upcoming innings based on the previous performance of the team. Man of the match which indicates the best player of the match is determined by the neural network in our framework. According to the

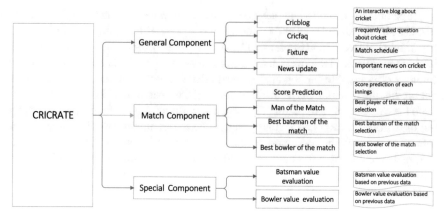

Fig. 1 Features of the CRICRATE framework

performance of the current match, best batsman of the match and best bowler of the match are determined.

Special Component: In the special component, we predict the performance of the individual player (typically batsman and bowler) based on his previous record and current fitness. We also classify the players of a specific country according to their performances such as level 1 player, level 2 player, level 3 player, etc., where level 1 player is a first-class player.

3.1 Features of Proposed System

In this section, we describe some algorithms to implement some significant modules of our framework.

Special Component: Here, first, we describe the algorithm for evaluating batsman and then bower from their performance in the previous match.

Algorithm 1: Batsman evaluation

Input: Global average(G_{Avg}), strike rate(G_{Sr}), LTM_{Avg}, strike rate(LTM_{Sr})
Output: Evaluated Value(E_v) within range [0,1]

1   $I_G = G_{Avg} \times G_{Sr}$;
2   $I_{LTM} = LTM_{Avg} \times LTM_{Sr}$;
3   $I = Avg(I_G, I_{LTM})$;
4   $E_v = map(I, 0, 60000, 0, 10)$
5   **return** E_v

Batsman Evaluation: We can calculate an evaluated value for every batsman with the help of their record data. The value helps the system to define the batsman with good performance and bad performance by observing the evaluated value. The evaluated value is stored in the database and used for the future evaluation. Algorithm 1 depicts the batsman evaluation.

Algorithm 2: Bowler evaluation

Input: $G_{Avg}, G_{Ec}, LTM_{Avg}, LTM_{Ec}$
Output: Evaluated Value within range [0,1]
1   $I_G = (100 - G_{Avg}) \times (100 - G_{Ec})$;
2   $I_{LTM} = (100 - LTM_{Avg}) \times (100 - LTM_{Ec})$;
3   $I = Avg(I_G, I_{LTM})$;
4   $E_v = map(I, 0, 60000, 0, 10)$;
5   **return** E_v ;

In Algorithm 1, we use the function $batsmanValue$ that contains four parameters, i.e., G_{Avg}, G_{Sr}, LTM_{Avg}, LTM_{Sr} refers to global average, global strike rate, last ten matches average, and last ten matches strike rate, respectively. The I_G refers to global index at line 2 which is calculated from the product of global average and global strike rate. At line 3, the procedure of calculation of I_{LTM} (which is index of last ten matches) is the same as before except it is calculated from the last ten matches average and strike rate.

And the final Index (I) is the average of I_G and I_{LTM}. Finally, the **evaluated value** (E_v) is the value which is mapped within a range [0, 10] from the range [0, 60000].

Algorithm 3: Batsman Rating Point

Input: Bowler E_v, Run of the ball
Output: Batsman rating point
1   **while** $j \leq 300$ *and* $i \leq 10$ **do**
2   **if** Bt_i *is out* **then**
3   **if** $Bt_{iTR} \geq 50$ *and* $Bt_{iTR} < 100$ **then**
4      $\big|$ $RP_i = RP_i + (RP_i \times \frac{5}{100})$;
5   **end**
6   **else if** $Bt_{iTR} \geq 100$ *and* $Bt_{iTR} < 150$ **then**
7      $\big|$ $RP_i = RP_i + (RP_i \times \frac{10}{100})$;
8   **end**
9   **else if** $Bt_{iTR} \geq 150$ *and* $Bt_{iTR} < 200$ **then**
10      $\big|$ $RP_i = RP_i + (RP_i \times \frac{15}{100})$;
11   **end**
12   **else if** $Bt_{iTR} \geq 200$ **then**
13      $\big|$ $RP_i = RP_i + (RP_i \times \frac{20}{100})$;
14   **end**
15   **end**
16   **else**
17     $PP_j = R_j \times \frac{1}{36}$;
18     $PB_j = Bw_{cEv} \times R_j$;
19     $B_j = map(PB_j, 0, 60, 0, \frac{1}{36})$;
20     $RP_i = RP_i + (PP_j + B_j)$;
21   **end**
22    $j = j + 1$
23   **end**

The upper limit of mapping range is 60000 because the highest average of a batsman would be 100 and the highest strike rate would be 600. So the product would be 60000.

Batsman Evaluation: We reckon the performance of the bowlers in the same way we evaluate the performance of the batsman. Algorithm 2 depicts the Bowler Evaluation.

The evaluation of bowler is the same as Algorithm 1 except for some subtle differences. A bowler who has less bowling average is considered as a good bowler. So, we subtract a bowler average and economy (which is expressed as G_{Ec} and LTM_{Ec}) from 100. The remaining calculation is the same as Algorithm 1 described above.

Match Component: We calculate every player's a ball by ball rating value which finally helps to select the most valuable batsman, bowler and the man of the match. This rating is calculated from the information of the running match.

Algorithm 4: Bowler Rating Point

Input: Batsman E_v, Run given of the ball
Output: Bowler rating point
1 $i = cBw$;   // Current Bowler

2 **while** $j \leq 300$ **do**
3 $\quad$ $PP_j = (3 - R_j) \times \frac{1}{36}$;
4 $\quad$ $PB_j = Bt_{cE_v} \times (3 - R_j)$;

5 $\quad$ $B_j = \text{map}(PB_j, 0, 30, 0, \frac{1}{36})$;

6 $\quad$ $PR_i = PR_i + (PP_j + B_j)$;
7 **end**
8 **if** $Bw_{iTw} \geq 3$ *and* $Bw_{iTw} < 5$ **then**
9 $\quad$ $PR_i = PR_i + (PR_i \times \frac{5}{100})$;
10 **end**
11 **else if** $Bw_{iTw} \geq 5$ **then**
12 $\quad$ $PR_i = PR_i + (PR_i \times \frac{10}{100})$;
13 **end**

Batsman Rating Point: The batsman rating point calculation is illustrated in Algorithm 3. Here, we calculate the point for all 11 batsmen. The system generates a rating point for every ball faced by a batsman with the help of the run he gets from that ball. It also takes into account that the run has been produced based on which bowler, i.e., how valuable the bowler is. The **While** loop from lines 1 to 16 check if the match running or not, i.e., if the total ball is less than or equal 300 or wicket is less than or equal 10. The **if** condition at lines 2–10 check if the batsman is out or not. If OUT then check his run. If the run is between 50 and 99, then he will get 5% bonus of his total point and also get 10%, 15%, 20% for 100 to 149, 150 to 199, and 200 to more respectfully. If the batsman is not out and on strike, then he obtains a point for every ball he faces. The primary point of ball j (PP_j) is the product of Run of ball j (R_j) and $\frac{1}{36}$. We multiply it by $\frac{1}{36}$ because we want to bind the point for every over in between 0 and 1. So we know there are six balls in an over and the maximum run of a ball could be 6. So for 6 balls, there are at most 36 runs [6]. For this reason, we use $\frac{1}{36}$. Next, we calculate the bonus point of each ball concerning the current bowler's evaluation value (see at line 13). Where Bw_{cE_v} is the evaluated value of current bowler. Finally, we are adding the primary point and bonus point to obtain the final rating point player i at line 15.

Bowler Rating Point: The bowler rating point calculation (pictured at Algorithm 3) is the same as batsman rating point except for the calculation of primary point (PP_j) which is done by subtracting the run (R_j) from 3. Because, the fewer runs batsmen earn for a bowler, the better the bowler is. But the maximum run of a ball is 6. So we can do it by subtracting the run from 6. But if a bowler makes a dot ball, i.e., gives 0 runs of a ball then this value would be ($6 - 0 = 6$). But the occurrence of a dot is high. Counting dot is as valuable as a batsman hits a six or boundary. But the occurrence of hitting a six is very low. So it results in a mismatch with respect to batting rating point. So we see that if we subtract the run from 3, then that is more relevant.

Algorithm 5: Score prediction

Input: Statistics of Squad members of both team
Output: Predicted Score

1 **foreach** Bt_i in Batting team **do**
2 $\quad BtCFB_i = \dfrac{Avg_{Bt_i} \times 100}{SR_{Bt_i}}; TB = TB + BtCFB_i;$
3 $\quad$ **if** $TB \geq 300$ **then**
4 $\quad\quad |\quad TWF =$ the loop counter; Break;
5 $\quad$ **end**
6 **end**
7 **if** $TB \geq 300$ **then**
8 $\quad |\quad TB = 300; ;$ $\qquad\qquad$ // Number of balls can be faced by the batting team
9 **end**
10 $BwCTW_i = \dfrac{TBD_{Bw_i}}{Avg_{Bw_i}}; ;$ $\qquad\qquad$ // Bowler can take wicket
11 $TWT = \sum_{j=0}^{6} BwCTW_j; ;$ $\qquad\qquad$ // Assume there are 6 bowler
12 **if** $TWT < 10$ **then**
13 $\quad |\quad TBBwD = 300;$
14 **end**
15 **else**
16 $\quad |\quad TBBwD =$ Number of balls need to do;
17 **end**
18 $Avg_b = average(TB, TBBwD); Avg_w = average(TWF, TWT);$
19 **for** $k = 0$ *to* Avg_w **do**
20 $\quad |\quad sum = sum + BtCFB_k;$
21 **end**
22 $BoP = \dfrac{Avg_b - sum}{Avg_w};$
23 **for** $l = 0$ *to* Avg_w **do**
24 $\quad |\quad FPR = FPR + \dfrac{BtCFB_l + BoP}{SR_{Bt_l}};$
25 **end**

We assume that at the end of an innings if a bowler can take wickets more than two but less than five, then he will get additional 5% bonus of his existing rating point, and if he will get wickets more than four, then he gets 10% additional bonus of his existing rating point.

A **score predictor** technique is implemented in our framework. The algorithm is depicted in Algorithm 5. The score of the innings is measured based on the performance of each player in the team. We combine this approach with machine learning algorithms such as decision tree and fuzzy classification to obtain the ultimate predicted score.

First, we calculate all batsmen faced balls ($BtCFB_i$) and find the Total Ball (TB) faced by all batsmen. If the TB is greater than or equal 300, then we find the total wicket fallen to face 300 balls. This task is shown on lines 1–9 in Algorithm 5. Then, we find total balls that can be faced by the batting side and how many wickets fallen to face those balls. Next, count the total wicket was taken (TWT) by the bowling side and how many balls it takes to do that job depicts at Algorithm 5 at lines 11–14.

At line 18, we calculate the average of ball faced by batsmen predicted balls and bowler predicted balls. The average number of wickets predicts the falling rate of wicket from the batting side and the obtaining rate of wicket from the bowling side. We calculate the bonus or penalty (BoP) shown at line 22. Finally, the Final Predicted Run (FPR) is calculated at lines 23–25.

Man of the Match: To find out the man of the match, we use neural network trained by back propagation algorithm. To train the neural network, we use some features such as the age of the player, number of matches played by the player, balling rate, batting rate, the total number of runs in the match, total number of runs given to the batsmen, the output of the designed algorithms, etc.

4 Performance Analysis

For performance analysis, we evaluate team ODIs score played in a year from 2002 till 2016. The datasets used here are collected from ESPNCricinfo Statsguru [3]. We compare the predicted score with the actual score and visualize in Highchart [2].

As we observe that score in each inning depends on the various factors. So predicted score might rise or fall sharply. For example, score drastically fall from an innings to another or score can rise highly from an innings to another. But it is clear from statistics that falling rate of the score is more frequent than the rising rate.

In Fig. 2, we have visualized the comparison of all the ODI innings played by Australia in the year 2016. We can see that in some innings difference is too high as score has drastically fallen. Also for that reason score has risen too high. But for an average prediction, it is clear that score risen is considerable. In Fig. 3, we have shown the innings comparison for South Africa in the year 2016. In all cases, our predicted algorithms work as we described.

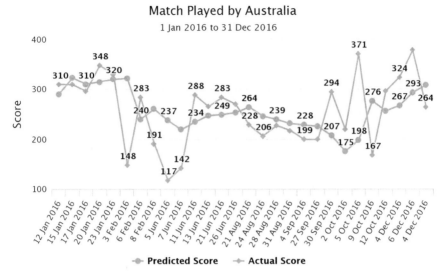

Fig. 2 Score comparison of Australia

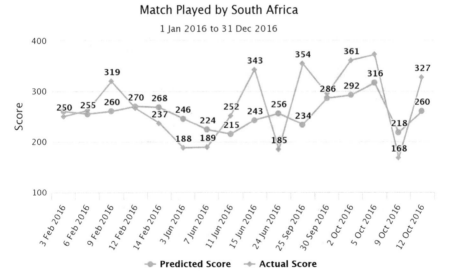

Fig. 3 Score comparison of South Africa

5 Conclusions

Our CRICRATE framework is one of the cricket matching conducting web applications. In this work, we design some algorithms to predict player performances. The application helps the cricket board to select the most potential players for an international match. In the future, we develop mobile apps by using our developed algorithm and plans to devise a more effective and efficient score prediction algorithm by using machine learning algorithm. Psychological aspects of a player influence his performance in the match. So, our future work is to incorporate the psychological factor of player to predict his score.

Acknowledgements This research was partially supported by Academic Innovation Fund, ICT Division, Government of the Peoples Republic of Bangladesh.

References

1. Website [espncrickinfo]. http://www.espncricinfo.com. Accessed 10 Sept 2017
2. Website [highchart]. https://www.highcharts.com. Accessed 25 Sept 2017
3. Website [espncrickinfostatsguruu]
4. Webstie [cricbuzz]. https://en.wikipedia.org/wiki/cricbuzz. Accessed 10 Sept 2017
5. Das, A., Parida, A.R., Srivastava, P.R.: ICC cricket world cup prediction model. In: Information Systems Design and Intelligent Applications, pp. 529–539. Springer (2016)
6. Fenske, M.: Playing the field. Text Perform. Q. **33**(3), 220–222 (2013)
7. Ghansela, S., Negi, A.: Behavior of search engines in popular queries. Indian J. Sci. Technol. **9**(38) (2016)
8. SCB GROUP, et al.: Cricbuzz Cricket Scores & News (2015)
9. Kampmark, B.: Australian cricket: the diminishing light. Sport Soc. **16**(1), 45–55 (2013)
10. Raj, K.A.A.D., Padma, P.: Application of association rule mining: a case study on team India. In: 2013 International Conference on Computer Communication and Informatics (ICCCI), pp. 1–6. IEEE (2013)
11. Shah, A., Jha, D., Vyas, J.: Winning and score predictor (wasp) tool
12. Singh, T., Singla, V., Bhatia, P.: Score and winning prediction in cricket through data mining. In: 2015 International Conference on Soft Computing Techniques and Implementations (ICSCTI), pp. 60–66. IEEE (2015)
13. Swartz, T.B., Gill, P.S., Muthukumarana, S.: Modelling and simulation for one-day cricket. Can. J. Stat. **37**(2), 143–160 (2009)

Predicting Factors of Students Dissatisfaction for Retention

Mohammad Aman Ullah, Mohammad Manjur Alam, Md. Mahiuddin and Mohammed Mahmudur Rahman

Abstract Education sector became challenging as well as competitive due to the huge availability of institutions worldwide. This increases students drop out at an alarming rate. Most of the dropout is due to dissatisfaction of the students. This paper investigates the reasons of students dissatisfaction from a feedback of a university student using Descriptive Statistics, Logistic Regression Analysis and some data mining techniques such as Naïve Bayes, Logistic Regression and Random Forest and found relationship exists between student dissatisfaction and student retention. Our study also found that, a high level of dissatisfaction of the students is mostly due to the cafeteria services and extra-curricular activities. They are moderately dissatisfied due to quality lecture and Lab Facilities. Our analysis has also shown that, overall, students are satisfied, but mostly they do not want to recommend others for this university. On the basis of the result, we then recommended some issues to be fulfilled for student retention.

Keywords Satisfaction · Student · Retention · Algorithms · SPSS

1 Introduction

Today's education system became like a consumer product. There are large numbers of institution selling the education and the marketplace is too competitive and challenging. The students have many options to drop out from the institution due to

M. A. Ullah · M. M. Alam (✉) · Md. Mahiuddin · M. M. Rahman
International Islamic University Chittagong, Chittagong 4203, Bangladesh
e-mail: manjuralam44@yahoo.com

M. A. Ullah
e-mail: ullah047@yahoo.com

Md. Mahiuddin
e-mail: mmuict@gmail.com

M. M. Rahman
e-mail: provaiiuc@gmail.com

© Springer Nature Singapore Pte Ltd. 2019
A. Abraham et al. (eds.), *Emerging Technologies in Data Mining and Information Security*, Advances in Intelligent Systems and Computing 755,
https://doi.org/10.1007/978-981-13-1951-8_45

availability of alternative. Therefore, every institution needs to find out some ways to deal with these challenges and competitions. Also, it is important to take regular feedback from the students and find any relationship exists between student satisfaction and student retention. If exists, the institution need to find the causes of student satisfaction. As the key elements of any educational institution are academic staffs and they are highly attached to the student service. So, it is equally important to find any relationship exists between student satisfactions and academic staffs. If exists, the institution need to find some ways to develop academic staffs. Both the relationships could be identified using statistical and data mining tools.

Data mining helps in predicting the causes of any occurrences from the given dataset. To help this analysis done, in this paper, we have collected the feedback of near 400 students from different departments of International Islamic University Chittagong (IIUC), Bangladesh; deeply investigated the students opinion regarding lecture quality of the teachers, advisors behavior, Lab facilities, availability of resources in library, medical service, campus security, easiness in course registration, food quality, extra curriculum activities, research opportunity, transport facility and intensity to suggest the university to others. Preprocess them to bring out a standard dataset of 12 variables. We have then analyzed those data using Descriptive Statistics, Logistic Regression Analysis and some data mining techniques such as Naïve Bayes, Logistic Regression and Random Forest and found high level of dissatisfaction of the students. Correlating different results of the analysis it is clear that, this dissatisfaction increases the drop out of the students. The total analysis was conducted using SPSS 20.0 software tool and Python programming Language.

The remaining work is organized as follows: Sect. 2 includes Literature Review, Sect. 3 represents details methodology. Whereas, Sect. 4 describes the results and Discussions and Sect. 5 includes Recommendations and Sect. 6 describe Conclusions.

2 Literature Review

A large number of researches were done on the student satisfaction issue. For instance, few researchers attempt to find out the factors that affect student satisfaction [1–3], some of them searched for the association between student satisfaction and other factors such as service quality [4–6], others dealt with the topics on student satisfaction relating to the learning environment [7] or student satisfaction and lecture notes [8]. Using data mining techniques, huge amount of research was conducted. Among them, Guo [9] used nine variables and two data mining algorithms such as linear regression and Multilayer Perceptron(MLP) neural network to predict students' satisfaction with respect to course, where MLP outperform linear regression.

A study for student retention using an engineering database of 39,277 students from nine different universities shows that high school GPA and good test score were significant predictors for engineering new student retention [10]. Herzog [11] carry out an investigation on student retention using forty variables, and some data mining algorithms such as decision trees and three backpropagation neural networks with a multinomial logistic regression model and found both the algorithm provides better accuracy and stronger analysis when predicting student retention using a large dataset. A study was conducted on the electrical engineering students of Eindhoven University of Technology to identify the causes that affect student retention. They used the decision trees, Bayesian, and random forests classifiers on data of period 2000–2009 to predict the causes, and found decision tree classifiers outperforming others [12].

Edin [13] has used supervised learning methods such as Naïve Bayes, Multilayer Perceptron, and C4.5 on the students' data, which are collected from university for the period of 2010–2011, in order to predict the performance growth of the students and teachers. They have used few attributes for predicting the performance and expected to be further extended with the use of more attributes and unsupervised learning methods. In [14], the authors used the two rule learners, decision tree, Naïve Bayes, and Nearest neighbor classifier classify the data of the National University of Bulgaria, in order to find the students performance and to predict the factors in enrolling the students in the University. As per them, though the algorithms do not perform well, but they were helpful in identifying student admission score and number of failure students in the first year.

An intelligent Mathematical model for analyzing student's expectation was proposed in [15]. Using Data Mining technique such as Classification, they classify and recognize the facilities hope by students from a college. In [16], they used support vector machines, discriminant analysis, decision tree algorithms, and artificial neural network classifier to predict instructor performance on student response course evaluation dataset and found the decision tree as best on the accuracy, specificity performance, precision, and metrics recall.

Ruba et al. (2011) created two neural network model to predict student retention at two levels in science and engineering programs. At first level, prospect students were predicted. In the second level, newly enrolled students were predicted. In the models a feed-forward backpropagation network were used with 70.1% accuracy [17]. A study done by Dorina (2012) for three consecutive years in a university to find the effect of data mining applications such as decision tree, a neural network and a Nearest Neighbor classifier on university management in enrolling and retention of new students [18]. Kovacic in 2010 performs the quantitative analysis of Open Polytechniques data, collected from 2006–2009 to predict students performance. The author applied the classification and regression tree methods on that data. Though the classification accuracy were not high, but they said to achieve three factors such as ethnicity, course program and course block are the key factors for separating successful and unsuccessful students based on enrolled data [19].

3   Data Collection, Methodology and Data Preprocessing

The primary sources of data were the feedback of different levels of students of International Islamic University Chittagong. Total 400 questionnaires were distributed randomly to the students during the period of spring semester 2017, of which 376 filled-in questionnaires were returned. It may be mentioned that while distributing the questionnaire, we have taken into account the students of different faculties of both the undergraduate and postgraduate students (such as Shariah, Business, Engineering, Law, Arts and Humanities). A cross-sectional research design was applied in this study. To meet the objectives of this study, data was collected by using structured questionnaires. The questionnaire was consists of the following questions: What the students think about teachers lecture quality? Are the students satisfied about their lab facilities? How easy to obtain the resources the student needed from the university library? How helpful is the service at the on campus medical center? What do the students think about campus security? How easy is it to register for courses at this university? How helpful is their academic advisors? How healthy is the food served at the varsity campus/cafeteria? How happy they are with the extra curriculum activities at this university? How likely the students recommend this university to others? Is there enough opportunity to do research at this university? Are the students satisfied about transport facility?

Questionnaires for each of the students were examined carefully. For in-depth analysis of the topic of this research work, feedbacks of the students were coded, editing, removed the missing values, summarized the details output, reduce the inconsistencies and finally analyzed the data set using the Statistical Package for Social Sciences (SPSS Windows version 20) and Python 3.5. The preprocessing of the data brings out a standard dataset of 12 variables. Both statistical (such as Descriptive Statistics, Logistic Regression Analysis) and data mining tools (Such as Naïve Bayes, Random Forest, and Logistic Regression) were used to assess the student dissatisfaction for student retention and development of academic staffs.

4   Experiment Results and Discussions

4.1   Descriptive Statistics Results

From among many factors, in this section we have only included few significant factors related to students dissatisfaction.

Based on the results of Tables 1 and 2, it is clear that, students are moderately satisfied with lecture quality and Lab facilities. Only 46.8% and 31.6% students are respectively satisfied and remaining are dissatisfied. Which cause problems in student retention.

Results in Table 3 indicate that 48.4% students were satisfied with the library service and only 20.7% of them were not satisfied. This is a positive sign for student

Table 1 What do you think about teacher's lecture quality?

		Frequency	Percent	Valid percent	Cumulative percent
Valid	Non effective	31	8.2	8.2	8.2
	Less effective	169	44.9	44.9	53.2
	Effective	148	39.4	39.4	92.6
	Very effective	28	7.4	7.4	100.0
	Total	376	100.0	100.0	

Table 2 Are you satisfied about your Lab facilities?

		Frequency	Percent	Valid percent	Cumulative percent
Valid	Not satisfied	119	31.6	31.6	31.6
	Less satisfied	136	36.2	36.2	67.8
	Satisfied	108	28.7	28.7	96.5
	Highly satisfied	13	3.5	3.5	100.0
	Total	376	100.0	100.0	

Table 3 How easy to obtain the resources you need from the university library?

		Frequency	Percent	Valid percent	Cumulative percent
Valid	Not so easy	78	20.7	20.7	20.7
	Somewhat easy	182	48.4	48.4	69.1
	Very easy	92	24.5	24.5	93.6
	Extremely easy	24	6.4	6.4	100.0
	Total	376	100.0	100.0	

retention of the university. In case of registration process, academic advising, medical services, and campus security of the university, majority of the students were satisfied with few dissatisfied. This helps retentions of the students largely.

It is clear from Table 4 that, the services provided by the university cafeteria were not satisfactory to the most of the students. 59 percent of the students were dissatisfied with their services and food. Also, Almost 50% percent of the students were dissatisfied with the extra curriculum activities of the university.

Table 5 indicates that, most of the students (52.1%) were satisfied about the transport facility of the university while others were dissatisfied.

It has been found that, no single entity has observed satisfaction and dissatisfaction. Therefore, overall rating of the satisfaction and dissatisfaction is important for predicting student retention.

Table 4 How healthy is the food service at the varsity campus/cafeteria?

		Frequency	Percent	Valid percent	Cumulative percent
Valid	Not so healthy	222	59.0	59.0	59.0
	Somewhat healthy	118	31.4	31.4	90.4
	Very healthy	31	8.2	8.2	98.7
	Extremely healthy	5	1.3	1.3	100.0
	Total	376	100.0	100.0	

Table 5 Are you satisfied about transport facility?

		Frequency	Percent	Valid Percent	Cumulative Percent
Valid	Not satisfied	39	10.4	10.4	10.4
	Less satisfied	80	21.3	21.3	31.6
	Satisfied	196	52.1	52.1	83.8
	Highly satisfied	61	16.2	16.2	100.0
	Total	376	100.0	100.0	

Table 6 Overall are you satisfied with academic system at this university?

		Frequency	Percent	Valid percent	Cumulative percent
Valid	Not satisfied	87	23.1	23.1	23.1
	Less satisfied	95	25.3	25.3	48.4
	Satisfied	132	35.1	35.1	83.5
	Highly satisfied	62	16.5	16.5	100.0
	Total	376	100.0	100.0	

From Table 6 and Fig. 1 it is important to mention that 51.6% students were satisfied with IIUC academic system while others were dissatisfied. But, it is very important issue to be consider that, overall 52% of the students does not want to recommend this university to the others, which is a huge challenge in student retention.

4.2 Binary Logistic Regression Analysis

The binary logistic regression is a form of regression used when a dependent variable takes only two values (e.g., Study outcome with two values: Satisfaction or Not

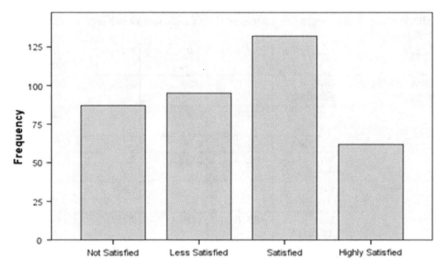

Fig. 1 Overall satisfaction chart

Satisfaction). In the logistic regression analysis, we have considered 11 predictors variables. Among these variables only three variables, Teachers Lecture Quality, Lab Facility, and recommendation of University to Others have a significant effect on Students overall satisfaction as shown in Table 7. On the other hand, Advising, Extra Curriculum, Transport Facility, Canteen Facility, Course Registration Facility, Campus Security, Medical Facility, and Access to Library are insignificant and thus does not include in Table 7.

The odds ratio is used for prediction of estimated logistic regression. The Odds ratio column contains predicted changes in odds for a unit increase in the corresponding independent variable. Odds ratios less that 1 corresponds to decreases in odds and odds ratios greater than 1 correspond to increases in odds. Odds ratios close to 1 indicate that unit changes in that independent variable do not affect the dependent variable. It has been observed from the logistic regression analysis that by improving teachers lecture quality and lab facility students drop out will be decreased. Thus, increase the recommendation to new students for admission to this university.

4.3   Analyzing with Data Mining Algorithms

We have also implemented some data mining techniques in python to find the satisfaction and dissatisfaction of the students and got nearby same results as statistical and logistic regression. Figure 2 and Table 8 shows the corresponding accuracy, precision, recall and F-score values, where, logistic regression outperform other classifiers. So, we may also recommend this classifier as best classifier for this sort of analysis.

Table 7 Binary logistic regression analysis of student satisfaction and retention with different co-variates

Variable	Category	Co-efficient (β)	S.E.	P-value	Odds ratio [Exp(β)]
Teachers lecture quality***	(Not effective)			0.023	
	Less effective	0.779	0.903	0.388	2.180
	Effective	−0.253	0.784	0.747	0.777
	Very effective	0.695	0.765	0.364	2.004
Lab facility***	(Not satisfied)			0.009	
	Less satisfied	−0.330	1.068	0.757	0.719
	Satisfied	−0.134	1.045	0.898	0.875
	Highly satisfied	1.306	1.047	0.212	3.690
Recommendation to others**	(Not so like)			0.025	
	Somewhat like	−0.929	0.676	0.169	0.395
	Very like	−0.077	0.615	0.901	0.926
	Extremely like	0.626	0.604	0.300	1.870

Reference category is marked by parenthesis, *p < 0.10, **p < 0.05, ***p < 0.01

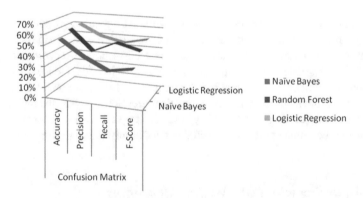

Fig. 2 Overall performance of the classifiers

Table 8 Overall performance of the classifiers

Algorithms	Confusion matrix			
	Accuracy (%)	Precision (%)	Recall (%)	F-Score (%)
Naïve Bayes	55	4	29	33
Random forest	60	40	50	44
Logistic regression	61	50	45	50

5 Recommendations

The following recommendations are given on the basis of analysis for the student retention:

1. Need to appoint the quality teachers with a teaching and practical experience
2. Increase the lab facility
3. Arrange more extra curriculum activities
4. Academic curriculum should be more focused based on the current need of the job market
5. Lecture period must be maintained strictly
6. Teaching should be made interactive through grouping the works of the students
7. Every teacher should follow a lecture plan
8. Avoid Guest teachers
9. Spoiled equipment's should be replaced in each semester
10. Increase canteen facilities availing hygienic and healthy foods
11. Need to increase Research facility
12. Need to increase books in the library
13. Enhance medical facility.

The following recommendations are given on the basis of analysis for the Academic Staff development:

1. Continuous observation is necessary
2. Train the teachers with more practical orientation
3. Motivating for Ph.D.
4. Increase Research facility for the teachers.

6 Conclusions and Future Work

Student satisfaction is the main concern to the authority of the institutions. Several attractive packages are offered to satisfy students. The reason is to prevent the students from leaving the institute. The institutions are also trying to find any other causes exist for students' dissatisfaction such as quality of Academic staff. Therefore, this paper emphasis on those matter and applied descriptive statistics, Logistic Regression and some data mining algorithms on our dataset and found there are close relation between student retention and Dissatisfaction. Also, found that student satisfaction come from the proper receiving of the lecture, thus development of academic staff is too important. In the future, there would like to analyze this data set using many more data mining techniques.

References

1. Manzoor, H.: Measuring student satisfaction in public and private universities. Pak. Global J. Manag. Bus. Res. Interdiscip. **13**(3) (2013)
2. Tessema, M.T., Ready, K., Yu, W.-Y.: Factors affecting college students' satisfaction with major curriculum: evidence from nine years of data. Int. J. Humanit. Social Sci. **2**, 34–44 (2012)
3. Negricea, C.I., Edu, T., Avram, E.M.: Establishing influence of specific academic quality on student satisfaction. In: Procedia—Social and Behavioral Sciences, vol. 116, pp. 4430–4435 (2014)
4. Yunus, N.K.Y., Ishak, S., Razak, A.Z.A.A.: Motivation, empowerment, service quality and polytechnic students' level of satisfaction in Malaysia. Int. J. Bus. Social Sci. **1**, 120–128 (2010)
5. Wei, C.C., Ramalu, S.S.: Students satisfaction towards the university: does service quality matters? Int. J. Educ. **3**(2), (2011)
6. Hanaysha, J., Abdullah, H.H., Warokka, A.: Service quality and students' satisfaction at higher learning institutions: the competing dimensions of Malaysian Universities' competitiveness. J. Southeast Asian Res. **1**, 1–9 (2011)
7. Stokes, S.P.: Satisfaction of college students with the digital learning environment: do learners' temperaments make a difference? Internet High. Educ. **4**, 31–44 (2001)
8. Macedo-Rouet, M., Ney, M., Charles, S., Lallich-Boidin, G.: Students' performance and satisfaction with Web vs. paper-based practice quizzes and lecture notes. Comput. Educ. **53**(2), 375–384 (2009)
9. Guo, W.W.: Incorporating statistical and neural network approaches for student course satisfaction analysis and prediction. Expert Syst. Appl. **37**(4), 3358–3365 (2010)
10. Zhang, G. et al.: Identifying factors influencing engineering student graduation and retention: a longitudinal and cross-institutional study (2002)
11. Herzog, S.: Estimating student retention and degree-completion time: decision trees and neural networks vis-à-vis regression. In: New Directions for Institutional Research, p. 17, (2006)
12. Dekker, G. et al.: Predicting Students Drop Out: A Case Study. pp. 41–50 (2009)
13. Osmanbegović, E., Suljić, M.: Data mining approach for predicting student performance. Econ. Rev. **10**, 3–12 (2012)
14. Kabakchieva, D.: Predicting student performance by using data mining methods for classification. Cybern. Inf. Technol. **13**, 61–72 (2013)
15. Radha, D., et al.: A novel approach to analyze students' expectation from colleges using data mining technique. Int. J. Comput. Appl. **137**(5), 25–28 (2016)
16. Agaoglu, M.: Predicting instructor performance using data mining techniques in higher education. IEEE Access **4**, 2379–2387 (2016)
17. Alkhasawneh, R. et al.: Modeling student retention in science and engineering disciplines using neural networks. In: 2011 IEEE Global Engineering Education Conference (EDUCON), pp. 660–663 (2011)
18. Kabakchieva, D.: Student performance prediction by using data mining classification algorithms. Int. J. Comput. Sci. Manag. Res. **1**, 686–690 (2012)
19. Kovacic, Z.: Predicting student success by mining enrolment data. Res. High. Educ. J. (2012)

Establishing the Correlation Relationship Between Size of Code and New Functionalities Using Regression Line Equation

Kanupriya Kashyap, Abdul Wahid and Vikrant Shokeen

Abstract With the rise in the usage of Software and huge upsurge of Software products globally it has become a great deal to develop and deliver Software products which can meet the requirements imposed by the real world. Various Software quality attributes such as reliability, interoperability, scalability, maintenance, etc., has widely been discussed in the research community but we see that very crucial quality attribute, like expandability has not received adequate attention. This can be stated as the main reason behind the failure of 70% of incapacitated industrial projects, incapability of deploying a qualitative product. So this is being taken as the challenge in order to overcome this issue and take approaches which are unprecedented and divergent from all the traditional approximations.

Keywords Expandability · Correlation · Linear regression · Line of code
Functionalities

1 Introduction

Software has become a critical part of our systems in the most important fields of today's society, from transportation and communication to financial and medical applications [1]. If we look at the Software development world, we will find that various baselines are being set by different researchers for the software development but only some of the companies can utilize the capabilities of doing that accordingly,

K. Kashyap (✉) · A. Wahid
National Institute of Technology Kurukshetra, Kurukshetra 136119, Haryana, India
e-mail: kanupriya_kashyap@yahoo.com

A. Wahid
e-mail: awahid.nitp@gmail.com

V. Shokeen
C-DAC, Noida 201309, Uttar Pradesh, India
e-mail: shokeen18@gmail.com

© Springer Nature Singapore Pte Ltd. 2019
A. Abraham et al. (eds.), *Emerging Technologies in Data Mining and Information Security*, Advances in Intelligent Systems and Computing 755,
https://doi.org/10.1007/978-981-13-1951-8_46

511

usually by CMM level companies and thus are successful in producing a quality product at the end of the day [2].

It has been strongly felt that expandability should be evaluated along with other quality measures. But it does not mean that other attributes are not worth measuring. Its just that measuring Expandability will thereby help us people in knowing the reusability of the Software. In this paper we will determine the Software Quality Attribute "Expandability".

Before discussing the approach behind this first of all we should know what basically "Expandability" is. Expandability is to check how we can use a code or software with minimum alteration is and at the end getting a new functionality from our previous code [3, 4]. This will thereby use the concept of reusability. The less the alteration the more expandable the code is or in other terms with less effort we can modify the existing code to incorporate the new functionalities into the existing software system. It has two major benefits; the less effort means low development cost as well time to develop the software is very less. Human productivity to develop the code also increases manifolds.

2 Our Approach

A proposed indicator of quality expandability of software product in terms of degree of alteration is given in Table 1.

Degree of Alteration* is directly proportional to the amount of code modified to incorporate a new functionality. That is if less amount of code is modified to incorporate functionality more reusable it is and thus more Expandable the product is.

Following Table 2 showing the mean of above data of ten different projects of varying size is shown below. Date has been collected of ten different projects in order to find the relation between KLOC and Correlation. KLOC is generally used to measure thousands of LOC.

Small theory about correlation: Correlation is an important indicator of the relationship between two variables. The value of correlation ranges from −1 to +1. If the value of correlation between two variables is 0, it means there is no relationship

Table 1 Quality criteria expandability of software product	S. no	Degree of alteration* (%)	Quality criteria expandability of software product
	1	0–10	Excellent
	2	10–20	Very good
	3	20–30	Good
	4	30–100	Discard

Table 2 Mean

Mean of new functionalities	1	2	3	4	5	6	7	8	9	10
KLOC <=1	7	7	6	6	9	2	4	1	1	9
KLOC <=3	6	5	3	4	4	2	3	6	6	4
KLOC <=7	6	6	5	2	2	2	2	2	2	2
KLOC <=10	7	7	4	3	2	1	2	9	7	2
KLOC <=50	3	2	2	2	1	1	2	1	1	1
Correlation coeff.	-0.90	-0.84	-0.73	-0.58	-0.57	-0.54	-0.51	-0.32	-0.37	-0.57

between the two variables or in other words both the variables are independent to each other [5, 6]. A +ve value of correlation, i.e., a value greater then 0 and up to +1 indicates a +ve correlation meaning both variables move in same direction. A −ve value of correlation, i.e., value greater than −1 and up to 0 indicates a −ve correlation meaning both variables move in opposite direction.

Few definitions and propositions used

- Definition 1: $E(X) =$ the expected value of a random variable X is the mean or average value of X.
- Definition 2: $Var(X) =$ the variance of $X = E([X − E(X)]^2)$.
- Definition 3: $Stdev(X) =$ the standard deviation of $X = \sqrt{Var(X)}$.
- Definition 4: $Cov(X; Y) =$ the covariance of X and

$$Y = E([X − E(X)][Y − E(Y)]).$$

- Definition 5: $Cor(X; Y) =$ the correlation coefficient of X and

$$Y = Cov(X; Y)/Stdev(X)Stdev(Y)$$

Mean value of code modified in percentage and the values are given as in Table 2. The Values given in Table 2 are the values based on a 10 point scale, for example if KLOC <=1 and number of Functionalities to be modified is 3 than 6 out of 10 code need to be modified i.e. 60% of the code will get modified for size of code 1KLOC and to add three new Functionalities. Using these mean values and applying the above mentioned Formula for finding Correlation, Correlation Coefficient is calculated here and written in the last column.

Here, Correlation coefficient is negative that means X and Y holds an inverse relationship, which means with the increase in the size of the project alteration in the code would be less and vice versa.

Now we will be making Linear Regression for each column. This will help in clearly understanding the above stated relationship and concept. Simple linear regression is a method that describes relationship between two variables with the help of an equation of straight line, called line of best fit, that models this relationship. Here we will be taking X-axis as %age change in Functionalities or Functionalities modified where y-axis will denote Line of code (Figs. 1, 2, 3, 4).

Linear Regression Formula

We are using the following formula to derive the equation for the line of best fit: $y = a + bx$, where $b = \frac{\sum_{i=1}^{n} xi\, yi - n\bar{x}\,\bar{y}}{\sum_{i=1}^{n} xi^2 - n\bar{x}^2}$ and $a = \bar{y} - b\bar{x}$

$$\text{Regression line equation } y = 7.1024 − 0.1276x$$

$$\text{Regression line equation : } y = 6.9409 − 0.1510x$$

$$\text{Regression line equation : } y = 4.8085 − 0.0569x$$

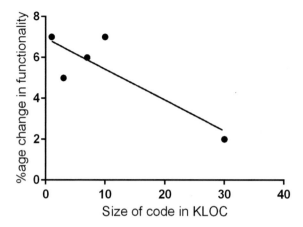

Fig. 1 Linear regression 1st for column

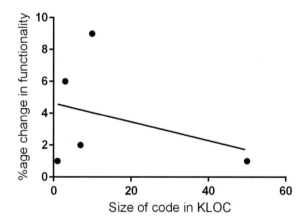

Fig. 2 Linear regression for 2nd column

$$\text{Regression line equation}: y = 4.0829 - 0.0480x$$

In this way we can make Scatter graphs for other also (Fig. 5).

3 Conclusion

However, comprehensive dissection were carried out with the aim of scrutinizing the quality of the software but in this fast changing environment it is still kind of baseless to measure Line of Code by itself without any correlation to other, more important

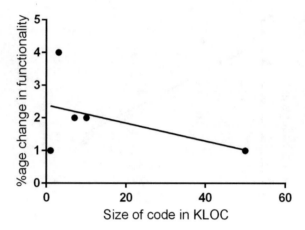

Fig. 3 Linear regression for 3rd column

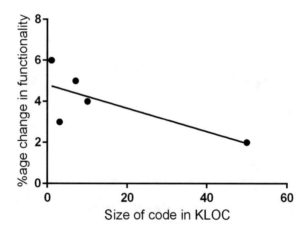

Fig. 4 Linear regression for 4th column

aspects of a project. Rate of change versus Rate of functionality are inversely proportional, i.e., all the Regressions above illustrate the exact relationship between the
two, the line of code and the Functionalities modified. So it can be concluded that
with the increase in the number of line of code, there is less need of changing the code
in order to add a new functionality. So instead of running behind different attributes,
one should take the expandability factor into consideration.

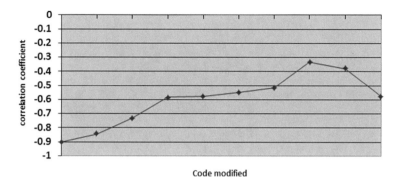

Fig. 5 Relationship between the code modification and the correlation coefficient

References

1. da Silva, L.A.F., e Abreu, F.B.: Exploring and overcoming major challenges in IT infrastructures faced by IT executives. In: Fourth International Conference on Software Engineering Advances, 2009. ICSEA'09, pp. 576–581. IEEE (Sept 2009)
2. Pressman, R.S.: Software Engineering: A Practitioner's Approach. Palgrave Macmillan (2005)
3. Betta, G., Cariglione, D., Pietrosanto, A., Sommella, P.: A statistical approach for improving the performance of a testing methodology for measurement software. IEEE Trans. Instrum. Meas. **57**(6), 1118–1126 (2008)
4. Dolado, J.J.: A validation of the component-based method for software size estimation. IEEE Trans. Softw. Eng. **26**(10), 1006–1021 (2000)
5. Back, T., Hammel, U., Schwefel, H.P.: Evolutionary computation: comments on the history and current state. IEEE Trans. Evol. Comput. **1**(1), 3–17 (1997)
6. Hakuta, M., Tone, F., Ohminami, M.: A software size estimation model and its evaluation. J. Syst. Softw. **37**(3), 253–263 (1997)

Cassandra—A Distributed Database System: An Overview

Abdul Wahid and Kanupriya Kashyap

Abstract In big data environment, apache Cassandra is a distributed database which offers very high availability. It is an open source database system and is designed to manage large transactional data across various server globally. Main feature of Cassandra is to provide high availability and very high fault tolerance, decentralized database system with zero downtime. A traditional relational database (RDBMSs) is used to storing data for various applications from many years, but some changes are required because application must be scale to levels that were unimaginable. But only scaling is not the main concern of changes, companies are also requires such type of applications that always available and running fast where RDBMS database fail. Apache Cassandra is a fully distributed database that has such type of architecture where it handles extreme data velocity with highly availability, scalability and recovers from fault tolerance easily. In Cassandra architecture, there is no master node to handle all the nodes in the ring or network. The data distribution among nodes in this architecture is in equal probation. Cassandra creates such type of environment where an entire datacenter can lose but still perform as if nothing happened. This paper provides a brief idea about Cassandra.

Keywords Distributed database · Cassandra · Decentralized · Fault tolerant

A. Wahid (✉)
Department of Computer Science and Engineering, National Institute
of Technology Patna, Patna 800005, Bihar, India
e-mail: awahid.nitp@gmail.com

K. Kashyap
Department of Computer Engineering, National Institute
of Technology Kurukshetra, Kurukshetra 136119, Haryana, India
e-mail: kanupriya_kashyap@yahoo.com

© Springer Nature Singapore Pte Ltd. 2019
A. Abraham et al. (eds.), *Emerging Technologies in Data Mining and Information
Security*, Advances in Intelligent Systems and Computing 755,
https://doi.org/10.1007/978-981-13-1951-8_47

1 Introduction

Distributed database is a collection of database which are interlinked over geographically large area. In distributed database, there is a centralized server that handles the
whole system. But Cassandra is a distributed storage system, which is decentralized.
Apache Cassandra [1, 2] is a distributed database for handling very huge data (in
petabytes) of structured type across thousands of servers, offers high availability as
well as very high fault tolerance with zero downtime. The main characteristic of
Cassandra is to support an infrastructure which involve thousands of nodes (possibly
spread different data centers of data set clusters). On a very large scale of data management, the occurrence of failure of different type (size) of components happens
continuously (Fig. 1).

The wide adoption of Cassandra in big data environment is due to its high fault
tolerance, user friendly, Cassandra query language [3], proficient and versatile data
model. It does not provide full support to relational data model. It highly relies on
application queries as per need of application run. The direct application of traditional
relational database model in Cassandra can't be possible because it does not support
many sql construct (query).

Some points about Cassandra

- Common data can be stored in multiple Cassandra table.
- Data duplication at various table to provide different query strategies.

The modeling technique used in apache Cassandra is opposite of the traditional
normalization technique to minimize data redundancy.

Fig. 1 A distributed data
base system

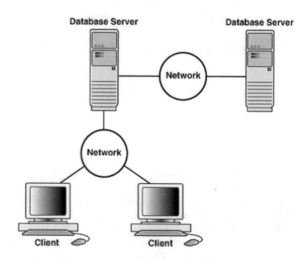

2 Data Model

Cassandra database is distributed over many nodes that work together to perform task. To handle failure, every node contains a replica; if any node goes down or fail due to some reason (like power loss, hardware failure etc.), replica works as alternate for that node. The arrangement of the nodes in Cassandra usually occurs in ring format (Figs. 2 and 3).

A. Cassandra Column

Cassandra column is a fundamental unit of CQL Table which can be view as partition of table vertically, and each column consist a name and value. We can say column is name value pair. (Such as author = "RAJ") (Fig. 4).

B. Cassandra Row

It can be viewed as the aggregation of Cassandra columns with different label values (names) (Fig. 5).

E.g.:-"Second Foundation"->{
Employee = "Raj",
Joining_Date = "......"
tag1 = "manager", tag2 = "Raj"
}

C. Keyspace

Keyspace can be viewed as a top-level namespace which is used to represent the database schema in Cassandra. Keyspace has three important attribute

Fig. 2 The arrangement of nodes in a Cassandra

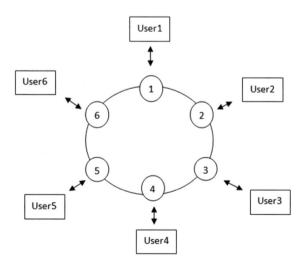

Fig. 3 Cassandra data model

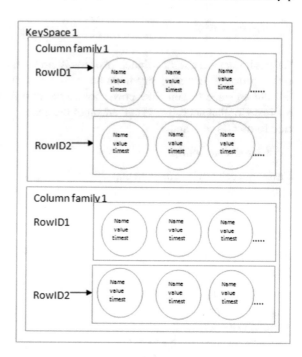

Fig. 4 Cassandra column

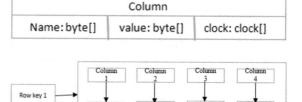

Fig. 5 Cassandra row

- Replication factor—In Cluster of nodes, the number of machine involve which keep the same replicated data is termed as Replication factor.
- Replica placement strategy—The methodology of keeping replicas in the ring of nodes are
- Simple strategy (rack unaware strategy)
- Old network topology strategy (rack-aware) and Network topology strategy (Data center-shared) (Fig. 6).

    e.g., CREATE KEYSPACE Keyspace name
    WITH replication = {'class': 'SimpleStrategy', 'replication factor': 3};

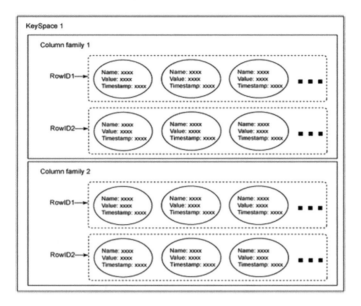

Fig. 6 Cassandra keyspace

D. **Column families**

Keyspace contains one or more column families. A column family can be defined as a container of a collection of rows. Where each row exhibits columns in ordered way. It also represents the structure of data. Each keyspace has at least one or many column families.

3 Cassandra Architecture

Figure 7, shows the architecture of a Cassandra cluster [4]. The first observation is that Cassandra is a distributed system. Cassandra consists of multiple nodes, and it distributes the data. Cassandra uses hashing to store data at various nodes. Cassandra uses a specific hash algorithm that is used to generate the hash-keys for each data item stored in Cassandra (e.g., column name, row ID, etc.). The various nodes in cluster assigned with different hash values (among all possible range of hash values). These values also termed as Keyspace values. Now Cassandra assigns each data item to the various nodes. The storing and managing of the data item is carried out by these nodes.

The Cassandra architecture provides the following properties:

- **Transparent Elasticity**: We can add or remove any node in the cluster/ring easily. Cassandra distributes data among its nodes transparently to the users [5, 6]. Any

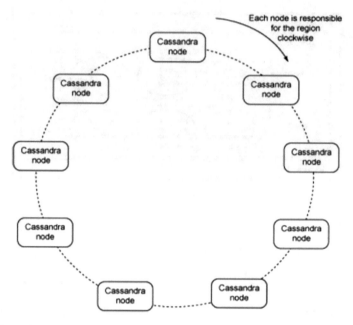

Fig. 7 Cassandra architecture

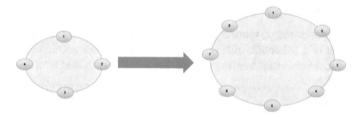

Fig. 8 Transparent elasticity

node can accept any request (read, write, or delete) and route it to the correct node even if the data is not stored in that node (Fig. 8).

- **Transparent Scalability**: When we add new node into the cluster, it assigned with a token such that it can alleviate a heavily loaded node. Its performance increase as increase the number of nodes in the cluster (Fig. 9).
- **High Availability**: Cassandra provides high availability. If a node in the cluster goes down or fail due to some reason, then other node take responsibility of that node and provides service with no delay (Fig. 10).
- **Data Replication**: The use of replication technique in Cassandra provides very high availability. The data spread at various nodes is replicated at N hosts. Where N is the replication factor. Cassandra provides various replication policy:

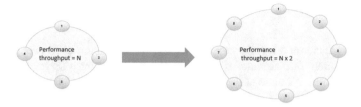

Fig. 9 Transparent scalability

Fig. 10 High availability

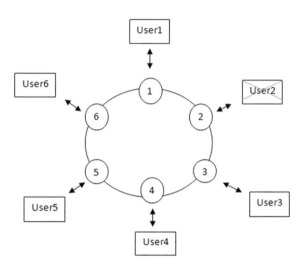

- Rack Unaware/Simple strategy
- Network topology strategy (With in a data center)
- Datacenter Aware.

(i) In Rack unaware replication/simple strategy kept the first replica on a node determined by the practitioner. On the next nodes additional replicas are kept in ring in clockwise regardless of topology.

(ii) Network topology strategy: As the name indicates, this strategy is aware of the network topology (location of nodes in racks, data centers, etc.) and is much intelligent than Simple Strategy [7]. This strategy is a must if your Cassandra cluster spans multiple data centers and lets you specify how many replicas you want per data center. It tries to distribute data among racks to minimize failures. That is, when choosing nodes to store replicas, it will try to find a node on a different rack.

4 Conclusion

Distributed database is a collection of database which is interlinked over geographically large area. In distributed database, there is a centralized server that handles the whole system. But one of the applications of Distributed database is Cassandra, which is decentralized. Cassandra is such type of distributed system that provides high scalability, high performance, and wide applicability with no single point of failure. Due to this, it is used by Facebook, Twitter, etc. Some future work is distributed transaction, Compression Support in Cassandra.

References

1. http://cassandra.apache.org/. Last accessed on 20 Dec 2015
2. http://planetcassandra.org/. Last accessed on 20 Dec 2015
3. Wang, G., Tang, J.: The NoSQL principles and basic application of cassandra model. In: 2012 International Conference on Computer Science and Service System (CSSS), pp. 1332–1335. IEEE, Augt 2012
4. Dean, J., Ghemawat, S.: MapReduce: simplified data processing on large clusters. Commun. ACM 51(1), 107–113 (2008)
5. Bagade, P., Chandra, A., Dhende, A.B.: Designing performance monitoring tool for NoSQL Cassandra distributed database. In: 2012 International Conference on Education and e-Learning Innovations (ICEELI), pp. 1–5. IEEE, July 2012
6. Jiang, W., Zhang, L., Qiang, W., Jin, H., Peng, Y.: MyStore: a high available distributed storage system for unstructured data. In: 2012 IEEE 14th International Conference on High Performance Computing and Communication and 2012 IEEE 9th International Conference on Embedded Software and Systems (HPCC-ICESS), pp. 233–240. IEEE, June 2012
7. Terry, D.B., Theimer, M.M., Petersen, K., Demers, A.J., Spreitzer, M.J., Hauser, C.H.: Managing update conflicts in Bayou, a weakly connected replicated storage system. In: ACM SIGOPS Operating Systems Review, vol. 29, no. 5, pp. 172–182. ACM, Dec 1995

Part V
Computational Science

Subcellular Localization of Gram-Negative Proteins Using Label Powerset Encoding

Hasnaeen Ferdous, Raihan Uddin and Swakkhar Shatabda

Abstract Bacterial proteins play an important role in cell biology due to their importance in drug design and antibiotics research. The localization of bacterial proteins is very important since the function of a protein is closely linked with its location. A single gram-negative bacteria proteins can be located in multiple locations in a protein. Prediction of subcellular locations of gram-negative bacteria proteins is thus more difficult. In this paper, we propose a novel method for subcellular localization of gram-negative bacteria. Our method uses label powerset encoding scheme for the associated multi-label classification problem. Using a set of effective features also used in the literature our encoding significantly improves over the traditional approaches on several base classifiers. Our method was tested using a standard benchmark dataset and showed promising results.

Keywords Supervised learning · Classification problem · Label encoding
Protein subcellular localization

1 Introduction

Locations of proteins in a cell are closely related to their functions within a cell. Subcellular localization of proteins is very important for the knowledge of metabolic pathways and signaling biological processes within the cell. Bacterial proteins can be broadly categorized into two types: gram positive and gram negative. Gram-positive bacteria take purple color during the gram stain test. On the contrary, due to the

H. Ferdous · R. Uddin · S. Shatabda (✉)
Department of Computer Science and Engineering, United International University,
Madani Avenue, Satarkul, Badda, Dhaka 1212, Bangladesh
e-mail: swakkhar@cse.uiu.ac.bd

H. Ferdous
e-mail: mhfnoyon@gmail.com

R. Uddin
e-mail: raihanuddin21@gmail.com

© Springer Nature Singapore Pte Ltd. 2019   529
A. Abraham et al. (eds.), *Emerging Technologies in Data Mining and Information Security*, Advances in Intelligent Systems and Computing 755,
https://doi.org/10.1007/978-981-13-1951-8_48

thinner peptidoglycan layer of gram-negative bacteria, they take up the counterstain and appear red or pink. In vitro localization methods like fluorescent microscopy [1] are very time-consuming and expensive. This is why the computational approaches are becoming very popular to predict the subcellular localization of bacterial proteins. Proteins are located in various locations within a cell. Supervised learning methods used for protein subcellular localization defines the problem as a multi-class classification problem. Many supervised learning methods have been proposed in the literature to handle protein subcellular localization problem. Most successful methods were as follows: Support Vector Machines (SVM) [2], Artificial Neural Networks (ANN) [3], naive Bayesian classifiers [4], decision tree [5], and ensemble of classifiers [6].

An additional difficulty to this problem is added by the fact that a single protein can be located at multiple locations which makes the problem a multi-label classification problem [7]. Traditionally, the researchers employ binary relevance encoding to handle the multi-label classification problem of protein subcellular localization. One of the main drawbacks of this method is they have learned multiple classifiers either for each label or for each nth location [2, 8, 9]. This increases the time complexity of the training phase.

In this paper, we propose a label powerset-based encoding for protein subcellular localization of gram-negative bacteria proteins. Using this encoding scheme, our method learns only a single model compared to multiple models used in traditional methods. To the best of our knowledge, this is the first application of label powerset encoding for prediction of gram-negative protein subcellular localization problem. We tested the effectiveness of our method using different classifiers using a standard set of features. On a standard benchmark dataset, our method was able to significantly improve over the state-of-the-art prediction methods for gram-negative protein subcellular localization.

2   Related Work

A number of successful methods and tools are available in the literature to predict subcellular localization of bacterial proteins or proteins in general. They are applied on a wide range of proteins: eukaryotic, human, plant, gram positive, gram negative, and virus proteins [10]. Dehzangi et al. [11] used Support Vector Machines (SVM) with evolutionary information as features. In a subsequent work [9], they used rotation forest and physicochemical features to improve over their work. Sharma et al. [2] used normalization-based profile features to solve the problem using SVM classifiers. Saini et al. [8] extracted probabilistic profile-based features with linear interpolation smoothing model to predict gram-positive and gram-negative bacteria proteins.

Most of these methods use binary relevance method for addressing the multi-label classification problem. In binary relevance method, they used to learn a multiple number of classifiers or classifier chains [12] for each different location [13] and then predict the multiple locations simultaneously. One of the major drawbacks of this method is the necessity of learning multiple learners. In order to solve this problem label space transformation or label powerset encoding was proposed in the literature [13]. In this paper, we apply this transformation for the first time to predict subcellular localization of gram-negative bacteria proteins.

3   Methods and Materials

In this section, we describe the details of the method and materials used in this paper. A system diagram of our proposed model is shown in Fig. 1. Our system starts by fetching the protein sequences in the dataset and feeding them into PSI-BLAST [14] software to fetch Position-Specific Scoring Matrix (PSSM) files using the nr database. PSSM files are then fed to the SPIDER [15] software to generate secondary structure-related information prediction to generate SPD files. SPD and PSSM files are then used to generate features for classification using a feature extraction procedure. Extracted features and locations (multiple or single) of the proteins are then fed into a label space transformation module that converts the label space into a new one. The training dataset with newly transformed labels are then fed into a classifier to learn a model, which stored later can be used to find locations for any query sequence.

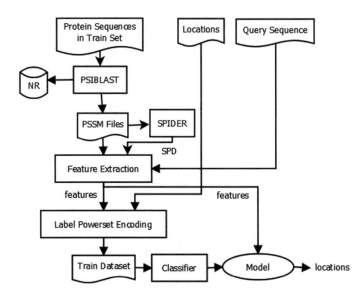

Fig. 1 System diagram

3.1 Benchmark Dataset

In this paper, we have used a standard benchmark dataset proposed in [16] and previously used in [2, 8, 9, 11]. There are in total 1392 different proteins in this gram-negative bacteria protein dataset. However, among these 1392 proteins only 64 have multiple locations and 1328 are only single location proteins. In total, there are 1456 labeled protein sequences considering the single locations. This benchmark dataset is available for download and use from: http://www.csbio.sjtu.edu.cn/bioinf/Gneg-multi/.

3.2 Feature Extraction Method

We use evolutionary information fetched by PSSM files to extract features. PSSM files contain substitution probabilities of each amino acid residue at each position of the given protein sequence. It is a matrix, P, of dimension $L \times 20$, where L is the length of the protein. First, this matrix is normalized using a method similar to that described in [2] and found to be effective for subcellular localization previously. Let us call this matrix N with same dimension as P. Now, the features are extracted from this matrix. This section gives a brief description of the features used in this paper.

PSSM Composition (PSSM-C): Composition of the PSSM matrix is the simple normalized sum of the columns. It can be formally defined as below:

$$PSSM - C(j) = \frac{1}{L} \sum_{i=1}^{L} N_{i,j}$$

PSSM Segmented Distribution (PSSM-SD): Segmented distribution of the normalized PSSM matrix is calculated columnwise. For column, j set of features are extracted as I_j. This is formally defined as below:

$$I_j^+ = \{m : \sum_{i=1}^{m} N_{i,j} \geq T_j \times F_p\}$$

Note that, T_j is the total sum of the column j values and F_p are the percentage of the total sum distributed in the column as partial sums. The values of F_p are in the range of 10, 20, 30, 40, and 50%. This is the partial sum from the top. Another set is defined as from bottom:

$$I_j^- = \{m : \sum_{i=L}^{m} N_{i,j} \geq T_j \times F_p\}$$

The total distribution set I_j is the union of these two sets $I_j^+ \cup I_j^-$. These features were first used in [11].

PSSM Segmented Auto-Covariance (PSSM-SAC): PSSM segmented auto-covariance is calculated based on the segmented distributions found from columnwise values. It can be formally defines as below:

$$PSSM - SAC(j, k, l) = \frac{1}{I_j^l} \sum_{i=1}^{I_j^l - l} (N_{i,j} - N_{ave,j}) \times (N_{i+k,j} - N_{ave,j})$$

Here, $N_{ave,j}$ is the average of substitution probabilities for column j of the normalized PSSM matrix. Value of k is in the range of $1, \ldots, K_p$. We used $K_p = 10$ in this paper. This is called distance factor. Here, the value of l is iterated through the distribution set I_j. These features were first used in [11].

PSSM Auto-covariance (PSSM-AC): This feature is calculated as the global PSSM matrix auto-covariance coefficient. This is depended on a distance factor called K_p and was introduced in [11]. Values of K_p is in the range $1, \ldots, 10$. This is formally defined as

$$PSSM - AC(j, k) = \frac{1}{L - k} \sum_{i=1}^{L-k} (N_{i,j} - N_{ave,j}) \times (N_{i+k,j} - N_{ave,j})$$

Apart from the features generated from the PSSM files we also use other features that are generated from SPD files. SPD files contain secondary structure-related information. We use three important information produced as matrix by the SPIDER software as output in SPD file: Accessible Surface Area (ASA), Torsional Angles (TA), and Probabilities of Motifs (PM). These three information are encoded as matrix of different dimensions. ASA is a $L \times 1$ dimension matrix, TA is a $L \times 8$-dimensional matrix where each of four different angles is first converted into radian angles, and then sine and cosine are taken for each of them to build the matrix. Lastly, the probability of secondary motifs is a $L \times 3$ matrix denoting probability of alpha (α) helix, beta (β) sheet, or beta coils at each random position. Now, we describe the other features used in this paper.

1-lead bigram of ASA: 1-lead bigram features for ASA are defined as below:

$$ASA - 1LB = \frac{1}{L} \sum_{i=0}^{L-2} ASA_i \times ASA_{i+2}$$

Torsional Angles Composition: Torsional angles composition is similar to PSSM composition and defined as below:

$$TA - C(j) = \frac{1}{L} \sum_{i=0}^{L} TA_{i,j}$$

These features are calculated for each column of the respective matrix.

Auto-Covariances of Probabilities: Auto-covariance of probability matrix is also calculated depending on a distance factor K_p. It is formally defined as below:

$$PM - AC(j, k) = \frac{1}{L} \sum_{i=0}^{L-k} PM_{i,j} PM_{i+k,j}$$

3.3   Label Transformation Method

We use the label transformation method using label powerset encoding. Apart from label powerset, another successful method which is traditionally used in protein subcellular localization is binary relevance method. Let us suppose that for the multi-label classification problem, the set of different labels is $\mathscr{L} = \{1, 2, 3, \ldots, K\}$. We define a label, $\mathscr{Y} \subseteq \mathscr{L}$. Now, the training set for any multi-label classification becomes, a set $\mathbb{S} = \{(\mathbf{x}_i, \mathscr{Y}_i) : \mathbf{x}_i \in \mathbb{R}^n, 1 \leq i \leq m\}$. Here, m is the number of total training instances or individual unique proteins and n is the dimensionality of the feature space. In binary relevance method, multiple learners are trained using the training dataset $\mathbb{S}$, each for labels, $l \in \mathscr{L}$. Predictions from each of these binary classifiers are then merged together in a vector $\{\hat{y}^1(\mathbf{x}, \hat{y}^2(\mathbf{x}, \ldots, \hat{y}^K(\mathbf{x})\}$. Elements of this vector are either 0 or 1 and the predicted labels are decided from this vector. However, in case of label powerset encoding, preprocessing is required on the label space. The pseudocode of the label powerset encoding is given in Algorithm 1.

Algorithm 1: Label Powerset

1 **preprocess:**
2 let B a bijective function
3 $\mathbb{P} = \{1, 2, 3, \ldots, 2^K\}$ learn B by mapping each label $\mathscr{Y}_i \in \mathbb{S}$ to a hyper-label $y_i \in \mathbb{P}$
4 $\mathbb{T} = \text{transform}(\mathbb{S}, B)$
5 **train:**
6 learn a single multi-class classifier $h(\mathbf{x})$ from $\mathbb{T}$
7 **predict:**
8 for each $\mathbf{x}$, return $B^{-1}(h(\mathbf{x}))$

Label powerset encoding transforms each multi-label of single labels to a hyper-label. This transformation maps each previous label to the power super set of the possible labels. A single base classifier is learned for the transformed training dataset and while prediction after the labels are decided by the classifier they are transformed back to the original labels. Note that compared to the binary relevance, here only a single learner has to be trained and hence reduces the training time of the dataset.

3.4 Classification Algorithms

We tested the effect of our proposed method on three base classifiers: Decision Tree (DT), Random Forest (RF), and Support Vector Machines (SVM). Among these three classifiers, SVM performed best. In this section, we provide a brief discussion of these classifiers.

Decision Tree: Decision tree classification [17] is based on selecting attributes as decision nodes in each level of the tree where instances are divided based on the attribute value or decision. Generally, nodes are mapped to decisions based on the values of the attribute that can discriminate the instances best. As discriminatory information, Gini impurity, entropy, and information gain are widely used. For a better generalization on the training dataset, often trees are pruned.

Random Forest: Random forest classifier [18] is an ensemble classifier with decision trees as base classifiers. It is an application of bootstrap aggregating or bagging technique that first samples the original dataset into a number of datasets by a sampling with replacement technique. This bootstrap method of sampling can radically reduce noise or outliers in the data. After creating K samples from the original dataset, K decision trees are learned on each dataset, and the decision of these bag of classifiers is combined by taking voting on the predictions made by them. The decision trees learned by the random forest algorithm are random in nature. In each iteration, features are selected randomly and based on the selected features only the trees are learned. The underlying decision tree algorithm uses the discriminatory information to build the tree structure of the data for classification.

Support Vector Machines: Support vector machines [19] are classifiers that try to separate the different classes in the dataset using a hyperplane learned from the training data that maximizes the separation between the borderline instances. It is also known as the maximum margin classifier. SVMs generally try to optimize a multiplier function that goes as follows:

$$L = \underset{\alpha}{argmax} \sum_j \alpha_j - \frac{1}{2} \sum_{j,k} \alpha_j \alpha_k y_j y_k \phi(\mathbf{x}_j.\mathbf{x}_k)$$

The prediction of an SVM classifier is defined as below:

$$h(\mathbf{x}) = sign(\sum_j \alpha_j y_j (\mathbf{x}.\mathbf{x}_j) - b)$$

Here, the transformation of the data points by the function ϕ could be linear, polynomial or any other kernel functions. Multi-class SVMs are extension of binary SVMs with an appropriate function (e.g., *softmax*) to approximate a multinoulli distribution. The parameters used for SVM in our experiments were as follows: gamma $(\gamma) = 0.05$ and $C = 3000$, and radial basis function (RBF) kernel was used.

3.5 Performance Evaluation

A number of sampling methods have been used in the literature for comparison of different prediction methods [20]. Among them are percentage split, k-fold cross-validation, and jackknife tests. In most of the papers of protein subcellular localization, tenfold cross-validation of the learners has been widely applied [2, 8, 9, 11] and also suggested in [21]. In this paper, we also adopt tenfold cross-validation to validate our method and results with those of the state-of-the-art methods.

Accuracy for simple binary or multi-class classification problems is calculated by taking the percentage of true positives for each class to the total number of instances. However, such metrics could be misleading for classification of multi-label classification problems. Here, one should give importance to the accurate prediction of all multiple locations simultaneously. We define *absolute accuracy* for this purpose. Absolute accuracy was previously used in [2]. Absolute accuracy can be formally defined as below:

$$\text{absolute accuracy} = \frac{1}{N_{dif}} \sum_{i=1}^{N_{dif}} C_i \tag{1}$$

Here, N_{dif} is the total number of protein sequences in the dataset and $C_i = 1$ if all the locations of a protein are predicted correctly and otherwise, 0. Note that typical evaluation metrics like accuracy, sensitivity, specificity, and others are only suitable for binary or multi-class classification problems and are not recommended for multi-label classification problem.

4 Results and Discussion

All the experiments done in this paper are done using Weka [22]. All the algorithms except our preprocessing part for label encoding were done using Weka implementation. All the experiments were run five times. In this section, we describe the details of the experimental results. After the preprocessing of data, eight locations and their combinations were transformed into 20 different labels.

4.1 Effect of Using Label Powerset Encoding

The first set of experiments were done on the gram-negative bacteria protein to show the effectiveness of the label powerset encoding with that of binary relevance. We tested our method with three base classifiers: SVM, decision tree J48 algorithm,

Table 1 Performance comparison of binary relevance and label powerset encoding on gram-negative bacteria

		J48	SVM	RF
Binary relevance	Avg	42.14	71.74	55.16
	Max	45.71	78.82	62.16
	St. Dev	3.49	3.49	3.49
Label powerset	Avg	**60.84**	**82.09**	**69.97**
	Max	**64.87**	**83.65**	**72.21**
	St. Dev	3.18	1.14	2.23

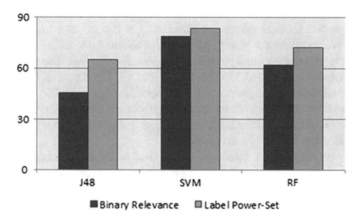

Fig. 2 Bar plot showing the absolute accuracy achieved by different classifiers using label powerset encoding compared to binary relevance

and random forest algorithm. For each of these classifiers, we used the same set of hyperparameters and run each of them five times for binary relevance and label powerset encoding. Average and maximum absolute accuracy and standard deviation for each of the classifiers using two different schemes are given in Table 1.

Best values in Table 1 are shown in bold-faced fonts. From the reported results, it is clear that for all three classifiers label powerset method is achieving much higher absolute accuracy in comparison to the binary relevance method. The trend is similar for both in terms of maximum accuracy and average accuracy. We also plot the bars in Fig. 2 to show a clear comparison. We also note that the standard deviation in the results is somehow lower in case of label powerset encoding. Also note that among all the classifiers, Support Vector Machines (SVM) with rbf kernel performed the best and achieved superior performance. Thus, we select SVM as a classifier for our proposed method.

Table 2 Comparison of the absolute accuracy of our method with other state-of-the-art methods

Method name	Reference	Absolute accuracy (%)
Pacharawongsakda et al.	[16]	73.2
Dehzangi et al.	[11]	76.6
Dehzangi et al.	[9]	79.6
Our method	This paper	**83.65**

4.2 Comparison with Other Methods

We also compare the performance of our method to that of other predictors in the literature. We compare our results with three other methods in the literature that predict gram-negative bacterial proteins subcellular localization. These methods are from Pacharawongsakda et al. [16], Dehzangi et al. [9, 11]. These all three methods use binary relevance on the protein labels and tenfold cross-validation. The reported accuracy for these methods is taken from their papers and shown in Table 2. It is clear from the results shown in Table 2 that our method is able to produce superior results compared to these state-of-the-art predictors.

One of the potential drawbacks of our method is the hyper-labeling. It solely depends on the success of this hyper-labeling. Now, if any combinations of protein locations are missing in the training data but are present in the test data might hamper the performance of the label powerset encoding. We believe our method is more efficient and effective in datasets like gram negative and others where the number of different locations is higher in number. It is often noticed that the combinations of locations follow a specific pattern. However, in comparison of training time label powerset seems to be a definite winner.

5 Conclusion

Subcellular localization of gram-negative bacteria is a very important problem to solve in cell biology. One of the major challenges to solve this problem is the nature of the proteins that make them located in multiple locations simultaneously. Thus, the problems belong to the category of multi-label classification. Traditional approaches in prediction of gram-negative protein locations binary relevance are used. In this paper, we proposed to use label space transformation using label powerset encoding. Our method was able to produce significantly improved results on a standard benchmark dataset and also tested of a number of classifiers shown promising results. We further wish to test our method on other multi-label datasets like plant or human datasets. We also wish to develop a web-based tool for the biologists so that they can use it for practical purposes. We also believe that suitable feature selection technique and other features could result in enhancement of the proposed method.

References

1. Gunsolus, I.L., Hu, D., Mihai, C., Lohse, S.E., Lee, C.s., Torelli, M.D., Hamers, R.J., Murhpy, C.J., Orr, G., Haynes, C.L.: Facile method to stain the bacterial cell surface for super-resolution fluorescence microscopy. Analyst **139**(12), 3174–3178 (2014)
2. Sharma, R., Dehzangi, A., Lyons, J., Paliwal, K., Tsunoda, T., Sharma, A.: Predict gram-positive and gram-negative subcellular localization via incorporating evolutionary information and physicochemical features into chou's general pseaac. IEEE Trans. Nanobiosci. **14**(8), 915–926 (2015)
3. Emanuelsson, O., Nielsen, H., Brunak, S., Von Heijne, G.: Predicting subcellular localization of proteins based on their n-terminal amino acid sequence. J. Mol. Biol. **300**(4), 1005–1016 (2000)
4. Lu, Z., Szafron, D., Greiner, R., Lu, P., Wishart, D.S., Poulin, B., Anvik, J., Macdonell, C., Eisner, R.: Predicting subcellular localization of proteins using machine-learned classifiers. Bioinformatics **20**(4), 547–556 (2004)
5. Shen, Y.Q., Burger, G.: 'Unite and conquer': enhanced prediction of protein subcellular localization by integrating multiple specialized tools. BMC Bioinform. **8**(1), 420 (2007)
6. Chou, K.C., Shen, H.B.: Cell-ploc: a package of web servers for predicting subcellular localization of proteins in various organisms. Nat. Protoc. **3**(2), 153 (2008)
7. Tsoumakas, G., Katakis, I.: Multi-label classification: an overview. Int. J. Data Warehous. Min. **3**(3) (2006)
8. Saini, H., Raicar, G., Dehzangi, A., Lal, S., Sharma, A.: Subcellular localization for gram positive and gram negative bacterial proteins using linear interpolation smoothing model. J. Theor. Biol. **386**, 25–33 (2015)
9. Dehzangi, A., Sohrabi, S., Heffernan, R., Sharma, A., Lyons, J., Paliwal, K., Sattar, A.: Gram-positive and gram-negative subcellular localization using rotation forest and physicochemical-based features. BMC Bioinform. **16**(4), S1 (2015)
10. Chou, K.C.: Impacts of bioinformatics to medicinal chemistry. Med. Chem. **11**(3), 218–234 (2015)
11. Dehzangi, A., Heffernan, R., Sharma, A., Lyons, J., Paliwal, K., Sattar, A.: Gram-positive and gram-negative protein subcellular localization by incorporating evolutionary-based descriptors into chou's general pseaac. J. Theor. Biol. **364**, 284–294 (2015)
12. Read, J., Pfahringer, B., Holmes, G., Frank, E.: Classifier chains for multi-label classification. Mach. Learn. Knowl. Discov. Databases, 254–269 (2009)
13. Tai, F., Lin, H.T.: Multilabel classification with principal label space transformation. Neural Comput. **24**(9), 2508–2542 (2012)
14. Altschul, S.F., Madden, T.L., Schäffer, A.A., Zhang, J., Zhang, Z., Miller, W., Lipman, D.J.: Gapped blast and psi-blast: a new generation of protein database search programs. Nucleic Acids Res. **25**(17), 3389–3402 (1997)
15. Heffernan, R., Yang, Y., Paliwal, K., Zhou, Y.: Capturing non-local interactions by long short term memory bidirectional recurrent neural networks for improving prediction of protein secondary structure, backbone angles, contact numbers, and solvent accessibility. Bioinformatics **33**(18), 2842–2849 (2017)
16. Pacharawongsakda, E., Theeramunkong, T.: Predict subcellular locations of singleplex and multiplex proteins by semi-supervised learning and dimension-reducing general mode of chou's pseaac. IEEE Trans. Nanobiosci. **12**(4), 311–320 (2013)

17. Dobra, A.: Decision tree classification. In: Encyclopedia of Database Systems. Springer, pp. 765–769 (2009)
18. Breiman, L.: Random forests. Mach. Learn. **45**(1), 5–32 (2001)
19. Cortes, C., Vapnik, V.: Support-vector networks. Mach. Learn. **20**(3), 273–297 (1995)
20. Efron, B., Gong, G.: A leisurely look at the bootstrap, the jackknife, and cross-validation. Am. Statist. **37**(1), 36–48 (1983)
21. Chou, K.C.: Some remarks on protein attribute prediction and pseudo amino acid composition. J. Theor. Biol. **273**(1), 236–247 (2011)
22. Garner, S.R., et al.: Weka: the waikato environment for knowledge analysis. In: Proceedings of the New Zealand Computer Science Research Students Conference, pp. 57–64 (1995)

Impact of Thermophoretic MHD Visco-Elastic Fluid Flow Past a Wedge with Heat Source and Chemical Reaction

Bibhash Deka and Rita Choudhury

Abstract An analysis is carried out to investigate the boundary layer flow of an electrically conducting visco-elastic fluid past a wedge in presence of thermophoresis, heat source, chemical reaction and magnetic field with heat and mass transfer. The problem has been solved by the application of steepest descent method used by Meksyn. Analytical expressions for the velocity, temperature, concentration, shearing stress, Nusselt number and Sherwood number have been obtained and illustrated graphically to observe the effects of visco-elastic parameter with the combination of various values of pertinent flow parameters involved in the solution. The relevancy of this model has been observed in various chemical and industrial processes.

Keywords Visco-elastic · Boundary layer · MHD · Wedge · Thermophoresis
Chemical reaction

1 Introduction

The study of steady and unsteady motions in visco-elastic second-order fluids are particularly useful in chemical and process industries because they encounter both viscous and elastic characteristics exhibited by most polymers and biological liquids. Due to complicated rheological equations involved in describing the dynamical behaviour of these fluids the investigation in this field is limited but interesting.

The term boundary layer in general refers to a thin layer of fluid next to the solid wall in high Reynolds number. Many processes of practical importance viz. viscous diffusion, heat, and mass transfer are controlled by the boundary layer. In the study of non-Newtonian fluid flow mechanics, the boundary layer phenomenon is relevant to a number of engineering fields viz. manufacturing processes such as drawing of copper

B. Deka · R. Choudhury (✉)
Department of Mathematics, Gauhati University, Guwahati 781014, Assam, India
e-mail: rchoudhury66@yahoo.in

B. Deka
e-mail: bibhashdeka66@gmail.com

© Springer Nature Singapore Pte Ltd. 2019   541
A. Abraham et al. (eds.), *Emerging Technologies in Data Mining and Information Security*, Advances in Intelligent Systems and Computing 755,
https://doi.org/10.1007/978-981-13-1951-8_49

wires, polymer extrusion, hot rolling, metal extrusion, glass fibre, etc. Watanabe [1] has presented the thermal boundary layer forced flow over a wedge in the presence of uniform injection or suction. Falkner–Skan equation for flow past a moving wedge with suction or injection has been analysed by Ishak et al. [2]. Yih [3] has studied the force convection flow about a wedge with uniform blowing/suction. Hydromagnetic flow of viscous incompressible fluid past a wedge with permeable surface has been presented by Mahmood et al. [4]. Srivastava and Usha [5] have investigated the laminar boundary layer on a moving continuous flat surface in presence of suction and magnetic field by using boundary value problem. The boundary layer flow over a plate affected by magnetic field has been analysed by Murthy and Sapre [6]. Meksyn [7] has used steepest descent method to find the value of arbitrary constants arising from the boundary condition at infinity in the case where the similarity solutions exist. Unsteady boundary layer flow past a wedge in the presence of magnetic field has been presented by Satter [8]. Cheng and Lin [9] also have significant subscription in the boundary layer flow over a wedge.

Thermophoresis is the phenomenon in which micron sized particles suspended in a non-isothermal fluid acquired a velocity relative to the fluid as in the direction of diminishing temperature. Such type of phenomenon has wide range of applications in aerosol reactors, drug discovery process, heat exchange fouling etc. In optical fibre fusion, thermophoresis is also recognized as the mass transfer procedure as utilized in the Modified Chemical Vapour Deposition. Thermophoresis of aerosol particles in laminar boundary layer on flat plate has been analysed by Goren [10]. Chamkha and Pop [11] have presented the effect of thermophoresis particle deposition in free convection boundary layer from a vertical flat plate embedded in a porous medium. Also, the authors viz. Selim et al. [12], Postelnicu [13], Bakier and Gorla [14], Zueco et al. [15], Noor et al. [16], Kundu et al. [17], Das [18], etc., have remarkable contribution in this area.

In this study, we have analysed the boundary layer flow past a wedge for visco-elastic fluid (non-Newtonian) in the presence of heat and mass transfer with other physical properties as mentioned above. It is well known that non-Newtonian fluids behave differently under the action of stress imposed on them. As a result, different types of non-Newtonian fluids have emerged with different rheological properties to describe the dynamical behaviour of these fluids. The second-grade fluid model based on the postulate of gradually fading memory deduced by Coleman and Noll [19] is defined as

$$\sigma = -pI + \mu_{\{1\}}A_{\{1\}} + \mu_{\{2\}}A_{\{2\}} + \mu_{\{3\}}A_{\{1\}}^2 \tag{1.1}$$

where σ is the stress tensor, p is the indeterminate pressure and $\mu_{\{1\}}$, $\mu_{\{2\}}$ and $\mu_{\{3\}}$ are material constants known as viscosity, elasticity and cross viscosity respectively. $A_{\{i\}}(i = 1, 2)$ are the kinematic Rivlin-Ericksen tensors. It is noticed from thermodynamic consideration that the constants $\mu_{\{1\}}$ and $\mu_{\{3\}}$ are positive and $\mu_{\{2\}}$ is negative (Coleman and Markovitz [20]). It has already been reported that the solution of polyisobutylene in cetane at 30 °C simulate a second-grade fluid and the material constants for the solutions of various concentrations have been determined by Markovitz.

2 Mathematical Formulation

We consider a steady two-dimensional incompressible flow of a visco-elastic fluid past a wedge under the influence of a magnetic field of uniform strength $B[x]$ in the presence of thermophoresis, heat source and a first-order chemical reaction with heat and mass transfer. The magnetic field acts in the transverse direction of the wedge. The magnetic Reynolds number is assumed to be small enough so that the induced magnetic field can be neglected in comparison with the applied magnetic field. The fluid flow is assumed to be in the x-direction along the surface of the wedge and y-axis is normal to it. With the consideration of above presumption, the governing equations of motion are:

$$\frac{\partial u}{\partial x} + \frac{\partial v}{\partial y} = 0 \tag{2.1}$$

$$u\frac{\partial u}{\partial x} + v\frac{\partial u}{\partial y} = v_{[1]}\frac{\partial^2 u}{\partial y^2} + v_{[2]}\left\{\frac{\partial u}{\partial x}\frac{\partial^2 u}{\partial y^2} + u\frac{\partial^3 u}{\partial x \partial y^2} + v\frac{\partial^3 u}{\partial y^3} + \frac{\partial u}{\partial y}\frac{\partial^2 v}{\partial y^2}\right\} + \frac{\sigma^* B^2}{\rho}(U - u) + U\frac{dU}{dx} \tag{2.2}$$

$$u\frac{\partial T}{\partial x} + v\frac{\partial T}{\partial y} = \frac{k_{[t]}}{\rho C_{[p]}}\frac{\partial^2 T}{\partial y^2} + \frac{v_{[1]}}{C_{[p]}}\left(\frac{\partial u}{\partial y}\right)^2 + \frac{v_{[2]}}{C_{[p]}}\left(u\frac{\partial u}{\partial y}\frac{\partial^2 u}{\partial x \partial y} + v\frac{\partial u}{\partial y}\frac{\partial^2 u}{\partial y^2}\right) + \frac{Q}{\rho C_{[p]}}(T - T_{[\infty]}) \tag{2.3}$$

$$u\frac{\partial C}{\partial x} + v\frac{\partial C}{\partial y} = D\frac{\partial^2 C}{\partial y^2} - \frac{\partial}{\partial y}\left\{V_{[T]}(C - C_{[\infty]})\right\} - k_{[c]}(C - C_{[\infty]}) \tag{2.4}$$

The boundary conditions on u, v, T and C are given as follows:

$$u = 0,\, v = 0,\, T = T_{[w]},\, C = C_{[w]} \text{ at } y = 0 \tag{2.5}$$

$$u \to U[x],\, T \to T_{[\infty]},\, C \to C_{[\infty]} \text{ in } y \to \infty, \tag{2.6}$$

where u and v are the velocity components along x and y axes respectively, $U[x]$ is the velocity of the potential flow, ρ is the fluid density, σ^* is the electric conductivity, $C_{[p]}$ is the specific heat, $k_{[t]}$ is the thermal conductivity, T is dimensional temperature, C is dimensional species concentration, D is the coefficient of chemical molecular diffusivity, $k_{[c]}$ is the rate of chemical reaction, Q is the heat generation coefficient, $v_{[1]}$ is the kinematic viscosity, $v_{[2]}$ is the visco-elasticity, $V_{[T]}\left(= -\frac{k'v_{[1]}}{T_{[r]}}\frac{\partial T}{\partial y}\right)$ is the thermophoretic velocity, where k' is the thermophoretic coefficient and $T_{[r]}$ is the reference temperature, $T_{[w]}$ and $T_{[\infty]}$ are wedge and ambient temperature respectively, $C_{[w]}$ and $C_{[\infty]}$ are wedge and ambient concentration respectively.

We consider a stream function $\psi[x, y]$ such that $u = \frac{\partial \psi}{\partial y}$, $v = -\frac{\partial \psi}{\partial x}$ which satisfies the Eq. (2.1).

Following Schlichting [21], we set $U[x] = ax^m$, where a and m are constants, $B[x] = B_{[0]}x^{\frac{m-1}{2}}$, $\eta[x, y] = \sqrt{\frac{(1+m)U[x]}{2v_{[1]}x}}y$, $\psi[x, y] = \sqrt{\frac{2v_{[1]}xU[x]}{(1+m)}}F[\eta]$ and put $[\eta] = \frac{T-T_{[\infty]}}{T_{[w]}-T_{[\infty]}}$, $\phi[\eta] = \frac{C-C_{[\infty]}}{C_{[w]}-C_{[\infty]}}$, $Sc = \frac{v_{[1]}}{D}$, $Pr = \frac{\rho c_{[p]}v_{[1]}}{k_{[t]}}$, $E = \frac{U^2}{C_{[p]}(T_{[w]}-T_{[\infty]})}$, $S = \frac{2Qx}{\rho c_{[p]}(m+1)U}$, $Kc = \frac{2xk_{[c]}}{U(m+1)}$.

Then the transformed forms of Eqs. (2.2)–(2.4) are given by

$$F''' + FF'' = M(F' - 1) + \beta\left(F'^2 - 1\right) + \alpha_1\left\{(1 - 3m)F'F''' + \frac{(m+1)}{2}FF'''' + \frac{(3m-1)}{2}F''^2\right\} \tag{2.7}$$

$$\theta'' + PrF\theta' = Pr\left[\alpha_1\left\{\frac{(1-3m)}{2}F'F''^2 + \frac{(m+1)}{2}FF''F'''\right\} - F''^2\right]E - \theta SPr \tag{2.8}$$

$$\phi'' + ScF\phi' = Sc(\phi'\theta' + \phi\theta'')\tau + \phi Kc \tag{2.9}$$

where $M = \dfrac{2\sigma^* B_{[0]}^2}{a\rho(1+m)}$, $\beta = \dfrac{2m}{m+1}$, $\alpha_1 = \dfrac{\nu_{[2]}U}{x\nu_{[1]}}$ and $\tau = -\dfrac{k'\left(T_{[w]} - T_{[\infty]}\right)}{T_{[r]}}$

The corresponding boundary conditions are

$$F[0] = 0, F'[0] = 0, \theta[0] = 1, \phi[0] = 1 \text{ at } \eta = 0 \tag{2.10}$$

$$F'[\eta] \to 1, \theta[\eta] \to 0, \phi[\eta] \to 0 \text{ in } \eta \to \infty \tag{2.11}$$

where $F[\eta]$ is the non-dimensional stream function, η is the transformed co-ordinate, M is the magnetic parameter, α_1 is the visco-elastic parameter, $\theta[\eta]$ is the dimensionless temperature, $\phi[\eta]$ is the dimensionless species concentration, Sc is the Schmidt number, Pr is the prandtl number, E is the Eckert number, S is the heat source parameter, Kc is the chemical reaction parameter, β is the wedge angle parameter that corresponds to the total angle $\Psi = \beta\pi$ and τ is the thermophoretic parameter.

3 Method of Solution

The Eqs. (2.7)–(2.9) subject to the boundary conditions (2.10) and (2.11) can be solved by any numerical method but we have solved these equations by the application of steepest descent method used by Meksyn. For this purpose, we express the functions $F[\eta]$, $\theta[\eta]$ and $\phi[\eta]$ in power series of η as,

$$F[\eta] = \sum_{i=2}^{\infty} \frac{B_i\eta^i}{i!} \tag{3.1}$$

$$\theta[\eta] = 1 + \sum_{j=1}^{\infty} \frac{C_j\eta^j}{j!} \tag{3.2}$$

$$\phi[\eta] = 1 + \sum_{k=1}^{\infty} \frac{D_k\eta^k}{k!} \tag{3.3}$$

The forms (3.1)–(3.3) of $F[\eta]$, $\theta[\eta]$ and $\phi[\eta]$ respectively satisfy the boundary conditions (2.10). Substituting the expressions of $F[\eta]$, $\theta[\eta]$ and $\phi[\eta]$ from (3.1)–(3.3) into Eqs. (2.7)–(2.9) and equating the coefficients of different powers of η to zero, we obtain the constants B_i, C_j and D_k ($i = 2, 3, 4 \ldots$; $j = 1, 2, 3 \ldots$; $k = 1, 2, 3 \ldots$). We will first determine the values of B_2, C_1 and D_1. We write the Eqs. (2.7)–(2.9) in the following forms,

$$F''' + FF'' = X[\eta] \tag{3.4}$$

$$\theta'' + PrF\theta' = Y[\eta] \tag{3.5}$$

$$\phi'' + ScF\phi' = Z[\eta], \tag{3.6}$$

where $X[\eta]$, $Y[\eta]$ and $Z[\eta]$ are the right hand sides of Eqs. (2.7)–(2.9) respectively. Now, integrating twice the Eqs. (3.4)–(3.6), we get,

$$F'[\eta] = \int_0^\eta e^{-L[\eta]} G[\eta] d\eta \tag{3.7}$$

$$\theta[\eta] = 1 + \int_0^\eta e^{-R[\eta]} H[\eta] d\eta \tag{3.8}$$

$$\phi[\eta] = 1 + \int_0^\eta e^{-N[\eta]} I[\eta] d\eta, \tag{3.9}$$

where,

$$L[\eta] = \int_0^\eta F[\eta] d\eta \tag{3.10}$$

$$R[\eta] = Pr \int_0^\eta F[\eta] d\eta \tag{3.11}$$

$$N[\eta] = Sc \int_0^\eta F[\eta] d\eta \tag{3.12}$$

$$G[\eta] = B_2 + \int_0^\eta e^{L[\eta]} X[\eta] d\eta \tag{3.13}$$

$$H[\eta] = C_1 + \int_0^\eta e^{R[\eta]} Y[\eta] d\eta \tag{3.14}$$

$$I[\eta] = D_1 + \int_0^\eta e^{N[\eta]} Z[\eta] d\eta \tag{3.15}$$

Now, taking $\eta \to \infty$ in the Eqs. (3.7)–(3.9), we get

$$\int_0^\infty e^{-L[\eta]} G[\eta] d\eta = 1 \tag{3.16}$$

$$\int_0^\infty e^{-R[\eta]} H[\eta] d\eta = -1 \tag{3.17}$$

$$\int_0^\infty e^{-N[\eta]} I[\eta] d\eta = -1 \tag{3.18}$$

These integrals can be solved asymptotically by Laplace's method. Putting $L = R = N = \xi$, transforming the Eqs. (3.16)–(3.18) to the variable ξ and integrating in the gamma functions, we get

$$1 = \sum_{m=0}^{\infty} r_m \Gamma_{\frac{(m+1)}{3}} \tag{3.19}$$

$$-1 = \sum_{m=0}^{\infty} a_m \Gamma_{\frac{(m+1)}{3}} \tag{3.20}$$

$$-1 = \sum_{m=0}^{\infty} b_m \Gamma_{\frac{(m+1)}{3}} \tag{3.21}$$

Solving Eqs. (3.19)–(3.21) using MATLAB, the values of B_2, C_1 and D_1 have been determined for different values of other flow parameters. After knowing the values of B_2, C_1 and D_1, we can easily find the values of $B_3, B_4, B_5, \ldots, C_2, C_3, C_4, \ldots,$ $D_2, D_3, D_4, \ldots,$ etc. Now using these values in Eqs. (3.1)–(3.3), we get the expressions for $F[\eta]$, $\theta[\eta]$ and $\phi[\eta]$ respectively. The constants are not presented here due to sake of brevity.

4 Results and Discussion

The velocity components along x and y axes are given by,

$$u = U[x]F'[\eta]$$

$$v = \frac{a}{2}(1-m)yF'[\eta]x^{m-1} - F[\eta]x^{\frac{(m-1)}{2}}\sqrt{\frac{(m+1)a\nu_{[1]}}{2}}$$

Knowing the velocity $F'[\eta]$, temperature $\theta[\eta]$ and concentration $\phi[\eta]$ fields, we obtain some important flow characteristics of the problem viz. wedge shear stress, local heat flux and mass flux.

The non-dimensional shearing stress τ_0 at the wedge $\eta = 0$ is given by,

$$\tau_0 = \frac{\sigma_{xy}}{\rho U \left\{ \frac{U(1+m)\nu_{[1]}}{2x} \right\}^{\frac{1}{2}}} = \frac{d^2 F}{d\eta^2} \bigg]_{\eta=0} \tag{4.1}$$

The non-dimensional heat flux at the wedge $\eta = 0$, in terms of Nusselt number Nu is given by,

Fig. 1 Variation of $F'[\eta]$
against η for $M=2$ and $\beta = 1$

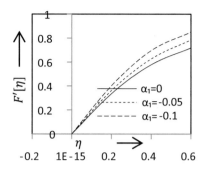

$$Nu = \frac{d\theta}{d\eta}\bigg]_{\eta=0} \tag{4.2}$$

The non-dimensional mass flux at the wedge $\eta = 0$, in terms of Sherwood number Sh is given by,

$$Sh = \frac{d\phi}{d\eta}\bigg]_{\eta=0} \tag{4.3}$$

In order to get an insight into the physical situation of the problem, the numerical computations have been carried out for velocity, temperature and concentration fields. Throughout the computations, we employ different values of the visco-elastic parameter $(-0.1 \le \alpha_1 \le 0)$, magnetic parameter $(0 < M \le 4)$, wedge angle parameter $(0 < \beta \le 1.6)$, heat source parameter $(0 < S \le 5)$, thermophoretic parameter $(0 < \tau \le 3.5)$ and chemical reaction parameter $(0 < Kc \le 3)$ with fixed value of Schmidt number $Sc = 0.3$, Eckert number $E = 0.01$ and Prandtl number $Pr = 2.5$. The non-zero values of α_1 characterize the visco-elastic fluid and $\alpha_1 = 0$ represents the character of Newtonian fluid flow phenomenon.

Figures 1, 2 and 3 demonstrate the variation of fluid velocity $F'[\eta]$ against η with various values of other flow parameters. The graphs reveal that the fluid velocity accelerates with the growth of the absolute values of the visco-elastic parameter α_1 ($\alpha_1 = 0, -0.05, -0.1$) in comparison with Newtonian fluid flow phenomenon. The growth of magnetic parameter M decelerates the fluid velocity in both system of fluids (Figs. 1 and 2) but an opposite trend is demonstrated during the growing behaviour of wedge angle parameter β (Figs. 1 and 3). The presence of transverse magnetic field yields Lorentz force which acts as a retarding force on the velocity profile, as a result fluid velocity decreases.

The variation of temperature $\theta[\eta]$ and concentration $\phi[\eta]$ against η are illustrated by the Figs. 4, 5, 6, 7, 8 and 9. The figures depict that the temperature and concentration of the fluid gradually drops away from the wedge for both Newtonian and visco-elastic fluids. Also, temperature and concentration of the fluid decelerate with the growth of the absolute values of the visco-elastic parameter α_1 ($\alpha_1 = 0, -0.05, -0.1$) in comparison with Newtonian fluid flow phenomenon. The rising values of

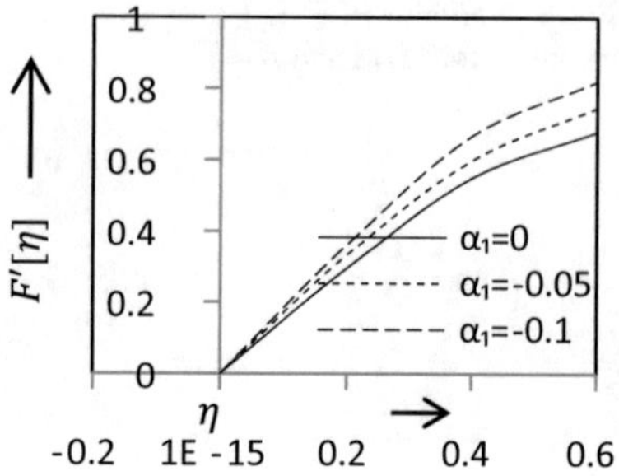

Fig. 2 Variation of $F'[\eta]$ against η for $M = 4$ and $\beta = 1$

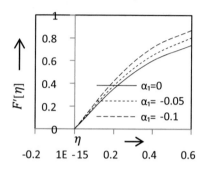

Fig. 3 Variation of $F'[\eta]$ against η for $M = 2$ and $\beta = 1.6$

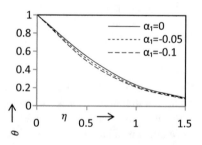

Fig. 4 Variation of $\theta[\eta]$ against η for $M = 2$, $S = 3$, $Pr = 2.5$ and $\beta = 1$

magnetic parameter M (Figs. 4 and 5) depict an accelerating trend of the fluid temperature but an opposite behaviour is demonstrated during the growing behaviour of heat source parameter S (Figs. 4 and 6). With the increasing value of M, the thickness of the thermal boundary layer increases because of the retarding force exerted by the magnetic field on fluid flow region and as a result temperature profile increases. The rising values of thermophoretic parameter τ (Figs. 7 and 8) depict a decelerating trend of the species concentration. This is true only for least values of Schmidt number for which the Brownian diffusion effect is excessive compared to the convection effect. However, for large values of Sc the diffusion effect is inferior in comparison with the convection effect. Hence the thermophoretic parameter τ is expected to reform the concentration boundary layer considerably. The similar result is obtained during the growing behaviour of chemical reaction parameter Kc (Figs. 7 and 9).

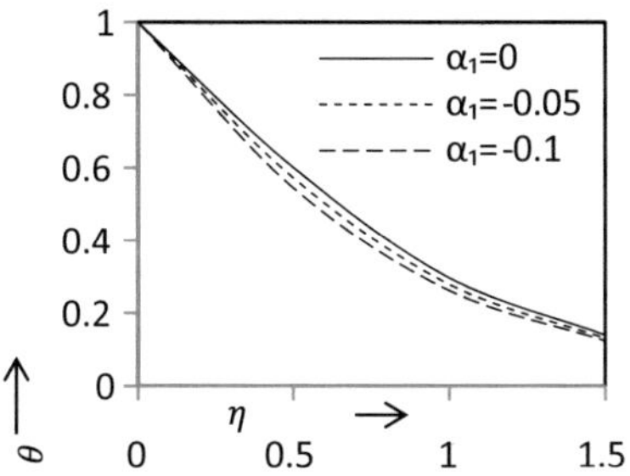

Fig. 5 Variation of $\theta[\eta]$ against η for $M = 4$, $S = 3$, $Pr = 2.5$ and $\beta = 1$

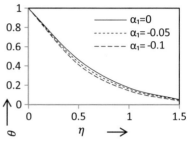

Fig. 6 Variation of $\theta[\eta]$ against η for $M = 2$, $S = 5$, $Pr = 2.5$ and $\beta = 1$

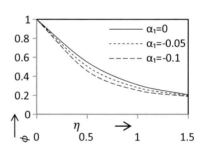

Fig. 7 Variation $\phi[\eta]$ against η for $\tau = 1.5$, $Kc = 1, M = 2, S = 3, Pr = 2.5$, $Sc = 0.5$ and $\beta = 1$

Fig. 8 Variation ϕ against η for $= 3.5$, $Kc = 1, M = 2$, $S = 3$, $Pr = 2.5$, $Sc = 0.5$ and $\beta = 1$

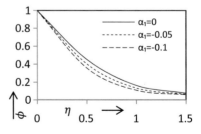

Figures 10 and 11 represent the variation of shearing stress τ_0 against M and β respectively. It is observed that the shearing stress against M increases with the growth of absolute value of visco-elastic parameter α_1 but an opposite trend is found against β. Variation of Nusselt number Nu against M and S are revealed by the Figs. 12 and 13 respectively. Again, it is seen that with the enhancement of absolute value of

Fig. 9 Variation $\varphi[\eta]$
against η for $\tau = 1.5$,
$Kc = 3, M = 2, S = 3, Pr = 2.5$,
$Sc = 0.5$ and $\beta = 1$

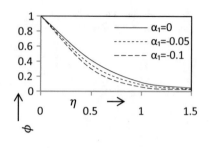

Fig. 10 Variation of τ_0
against M τ_0 for $\beta = 1$

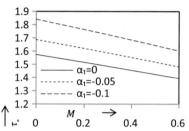

Fig. 11 Variation of τ_0
against β for $M = 2$

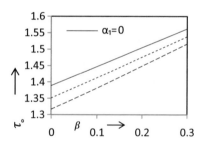

visco-elastic parameter, the Nusselt number shows a growing behaviour against M but opposite behaviour is observed in case of heat source parameter S. Figures 14 and 15 demonstrate the variation of Sherwood number Sh against thermophoretic parameter τ and chemical reaction parameter Kc respectively. It is noticed that the Sherwood number has decelerating trend in both visco-elastic and simple fluid cases. Also, with the growth of absolute value of visco-elastic parameter α_1 diminishes the Sherwood number in compared to Newtonian fluid flow phenomenon.

5 Conclusion

The study leads to the following conclusions:

- The visco-elastic parameter significantly affects the velocity field at each point of the fluid flow region.

Fig. 12 Variation of Nu against M for $S = 3$, $Pr = 2.5$ and $\beta = 1$

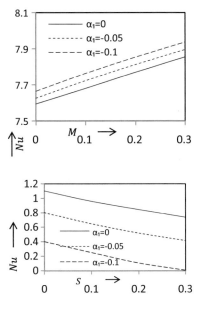

Fig. 13 Variation of Nu against S for $M = 2$, $Pr = 2.5$ and $\beta = 1$

Fig. 14 Variation of Sh against τ for $Kc = 1$, $M = 2$, $S = 3$, $Pr = 2.5$, $Sc = 0.5$ and $\beta = 1$

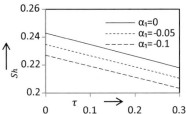

Fig. 15 Variation of Sh against Kc for $\tau = 1.5$, $M = 2$, $S = 3$, $Pr = 2.5$, $Sc = 0.5$ and $\beta = 1$

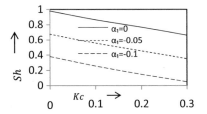

- The fluid velocity is diminished by the enhancement of magnetic parameter and this shows conformity with physical impact of magnetic parameter.
- The temperature and concentration fields are affected remarkably by visco-elasticity.
- The variation of shearing stress in the flow region against the pertinent flow parameter in presence of visco-elasticity is influential.

- The Nusselt number and Sherwood number play vital roles in fluid flow phenomenon. The effect of visco-elasticity is considerable in both the cases.

References

1. Watanabe, T.: Thermal boundary layer over a wedge with uniform suction or injection in forced flow. Acta Mech. **83**, 119–126 (1990)
2. Ishak, A., Nazar, R., Pop, I.: Falkner–Skan equation for flow past a moving wedge with suction or injection. J. Appl. Math. Comput. **25**, 67–83 (2007)
3. Yih, K.A.: Uniform suction/blowing effect on forced convection about a wedge: uniform heat flux. Acta Mech. **128**, 173–181 (1998)
4. Mahmood, M., Asghar, S., Hossain, M.A.: Hydromagnetic flow of viscous incompressible fluid past a wedge with permeable surface. J. Appl. Math. Mech. **89**, 174–188 (2009)
5. Shrivastava, U.N., Usha, S.: Magneto-fluid dynamic boundary layer on a moving continuous flat surface. Indian J. Pure Appl. Math. **18**, 741–751(1987)
6. Murthy, S.N., Sapre, M.: Effect of magnetic field on laminar boundary layer flow on a flat plate. Indian J. Pure Appl. Math. **22**, 601–609 (1991)
7. Meksyn, D.: New Method in Laminar Boundary Layer Theory. Pergamon Press (1961)
8. Sattar, M.A.: A local similarity transformation for the unsteady two-dimensional hydrodynamic boundary layer equations of a flow past a wedge. Int. J. Appl. Math. Mech. **7**, 15–28 (2011)
9. Cheng, W.T., Lin, H.T.: Non-similarity solution and correlation of transient heat transfer in laminar boundary layer flow over a wedge. Int. J. Eng. Sci. **40**, 531–538 (2002)
10. Goren, S.L.: Thermophoresis of aerosol particles in laminar boundary layer on flat plate. J. Colloid Interface Sci. **61**, 77–85 (1977)
11. Chamkha, J.A., Pop, I.: Effect of thermophoresis particle deposition in free-convection boundary layer from a vertical flat plate embedded in a porous medium. Int. Commum Heat Mass Transf. **31**, 421–430 (2004)
12. Selim, A., Hossain, M.A., Das, R.: The effect of surface mass transfer on mixed convection flow past a heated vertical flat permeable plate with thermophoresis. Int. J. Therm. Sci. **42**, 973–982 (2003)
13. Postelnicu, A.: Effects of thermophoresis particle deposition in free convection boundary layer from a horizontal flat plate embedded in a porous medium. Int. J. Heat Mass Transf. **50**, 2981–2985 (2007)
14. Bakier, A.Y., Gorla, R.S.R.: Effects of thermophoresis and radiation on laminar flow along a semi-infinite vertical plate. Heat Mass Trans. **47**, 419–425 (2011)
15. Zueco, J., Beg, O.A., Takhar, H.S., Prasad, V.R.: Thermophoretic hydromagnetic dissipative heat and mass transfer with lateral mass flux, heat source, Ohmic heating and thermal conductivity effects: network simulation numerical study. Appl. Therm. Eng. **29**, 2808–2815 (2009)
16. Noor, N.F.M., Abbasbandy, S., Hashim, I.: Heat and mass transfer of thermophoretic MHD flow over an inclined radiate isothermal permeable surface in the presence of heat source/sink. Int. J. Heat Mass Trans. **55**, 2122–2128 (2012)
17. Kundu, P.K., Das, K., jana, S.: Combined effects of thermophoresis and chemical reaction on magnetohydrodynamics mixed convection flow. J. Thermophys. Heat Trans. **27**, 741–747 (2013)
18. Das, K.: Influence of thermophoresis and chemical reaction on MHD micropolar fluid flow with variable fluid properties. Int. J. Heat Mass Trans. **55**, 7166–7174 (2012)
19. Coleman, B.D., Noll, W.: An application theorem for functional with applications in continuum mechanics. Arch. Ration Mech. Anal. **6**, 350–360 (1960)
20. Coleman, B.D., Markovitz, H.: Incompressible second-order fluids. Adv. Appl. Mech. **8**, 69–101 (1964)
21. Schlichting, H.: Boundary Layer Theory. McGraw Hill, New York (1968)

Slip Effects on Heat and Mass Transfer in MHD Visco-Elastic Fluid Flow Through a Porous Channel

Bamdeb Dey and Rita Choudhury

Abstract The problem of MHD flow of visco-elastic fluid through a horizontal channel has been analysed in the presence of heat and mass transfer. The fluid is subjected to a transverse magnetic field and the slip velocity at the lower wall of the channel taken into consideration. A mathematical model (Walters liquid model B′) has been analyzed using appropriate mathematical techniques. Expressions for velocity, temperature, concentration, wall shear stress and rate of heat and mass transfer have been obtained. Variations of the quantities with different parameters are computed by using MATLAB software. The results are discussed graphically to measure the impact of visco-elasticity.

Keywords Visco-elastic · MHD · Slip velocity · Free convection · Shear stress

Nomenclature

L	The mean-free path
m_1	The Maxwell's reflexion
B_0	Transverse magnetic field
D_m	Co-efficient of mass diffusivity
D	The mean width of the channel
p'	Pressure
T'	Temperature distribution
x, y	Cartesian co-ordinate
u, v	Velocity components along x and y axis
K	Co-efficient of thermal conductivity
T_1', T_2'	Wall temperatures

B. Dey · R. Choudhury (✉)
Department of Mathematics, Gauhati University, Guwahati 781014, Assam, India
e-mail: rchoudhury66@yahoo.in

B. Dey
e-mail: bamdebdey88@gmail.com

© Springer Nature Singapore Pte Ltd. 2019
A. Abraham et al. (eds.), *Emerging Technologies in Data Mining and Information Security*, Advances in Intelligent Systems and Computing 755,
https://doi.org/10.1007/978-981-13-1951-8_50

C_1', C_2' Wall concentrations

K_1 Visco-elastic parameter $\left(\frac{K_0 V_0}{\eta_0 d} \right)$

Dimensionless Parameters

M Magnetic Parameter $\left(\frac{\sigma B_0^2 \mu' d^2}{\mu_0} \right)$

P_r Prandtl number $\left(\frac{\mu C_p}{K} \right)$

S Suction parameter $\left(\frac{v_0 d}{v} \right)$

S_c Schmidth number $\left(\frac{v_0 d}{Dm} \right)$

Greek Symbols

$\in$ The amplitude parameter

λ Frequency parameter $(K_0 d)$

υ Kinematic viscosity $\left(\frac{\mu'}{\rho} \right)$

ρ Density

σ Electric conductivity

θ Dimensionless temperature

ϕ Dimensionless concentration

1 Introduction

In recent times, the study of non-Newtonian fluids have attracted very much attention because of their extensive applications in different fields in engineering and indus-try, especially in extraction of crude oil from petroleum products. In the category of non-Newtonian fluids, visco-elastic fluid has distinct features. The study of visco-elastic fluid flow through porous medium finds applications in agricultural engi-neering technology and geophysics. The applications of magneto-hydrodynamics in viscous fluid theory are noticed in many engineering and technological fields. Sparrow and Cess [1], Drake [2], Ram and Singh [3]. Makinde [4], Singh [5], Al-Hadhrami [6], Adesanya and Ghadeyan [7], Kishnambol [8], Sanyal and Sanyal [9], Chauhan and Priyanka [10], Falade et al. [11], Raju and Venkataraman [12], Om Prakash et al. [13], Mohamed et al. [14] have remarkable contributions in this field for analyzing the characteristics of viscous fluid with different geometries. The visco-elastic fluid behave differently due to the presence of both viscosity and elastic properties. As a result, the difficulties in Newtonian fluid may be removed using ana-lytical and numerical methods applicable in different models of non-Newtonian fluid with various rheological properties. Related to this area, the works of some authors viz. Choudhury and Dey [15], Choudhury and Das [16], Choudhury and Dhar [17], Shekhar and Reddy [18], etc., are worth mentioning. Motivated by the above cited works, here we have studied the features of a visco-elastic fluid characterized by Walters liquid model (B') [19, 20] through non-isothermal parallel flat wall and a long wavy wall in presence of heat and mass transfer under the influence of uniform magnetic field.

2 Formulation of the Problem and Basic Equations

We consider the steady, visco-elastic, incompressible MHD fluid flow through a non-isothermal parallel flat wall and a long wavy wall. The x'-axis and y'-axis are taken along the parallel flat wall and normal to the wall respectively. The expression $y = d + \epsilon \cos kx$ represents the wavy wall and $y = 0$ depicts the flat wall. A uniform magnetic field is applied in the direction to the walls. The wavy and flat walls are maintained at constant temperatures T_1' and T_2' respectively. The governing equations are given by

$$\frac{\partial u'}{\partial x'} + \frac{\partial v'}{\partial y'} = 0 \tag{2.1}$$

$$\begin{aligned}
\rho \left(u' \frac{\partial u'}{\partial x'} + v' \frac{\partial u'}{\partial y'} \right) = &-\frac{\partial p'}{\partial x'} + \eta_0 \left(\frac{\partial^2 u'}{\partial x'^2} + \frac{\partial^2 u'}{\partial y'^2} \right) - k_0 \left(u' \frac{\partial^3 u'}{\partial x'^3} + v' \frac{\partial^3 u'}{\partial y'^3} + 2v' \frac{\partial^3 u'}{\partial x'^2 \partial y'} + v' \frac{\partial^3 v'}{\partial x' \partial y'^2} \right. \\
&+ u' \frac{\partial^3 u'}{\partial x' \partial h'^2} - 6 \frac{\partial u'}{\partial x'} \cdot \frac{\partial^2 u'}{\partial x'^2} + 6 \frac{\partial u'}{\partial y'} \cdot \frac{\partial^2 u'}{\partial x' \partial y'} - 2 \frac{\partial u'}{\partial y'} \cdot \frac{\partial^2 v'}{\partial x'^2} + 2 \frac{\partial u'}{\partial y'} \cdot \frac{\partial^2 u'}{\partial x'^2} \\
&\left. + \frac{\partial u'}{\partial y'} \cdot \frac{\partial^2 v'}{\partial x'^2} - \frac{\partial^2 u'}{\partial h'^2 \partial x'} \cdot \frac{\partial v'}{\partial x'} - 2 \frac{\partial u'}{\partial x'} \cdot \frac{\partial^2 v'}{\partial y' \partial x'} + 2 \frac{\partial u'}{\partial x'} \cdot \frac{\partial^2 u'}{\partial y'^2} \right) - \sigma B_0^2 \mu' u'
\end{aligned} \tag{2.2}$$

$$\begin{aligned}
\rho \left(u' \frac{\partial v'}{\partial x'} + v' \frac{\partial v'}{\partial y'} \right) = &-\frac{\partial p'}{\partial y'} + \eta_0 \left(\frac{\partial^2 v'}{\partial x'^2} + \frac{\partial^2 v'}{\partial y'^2} \right) \\
&- k_0 \left(u' \frac{\partial^3 u'}{\partial x'^2 \partial y'} + 2u' \frac{\partial^3 v'}{\partial x' \partial y'^2} + u' \frac{\partial^3 u'}{\partial x'^3} \right. \\
&+ v' \frac{\partial^3 v'}{\partial y'^3} - v' \frac{\partial^3 u'}{\partial x'^3} - 2 \frac{\partial u'}{\partial x'} \cdot \frac{\partial^2 v'}{\partial x'^2} - \frac{\partial v'}{\partial x'^2} - \frac{\partial v'}{\partial x'} \cdot \frac{\partial^2 u'}{\partial y'^2} \\
&\left. + 2 \frac{\partial^2 u'}{\partial x'^2} \cdot \frac{\partial u'}{\partial y'} - \frac{\partial^2 v'}{\partial x'^2} \cdot \frac{\partial u'}{\partial y'} - \frac{\partial^2 v'}{\partial x'^2} \cdot \frac{\partial v'}{\partial y'} + \frac{\partial v'}{\partial x'} \cdot \frac{\partial^2 u'}{\partial x'^2} \right) - 3 \frac{\partial v'}{\partial y'} \cdot \frac{\partial^2 v'}{\partial y'^2}
\end{aligned} \tag{2.3}$$

$$\rho C_P \left(u' \frac{\partial T'}{\partial x'} + v' \frac{\partial T'}{\partial y'} \right) = K' \left(\frac{\partial^2 T'}{\partial x'^2} + \frac{\partial^2 T'}{\partial y'^2} \right) \tag{2.4}$$

$$u' \left(\frac{\partial C'}{\partial x'} + v' \frac{\partial C'}{\partial y'} \right) = D_m \left(\frac{\partial^2 C'}{\partial x'^2} + \frac{\partial^2 C'}{\partial y'^2} \right) \tag{2.5}$$

The relevant boundary conditions are as follows:

$$y' = 0; \ u' = L_1 \frac{\partial u'}{\partial y'}, \ v' = -v_0, \ T' = T'_1, \ C' = C'_1$$

$$y' = d + \epsilon \cos k_0 x'; \ u' = -L_1 \frac{\partial u'}{\partial y'}, \ v' = 0, \ T' = T'_2, \ C' = C'_2$$ \qquad (2.6)

where, $L_1 = \frac{2-m_1}{m_1} L$

As the flat-wall is infinite in length, so

$$\frac{\partial u'}{\partial x'} = 0 \qquad (2.7)$$

Then Eq. (2.1) reduces to

$$\frac{\partial v'}{\partial y'} = 0, \qquad (2.8)$$

which gives $v' = -v_0$ with the help of Eq. (2.6)

We introduce following dimensionless quantities

$$x' = \frac{x'}{d}, \ y' = \frac{y'}{d}, u = \frac{u'}{v_0}, \ v = \frac{v'}{v_0}, \ \theta = \frac{T'-T'_1}{T'_2-T'_1}, \ \varphi = \frac{C'-C'_1}{C'_2-C'_1},$$

$$p = \frac{p'd}{\eta_0 v_0}, \ S = \frac{V_0 d}{\nu}, \ M = \frac{\sigma B_0^2 d^2 \mu'}{\eta_0}, \ P_r = \frac{\eta_0 C_p}{K'}, \ S_c = \frac{V_0 d}{D_m}$$ \qquad (2.9)

where S, M, P_r, S_c are the respective suction parameter, magnetic parameter, Prandtl number and schmidt number.

The dimensionless governing equations are:

$$\frac{d^2 u}{dy^2} + S \frac{du}{dy} - M^2 u + k_1 \frac{d^3 u}{dy^3} = \frac{dp}{dx} \qquad (2.10)$$

$$\frac{dp}{dy} = 0 \qquad (2.11)$$

$$\frac{\partial^2 \theta}{\partial y^2} + P_r S \frac{\partial \theta}{\partial y} = 0 \qquad (2.12)$$

$$\frac{\partial^2 \varphi}{\partial y^2} + S_c \frac{\partial \varphi}{\partial y} = 0 \qquad (2.13)$$

with relevant boundary conditions:

$$y = 0; \ u = \gamma u', \ v = -1, \ \theta = 0, \ \varphi = 0$$

$$y = h; \ u = -\gamma u', \ v = 0, \ \theta = 1, \ \varphi = 1$$ \qquad (2.14)

Now Eq. (2.11) shows that p is independent of y, so we assume that $\frac{dp}{dy}$ = constant.

3 Method of Solution

To solve the system of differential Eqs. (2.10)–(2.13) under boundary conditions (2.14), the physical quantity u may be expanded in power of visco-elastic parameter k_1(as $k_1 \ll 1$ due to small shear rate).

$$u = u_0 + k_1 u_1 + 0\left(k_1^2\right) \tag{2.15}$$

The corresponding boundary conditions are

$$\left. \begin{array}{l} y = 0; \quad u_0 = \gamma \frac{\partial u_0}{\partial y}, \quad u_1 = \gamma \frac{\partial u_1}{\partial y} \\ y = h; \quad u_0 = -\gamma \frac{\partial u_0}{\partial y}, \quad u_1 = -\gamma \frac{\partial u_1}{\partial y} \end{array} \right\} \tag{2.16}$$

Substituting (2.15) in Eq. (2.10), thereafter equating the co-efficients of like powers of k_1 and neglecting the higher order terms, we get

Zeroth-order equation:

$$u_0''' + Su_0' - M^2 u_0 = C \text{ where } C = \frac{dp}{dx} \tag{2.17}$$

First-order equation:

$$u_1''' + Su_1' - M^2 u_1 + u_0'''' = 0 \tag{2.18}$$

Solving (2.17) and (2.18) under the boundary conditions (2.16), we obtain the expression for velocity as follows:

$$u = a_{13} e^{a_1 y} + a_{14} e^{-a_2 y} - \frac{C}{M^2}$$

Solving the Eqs. (2.12) and (2.13) under the boundary conditions (2.14) the temperature and concentration are acquired as

$$\theta = \frac{e^{-P_r S y} - 1}{e^{-P_r S h} - 1}, \quad \phi = \frac{e^{-S_c y} - 1}{e^{-S_c h} - 1}$$

The shearing stress at any point is given by

$$\sigma = \frac{\partial u}{\partial y} - k_1 \frac{\partial^2 u}{\partial y^2}$$

The dimensionless shearing stress at the flat wall $y = 0$ and the wavy wall $y = h$ are given respectively as

$$\sigma_0 = [\sigma]_{y=0} \text{ and } \sigma_1 = [\sigma]_{y=h}$$

The heat transfer co-efficient characterized by Nusselt number (N_u) in dimensionless form is: $N_u = -k \left(\frac{T_2 - T_1}{d} \right) \frac{\partial \theta}{\partial y}$

Nusselt number at the flat wall $y = 0$ and the wavy wall $y = h$ are given by $N_{u_0} = [N_u]_{y=0}$, $N_{u_1} = [N_u]_{y=h}$ respectively.

The dimensionless mass transfer number corresponding to the Sherwood number is:

$$S_h = \frac{\partial \phi}{\partial y}$$

Dimensionless Sherwood number at the flat wall $y = 0$ and at the wavy wall $y = h$ are given by

$$Sh_0 = \left[\frac{\partial \phi}{\partial y} \right]_{y=0}, \quad Sh_1 = \left[\frac{\partial \phi}{\partial y} \right]_{y=h} \quad \text{respectively.}$$

The values of the constants are obtained but not presented here due to brevity.

4 Results and Discussion

We analyse the effects of visco-elastic parameter k_1 on the MHD visco-elastic fluid flow in a horizontal channel in presence of heat and mass transfer. The simple Newtonian and visco-elastic fluid flow characteristics are illustrated by zero and non-zero values of k_1 respectively. The effects of various physical parameters on these flow quantities are studied through graphs. The velocity distribution is depicted through Figs. 1, 2, 3, 4 and 5. Figures 6, 7, 8 and 9 illustrate the changes in shearing stress with respect to different parameters for flat and wavy plates. Figures 10 and 11 represent the temperature and concentration profiles respectively. The value $k_1 = 0$ reveals the Newtonian fluid flow phenomenon, whereas the non-zero values of k_1 characterizes the visco-elastic fluid. Throughout the numerical calculations, we consider $P_r = 2$, $M = 5$, $S_c = 2$, $\gamma = 0.15$, $S = 0.5$, $c = 1$, $h = 1$ unless otherwise stated. The slip parameter (γ), the Prandtl number (P_r,) the suction parameter (S) and the magnetic parameter (M) play important roles in graphical illustrations. The velocity distribution u against y for various flow parameter reveals that the increasing trend of visco-elasticity subdues the fluid flow (Fig. 1). The growth of the slip parameter γ (Figs. 1 and 2) accelerates the fluid velocity in both cases. The enhancement of Prandtl number does not alter the flow curves in Newtonian and visco-elastic fluid flows (Figs. 1 and 3) significantly. The increase of magnetic parameter M diminishes the fluid velocity and it clearly agrees with the physical situation (Figs. 1 and 4), that the fluid velocity is retarded for the formation of Lorentz force. The nature of fluid flow for different values of suction parameter S is shown in Figs. 1 and 5. Here the fluid velocity decreases continuously for non-Newtonian fluid flow but in case of Newtonian fluid flow velocity, the velocity first decreases

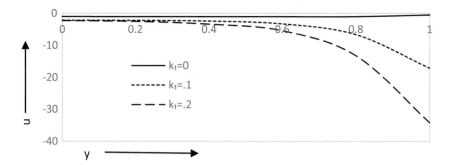

Fig. 1 Fluid velocity u against y for M = 5, S = 0.5, γ = 0.15, Pr = 2, Sc = 2, c = 1, h = 1

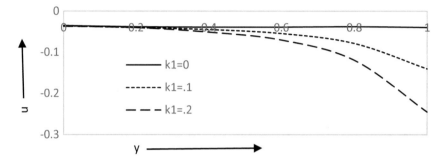

Fig. 2 Fluid velocity u against y for M = 5, S = 0.5, γ = 1.5, Pr = 2, Sc = 2, c = 1, h = 1

and then gradually enhances. The variation of shearing stress σ against the magnetic parameter M and suction parameter S are illustrated graphically in Figs. 6, 7, 8 and 9 in presence of other flow parameters. Figures 6 and 7 represent the behaviour of shearing stress against the magnetic parameter M at the flat plate y = 0 and the wavy plate y = h respectively. The magnitude of shearing stress first increases but then gradually decreases with the increase of the magnetic parameter M in both the walls. The growth of suction parameter S on shearing stress reveals that the shearing stress decreases in both Newtonian and non-Newtonian fluids in both the walls. Figures 10 and 11 demonstrate the temperature and concentration profiles, and consequently it can be remarked that the Nusselt number and the Sherhood number which character-ize the rate of heat transfer and the rate mass transfer respectively are not significantly affected by the change in visco-elastic parameter due to preventing effect of visco-elasticity. As a result, the elasticity parameter keeps on reducing regardless of its character and a visco-elastic fluid acts as a viscous fluid.

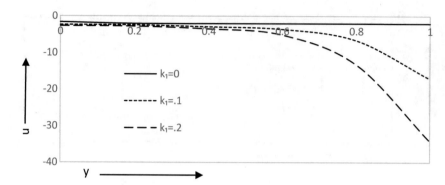

Fig. 3 Fluid velocity u against y for M = 5, S = 0.5, γ = 0.15, Pr = 4, Sc = 2, c = 1, h = 1

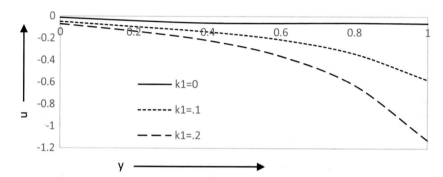

Fig. 4 Fluid velocity u against y for M = 3.5, S = 0.5, γ = 0.15, Pr = 4, S_c = 2, c = 1, h = 1

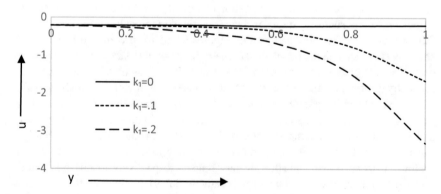

Fig. 5 Fluid velocity u against y for M = 5, S = 2, γ = 0.15, Pr = 2, Sc = 2, c = 1, h = 1

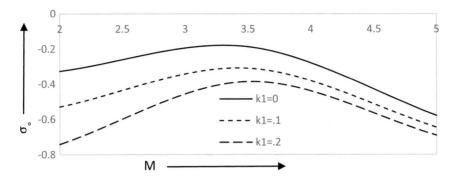

Fig. 6 Shearing stress at y = 0 against Magnetic parameter M

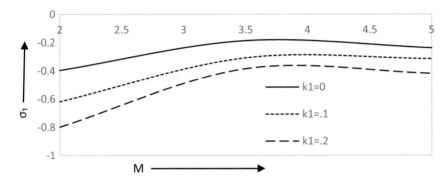

Fig. 7 Shearing stress at y = h against Magnetic parameter M

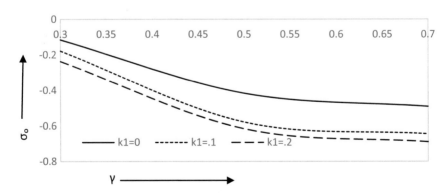

Fig. 8 Shearing stress at y = 0 against slip parameter γ

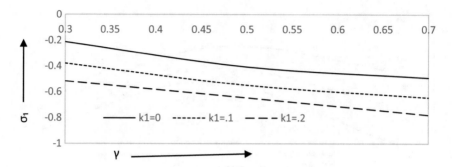

Fig. 9 Shearing stress at y = h against slip parameter γ

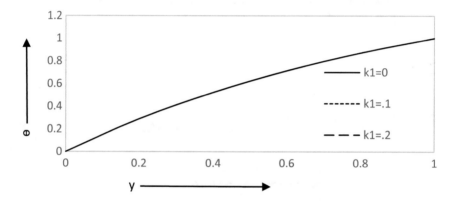

Fig. 10 Temperature profile against y

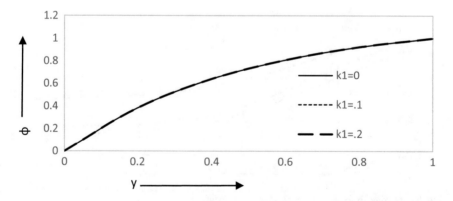

Fig. 11 Concentration profile against y

5 Conclusion

In this paper, we have studied the effects of visco-elastic parameter on the MHD flow in a horizontal channel with heat and mass transfer. From the present investigation, we make the following conclusions:

(i)   With the growth of visco-elastic parameter the fluid velocity decreases in the fluid flow region.

(ii) The fluid velocity shows a decreasing trend for both Newtonian and visco-elastic fluids for different flow parameters.

(iii) The shearing is visibly affected at both the plates with the variation in visco-elastic parameter.

(iv) The Nusselt number and the Sherwood number are not significantly affected at both the plates with the change of visco-elastic parameter due to restraining effect cavorted by the elasticity of the fluid.

References

1. Sparrow, E.M., Cess, R.D.: The effect of a magnetic field on free convection heat transfer. Int. J. Heat Mass Transf. **3**(4), 267–274 (1961)
2. Drake, D.G.: Flow in a channel due to periodic pressure gradient, quart. J. Mech. Appl. Math. **18**(1) (1965)
3. Ram, P.C., Singh, C.B.: Unsteady MHD fluid flow through a channel. J. Sci. Res. **28**(2) (1978)
4. Makinde, O.D., Mhone, P.Y.: Heat transfer to MHD oscillatory flow in a channel filled with porous medium. Ramanian J. Phys. **50**, 931–938 (2005)
5. Singh, C.B.: MHD steady flow of liquid between two parallel plates. In: Proceeding of the First Conference of Kenya Mathematical Society, pp. 24–26 (1993)
6. Al-Hadhrami, A.K., Elliot, L., Ingham, M.D., Wen, X.: Flow through horizontal channels of porous materials. Inter. Energy Res. **27**, 875–899 (2003)
7. Adesanya, O., Gbadeyan, J.A.: Adomian decomposition approach to steady visco-elastic fluid flow with slip through a planner channel. Int. J. Nonlinear Sci., 986–94 (2010)
8. Ganesh, S., Krishnambal, S.: Unsteady MHD Stokes flow of viscous fluid between two parallel porous plates. J. Appl. Sci. **7**, 374–379 (2007)
9. Sanyal, D.C., Sanyal, M.K.: Hydromagnetic slip flow with heat transfer is an inclined channel. Czechoslov. J. Phys. **39**(5), 529–536 (1989)
10. Chauhan, D.S., Rastogi, P.: Heat transfer effects on rotating MHD Couette flow in a channel partially filled by a porous medium will hall current. J. Appl. Sci. Eng. **15**(3), 281–290 (2012)
11. Falade, J.A., Ukaegbu, J.C., Egere, A.C., Adesanya, S.O.: MHD oscillatory flow through a porous channel saturated with porous medium. Alex. Eng. J. **56**, 147–152 (2017)
12. Raju, K.V.S., Venkataramana, S.: MHD convective flow through porous medium in a horizontal channel with insulated and impermeable bottom wall in the presence of viscous dissipation and Joule heating. Ain Shams Eng. J. **5**(2), 1–8 (2014)
13. Prakash, O.M., Makinde, O.D., Kumar, D., Dwivedi, Y.K.: Heat transfer to MHD oscillatory dusty fluid flow in a channel filled with a porous medium. Indian Acad. Sci. **40**, part-4, 1273–1282 (2015)
14. Mohamed Ismail, A., Ganesh, S., Kirubhashankar, C.K.: Unsteady MHD flow between two parallel plates through porous medium with one plate moving uniformly and the other plate at rest with uniform suction. Int. J. Sci. Eng. Tech. Res. **3**(1), 6–10 (2014)

15. Choudhury, R., Dey, D.: Free convective visco-elastic flow with heat and mass transfer through a porous medium with periodic permeability. Int. J. Heat Mass Transf. **53**, 1666–1672 (2010)
16. Choudhury, R., Das, S.: Visco-elastic MHD free convectine flow through porous media in presence of radiation and chemical reaction with heat and mass transfer. J. Appl. Fluid Mech. **7**(4), 603–609 (2014)
17. Choudhury, R., Dhar, P.: Mixed convective MHD oscillatory flow of a visco-elastic fluid flow with heat source and variable suction through porous media. Int. J. Math. Stast. **16**(1), 55–70 (2015)
18. Sekhar, D.V., Reddy, G.V.: Chemical reaction effects on MHD unsteady free convective Walter's memory flow with constant suction and heat sink. Adv. Appl. Sci. Res. **3**(4), 2141–2150 (2012)
19. Walters, K.: The solutions of flow problems in the case of material with memories. J. Mec. **1**, 473–478 (1962)
20. Walters, K.: Non-Newtonian effects in some elastico-viscous liquids whose behaviour at small rates of shear in characterized by a general linear equation of state. Q. J. Mec. Appl. Math. **15**, 63–76 (1962)

A Channel Selection Method
for Epileptic EEG Signals

Satarupa Chakrabarti, Aleena Swetapadma and Prasant Kumar Pattnaik

Abstract Epilepsy is a disorder of the central nervous system in which a considerably large number of neurons at a certain instance of time show abnormal electrical activity. EEG or electroencephalogram signals plays a significant role in the diagnosis of epilepsy. Worldwide roughly 50 million people are affected by epilepsy that includes patients from all walks of life. In this paper we have studied the performance of artificial neural network (ANN) on EEG signals (CHB-MIT database) by applying principal component analysis (PCA) for selection of channels. The results reflect the performance of the neural network in different configuration. Out of the 23 channels considered, after using PCA, the highest accuracy of 86.7% is achieved with 18 channels. With the reduction of channels the accuracy decreased simultaneously. The study also brings forth the shortcomings as well as determines the area in this domain that holds prospective for future scope of work.

1 Introduction

Epilepsy is a chronic disorder of the central nervous system that predisposes individuals to experiencing recurrent seizures. A seizure is a sudden, transient aberration in the brain's electrical activity that produces disruptive symptoms. These symptoms range between a lapse in attention, a sensory hallucination, or a whole-body convulsion. Epilepsy is not a single disease, but a family of syndromes that share the feature of recurrent seizures [6]. In the current scenario of disease and disorders in the world population, epilepsy stands out to be the third commonest neurological disorder that

S. Chakrabarti · A. Swetapadma (✉) · P. K. Pattnaik
KIIT University, Bhubaneswar 751024, Odisha, India
e-mail: aleena.swetapadmafcs@kiit.ac.in

S. Chakrabarti
e-mail: chakrabartisatarupa@gmail.com

P. K. Pattnaik
e-mail: patnaikprasantfcs@kiit.ac.in

© Springer Nature Singapore Pte Ltd. 2019
A. Abraham et al. (eds.), *Emerging Technologies in Data Mining and Information Security*, Advances in Intelligent Systems and Computing 755,
https://doi.org/10.1007/978-981-13-1951-8_51

is affecting millions of people across the continents. Two recurrent seizure attacks are considered to be the general manifestation of epilepsy [13].

Depending on their characteristics and features, classification of epilepsy can be done in two distinct ways. One based on the origin or location of the seizure such as Temporal, frontal, parietal or occipital lobes of the brain, where the seizure manifests [5]. And the other classification based on the type of seizures, focal or partial seizures and generalized seizures. Researchers over the years have painstakingly dealt with epilepsy and its accompanying unpredictable abnormal electrical discharges. The earliest of work can be traced back to 1982 by Gotman who developed patient non-specific detectors. In [17], intra-cranial EEG (iEEG) signals were used and band pass filters were used having the range of frequency in between 0.5 and 100 Hz to remove noise and artifacts. The line noise was recorded around the frequency of 50 Hz was removed using notch filters. In [10], wavelet filters were used in the study so that the signals were restricted under 60 Hz. Different other techniques had been used for artifacts removal such as ICA (independent component analysis) while for segregating electroencephalogram (EEG) data blind source separation (BSS) was used. The amplitude as feature was used in [1] so that an input vector was built which was used in the artificial neural network (ANN). Regularity and synchronicity were the two other parameters used to analyze the similarity between the signals.

In [3], a study regarding the use of time frequency analysis for classification of EEG segments was presented for epilepsy along with comparison of different other methods based on EEG signals. The different types of classification problem considered for evaluation are Naïve Bayes, k-nearest neighbor (KNN) classifier, decision tree and logistic regression. In [8], a framework based on discrete wavelet transformation (DWT) was presented. A set of 14 combinations of Naïve Bayes and k-nearest neighbor classification were used to analyze the performance of data that was obtained after performing discrete wavelet transformation (DWT). In [15], an algorithm that would predict seizure using nonlinear features had been elaborated called entropy and approximate entropy. The prediction time varied between 5 and 60 min and the false detection rate was low. In [11], multimodal signal processing was used in order to detect seizures. Wavelet transformation and SVM was chosen for the study and it showed an accuracy of 93%.

In [9], a much newer method called GLCM for obtaining distinct characteristics from EEG signals had been mentioned. The features were then given in SVM for classification. The mentioned methodology gave accuracy, sensitivity and specificity in the range of 90%. In [4], signal processing of EEG recordings in order to get a hold about epilepsy and its occurrence was discussed. The authors proposed fixed cluster size for extracting the features that would further help in classification. In [16], general epileptic seizure detection and prediction procedure for obtaining features of ictal and inter ictal recordings using different transformations and decompositions were explored and analyzed.

In this work a comparison is done using different configuration of the neural network and also by applying principal component analysis on the same classifier. The remaining of the paper is organized in the following manner: general idea about principal component analysis which is followed by ANN classifier. Performance of the proposed method using the classifier is studied in the next following section which is finally followed by conclusion.

2 Principal Component Analysis

Principal component analysis (PCA) is a tool used in data analytics that helps in reducing dimension for a large set of inputs while still keeping most of the needed information. A mathematical operation that converts correlated variables into uncorrelated variables that are smaller in number and known as principal components. The application of PCA ranges from neuroscience to computer graphics. Principal component analysis shows a way to reduce large complex datasets to smaller dimension with relevant information intact [18]. Generally PCA is calculated using a square symmetric matrix. The matrix can either be a pure sum of squares and cross products or a correlation matrix which comes into the scenario when variance differs to a great extent. The geometry of PCA is seen as rotation of the co-ordinate axes of the original variables into new axes (orthogonal) that are termed as principal axes. The first principal axis is defined by the variation (maximum) of the projected points and also known as the best fit line. The values that correspond to this particular direction are known as the principal scores. The next consecutive lines are established using the criteria that they are orthogonal to the previous axes. The main objective of principal component analysis is reduction of attribute space thus it is a data compression method [7]. The selection of the subset depends on the variables that have highest correlation with the principal components. The main aim of PCA is to extract maximum variance from the input. Principal component analysis is supposed to reflect both unique as well as common variance of the input. The approach is variance focused and that seeks to represent the correlation as well as the total variance. The Eigen value is used for the measurement of variance for all variables in respect to a particular factor. Eigen values give the idea about the variation in the whole input sample that is considered for each factor.

$$Linear\ Features: Z = WX \tag{1}$$

3 Artificial Neural Network

The most substantial presence of neural network is in the human brain that records the presence of millions of neurons that are capable and responsible for the functioning of the human body. Artificial neural network is a representation of a model that closely resembles the human nervous system but stands on the foundation of mathematical equations and derivations [2]. The key element in the neural network architecture is the neurons or specifically the artificial neurons, sometimes called nodes. The synapses present in the actual nervous system is represented as effect of the different input signals with connection weights, transfer function presents the nonlinear features of the neurons. The impulse of the neuron is calculated as the sum of input (weighted) that gets modified with the help of the transfer function.

The effectiveness of the learning of neurons is reached by changing and modifying the weights in accordance to the specific learning algorithm that is selected [14]. The output of the neuron is given by the following relationship in (2)

$$O = f(net) = f\left(\sum_{j=1}^{n} w_j x_j\right),\qquad(2)$$

where the parameters represent the transfer or activation function f(net) and are obtained as the scalar product of input (x_j) and weight (w_j). The net input of the method can be given as in (3),

$$net = w^T = w_1 x_1 + \cdots + w_n x_n \qquad(3)$$

Here T represents transpose matrix and in a simple scenario O is evaluated as in (4),

$$O = f(net) = \begin{cases} 1 \ if \ w^T x \geq \theta \\ 0 \ otherwise \end{cases} \qquad(4)$$

4 Proposed Method

4.1 Data Acquisition

The dataset that was considered for this work consist of scalp EEG recordings of 10 pediatric patients between the ages of 1.5 years to 15 years suffering from intractable seizure. The dataset belonged to the patients from Children's Hospital Boston (CHB) who was monitored following withdrawal of anti-seizure medicines so that seizures could be assessed. Each file contained one to maximum of four seizures that were recorded using bipolar placement of electrodes according to the 10–20 international electrode placements and were sampled at 256 Hz at 16-bit resolution. The onset of seizure had aparoxysmal burst of 4 Hz that was present in right frontal, fronto-central channels, as well as in central and posterior channels.

The location or type of seizure- generalized, partial or lateral was present in the recordings. The EEG recording was performed in clinical environments and the patients were constantly monitored for several hours over several days. Artifacts were present along with seizure and non-seizure activities such as eye movements, head and body movements. Sleep spindles were also present in the recordings.

The database from where the EEG recordings were used is contained in CHB-MIT database that is available for download from Physionet website [12].

4.2 Principal Component Analysis and Channel Selection

Principal component analysis (PCA) helps in dimension reduction for a large set of inputs while keeping the required information intact. Principal component analysis helps in reducing large complex datasets into smaller sets with necessary information intact. In this work PCA has been used to select channels from the recorded EEG dataset. The dataset that has been considered for the work consist of 23 channels, thus containing huge amount of recorded data. Depending on the contribution, the number of channels have been selected that ranges from minimum of three channels to a maximum of nineteen channels and the reduced dataset has been used as an input for the classifier for further evaluation.

4.3 Detection of Epilepsy Using ANN Classifier

Prior to the detection framework either wavelet decomposition or principal component analysis is used to extract the desirable features that would help in identifying a non-seizure signal from a seizure recording. The obtained features are used for classification and decision making. The main goal of this step is to create boundaries between the data, label the classes and maintain a threshold value for the features. In this work PCA has been used for feature extraction and ANN is used as classifier to detect epileptic and normal signals. The performance of the method is shown in Tables below for all types of wavelets that has been used and also for PCA. The optimal neural network obtained is a 4-layered network with 20 neurons in hidden layer and tan-sig transfer function. Figure 1 shows the final neural network obtained for the proposed method. The performance of the proposed method is discussed in detail in result section.

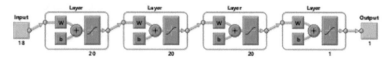

Fig. 1 Final neural network obtained

Table 1 Performance with all channels

Network	Transfer function	Error goal	Accuracy (%)
10-1	Tan-sig	0.1	75.3
20-1	Tan-sig	0.1	78.4
10-10-1	Tan-sig	0.1	80.1
20-10-1	Tan-sig	0.1	83.3
20-20-1	Tan-sig	0.1	85.6

5 Proposed Method

Performance of the proposed method is evaluated under two categories, one using the whole dataset in different configuration of the artificial neural network and the other using principal component analysis. Performance measures used in this work are %accuracy, correct classification rate and misclassification rate. Results of the proposed PCA and ANN based method are discussed in the subsection below.

5.1 Performance with All Channels

Performance of the method is evaluated considering all EEG channels. The results are shown in Table 1 for various network obtained. The accuracy of the proposed method is highest with the network with 20 neurons in hidden layers and tan-sig transfer function. The performance of the method may change after selecting appropriate number of channels which is discussed in next section.

5.2 Performance Varying Channels

Performance of the method is evaluated considering varying EEG channels to select the appropriate number of channels required. The results for various selected channels are shown in Table 2. From Table 2 it can be observed that the method is having highest accuracy with 18 channels. Confusion matrix obtained using 18 channels is shown in Fig. 2. Hence 18 channels are chosen as appropriate using PCA and ANN for epilepsy detection.

Table 2 Performance with varying channels

Channels	Network	Transfer function	Error goal	Accuracy (%)
23	20-20-20-1	Tan-sig	0.1	85.6
22	20-20-20-1	Tan-sig	0.1	86.1
21	20-20-20-1	Tan-sig	0.1	85.7
18	20-20-20-1	Tan-sig	0.1	86.7
12	20-20-20-1	Tan-sig	0.1	83.4
10	20-20-20-1	Tan-sig	0.1	82.7
8	20-20-20-1	Tan-sig	0.1	81.7
6	20-20-20-1	Tan-sig	0.1	80.5
4	20-20-20-1	Tan-sig	0.1	78.5

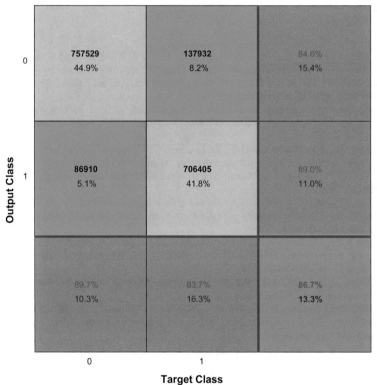

Fig. 2 Confusion matrix obtained

6 Conclusion

In this proposed method we have worked with raw EEG signals and ANN for the purpose of classifying seizure and non-seizure data. The classifier has been used on both the raw signal as well as to data after applying PCA. The results reflect that with the reduction in the number of channels the performance of the classifier surged at a particular number of channels and then gradually decreased. The best accuracy of 86.7% is reached with 18 channels. Therefore, this work gives us future scope of using and exploring different other techniques to compress and reduce dimension with relevant information still intact so that better classification accuracy can be achieved and also a robust method is devised that would have a positive impact in the real-world scenario.

References

1. Aarabi, A., Fazel-Rezai, R., Aghakhani, Y.: A fuzzy rule-based system for epileptic seizure detection in intracranial EEG. Clin. Neurophysiol. **120**(9), 1648–1657 (2009)
2. Abraham, A.: Artificial neural networks. In: Sydenham, P.H. (ed.) Handbook of Measuring System Design. Wiley, New York (2005)
3. Adeli, H., Ghosh-Dastidar, S., Dadmehr, N.: A wavelet-chaos methodology for analysis of EEGs and EEG subbands to detect seizure and epilepsy. IEEE Trans. Biomed. Eng. **54**(2), 205–211 (2007)
4. Bizopoulos, P.A., Tsalikakis, D.G., Tzallas, A.T.: EEG epileptic seizure detection using k-means clustering and marginal spectrum based on ensemble empirical mode decomposition. In: IEEE 13th International Conference on Bioinformatics and Bioengineering (BIBE), Chania, Greece (2013)
5. Gajic, D., Djurovic, Z., Di Gennaro, S., et al.: Classification of EEG signals for detection of epileptic seizures based on wavelets and statistical pattern recognition. Biomed Eng Appl Basis Commun **26**(02) (2014)
6. Giannakakis, G., Sakkalis, V., Pediaditis, M., et al.: Methods for seizure detection and prediction: an overview. In: Modern Electroencephalographic Assessment Techniques, 131–157. Humana Press, New York (2014)
7. Jolliffe, I.T.: Principal component analysis and factor analysis. In: Principal Component Analysis, 2nd edn. Springer, New York (2013)
8. Minasyan, G.R., Chatten, J.B., Chatten, M.J., et al.: Patient-specific early seizure detection from scalp EEG. J. Clin. Neurophys. **27**(3), 163 (official publication of the American Electroencephalographic Society)
9. Nanthini, B.S., Santhi, B.: Seizure detection using SVM classifier on EEG signal. J. Appl. Sci. **14**(14), 1658–1661 (2014)
10. Orosco, L., Correa, A.G., Laciar, E.: A survey of performance and techniques for automatic epilepsy detection. J. Med. Biol. Eng. **33**(6), 526–537 (2013)
11. Parvez, M.Z., Paul, M.: Epileptic seizure detection by analyzing EEG signals using different transformation techniques. Neurocomputing **145**, 190–200 (2014)
12. Physionet CHB-MIT Scalp EEG Database: Boston. https://physionet.org/pn6/chbmit (2010)
13. Ramgopal, S., Thome-Souza, S., Jackson, M., et al.: Seizure detection, seizure prediction, and closed-loop warning systems in epilepsy. Epilepsy Behav. **37**, 291–307 (2014)
14. Rojas, R.: Neural Networks: A Systematic Introduction. Springer Science & Business Media, Netherlands (2013)

15. Saab, M.E., Gotman, J.: A system to detect the onset of epileptic seizures in scalp EEG. Clin. Neurophysiol. **116**(2), 427–442 (2005)
16. Sharmila, A., Geethanjali, P.: DWT based detection of epileptic seizure from EEG signals using naive Bayes and k-NN classifiers. IEEE Access **4**, 7716–7727 (2016)
17. Shoeb, A.H.: Application of machine learning to epileptic seizure onset detection and treatment. Dissertation, Massachusetts Institute of Technology (2009)
18. Smith, L.I.: A Tutorial on Principal Components Analysis. Cornell University, USA (2002)

Visco-Elastic Effects on Nano-fluid Flow in a Rotating System in Presence of Hall Current Effect

Debasish Dey and Ashim Jyoti Baruah

Abstract An unsteady free convective flow of visco-elastic fluid though a porous medium has been investigated in presence of Hall effects. Constitutive equation of visco-elastic fluid is governed by Walters liquid model for short relaxation memories. The medium is rotating with a constant angular velocity. A magnetic field of uniform strength is applied along the transverse direction to the surface. Conservation laws of mass, momentum, energy and species concentration are formed mathematically using suitable approximations. Governing equations of motion are solved analytically using perturbation scheme. Results are discussed graphically and numerically for various values of flow parameters involved in the solution.

Keywords Angular acceleration · Visco-elastic · Prandtl number · Volume fraction · Nano-fluid

1 Introduction

Fluid flow problems governed by visco-elastic fluid model have been paying attention to researchers due to its uses in engineering science, industries and medical science dealing with blood flow problems, etc. The motion of visco-elastic materials under the action of force generates a concept termed as "memory" of fluid. It differs the visco-elastic materials from purely elastic bodies. Walters has derived the constitutive equation of visco-elastic memory fluid flows in [21, 22]. Several authors in their research works [2, 3, 7–11] have used the constitutive equation given by Walters or Oldroyd in the analysis of visco-elastic fluid flow problems through various flow

D. Dey
Department of Mathematics, Dibrugarh University, Dibrugarh 786004, India
e-mail: debasish41092@gmail.com

A. J. Baruah (✉)
Department of Mathematics, Namrup College, Namrup 786623, India
e-mail: ashimjyotibaruah1@gmail.com

© Springer Nature Singapore Pte Ltd. 2019
A. Abraham et al. (eds.), *Emerging Technologies in Data Mining and Information Security*, Advances in Intelligent Systems and Computing 755,
https://doi.org/10.1007/978-981-13-1951-8_52

configurations. Visco-elastic fluid flow through porous medium is used in petroleum technology for flow of oil through porous rocks, in chemical engineering and in drug permeation through human skin.

Inclusion of nano-sized particles in the base fluid enhances the thermal conductivity of fluid. Nano-fluid flow guided by convective heat transfer is used in various technologies related to cooling or heating, solar energy, nuclear reactors etc. The concept of addition of nano-sized solid particles in base fluid was developed by Choi in [1] during the research on cooling technologies.

A significant enhancement in thermal conductivity of ethylene glycol is noticed during the inclusion of cupric oxide is seen in [12]. Growth of heat transfer in the natural convection flow through porous medium has been noticed in [20]. Analysis from work in [13] has concluded that the addition of nano-particles leads to an significant effect on unsteady fluid flow in a rotating system. Eiyad and Abu-Nanda studied the combined effects of viscosity and thermal conductivity on the flow of water-aluminium oxide nano-fluid in [15]. Yasin et al. [23] have investigated flow problems oa various nano-fluids in the presence of volume fraction. Sheikholeslami et al. [18] have studied the hydro-magnetic flow with heat transfer over a stretching sheet in a rotating system. Boundary layer flow of a nano-fluid past a permeable stretching/shrinking sheet has been studied in [14]. Das [4, 6] have analysed flow problems of nano-fluid in a rotating frame. Nano-fluid flow and heat transfer past a vertical stretching sheet in presence of non-uniform heat source or sink was formulated in [5]. Takhar et al. [19] have studied the hydro-magnetic fluid flow over a moving plate in a rotating system with hall current. Hall current effects on hydro-magnetic convective flow of elastico- viscous fluid in a rotating porous channel with thermal radiation have been discussed in [16].

The purpose of this work is to find out the effects of visco-elasticity on memory fluid flow in rotating system containing nano-sized particles with Hall current. Equations are solved analytically using perturbation technique and MATLAB programming has been used to plot the graphs.

2 Mathematical Formulations

We consider an unsteady flow of visco-elastic fluid guided by Walters liquid model for short relaxation memories containing nano-sized particles though a porous medium with Hall effects. The medium is rotating with a constant angular velocity. To stabilize the system, a magnetic field is applied along the transverse direction to the surface. The density difference creates the free convection and using Boussinesq approximation, free convection terms are given by $\frac{\rho}{\rho_{nf}} g \beta_{nf} (T' - T_\infty) + \frac{\rho}{\rho_{nf}} g \beta_{nf}^* (C' - C_\infty)$. The system is rotating with an angular velocity $\vec{\Omega} = (0, 0, \Omega)$. For smaller order of magnitude of angular velocity, the centripetal acceleration $|\mathbf{\Omega x}(\mathbf{\Omega x r})| \sim \Omega^2 d$ is neglected. Let (u', v') be velocity components along x' and y' axes respectively. Following Reddy et al. [17], the effective density, thermal diffusivity, heat capacity, thermal conductiv-

ity, thermal expansion coefficient, electrical conductivity, effective viscosity of nano-fluid is given by $\rho_{nf} = (1 - Q)\rho_f + Q\rho_s$, $\alpha_{nf} = \frac{k_{nf}}{(\rho C_p)_{nf}}$, $(\rho C_p)_{nf} = (1 - Q)(\rho C_p)_f +$
$Q(\rho C_p)_s$, $\frac{k_{nf}}{k_f} = \frac{(k_s + (n_p - 1)k_s) - Q(n_p - 1)(k_f - k_s)}{(k_s + (n_p - 1)k_s) + Q(k_f - k_s)}$, $(\rho\beta)_{nf} = (1 - Q)(\rho\beta)_f + Q(\rho\beta)_s$, $\sigma_{nf} =$
$\sigma_f \left[1 + \frac{3(\sigma - 1)Q}{(\sigma + 2) - (\sigma - 1)Q}\right]$, $\mu_{nf} = (1 - Q)^{-2.5}\mu_f$, Q volume fraction, ρ_f & ρ_s are densities of fluid and nano-particles, C_p specific heat capacity, k_f, k_s are thermal conductivities of fluid and nano-particles respectively, β coefficient of volume expansion, σ_f electrical conductivity of fluid, μ_f viscosity of fluid.

$$\frac{\partial u'}{\partial t'} + w'\frac{\partial u'}{\partial z'} - 2\Omega'v' = \gamma_{nf}\frac{\partial^2 u'}{\partial z'^2} - k_0/\rho_{nf}\left[\frac{\partial^3 u'}{\partial t'\partial z'^2} + w'\frac{\partial^3 u'}{\partial z'^3}\right] + \frac{\rho}{\rho_{nf}}g\beta_{nf}(T' - T_\infty)$$

$$+ \frac{\rho}{\rho_{nf}}g\beta_{nf}^*(C' - C_\infty) - \frac{\mu_f u'}{\rho_{nf}k} + \frac{\sigma_{nf}\beta_0^2(mv' - u')}{\rho_{nf}(1 + m^2)} \tag{1}$$

$$\frac{\partial v'}{\partial t'} + w'\frac{\partial v'}{\partial z'} + 2\Omega'v' = \gamma_{nf}\frac{\partial^2 v'}{\partial z'^2} - k_0/\rho_{nf}\left[\frac{\partial^3 v'}{\partial t'\partial z'^2} + w'\frac{\partial^3 v'}{\partial z'^3}\right] - \frac{\mu_f v'}{\rho_{nf}k} - \frac{\sigma_{nf}\beta_0^2(mu' + v')}{\rho_{nf}(1 + m^2)} \tag{2}$$

$$\frac{\partial T'}{\partial t'} + w'\frac{\partial T'}{\partial z'} = \alpha_{nf}\frac{\partial^2 T'}{\partial z'^2} + \frac{Q}{(\rho c_p)_{nf}}(T' - T_\infty) - \frac{1}{(\rho c_p)_{nf}}\left[\frac{-16T_\alpha^3\sigma}{3K^*} \cdot \frac{\partial^2 T'}{\partial z'^2}\right] \tag{3}$$

$$\frac{\partial C'}{\partial t'} + w'\frac{\partial C'}{\partial z'} = D_B\frac{\partial^2 C'}{\partial z'^2} + k_p(C' - C_\infty) \tag{4}$$

The boundary conditions of the problem are given by Eq. (5) as follows:

$$\left.\begin{array}{c} t' = 0: u' = v' = 0, T' = T_\infty, C' = C_\infty \forall z, \\ t' > 0, z' = 0: u' = U_r\left(1 + \frac{\varepsilon}{2}\left(e^{in't'} + e^{-in't'}\right)\right), v' = 0, -k_f\frac{\partial T'}{\partial z'} = h_f(T_w - T), \\ -D_B\frac{\partial C'}{\partial z'} = h_s(C_w - C') \\ t' > 0, z' \to \infty: u' \to 0, \quad v' \to 0, \quad T' \to T_\infty, \quad C' \to C_\infty \end{array}\right\} \tag{5}$$

3 Method of Solution

We introduce the following non-dimensional variables into the governing Eqs. (1, 2, 3, 4),

$$u = \frac{u'}{U_r}, v = \frac{v'}{U_r}, z = \frac{z'U_r}{\gamma_f}, t = \frac{t'U_r^2}{\gamma_f}, \eta = \frac{n'\gamma_f}{U_r^2}, \theta = \frac{(T' - T_\infty)}{(T_w - T_\infty)}, \Psi = \frac{(C' - C_\infty)}{(C_w - C_\infty)} \tag{6}$$

The dimensionless equations are:

$$\frac{\partial u}{\partial t} - S\frac{\partial u}{\partial z} - Rv = \frac{1}{(1 - \emptyset)^{2.5}}\frac{1}{J_1}\frac{\partial^2 u}{\partial z^2} - e\frac{1}{J_1}\left[\frac{\partial^3 u}{\partial t\partial z^2} - S\frac{\partial^3 u}{\partial z^3}\right] + \frac{g\beta_f\gamma_f}{U_r^3}(T_w - T_\infty)\theta\frac{J_2}{J_1}$$
$$+ \frac{g\beta_f^*\gamma_f}{U_r^3}(C_w - C_\infty)\emptyset\frac{J_3}{J_1} - \frac{1}{k}\frac{1}{J_1}u + \frac{J_4}{J_1}M\frac{(mv - u)}{1 + m^2}, \tag{7}$$

$$\frac{\partial v}{\partial t} - S\frac{\partial v}{\partial z} - Ru = \frac{1}{(1 - \emptyset)^{2.5}}\frac{1}{J_1}\frac{\partial^2 v}{\partial z^2} - e\frac{1}{J_1}\left[\frac{\partial^3 v}{\partial t\partial z^2} - S\frac{\partial^3 v}{\partial z^3}\right] - \frac{1}{k}\frac{1}{J_1}v - \frac{J_4}{J_1}M\frac{(mu + v)}{1 + m^2}, \tag{8}$$

$$J_5\left[\frac{\partial\theta}{\partial t} - S\frac{\partial\theta}{\partial z}\right] = \frac{1}{PrJ_5}\left[\frac{(k_s + (n_p - 1)k_s) - \emptyset(n_p - 1)(k_f - k_s)}{(k_s + (n_p - 1)k_s) + \emptyset(k_f - k_s)}\right]\frac{\partial^2\theta}{\partial z^2} + Q_H\theta + NJ_5\frac{\partial^2\theta}{\partial z^2}, \quad (9)$$

$$\frac{\partial\Psi}{\partial t} - S\frac{\partial\Psi}{\partial z} = \frac{1}{Sc}\frac{\partial^2\Psi}{\partial z^2} + K_r\Psi, \quad (10)$$

To solve (7) and (8), let us define a new function F as (u + iv), and combining them we get

$$\frac{\partial F}{\partial t} - S\frac{\partial F}{\partial z} + iRF = \frac{1}{J_1(1-\emptyset)^{2.5}}\frac{\partial^2 F}{\partial z^2} - \frac{e}{J_1}\left[\frac{\partial^3 F}{\partial t\partial z^2} - S\frac{\partial^3 F}{\partial z^3}\right]$$
$$- \frac{F}{kJ_1} + \frac{J_4 MF}{J_1}[im - 1] + \frac{G_r\theta J_2 + G_m\emptyset J_3}{J_1}, \quad (11)$$

where, the dimensionless parameters appeared in the above Eqs. (7, 8, 9, 10, 11) are given as follows:

$M = \frac{\sigma_f B_0^2 \mu_f}{U_r^2 \rho_f^2}$, Hartmann number, $e = \frac{k_0 U_r^2}{\rho_r \gamma_f}$, visco-elastic parameter, $S = \frac{\omega o}{Ur}$ suction parameter, $N = \frac{16T\alpha 3\sigma}{3k\gamma f}$ radiation parameter, $QH = \frac{Q\gamma f\alpha f}{U_r^2 kf}$ heat source/sink parameter, $R = 2\left(\frac{\Omega\gamma f}{Ur2}\right)$ rotational parameter, $Gr = \frac{g\beta_f\gamma_f}{U_r^3}(T_w - T_\infty)$ Grashoff number for heat transfer, $Gm = \frac{g\beta_f^*\gamma_f}{U_r^3}(C_w - C_\infty)$ Grashoff number for mass transfer.

Dimensionless boundary conditions are given as follows:

$$\left.\begin{array}{l} t = 0 : u = v = 0, \theta = 0, \ \psi = 0\forall z, \\ t > 0, z = 0 : u = \left(1 + \frac{\varepsilon}{2}\left(e^{int} + e^{-int}\right)\right), v = 0, \theta'(z) = -N_c(1-\theta), -N_d(1-\Psi) \\ t > 0, z \to \infty : u \to 0, \quad v \to 0, \quad \theta \to 0, \quad \psi \to 0 \end{array}\right\} \quad (12)$$

To solve (11), (9) and (10), we express F, θ, Ψ (following Purkayastha and Choudhury [16] and Reddy et al. [17]) as

$$F = F_0 + \frac{\varepsilon}{2}\left[e^{int}F_1 + e^{-int}F_2\right], \theta = \theta_0 + \frac{\varepsilon}{2}\left[e^{int}\theta_1 + e^{-int}\theta_2\right], \psi = \Psi_0 + \frac{\varepsilon}{2}\left[e^{int}\Psi_1 + e^{-int}\Psi_2\right] \quad (13)$$

Substituting the above Eqs. (13) in Eqs. (11, 9, 10), and equating the terms containing e^{int}, e^{-int} and constants terms on both sides, we get the following equations.

$$eS\frac{\partial^3 F_0}{\partial^3 z} + B_2\frac{\partial^2 F_0}{\partial^2 z} + SJ_1\frac{\partial F_0}{\partial z} = A_2 F_0 + G_r\frac{J_2}{J_1}\theta_0 + G_m\frac{J_3}{J_1}\Psi_0 \quad (14)$$

$$eS\frac{\partial^3 F_1}{\partial^3 z} + A\frac{\partial^2 F_1}{\partial^2 z} + SJ_1\frac{\partial F_1}{\partial z} = BF_1 + G_r\frac{J_2}{J_1}\theta_1 + G_m\frac{J_3}{J_1}\Psi_1 \quad (15)$$

$$eS\frac{\partial^3 F_2}{\partial^3 z} - L\frac{\partial^2 F_2}{\partial^2 z} + SJ_1\frac{\partial F_2}{\partial z} = MF_2 + G_r\frac{J_2}{J_1}\theta_2 + G_m\frac{J_3}{J_1}\Psi_2 \quad (16)$$

$$B_1\frac{\partial^2\theta_0}{\partial z^2} + SJ_5\frac{\partial\theta_0}{\partial z} + Q_H\theta_0 = 0 \quad (17)$$

$$B_1 \frac{\partial^2 \theta_1}{\partial z^2} + SJ_5 \frac{\partial \theta_1}{\partial z} - A_{11}\theta_1 = 0 \tag{18}$$

$$B_1 \frac{\partial^2 \theta_2}{\partial z^2} + SJ_5 \frac{\partial \theta_2}{\partial z} + B_{11}\theta_2 = 0 \tag{19}$$

$$\frac{\partial^2 \Psi_0}{\partial z^2} + SS_c \frac{\partial \Psi_0}{\partial z} + C_{22}\Psi_0 = 0 \tag{20}$$

$$\frac{\partial^2 \Psi_1}{\partial z^2} + SS_c \frac{\partial \Psi_1}{\partial z} + A_{22}\Psi_1 = 0 \tag{21}$$

$$\frac{\partial^2 \Psi_2}{\partial z^2} + SS_c \frac{\partial \Psi_2}{\partial z} + B_{22}\Psi_2 = 0 \tag{22}$$

For solving the Eqs. (14, 15, 16, 17, 18, 19, 20, 21, 22), we use the following boundary conditions:

$$z = 0 : F_0 = F_1 = F_2 = 1, \theta_0' = -N_c(1 - \theta_0), \theta_1' = N_c\theta_1, \theta_2' = N_c\theta_2,$$
$$\left.\Psi_0' = -N_d(1 - \Psi_0), \Psi_1' = N_d\Psi_1, \Psi_2' = N_d\Psi_2 \right\} \tag{23}$$
$$z \to \infty, F_0 = F_1 = 0, F_2 = \theta_0 = \theta_1 = \theta_2 = \Psi_0 = \Psi_1 = \Psi_2 \to 0$$

Ordinary differential equations are solved using above mentioned boundary conditions (23) and the solutions are given as follows:

$$F = e^{\beta_9 z} + \frac{\varepsilon}{2}\left[e^{int}\left\{(1 - s_4 - s_6)e^{\beta_9 z} + s_4 e^{\beta_3 z} + s_6 e^{\beta_6 z}\right\}\right.$$
$$\left. + e^{-int}\left\{(1 - s_{10} - s_{12})e^{\beta_2 z} + s_{10}e^{\beta_4 z} + s_{12}e^{\beta_7 z}\right\}\right] \tag{24}$$

$$\theta = \frac{\varepsilon}{2}\left[e^{int}e^{\beta_3 z} + e^{-int}e^{\beta_4 z}\right] \tag{25}$$

$$\psi = \frac{\varepsilon}{2}\left[e^{int}e^{\beta_6 z} + e^{-int}e^{\beta_7 z}\right] \tag{26}$$

The constants appeared in the equations or solutions (24, 25, 26) are not given are for the sake of brevity.

4 Results and Discussions

Unsteady free convective flow problem of visco-elastic fluid in rotating system in presence of Hall effects has been studied analytically. Influences of the non-dimensional governing parameters, namely Visco-elastic (e), Volume fraction of Nano-particles (Q), Suction parameter (S), Heat Source parameter (Q_H), Chemical Reaction parameter (K_r), Schmidt number (Sc) and Prandtl number (Pr) on velocity, temperature and concentration profiles have been shown graphically. We have considered Q = 0.15, e = 0.1, S = 2, $Q_H = -5$, $K_r = 0.5$, Sc = 0.6, Pr = 10 for graphical results. MATLAB programming has been used to plot the graphs.

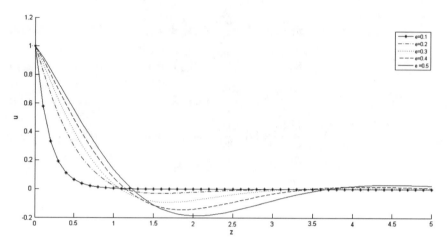

Fig. 1 Velocity profile against z when Q=0.15, S=2, $Q_H = -5$, $K_r = 0.5$, Sc=0.6, Pr=10

Figures 1 and 2 represent the velocity profile against the displacement variable z for various values of visco-elastic parameter (e) and volume fraction of nano-particles (Q). It is revealed that the visco-elasticity has a positive effect on fluid motion. It may be interpreted that during the growth in visco-elasticity, fluid flow gains the momentum and as a consequence speed increases. The volume fraction of nano-particles has a retarding effect on the fluid motion (Fig. 2). Physically, this is because of the fact that fluid flow experiences more obstruction with the increment of volume fraction. This encourages to select the parameters in such a way that the speed should not be very high (motion may be turbulent) or very low (more energies will be required to guide the motion). Fluid experiences the effect of adverse pressure gradient (back flow region) in the region 1<z<3.5 for e (visco-elasticity)>0.1.

Figure 3 shows the influences of volume fraction of nano-particles (Q) on the temperature profile of fluid flow. It is noticed that enhancement of volume fraction reduces the speed of fluid motion. As a result kinetic energy will be reduced due to slow speed. But, to maintain the conservation of energy, temperature should be increased in the form of thermal energy. Figure 4 shows the effect of heat source parameter (Q_H) on the temperature profile of governing fluid motion. It may be interpreted that temperature of fluid is enhanced by external heat agent.

Figures 5 and 6 reveal the influences of Schmidt number (Sc) and Chemical reaction parameter (K_r) on the concentration profile of fluid motion. Schmidt number characterizes the simultaneous effect of momentum and mass diffusions. Growth in Schmidt number reduces the concentration level in fluid motion. Effect of chemical reaction parameter is prominent in viscous dominant region (neighbourhood of the surface) and concentration experience a falling trend with the chemical reaction parameter. Figure 7 shows the effect of suction parameter on the concentration profile of the fluid motion. It can be interpreted that the suction parameter has a negative impact on the concentration of the fluid motion.

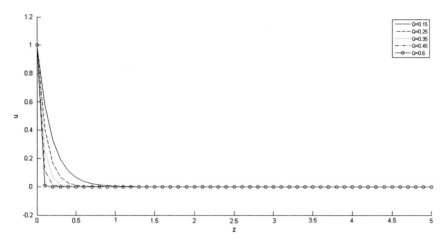

Fig. 2 Velocity profiles against z when e = 0.1, S = 2, $Q_H = -5$, $K_r = 0.5$, Sc = 0.6, Pr = 10

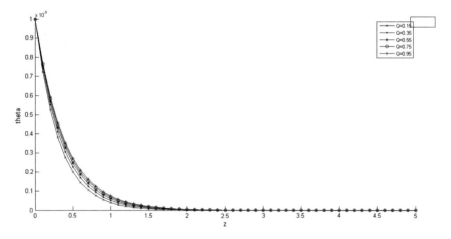

Fig. 3 Temperature profiles against z when e = 0.1, S = 2, Q = 0.15, $K_r = 0.5$, Sc = 0.6, Pr = 10

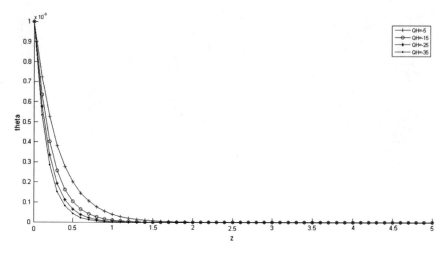

Fig. 4 Temperature profiles against z when e = 0.1, S = 2, $Q_H = -5$, $K_r = 0.5$, Sc = 0.6, Pr = 10

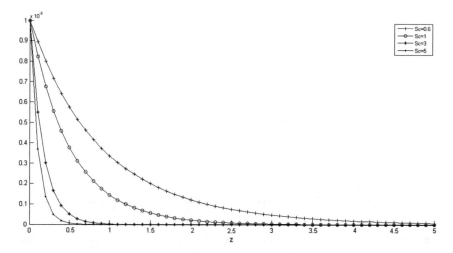

Fig. 5 Concentration profiles against z when Q = 0.15, e = 0.1, S = 2, $Q_H = -5$, $K_r = 0.5$, Pr = 1

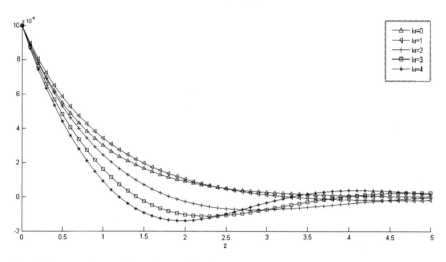

Fig. 6 Concentration profiles against z when Q = 0.15, e = 0.1, $Q_H = -5$, S = 2, Sc = 0.6, Pr = 10

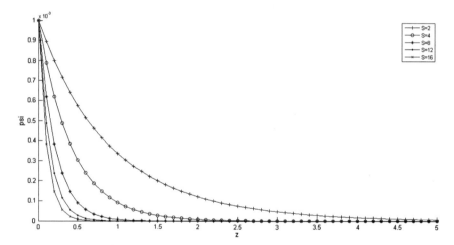

Fig. 7 Concentration profiles against z when Q = 0.15, e = 0.1, $Q_H = -5$, Kr = 0.5, Sc = 0.6, Pr = 10

References

1. Choi, U.S.: Enhancing thermal conductivity of fluids with nanoparticles. In: Proceedings of the ASME International Mechanical Engineering Congress and Expositions, vol. 66, pp. 99–105 (1995)
2. Choudhury, R., Dey, D.: Free convective viscoelastic flow with heat and mass transfer through a porous medium with periodic permeability. Int. J. Heat Mass Transf. **53**, 1666–1672 (2010)
3. Choudhury, R., Dey, D.: Free convective elastico-viscous fluid flow with heat and mass transfer past an inclined plate in slip flow regime. Latin Am. Appl. Res. **42**(4), 327–332 (2012)
4. Das, K.: Flow and heat transfer characteristics of nanofluids in a rotating frame. Alex. Eng. J. **53**, 757–766 (2014)
5. Das, S., Jana, R.N., Makinde, O.D.: MHD boundary layer slip flow and heat transfer of nanofluid past a vertical stretching sheet with non-uniform heat generation/absorption. Int. J. Nano Sci. **13**, 1450019 (2014)
6. Das, S., Mandal, H.K., Jana, R.N., Makinde, O.D.: Magneto-nanofluid flow past an impulsively started porous flat plate in a rotating frame. J. Nanofluid. **4**, 167–175 (2015)
7. Dey, D.: Dusty hydromagnetic oldroyd fluid flow in a horizontal channel with volume fraction and energy dissipation. Int. J. Heat Tech. **34**(3), 415–422 (2016)
8. Dey, D.: Hydromagnetic oldroyd fluid flow past a flat surface with density and electrical conductivity. Latin Am. Appl. Res. **47**(2), 41–45 (2017)
9. Dey, D., Baruah, A.J.: Stratified visco-elastic fluid flow past a flat surface with energy dissipation: an analytical approach. Far East J. Appl. Math. **96**(5), 267–278 (2017)
10. Dey, D., Khound, A.S.: Relaxation and retardation effects on free convective visco-elastic fluid flow past an oscillating plate. Int. J. Comp. Appl. **144**(9), 34–40 (2016)
11. Dey, D., Khound, A.S.: Hall current effects on binary mixture flow of Oldroyd-B fluid through a porous channel. Int. J. Heat Tech. **34**(4), 687–693 (2016)
12. Eastman, J.A., Choi, S.U.S., Yu, W., Thompson, L.J.: Anomalously increased effective thermal conductivities of ethylene glycol based nano fluids containing copper nano particles. Appl. Phys. Lett. **78**, 718–720 (2001)
13. Hamad, M.A.A., Pop, I.: Unsteady MHD free convection flow past a vertical permeable flat plate in a rotating frame of reference with constant heat source in a nano fluid. Heat Mass Transf. **47**, 1517–1524 (2011)
14. Khairy, Z., Ishak, A., Pop, I.: Boundary layer flow and heat transfer over a nonlinearly permeable stretching/shrinking sheet in a nanofluid. Sci. Rep. 4 (2014) http://dx.doi.org/10.1038/srep04 404
15. Nada, E.A.: Effects of variable viscosity and thermal conductivity of Al2O3–water nanofluid on heat transfer enhancement in natural convection. Int. J. Heat Fluid Flow **30**, 668–679 (2009)
16. Purkayastha, S., Choudhury, R.: Hall current and thermal radiation effect on mhd convective flow of an elastico- viscous fluid in a rotating porous channel. WSEAS Trans. Appl. Theor. Mech. **9**, 196–205 (2014)
17. Reddy, J.V.R., Sugunamma, V., Sandeep, N., Sulochana, C.: Influence of chemical reaction, radiation and rotation on mhd nanofluid flow past a permeable flat plate in porous medium. J. Niger. Math. Soc. **35**, 48–65 (2016)
18. Sheikholeslami, M., Ashorynejad, H.R., Domairry, G., Hashim, I.: Flow and heat transfer of Cu–Water nanofluid between stretching sheet and a porous surface in rotating system. J. Appl. Math. 421320 (2012)
19. Takhar, H.S., Chamkha, A.J., Nath, G.: MHD flow over a moving plate in a rotating fluid with magnetic field, hall currents and free stream velocity. Int. J. Eng. Sci. **40**, 1511–1527 (2002)

20. Uddin, Z., Harmand, S.: Natural convection heat transfer flow of nano fluids along a vertical plate embedded in porous medium. Nano scale Res Lett **8**, 64–73 (2013)
21. Walters, K.: The motion of an elastico-viscous liquid contained between co-axial cylinders. Q. J. Mech. Appl. Math. **13**, 444–461 (1960)
22. Walters, K.: The solutions of flow problems in the case of materials with memories. J. Mech. **1**, 473–478 (1962)
23. Yasin, M.H.M., Arifin, N.M., Nazar, R., Ismail, F., Pop, I.: Mixed convection boundary layer flow embedded in thermally stratified porous medium saturated by a nano fluid. Adv. Mech. Eng. 121943 (2013) http://dx.doi.org/10.1155/2013/121943

RNA Structure Prediction Using Chemical Reaction Optimization

Md. Rayhanul Kabir, Fatema Tuz Zahra and Md. Rafiqul Islam

Abstract RNA structure prediction (RSP) is an NP-complete problem which has been implemented by various algorithms, e.g., exact and metaheuristics. Here, we have proposed a method to solve RSP problem, which is founded on chemical reaction optimization (CRO), a metaheuristic approach. At first, we have developed a method to generate the solution, and the next four reaction operators of CRO are redesigned and implemented to expand the population all over the solution space. To find the best results from solution space, the values of the fundamental parameters of CRO have assigned accurately. The minimum free energy of RNA structure is computed through INN-HB model to evaluate the fitness of structure. Ten different RNA sequences have been taken for the experiment, and the experimental results of the proposed method are compared with three previous works, RNAPredict, SARNA-Predict, and COIN, to verify the performance of the proposed method.

Keywords Chemical reaction optimization · RNA structure prediction
Ribonucleotide sequences · Minimum free energy

1 Introduction

Ribonucleic acid (RNA) is a macromolecule which works as a mediator between protein synthesis and DNA. This single-stranded macromolecule has four nucleotides: guanine, cytosine, adenine, and uracil. The RNA has the capacity to bend backward and build hydrogen bond between the corresponding nucleotides. There are two types

Md. R. Kabir (✉) · F. T. Zahra · Md. R. Islam
Computer Science and Engineering Discipline, Khulna University,
Khulna 9208, Bangladesh
e-mail: nabid.ku.cse@gmail.com

F. T. Zahra
e-mail: fatema130210@gmail.com

Md. R. Islam
e-mail: dmri1978@gmail.com

© Springer Nature Singapore Pte Ltd. 2019
A. Abraham et al. (eds.), *Emerging Technologies in Data Mining and Information Security*, Advances in Intelligent Systems and Computing 755,
https://doi.org/10.1007/978-981-13-1951-8_53

of base pairs: A-U, G-C and temporary wobble base pair: G-U can be produced by the folding of RNA. The primary RNA structure is the chain of nucleotides or bases and representation of a set of base pairs called secondary structure. The RNA represented in a 3-D way is known as a tertiary structure. In recent times, RNA secondary structure prediction has grown into one of the focal points among the researchers. The problem RSP is to construct potential stems and takes up some of the stems to form a structure of RNA. The challenge of the problem is to find out the more stable structure among all the possible structures. Several methods were introduced to solve this problem. Tough and expensive processes are also applied to predict RNA structure such as NMR and X-ray crystallography [1]. Due to the disadvantages of these processes, it is obvious to implement an easy and computational process to predict RNA structure. The RSP problem is the technique of predicting the RNA secondary structure from a particular nucleotide sequence. If an RNA sequence is $X = x_1 x_2 \ldots x_n$, X is a set of letters, which consists of letters from $\sum = \{a, u, g, c\}$. A pair $\{a, b\}$ is canonical base pair if $\{a, b\} = \{a, u\}$ or $\{u, a\}$ and $\{a, b\} = \{g, c\}$ or $\{c, g\}$. Other pairs from the alphabet like $\{a, g\}$, $\{a, c\}$, $\{c, u\}$ are not said to be canonical base pairs [2]. The target of the proposed work is to search the possible stacking base pairs as well as the helices from a primary sequence and after that selecting the more stable structure by evaluating the structure. The stability of the structure can be measured by calculating minimum free energy. The structure with the lowest free energy is said to be the most steady and stable structure. In this method, we have built an objective function (shown in Eq. 1) which is based on individual nearest-neighbor hydrogen bond model (INN-HB), a group of thermodynamic models [3]. The free energy of every helix is calculated by Eq. (2). In Eq. (2), ΔG°_{37init} is used for the entropy loss when the initiation occurs and the first base pair is formed. If it is a self-complementary duplex, $\Delta G^\circ_{37sym} = 0.43$ kcal/mol; otherwise, none is applied. $\sum \left[\Delta G^\circ_{37NN} \right]$ means the total summation of all base pairs shaped by a helix reading from the left side to the right. When a single AU or GU pair is found at the tail of a helix, $\Delta G^\circ_{37AU/GUend}(per\, AU/GU\, end) + \Delta G^\circ_{37sym}$ is applied once.

$$F = min\{\Delta G_k\};\ 1 \leq k \leq m;\ m = no\ of\ secondary\ structure \\ for\ one\ sequence \tag{1}$$

$$\Delta G^\circ_{37} = \Delta G^\circ_{37init} + \sum \left[\Delta G^\circ_{37NN} \right] + \Delta G^\circ_{37AU/GUend}(per\, AU/GU\, end) \\ + \Delta G^\circ_{37sym} \tag{2}$$

Many methods were used to determine RNA structure. DP-based method mfold [4] was implemented to solve RSP problem without pseudoknots. A pseudoknot in RNA has two loops of the stem, where half portion of a stem is crossed with another stem. It is a tough task to compute the structure with pseudoknot, and it is also tedious [2]. Ant colony optimization [5] was also implemented to find out the solution to this problem in which all the stems are searched by brute-force algorithm. The method shows that it is good for the shorter RNA sequences. In this paper, we have used CRO (Chemical reaction optimization) algorithm for this optimization problem. Over recent years, CRO has become very popular by solving various optimization problems, such as cognitive radio spectrum allocation problem [6], quadratic assignment problem [7], and network coding optimization [8]. For the implementation, the four reaction operators are reconstructed. These operators find the global optimal points among the solutions. For the calculation of free energy, we followed the INN-HB model which is convenient to calculate. For testing, we have taken ten shorter and longer RNA sequences whose detailed information is also given. Prediction accuracy is shown by computing the predicted base pairs which are true positive (TP), the number of predicted base pairs which are false negative (FN), the number of predicted base pairs which are false positive (FP), sensitivity, specificity, and f-measure. The experiment results are also compared with three methods: RNAPredict (GA) [9], SARNA-Predict (SA) [10], and COIN [11]. From the comparison, it is proved that RNA structure prediction using CRO is better for all sequences than the mentioned three methods.

2   Related Works

Mfold, a mechanism for RNA and DNA secondary structure prediction, was implemented by Zuker [4]. It is a web server that estimates multiple foldings of RNA and DNA sequences. The server was established on dynamic programming employing thermodynamic methods. The primary sequence along with sequence name is entered as input in the server. Some constraint information can also be entered which is optional and used if any user wants to force or prevent any base pair to build. Some folding parameters are initialized at the starting of the process and the temperature of folding is fixed at 37 °C. After the submission of the input, the output is presented in energy dot plot matrix [10]. In the upper triangular part of the matrix, each dot represents base pair in that row and column. To differing the suboptimality, particular colors are used. The red dots describe the optimal base pairs. Besides, black dots are the least possible base pairs. The free energy of the structure can be calculated at the structure display of the server.

RNAPredict was invented by Wiese [9] which is established by genetic algorithm. First, the evaluation of two thermodynamics model has been shown in RNAPredict, Individual Nearest-Neighbor (INN), and Individual Nearest-Neighbor Hydrogen Bond (INN-HB). RNAPredict was implemented by the standard generational evolutionary algorithm (EA) in which at every generation the population of solutions has relatively lowest free energy than earlier generation. The algorithm activates by initializing population and ends when it meets the stopping condition, the number of generation. To create a population, at first, all feasible base pairs are traced and then by stacking the base pairs all helices are built and added to a set of helices. In every generation, crossover occurs in which parts of two parent solutions are made up to create child solution. Mutation does the random changes in population. And there is a fitness function to score the solution to make a decision to choose good solutions and omit the others. For the experiment, RNAPredict took 19 RNA sequences. They calculated the total of true positive (TP), false negative (FN), and false positive (FP) base pairs. The sensitivity, specificity, and f-measure are also estimated. They compared the results of these parameters with DP-based mfold [4] to show the accuracy of the structure prediction. They have shown that RNAPredict performs better than mfold, and when comparing with the suboptimal structure built by mfold, it shows similar performance.

Simulated annealing-based method SARNA-Predict was introduced by Schmitz and Seger [12], which was established on the concept of the heating element and cooling down until it minimizes its faults. In RNA prediction, when the temperature is high, more mutations occur as well as more structures are received. Whereas the temperature falls down, the lesser mutation occurs and the secondary structure merges into low-energy structure. This method is configured by some design elements. One of them is state presentation that puts the parameters into code. Comparison occurs by mutation function. Another element is the evaluation function which follows the INN-HB model. Taking or omitting new structure depends on the decision mechanism. Annealing mentions the method in which the temperature falls down in time. Geometric annealing schedule and adaptive annealing schedule were implemented in SARNA-Predict. For testifying, 13 RNA sequences were used. They have calculated TP, FP, FN, specificity, sensitivity, and f-measure to evaluate the predicted structure. The algorithm selects inversion mutation operator over swap mutation operator while using the geometric annealing.

Evolutionary algorithm-based coincidence algorithm was used in RNA structure prediction by Srikamdee [11]. For the generation of the helices, the method first searches for all feasible canonical base pair for a distinct RNA sequence. By a probability matrix, all the possible helices are calculated and added to a set. The information of every helix is recorded. After generation the initial solutions, the method calculates the minimum free energy by the thermodynamic model (INN-HB). When the evaluation of the solution is completed by the thermodynamic model, the algorithm selects the candidate solutions. After that, the generation matrix is renewed by the recorded information of good solutions. The algorithm took ten RNA sequences for the experiment and they estimated the value of the TP, FP, FN, specificity, sensitivity, and F-measure and compared the results with the SARNA-Predict [10] and RNAPredict [9]. From the experimental data, it is shown that it gives better output than the other two methods.

3 Proposed Method Using Chemical Reaction Optimization

To solve the RNA structure prediction problem, we have proposed an algorithm based on the characteristics of chemical reaction, chemical reaction optimization (CRO) algorithm. Both local search and the global search can be done in CRO. The main convenience of the chemical reaction optimization algorithm is the high flexibility of designing the reaction operators which helps CRO to suit for any optimization problem not depending on the complexity. There are three independent stages in the functionality of chemical reaction optimization algorithm. These are initialization, iteration, and final stage [9]. We generate a certain number of population for the RNA structure prediction problem in the opening stage. There are some parameters needed to be initialized at this stage. These are InitialKE, two threshold limits (α and β), KELossRate, MoleColl, and POPSize. A chemical reaction can be two different types. Either it is unimolecular reaction or intermolecular reaction. Synthesis and intermolecular ineffective collision are for the intermolecular reaction. On the other hand, both decomposition and on-wall ineffective collision are for the unimolecular reaction. Two threshold limits (α and β) determine the selection criteria of decomposition and synthesis operator. For every iteration, only single reaction operator is chosen among them in the iteration stage. Selecting criteria between unimolecular and intermolecular is done by comparing the MoleColl with a random value (t). After completing each iteration, we check the termination condition and if it fulfills, the iteration stage ends. After ending of the iteration stage, we have the desired solution, which is the best among the population.

3.1 Solution Generation

Solution generation is the first stage of chemical reaction optimization problem. The primary structure of an RNA sequence is the input of this stage. At first, we take a matrix. A specific row and column are assigned with the input RNA sequence where each index contains a nucleotide of the given RNA sequence. Then, we fill the upper triangular indices of the matrix with the value 1 if it creates a Watson–Crick base pair (AU, GC) or Wobble pair (GU). Otherwise, we skip that index. We do this only on the upper triangular matrix to avoid the repetition. When the assigning of 1s is completed, we check for the diagonals with 1s in the matrix. In Fig. 1, the diagonals with 1s are showed using ellipses. If the number of 1s is more than two along with a particular diagonal, we track the diagonal using ellipse and store this in the information table. In the information table, we have the information of all helices along with its top index, bottom index, and count. At the end of this task, we have the indices of stem list. We store different solutions by doing permutation of the indices. A "molecule" class with some attributes is created here. NumHit, ω, PE, and KE are the attributes which are initialized here. Using Eq. (2), we calculate the PE of every solution.

3.2 Reaction Operators

We have designed four reaction operators for our proposed method which are described below.

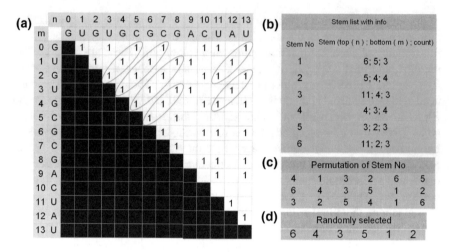

Fig. 1 **a** Matrix that represents the primary sequence in a distinct row and column with index n and m from 0 to 13. **b** Stem list with info table consisting of every stem information. **c** Some values of permutation of the stem no. found on the info table. **d** A randomly selected solution

Fig. 2 On-wall ineffective collision

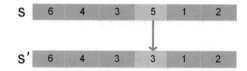

3.2.1 On-Wall Ineffective Collision

On-wall ineffective reaction is applied to next door neighbor search. In this collision, the resultant solution is slightly different from the initial solution and the chemical structure of the new solution remains the same. At first, a solution is randomly selected. In Fig. 2, s is the solution and s′ is the new solution generated from s. A random integer value is chosen within the solution length and one random position in the solution s is selected. The randomly chosen integer value is summed with the value of the selected position of solution s and added to the new solution s′, if it is less than or equal to the solution length. Otherwise, the selected integer value is directly added to the new solution. Rest of the indices of s′ are filled with the corresponding value of s.

3.2.2 Decomposition

In decomposition, the reaction helps to find the solution from a different arena of the search space. There happens a huge diversity in the solution. Here, a solution is chosen randomly. Then this solution is divided into two equal parts. In Fig. 3, the

Fig. 3 Decomposition

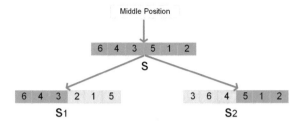

solution s is chosen arbitrarily. First bisected part of the solution s is assigned directly to the first half of the new solution s1 and second bisected part of the solution s is assigned directly to the second half of the new solution s2. Rest of the indices of s1 and s2 are filled with the random values between 1 to the length of s.

3.2.3 Intermolecular Ineffective Collision

In this reaction operator, two solutions are chosen randomly to generate two different solutions. In Fig. 4, two solutions s1 and s2 are selected randomly from the population. Two points are selected dividing the solution s1 and s2 into three different parts. Now, the odd parts of s1 and the even part of s2 create a new solution s1′. On the other hand, the odd parts of s2 and the even part of s1 create a new solution s2′.

3.2.4 Synthesis

Synthesis is another reaction operator when the reaction is intermolecular. Synthesis operator is just the opposite of decomposition. It creates a single solution from two distinct solutions. In Fig. 5, two solutions s1 and s2 are chosen randomly to create a new solution s′. Selection of two distinct solutions randomly is the first task of this operator. The second task is to take a random value between 0 and 1. If the value is

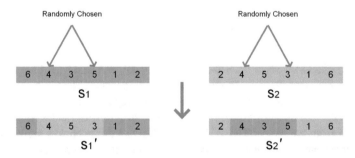

Fig. 4 Intermolecular ineffective collision

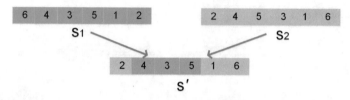

Fig. 5 Synthesis

Table 1 Parameters of CRO
for RSP problems with their
initial values

Symbol	Value
POPSize	200
KELossRate	0.8
MoleColl	0.3
InitialKE	4
α	1
β	8
Buffer	0
NumHit	0

more than 0.5, s2 is chosen to take the value and add to the new solution s′. Otherwise,
s1 is chosen to add the value to the new solution s′.

3.3 Parameter Settings

The effectiveness and efficiency of the CRO algorithm depend on the parameters.
These parameters help to find the desired best solution from the search space. These
parameters are needed to be initialized with proper values. In Table 1, we have shown
the parameters along with their initial value.

4 Experimental Results

The results of RNAPredict [9], SARNA-Predict [10], and COIN [11] are compared
with our proposed method, RNA structure prediction using CRO algorithm. We use
GA, SA, COIN, and CRO in Tables 3, 4, and Fig. 6 to represent RNAPredict, SARNA-
Predict, COIN and our proposed method, respectively. Ten different RNA sequences
were taken from the RNA STRAND v2.0 [13] to test our proposed method. These
ten sequences were collected from the website link (http://www.rnasoft.ca/strand).
In Table 1, parameters of CRO are given with the symbols and their corresponding
values.

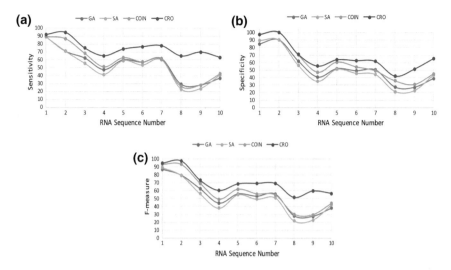

Fig. 6 Comparison of the experimental results of the proposed method with RNAPredict (GA) [9], SARNA-Predict (SA) [10], and COIN [11] in terms of **a** Sensitivity, **b** specificity, and **c** F-measure, respectively

Every RNA sequence detail is shown in Table 2 where the first column is the sequence number which is maintained properly on the other tables, the second column contains the RNA sequence name, the third column contains the RNA class, the fourth column is the accession number of the RNA sequence, the length of each sequence is given in the fifth column, and the number of base pairs of each sequence is given in the last column. The anticipated structure accuracy is obtained by comparing the predicted structure with the known structure. FP, FN, TP, sensitivity, specificity, and F-measure are the parameters for the performance measurement. TP is the predicted base pair which is true positive, FN is the predicted base pair which is false negative, and FP is the predicted base pair which is false positive.

In Table 3, we show the comparison of GA, SA, and COIN with our proposed method in terms of TP, FP, and FN. For every sequence, CRO shows better results in terms of TP and FN. For some sequences, the result of FP of CRO is slightly greater than COIN. Since the values of FN of CRO are less for all sequences than the three methods, the structures are more proper and error free. As the prediction of true base pair of CRO is more accurate than other methods, the prediction of false negative base pair of CRO is less than other methods. In Table 4, we show the comparison of GA, SA, and COIN with our proposed method in terms of sensitivity, specificity, and F-measure. In both tables, the best results are shown in bold. Sensitivity is the ratio of true positive base pair predicted and the total positive base pair in the known structure. So, the higher sensitivity means the higher accuracy of prediction. Specificity is the ratio of true positive base pair predicted and the summation of true positive base pair predicted and false positive base pair predicted. That means, if the availability of false positive base pair in the prediction is null, the accuracy of

Table 2 RNA sequence detail taken from the RNA STRAND V2.0

Seq No.	Sequence	RNA Class	Accession No.	Length (nt.)	#Base pair
1	S. cerevisiae	5s rRNA	X67579	118	37
2	H. marismortui	5s rRNA	AF034620	122	38
3	M. anisopliae(3)	Group I intron, 23S RNA	AF197120	394	120
4	M. anisopliae(2)	Group I intron, 23S RNA	AF197122	456	115
5	A. lagunensis	Group I intron, 16S RNA	U40258	468	113
6	H. rubra	Group I intron, 16S RNA	L19345	543	141
7	A. griffini	Group I intron, 16S RNA	U02540	556	131
8	C. elegans	16S RNA	X54252	697	189
9	D. virilis	16S RNA	X05914	784	233
10	X. laevis	16S RNA	M27605	945	251

Table 3 The proposed method compared with the RNAPredict (GA) [9], SARNA-Predict (SA) [10], and COIN [11] method in terms of TP, FP, and FN

Seq No.	TP				FP				FN			
	GA	SA	COIN	CRO	GA	SA	COIN	CRO	GA	SA	COIN	CRO
1	33	33	33	**34**	6	4	**1**	1	4	4	4	**3**
2	27	27	33	**36**	3	3	**0**	0	11	11	5	**1**
3	75	67	82	**90**	46	52	**35**	37	45	53	38	**30**
4	55	48	59	**75**	80	90	66	**60**	60	67	56	**40**
5	68	67	71	**83**	63	64	**46**	47	45	46	42	**30**
6	79	74	81	**108**	82	88	68	**64**	59	64	60	**33**
7	81	79	80	**102**	80	100	83	**63**	50	52	51	**29**
8	55	43	50	**123**	147	160	**91**	167	134	146	139	**66**
9	65	55	66	**163**	177	191	**149**	152	168	178	167	**70**
10	93	103	107	**161**	147	133	**129**	153	158	148	144	**93**
Avg	63.1	59.6	66.2	**97.5**	83.1	88.5	**66.8**	74.4	73.4	76.9	70.6	**39.5**

Table 4 The proposed method compared with the RNAPredict (GA) [9], SARNA-Predict (SA) [10], and COIN [11] method in terms of sensitivity, specificity, and F-measure

Seq No.	Sensitivity				Specificity				F-measure			
	GA	SA	COIN	CRO	GA	SA	COIN	CRO	GA	SA	COIN	CRO
1	89.2	89.2	89.2	**91.9**	84.6	89.2	**97.1**	**97.1**	86.8	89.2	93.0	**94.4**
2	71.1	71.1	86.8	**94.7**	90.0	90.0	**100**	**100**	79.4	79.4	93.0	**97.3**
3	62.5	55.8	68.3	**75**	62.0	56.3	70.1	**70.9**	62.2	56.1	69.2	**72.9**
4	47.8	41.7	51.3	**65.2**	40.7	34.8	47.2	**55.6**	44.0	37.9	49.2	**60**
5	60.2	59.3	62.8	**73.5**	51.9	51.1	60.7	**63.9**	55.7	54.9	61.7	**68.3**
6	57.2	53.6	57.4	**76.6**	49.1	45.7	54.4	**62.8**	52.8	49.3	55.9	**69**
7	61.8	60.3	61.1	**77.9**	50.3	44.1	49.1	**61.8**	55.5	51.0	54.4	**68.9**
8	29.1	22.8	26.5	**65.0**	27.2	21.2	35.5	**42.4**	28.1	21.9	30.3	**51.3**
9	27.9	23.6	28.3	**69.9**	26.9	22.4	30.7	**51.7**	27.4	23.0	29.5	**59.4**
10	37.1	41.0	42.6	**63.3**	38.8	43.6	45.3	**51.2**	37.9	42.3	43.9	**56.6**
Avg	54.4	51.8	57.4	**75.3**	52.1	49.8	58.9	**65.7**	52.9	50.5	58.0	**69.8**

prediction is maximum. It also shows that the accuracy of prediction is proportional to the specificity. F-measure is the ratio of two times multiplication of sensitivity, specificity, and the summation of specificity and sensitivity. As there exist both specificity and sensitivity in the term F-measure, so, higher F-measure means higher specificity and sensitivity likewise higher accuracy. Our method, CRO outperforms all method for every sequence in terms of sensitivity and F-measure. In Table 4, we have also calculated the average of the results of sensitivity, specificity, and F-measure. From the averages, it is verified that CRO is better than others. Since, the accuracy of prediction is proportional to the sensitivity, specificity, and F-measure, we can say that CRO predicts the best structures than all other methods for all cases. Figure 6 shows the line graph representing the comparison of the results of the sensitivity, specificity, and F-measure of the proposed method (represented by CRO) with three methods GA, SA, and COIN. Here, in all figure (a), (b), and (c) the x-axis represents the sequence number of RNA which is given sequentially in Table 2. The y-axis of Figures (a), (b), and (c) represents sensitivity, specificity, and F-measure, respectively, measured in percentage. All the three graphs prove the excellent performance of the proposed method in predicting RNA structure. Here, we have examined our proposed method with only ten sequences. In the future, we will apply our proposed method to more long and short sequences.

5 Conclusion

RNA has a significant part in molecular biology, its structure prediction is essential in fulfilling the role. Due to the significance of RNA, the structure prediction has attained much popularity among the researchers. Here, we applied chemical reaction optimization to predict the structure of RNA. CRO has successfully solved various optimization problems. It is also proven that the proposed work gives more stable RNA structure because of the CRO algorithm. It outperforms effectively than all other methods described in this paper. The most challenging part of the implementation was to design the reaction operators. However, it was completed efficiently and works perfectly. Our future attempts aim at the accuracy of the result for both short and long sequences.

References

1. Wang, G., Zhang, W., Ning, Q., Chen, H.l.: A novel framework based on ACO and PSO for RNA secondary structure prediction. Math. Probl. Eng. (2013)
2. Mizuno, H., Sundaralingam, M.: Stacking of crick wobble pair and Watson-Crick pair: stability rules of gu pairs at ends of helical stems in trnas and the relation to codon-anticodon wobble interaction. Nucleic Acids Res. 5(11), 4451–4462 (1978)
3. Neethling, M., Engelbrecht, A.P.: Determining RNA secondary structure using set-based particle swarm optimization. In: IEEE Congress on Evolutionary Computation, 2006. CEC 2006, pp. 1670–1677. IEEE (2006)
4. Zuker, M.: Mfold web server for nucleic acid folding and hybridization prediction. Nucleic Acids Res. 31(13), 3406–3415 (2003)
5. McMellan, N.: RNA secondary structure prediction using ant colony optimisation. Master thesis, University of Edinburgh, School of Informatics (2006)
6. Lam, A.Y., Li, V.O.: Chemical reaction optimization for cognitive radio spectrum allocation. In: Global Telecommunications Conference (GLOBECOM 2010), 2010 IEEE, pp. 1–5. IEEE (2010)
7. Lam, A.Y., Li, V.O.: Chemical-reaction-inspired metaheuristic for optimization. IEEE Trans. Evol. Comput. 14(3), 381–399 (2010)
8. Pan, B., Lam, A.Y., Li, V.O.: Network coding optimization based on chemical reaction optimization. In: Global Telecommunications Conference (GLOBECOM 2011), 2011 IEEE, pp. 1–5. IEEE (2011)
9. Wiese, K.C., Deschenes, A.A., Hendriks, A.G.: Rnapredictan evolutionary algorithm for RNA secondary structure prediction. IEEE/ACM Trans. Comput. Biol. Bioinform. (TCBB) 5(1), 25–41 (2008)
10. Grypma, P., Tsang, H.H.: SARNA-predict: using adaptive annealing schedule and inversion mutation operator for RNA secondary structure prediction. In: 2014 IEEE Symposium on Computational Intelligence in Multi-Criteria Decision-Making (MCDM), pp. 150–156. IEEE (2014)
11. Srikamdee, S., Wattanapornprom, W., Chongstitvatana, P.: RNA secondary structure prediction with coincidence algorithm. In: 2016 16th International Symposium on Communications and Information Technologies (ISCIT), pp. 686–690. IEEE (2016)
12. Schmitz, M., Steger, G.: Description of RNA folding by "simulated annealing". J. Mol. Biol. 255(1), 254–266 (1996)
13. Andronescu, M., Bereg, V., Hoos, H.H., Condon, A.: RNA strand: the RNA secondary structure and statistical analysis database. BMC Bioinform. 9(1), 340 (2008)

Mapping of Flow Visualization and Heat Transfer Analysis Over Roughened Plate Inside Rectangular Duct

Anup Kumar and Apurba Layek

Abstract In recent decades, liquid crystals thermography (LCT) technique which is optical and inexpensive technique for visualizing surface temperatures distribution and measuring heat transfer coefficients. The optical technique is build with the help of thermo-chromic liquid crystals enforced to the test surface. The main objective of this study is to visualize temperature fields through image processing for certain set of working parameters and to study of heat transfer characteristics with turbulators mounted over flat plate in a rectangular channel. The working geometric parameter for the roughened rectangular air duct are angle of attack of fluid flow (α) with the range of 30°–90°, relative pitch ratio (P/e) with range of 7–9, relative twist ratio (Y/e) with of range 3–5 and Reynolds number (Re) with the range of 14000–21000. Predictions of Nusselt number with roughened surface is compared with the smooth duct with same condition.

Keywords Solar air heater · Heat transfer · Liquid crystal thermography
Image processing

Nomenclature

L	Length of the duct, m
W	Width of the duct, m
D	Depth of the duct, m
P/e	Relative pitch ratio (dimensionless)
Y/e	Relative twist ratio (dimensionless)
α	Angle of attack (degrees)

A. Kumar (✉) · A. Layek
Department of Mechanical Engineering, National Institute
of Technology Durgapur, Durgapur 713209, West Bengal, India
e-mail: anupkr335@gmail.com

A. Layek
e-mail: apurba_layek@yahoo.co.in

© Springer Nature Singapore Pte Ltd. 2019
A. Abraham et al. (eds.), *Emerging Technologies in Data Mining and Information Security*, Advances in Intelligent Systems and Computing 755,
https://doi.org/10.1007/978-981-13-1951-8_54

H   Hue (dimensionless)
T_{fm} Mean temperature of air (°C)
T_{LCT} Temperature of LCT sheet/absorber plate (°C)
D_h Hydraulic diameter (m)
k   Thermal conductivity of air (W/m°C)
V   Wind velocity of air (m/sec)
υ Kinematic viscosity (m^2/s)
h   Convective heat transfer coefficient (W/m^2°C)
Q   Convective heat flux (W)
Nu   Nusselt number (dimensionless)
Re   Reynolds number (dimensionless)

1   Introduction

In last four decades, liquid crystals thermo-chromic technique (LCT) have been successfully applied in heat transfer, flow visualization and fluid mechanics studies. Liquid crystal sheets consist of thin coatings of some cholesteric and chiral-nematic liquid which are utilized to obtain definite colours at specific temperatures for steady or transient process [1, 2]. Liquid crystal thermography technique allows instantaneous measurement of velocity fields and temperature profile.

In recent years, several researchers have used Liquid Crystal Thermography for various purposes. Different researchers such as Copper et al. [1], Baughn et al. [2] have implemented LCT for convective heat transfer problem and Lee et al. [3] studied behaviour and structure of jet impingement cooling using LCT. Detailed analysis of twelve different shaped vortex generators in a square duct and investigation of boiling heat transfer are found in the work of Liou et al. [4] and Kenning [5] respectively. Tanda [6], Gao et al. [7] and Cavallero [8] studied the effect of heat transfer and pressure drop in a rectangular channel with turbolators. Tariq et al. [9] had experimentally examined heat transfer distribution over flat and ribbed surface by liquid crystal thermography technique. Malay et al. [10] had tested liquid crystal technique to analyzed convective heat transfer coefficient in transient state.

In the present study, flow visualization techniques allow an automatic quantification of surface temperature distribution which is need to be analyzed and examined. Here, experimental work is carried with selected problems for different set of roughness geometric configurations. The experimental procedure includes modelling of different geometric parameters which are angle of attack (α), relative pitch ratio (P/e), relative twist ratio (Y/e) and flow operating parameter termed as Reynolds number.

2 Experimental Set-up

2.1 Rib Geometry

The general geometry of the twisted rib roughness on absorber plate is shown in Fig. 1. The parameters are been expressed in the form of the following dimensionless roughness parameters and their range are given in Table 1., and 'e' is the height of the rib.

2.2 Experimental Set-up

The schematic diagram of experimental set-up is shown in Fig. 2. The rectangular channel test section is designed according to ASHARE Standard [11]. It consists of heater, entry section, test section, exit section, blower, illuminating system, Camera, data logger and computer. The length of inlet and exit section are consisting minimum length of $5\sqrt{WD}$ and $2.5\sqrt{WD}$ respectively and the length of test section test is 20 times of hydraulic diameter [11]. A uniform heat flux is supplied to 3 mm Aluminium plate placed inside the test section and the plate temperature is monitored by high-accuracy thermocouples connected to data logger.

Fig. 1 Position of LCT Sheet and Rib

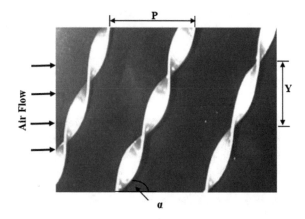

Table 1 Desire set of geometric parameters

Parameters/Factors	Levels		
Relative pitch ratio	7	8	9
Angle of attack	30°	60°	90°
Relative twist ratio	3	4	5
Reynolds number	14000	17500	21000

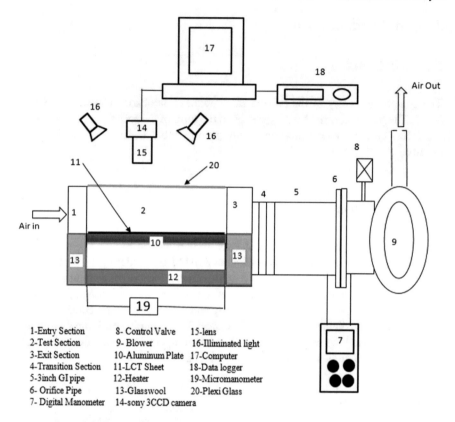

Fig. 2 Schematic Diagram of experimental set-up

1-Entry Section
2-Test Section
3-Exit Section
4-Transition Section
5-3inch GI pipe
6- Orifice Pipe
7- Digital Manometer

8- Control Valve
9- Blower
10-Aluminum Plate
11-LCT Sheet
12-Heater
13-Glasswool
14-sony 3CCD camera

15-lens
16-Illuminated light
17-Computer
18-Data logger
19-Micromanometer
20-Plexi Glass

2.3 Image Capturing

A 3CCD Sony camera is used to capture images of colour distribution of the LCT (test zone). Illuminating Light sources 4 halogen lamps (each of 35 W power, without UV radiation) are used for the test zone. Image is captured through IC capture 2.3 software in the form jpeg format. Important point is to be noted the activation of colour sighting of LCT sheet dependent on light source and viewing angle [12]. Therefore, camera should not be distributed and lies at same position throughout experiments in all respects.

2.4 Image Processing

With this optical technique, an image of the test surface is acquired and image processing coding is used in order to convert the colour from the RGB domain into

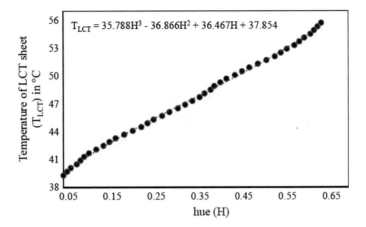

$T_{LCT} = 35.788H^3 - 36.866H^2 + 36.467H + 37.854$

Fig. 3 Calibration curve for LCT sheet

the HSI (hue, saturation and intensity) domain by using Eqs. (1)–(5). Among these parameters, only hue is retained, other quantities (S and I) need not be computed since they show small variation with temperature.

$$R = Max; H = \frac{G - B}{6[R - min\{R, G, B\}]} \tag{1}$$

$$G = Max; H = \frac{2 + B - R}{6[G - min\{R, G, B\}]} \tag{2}$$

$$B = Max; H = \frac{4 + R - G}{6[G - min\{R, G, B\}]} \tag{3}$$

Saturation values can be obtained by,

$$S = 1 - \left[\frac{Min(R, G, B)}{I}\right] \tag{4}$$

Intensity of any colour at particular point can be obtained by,

$$I = \frac{R + G + B}{3} \tag{5}$$

2.5 Calibration of LCT Sheet

The LCT sheet (specification R40C5 W) is used for temperature visualization where colour change is reversible and repeatable [12]. Therefore, LCT sheet requires proper care to calibrated the sheet accurately. The colour transformation for the sheets

Fig. 4 LCT images captured
at Re of 21000 for different
(P/e)

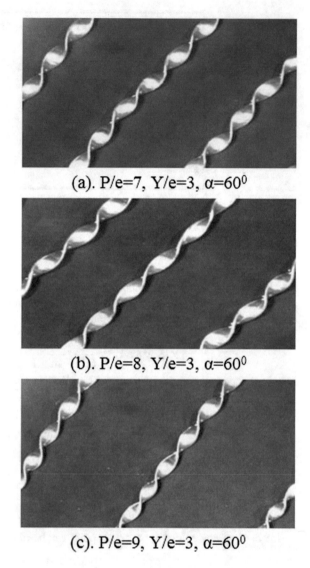

(a). P/e=7, Y/e=3, α=60⁰

(b). P/e=8, Y/e=3, α=60⁰

(c). P/e=9, Y/e=3, α=60⁰

dawning red colour at 40 °C and changes to green and then to blue with the increase
in temperature and turning achromatic again at temperature more than 55 °C. The
results shows, sheet is activated within desire range of the temperature and appears
colourless below and above of the active range.

A calibration curve shown in Fig. 3. represents temperature distribution along the
plate and hue and relation between two is given by Eq. (6).

$$T_{LCT} = 35.7881H^3 - 36.866H^2 + 36.46H + 37.854 \qquad (6)$$

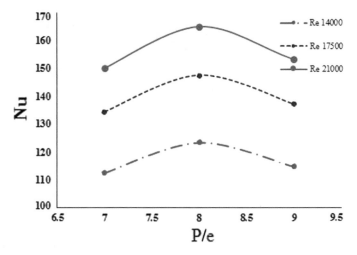

Fig. 5 effect of (P/e)

2.6 Data Reduction

The captured images are processing through MATLAB coding and heat transfer characteristic is determined by following Eqs. (7–9).

$$Re = \frac{VD_h}{\upsilon} \tag{7}$$

$$h = \frac{Q}{(T_{LCT} - T_{fm})} \tag{8}$$

$$Nu = \frac{hD_h}{k} \tag{9}$$

3 Result and Discussion

3.1 Effect of Relative Pitch Ratio (P/e)

Figure 4, demonstrates flow visualization through liquid crystal for the arrangements at angle of attack 60°, twist ratio 3 and at various relative pitch ratio of 7, 8 and 9. The flow is from left to right. For such images, the LCT sheets indicates colour throughout the entire temperature field and thus helps to determinate the local value of the heat transfer coefficient for each position in the inter-rib region. At relative roughness pitch of P/e-7, shown in Fig. 4a flow reattachment does not occur properly as, inter-

Fig. 6 LCT images captured at Re of 21000 for different (α)

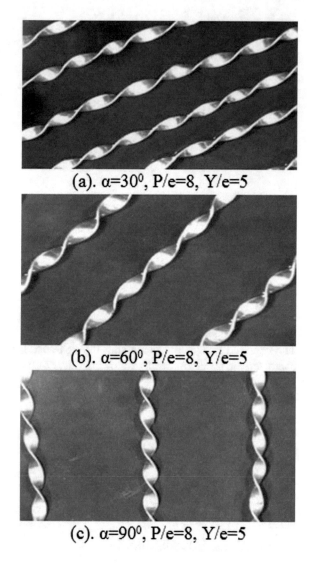

(a). $\alpha=30^0$, P/e=8, Y/e=5

(b). $\alpha=60^0$, P/e=8, Y/e=5

(c). $\alpha=90^0$, P/e=8, Y/e=5

rib region consists of green colour corresponding to higher value of the surface temperature.

Further, increase in P/e to 8 in Fig. 4b, red colour appearance between inter-rib region indicates appearance of flow reattach occur over the heated surface plate and even throughout the wall region. As P/e is further increased to 9 (Fig. 4c), the gradually decrease of red colour and increase of green colour intensity, indicates that a reattach of the flow between ribs occur less than previous arrangement. Figure 5, shows the effect. of relative pitch ratio on average Nusselt number and observed the maximum heat transfer at P/e 8 for all values of Re.

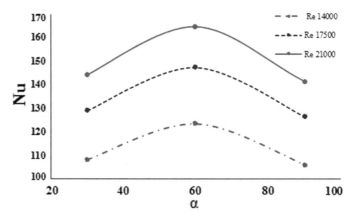

Fig. 7 Effect of (α)

3.2 Effect of Angle of Attack (α)

The angle of attack helps to generate secondary flow resulting enhancement in heat transfer characteristic. Figure 6, shows flow visualization for the arrangements at twist ratio of 3, relative roughness pitch of 8 and varying angle of attack at 30°, 60° and 90°.

As in Fig. 6a and c, the angle of attack at 30° and 90°, inter-rib region consists of black and green colour respectively depicts higher surface temperature as no generation of secondary flow. However, at angle of attack 60° Fig. 6b, shows flow reattach over the surface plate as presence red colour over the region. These conclude the maximum secondary flow ensues at 60°. The effect of angle of attack on average Nusselt number is shown in Fig. 7, and observed maximum Nusselt number obtained at α = 60° for all values of Reynolds number.

3.3 Effect of Relative Twist Ratio (Y/e)

Figure 8, exhibits flow visualization for the arrangements at α = 60°, P/e = 8 and various twist ratio of 3, 4 and 5. For arrangement for Y/e = 3, such images (Fig. 8.a) shows that the flow reattachment and jet formation which influences to break sub-laminar region over the heated plates properly as inter-rib region consists of red colour corresponding to lower value of the surface temperature.

Fig. 8 LCT images captured
at Re 21000 for different
(Y/e)

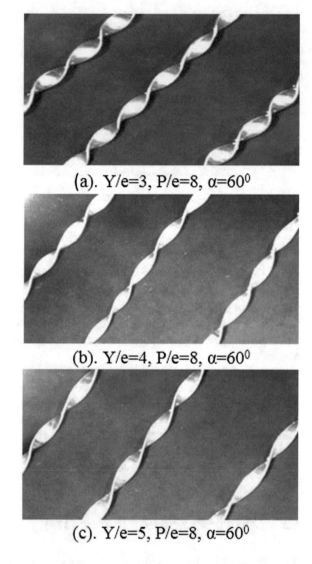

(a). Y/e=3, P/e=8, α=60⁰

(b). Y/e=4, P/e=8, α=60⁰

(c). Y/e=5, P/e=8, α=60⁰

Further, increase of Y/e to 4 (Fig. 8b) and 5 (Fig. 8c), resulting the appearance of blue and black colour indicate that increase in relative twist ratio does not permit sufficient flow reattach and jet formation over the surface plate. Figure 9, shows the effect of relative twist ratio on average Nusselt number and has been concluded the maximum Nusselt number obtained at Y/e = 3 for all values of Re and decrease with increase in twist ratio

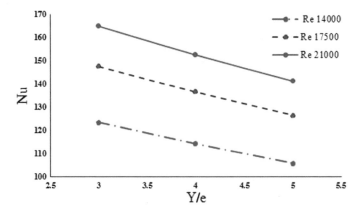

Fig. 9 Effect of (Y/e)

4 Conclusion

In present study, the method of flow visualization and heat transfer measurement technique has been demonstrated in a rectangular channel with roughened rib over absorber plate. In an order the means of visualization, liquid crystal thermography indicates reattachment regions followed by a slight reduction in the stream wise direction. Temperature distribution over the heated plates with various geometric configurations is analyzed and also results the fluctuation in heat transfer coefficients due to vortex shedding. Several configurations are experimentally investigated to obtain the optimal roughness geometric parameters. The experimental and image processing approach shows the maximum heat transfer obtained with geometric configuration at P/e = 8, Y/e = 3, α = 60°. Thus, liquid crystals have been successfully applied to flow visualization and heat transfer.

Acknowledgements The authors gratefully acknowledge the financial support of the Science and Engineering Research Board (SERB), Department of Science and Technology (DST), Govt. of India for initiating this research activity in Mechanical Engineering Department at National Institute of Technology Durgapur, India—Reference SERB-DST Grant: SB/EMEQ-314/2013; dated: 08/07/2013.

References

1. Copper, T. E., Filed, R. J., Meyer, J. F.: Liquid crytal thermography and its application to the study of convective heat transfer, transactions of the ASME, pp 442–450 (1975)
2. Baughn, J.W.: Liquid crystal methods for studying turbulent heat transfer. Int. J. Heat Fluid Flow **16**, 365–375 (1995)
3. Lee, K.C., Yianneskis, M.: A liquid crystal thermo graphic technique for the measurement of mixing characteristics in stirred vessel. Trans. Icheme. **75**(A), 746–754 (1997)

4. Liou, T.M., Chen, C.C., Tsai, T.W.: Heat transfer and fluid flow in a square duct with 12 different shaped vortex generators. J. Heat Transf. **122**, 327–335 (2000)
5. Kenning, D.B.R., Konno, T., Wienecke, M.: Investigation of boiling heat transfer by liquid crystal Thermography. Exp. Therm. Fluid Sci. **25**, 219–229 (2001)
6. Tanda, G.: Heat transfer and pressure drop in a rectangular channel with diamond shaped elements. Int. J. Heat Mass Transf. **44**, 3529–3541 (2001)
7. Gao, X., Sunden, B.: Heat transfer and pressure drop measurement in rib roughened rectangular ducts. Exp. Therm. Fluid Sci. **24**, 25–34 (2001)
8. Cavallero, D., Tanda, G.: An experimental investigation of forced convection heat transfer in channels with rib turbulators by means of liquid crystal thermography. Exp. Therm. Fluid Sci. **26**, 115–121 (2002)
9. Tariq, A., Swain, S.K., Panigrahi, P.K.: An experimental study of convective heat transfer from flat and ribbed surface. Indian J. Eng. Mater. Sci. **9**, 464–471 (2002)
10. Malay K.D., Tariq, A., Panigrahi, P.K., Muralidhar, K.: Estimation of convective heat transfer coefficient from transient liquid crystal data using an inverse technique. Inverse Probl. Sci. Eng.1–23(2004)
11. ASHRAE Standards: Methods of testing to determine the thermal performance of solar collectors, pp. 93–77. New York
12. Stasieka, J., Stasieka, A., Jewartowskia, M., Collins, M.W.: Liquid crystal thermography and true-colour digital image processing. Opt. Laser Tech. **38**, 243–256 (2006)

Impact of Oblique Magnetic Flux upon Second-Grade Fluid Flow with Dissipation

Bandita Das and Rita Choudhury

Abstract The analysis endeavors the effects of boundary-layer approximation on unsteady second-grade fluid flows past an erect plate in porous medium. The fluid is influenced by inclined magnetic flux, Joule effect is contemplated into exposition. The multi-parameter perturbation procedure is enrolled to solve the flow problem. Some indispensable flow features are emphasized graphically to acquire physical perception of the problem.

Keywords MHD · Visco-elasticity · Joule effect

1 Introduction

The boundary-layer flow problems of free-convection with existence of magnetic flux in porous medium have been scrutinized by numerous analyzers due to their heterogeneous relevancies in multiple engineering and technological extents. The efficacy of radiation and dissipation also cavort consequential involvement in these kinds of flow problems. Furthermore, Porous medium is contemplated to serviceable in controlling the free-convection which would differently transpire potency on a vertical heated surface. Some works of the scientists are intimated here. Huges and Yong [1], Cogley [2], Singh and Gorla [3], Singh [4], Sandeep and Sugunamma [5], Alizadeh and Rahmdel [6], Sharma and Chaturvedi [7], Seddeek [8], Chen [9], Sharma and Singh [10], Sharma et al. [11], Umamaheswar et al. [12], Soundalgekar 13], Labropulu et al. [14], Alcocer and Singh. [15], Khan and Sanjayanand [16], Ghosh and Sana [17, 18], Choudhury and Dey [19], Choudhury and Das [20, 21]

B. Das
Department of Mathematics, B. Borooah College, Guwahati 781003, Assam, India
e-mail: banditadas1234@gmail.com

R. Choudhury (✉)
Department of Mathematics, Gauhati University, Guwahati 7810014, Assam, India
e-mail: rchoudhury66@yahoo.in

© Springer Nature Singapore Pte Ltd. 2019
A. Abraham et al. (eds.), *Emerging Technologies in Data Mining and Information Security*, Advances in Intelligent Systems and Computing 755,
https://doi.org/10.1007/978-981-13-1951-8_55

Choudhury and Dhar [22] etc. who have sensational involvement in solving flow problems of various geometries in both Newtonian non-Newtonian cases.

In this study, a subclass of non-Newtonian fluid called visco-elastic fluid (characterized by Second-grade model) is appraised. The fluid flow is unsteady, radiative, dissipative and influenced by inclined magnetic flux. The flow of fluid past a vertical plate in the porous medium.

2 Mathematical Formulation

We contemplate an unsteady two-dimensional, free convective, boundary-layer flow of second-grade fluid past a vertical plate. The medium is taken as porous. The $\bar{x}$-axis is taken along the vertical plate and $\bar{y}$-axis is normal to it. Let $\bar{u}$ and $\bar{v}$ be the components of velocity along $\bar{x}$ and $\bar{y}$ directions respectively. The influence of oblique magnetic flux is taken into account with thermal radiation We adjudge

$$\frac{\partial q_r}{\partial y} = 4\bar{I}(\bar{T} - T_\infty) \tag{1}$$

The dimensional governing equations are:

$$\rho \frac{\partial \bar{u}}{\partial \bar{t}} = \mu_1 \frac{\partial^2 \bar{u}}{\partial \bar{y}^2} + \mu_2 \frac{\partial^3 \bar{u}}{\partial \bar{y}^2 \partial \bar{t}} + g\beta\rho(\bar{T} - T_\infty) - \left(\frac{\mu_1}{K^*} + \sigma B_0^2 \sin^2 \varphi\right)\bar{u} \tag{2}$$

$$\rho C_P \frac{\partial \bar{T}}{\partial \bar{t}} = k\frac{\partial^2 \bar{T}}{\partial \bar{y}^2} + \mu_1\left(\frac{\partial \bar{u}}{\partial \bar{t}}\right)^2 + \mu_2\left(\frac{\partial \bar{u}}{\partial \bar{t}}\frac{\partial^2 \bar{u}}{\partial \bar{y}\partial \bar{t}}\right) + \bar{Q}(\bar{T} - T_\infty) - 4I(\bar{T} - T_\infty) + \sigma B_0^2 \bar{u}^2 \tag{3}$$

with suitable boundary conditions:

$$\bar{y} = 0 : \bar{u} = U\left(1 + \varepsilon e^{i\bar{\omega}\bar{t}}\right), \frac{\partial \bar{T}}{\partial \bar{y}} = -\frac{q}{k}$$

$$\bar{y} \to \infty : \bar{u} = 0, \bar{T} = T_\infty \tag{4}$$

where the mean velocity is denoted by U, $\varepsilon << 1$. The symbols $\rho, \bar{t}, \mu_1$, g, $\beta, \bar{T}$, $T_\infty, \sigma, B_0, \nu_1, K^*, k, C_p, \bar{Q}$ have their appropriate elucidations.

Inserting the dimensionless quantities

$$y = \frac{U\bar{y}}{\nu_1}, u = \frac{\bar{u}}{U}, t = \frac{U^2\bar{t}}{\nu_1}, \omega = \frac{\nu_1\bar{\omega}}{U^2}, \theta = \frac{kU(\bar{T} - T_\infty)}{q\nu_1}, Gr = \frac{g\beta\nu_1^2 q}{kU^4}, K = \frac{K^*U^2}{\nu_1^2}, M = \frac{\sigma B_0^2 \nu_1}{\rho U^2},$$

$$Pr = \frac{\mu C_p}{k}, R = \frac{4I^*\nu_1}{\rho C_p U^2}, Ec = \frac{kU^3}{\rho C_p q}, Q = \frac{\bar{Q}\nu_1}{\rho C_p U^2} \tag{5}$$

into the Eqs. (2) and (3), we get

$$\frac{\partial u}{\partial t} = \frac{\partial^2 u}{\partial y^2} + \alpha_1 \frac{\partial^3 u}{\partial y^2 \partial t} + Gr\theta - \left(M \sin^2 \varphi + \frac{1}{k} \right) u \tag{6}$$

$$\frac{\partial \theta}{\partial t} = \frac{1}{Pr} \frac{\partial^2 \theta}{\partial y^2} - (R - Q)\theta + Ec \left(\frac{\partial u}{\partial y} \right)^2 + Ec\alpha_1 \left(\frac{\partial u}{\partial y} \frac{\partial^2 u}{\partial y \partial t} \right) + EcMu^2, \tag{7}$$

where α_1 is the miniature visco-elastic parameter.

The mutated form of (4) is

$$y = 0 : u = 1 + \varepsilon e^{i\omega t}, \frac{\partial \theta}{\partial y} = -1$$

$$y \to \infty : u = 0, \theta = 0 \tag{8}$$

Here Ec, Gr, R, M, Pr are the eloquent flow parameters.

3 Method of Solution

Let us appraise

$$u = f_0 + \varepsilon e^{i\omega t} f_1 \tag{9}$$

$$\theta = \theta_0 + \varepsilon e^{i\omega t} \theta_1 \tag{10}$$

where u and θ be the solutions of the Eqs. (6) and (7) respectively.

We substitute the expressions (9) and (10) in the dimensionless Eqs. (6) and (7). After that the coefficients of ε^n $(n = 0, 1, 2 \ldots)$ are rearranged in both the equations with the neglect of term containing $\varepsilon \geq 2$ to obtain.

Zeroth-order equations:

$$f_0'' - \left(M \sin^2 \varphi + \frac{1}{k} \right) f_0 = -GrT_0 \tag{11}$$

$$T_0'' - Pr (R - Q)T_0 + Ec \ Pr \left(f_0' \right)^2 + Ec \ Pr \ Mf_0^2 = 0 \tag{12}$$

First-order equations:

$$f_1'' + \alpha_1 i\omega f_1'' - \left(M \sin^2 \varphi + \frac{1}{k} + i\omega \right) f_1 = -GrT_1 \tag{13}$$

$$T_1'' - Pr(R - Q + i\omega)T_1 + 2Ec \ Pr \ f_0' f_1' + Ec\alpha_1 \ Pr \ f_0' f_1' + Ec \ Pr \ Mf_1^2 = 0 \tag{14}$$

The metamorphosed boundary conditions are:

$$y = 0 : f_0 = 1, f_1 = 1, (T_0)_y = -1, (T_1)_y = 0$$

$$y \to \infty : f_0 = 0, f_1 = 0, T_0 = 0, T_1 = 0 \tag{15}$$

where the suffixes denote partial differention w.r.t. y

To solve the Eqs. (11) to (14), the Eckert number Ec is taken as perturbation parameter as $E \ll 1$.

We write,

$$f_0 = f_{00} + Ec f_{01} + O(Ec^2) \tag{16}$$

$$f_1 = f_{10} + Ec f_{11} + O(Ec^2) \tag{17}$$

$$T_0 = T_{00} + Ec T_{01} + O(Ec^2) \tag{18}$$

$$T_1 = T_{10} + Ec T_{11} + O(Ec^2) \tag{19}$$

and proceeding as above we have,

Zeroth-order equations:

$$f_{00}'' - \left(M \sin^2 \varphi + \frac{1}{k} \right) f_{00} = -Gr T_{00} \tag{20}$$

$$f_{10}'' + \alpha_1 i \omega f_{10}'' - \left(M \sin^2 \varphi + \frac{1}{k} + i\omega \right) f_{10} = -Gr T_{10} \tag{21}$$

$$T_{00}'' - \Pr(R - Q) T_{00} = 0 \tag{22}$$

$$T_{10}'' - \Pr(R - Q + i\omega) T_{10} = 0 \tag{23}$$

First-order equations:

$$f_{01}'' - \left(M \sin^2 \varphi + \frac{1}{k} \right) f_{01} = -Gr T_{01} \tag{24}$$

$$f_{11}'' + \alpha_1 i \omega f_{11}'' - \left(M \sin^2 \varphi + \frac{1}{k} + i\omega \right) f_{11} = -Gr T_{11} \tag{25}$$

$$T_{01}'' - \Pr(R - Q) T_{01} + \Pr(f_{00}')^2 + \Pr M f_{00}^2 = 0 \tag{26}$$

$$T_{11}'' - \Pr(R - Q + i\omega) T_{11} + 2\Pr f_{00}' f_{10}' + \alpha_{10} \Pr f_{00}' f_{10}' + \Pr M f_{10}^2 = 0 \tag{27}$$

With boundary conditions

$$y = 0: f_{00} = 1, \ f_{01} = 0, \ f_{10} = 1, \ f_{11} = 0, \ (T_{00})_y = -1, \ (T_{01})_y = 0,$$
$$(T_{10})_y = 0, \ (T_{11})_y = 0$$
$$y \to \infty: f_{00} = 0, \ f_{01} = 0, \ f_{10} = 0, \ f_{11} = 0, \ T_{00} = 0, \ T_{01} = 0, \ T_{10} = 0, \ T_{11} = 0 \tag{28}$$

To solve (21) and (25), α_1 is considered as perturbation parameter as α_1 is very very small for small shear rate and hence we take

$$f_{10} = f_{100} + \alpha_1 f_{101} \tag{29}$$

$$f_{11} = f_{110} + \alpha_1 f_{111} \tag{30}$$

Substitution of (29) and (30) in the Eqs. (21) and (25) yield the following equations:

$$f_{100}'' - \left(M \sin^2 \varphi + \frac{1}{k} + i\omega \right) f_{100} = -Gr T_{10} \tag{31}$$

$$f_{101}'' + i\omega f_{100}'' - \left(M \sin^2 \varphi + \frac{1}{k} + i\omega \right) f_{101} = 0 \tag{32}$$

$$f_{110}'' - \left(M \sin^2 \varphi + \frac{1}{k} + i\omega \right) f_{110} = -Gr T_{11} \tag{33}$$

$$f_{111}'' + i\omega f_{110}'' - \left(M \sin^2 \varphi + \frac{1}{k} + i\omega \right) f_{111} = 0 \tag{34}$$

with the conditions

$$y = 0 : f_{100} = 1, \ f_{101} = 0, \ f_{110} = 0, \ f_{111} = 0$$
$$y \to \infty : f_{100} \to 0, \ f_{101} \to 0, \ f_{110} \to 0, \ f_{111} \to 0 \tag{35}$$

Now, the governing equations are solved in appropriate way to obtain f_0, T_0, f_1, T_1, etc., and then substituting them in the expression (9) and (10), it is feasible to write the following articulation:

$$u = \left(A_{41} e^{-m_2 y} + A_{42} e^{-m_1 y} + A_{43} e^{-2m_2 y} + A_{44} e^{-(m_1+m_2)y} + A_{45} e^{-2m_1 y} \right)$$
$$+ \varepsilon e^{i\omega t} \left((e^{-m_4 y} + \alpha_1 A_{13} y e^{-m_4 y}) + Ec \left(A_{46} e^{-m_4 y} + A_{47} e^{-m_3 y} + A_{48} e^{-(m_2+m_4)y} + A_{49} y e^{-(m_2+m_4)y} \right. \right.$$
$$\left. \left. A_{50} e^{-(m_1+m_4)y} + A_{51} y e^{-(m_1+m_4)y} + A_{52} e^{-2m_4 y} + A_{53} y e^{-2m_4 y} + A_{54} y^2 e^{-2m_4 y} + A_{55} y e^{-m_4 y} \right) \right) \tag{36}$$

$$\theta = \left(A_1 e^{-m_1 y} + Ec \left(A_7 e^{-m_1 y} + A_4 e^{-2m_2 y} + A_5 e^{-(m_1+m_2)y} + A_6 e^{-2m_1 y} \right) \right)$$
$$+ \varepsilon e^{i\omega t} \left(Ec (A_{21} e^{-m_3 y} + A_{14} e^{-(m_{21}+m_4)y} + A_{15} y e^{-(m_{21}+m_4)y} + A_{16} e^{-(m_1+m_4)y} \right.$$
$$\left. + A_{17} y e^{-(m_1+m_4)y} + A_{18} e^{-2m_4 y} + A_{19} y e^{-2m_4 y} + A_{20} y^2 e^{-2m_4 y}) \right) \tag{37}$$

The dimensionless forms of shearing stress and the heat flux at the plate are manifested as

$$\sigma = A_{56} + \varepsilon e^{i\omega t} A_{57} + \alpha_1 \varepsilon e^{i\omega t} A_{58} \tag{38}$$

$$Nu = A_{59} + \varepsilon e^{i\omega t} A_{60} \tag{39}$$

respectively.

The constants are determined but not comprised here due to abridgment.

4   Results and Discussion

It is to be noted that the fluid is highly visco-elastic for smaller values of α_1. The result corresponding to the value $\alpha_1 = 0$ describes the behavior of Newtonian viscous fluid which is free from elastic effect. Allthrough the computation, we reckon the fixed values for Ec $= 0.2$, $\varepsilon = 0.2$, $\omega = 1$, $\omega t = \pi/2$, Q $= 0.2$ $\varphi = \frac{\pi}{2}$, K $= 0.4$.

The velocity of fluid versus the dispacementl are portrayed by the Figs. 1, 2, 3, 4 and 5. It is observed that the fluid velocity accelerates in the vicinity of the plate but decelerates apart from the same for both types of fluid with the magnification form of $|\alpha_1|$, the fluid velocity ascends in compared to classical form of Newtonian fluids.

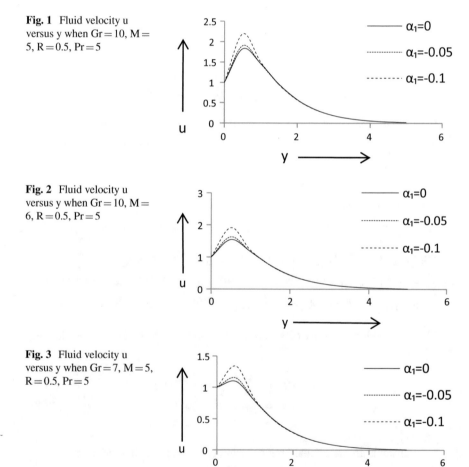

Fig. 1 Fluid velocity u versus y when Gr $= 10$, M $= 5$, R $= 0.5$, Pr $= 5$

Fig. 2 Fluid velocity u versus y when Gr $= 10$, M $= 6$, R $= 0.5$, Pr $= 5$

Fig. 3 Fluid velocity u versus y when Gr $= 7$, M $= 5$, R $= 0.5$, Pr $= 5$

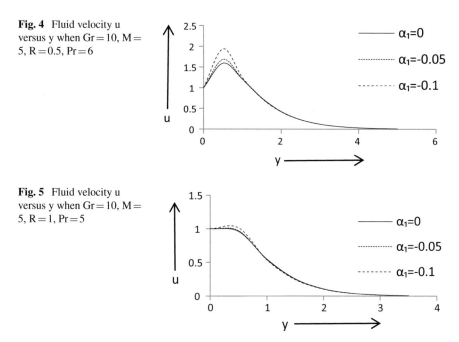

Fig. 4 Fluid velocity u versus y when $Gr = 10$, $M = 5$, $R = 0.5$, $Pr = 6$

Fig. 5 Fluid velocity u versus y when $Gr = 10$, $M = 5$, $R = 1$, $Pr = 5$

The Figs. 1 and 2 infer that the velocity of fluid shrinks with the inflation of the magnetic parameter. It occurs for the formation of Lorendz force generated by transverse magnetic field which has a propensity to retard fluid motion. The Grashof number explicates the proportion of buoyancy forces to viscous forces. Figures 1 and 3 hypothesize that the increase of buoyancy forces in comparison with viscous forces induced by other fixed flow parameters speed up the fluid velocity in the entire fluid flow region. Figures 1 and 4 represent effects of elasticity on the velocity of fluid with the enhancement of Prandtl number. Furthermore, the fluid velocity diminishes with the enhancement of radiation parameter (Figs. 1and 5).

Figure 6 demonstrates the temperature profile θ against the displacement y. The figure indicates that the temperature profile decelerates away from the plate in the fluid flow region. Consequently, it appears from the figure that the visco-elastic fluid neither grow nor decay as quickly as Newtonian viscous fluid due to the restraining effect of the fluid elasticity.

As a result, the effect of the elastic parameter α_1 on the fluid motion keeps on diminishing irrespective of its nature and a visco-elastic fluid behaves as a viscous fluid.

From practical attitude, it is very important to highlight the effect of shear tress and consequently the viscous drag of fluid. The fluctuations of the shear stress against the pertinent flow parameters are depicted in Figs. 7, 8, 9 and 10. In all the expositions,

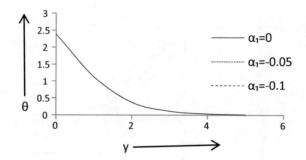

Fig. 6 Fluid temperature θ when Gr $= 10$, M $= 5$, R $= 0.5$, Pr $= 5$

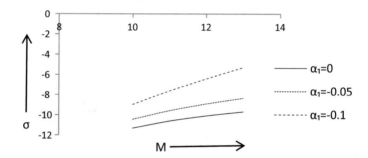

Fig. 7 Shearing stress versus Magnetic parameter M

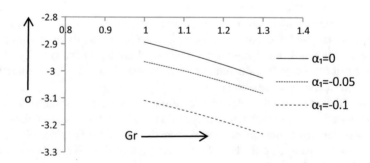

Fig. 8 Shearing stress versus Garshof number Gr

the effects of visco-elasticity is noteworthy and comparable with Newtonian fluids. The visco-elastic parameter has no sizeable execution on Nusselt number.

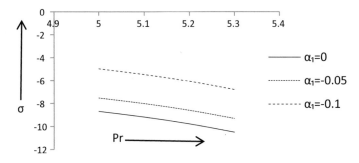

Fig. 9 Shearing stress versus Prandtl number Pr

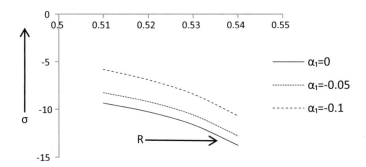

Fig. 10 Sheraing stress versus Radiation parameter R

5 Conclusions

Based on present analysis, the key physiognomics are accentuated below:

- The velocity profiles divulge the exactitude of boundary conditions elucidated for the flow problem.
- With the magnification of the electromagnetic force in compared to viscous force, the velocity lessens and this consequence unveils conformity with physical circumstances.
- Temperature profile is not affected by visco-elasticity due to the restraining outcome of the fluid elasticity.
- The fluctuation of the shearing stress due to vartion of the pertinent flow parameters induced by Visco-elasticity is ascertainable.
- Visco-elasticity has no consequential efficacy on Nusselt number.

References

1. Huges, W.F., Yong, F.J.: The electro-Magneto-Dynamics of fluids. Wiley, New York, USA(1966)
2. Cogley, A.C., Vincenti, W.G., Gilles, E.S.: Differential approximations for radiative heat transfer in a nonlinear equations-grey gas near equilibrium. Am. Inst. Aeronaut. Astronaut. **6**, 551–553 (1968)
3. Singh, A.K., Gorla, R.S.R.: Free convective heat and mass transfer with Hall current, Joule heating and thermal diffusion. Heat Mass Transf. **45**, 1341–1349 (2009)
4. Sharma, P.R., Singh, G.: Effects of variable thermal conductivity, viscous dissipation on steady MHD natural convection flow of low Prandtl fluid on an inclined porous plate with Ohmic heating. Meccanica **45**, 237–247 (2010)
5. Sandeep, N., Sugunamma, V.: Effect of inclined magnetic field on unsteady free convective flow of dissipative fluid past a vertical plate. Open J. Adv. Eng. Tech. **1**(1), 6–23 (2013)
6. Alizadeh, R., Rahmdel, K.: MHD free convection flow of a dissipative fluid over a vertical porous plate in porous media. Adv. Appl. Sci. Res. **5**(4), 31–42 (2014)
7. Sharma, P.R., Chaturvedi, R.: Unsteady flow and heat transfer along a plane wall with variable suction and free stream. Bull. Allahabad Math. Soc. **20**, 51–61 (2005)
8. Seddeek, M.A.: Effects of radiation and variable viscosity on a MHD free convection flow past a semi infinite flat plate with an aligned magnetic field in the case of unsteady flow. Int. J. Heat Mass Transf. **45**, 931–935 (2002)
9. Chen, C.H.: Combined heat and mass transfer in MHD free convection from a vertical surface with ohmic heating and viscous dissipation. Int. J. Eng. Sci. **42**, 699–713 (2004)
10. Sharma, P.R., Singh, G.: Unsteady MHD free convective flow and heat transfer along a vertical porous plate with variable suction and internal heat generation. Int. J. Appl. Math. Mech. **4**, 01–08 (2008)
11. Sharma, P.R., Sharma, P., Navin, K.: Radiation effects on unsteady MHD free convective flow with hall current and mass transfer through viscous incompressible fluid past a vertical porous plate immersed in porous medium with heat source/sink. J. Int. Acad. Phys. Sci. **13**, 231–252 (2009)
12. Umamaheswar, M., Varma, S.V.K., Raju, M.C.: Unsteady MHD free convective visco-elastic fluid flow bounded by an infinite inclined porous plate in the presence of heat source, viscous dissipation and ohmic heating. Int. J. Adv. Sci. **61**, 39–52 (2013)
13. Soundalgekar, V.M.: Flow of an elastic-viscous fluid past an oscillating plate, Czeehoslovak J. Phys. B **28**(11), 1217–1220 (1978)
14. Labropulu, F., Dorrepaal, J.M., Chanda, O.P.: Visco-elastic fluid flow impinging on a wall with suction or blowing. Mech. Res. Commun. **20**(2), 143–153 (1993)
15. Alcocer, F.J., Singh, P.: Permeability of periodic arrays of cylinders for visco-elastic flows. Phys. Fluids **14**(7), 2578–2581 (2002)
16. Khan, S.K., Sanjayanand, E.: Visco-elastic boundary layer MHD flow through a porous medium over a porous quadratic stretching sheet. Arch. Mech. **56**(3), 191–204 (2004)
17. Ghosh, A.K., Sana, P.: On hydromagnetic flow of an oldroyed-B fluid near a pulsating plate. Acta Astronuticca **7**(016), 1–9 (2008)
18. Ghosh, A.K., Sana, P.: On hydromagnetic rotating flow of an oldroyd- B fluid near an oscillating plate **60**, 1135–1155 (2009)
19. Choudhury, R., Dey, D.: Free convective visco-elastic flow with heat and mass transfer through a porous medium with periodic permeability. Int. J. Heat Mass Transf. **53**, 1666–1672 (2010)
20. Chodhury, R., Das, U.J.: Heat transfer to MHD oscillatory visco-elastic flow in a channel filled with porous medium. Phys. Res. Int., Ar ID87953, 5 p (2012)
21. Choudhury, R., Das, S.K.: Visco-elastic MHD fluid flow over a vertical plate with dufour and Soret effects. Int. J. Sci. Eng. Res. **4**(7), 11–17 (2013)
22. Choudhury, R., Dhar, P.: Ion slip effect on visco-elastic fluid flow past an impulsively started infinite vertical plate embedded in a porous medium with chemical reaction, Int. Scholarly Res. Not., Ar ID 481308, 10 p (2014)

Analyzing the Complexity of Loop Shifting for Optimization of Matrix-Multiplication Process for System Having One Level Cache

Yogesh Singh Rathore, Dharminder Kumar and Kavita Saxena

Abstract Although the use of Matrix-Multiplication is very extensive in research. In our paper it is being used for optimization in case of level one cache memory. Initially Open-MP (OMP) is used, followed by various optimizing techniques. Finally the results of Matrix-Multiplication are verified by the mathematical proof in terms of complexity.

Keywords Open-MP · Cache · Optimization · Cache conscious · Nested loops

1 Introduction

This research paper an effort to optimize the Matrix-Multiplication algorithm by various possible permutations and combinations of optimization techniques. Process of optimization involves partitioning of code into Basic blocks [1]. Each basic block has an entry point for the block known as leader of the block. To find the leader of the block, there is a set of rules, according to which: First statement is the initiator of a code, the statement which is an objective for unconditional or conditional transfer of control is a initiator, the statement which instantly chase the conditional transfer of control is a leader [2]. Leaders play a major role in identifying the basic block. Each leader is the first statement of a basic block that continues till the next leader minus statement or till the end of the program. A statement that is not in the basic block is non-reachable and can be eliminated, if desired [3]. Some basic block provided with the language is IF-THEN-ELSE, SWITCH, FOR, REPEAT-UNTIL and others are

Y. S. Rathore (✉) · K. Saxena
Mewar University, Chittorgarh, Rajasthan, India
e-mail: ysrathore@amity.edu

K. Saxena
e-mail: dr_ksaxena@gmail.com

D. Kumar
Guru Jambheshwar University of Science & Technology, Hissar, Haryana, India
e-mail: dr.dk.kumar.02@gmail.com

© Springer Nature Singapore Pte Ltd. 2019
A. Abraham et al. (eds.), *Emerging Technologies in Data Mining and Information Security*, Advances in Intelligent Systems and Computing 755,
https://doi.org/10.1007/978-981-13-1951-8_56

user defined. Loops form the part of the program where the control stays for larger durations, so it becomes very important to optimize loops. Some of the well-known techniques for loop optimization are Invariant code, Induction analysis, loop Expansion/Unrolling, loop Jamming/Fusion, loop Shifting, Strength Reduction and Dead Code Elimination [4]. A section of code that resides inside the loop and computes the same value at each iteration is known as loop invariant code. The code of this kind can be advanced outside the curl and protects repeated processing in each iteration of the code. A variable is said to be an induction variable if the value of the variable is altered within the loop, by a loop invariant value [5]. Loop Expansion/Unrolling is the concept of replacing the code inside the loop, a number of times. The number of times depends upon the subscript of the loop. In case of unrolled function only four statements are executed for complete execution, whereas in the case of normal function more than double the number iterations are done to get the result [6]. Loop fusion/Jamming is used when there are more than 2(two) loops having same number of iterations. In this technique a loop is designed in such a way that all the statements are placed in a single loop with the same number of iterations as both were having. Dead Code is a simple or compound statement that is never executed or if executed then it has no effect on the output.

So in the interest of optimization it is good to eliminate the dead code. The practical approach of this knowledge is shown and implemented in this paper [7].

2   Related Work

The variety of processors like single, dual and multi-core are utilized by author to run the Matrix-Multiplication algorithm for different sizes between 300×300 to 2000×2000 in steps of 300. Assess the mean value of the process duration in seconds for each set of algorithms. Afterwards author executed the identical procedure on the identical sets of the sizes for multiplication of matrix without and with OMP. Which is shown in Table 1 and corresponding Figure is presented as Fig. 1. Moreover, seeking to pursue the process duration in seconds. To obtain the time (duration) clock function is being utilized. For the computation of implementation of duration of time for specimen code, Compiler of language 'C' having 'time' header file is very obliging. The author utilizes the clock function for obtaining time, to do so the time header file attached. The obtained time 't' will be distributed by acceptable value to obtain it into seconds. Furthermore, we attempt to interpret the run-time in seconds for every composition. Likewise, the chunk's reading and multiplication is utilized in OMP. By utilizing OMP we can settle variety of threads to an algorithm relying on the number of cores of the machine. To obtain the superior parallelism, the number of threads should equate the number of cores [8]. For the machine under study, gives better results in case of OMP as compared with non-OMP with the gradual increase in size of Matrix-Multiplication. The composition for the three loops for Matrix-Multiplication is given as $3! = 3*2*1$ which equates to 6. If the loops considered as IJK then the composition should be IJK, IKJ, JIK, JKI, KI, and KIJ. Our next step is to find the mean value

Table 1 Matrix-multiplication with and without OMP

Matrix size	Without OMP	With OMP
300 × 300	0.9292	0.3192
600 × 600	3.6611	1.3183
900 × 900	11.3682	3.4266
1200 × 1200	24.4960	7.5200
1500 × 1500	30.6200	9.4000
1800 × 1800	140.5824	39.6156
2000 × 2000	156.2027	44.0173

Fig. 1 Matrix-multiplication with and without OMP

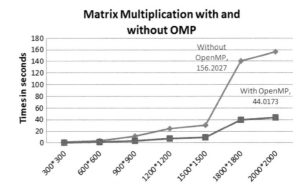

Table 2 All combinations of loops

Matrix size	For loop IJK	For loop IKJ	For loop JIK	For loop JKI	For loop KIJ	For loop KJI
300 × 300	0.7020	0.8040	0.7080	0.7260	0.6960	0.6660
600 × 600	2.9570	3.1114	3.0770	3.1114	3.1630	2.8800
900 × 900	10.7910	10.8090	10.7370	10.8720	10.7730	10.7370
1200 × 1200	18.6900	19.7300	19.4200	19.4100	19.1000	19.4200
1500 × 1500	23.3625	24.6625	24.2750	24.2625	23.8750	24.2750
1800 × 1800	78.7950	80.0000	78.0000	77.8800	79.0700	79.8200
2000 × 2000	87.5500	88.8900	86.6700	86.5300	87.8600	88.6900

of combinations, in seconds, for the set with OMP and compare the results with Non-OMP set. On executing the identical procedure on all the sets of the different sizes for multiplication of matrix with OMP plus other optimization techniques and without OMP, the variation in the execution time attained for an algorithm with and without OMP is considerably large. We can infer that with the increase in size of input algorithms [9] the time increases rapidly. This is shown in Tables 2, 3 and Fig. 2.

Table 3 Comparison of optimized and non-optimized

Matrix size	Av. of pptimized.	For loop IJK (plane)
300 × 300	0.7170	0.9240
600 × 600	3.0500	3.6600
900 × 900	10.7865	11.3400
1200 × 1200	19.3000	24.4000
1500 × 1500	24.1200	30.5000
1800 × 1800	78.9300	140.4000
2000 × 2000	87.7000	156.0000

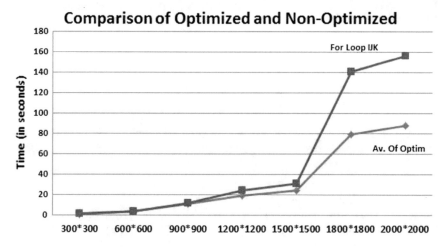

Fig. 2 Comparison of optimized and non-optimized

Table 4 Matrix pattern (3 × 3) each

2	1	1
1	2	3
3	2	1

1	2	3
3	2	1
2	1	1

3 Data Set Used

To study the pattern we have used two matrix with the given data sets and are shown in Table 4.

4 Methodology

A sample values are taken into consideration and all the possible combinations are explored. Their pattern of accessing the various locations is given in Fig. 3, which accesses in terms of spatial/temporal concepts and tried to find out the execution time in both the cases, that is in normal case and optimized case [10]. In each case the result has been evaluated in terms of execution time. After getting the access pattern for all the matrix of NXN matrix, it is quite necessary to know that whether there is any benefit in terms of cache access [11]. Let us consider the matrix of NXN order, the matrix are A, B and C. Multiplication of A and B is done and the result is to be stored in C matrix. Storage of elements in the memory is considered to be row major [12]. We will examine the case of systems having L1 cache only, with S words of the cache line with block size W. Access for matrix A is row wise for the ijk order. Size of matrix is considerable large as compared to the cache line W. If we divide the entire row of matrix A of $N \times N$ order into the W size cache line then we have N/W chunks for first row of matrix A [13]. N/W chunks of matrix A are to be multiplied to the column of matrix B. N/W gives the number is read Miss for the matrix A, for the initial row of matrix A which to be propagated with the initial column of B matrix. This matrix A's row is to be propagated with N columns of B matrix. So read Miss for a isolated matrix A's row with the entire B matrix is given by N^2/W. There is N rows in the matrix A so the total read Miss for the matrix A is given by $N*N^2/W$, that is N^3/W. Now let us see matrix B, access for matrix B is column wise for the IJK order. Size of matrix is considerable large as compared to the cache line W. If we divide the entire column of matrix B of NXN order into the W size cache line then we have N/W chunks for first column of matrix B. Here important point to be noted is that the storage of matrix is to be considered to be row wise so to access the first element of the first column of matrix B, the entire row has to be cached [14]. This constitutes the first miss. To access the second element of the first column for matrix B, the second row is to be cached. This is second miss. So to access the first column of N elements of the matrix B, there will be N read miss. For accessing N column of matrix B there will be $N*N = N^2$ read miss. So now we have N^2 miss for matrix B of N*N order for first row of matrix A. If we consider N rows for the matrix A then it would be $N*N^2 = N^3$. The outcome of propagation of A matrix & B matrix is to be reserved in C matrix. Access for matrix A is row wise for the IJK order [15]. Size of matrix C will be considerably large as compared due to the large size of matrix A and B. cache line is of size W. N/W miss for the first row of A matrix. N row of A matrix will have $N*N/W$ miss $= N^2/W$ miss for matrix A. The total cache miss for all the three matrix for C, A and B is

$$(N^2/W + N^3/W + N^3)/3 = N^3/W(1/N + 1 + W) \qquad (1)$$

The order ikj has access matrix, for all matrix namely A, B and C row wise [16]. also the storage in the memory is row wise. N2/W is the read for the matrix A. N/W*N*N = N^3/W miss for B matrix. Similarly Miss for C matrix will be N^3/W.

Fig. 3 The pattern of accessing the various locations

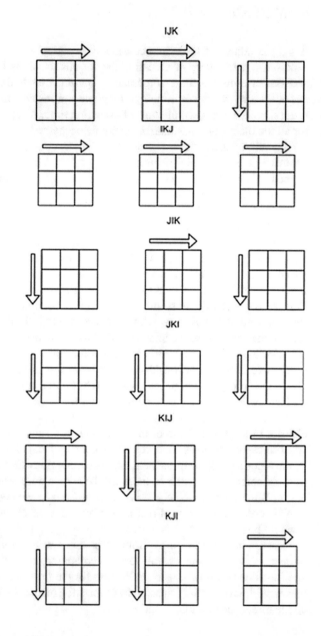

Total cache miss for order ikj is given by

$$(N^3/W + N^2/W + N^3/W)/3 = N^3/W(2 + 1/N) \qquad (2)$$

"ijk/ikj ∼

ijk/ikj ∼ (1 + w)/2, when n is large.

w = 4(32 − byte cache line, double precision data)

    −ratio ∼ 2.5.

w = 8(64 − byte cache line, double precision data)

    −ratio ∼ 4.5.

w = 16 (64 − byte cache line, integer data)

    −ratio ∼ 8.5."[17]

5   Implementation

Now profiling of cache miss under Ubuntu Linux using Perf Events, present in the kernel is performed. An event can have sub events. For counting the events there are some filters provided. The perf command in our experiment calculates cycles, instruction, L1d cache load and misses and has general syntax as:

perf stat –repeat 1 –e cycles : u –e instructions: u –e L1-dcache-loads: u –e L1–dcache- load-misses: u ./a.out

The perf command available in the kernel of Ubuntu operating system has been used to evaluate the number of instructions per cycle and execution time of set of Matrix-Multiplication. We executed the normal code and the optimized code with the perf command, for various sets of matrix and read the output. Results for 'Estimation of Execution of Instructions Per Cycle' are given in Table 5 and hence the Fig. 4. It clearly shows that execution of instructions per cycle in optimized case requires less time than in non-optimized case.

Simultaneously we find execution time for all size of matrix, which are given in the Table 6 and hence the Fig. 5.

Along X-axis the various sizes of matrices for optimized and non-optimized are taken. Along Y-axis the execution time of Matrix-Multiplication in seconds is taken. The execution time for the set of optimized and non-optimized from 300 X300 till

Table 5 Estimation of execution of instructions per cycle

Matrix size	Non optimized	Optimized
300 × 300	1.3800	1.4000
600 × 600	1.1700	1.1700
900 × 900	0.8600	0.9300
1200 × 1200	0.8900	1.0300
1500 × 1500	0.8900	1.0300
1800 × 1800	0.8900	1.0300
2000 × 2000	0.9000	1.0300

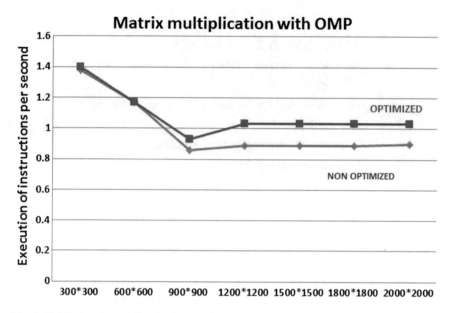

Fig. 4 Estimation of execution of instructions Per Cycle

Table 6 Estimation of execution time of program

Matrix size	Non-optimized	Optimized
300 × 300	0.2070	0.2040
600 × 600	1.8780	1.8790
900 × 900	8.5680	7.9630
1200 × 1200	19.5890	17.0210
1500 × 1500	38.2980	33.1410
1800 × 1800	38.2980	33.1410
2000 × 2000	89.7520	78.3260

900 × 900 remains the same approximately, but there is an increase in the difference of execution time for the matrix size 900 × 900 to 2000 × 2000 is considerable.

6 Result

Evaluating the performance of Matrix-Multiplication in combination with OMP and non-OMP results are promising. Further all the possible combinations of the loop shifting and other techniques are applied on the set with OMP and comparison is done with the normal Matrix-Multiplication, still the results are positive. The same has been verified by calculating the complexity of the system in reference to one level

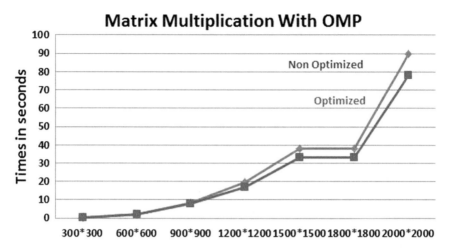

Fig. 5 Estimation of execution time of program

cache. This sequence supports cache for optimization that implies this sequence lies under cache conscious data structures. The sequence is useful because it is an addition to the optimization done at compiler level.

References

1. Compunity. The community of OpenMP users, researchers tool developers and provider website (2006). http://www.compunity.org/
2. Hammond, L., Nayfeh, B.A., Olukotun, K.: A single-chip multiprocessor. Computer **30**(9), 79–85 (1997)
3. Jerraya, A., Tenhunen, H., Wolf, W.: Guest editors' introduction: multiprocessor systems-on-chip. Computer **l38**(7), 36–40 (2005)
4. Ayguade, E., Copty, N., Duran, A., Hoeflinger, J., Lin, Y., Zhang, G.: A proposal for task parallelism in OpenMP. In: Proceedings of the 3rd International Workshop on OpenMP, June 2006
5. Zhong, H., Lieberman, S.A., Mahlke, S.A.: Extending multicore architecture to exploit hybrid parallelism in single thread applications. HPCA 25–36 (2007)
6. Dally, W.J., Lacy, S.: VLSI architecture: past, present, and future. In: ARVLSI '99: Proceedings of the 20th Anniversary Conference on Advanced research in VLSI, p. 232. IEEE Computer Society, Washington, DC, USA (1999)
7. Kumar, S., Hughes, C.J., Nguyen, A.: Carbon: architectural support for fine-grained parallelism on chip multiprocessors. In: ISCA '07: Proceedings of the 34th Annual International Symposium on Computer Architecture, pp. 162–173. ACM, New York, NY, USA (2007)
8. IOSR Journal of Engineering (IOSRJEN) www.iosrjen.org ISSN (e): 2250–3021, ISSN (p): 2278–8719 Vol. 04, Issue 01 (January. 2014), ||V3|| PP 56–59
9. IOSR Journal of Engineering (IOSRJEN) www.iosrjen.org ISSN (e): 2250–3021, ISSN (p): 2278–8719 Vol. 04, Issue 03 (March. 2014), ||V1|| PP 19–22
10. https://www.google.co.in/search?sitesearch=www.tutorialspoint.com&q=optimisation&cof=FORID:11&ie=ISO-8859-1&gws_rd=cr&dcr=0&ei=OlEmWoiqL4bgjwTkpZHwDQ

11. Bousias, K., Hasasneh, N., Jesshope, C.: Instruction level parallelism through micro threading-a scalable approach to chip multiprocessors. Comput. J. **49**(2), 211–233 (2006)
12. Rodrigues, A., Murphy, R., Kogge, P., Underwood, K.: Characterising a new class of threads in scientific applications for high end super computers. In: ICS '04: proceedings of the 18th Annual International Conference on Super Computing, pp. 164–174. ACM, New York, NY, USA (2004)
13. Frigo, M., Leiserson, C.E., Randall, K.H.: The implementation of the Clik-5 multithreaded language. SIGPLAN Not. **33**(5), 212–223 (1998)
14. Buluc, A., Gilbert, J.R.: Challenges and advances in parallel sparce Matrix-Matrix-Multiplication. In: Proceedings of 37th International Conference on Parallel Processing ICPP'08, Portland, Sept 2008
15. Alonso, P., Reddy, R., Lastovetsky, A.: Experimental study of six different implementations of parallel Matrix-Multiplications on Hetrogenous Computational Clusters of Multi-core processors. In: Proceedings of Parallel, Distributed and Network Based Processing (PDP), Pisa, Feb. 2010
16. Ohshima, S., Kise, K., Katagiri, T., Yuba, T.: Parallel processing of matrix-multiplication in a CPU and GPU heterogeneous environment. High Perform. Comput. Computat. Sci.-VECPAR (2006)
17. Gorder, P.F.: Multi-Core processors for science and engineering. Comput. Sci. Engg. 9(2), 3–7 (2007). https://doi.org/10.1109/mcse.2007.35

Modeling Compensation of Data Science Professionals in BRIC Nations

M. J. Smibi and Vivek Menon

Abstract This paper proposes a model for predicting the compensation of data science professionals in BRIC nations based on the worldwide Data Science Survey conducted by Kaggle in 2017. In this paper, we have used the Rosling's approach to adjust the compensation amount in BRIC currencies with respect to Purchasing Power Parity (PPP) units. Exploratory data analysis is used to identify the factors that influence the compensation amount, and an XGBoost algorithm is employed to predict the compensation. We evaluate the performance of the model by generating the Root Mean Squared Log Error (RMSLE) score. The results indicate a robust prediction using the XGBoost algorithm.

1 Introduction

With rapid digitization of economies, data has replaced oil as one of the most valuable resources in the world [1], leading to the emergence of a range of new roles and redefinition of associated skill sets, revolving around the analysis of data, and its organization. According to the Glassdoor's 50 Best Jobs in America report 2017 [2], "Data Scientist" continues to be ranked as the best job across every industry, given the lucrative earnings potential, exciting career opportunities, and the sizeable number job openings in this profession. According to an IBM report [3], the demand for data scientists will increase by 28% by 2020, with employers willing to pay premium salaries for qualified candidates.

Kaggle is an online platform for data science and machine learning enthusiasts to connect, learn, find, and explore data, as well as compete in data science and

M. J. Smibi
Department of Management, Amrita Vishwa Vidyapeetham, Kochi 682041, Kerala, India
e-mail: smibimj@gmail.com

V. Menon (✉)
Department of Computer Science and Engineering, Amrita Vishwa Vidyapeetham,
Kollam 690525, Kerala, India
e-mail: vivekmenon@am.amrita.edu

© Springer Nature Singapore Pte Ltd. 2019
A. Abraham et al. (eds.), *Emerging Technologies in Data Mining and Information Security*, Advances in Intelligent Systems and Computing 755,
https://doi.org/10.1007/978-981-13-1951-8_57

machine learning challenges. As the largest active community of data scientists and machine learning engineers, Kagglers span over 194 countries and come from a wide variety of backgrounds, including fields such as computer science, biology, computer vision, medicine, and even glaciology. In 2017, Kaggle conducted an industry-wide survey to understand the state of data science and machine learning [4]. Around 16,716 Kagglers from 171 countries responded to the study that collected a wide range of data totaling 228 features. The features range from primary demographic factors (age, gender, education level, etc.) to domain-specific data on data science and machine learning (time spent in analyzing data, frequently used data analysis tool, job function, etc.).

BRICS (Brazil, Russia India, China, and South Africa) grouping has grown into a vibrant platform for development and mutual cooperation among the major emerging markets and developing countries in Asia, Africa, Europe, and America. Together, BRICS accounts for 26.46% of world land area and 42.58% of world population. Additionally, BRICS generated 22.53% of the world GDP in 2015 and contributed more than 50% of world economic growth during the last 10 years. Given the increasing importance of the BRICS, various studies have been undertaken to focus on the opportunities and challenges within this crucial group of countries [5–7].

Regression models tend to remain a primary choice for modeling salary and compensation trends worldwide [8–10]. While the traditional focus of most of these regression studies has been on explanatory modeling, the primary focus of our paper is on predictive modeling [11] of the compensation of the data science professionals in BRIC nations (Brazil, Russia, India, and China). The worldwide Data Science Survey conducted by Kaggle in 2017 forms the basis for our study. As the respondents from South Africa were limited, we focus only on the BRIC nations in this study.

2 Exploratory Data Analysis

The Kaggle community released the Kaggle survey 2017 dataset along with a report describing the basic features of the dataset. This initial report is an excellent starting point and provides useful insights about the average age, level of education, most preferred data analysis tool, etc. of the respondents. Additional insights from exploratory data analysis performed by other Kagglers, available in their public kernels [12], provided a better understanding of this dataset and helped to fine-tune the proposed model.

From the original Kaggle dataset with 16,716 records, we extract the subset of BRIC nations, which contains 4,218 records. Exploratory data analysis is performed on this subset to gather insights about the BRIC nations. These insights help to identify the factors that influence the compensation amount. Majority of the respondents in the BRIC dataset are from India as can be seen in Fig. 1a.

Females constitute only 12.36% of the total respondents from BRIC nations, and a majority of the respondents are male. From Fig. 1b we can see that the highest numbers of male respondents are from Brazil. Around 91.7% of the respondents

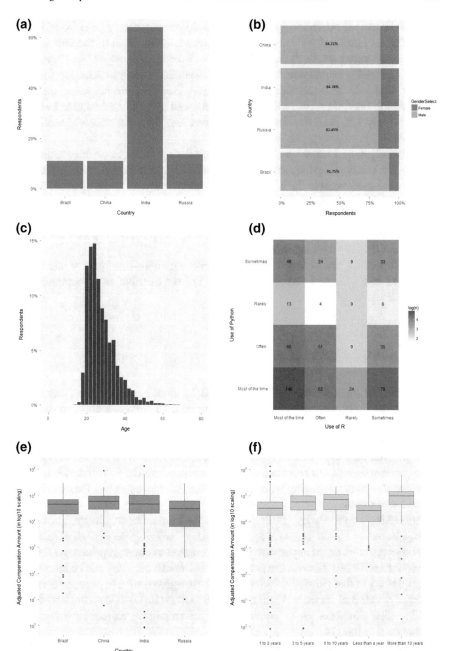

Fig. 1 Exploratory data analysis: Kaggle data of BRIC nations

from BRIC nations are in the age group of 20–40 years as can be seen from Fig. 1c. Only 10.6% of professionals are Doctoral degree holders. Most of the data science professionals have Bachelor's degree or a Master's degree. All respondents belonging to BRIC nations are code writers, i.e., they write code to analyze data. Job function plays a vital role in determining compensation. Building prototypes is the most rewarded job function across BRIC nations followed by building Machine Learning services. Only 7.87% of the data science professionals have a tenure of more than 10 years.

Tools and programming languages used at work vary from one country to another. While AWS and Hadoop users receive the highest compensation in India and Russia, Tableau tool users receive the highest compensation in Brazil and China. MATLAB users are the lowest paid as per this survey. Naive Bayes appears to be the least popular method used at work. Python and R are the most commonly used data analysis tools among the data science professionals and a majority of respondents prefer using both Python and R as can be seen in Fig. 1d.

Boxplots of log-transformed adjusted compensation against country and tenure are plotted in Fig. 1e and f to get a better idea of the distribution of compensation data in the BRIC dataset.

2.1 Adjusted Compensation

Direct comparison of compensation across BRIC nations is a challenging task, as the cost of living is not the same across these countries. To provide a uniform basis for comparison, we utilize the Rosling's approach [13] for standardizing the compensation across different countries. In this method, the typical elements of variation such as the population size of a country, the relative standard of living, or the period over which certain figures are reported, are adjusted by normalization. Individual compensation values are adjusting with respect to the Purchasing Power Parity (PPP) of respective currencies.

Purchasing Power Parity (PPP) compares the currencies of different countries through a market basket of goods approach. After factoring the exchange rate, two currencies are considered at par when a market basket of goods is priced the same in both countries. In other words, the purchasing power of a currency can be interpreted as the quantity of that currency needed to purchase a given unit of a good, or common basket of goods and services. Keeping the US Dollar (USD) as the benchmark, the PPP in effect translates as an equivalent of the respective national currency per international dollar [14].

The currency and Purchasing Power Parity (PPP) of BRIC nations are shown below in Table 1 [14].

Table 1 PPP of BRIC nations

Country	Currency	PPP
Brazil	BRL	2.05
Russia	RUB	25.13
India	INR	17.70
China	CNY	3.51

Thus, the adjusted compensation amount in US Dollars (USD) for the BRIC nations is calculated according to the formula given below:

$$Adjusted\,Compensation = \frac{Compensation\,Amount * Market\,Exchange\,Rate}{PPP * Market\,Exchange\,Rate}$$

(1)

The initial step is to convert the compensation amount of respondents into USD using the Market Exchange Rate (MER), available in the Kaggle dataset. The compensation amount of the respondents in USD is subsequently normalized using the PPP, also adjusted to USD using the MER, as illustrated in Eq. 1.

2.2 Preprocessing

The compensation amount is adjusted using the approach discussed above. One of the main issue with the dataset is the presence of missing fields. Outliers in the compensation range of BRIC nations are removed. For the numerical attributes in the dataset, median values as central tendency are used for filling the missing values.

3 Prediction Model

3.1 Feature Selection

The feature selection process involves selecting important attributes from the dataset, specific to BRIC nations. We identify 23 independent attributes to build a model for predicting compensation, with adjusted compensation as the target variable. Table 2 lists the independent numerical attributes, while Table 3 lists the independent categorical attributes. We transform the categorical attributes into dummy variables for incorporating them into the prediction model.

Table 2 Numerical attributes

Attributes	Description
Age	Age of the respondent
TimeGatheringData	% of time devoted to gathering data
TimeModelBuilding	% of time devoted to model building
TimeProduction	% of time devoted to production
TimeVisualization	% of time devoted to visualizing data
TimeFindingInsights	% time devoted to finding insights

Table 3 Categorical attributes

Attributes	Description
Country	Country in which the respondent lives in
GenderSelect	Gender of the respondent
FormalEducation	Level of formal education
MajorSelect	Under graduation major
ParentsEducation	Education level of parents
EmploymentStatus	Current employment status
CodeWriter	Whether the respondent writes code to analyze data
Tenure	How long the respondent has been writing code to analyze data
JobFunctionSelect	Functions of the respondent's job
MATLABUsers	Whether the respondent uses MATLAB
AWSUsers	Whether the respondent uses AWS
HadoopUsers	Whether the respondent uses Hadoop
SparkUsers	Whether the respondent uses Spark
NaiveBayesUsers	Whether the respondent uses naive Bayes
PythonVsR	Whether the respondent uses Python or R or both
RecommenderSystemUsers	Whether the respondent uses recommender systems
DataVisualizationUsers	Whether the respondent uses data visualization

3.2 Prediction

Though multiple data mining techniques are available for prediction tasks involving continuous/categorical predictors and continuous response [15], we have used the XGBRegressor function of the XGBoost algorithm [16] to model adjusted compensation as a function of 23 independent predictor attributes listed across Tables 2 and 3. The preprocessed dataset consists of 833 rows and 71 columns with no missing data, is partitioned into training and test sets. The training set consists of 666 data points randomly drawn from the dataset.

XGBoost is a highly sophisticated algorithm used for predictive modeling and is capable of dealing with all sorts of irregularities in the data [17]. It has high flexibility and allows the user to run cross-validation at each iteration of the boosting process. The prediction model is applied to the test data with 167 data points.

3.3   Performance Evaluation

The performance of the prediction model is evaluated based on the Root Mean Squared Logarithmic Error (RMSLE) metric [18]. Effectively, RMSLE metric is the Root Mean Square Error (RMSE) of the log-transformed predicted and target values [19]. In other words, it is the standard deviation of magnitude of the predictions compared to the magnitude of the actual measurements, on a log-scale. RMSLE score is used to avoid penalizing the huge differences in the predicted and actual values of the compensation amount, the response variable in our study, especially when both predicted and actual values are huge. On applying the XGBoost model, we obtain an RMSLE score of 0.1158, which translates to a standard deviation of $e^{0.1158}$ or 1.12. This means the predicted values are well within a small range of 1.12 times greater or lesser than the actual measurements, thus indicating a robust prediction.

4   Conclusion

An XGBoost algorithm is tested to predict the compensation of the Kaggle users belonging to the BRIC nations based on the worldwide survey conducted by Kaggle in 2017. The compensation amount is converted to PPP dollars for uniform comparison across different countries. In the PPP approach, a basket of similar products is used for price comparison across different countries, though this may not always be reasonable, considering the geographical factors that influence the production/availability of such products in some countries. The model generated an RMSLE score of 0.1158, which translates to a standard deviation of $e^{0.1158}$ or 1.12, indicating a robust prediction. Improvements in feature selection process as well as better handling of missing data could further enhance the overall quality of the training data and help to further improve the results obtained from the XGBoost algorithm.

References

1. Economist: The world's most valuable resource is no longer oil, but data (2017). https://www.economist.com/news/leaders/21721656-data-economy-demands-new-approach-antitrust-rules-worlds-most-valuable-resource
2. Best Jobs in America (2017). https://www.glassdoor.com/List/Best-Jobs-in-America-LST_KQ0,20.htm
3. Columbus, L.: IBM predicts demand for data scientists will soar 28% by 2020 (2017). https://www.forbes.com/sites/louiscolumbus/2017/05/13/ibm-predicts-demand-for-data-scientists-will-soar-28-by-2020/
4. Kaggle, M.L.: Data science survey (2017). https://www.kaggle.com/kaggle/kaggle-survey-2017
5. Carnoy, M., et al.: University expansion in a changing global economy: triumph of the BRICS (2013)
6. Vizgunov, A., Glotov, A., Pardalos, P.M.: Comparative analysis of the BRIC countries stock markets using network approach. In: Proceedings in Mathematics and Statistics, pp. 191–201 (2013)
7. Mazzioni, S., et al.: The relationship between intangibility and economic performance: study with companies traded in Brazil, Russia, India, China and South africa (BRICS). In: Advances in Scientific and Applied Accounting, pp. 122–148 (2014)
8. Scott, E.: Higher Education Salary Evaluation Kit. American Association of University Professors (1977)
9. Moore, N.: Faculty salary equity: issues in regression model selection. Res. High. Educ. **34**, 107–126 (1993)
10. Billard, L.: Study of salary differentials by gender and discipline. Stat. Public Policy **4**, 1–14 (2017)
11. Shmueli, G.: To explain or to predict. Stat. Sci. **25**(3), 289–310 (2010)
12. Jabri, M.: Salary and purchasing power parity (2017). https://www.kaggle.com/mhajabri/salary-and-purchasing-power-parity
13. Hirst, T., Rosling, H.: How to compare income across countries (2015). http://www.open.edu/openlearn/science-maths-technology/mathematics-and-statistics/how-compare-income-across-countries
14. Implied PPP conversion rate (2017). http://www.imf.org/external/datamapper/PPPEX@WEO/OEMDC/ADVEC/WEOWORLD/IND?year=2017
15. Shmueli, G., Bruce, P.C., Patel, N.R.: Data mining for business analytics: concepts, techniques, and applications with XLMiner (2016)
16. Python API Reference—xgboost 0.6 Documentation. http://xgboost.readthedocs.io/en/latest/python/python_api.html
17. Jain A.: Complete guide to parameter tuning in XGBoost (with codes in Python) (2016). https://www.analyticsvidhya.com/blog/2016/03/complete-guide-parameter-tuning-xgboost-with-codes-python/
18. Kaggle Forums (2014). https://www.kaggle.com/general/9933
19. Pentreath, N., Ghotra, M.S., Dua R.: Machine Learning with Spark, 2nd edn (2017)

Horizontal Scaling Enhancement for Optimized Big Data Processing

Chandrima Roy, Kashyap Barua, Sandeep Agarwal, Manjusha Pandey and Siddharth Swarup Rautaray

Abstract Big Data, as we all know, is becoming a new technological trend in the industries, in science and even businesses. Indefinite data scalability allows organizations to process huge amounts of data in parallel, assisting dramatically decrease the amount of time it takes to manage several amount of work, optimize hardware resource usage and permit the extreme quantity of data per node to be handled. Optimization is to done to attain the finest strategy relative to a set of selected constraints which include maximizing factors such as efficiency, productivity, reliability, strength, and utilization. When the current system becomes insufficient, instead of upgrading it by adding more components to the existing structure you just add more computers to a cluster. This research discusses a hierarchical architecture of Hadoop Nodes namely Name nodes and Data nodes and mainly focuses on the optimization of Data Node by distributing some of its work load to Name Node.

Keywords Big data · Hadoop · Optimization · Scalability · Horizontal scaling

C. Roy (✉) · K. Barua · S. Agarwal · M. Pandey · S. S. Rautaray
School of Computer Engineering, KIIT, Deemed to be University,
Bhubaneswar 751024, Odisha, India
e-mail: Chandrima.roy.1914@gmail.com

K. Barua
e-mail: kashyapbarua@gmail.com

S. Agarwal
e-mail: sandygarg65@gmail.com

M. Pandey
e-mail: manjushafcs@kiit.ac.in

S. S. Rautaray
e-mail: siddharthfcs@kiit.ac.in

© Springer Nature Singapore Pte Ltd. 2019
A. Abraham et al. (eds.), *Emerging Technologies in Data Mining and Information Security*, Advances in Intelligent Systems and Computing 755,
https://doi.org/10.1007/978-981-13-1951-8_58

639

1 Introduction

The world is changing rapidly with time it is growing towards the hi-tech era where technology is playing a great role. As the world is rapidly pacing into hi-tech era one thing has also grown tremendously within a few years that is data. Today data has rose from terabytes to zettabytes of data. As data is growing tremendously in a rapid velocity and different varieties and structure, efficient [1] tools are needed to organize these huge data known as big data. Some of the tools are already playing vital roles in efficiently working with big data, one such tool is hadoop. As hadoop is a new technology there are still some areas in hadoop which can be better optimized to increase the efficiency and throughput of hadoop ecosystem.

Optimization is the act of design and developing systems in such a way that it can take greatest advantage of the available resources. Optimization of applications can be done to take advantage from the huge amount of memory [2] space present on a specific computer, or the hardware speed, or the processor being used. Finding the alternative way to achieve the highest performance in cost effective manner and under given limitation by mostly utilizing the desired factors. In comparison, maximization is the process of attaining highest or maximum performance deprived of cost or expenditure. Practice of optimization is restricted when there is shortage of information and lack of time to estimate what information is obtainable.

"Horizontal scaling" is one of the important aspects of Big Data technologies. Hadoop arranges computers in the structure of a tree topology which can be developed to any size; the biggest ones presently in use process many petabytes through thousands of nodes. Fault tolerance [3] is the outcome of horizontal scaling. Big Data structures take it as a fact of life, and handle them automatically.

1.1 Hadoop Cluster

The major components used by name node to allocate location of block on different data nodes are depends upon the Nearest Location, Data Redundancy and Network Traffic (If any of the Node has more usage or more traffic name node will divert the data to different data node which is second nearby) are depicted in below Fig. 1.

NameNode has Meta data. DataNode will constantly send a heartbeat to Name node in this way Name node understands that Data node [4] is working, if in case (due to any reason) Data node stops sending the heartbeat to the Name node, then name node will come to know that that particular Data node is down and then make

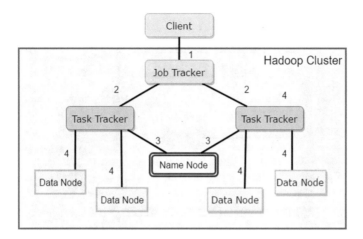

Fig. 1 Task distribution on Hadoop cluster

sure that the Blocks in that Date node get replicated in another node and if in case the node which stopped sending [5] the heartbeat again started to send its heartbeat then Name node will balance the replication factor again. In this way, Name node handles the data node failure in Hadoop HDFS.

Task Tracker is a slave node daemon service for running the tasks that job tracker assigns for each node. One Task Tracker used for processing tasks (map, shuffle, sort, reduce) received from Job Tracked. A Task Tracker accepts tasks such as map, reduce and shuffle operations from a Job Tracker. Every Task Tracker is designed with a set of slots; that means the number of tasks it can accept.

1.2 Dead Node Detection

The Task Trackers send out heartbeat messages to the Job Tracker, typically every few minutes, to assure the Job Tracker that it is still alive. These messages also notify the Job Tracker of the number of available slots, so the Job Tracker can stay up to date with where in the cluster work can be delegated [6]. DataNode will constantly send a heartbeat to Name node in this way Name node understands that Data node is working, if in case (due to any reason) Data node stops sending the heartbeat to the Name node , then name node will come to know that that particular Data node is

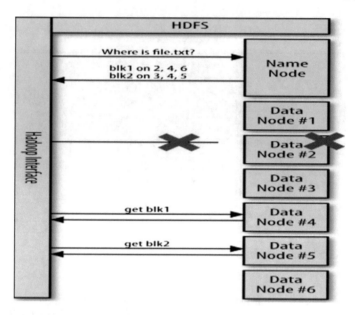

Fig. 2 Data node detection

down and then make sure that the Blocks in that Date node get replicated in another node and if in case the node which stopped sending the heartbeat again started to send its heartbeat then Name node will balance the replication factor again.

As we can see in Fig. 2, data blocks are replicated in several data nodes. Among them the data node #2 is not responding [7], which contained the block. After some time the block goes to the dead state.

2 Problem Definition

Historical data is digital evidence drawing activity, circumstances and developments in a company's past. Data analysis forecasts future trends, performances, or actions based on historical data. Previously number of historical Data [8] is limited but day by day it is increasing. To handle such amount of vast data we need data nodes which can perform action on that data. As number of data node in a cluster is also limited, all the data cannot be processed at a time. Suppose 260 MB of data is present, for that we need 5 data nodes. As a single data node can handle up to 64 MB of data. Now 64 * 4 = 256 MB of data is now handled by 4 data nodes, still some small amount of data (260 – 256 = 4 MB) is left, for which another data node is needed to complete the execution. But that last data node is not utilized fully. It can handle more (64 – 4 = 60 MB) data. Instead of using the 5th data node the smallest amount of data could have been analyzed by the name node. A framework is proposed to reduce the work load from data node by scaling the name node horizontally.

2.1 Name Node Taking Over the Function of Dead Data Node

NumLiveDataNodes and NumDeadDataNodes together track the number of Data Nodes in the cluster. Generally, number of live Data Nodes and number of Data Node allocated for the cluster are equal. DataNode is marked as "stale", when the name node does not hear from a DataNode for 30 s. If the data node fails to connect with the name node for 10 min, the DataNode is marked "dead." The death of a data node causes a flood of network activity, as the dead data nodes contains replicas of blocks. Though the loss of a single data node may not affect performance [9] much, losing several Data Nodes could result in data loss. Rather waiting for another data node to free and continue the incomplete task of that dead node, that small amount of task can be performed by the name node itself. As name node has the information of the replicated blocks of that dead node, it is easy for name node to continue execution from the left state of the incomplete task.

3 Methodology

Hadoop accomplishes its operations with the help of the MapReduce model, which comprises two functions namely, a mapper and a reducer. The Mapper function is responsible for mapping the computational subtasks to different nodes and also responsible for task distribution, load balancing and managing the failure recovery. The Reducer function takes the responsibility [10] of reducing the responses from the compute nodes to a single result. The reduce component has the responsibility to aggregate all the elements together after the completion of the distributed computation.

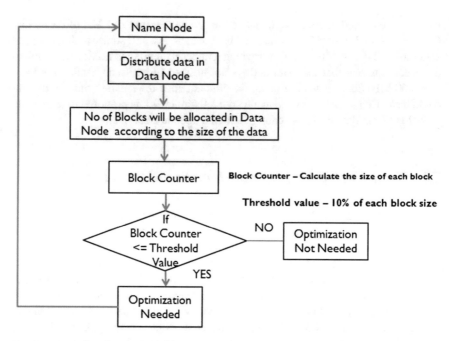

Fig. 3 Flowchart of the proposed framework

The flowchart in the Fig. 3 represent the sequence of steps of how data flows through the MapReduce process and how MapReduce handles and processes data. Also exploring the mapping and reducing steps. A counter variable is present to count the number of Data Node used in a certain execution. The Threshold is the optimum number of Data nodes that can be connected to the Name node for optimized processing and perfect load balancing. The result of the checking decides whether optimization is needed or not. That means if the counter value is greater than the threshold value then only optimization is needed.

```
Algorithm 1:
Input:
JobConf-> Job configuration object for a specific
tracker.
JobId-> Represents the unique identifier for a Job.
JobTrackAddr-> Get job status for all jobs in
the cluster and report it to JobTrackAddr.
TaskTrackerName-> Name of the Task tracker
TaskID-> Represents the unique identifier for
a Map or Reduce Task.
TaskReporter-> Get task status for all tasks
Key-> file name, line number.
value -> Contents of the line
N(D_N)-> Number of Data Nodes present in the execution
(T_DN)-> Task Assign to each data node
N(T_DN)-> Total task assign
J(N_N) -> A Job assign to Name Node
Step 1:- f_T -> { □ T_x ∈ J(N_N)
        Let T_x= {T_1,T_2,T_3,…,T_N} }
//Where f_T is the function which breaks the Job J(N_n)
to various task T_x that is a set of  T_1,T_2,T_3,…,T_N.
Each node represented by a count value.
Step 2:-
RunningJob-> addTaskToJob(JobID,TaskInProgress)
//Here JobTracker finds the best TaskTracker nodes to
execute tasks based on the data locality and JobID is
asked from JobTracker
TaskInProgress -> It maintains all the info for a
//Task that lives at   this TaskTracker and also
maintains the Task object, TaskStatus and the
TaskRunner.
Step 3:- taskId -> TaskInProgress
//This is a map of the tasks that have been assigned
to task trackers, but  that have not yet been seen
in a status report.
Step 4:- Thread.sleep(TASKTRACKER_EXPIRY_INTERVAL/3)
//Every 3 minutes checks for any overdue tasks.
Step 5:- TaskId--> Assign {getKey,getValue}
//Here Key /value pair is assigned to each task to
each data node.
Step 6:- addToMemoryManager --> {TaskID,JobConf}
//Memory has been alloted to each job along with its
individual task.
```

The above Algorithm 1 presents the task allotment process.

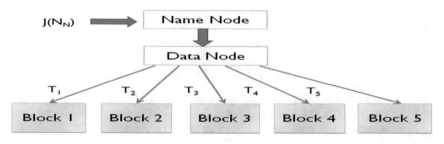

Fig. 4 Task distribution between name node and data node

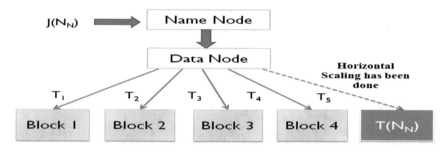

Fig. 5 Optimization of data node

```
Algorithm 2:
Step 7:- class Mapper
         method Map(string t; integer r)
         Emit(string t; pair (r; 1))
Step 8.1:- fTDN -> Assign { T1 -> B1
              T2 -> B2
              T3 -> B3
                 .
                 .
                 .
              TN -> BN}
fTDN is the function which assign each task to
each block.
```

The mapper emits an intermediate key-value pair for each word in a document and simply emit "1″ for each term it processes. f_{TDN} is the function which assign each task to each data node which is represented in Fig. 4. A counter variable is present to count the number of Data Node used in a certain execution [5] which is represent by TDNN. Threshold calculates the optimum number of data nodes that can be connected to the name node for optimized processing and perfect load balancing. At any instant of time it is found that the load on a particular name node has reached the threshold limit that means the counter value is greater than the threshold value, and then only optimization is needed. Figure 5 demonstrate that up to D4 all the data nodes have been fully utilized. In this scenario some small amount of remaining data has been left for which we need another data node. But that last node is not utilized fully. That's why name node enhancement comes to the picture. Here name node executes the remaining amount of data instead of the data node. Name node has been horizontally scale along with the previous data node to enhance the execution time and as well as memory. $T_{(NN)}$ indicate the virtual work done by name node.

```
Step 8.2.1:-Conditional statement
            C₁= {If T_DNN > Threshold}
    Here we are checking whether the T_DNN value is
greater than the threshold value or not.
            If C₁ == True
            f_o -> optimize(f_TDN)
            else
            Follow f_TDN
            T(N₅)-> N_N
    Leftover task assign to Name Node rather than
    using another block for execution.
```

```
Algorithm 3:
Step 9:- class Combiner
      method Combine(string t; pairs [(s1; c1);
   (s2; c2) : : :])
      sum <- 0
      cnt <- 0
      for all pair (s; c) 2 pairs [(s1; c1);
   (s2; c2) : : :] do
      sum <- sum + s
      cnt <- cnt + c
      Emit(string t; pair (sum; cnt))
Step 10:- class Reducer
      method Reduce(string t; pairs [(s1; c1);
   (s2; c2) : : :])
      sum <- 0
      cnt <- 0
      for all pair (s; c) 2 pairs [(s1; c1);
   (s2; c2) : : :] do
      sum <- sum + s
      cnt <- cnt + c
      ravg <- sum=cnt
      Emit(string t; integer ravg)
Step 11:- ravg -> TaskReporter
      Execution result after the reducer
      phase is stored in TaskReportAddress
Step 12:- TaskReporter -> JobTrackAddr
      Update the final execution result.
Output: When the Job Tracker receives notification
      from Task Tracker that the last task of
      the Job is complete;
      Status of job is changed to successful.
```

Combiners processed the output which is generated by the mappers. Combiner performs local aggregation [11] before map and reduce phase to cut down on the number of intermediate key-value pairs. Combiner produces a key-value pair for each word in the group. All values associated with the same intermediate key are executed by the reducer to generate output key-value pairs. The reducer emits final key-value pairs with the word as the key, and the count as the value. Execution result after the reducer phase is stored in TaskReportAddress. When the Job Tracker receives notification from Task Tracker that the last task of the job is complete, status of job is changed to successful.

4 Conclusions

Today, every minute, every second huge amount of data is produced. This vast quantity of data produced makes it very hard to store, manage and analyses it. The improvement of existing big data analysis tools has helped with handling this gigantic amount of data to a great extent. In this research some modification has been proposed on the existing name node and data node work process. Optimization [9] of data node is needed periodically with the increase of the huge amount of data, huge number of the data node also needed to process all the data. As it is time consuming and as well as memory consuming, a framework is proposed for ensuring effective data placement policies to speed up HDFS and reducing work load from data node by scaling the name node horizontally. The major idea behind the design is to produce more optimized and more scalable HDFS architecture.

References

1. Yadav, K., Pandey, M., Rautaray, S.S.: Feedback analysis using big data tools. In: International Conference on ICT in Business Industry & Government (ICTBIG). IEEE (2016)
2. Chakraborty, S. et al.: A proposal for high availability of HDFS architecture based on threshold limit and saturation limit of the namenode (2017)
3. Jena, B. et al.: Name node performance enlarging by aggregator based HADOOP framework. In: 2017 International Conference on I-SMAC (IoT in Social, Mobile, Analytics and Cloud)(I-SMAC). IEEE (2017)
4. Shvachko, K., et al.: The hadoop distributed file system. In: 2010 IEEE 26th Symposium on Mass Storage Systems and Technologies (MSST). IEEE (2010)
5. Jahani, Eaman, Cafarella, Michael J., Ré, Christopher: Automatic optimization for MapReduce programs. Proc. VLDB Endow. **4**(6), 385–396 (2011)
6. Lee, K.-H. et al.: Parallel data processing with MapReduce: a survey. ACM sIGMoD Record **40**(4), 11–20 (2012)
7. White, T.: Hadoop: The Definitive Guide. O'Reilly Media, Inc. (2012)
8. Kanaujia, P.K.M., Pandey, M., Rautaray, S.S.: Real time financial analysis using big data technologies. In: 2017 International Conference on I-SMAC (IoT in Social, Mobile, Analytics and Cloud)(I-SMAC). IEEE (2017)
9. Borthakur, Dhruba: The hadoop distributed file system: architecture and design. Hadoop Proj. Website **11**(2007), 21 (2007)
10. Jena, B. et al.: A survey work on optimization techniques utilizing map reduce framework. Hadoop Cluster. Int. J. Intell. Syst. Appl. **9**(4), 61 (2017)
11. Feng, D., Zhu, L., Zhang, L.: Review of hadoop performance optimization. In: 2016 2nd IEEE International Conference on Computer and Communications (ICCC). IEEE (2016)

Part VI
Cryptology

Implementation and Analysis of Cryptographic Ciphers in FPGA

V. G. Kiran Kumar and C. Shantharama Rai

1 Introduction

One of the defining trends in the digital era is the information about every aspect of our lives needs to be secured and can be achieved by various cryptographic algorithms and these algorithms have been extensively deployed in tiny computing devices. As the current research focus and challenge is on devising ciphers for sides of high end devices like desktop and servers and also the low constrained devices like the RFID, IoT, etc. This paper presents the implementation of few Lightweight cryptographic algorithms and conventional cryptographic algorithms [1–3]. PRESENT and SIMON is newly introduced lightweight cryptographic algorithm but both with a different architecture and with different operations [2, 4–9]. The lightweight ciphers like TEA, XTEA, and SIMON uses simple operations like shift and XOR, while PRESENT makes uses of substitution and permutation of bits [10–12]. These algorithms are designed for low constrained resources in terms of area power and time. BLOW-FISH, IDEA, and AES are conventional ciphers. IDEA makes use of XOR, Addition and multiplication modulo operations. The AES and BLOWFISH make the use of S-BOX and Permutation so as to achieve the confusion and diffusion operations. The conventional and lightweight ciphers are evaluated by the parameters—speed, cost, performance, and balanced efficiency in hardware implementation. Since the PRESENT cipher has a status of standard cipher it will be taken as the basis performance evaluation.

V. G. Kiran Kumar (✉)
Sahyadri College of Engineering and Management, Adyar, Mangalore 575009, India
e-mail: kiranvgk@gmail.com

C. Shantharama Rai
AJ Institute of Technology, Kuloor, Mangalore 575013, India
e-mail: csraicec@gmail.com

© Springer Nature Singapore Pte Ltd. 2019
A. Abraham et al. (eds.), *Emerging Technologies in Data Mining and Information Security*, Advances in Intelligent Systems and Computing 755,
https://doi.org/10.1007/978-981-13-1951-8_59

2   The Implementation of the Ciphers

This section will discuss briefly the algorithms implemented in this paper.

2.1   IDEA Algorithm

IDEA is a conventional, block cipher designed by Lai, Massey, and Murphy in 1991. (IDEA) is a symmetric key cipher, with 128-bit key length and 64-bit input data. The IDEA cipher uses mixing of three different elementary mathematical operations.

- Bitwise-XOR.
- Integer-Addition (Modulo-2^{16}).
- Integer-multiplication (Modulo-$2^{16} + 1$).

IDEA cipher for one round is shown in Fig. 1. The 64-bit input plaintext is divided into four sub-blocks P_1, P_2, P_3, and P_4, each of 16-bit, P_1, P_2, P_3, and P_4 forms the input to the first round. The four sub-keys K_1, K_2, K_3 and K_4 are added in each round, multiplied and XORed with 16-bit sub-keys. The second sub-block is swapped with third sub-block at the end of each round. The steps are performed for eight rounds., In the output transformation round, four sub-keys are combined with the four sub-blocks after the eighth round. The following are the sequence of steps for each round (All multiplication and addition are multiplication modulo and addition modulo operation):

1. Sub-block P_1 is multiplied with the first sub-key K_1.
2. Sub-block P_2 is added with the sub-key K_2.
3. Sub-block P_3 is added with the sub-key K_3.
4. Sub-block P4 is multiplied with the sub-key K_4.
5. The output of Step 1 is bitwise-XORed with output of Step 3.
6. The output of Step 2 is bitwise XORed with output of Step 4.
7. The output of Step 5 is multiplied with the sub-key K_5.
8. The outputs of Steps 6 and 7 will be added.
9. The output of Step 8 is multiplied with the sub-key K_6.
10. The outputs of Step 7 and Step 9 are added.
11. The output of Step 1 is are bitwise XORed with output of step 9.
12. The output of Step 3 is are bitwise XORed with output of step 9.
13. The output of Step 2 is are bitwise XORed with output of step 10.
14. The outputs of Steps 4 and 10 are bitwise XORed.

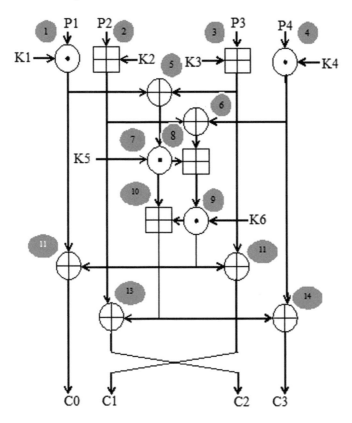

Fig. 1 Block diagram of IDEA encryption process

2.1.1   Key Generation

The 128 bit key is sub-divided into 16 bit sub-blocks, the first 96 bits are the six sub-keys K_1 to K_6 for the initial round. Thus, after the first round, 32 bits (i.e. 97–128) of the primary key are unused. In the second round the unused 32 bits of the primary key of the first round and another 64 bits required are generated from the primary key which is left shifted circularly by 25 bits. We then generate the remaining four sub-keys using the modified key in the same way as the first round keys were generated. Thus we repeat the sub-key generation for the remaining seven rounds.

2.2   *SIMON Cipher*

SIMON cipher designed by the NSA for high-performance in hardware for encryption for highly constrained devices; it is a Feistel block ciphers. SIMON 2n/ωn refers

Fig. 2 SIMON round
function

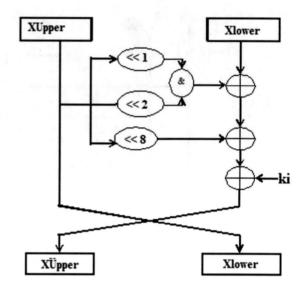

to SIMON cipher with plain text size of 2n and size of the key is ωn bits, where n is 16, 24, 32, 48, or 64 and ω is 2, 3, or 4.

SIMON $2n$ cipher performs three operations:

1. Bitwise-XOR
2. Bitwise-AND
3. Left bit rotation and Right bit Circular Shifts, Sj and S-j by j bits, respectively.

For $k \in GF(2)^n$, the Simon$2n$ round function is
$R_k : GF(2)^n \times GF(2)^n \to GF(2)^n \times GF(2)^n$ defined by

$$R(l, r, k_i) = ((Shl(1) \& Shl(8)) \oplus Shl(2) \oplus X_{lower} \oplus k_i, l)$$

Figure 2 shows the SIMON round function.
The sequence of steps for each round is as follows:

1. Shift Left by one bit.
2. Shift left by 8 bit
3. AND the output of step 1 with output of step 2
4. XOR the output of step 3 with Lower word
5. Shift left by 2 bit
6. XORing the output of step 4 with step 5.
7. XORing the output of step 6 with key K_i
8. Swap Lower and upper words.

In the SIMON key generation, the round keys will be generated from the primary key. Simon key expansion for m=4 is shown in Fig. 3. In the SIMON64/128 key generator, a 128 bit master key is used to generate 44, 32-bit sized round keys. For

Fig. 3 Simon key expansion

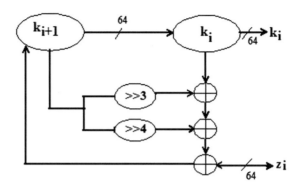

any given round i, the currently cached previous n round keys (where n is the key words parameter) is combined with a constant and a 1-bit round constant to generate the keys. The key expansion function utilizes the following operations:

- Bitwise XOR, denoted as $x \oplus y$.
- Right bitwise rotation, denoted as $S^{-y}(x)$ where y is the rotation count.

2.3   *PRESENT Algorithm*

PRESENT the block cipher was designed in 2007 by the Technical university of Denmark, The Orange Labs (France) and The Ruhr University Bochum (Germany). It is a SPN structure; it encrypts 64 bit plain text and uses 80 and 128 bit key. It consists of 32 rounds in which have 31 regular rounds and one final round. There are three transformations in each round, add Round Key, Substitution-box and the Permutation. These three transformations are operated on previous output called intermediate result. The round key is generated in the Key Schedule which used for round operation. The 64-bit plaintext P after 31 rounds of operation and XORed in the last round with the round key to get 64-bit cipher-text C. Permutation is simple bit transposition. Each bit position is changed and it is moved to corresponding position as provided in the P box. Substitution layer has 16 S-boxes. Each S-box has four bit inputs and four bit output. S-box performs nonlinear substitution so as to ensure avalanche effect. Figure 4 depicts PRESENT Encryption algorithm.
    The steps are as follows

1. Generate the round keys.
2. XOR the Round key with the state.
3. S-box substitution for the state.
4. Permutation layer for the state.
5. Perform steps 1 to 4 to be for 31 rounds.
6. For 32 round XOR addRoundkey.

Fig. 4 PRESENT
encryption algorithm

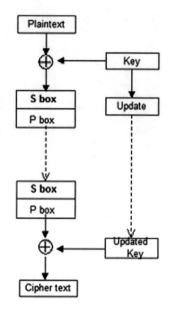

2.4   Tiny Encryption Algorithm (TEA)

The Tiny Encryption Algorithm (TEA), a Fiestel type cipher, developed by David
Wheeler and Roger Needham of the Cambridge Computer Laboratory in 1994. TEA
encrypts a plaintext block of 64 bits using a key of 128 bits. The algorithm performs
32 rounds. The TEA cipher partitions the 128 bit key into 4 sub-keys of 32 bits each,
namely K_0, K_1, K_2 and K_3. 64 bit data block is split into two 32 bit blocks. The 32
bit left side block is denoted as L and the 32 bit block on the right side is denoted as
R. A constant integer value 2^{32} is defined, the multiples of which are used in each
round. Figure 5 shows the block diagram of TEA Algorithm.

For every feistel round, R will perform simple different operations like left shift,
right shift and integer modulo addition. The addition operations performed are modulo 2^{32}.

The sequences of steps for each round are

1. R is left shifted by 4 and then sub-key K_0 is added.
2. The constant Delta is added to R.
3. R is shifted right by 5 and sub-key K_1 is added.

An XOR operation is performed on the output of the above three steps and the
result is added to L. It is then swapped, so that the result becomes R for the next
feistel round.

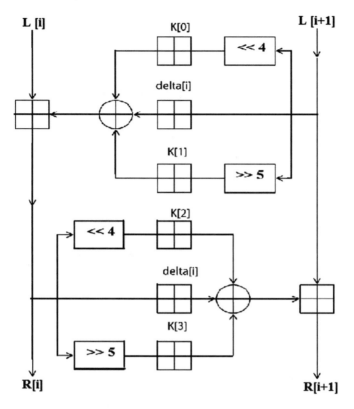

Fig. 5 Block diagram of tiny encryption algorithm

2.4.1 Key Generation Using LFSR

Figure 6 shows the LFSR, a shift register which generates the key sequences.

It designed by tapping the bits, the taps are XORed sequentially and then fed back into the leftmost bit as shown. The LFSR takes an initial seed value of 128 bits and by XORing the certain bits of the shift register value and drives the ciruitry.

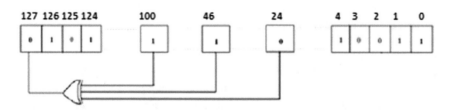

Fig. 6 Key generation using linear feedback shift register

Fig. 7 Block diagram of
XTEA algorithm

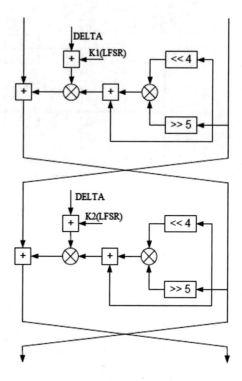

2.5   *EXTENDED Tiny Encryption Algorithm (XTEA)*

XTEA is a fiestel type cipher, developed by Wheeler and Needham in 1997. The
weaknesses and vulnerabilities of TEA led to the implement of some improvements
over TEA and called it Extensions of TEA (XTEA). A block of 64-bit plaintext
using a 128-bit key is encrypted. XTEA consists of 64 rounds, or 32 cycles. Figure 7
depicts the Block diagram of XTEA. The steps are

1. The 128 bit key is partitioned into four sub-blocks each of 32 bits namely K_0,
   K_1, K_2 and K_3
2. The 64 bit plain text is partitioned into two blocks of 32 bits each. The Left block
   denoted by L and right block denoted as R.
3. Left Shift R by 4 bits.
4. Right shift R by 5 bits.
5. Bitwise XOR operation of steps 3 and 4.
6. Perform Modulo addition 2^{32} operation with result of step 5 with R.
7. XOR the LFSR and Delta function with shifted data.
8. Perform steps 3 to 7 operations for 64 rounds to obtain the cipher text.

Fig. 8 Block diagram of Blowfish algorithm

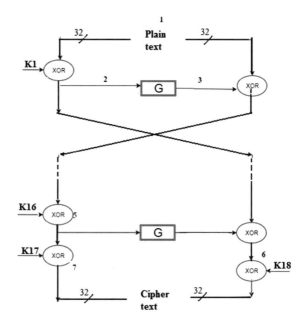

2.6 BLOWFISH Lightweight Cryptographic Algorithm

BLOWFISH encrypts a 64 bit Plaintext block, it is a fiestel block cipher, has 16 round operations with a key of variable size from 32 bits up to 448 bits. Figure 8 depicts the block diagram of BLOWFISH algorithm.

This algorithm is divided into two parts such as Key expansion and Data Encryption.

2.6.1 Key Expansion

These keys are pre-computed to encrypt and decrypt the data block.

1. The 448 bit key is expanded and stored in the p-array, there are 18, 32-bit sub-keys in each array. P1, P2,…, P18
2. There are Four, 32-bit S-BOXes with 256 entries in each S-BOX

$$
\begin{pmatrix}
S_{1,0}, S_{1,1}, \dots S_{1,255} \\
S_{2,0}, S_{2,1}, \dots S_{2,255} \\
S_{3,0}, S_{3,1}, \dots S_{3,255} \\
S_{4,0}, S_{4,1}, \dots S_{4,255}
\end{pmatrix}
$$

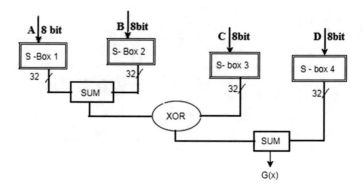

Fig. 9 The G function

The sub-keys are generated by the following steps:

1. Initialization of the P-array and the four S-boxes, in order, with a fixed string which consists of the hexadecimal digits of Pi. i.e., P1 $=0 \times$ 243f6a88, P2 $=0 \times$ 85a308d3, P3 $=0 \times$ 13198a2e, P4 $=0 \times$ 03707344 and so on.
2. P_1 is XORed with the K1; P_2 is XORed with K2, and so on. (K1 is first 32 bits of input key, K2 is the next 32 bits of the input key and so on). The cycle is repeated until the entire P-array has been XORed with key bits (Fig. 8).
3. Using the Blowfish algorithm and the sub-keys of step 1 and step 2 all the zero string are encrypted.
4. The output of step (3) replaces P_1 and P_2.
5. Using the modified sub-keys encrypt the output of step (3) is encrypted using the Blowfish cipher.
6. The output of step (5) replaces P_3 and P_4.

2.6.2 The G Function

Figure 9 shows G(L) function and is given as
    Split L into four 8 bit data A, B, C, D;
then G(L) is given by

$$\left(\left(\left(S_1, A + S_2, B\right) \bmod 2^{32}\right) XOR S_3, C\right) + S_4, D) \bmod 2^{32}$$

2.7 The AES Algorithm

Joan Daemen and Vincent Rijmen of Belgium proposed a Rijndael algorithm in 1997, which NIST in 2001 renamed it as standard AES (Advanced Encryption Standard). The AES algorithm consists of 10, 12, or 14 rounds for the key length is 128, 192,

Fig. 10 The AES algorithm

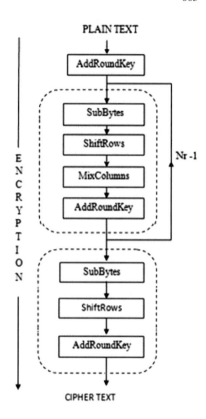

or 256 bits respectively to encrypt a plain text block of 128 bits. Figure 10 shows the AES Cipher.

The encryption process starts with an Add Round key process followed 9 rounds of four transformations and the 10th round with three transformations. The four transformations are as follows,

- Substitution Bytes
- Shift Rows
- Mix Columns
- Add Round Key

The Mix Columns transformation operation is not performed in the 10th round.

Table 1 Performance results of the ciphers

Algorithm	Key size	Plaintext block size	Rounds	Maximum frequency in MHz	Time delay ns	LUT's	FF's
TEA	128	64	32	147.29	6.789	758	149
PRESENT	80	64	32	607.4	1.646	217	149
XTEA	128	64	32	108.93	9.38	968	149
BLOWFISH	64	64	16	76.653	7.9	4477	1551
SIMON	128	128	68	114.116	14.317	92	30
IDEA	128	64	8.5	69.845	8.763	357	288
AES	128	128	10	15.04	66.446	23874	12137

Table 2 Area and power comparison

Algorithm	Power consumption in Watts	Area
TEA	0.0177	190659
PRESENT	0.01293	191249
XTEA	0.0002545	258965
BLOW FISH	0.0004983	372994
SIMON	0.000643	4733
IDEA	0.003236	547
AES	0.0001585	225358

Fig. 11 Power in Watts

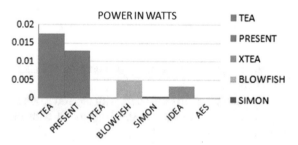

3 The Performance Results of the CIPHERS

We have implemented the above ciphers in VerilogHDL and synthesized Xilinx 14.2 the target device is xc2v40-5cs144 Table 1 shows the corresponding results.

The Power and Area report for the above ciphers has been generated by *Cadence ncsim* and Table 2 shows the corresponding results.

Figures 11 and 12 depicts power consumption and Area.

Fig. 12 Area in mm^2

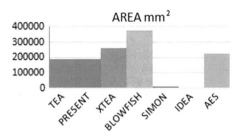

4 Conclusion

This paper has analyzed the ultra-lightweight block ciphers like PRESENT and SIMON with the block ciphers like the AES and other lightweight block ciphers. PRESENT and SIMON ciphers has proven the ability to resist differential and linear attacks. For low cost constrained devices SIMON is alternative compared to AES since it has AES-like structure. XTEA may not perform as fast as AES but well suited for low security and has ultra low power and has higher throughput. PRESENT encrypts faster than other constrained algorithms listed in the implementation. The future work would be to optimize the algorithms in terms of Area and power.

References

1. Eisenbarth, T., Kumar, S., Paar, C., Poschmann, A., Uhsadel, L.: A Survey of lightweight-cryptography implementations. IEEE Des. Test **24**(6), 522–33 (2007)
2. Tang,Z., Cie, J., Zhong, H., Yu, M.: A Random PRESENT encryption algorithm based on dynamic s box. Int. J. Secur. Appl. **10**(3) (2016)
3. Chaitra, B., Kiran Kumar, V.G., Shantharama, R.C.: A survey on various lightweight cryptographic algorithms on FPGA. IOSR J. Electron. Commun. Eng. **12**(1) (2017)
4. Beaulieu, R., Shors, D., Smith, J., Treatman-Clark, S., Weeks, B., Wingers, L.: The simon and speck of lightweight block ciphers. National Security Agency 9800 Savage Road, Fort Meade, MD 20755, USA, June 2013
5. Sbeiti, M., Silbermann, M., Poschmann, A., Paar, C.: Design space exploration of PRESENT implementations for FPGA's. Hortz gortz institute for IT security
6. PRESENT: an ultralightweight block cipher. In: Bogdanov, A., Knudsen, L.R., Leander, G., Paar, C., Poschmann, A., Robshaw, M.J.B., Securin, Y., Vikkelsoe, C. (eds.), Hortz gortz institute for IT security International Workshop on Cryptographic Hardware and Embedded Systems CHES 2007, pp. 450–466 (2007)
7. Ameli, R.: PRESENT cipher encryption IP core. Digital system lab, Iran
8. A hardware implementation of simon cryptography algorithm. In: Feizi, S., Ahmadi, A., Nemati A. (eds.), 4th International Conference on Computer and Knowledge Engineering (ICCKE) (2014)
9. Aysu, A., Ege, Gulcan, E., Schaumont, P.: SIMON says: break area records of block ciphers on FPGAs. IEEE Embedded Syst. Lett. **6**(2) (2014)
10. Thaduri, M., Yoo, S.M., Gaede, R.: An efficient implementation of IDEA encryption algorithm using VHDL. Elsevier (2004)

11. Nie, T., Zhang, T.: A Study of DES and Blowfish Encryption) Algorithm. 978-1-4244-4547-9/09/IEEE, TENCON (2009)
12. Michael, C.-J.L., Lin, Y.-L.: A VLSI implementation of blowfish encryption/decryption algorithm. In: IEEE Design Automation Conference (A" SP_DAC'00), pp. 1–2 (2000)
13. Liu, S., Gavrylyako, O., Bradford, P.: Implementing the TEA algorithm on sensors. In: Proceedings of the 42nd Annual Southeast Regional Conference, pp. 64–69 (2004)
14. Wheeler, D., Needham, R.: TEA, a tiny encryption algorithm. In: Proceedings Fast Software Encryption: Second International Workshop, Lecture Notes in Computer Science, vol. 1008, pp. 363–366, December 2004
15. Lai, X., Massey, J.: A proposal for a new block encryption standard. In: Proceedings Eurocrypt '90, (1990)
16. Kumar, S., Paar, C., poschmann, A.: Leif Uhsadel A survey of lightweight cryptographic implementations. Thomas eisenbarth. Design and test IC's for secure embedded computing. IEEE Des. Test of Comput. **24**(6) (2007)
17. Kiran Kumar, V., Mascarenhas, S.J., Kumar, S., Rakesh, V.: Design and implementation of Tiny encryption algorithm. J. Pais Kiran Kumar, V.G., et al.: Int. J. Eng. Res. Appl. **5**(6) (2015)

Attack Experiments on Elliptic Curves of Prime and Binary Fields

Ni Ni Hla and Tun Myat Aung

Abstract At the beginning the paper describes the basic properties of finite field arithmetic and elliptic curve arithmetic over prime and binary fields. Then it discusses the elliptic curve discrete logarithm problem and its properties. We study the Baby-Step, Giant-Step method, Pollard's rho method and Pohlig–Hellman method, known as general methods that can exploit the elliptic curve discrete logarithm problem, and describe in detail attack experiments using these methods over prime and binary fields. Finally, the paper discusses the expected running time of these attacks and suggests the strong elliptic curves that are not vulnerable to these attacks.

1 Introduction

Elliptic Curve Cryptosystem (ECC) is an alternative approach for implementing Public Key Cryptosystem (PKC) in which each entity connecting in the public communication channel generally has a couple of keys, a public key and a private key to perform cryptographic operations such as encryption, decryption, signing, verification, and authentication. The private key must be kept secret but the corresponding public key is distributed to all entities connecting in the public communication channel [1]. ECC can be applied for providing the security services: confidentiality, authentication, data integrity, non-repudiation, and authenticated key exchange.

These days, ECC becomes a major in the industry of information and network security technology. It substitutes other public key cryptosystems such as RSA and DSA. It becomes the industrial standard as a consequence of an increase in speed and a decrease in power consumption during implementation as a result of less memory usage and smaller key sizes. Its security depends on the complexity of solving the Elliptic Curve Discrete Logarithm Problem (ECDLP). Although the ECDLP is

N. N. Hla (✉) · T. M. Aung
University of Computer Studies, Yangon (UCSY), Yangon, Myanmar
e-mail: ni2hla@ucsy.edu.mm

T. M. Aung
e-mail: tma.mephi@gmail.com

© Springer Nature Singapore Pte Ltd. 2019
A. Abraham et al. (eds.), *Emerging Technologies in Data Mining and Information Security*, Advances in Intelligent Systems and Computing 755,
https://doi.org/10.1007/978-981-13-1951-8_60

thought to be a difficult problem, it has not stopped attackers attempting to attack on ECC. Several attacks have been created, experienced and analyzed by mathematicians over the years, to discover defects in ECC. Some attacks have done partially well, but others have not.

The idea of this paper is to apply the knowledge of the general methods of attacking the ECDLP in attempting to select powerful elliptic curves over prime and binary fields under large integer. The structure of this paper is as follows. The Sect. 2 includes finite field arithmetic over prime and binary fields and their properties. In Sect. 3, we discuss elliptic curve arithmetic over prime and binary fields, its geometric properties, the ECDLP and its properties. The Sect. 4 describes in details the general methods of attacking on the elliptic curve discrete logarithm problem. In Sect. 5 we discuss attack experiments on the ECDLP over prime and binary fields. Finally, in Sect. 6 we conclude our discussion by describing time complexity of the attacking methods and by suggesting powerful elliptic curves for best secure implementation of ECC.

2 Finite Field Arithmetic

A finite field, generally signified by F, is a field which consists of a finite number of elements. Finite Fields are applied to the rational number system, the real number system and the complex number system. They contain a set of elements together with two arithmetic operations: *addition* signified by the symbol $+$ and *multiplication* signified by the symbol, that satisfy the typical arithmetic properties:

- $(F, +)$ is an *additive* group with operation by $+$ and identity element by 0.
- $(F\backslash\{0\},)$ is a *multiplicative* group with operation by . and identity element by 1.
- Elements of finite group hold the distributive law: $(x+y) \cdot z = (x \cdot z) + (y \cdot z)$ for all $x, y, z \in F$

When the number of elements in the field is finite, then the field is said to be *finite* [2]. Galois open that the elements in the field to be finite and the number of elements should be p^m, where p is a prime number called the *characteristic* of the field and m is a positive integer. The finite fields are generally called *Galois fields* and also signified by $GF(p^m)$. When $m=1$, then the field $GF(p)$ is called a *prime field*. When $m \geq 2$, then the field $GF(p^m)$ is called an *extension field*. The number of elements in a finite field is called the *order* of the field. Any two fields are *isomorphic* when their orders are the same [3].

2.1 Field Operations

Finite field F performs two arithmetic operations, *addition* and *multiplication*. However, the *subtraction* of field elements is defined in the expressions of addition operation. For instance, let $x, y \in F$, $x - y$ is defined as $x + (-y)$, in this case $-y$ is called *additive inverse* of b such that $y+(-y)=0$. Correspondingly, the *division* of field elements is defined in the expression of multiplication operation. For instance, let x, $y \in$ F with $y \neq 0$, x/y is defined as $x \cdot y^{-1}$, in this case y^{-1} is called the *multiplicative inverse* of y such that $y \cdot y^{-1} = 1$ [2].

Prime Field. A finite field of prime order p is called *prime field* signified by $GF(p)$. It contains a set of integer elements modulo p, $\{0,1,2,..., p-1\}$ with additive and multiplicative groups performed modulo p. For any integer x, $x \bmod p$ refers to the integer remainder r that obtained upon dividing x by p. This operation is called *reduction modulo p*. In this case, the remainder r is the distinct integer element between 0 and $p-1$, i.e. $0 \leq r \leq p-1$ [2]. The arithmetic operations of elements over $GF(p)$ are performed as the following example (1).

Example (1). (*prime field* GF(23)) The elements of GF(23) are $\{0,1,2,...,23\}$. The following examples demonstrate for arithmetic operations of elements in GF(23).

- Addition: $20 + 10 \pmod{23} = 7$ since $30 \bmod 23 = 7$.
- Subtraction: $10 - 20 \pmod{23} = 13$ since $10 + (-20) \bmod 23 = 13$.
- Multiplication: $20 \cdot 10 \pmod{23} = 6$ since $200 \bmod 23 = 6$.
- Inversion: $10^{-1} \pmod{23} = 7$ since $10 \cdot 7 \bmod 23 = 1$.
- Division: $20/10 \pmod{23} = 2$ since $20. \ 10^{-1} \pmod{23}$ and $20. \ 7 \pmod{23} = 2$.

Binary Field. A finite field of order 2^m is called *binary field* signified by $GF(2^m)$. It also refers to the *finite field with characteristic-two*. The elements over $GF(2^m)$ can be constructed by applying a *polynomial basis representation* defined by the Eq. (1). In this case, the elements of $GF(2^m)$ are the binary representation polynomials with degree at most $m-1$.

$$GF(2^m) = a_{m-1}x^{m-1} + a_{m-2}x^{m-2} + \cdots + a_2x^2 + a_1x + a_0, a_i \in \{0, 1\}. \quad (1)$$

$f(x)$ is defined as an irreducible binary representation polynomial with degree m if $f(x)$ cannot be factored as a product of binary representation polynomials with degree less than m. Let $a(x)$ and $b(x)$ be elements over $GF(2^m)$. They are the binary representation polynomials with degree at most $m-1$. Addition of elements in binary field refers to the addition of binary representation polynomials, that is, $a(x) \oplus b(x)$. Multiplication of elements in $GF(2^m)$ refers to refers to the expression $a(x) \times b(x) \bmod f(x)$. Let $c(x) = a(x) \times b(x)$ and $c(x)$ be an binary representation polynomial with degree more than m. The result of the expression $c(x) \bmod f(x)$ refers to the unique remainder polynomial $r(x)$ with degree less than m that obtained upon the division of $c(x)$ by $f(x)$; this operation is called *reduction modulo f(x)*. Division of elements in $GF(2^m)$ refers to refers to the expression $a(x)/b(x) \bmod f(x)$. In this case, the division of elements in $GF(2^m)$ is calculated as the expression $a(x) \times b(x)^{-1} \bmod f(x)$ [2]. The arithmetic operations of elements over $GF(2^m)$ are performed as the following example (2).

Example (2). (*binary field* $GF(2^m)$) The elements of $GF(2^m)$ generated by the polynomial $f(x) = x^4 + x + 1$ are represented by 16 binary polynomials of degree at most 3 as shown in Table (1).

Table 1 Binary representation polynomials

Polynomial	Polynomial	Polynomial	Polynomial
0	x^2	x^3	$x^3 + x^2$
1	$x^2 + 1$	$x^3 + 1$	$x^3 + x^2 + 1$
x	$x^2 + x$	$x^3 + x$	$x^3 + x^2 + x$
$x + 1$	$x^2 + x + 1$	$x^3 + x + 1$	$x^3 + x^2 + x + 1$

The followings are some examples of arithmetic operations in $GF(2^4)$ with the elements generated by reduction polynomial $f(x) = x^4 + x + 1$.

- Addition: $(x^3 + 1) + (x + 1) = x^3 + x$ since $(x^3 + 1) \oplus (x + 1) = x^3 + x$.
- Subtraction: $(x^3 + 1) - (x + 1) = x^3 + x$ since $(x^3 + 1) \oplus (x + 1) = x^3 + x$.
- Multiplication $(x^3 + 1).(x + 1) = x^3$ since $(x^3 + 1) \times (x + 1) = x^4 + x^3 + x + 1$ and $(x^4 + x^3 + x + 1) \bmod f(x) = x^3$.
- Inversion: $(x + 1)^{-1} = x^3 + x^2 + x$ since $(x^3 + x^2 + x) \times (x + 1) \bmod f(x) = 1$.
- Division: $(x^3 + 1)/(x + 1) = x^2 + x + 1$ since $(x^3 + 1) \times (x + 1)^{-1} \bmod f(x) = x^2 + x + 1$

3   Elliptic Curve Arithmetic

The elliptic curve over finite field E(GF) is a cubic curve defined by the general Weierstrass equation: $y^2 + a_1 xy + a_3 y = x^3 + a_2 x^2 + a_4 x + a_6$ over GF where $a_i \in GF$ and GF is a finite field. We study elliptic curves over $GF(p)$ and $GF(2^m)$.

3.1   Elliptic Curve Arithmetic Over Prime Field -GF(P)

Elliptic curves are driven from the general Weierstrass equation. The elliptic curve E(GF(p)) over $GF(p)$ is determined by the Eq. (2) [4]:

$$y^2 = x^3 + ax + b, \tag{2}$$

where $p > 3$ is a prime and $a, b \in GF(p)$ satisfy that $4a^3 + 27b^2 \neq 0$. ($a_1 = a_2 = a_3 = 0$; $a_4 = a$ and $a_6 = b$ corresponding to the general Weierstrass equation)

Points on E(GF(p)). The elliptic curve E(GF(p)) over GF(p) belongs to a set of points together with a point at infinity signified by symbol O. In this case,$\{P = (x, y) | y^2 = x^3 + ax + b; x, y, a, b \in GF(p)\}$. Every point on the curve generally has its corresponding *inverse*. The inverse of a point (x, y) on E(GF(p)) is defined as $(x, -y)$. The number of points on the curve, including a point at infinity, is defined as its *order* #E. The points on E(GF(p)) are generated by using Algorithm (1).

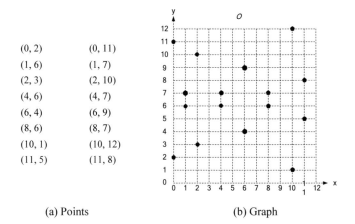

(0, 2)	(0, 11)
(1, 6)	(1, 7)
(2, 3)	(2, 10)
(4, 6)	(4, 7)
(6, 4)	(6, 9)
(8, 6)	(8, 7)
(10, 1)	(10, 12)
(11, 5)	(11, 8)

(a) Points   (b) Graph

Fig. 1 Points on $E : y^2 = x^3 + 5x + 4$

Algorithm (1). Generating the points on E(GF(p))
Input: a, b, p
Output: $P_i = (x_i, y_i)$
Begin
x = 0;
while(x < p){
$\quad w = (x^3 + ax + b) \bmod p$.
$\quad$ If(w is perfect square in Z_p) output $(x, \sqrt{w}), (x, -\sqrt{w})$
$\quad$ x = x + 1.
}
End

Example (3). Let p = 13 and consider the elliptic curve $E : y^2 = x^3 + 5x + 4$ over GF(13) where $a = 5$ and $b = 4$. Note that $4a^3 + 27b^2 = 500 + 432 = 932 \bmod 13 = 9$, so E is indeed an elliptic curve. The points on the curve and its graph are shown in Fig. (1a and b). The *order* of the elliptic curve $E : y^2 = x^3 + 5x + 4$ over GF(13) is 17.

Arithmetic Operations on E(GF(p)). Addition of two points on an elliptic curve E(GF(p)) applied the *chord-and-tangent rule* to find a third point on the curve. The addition operation with the points on E(GF(p)) generates a group with point at infinity O serving as its identity. It is the group of points on E(GF(p)) that is used in the construction of elliptic curve cryptosystems [5]. It is the best way to explain the point addition rule geometrically. Let $P = (x_1, y_1)$ and $Q = (x_2, y_2)$ be two distinct points on E(GF(p)). Assume that the point $R = (x_3, y_3)$ is obtained by *addition* of P and Q. This point addition is illustrated in Fig. (2a). The line connecting through P and Q intersects the elliptic curve at the point called -R. R is the reflection of -R with respect to the x-axis. Assume that *doubling* of P is $R = (x_3, y_3)$ in the case of $P = (x_1, y_1)$. This point doubling is illustrated in Fig. (2b). The tangent line drawing from point P intersects the elliptic curve at the point called -R. R is the reflection of -R with respect to the x-axis as in the case of addition.

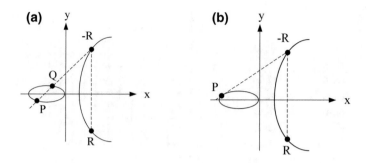

Fig. 2 a Addition. ($R = P + Q$). **b** Doubling. ($R = P + P$)

The geometric description open following algebraic methods for the addition of two points and the doubling of a point [4].

1. $P + O = O + P = P$ for all $P \in E(GF(p))$.
2. If $P = (x, y) \in E(GF(p))$, then $(x, y) + (x, -y) = o$ where the point $(x, -y)$ is signified by $(-P)$ that is called the *inverse* of P.
3. (*Point addition*). Let $P = (x_1, y_1) \in E(GF(p))$ and $Q = (x_2, y_2) \in E(GF(p))$, where $P \neq \pm Q$. Then $P + Q = (x_3, y_3)$, where $x_3 = \lambda^2 - x_1 - x_2$ and $y_3 = \lambda(x_1 - x_3) - y_1$. In this case, $\lambda = (y_2 - y_1)/(x_2 - x_1)$.
4. (*Point doubling*). Let $P = (x_1, y_1) \in E(GF(p))$, where $P \neq -P$. Then $2P = (x_3, y_3)$, where $x_3 = \lambda^2 - 2x_1$ and $y_3 = \lambda(x_1 - x_3) - y_1$. In this case, $\lambda = (3x_1^2 + a)/2y_1$.

Example (4). (*elliptic curve addition and doubling*) Let us consider the elliptic curve defined in Example (3).

- *Addition*. Let P=(1, 6) and Q=(4, 6). Then P+Q=(8, 7).
- *Doubling*. Let P=(1, 6). Then 2P=(10, 1).
- *Inverse*. Let P=(1, 6). Then –P=(1, 7).

3.2 Elliptic Curve Arithmetic Over Binary Field - GF(2 ^{M})

Elements over GF(2 ^{m}) must be firs generated by using a reduction polynomial f(x). These elements are applied to construct an elliptic curve E(GF(2 ^{m})) over GF(2 ^{m}). The curve E(GF(2 ^{m})) is determined by the Eq. (3) [4]:

$$y^2 + xy = x^3 + ax + b. \tag{3}$$

where $a, b \in GF(2^m)$ and $b \neq 0$.

Points on E(GF(2 ^{m})). The elliptic curve E(GF(2 ^{m})) over GF(2 ^{m}) belongs to a set of points together with a point at infinity signified by symbol O. In this case,

Table 2 Binary and polynomial representations for elements of $GF(2^4)$

Binary	Polynomial	Binary	Polynomial	Binary	Polynomial	Binary	Polynomial
0000	0	0100	x^2	1000	x^3	1100	$x^3 + x^2$
0001	1	0101	$x^2 + 1$	1001	$x^3 + 1$	1101	$x^3 + x^2 + 1$
0010	x	0110	$x^2 + x$	1010	$x^3 + x$	1110	$x^3 + x^2 + x$
0011	$x + 1$	0111	$x^2 + x + 1$	1011	$x^3 + x + 1$	1111	$x^3 + x^2 + x + 1$

$\{P = (x, y) | y^2 + xy = x^3 + ax + b; x, y, a, b \in GF(2^m)\}$. Every point on the curve has its corresponding *inverse*. The inverse of a point (x, y) on E(GF(2^m)) is defined as $(x, x \oplus y)$. The number of points on the curve, including a point at infinity, is generally called its *order* #E. The points on E(GF(2^m)) are generated by using Algorithm (2).

Algorithm (2). Generating the points on E(GF(2^m))
Input: a, b, f(x)
Output: $P_i = (x_i, y_i)$
Begin

$$x_i = \{0, 1, g^1, K, g^{m-2}\}$$

$$y_i = \{0, 1, g^1, K, g^{m-2}\}$$

for(i=0; i<2^m; i++){
    for(j=0; j < 2^m; j++){
    $w_1 = x_i^3 \oplus ax_i \oplus b$.

    $w_2 = y_j^2 \oplus x_i y_j$
    If $(w_1 = w_2)$ output (x_i, y_j), $(x_i, y_j \oplus x_i)$
    }
}
End

Example (5). Let $f(x) = x^4 + x + 1$ be the reduction polynomial. Then binary and polynomial representations for 16 elements of $GF(2^4)$ generated by the reduction polynomial $f(x) = x^4 + x + 1$ are shown in Table (2).

Table (3) shows the power representations of g and corresponding binary representations for elements of $GF(2^4)$ generated by the reduction polynomial $f(x) = x^4 + x + 1$. The element of g=(0010) is a generator of $GF(2^4)$ and its order is 15 ($2^4 - 1$).

The elliptic curve $E : y^2 + xy = x^3 + g^{11}x + g^{13}$ where $a = g^{11}$ and $b = g^{13}$ belongs to the points on the curve, as shown in Fig. (3). The points on the curve and its graph are shown in Fig. (3a and b). The *order* of the elliptic curve $E : y^2 + xy = x^3 + g^{11}x + g^{13}$ is 22.

Arithmetic Operations on E(GF(2^m)). Addition of two points on an elliptic curve E(GF(2^m)) also applied the *chord-and-tangent rule* to find a third point on the curve. The addition operation with the points on E(GF(2^m)) generates a group

Table 3 Power and binary representations of elements of GF(2^4)

Power	Binary	Power	Binary	Power	Binary	Power	Binary
g	0010	g^5	0110	g^9	1010	g^{13}	1101
g^2	0100	g^6	1100	g^{10}	0111	g^{14}	1001
g^3	1000	g^7	1011	g^{11}	1110	g^{15}	0001
g^4	0011	g^8	0101	g^{12}	1111		

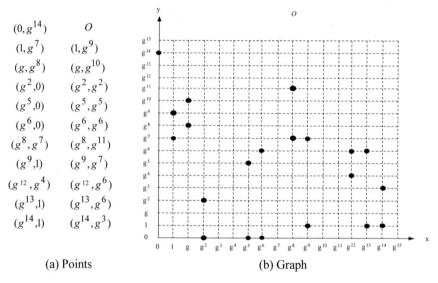

(a) Points   (b) Graph

Fig. 3 Points on $E : y^2 + xy = x^3 + g^{11}x + g^{13}$

with point at infinity O serving as its identity. It is the group of points on E(GF(2^m)) that is used in the construction of elliptic curve cryptosystems [5]. The followings are algebraic methods for the addition of two distinct points and the doubling of a point [4].

1. $P + O = O + P = P$ for all $P \in E(GF(2^m))$.
2. If $P = (x, y) \in E(GF(2^m))$, then $(x, y) + (x, x + y) = O$ where the point $(x, x+y)$ signified by $(-P)$ is called the *inverse* of P.
3. (*Point addition*). Let $P = (x_1, y_1) \in E(GF(2^m))$ and $Q = (x_2, y_2) \in E(GF(2^m))$, where $P \neq \pm Q$. Then $P + Q = (x_3, y_3)$, where $x_3 = \lambda^2 + \lambda + x_1 + x_2 + a$ and $y_3 = \lambda(x_1 + x_3) + x_3 + y_1$. In this case, $\lambda = (y_2 + y_1)/(x_2 + x_1)$.
4. (*Point doubling*). Let $P = (x_1, y_1) \in E(GF(2^m))$, where $P \neq -P$. Then $2P = (x_3, y_3)$, where $x_3 = \lambda^2 + \lambda + a$ and $y_3 = x_1^2 + \lambda x_3 + x_3$. In this case, $\lambda = x_1 + (y_1/x_1)$.

Example (6). (*elliptic curve addition and doubling*) Let's consider the elliptic curve defined in Example (5).

- *Addition*. Let $P = (g^2, g^2)$ and $Q = (g^6, g^6)$. Then $P + Q = (g^5, 0)$.
- *Doubling*. Let $P = (g^2, g^2)$. Then $2P = (g^{14}, 1)$.
- *Inverse*. Let $P = (g^2, g^2)$. Then $-P = (g^2, 0)$.

3.3 Elliptic Curve Discrete Logarithm Problem

The complexity of solving ECDLP determines the security of ECC. Let P and Q be the points on an elliptic curve such that $Q=kP$, where k is an integer number. k is called the discrete logarithm of Q to the base P. Known two points, P and Q, it is unable to compute k, when the group order of the points is enough large.

Point Multiplication. *Point Multiplication* is a major operation usually used in ECC. The scalar multiplication operation of a integer scalar k with a point P on the elliptic curve creates another point Q on this curve [1]. The point Q is gotten by performing *point addition and point doubling* operations according to bit sequence patterns of integer scalar k. The bit sequence patterns of integer k is shown as the Eq. (4)

$$k = k_{n-1}2^{n-1} + k_{n-2}2^{n-2} + n + k_1 + k_0, \tag{4}$$

where $k_{n-1} = 1$ and $k_i \in \{0, 1\}$, $i = 0, 1, 2, \ldots, n-1$. This operation is based on the *binary method* which scans the bit sequence patterns of k either from left-to-right or right-to-left [2]. The Algorithm-3 illustrates the scalar multiplication operation of a integer scalar k with a point P on the elliptic curve using binary method. This method can be applied for both elliptic curves over GF(p) and GF(2^m).

Algorthm (3). Scalar Multiplication of a Point
Input : point P and integer scalar k
Output : point Q such that Q = kP
Begin
$k_i \in \{0,1\}, i = 0,1,2,\ldots,n-1$
Q=P
For i = n-1 to 0 do
{
    Q = Point-Doubling of Q
    If k_i = 1 then
    Q = Point-Addition of P and Q
}
Return Q
End

The cost of this point multiplication method relies on the number of 1 s in bit sequence patterns of integer scalar k. The number of 1 s is called the *Hamming Weight* of the scalar. Generally, this method performs $(n - 1)$ *point doublings* and $(n - 1)/2$ *point additions*. For each bit .1., this method must perform both *point doubling* and *point addition*, if the bit is .0., this method performs only *point doubling* operation.

Consequently, making less the number of 1 s in bit sequence patterns of integer scalar k will increase the speed of scalar multiplication of a point on elliptic curve [6].

The Order of Point. Let $P \in E(GF(p))$. The *order* of point P is defined as the smallest positive integer value N such that $NP = O$. In this case, O is the identity of the group of points on the elliptic curve.

$$p + 1 - 2\sqrt{p} \le N \le p + 1 + 2\sqrt{p}. \tag{5}$$

All different values of N must be tried in the range defined in the Eq. (5) [7] and then check which value of N agrees this statement $NP = O$.

Example (7). Let E be the elliptic curve $E : y^2 = x^3 + 5x + 4$ over GF(13). The order of point $(1, 7)$ is 17. The Eq. (5) is solved as $13 + 1 - 2\sqrt{13} \le N \le 13 + 1 + 2\sqrt{13}$. All different values of N in the range, $7 \le N \le 21$, could be tried to find $N = 17$ such that $NP = O$. Therefore, $N = 17$.

4   General Methods of Attacking on ECDLP

The complexity of solving the Discrete Logarithm Problem (DLP) is deeply important for the security of PKC. PKC is constructed based on the assumption that the DLP is extremely difficult to compute; the more difficult it is, the more security it supports. Therefore, PKC is constructed on a larger group order under large integer in order to increase the complexity of solving the DLP.

General methods of attacking on the ECDLP can be classified into three groups as following [8]. These methods can solve the ECDLP under small integer.

1. Methods stand on random walks, such as the exhaustive search method and the Baby-Step, Giant-Step method,
2. Methods stand on random walks with special conditions, like Pollard's rho method and Pollard's lambda method, and
3. Methods stand on multiplicative groups, such as the Index Calculus method and Pohlig–Hellman method.

We studied the following general methods of attacking on the ECDLP.

4.1   Baby-Step, Giant-Step Method

Let $P, Q \in E$. Assume that we solve an integer scalar k such that $Q = [k]P$ and P has prime order N. At first, we must compute the order N of P. This method generally performs about $\sqrt{N}$ steps and requires about $\sqrt{N}$ storage. Therefore, this method only works well for memory storage size N. This method follows the procedure below [7].

1. Fix an integer m such that $m = \lceil \sqrt{N} \rceil$ and compute mP.
2. Compute and store a list of iP for $1 \le i \le m$.
3. Compute the points such that $Q - jmP$ for $j = 0, 1, \ldots$ until one of resulting points matches one from the stored list.
4. If $iP = Q - jmP$, then $Q = kP$ with $k \equiv i + jm \pmod{N}$.

The list of points iP are calculated by adding P to $(i - 1)P$. It is *the baby-step*. The list of points $Q - jmP$ are computed by adding $-mP$ to $Q - (j - 1)mP$. It is *the giant-step*. This method may generally perform about m steps to find a match and its time complexity is $O\left(\sqrt{N}\right)$ [7].

4.2 Pollard's Rho Method

Let $P, Q \in E$. Assume that we solve an integer scalar k such that $Q = [k]P$ where P has prime order N and $Q \in< P >$. This method generally find two different pair of integers: (a, b) and $\left(a', b'\right)$ modulo N such that $[a]P + [b]Q = [a']P + [b']Q$. This method follows the procedure below [7]:

1. Select $a, b \in [0, N - 1]$ uniformly at random.
2. Compute $[a]P + [b]Q$.
3. Store the triple $(a, b, [a]P + [b]Q)$.
4. Select new pairs $\left(a', b'\right)$ uniformly at random such that $(a, b) \ne \left(a', b'\right)$.
5. Compute $[a']P + [b']Q$.
6. Store the new triple $(a', b', [a']P + [b']Q)$.
7. Compute and Check the new triple against all previously stored triples until we find a pair $\left(a', b'\right)$ satisfied with the Eq. (6).
8. Compute $k \equiv (a - a')(b' - b)^{-1} \bmod N$.

$$[a]P + [b]Q = [a']P + [b']Q \tag{6}$$

The time complexity of this method is $O\left(\sqrt{\pi N / 2}\right)$ [7]. The diagram of the sequence of resulting points looks like the Greek letter ρ. Therefore, this method is called the Pollard-Rho method.

4.3 Pohlig–Hellman Method

Let $P, Q \in E$. Assume that we solve an integer scalar k such that $Q = [k]P$ where P has prime order N.

$$N = \prod_i q_i^{e_i} \tag{7}$$

The main idea of this method is as following:

- Compute the order N of P.
- Compute prime factorization of N that satisfied the Eq. (7).
- Compute $k(\bmod q_i^{e^i})$ for each i,
- Combine them to obtain k (mod N) using the Chinese Remainder theorem [9].

Let q be a prime, and let q^e be the exact power of q dividing N. This method defines k in its base q expansion as the Eq. (8).

$$k = k_0 + k_1 q + k_2 q^2 + n \tag{8}$$

where $0 \le k_i < q$. This method evaluates $k(\bmod q_i^e)$ by successively determining $k_0, k_1, k_2, n, k_{e-1}$. This method follows the procedure below [7]:

1. Compute $T = j \cdot \left(\frac{N}{q} \cdot P\right), 0 \le j \le \text{q-1}$.
2. Compute $\frac{N}{q} \cdot Q$. It is an element of $k_0 \left(\frac{N}{q} \cdot P\right)$ of T.
3. If $e = 1$, stop. Otherwise, continue.
4. Let $Q_1 = Q - k_0 P$.
5. Compute $\frac{N}{q^2} \cdot Q$. It is an element of $k_1 \left(\frac{N}{q^2} \cdot P\right)$ of T.
6. If $e = 2$, stop. Otherwise, continue. Assume that we have calculated: k_1, k_2, n, k_{r-1} and Q_1, Q_2, n, Q_{r-1}.
7. Let $Q_r = Q_{r-1} - k_{r-1} q^{r-1} P$.
8. Determine k_r such that $\frac{N}{q^{r+1}} \cdot Q_r = k_r \left(\frac{N}{q} \cdot P\right)$.
9. If $r = e - 1$, stop. Otherwise, return to step (7).

Then the method computes $k \equiv k_0 + k_1 q + n + k_{e-1} q_{e-1} (\bmod q_e)$. Therefore, early we find k_1. In the same way, the method produces $k_2, k_{3,n}$. We must stop after $r = e - 1$. The time complexity of this method is $O(\sqrt{q})$ [7]. In this case, q is the largest prime divisor of N. In practice this method becomes infeasible as a result of that N has a large prime divisor. Then it becomes difficult to make and store the list T to find matches.

5 Attack Experiments

We implemented well-known general common attacks such as Baby-Step Giant-Step method, Pollard's rho method and the Pohlig–Hellman method by using our implementations of finite field arithmetic operations [10] and elliptic curve arithmetic operations [11] under java BigInteger class.

5.1 Baby-Step Giant-Step Attack

Prime Field. Let an elliptic curve be $E : y^2 = x^3 + 5x + 4$ over $GF(13)$, $P = (0, 2)$ and $Q = (6, 4)$. Assume that we solve an integer scalar k such that $Q = [k]P$ by using *Baby-Step, Giant-Step method*. P has order 17. We first compute $m = |\sqrt{17}| = 5$. The points iP for $1 \leq i \leq 5$ are:

$$(0, 2), (4, 6), (10, 1), (6, 9), (8, 6).$$

We calculate $Q - jmP$ for $j = 0, 1, 2, 3, \ldots$ and obtain:

$$(6, 4), (11, 8), (10, 1), (4, 7), (1, 6)$$

at which point we stop since this third point matches $3P$. Since $j = 2$ yielded the match, we got:

$$(6, 4) = (3 + 2.5)P = 13P.$$

Therefore $k = 13$.

Binary Field. Let an elliptic curve be $E : y^2 + xy = x^3 + g^{11}x + g^{13}$ over $GF(2^4)$, $P = (g^9, 1)$ and $Q = (g^6, g^6)$. Assume that we solve an integer scalar k such that $Q = [k]P$ by using *Baby-Step, Giant-Step method*. P has order 11. We first compute $m = \lceil \sqrt{11} \rceil = 4$. The points iP for $1 \leq i \leq 4$ are:
$(g^9, 1), (g^{12}, g^4), (g^6, 0), (g^{14}, 1)$.
We calculate $Q - jmP$ for $j = 0, 1, 2, 3, 4, \ldots$ and obtain:
$(g^6, g^6), (g^{14}, 1), O, (g^{14}, g^3), (g^6, 0)$.
at which point we stop since this second point matches $4P$. Since $j = 1$ yielded the match, we got:

$$(g^6, g^6) = ((4 + 1.4) \bmod 11)P = 8P.$$

Therefore $k = 8$.

5.2 Pollard's Rho Attack

Prime Field. Let an elliptic curve be $E : y^2 = x^3 + 5x + 4$ over $GF(13)$, $P = (0, 2)$ and $Q = (6, 4)$. Assume that we solve an integer scalar k such that $Q = [k]P$ by using *Pollard's rho method*. The point P has prime order 17. We choose $a, b \in [0, 17]$ uniformly at random, compute $R = [a]P + [b]Q$ and keep the triple (a, b, R) in the memory until we meet an another triple (a', b', R') such that $R = R'$ or $R = -R'$. Table (4) shows computing data used for Pollard's rho attack on $E : y^2 = x^3 + 5x + 4$

Table 4 Data for Pollard's rho attack on $E : y^2 = x^3 + 5x + 4$ over $GF(13)$

[a]	[b]	R = [a]P + [b]Q
5	**12**	**(11, 8)**
3	8	(8, 6)
10	4	(2, 3)
6	11	(6, 4)
2	**7**	**(11, 8)**
1	15	(11, 5)

Table 5 Data for Pollard's rho attack on $E : y^2 + xy = x^3 + g^{11}x + g^{13}$ over $GF(2^4)$

[a]	[b]	R = [a]P + [b]Q
10	**5**	**$(g^{13}, 1)$**
8	3	(g^9, g^7)
4	10	(g^{14}, g^3)
5	6	(g^{12}, g^6)
7	**4**	**$(g^{13}, 1)$**
2	7	$(g^6, 0)$

over $GF(13)$. We have that $[5]P + [12]Q = [2]P + [7]Q$. Then $k = (5 - 2)(7 - 12)^{-1} \bmod 17$; $k = 3(-5)^{-1} \bmod 17$; $k = 3.10 \bmod 17$; Hence $k = 13$.

Binary Field. Let an elliptic curve be $E : y^2 + xy = x^3 + g^{11}x + g^{13}$ over $GF(2^4)$, $P = (g^9, 1)$ and $Q = (g^6, g^6)$. Assume that we solve an integer scalar k such that $Q = [k]P$ by using *Pollard's rho method*. The point P has prime order 11. We choose $a, b \in [0, 11]$ uniformly at random, compute $R = [a]P + [b]Q$ and keep the triple (a, b, R) in the memory until we meet an another triple (a', b', R') such that $R = R'$ or $R = -R'$. Table (5) shows computing data used for Pollard's rho attack on $E : y^2 + xy = x^3 + g^{11}x + g^{13}$ over $GF(2^4)$. We have that $[10]P + [5]Q = [7]P + [4]Q$. Then $k = (10 - 7)(4 - 5)^{-1} \bmod 11$; $k = 3(-1)^{-1} \bmod 11$; $k = 3.10 \bmod 11$; Hence $k = 8$.

5.3 Pohlig–Hellman Attack

Prime Field. Let an elliptic curve be $E : y^2 = x^3 + 77x + 28$ over $GF(157)$, $P = (9, 115)$ and $Q = (2, 70)$. Assume that we solve an integer scalar k such that $Q = [k]P$ by using *Pohlig–Hellman method*. The order N of point P is 162. The prime factorization of N is 2.3^4. We compute $k \bmod 2$, and $\bmod 81$, then recombine them to obtain $k \bmod 162$ using the Chinese Remainder Theorem.

k mod 2. We compute $T = \{(24, 0)\}$.
Since $\frac{N}{2}.Q = (24, 0) = 1.(\frac{N}{2}.P)$, we have $k_0 = 1$.
Therefore $k \equiv 1 \pmod 2$.

k mod 81. We compute $T = \{(57, 41), (5, 99), (57, 116), O\}$.

Since $\frac{N}{3}.Q = (57, 41) = 1.(\frac{N}{3}.P)$, we have $k_0 = 1$.

Therefore $Q_1 = Q - 1.P = (5, 99)$.

Since $\frac{N}{9}.Q_1 = O = 0.(\frac{N}{3}.P)$, we have $k_1 = 0$.

Therefore $Q_2 = Q_1 - 0.3.P = Q_1$.

Since $\frac{N}{27}.Q_2 = (57, 116) = 2.(\frac{N}{3}.P)$, we have $k_2 = 2$.

Therefore $Q_3 = Q_2 - 2.9.P = (57, 41)$.

Since $\frac{N}{81}.Q_3 = (57, 116) = 2.(\frac{N}{3}.P)$, we have $k_3 = 2$.

Therefore $k = 1 + 0.3 + 2.9 + 2.27 \equiv 73 (\bmod 81)$.

We now have the simultaneous congruence:

$k \equiv 1 \pmod 2$

$k \equiv 73 \pmod{81}$.

Then we obtain $k = 73$ using the Chinese Remainder theorem to recombine simultaneous congruences as following:

$$M_1 = 162/2 = 81.$$

$$y_1 = M_1^{-1} \bmod 2 = 1.$$

$$M_2 = 162/81 = 2.$$

$$y_2 = M_2^{-1} \bmod 81 = 41.$$

$$k = 1.(81).1 + 73.(2).41 (\bmod 162) = 73.$$

Binary Field. Let an elliptic curve be $E : y^2 + xy = x^3 + g^{11}x + g^{13}$ over $GF(2^4)$, $P = (g^2, g^2)$ and $Q = (g^6, g^6)$. Assume that we solve an integer scalar k such that $Q = [k]P$ by using *Pohlig–Hellman method*. The order N of point P is 22. The prime factorization of N is 2.11. We compute k mod 2, and mod 11, then recombine them to obtain k mod 22 using the Chinese Remainder Theorem.

k mod 2. We compute $T = \{O\}$.

Since $\frac{N}{2}.Q = O = 0.(\frac{N}{2}.P)$, we have $k_0 = 0$.

Therefore $k \equiv 0 (\bmod 2)$.

k mod 11. We compute $T = \{(g^{13}, g^6)\}$.

Since $\frac{N}{11}.Q = (g^{13}, g^6) = 4.(\frac{N}{11}.P)$, we have $k_0 = 4$.

Therefore $k \equiv 4 (\bmod 11)$.

We now have the simultaneous congruence:

$k \equiv 0 \pmod 2$

$k \equiv 4 \pmod{11}$.

Then we obtain $k = 4$ using the Chinese Remainder theorem to recombine simultaneous congruences as following:

Table 6 Time complexity

Attacks	Expected running time
Baby-Step Giant-Step	$O(\sqrt{N})$
Pollard's rho	$O(\sqrt{\pi N}/2)$
Pohlig–Hellman	$O(\sqrt{q})$

$$M_1 = 22/2 = 11.$$
$$y_1 = M_1^{-1} \mod 2 = 1.$$
$$M_2 = 22/11 = 2.$$
$$y_2 = M_2^{-1} \mod 11 = 6.$$
$$k = 0.(11).1 + 4.(2).6(\mod 22) = 4.$$

6 Conclusion

The security strong point of ECC relies on the complexity of solving ECDLP for a cryptanalyst to find the secret key k such that $Q = kP$. The Table (6) summarizes time complexity of general methods of attacking on ECDLP. Our research found that these attacking methods can solve ECDLP within the corresponding expected running time when the group order N of the elliptic curve is not enough large and its prime factorization is composed of smooth primes.

When implementing the ECC, the following several classes of elliptic curves should be applied in order to gain the maximum security level of the cryptosystems. The National Institute of Standards and Technology (NIST) issued several classic elliptic curves with larger key sizes for federal government use.

NIST recommends the 15 elliptic curves: five elliptic curves over $GF(p)$ where p equals 192, 224, 256, 384, and 521 bits and five elliptic curves over $GF(2^m)$ where m equals 163, 233, 283, 409, and 571. For each of the binary fields, one Koblitz curve is recommended [12]. Thus, NIST issue contains a total of five prime curves and ten binary curves. These curves should be selected for best security and implementation efficiency. The group order for each of these curves is enough large and has large prime factors. Therefore, these curves are resistant to the attacking methods we studied in the Sect. 4.

References

1. Anoop, M.S.: Elliptic Curve Cryptography. http://www.infosecwriters.com/Papers/Anoopms_ECC.pdf
2. Hankerson, D., Menezes, A., Vanstone, S.: Guide to Elliptic Curve Cryptography. Springer press (2004)
3. Lidl, R., Niederreiter, H.: Introduction to Finite Field Arithmetic and their Applications. Cambridge University Press (1986)
4. Behrouz, A.: Forouzan, Cryptography and Network Security. McGraw-Hill press, International Edition (2008)
5. Liao, H.-Z., Shen, Y.-Y.: On the elliptic curve digital signature algorithm. Tunghai Sci. **8** (2006)
6. Karthikeyan, E.: Survey of elliptic curve scalar multiplication algorithms. Int. J. Adv. Netw. Appl. **04**(02) (2012)
7. Washington, L.C.: Elliptic Curves: Number Theory and Cryptography. Discrete Mathematics and its Applications (Boca Raton). Chapman & Hall/CRC, Boca Raton, FL (2003)
8. Musson, M.: Attacking the elliptic curve discrete logarithm problem. Master Thesis of Science (Mathematics and Statistics) Acadia University (2006)
9. Rosen, K.H.: Discrete Mathematics and its Applications, Global Edition (2008)
10. Hla, N.N., Aung, T.M.: Implementation of finite field arithmetic operations for large prime and binary fields using java BigInteger class. Int. J. Eng. Res. Technol. (IJERT), **6**(08) (2017)
11. Aung, T.M., Hla, N.N.: Implementation of elliptic curve arithmetic operations for prime field and binary field using java BigInteger class. Int. J. Eng. Res. Technol. (IJERT), **6**(08) (2017)
12. Recommended Elliptic Curves for Federal Government Use, NIST (1999)

Gait-Based Authentication System

Vivek Kumar⑩, **Chirag Gupta**⑩ and **Vatsal Agarwal**

Abstract This paper deals with developing a method for classifying people based on data collected through their gait patterns. This paper presents a biometric user authentication based on a person's gait. This approach uses the accelerometer for capturing raw data to be evaluated on a machine learning model for identifying the person and various algorithms are used so as to obtain the highest precision.

1 Introduction

This paper presents gait as a robust means for human identification which is clinically proved to be highly subjective.

Biometrics is being used for a long time for identification purposes because of its robustness and easiness.

The conventional biometric identification techniques as fingerprint matching, face recognition require the user to log entry into a system and wait for confirmation—there is a delay involved in the process, whereas proposed system is unobstructed in nature, because mobile devices (mobile phones, PDAs, etc.) have inbuilt accelerometer sensors so continuous verification subject is possible. This makes it hold an edge over other identification methods. Also in comparison the equal error rate (EER) the gait system achieves lower EERs than fingerprint recognition [1] or two-dimensional face recognition [2]. This can also be user-friendly as the identification goes into background processes.

V. Kumar
Department of CSE, Moradabad Institute of Technology, Moradabad 244001, India
e-mail: contactviku@gmail.com

C. Gupta (✉) · V. Agarwal
Department of CSE, University of Petroleum & Energy Studies, Dehradun 248001, India
e-mail: chirag_gupta@stu.upes.ac.in

V. Agarwal
e-mail: vats14aug@gmail.com

© Springer Nature Singapore Pte Ltd. 2019
A. Abraham et al. (eds.), *Emerging Technologies in Data Mining and Information Security*, Advances in Intelligent Systems and Computing 755,
https://doi.org/10.1007/978-981-13-1951-8_61

Human gait corresponds to a specific style and way of moving legs, this, in turn, is synchronization between muscular, skeletal and neurological systems [3]. Therefore, gait features differ from individual to individual. From decades, Gait recognition has been studied as a behavioral biometric. Its implementation is categorized in three ways: Machine Vision Technology (WVT), Floor Sensor Technology (FST), and Wearable Sensor Technology (WST).

WST seems promising in all of them because of the fact it is newest and the sensor can be attached to subject's body. The subject can wear sensor in various positions as pockets hands or shoes. In Wearable Sensor Technology device's onboard sensors are put to use as—GPS, Accelerometer, Magnetometer, Compass, Proximity Sensor, etc. Thus, it helps to authenticate precisely without any accessory sensors.

In this paper, we propose identification through the onboard accelerometer present in smartphones.

The data is collected with an app and feature extraction is performed over the data, later the data is classified using the Decision Jungle algorithm: Decision Forest implementation of Microsoft Research [4].

2 Data Collection and Feature Extraction

In this unit, we will describe how we collect the accelerometer data and extract interesting features from the logs as to build predictive models for identification. The datasets are split into training and test datasets to obtain the robust model. The remaining section describes how the raw accelerometer data is being collected and transformed for this process.

2.1 Data Collection

People with age ranging from 18 to 40 (students and instructors) participated in the volunteer data collection program, both male and female participated and logged the data in a whole span of a week at the various time of the day, walking with a constant pace at a smooth surface. An upload utility provided to aggregate the data at a central repository [5].

The data is being collected with the help of Accelerometer Log android application available in Google Play Store with due permission. This application logs data from the device's accelerometer in three coordinate's axes—x, y, and z and stores them on device's external storage. The format used for logging accelerometer data is comma separated values (CSV).

The app starts logging upon launching and the device needs to be put in the following manner for effective logging of accelerometer data. The configuration of a mobile device for logging data effectively is illustrated in figure.

2.2 Feature Extraction

The unstructured data collected with the application needs to be prepared to be fed into the model. Prior to it, some parameters need to be calculated and stored. Various work done prior to this research handles data in the time domain, there also exists isolated research work that deals with this data in form of sliding window approach, under this technique selective, discrete time window is decided and data is processed as per time window, it helps in the way as it has compatibility with conventional classifier induction algorithms that do not work fine with time-series data.

We have calculated features for a single subject's log file at once because we have worked with individual isolated subject log files instead of an agglomerated database. Several features are computed with the dataset using custom R scripts which include:

- Average: The accelerometer sensor value is averaged for all of the axes—X, Y, and Z.
  Average Acceleration:

$$\left(\sum_{0}^{n} x, y, z \right) / n$$

- Standard Deviation: Standard deviation is calculated amongst axes—X, Y, and Z.

$$\sigma = \sqrt{\frac{1}{N} \sum_{i=1}^{N} (x_i - \mu)^2}$$

- Average Absolute Difference: Average absolute difference amongst the logged values and mean of same logged values across all axes.

$$\frac{1}{n^2} \sum_{i=1}^{n} \sum_{j=1}^{n} \left| yi - y^j \right|$$

- Average Resultant Acceleration: For all the logged accelerometer values, calculate the square root of the sum of the squares of the x, y, and z-axis values, and then their average is calculated.

$$\left(\frac{\sqrt{x^2 + y^2 + z^2}}{n} \right)$$

Legends:

x: Depicts acceleration for the axis along X
y: Depicts acceleration in the axis along Y
z: Depicts acceleration in the axis along Z
n: Total number of logged data elements
i: Single data element.

Every set of data is being tagged with the initials of the person logging the data so as to facilitate authentication at the end. These extracted features are being added to the current dataset which is being passed for evaluating the classification model.

The whole dataset is being converted to a comma separated file (CSV) for evaluation. The computation model is implemented with the help of the cloud-based platform—Microsoft Azure Machine Learning. The tagged dataset upon evaluation will contain a field—Scored Labels which is used to decide final result.

3   Classifier Algorithm

The classifier algorithm used for this task is Microsoft Research's implementation of Decision Forest—Decision Jungles. The Problem with decision forest is that when sufficient data is there, exponential growth in nodes in decision tree is being observed which is being addressed by Decision Jungles.

This algorithm deals with decision directed acyclic graphs (DAGs). A DAG may have a number of paths, whereas decision trees offer a single path to each node. The minimization of the objective function is done by splitting of node and merging of the DAG's node. This makes the DAGs achieve a considerable less memory footprint compared to other algorithms as decision forest and other baseline methods too. Generalization is also improved with the help of decision jungle as compared to other algorithms. The calculation of an optimal decision tree remains an NP-hard problem, but DAGs help in making effective predictions. Sometimes, learning optimal decision trees leads to overfitting as asserted by some of the researchers [6]. There are certain measures with the help of which complexity of the model can be reduced with the help of some stopping criteria, viz.

- Capping the depth of tree
- Post hoc pruning of DAGs.

In this algorithm, learning is known as an energy minimization problem. The methods which lead to the minimization of the energy objective include optimization of a split function at nodes and placing the branches arising from parent nodes, they lead to optimal DAGs. Both rooted binary DAGs and decision trees correspond to an entirely different architecture as asserted by Platt et al. [7]. He introduced them as a procedure to merging binary classifiers and goals with multiple classes. We can generalize a binary DAG (rooted) as:

- Node with single root, having in degree as 0
- Cluster of split nodes, having in degree ≥ 1 and out degree 2.

Suppose we have N—class classification problem, this strategy does not expect to have N DAG leaves. It means that each leaf has relation with a distribution within empirical class.

Each of the rooted decision DAG, in this algorithm, is trained in a jungle independently. The training method works by consecutively increasing the level one by one each time. The minimization function's critical parameters and features along with structural branching are fed to the algorithm at every level. The predictions that are made help in minimizing the objective function with the help of child nodes that emerges from other nodes whose major parameters are already being learned.

4 Results

This experiment involved building a predictive model by comparing all the algorithms and using best of them, it predicts the probability for each of the subjects to be the person under interrogation. The results hence obtained by using the said algorithm is described below in Table 1.

The results of model evaluation for a small set of subjects is illustrated in Fig. 1. The predicted class for subject1 is correct for 97.6% of the time, whereas 94.8% for subject2. Increasing data samples for other users may increase their accuracy. Overall precision is decided as:

Number of Correct Predictions/Total Number of predictions

The overall accuracy obtained is 87.31%, while 95.65% of average accuracy.

5 Conclusion and Future Work

Gait as a means of identification is explored and found to be reliable in this paper. The available commercial smartphone embedded accelerometer was able to carry out identification with modest accuracy.

Table 1 Prediction results

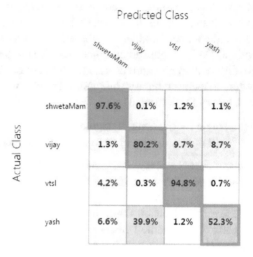

Fig. 1 Classification parameters

Overall Accuracy	0.873114
Average Accuracy	0.936557
Micro averaged precision	0.873114
Macro averaged precision	0.835511
Micro averaged recall	0.873114
Macro averaged recall	0.812187

This method of authentication seems of high value than other systems as the whole process is unobstructed—i.e., it does not require the subject to intervene in the authentication process, it's seamless and is more user-friendly, the authentication is being observed on-the-go. It can also be implemented with the help of embedded sensors—accelerometer, for that embedded devices that can evaluate the model on board needs to be procured.

The results hence achieved seems promising and the underlying algorithm used is robust and precise with fewer memory footprints. We are planning to improve the experiment by increasing our database, and as of current, we logged ~ 10 datasets from each subject, which we aim to increase. Instead of our own database, we will try to evaluate our model on various online available data sources. More experimentation is needed regarding the placement of the device, instead of placing it indefinite position and configuration we will try this placement to be natural so as to increase of user-friendliness to a level up. There are several important types of research conducted previously on the attack resistance of gait recognition with the help of biometrics. These types of studies need to be carried with the prospect of mobile phones onboard accelerometer.

References

1. FVC2006: the fourth international fingerprint verification competition. http://bias.csr.unibo.it/f vc2006/results/O_res_db2_a.asp (2017). Accessed 28 Mar 2017
2. Abeni, P., Baltatu, P., D'Alessandro, R.: A face recognition system for mobile phones. ISSE 2006 Secur. Electro. Busin. Process. 211–217, Oct 2006
3. Fish, D.J., Nielsen, J.: Clinical assessment of human gait. J. Prosthet. Orthot. **2**, Apr 1993
4. Shotton, J., Sharp, T., Kohli, P.: Decision Jungles: Compact and Rich Models for Classification
5. Gbas data upload utility. https://script.google.com/macros/s/AKfycbxEiWElCj9-CpHwsnGP9 HfrcIuaZgGOMfquRidZQmbhkJ36NdM/exec
6. Murthy, K.V.S., Salzberg, S.L.: On growing better decision trees from data. Ph.D. thesis, John Hopkins University (1995)
7. Platt, J.C., Cristianini, N., Shawe-Taylor, J.: Large margin DAGs for multiclass classification. In: Proceedings of NIPS 547–553 (2000)

Pen-Drive Based Password Management System for Online Accounts

Samruddhi Patil, Kumud Wasnik and Sudhir Bagade

Abstract This paper presents a system that manages user's password using pen-drive as a key for storing and protecting it from attackers, hackers, and from loss. Nowadays users are using single or easy password for all website accounts because they can be easy to remember but it increases chances of attack, also user sometime uses single password for all accounts. Instead of remembering or using same passwords, in this system we use pen-drive to store password and login to online accounts. Pen-drive is popular for storage of different files like audio, text, data, and video. In our password management system, passwords are stored in a password-protected file which is saved in Pen-drive and retrieved by application. The application stores the password and corresponding URL temporarily on PC and use till pen-drive is connected to the PC. Our system can solve the password memorability problem for user and also protect password from various attacks. Our System addresses the security issues like Confidentiality, Integrity and Authentication because of two-level protections. In first level, password is stored on users own pen-drive and in second level, pen-drive is protected by password.

Keywords Pen-drive · Password manager · Online account · Security
Password · URL

1 Introduction

Currently, many users are likely tended to use a password which is easily remembered or single password for all website. These are insecure and can leak easily and also

S. Patil (✉) · K. Wasnik · S. Bagade
UMIT, SNDT Women's University, Mumbai 400049, India
e-mail: patil.samruddhi01@gmail.com

K. Wasnik
e-mail: kumudwasnik@gmail.com

S. Bagade
e-mail: bsudhiran@ieee.org

© Springer Nature Singapore Pte Ltd. 2019　　　　　　　　　　　　　　693
A. Abraham et al. (eds.), *Emerging Technologies in Data Mining and Information
Security*, Advances in Intelligent Systems and Computing 755,
https://doi.org/10.1007/978-981-13-1951-8_62

use different passwords to the different website are difficult to remember. Due to this, there can be a high risk of security and confidentiality of user's account.

Nowadays many peoples are using Internet for shopping, banking, and paying bills. To access all this services user needs to register and open account to access this web services [1]. Most of the website basically use text password-protected security for users account, but for a single user, it is more difficult to remember all these passwords. Sometime it may happen that user open account after a long time or user need account sometime only, so it is more difficult and hectic to remember the password. Some user to uses one password for all account or there are chances that user can select an easy password which can be easy to remember or use the same password all over the internet, this can cause password guessing attack and a chance to leakage of passwords [1–3].

We usually take passwords for granted, but they are often the only defense against someone getting their hands on our personal information, including financial information, health data, and private documents [1].

Generally, most of the users write down the password in a diary or in a text file using encryption or other protection methods. The main aim of the password is to protect user's personal account information and their confidentiality, but if the password gets leaks or hacked by hackers then the hacker can misuse their account or sell them for personal purpose. There are two modes of password manager: the one is to store user's passwords in local file [4], and the other one is to store them "in the cloud" by some providers [5]. Some browser also saves the password in cookies but they are not safe [6, 7].

Hackers crack or break the passwords in a number of ways. Hence, there is a high-security risk and confidentiality risk of user accounts.

Existing research [2], about the password, shows that most of the user's use password which is less than nine characters and which can be attacked by most of the attacker easily. The guessing attack is more common attack among them. The user also seems to make little of rules and advises concerning the security. Another problem is that sometimes most of the users reuse the same or identical password for the multiple online accounts so to prevent this; the system used pen-drive to store and login to account. Pen-drive is mostly like for its storage capacity and transferring files in a fraction of seconds. Pen-drive is used to store passwords for user and application help user to retrieve and fill password to the login form. Our research focuses on most identification, security and authentication of user's account using pen-drive in which passwords are stored in existing password-protected system or password manager [4, 8].

The rest of the paper is organized as follows. In Sect. 2 we are going to discuss related work. Section 3 gives detail methodology of this system. Section 4 contains Usage overview. Section 5 contains the implementation of the system; Sect. 6 contains experimental result and discussion. Section 7 contains a conclusion.

2   Related Work

The most of the password manager's databases are stored on local machine or in the cloud. Related work is that of [9], they use mobile as a storage device for a password. It gives solution for memorability problem of the password in authentication system using a mobile-based password manager. In this system, it creates a faux password using Transformed-Based Algorithm and the Modified Levenshtein Distance. It uses Decentralized File Format architecture to distribute credential information into different files for password storage.

Another related work is studied in [10]. The Security Enhanced Secret Storage Scheme, In this when the client wants to access any account or any secure data, it will create and starts a session with server and information related master key is stored locally. So the user needs to remember master key for secure authentication. If the authentication is successful, the server will give access and send its partial data to the authenticated user.

In [7], the author used to store the password on cloud and on local devices. This research of password stored in the cloud and local Machine gives the security and confidentiality analysis of two password manager which mostly used. LastPass and RoboForm are both browsers and cloud-based password manager. They both save password on a local drive-in cache file or text file and also on the cloud. The recent technology, browsers are also storing the password on the local machine or on the cloud.

Google chrome stores password or related information on Google server to allow synchronization between different devices [11]. Chrome support pages claim that password are stored in encrypted format but a user with access to the database can recover all its content and make arbitrary modifications [7, 8, 11]. Mozilla Firefox stores login details or related information in an SQLite database. Users can choose a master password that is used for encrypting the database content. URLs are mainly stored unencrypted regardless of whether a master password is used or not [7, 8].

3   The Methodology

In this section, we will discuss login system methodology and its objectives which give solution for managing passwords using pen-drive. The conventional ways to remember passwords by humans is to memorize them. The chances of forgetting passwords are increases if the password is not used frequently. Hence, our objectives are listed as follows.

1. To protect web users from various attack.
2. Give easy authentication using a different password.
3. To provide an easy-to-use system with roaming capability.
4. The password does not require to be remembered.

5. The password is only valid if it is coming from particular pen-drive. So that password is safe with the owner.

The flow chart of login system is drawn in Fig. 1, which shows the management of passwords using pen-drive. This flowchart is explained with the following steps.

1. Application start.
2. Application detect pen-drive and check pin of pen-drive.
3. Then find password file in pen-drive.
4. If file found then read the file from pen-drive else display an error message.
5. Ask password to open the file from pen-drive.
6. If password matches display a list of login details else display an error message.
7. Copy all data from file and store in a temporary location.
8. After data is store application communicate with the browser through trust path.
9. For login to the website, user click on required login detail.
10. URL gets open in the browser and automatically fill login id and password for respective URL.
11. It will check the status of pen-drive every 30 s. if the status is removed then it will delete all data password from the application.
12. The application also gives the option to save backup and restore the file if by chance they lost pen-drive.

4 Usage Overview

First, the user inserts pen-drive in the USB port and run application which then detects the pen-drive and for opening password file from pen-drive one has to enter register password. Then the user can access login details which are stored in pen-drive. All the registered user's id, password and URL display on the application.

When a user registers a new account on any website that user has to save that login details in pen-drive through the application. The application will save information of URL of website, username, and password in encrypted format in pen-drive.

When the user wants to log into the website, the user will click on login details from a file and it will automatically open a webpage and fill login details for the user.

All websites login details are stored in pen-drive so when user wants to access the website from other computers then the user needs to install our application onto the respective machine.

5 Application Methodology

In Fig. 2, we have shown the architecture of our system which consists of three phases, namely pen-drive, application, and browser.

The phases of the proposed system in Fig. 2 are discussed as follows.

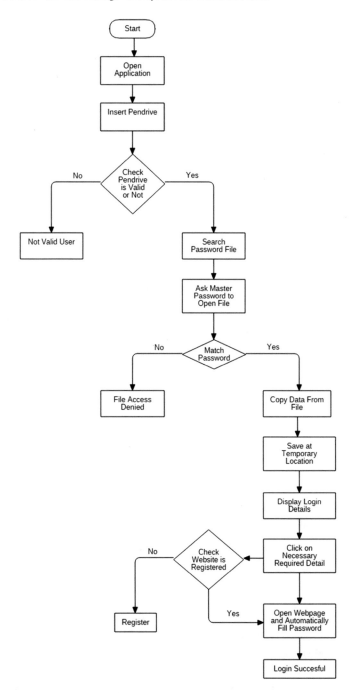

Fig. 1 Flow chart of pen-drive based password management system for online accounts

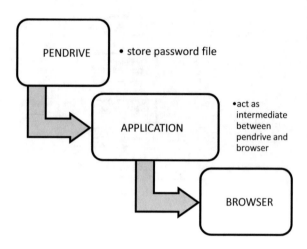

Fig. 2 The architecture of proposed system

5.1 Application

1. The application will act as an intermediate between pen-drive and browser.
2. After inserting pen-drive application will detect it and use a pin for a trusted path between pen-drive and application and browser [12].
3. Then it will find password file and copy data from it and store temporarily in pc. Then it will decrypt data.
4. The password is stored in word document file which is encrypted and password-protected.
5. While it checks browser parse that browser and take username and URL of the website, search it and automatically fill login form for the user.
6. If the user is not registered that website then asks for registration and save all login details and update file in pen-drive.
7. If loss of a file or the pen-drive user can export file from his trusted pc with encrypted by pin as a backup file. The user can import backup file to new pen-drive and no one can access that file without application pin.

5.2 Browser

1. The browser will get open when the user clicks on login details. Application securely passes all login details to the browser with a secure path between browser and application.

5.3 Pen-Drive

1. Pen-drive is most important part of this system; it is used to store user information in an encrypted format that can only use by application with authentication of a pin.
2. Login details stored in a word document which is encrypted and password-protected.
3. Without the correct password, it will not give access to file.
4. Application copies all data from pen-drive and store temporally on pc.

6 Implementation

We have implemented our system using following steps.

Step 1: Registering login details to pen-drive.

In this module, the user fills information about login details which can be taken when the user is registering on the website automatically by parsing register form or manually can be filled by the user.

In this step, we fetch login details from a webpage and fill in the registering form. The user can manually fill details but URL of the website is fetched only by the system. Thus we get details from the user are USERNAME and PASSWORD. Then all the data is saved in pen-drive by detecting pen-drive. All data saved in the file is in an encrypted format.

Step 2: Detection of pen-drive.

In this Step, pen-drive is connected or not is checked. It accesses all file directory of the operating system (here we are accessing windows 8.1). The application keeps track of drives if any additional drive is found then, it will assume that external device is connected and then it checks that pen-drive is users or not.

It will give all list of an external device connected to the system.

Step 3: Identify password file and copy data from a file.

After pen-drive is selected it will search for password file if the file is found then ask the user for a password if it matches then the only file can be accessible and on file File read(), File write() operation is performed. Then application read the file from the pen-drive line by line. And store in array format temporarily. If they did not detect any pen-drive or pen-drive disconnected then temporarily data is erased by the application. So it keeps track of pen-drive. We use a word document to store login details with encrypted by password.

Step 4: Parse webpage and autofill login details.

This is the last section where login details are automatically filled in the login page. When click on required login detail URL get open in the browser and automatically fill login id and password.

Figure 1 shows the flow chart diagram of the system.

7 Results and Discussion

We have implemented our application which read and write a file of pen-drive and also connects with the browser. We developed an application in c# and .net framework programing language.

7.1 Login Page

As Fig. 3, Shows Login page contain function buttons first button shows (SHOW LOGIN DETAILS) all saved login details from pen-drive. 2nd button (NEW ENTRY) is called for entry or new register website and remaining all two buttons are for backup and restore of the file.

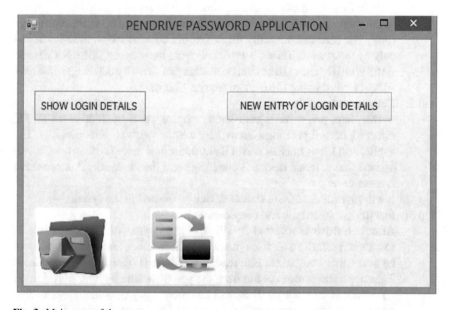

Fig. 3 Main page of the system

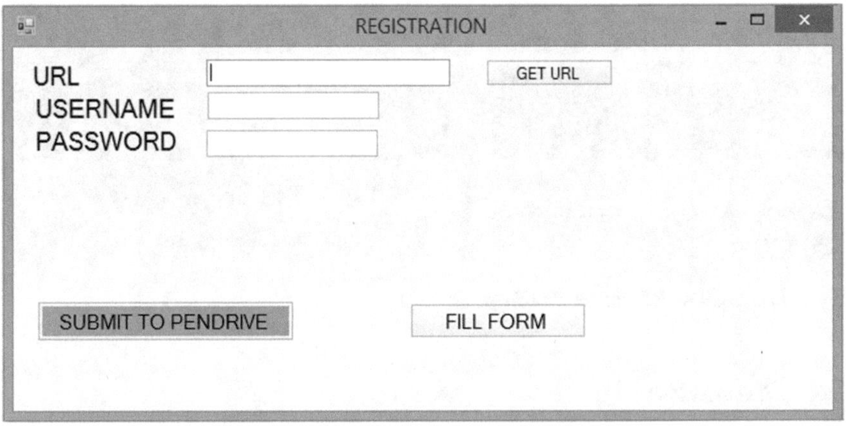

Fig. 4 Registration page of the system

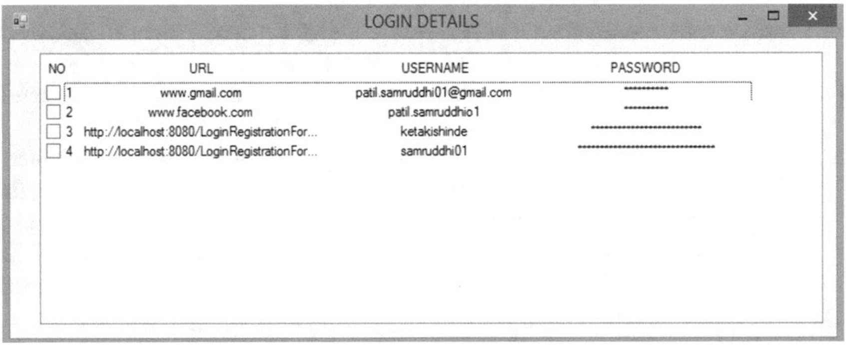

Fig. 5 Shows all login details from pen-drive

7.2 Registration Page

As Fig. 4 shows this is register form for new registration of website or web application. It collects all details from the user and fills in register page of the website and simultaneously saves in pen-drive.

7.3 All Login Details

Figure 5, shows all login details of register website like URL, username, password.

After clicking on an individual record, it will go to directly that URL and autofill login details. Figure 6 shows Facebook login page after clicking on the URL or username.

Fig. 6 Facebook with login details fill

7.4 Discussion

1. If pen-drive is stolen: we provide backup and restore function in the application, the user can create a backup file and store somewhere in the safe place.
2. Users do not need to remember password: this system saves users all password in the password-protected file so the user doesn't need to remember them.
3. Does login works for all the registered sites?: yes -it will work for all registered sites.
4. Does it address the security issues?: yes- Our system solves issues of Confidentiality, Integrity, and Authentication by using two-level password protection for the registered URL. First, the password and corresponding URL is stored in a file on pen-drive. This password file is again protected by the password which is known only to the owner of the pen-drive. Thus our system addresses the above-stated security issues.

Further, there is still need to do research on lost pen-drive. Currently, we give the backup option. We can also increase the security of pen-drive by adding a biometric feature like a fingerprint scanner. We can create passwords for user combining user's text password and pen-drive key so that user cannot use the online account without pen-drive.

7.5 Comparative Study

In this section, the comparison of some existing method or some technique of storing password and authentication is given. Mobile-based [9], local drive and cloud-based password manager [7] which proposes a system for storing password. Mobile devices are one of the insecure devices which have lots of application that they can steal credential information from mobile. There are many local and cloud-based password manager which stored password [13].

Cloud-based password managers are secured but most of the time they are accessible to its owner. The locally stored password has chances of attack to steal information. But in our proposed system we store the password in pen-drive which has not any other owner. Credential information is safe with the user and it's not connected to any Internet, so no chances of an attack. Even someone steals the pen-drive the password file is password-protected so no one will able the read the content. Hence, as compared to other existing systems our system is secure and personal to the owner only.

8 Conclusion

In this paper, we demonstrated the benefit of managing username and password using our system based on pen-drive. The system able to protects user credential information such as username, password, etc. It solves the problem of remembrance of passwords. The system automatically accesses the pen-drive and asks for the password before showing the user details. Further, the user can carry pen-drive anywhere so he/she can access its login details from everywhere and use pen-drive as login key to the registered URL. Even if somebody steals the pen- drive or loss it in person, the password cannot be retrieved as pen-drive is protected by a password. Thus, our system solves the issues of Confidentiality, Integrity, and Authentication by using two-level password protection.

In future, we are focusing on the creation of password by using the pen-drive serial number and user's text password.

References

1. Wasnik, K., Patil, S.: A survey on existing password storage methods and their security. Int. J. Sci. Res. 2319–7064 (2017). ISSN (Online)
2. Ma, W., Campbell, J., Tran, D., Kleeman, D.: A conceptual framework for assessing password quality. Int. J. Comput. Sci. Netw. Secur. **7**(1), 179–185 (2007)
3. Gaw, S., Felten, E.W.: Password management strategies for onlineaccounts. In Proceedings of the 2nd Symposium on Usable Privacy and Security (SOUPS),pp. 44–55. New York, NY, USA, ACM, July 2006
4. RoboForm Password Manager. http://www.roboform.com/
5. LastPass, online password manager, may have been hacked. http://www.pcworld.com/article/227223/LastPass_Online_Password_Manager_May_Have_Been_Hacked.html
6. Herzberg, A. Jbara, A.: Security and identification indicators for browsers against spoofing and phishing attacks. ACM Trans. Internet Technol **8**(1–36) (2008). New York, NY, USA
7. Zhao, R., Yue, C., Sun, K.: A security analysis of two commercial browser and cloud-based password managers. BioMedCom (2013)
8. Gasti, P., Rasmussen, K.P.: On the security of password manager database formats
9. Agholor, S., Sodiya, A.S., Akinwale, A. T., Adeniran, 0. J.: A secured mobile-based password manager. ISBN: 978-1-4673-7504-7©2016 IEEE

10. Fang, H., Aiqun, H., Shi, L., Li, T.: SESS: a security-enhanced secret storage scheme for password managers. 978-1-4673-7687-7/15/©2015 IEEE
11. Google. Protect our synced data. http://support.google.com/chrome/bin/answer.py?hl=en&answer=118103
12. Ye, Z., Smith, S., Anthony, D.: "Trusted paths for browsers", ACM Trans. Inf. Syst. Secur. 3, 153–186 (2005)
13. Fukumitsu, M., Hasegawa, S., Iwazaki, S.-Y.: A proposal of a password manager satisfying security and usability. In: 2016 IEEE 30th International Conference on Advanced Information Networking and Applications (2016)
14. LastPass Password Manager. https://lastpass.com/

A Comparative Analysis of Symmetric Lightweight Block Ciphers

Asmita Poojari and H. R. Nagesh

Abstract With the advancements in technology the requirement of security systems to protect data are becoming crucial. The shift from high end devices such as desktop computers, servers and laptops to small constrained devices such as RFID tags, smart cards, wireless sensor nodes and all those devices related to Internet of Things (IOT) brings a wide range of new security challenges. The major concern of such devices is the storage, access and transmission of critical data. Lightweight cryptography mainly aims for the security of such resource constrained devices. The existing conventional cryptographic algorithms cannot be applied to these resource constrained devices and hence the need for new lightweight cryptographic algorithms. A comparative analysis of the various symmetric lightweight block ciphers applicable to such devices is provided in this paper.

Keywords Lightweight cryptography · IOT · RFID

1 Introduction

IOT is an emerging technology that enables the physical devices to interact with each other. For this interaction to take place technologies like sensors or RFID are used. RFID is used to identify the things in IOT and track the status of things whereas sensors are used to sense and collect data. The IOT devices have very limited memory, computing power and battery supply and thus the need for lightweight cryptography which can be applied to such devices. While designing the lightweight cryptographic

A. Poojari (✉)
NMAMIT, Nitte, Karkala 574110, India
e-mail: asmitapoojari@nitte.edu.in

H. R. Nagesh
AJ Institute of Technology, Kuloor, Mangalore 575013, India
e-mail: nageshhrcs@rediffmail.com

© Springer Nature Singapore Pte Ltd. 2019
A. Abraham et al. (eds.), *Emerging Technologies in Data Mining and Information Security*, Advances in Intelligent Systems and Computing 755,
https://doi.org/10.1007/978-981-13-1951-8_63

algorithm the trade-offs between security, cost and performance should be taken into consideration. The lightweight ciphers can be categorized in terms of hardware versus software designs, block versus stream ciphers and symmetric versus asymmetric ciphers.

2 Hardware Versus Software Implementation

Lightweight cryptographic algorithms can be classified in terms of either hardware implementation or software implementation. The main design goal in hardware implementation is that of small gate equivalent i.e. the reduction of the logic gates that are required. This reduces the cost and power consumption. The other design constraints are energy consumption and power consumption. The main design goals for software implementations are memory consumption, processing power and throughput.

3 Symmetric Versus Asymmetric Cryptography

Most of the lightweight cryptographic algorithms developed so far fall under symmetric cryptographic systems wherein the same key is used for encryption as well as decryption. The algorithms used for survey in this paper are symmetric lightweight ciphers. Asymmetric lightweight ciphers wherein the key used for encryption and decryption process are different are usually used with devices with powerful hardware. Some of the lightweight asymmetric cryptographic algorithms are ECC [1]/HECC [2] based on Elliptic curve Cryptography and ElGamal [2] based on the Discrete Logarithm Problem in Finite fields (DLP). ECC and HECC present lower computational requirement. Also they provide shorter keys. ECC also provides software implementations. Some of the software implementations are TinyECC and WMECC. HECC uses the concept of elliptic curves and performs better than ECC. Other Asymmetric cryptosystems are Merkle signature scheme (MSS) [3] based on hash functions, NTRU [4] based on lattice based cryptography, Multivariate quadratic cryptography based on multivariable quadratic equations over finite fields.

4 Block Cipher Versus Stream Cipher

Many lightweight block and stream ciphers have been invented over the years. In block ciphers a block of plaintext along with a key is used as input to give a ciphertext whereas in stream cipher string of bits of plaintext along with a key is used as input to give bit strings of ciphertext. We study some of the lightweight block ciphers in this paper.

4.1 HIGHT

HIGHT [5] was developed in 2006. It has block size of 64 bits, 128 bits key size and 32 rounds. It makes use of Fiestel network. Fiestel network consists of XOR operations combined with left and right rotations. It provides low resource hardware implementation and can be used in devices such as RFID tag and tiny ubiquitous devices. It can be implemented with 3048 gates on 0.255 μm technology. It is hardware oriented cipher with simple operations such as XOR, addition mod^8 and left bitwise rotation. The steps are as follows:

- Initial transformation on the plaintext together using input whitening keys. The whitening keys are generated using the 128-bit master keys. The output of the initial transformation is the input for the first round.
- 32 byte oriented round function using modular addition, XOR and linear sub-round functions i.e. left bitwise rotation.
- A final transformation to the output of the last round using four whitening keys bytes.

4.2 PRESENT

PRESENT [6] is a ultra-lightweight block cipher with a block length of 64 bits and two key lengths of 80 and 128 bits. It has SP-networks and consists of 31 rounds. It is dedicated to be implemented in hardware. Each round has following operations:

- Subkey addition which consists of simple XOR function.
- Substitution using 16 4-bits S-boxes for the confusion layer.
- For diffusion, permutation consisting of bitwise permutation is applied.

4.3 PRINTcipher-48 and PRINTcipher-96

PRINTcipher-48 and PRINTcipher-96 [7] are lightweight block ciphers that are used for IC printing technology. It has the structure of SP-network. PRINTcipher-48 uses 48-bit blocks as input plaintext and 80-bit key.It consists of 48 rounds while PRINTcipher-96 uses 96-bit blocks as input and 160-bit key and consists of 96 rounds. Each round consists of five steps:

- The current state of the cipher is XORed with the round key.
- The cipher is shuffled using linear diffusion which is simple bit permutation.
- The rightmost 6 bits of the cipher is XORed with round constant.
- The bits are then permuted using key-dependent permutation.
- Finally the cipher is mixed using S-box.

4.4   TWINE

TWINE [8] is suitable for extremely small hardware devices. It has 64 bits block size and 36 rounds. It has two variants that differ with respect to the key size- TWINE-80 and TWINE-128 with key size of 80 bits and 128 bits respectively. It has a Fiestel structure. The steps are as follows:

- In data processing part 36, 32-bit round keys are applied to the 64-bit plaintext with each round comprising of F-function which is XORing of plaintext with sub key and applying 4 * 4 s-box.
- In key scheduling the input key uses thirty five 6-bit constants to produce a 36 32-bit round keys.

4.5   LBLOCK

LBLOCK [9] is a lightweight block cipher developed in 2011 with block size of 64 bit and key size of 80 bit with 32 rounds. It has a Fiestel structure. S-box includes S-box layer and permutation layer. It uses 10 different 4 * 4 bits S-box. Various attacks were applied on LBLOCK such as differential attack, biclique attack and zero correlation linear attack. Each round function is composed of the following operations:

- Subkey addition.
- For confusion eight 4-bit S-boxes applied in parallel.
- For diffusion eight 4-bit permutations are applied.

4.6   Piccolo

Piccolo [10] is a very recent cipher which is a 64 bits block cipher. The key size is 80 and 128 bits. It makes use of Fiestel structure. The working of Piccolo is as follows:

- In data processing part, 64 bit plaintext, four 16-bit whitening keys and 2r 16 bit round keys are used where r is the number of rounds used for encryption. The whitening keys and round keys are generated by the key scheduling of 80 and 128-bit keys.
- In each round, output of previous stage is permuted (shuffled on words of 8 bits) and is given as input to the next stage.
- In key scheduling part, the input key is divided into five 16-bit sub-keys for 80-bit key size and eight 16-bit sub-keys for 128-bit key size.

4.7 Simon

Developed in 2013, Simon [11] cipher uses a fiestel structure. The round function consisting of the following functions:

- bitwise XOR
- bitwise AND
- left circular shifts.

It can be optimized for both software and hardware implementations but performs well with hardware implementation. The block size and key size range goes from 32 bit block with 64 bit key size to 128 bit block with 256 bit key. Each round takes as input the round key and two n-bit intermediate ciphertext words.

4.8 SPECK

The SPECK [11] cipher also makes use of fiestel structure with each round consisting of following functions:

- Bitwise XOR
- Addition modulo 2^n
- Left and Right circular shifts

It can be optimized for both software and hardware implementations but performs well with software implementation. The block size and key size range goes from 32 bit block with 64 bit key size to 128 bit block with 256 bit key. The number of rounds can range from 22 to 34 depending on the block size and key size selected.

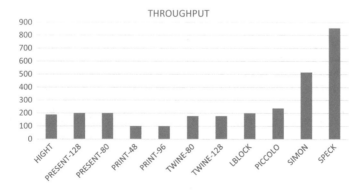

Fig. 1 Comparative analysis in terms of throughput (kilobits per second)

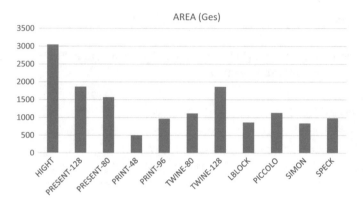

Fig. 2 Comparative analysis in terms of area (GEs)

Table 1 Comparative analysis of the block ciphers

Algorithm	Block size	Key size	Structure	Rounds	Throughput (kbps)	Area (GEs)
HIGHT	64	128	Fiestel	32	188.2	3048
PRESENT-128	64	128	SPN	31	200	1886
PRESENT-80	64	80	SPN	31	200	1570
PRINT-48	48	80	SPN	48	100	503
PRINT-96	96	160	SPN	96	100	967
TWINE-80	64	80	Fiestel	36	178	1116
TWINE-128	64	128	Fiestel	36	178	1866
LBLOCK	64	80	Fiestel	32	200	866
PICCOLO	64	80/128	Fiestel	25/31	237	1135
SIMON	32–128	64–256	Fiestel	32–72	515	838
SPECK	32–128	64–256	Fiestel	22–34	855	984

5   Conclusion

Internet of things has become part of our day-to-day life due to advancements in the technology but the most crucial part to be considered is the lack of security. This paper evaluated the performance of the lightweight block ciphers in terms of throughput and area (gate equivalence) with the comparison shown in Table 1. Also a comparative analysis in terms of throughput and area is shown (Figs. 1 and 2). The evaluation conducted in terms of throughtput metric showed that SPECK and SIMON block ciphers perform better than rest of the ciphers. The evaluation conducted in terms of cost and with GE acting as the metric determined SPECK and SIMON as the best ciphers in this respect. This comparison lays a platform to propose a novel lightweight algorithm with improved security in terms of better throughtput, reduced area, small key size and block size.

References

1. Roman, R., Alcaraz, C., Lopez, J.: A survey of cryptographic primitives and implementations for hardware-constrained sensor network nodes. J. Mob. Netw. Appl. **12**(4), pp. 231–244 (2007)
2. Oren, Y., Feldhofer, M.: WIPR a low-resource public-key identification scheme for RFID tags and sensor nodes. In: Basin, D.A., Capkun, S., Lee, W. (eds.) WISEC, pp. 59–68. ACM (2009)
3. Shen, X., Du, Z., Chen, R.: Research on NTRU algorithm for mobile java security. In: International Conference on Scalable Computing and Communications. The Eighth International Conference on Embedded Computing 2009, SCALCOMEMBEDDEDCOM'09, pp. 366–369 (2009)
4. Howgrave-Graham, N., Silverman, J.H., Whyte, W.: Choosing parameter sets for NTRUEncrypt with NAEP and SVES-3. In: Menezes, A. (ed.) Proceedings of the 2005 international conference on Topics in Cryptology (CT-RSA'05), pp. 118–135. Springer, Berlin, Heidelberg (2005)
5. Hong, D., Sung, J., Hong, S, Lim, J., Lee, S., Koo, B.S. et al.: A new block cipher suitable for low-resource device. In: Cryptographic Hardware and Embedded Systems—CHES 2006, LNCS 4249, pp. 46–59. Springer (2006)
6. Nakahara, J., Sepehrdad, P., Zhang, B., Wang, M.: Linear (Hull) and algebraic cryptanalysis of the block cipher PRESENT. In: Cryptology and Network Security—CANS 2009, LNCS 5888, pp. 58–75. Springer (2009)
7. Knudsen, L., Leander, G., Poschmann, A., Robshaw, M.: Printcipher: a block cipher for icprinting. Cryptographic Hardware and Embedded System. Springer **6225**, 16–32 (2010)
8. Suzaki, T., Minematsu, K., Morioka, S., Kobayashi, E.: TWINE: a lightweight block cipher for multiple platforms. In: Knudsen, L.R., Wu, H. (eds.) Selected Areas in Cryptography, volume 7707 of Lecture Notes in Computer Science, pp. 339–354. Springer, Berlin, Heidelberg (2013)
9. Wu, W., Zhang, L.: Lblock: a lightweight block cipher. In: Applied Cryptography and Network Security ACNS 2011, volume 6715 of LNCS, pp. 327–344. Springer (2011)
10. Shibutani, K., Isobe, T., Hiwatari, H., Mitsuda, A., Akishita, T., Shirai, T.: Piccolo: an ultra-lightweight blockcipher. In: Cryptographic Hardware and Embedded Systems—CHES 2011, vol. 6917, LNCS. Springer (2011)
11. Beaulieu, R., Shors, D., Smith, J., TreatmanClark, S., Weeks, B., Wingers, L.: The SIMON and SPECK families of lightweight block ciphers. IACR Cryptology ePrint Archive, vol. 2013, p. 404 (2013)

A Novel Approach to Generate Symmetric Key in Cryptography Using Genetic Algorithm (GA)

Chukhu Chunka, Rajat Subhra Goswami and Subhasish Banerjee

Abstract Purpose of computer network is to share the information and provide the secure services. Due to publically in nature of computer network opens the possibility of hacking and stealing the confidential information by the attackers. To maintain confidentiality, integrity and to defend interception, fabrication, and modification of data become a burning issue. In this regard, many mechanisms have been proposed by researchers, among which Automatic Variable Key (AVK) is a novel approach. But in AVK initial key is distributed through Rivest-Shamir-Adleman (RSA). Thus, to surmount this initial distribution of key, we have proposed a new technique using Artificial Intelligence where the initial key is distributed to both the parties through fitness function of GA. To validate the proposed scheme, National Institute of Standards Technology (NIST) statistical tools is used to check the randomness among the auto-generated keys and is compared with existing related schemes. The Standard Deviation of Hamming distance is calculated for three different experiments and values obtained are 8.05, 6.44 and 7.05 which shows improvement in performance as compared to similar existing methods.

Keywords Genetic algorithm · Initial key · AVK · Artificial intelligence

1 Introduction

Since the twenty-first era is kenned as technology world, all the time's online transaction, e-commerce, and e-business consummately depend on how reliably and correctly the stream of data is transmitted over the network. The flow of data conven-

C. Chunka (✉) · R. S. Goswami · S. Banerjee
Department of Computer Science & Engineering, National Institute of Technology,
Yupia 791110, Arunachal Pradesh, India
e-mail: chukhuchunka20@gmail.com

R. S. Goswami
e-mail: rajat.nitap20@gmail.com

S. Banerjee
e-mail: subhasishism@gmail.com

© Springer Nature Singapore Pte Ltd. 2019　　　　　　　　　　　　　713
A. Abraham et al. (eds.), *Emerging Technologies in Data Mining and Information Security*, Advances in Intelligent Systems and Computing 755,
https://doi.org/10.1007/978-981-13-1951-8_64

tionally meets certain problems like modification, damage, loss of data and exposure of data to illicit persons. In this regard, the security becomes a challenging issue among researchers [1–15]. As per literature survey, the key size is of 8 bits fixed and the probability of frequency attack, brute force attack and cryptanalysis attack may increase during the transmission of data. Later, Dutta et al. [9] modified the technique in which auto-generated keys and size of keys differ which ensures higher security. In Addition to a new technique, it has also been developed by Sania et al. [16] to improve the key efficiency and to generate the best fit key in cryptography using Genetic Algorithm (GA), in which chromosome/key sequence is non-repeating. The key size is increased to 100 bits, the larger the key size more difficult for the attacker to decrypt the original key. To enhance security, we increase the key size of symmetric key based on cryptosystem. In present work, the initial key is shared between both the parties, i.e., sender and receiver using the fitness function of GA. The selection of the initial key in the scheme is based on fitness function. For the next session apply AVK [3, 4] to generate dynamic keys. The generated key using fitness function of GA is postulated to be the best key/initial key for encryption, which is more arbitrary in nature. To validate the truthfulness about the randomness of key generation method, the Hamming distance is defined as difference made between the keys or difference made in two key bits. To choose the best chromosome/key depends upon the fitness value of the population. In this consequence, the GA is vastly utilized in the key generation techniques in cryptography [17, 16, 18, 19, 20, 5]. Outline of the manuscript is as follows. In Sect. 2 introduction to GA is given; Sect. 3 includes proposed work; Sect. 4 experimented results are discussed; Sect. 5 for complexity analysis; Sect. 6 summarizes performance analysis; in Sect. 7 and Sect. 8, we have discussed randomness verification and security analysis respectively and followed by the conclusion.

2   Genetic Algorithm

GA is computerized search and optimization technique based on mechanics of genetic and natural selection [16]. GA is very different from other traditional optimization methods. Figure 1 shows evaluations of the natural process, how the GA goes through the process until the fitness value of each chromosome is satisfied. The algorithm starts with a set of solutions called population and the individual chromosome with highest fitness value is selected for crossover and mutation process. The individual with highest fitness value will be in the new population.

- GA starts with initial population i.e. parents and transfers to a new population, i.e., offspring.
- Selection is the first operator practiced on the initial population. The chromosome is selected from the population by applying the fitness values on individual chromosomes. Among the chromosomes, the best ones will survive and produce new offspring.

Fig. 1 Basic genetic
algorithm process

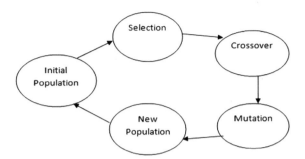

- Two parents' chromosomes are selected to perform crossover operation. The new offspring chromosomes consist of some features of first parent chromosome and the rest over from the second parent chromosome such as the given example.
- Mutation operator involves in modifying a bit of string chromosomes to maintain genetic diversity from one generation to new generation of the population. These changes are mainly caused by errors in copying genes from parents' chromosome [21] such as the given example.

Parent chromosome 1:	ABCD*EFGH*	Parent chromosome 2:	IJKL*MNOP*
Offspring1:	ABCD*MNOP*	Offspring2:	IJKL*EFGH*

Parent chromosome 1:	*Q*RSTUV*WX*	Parent chromosome 2:	*YZ*ABCDEF
Offspring1:	*G*RSTUV*HX*	Offspring2:	*I*ZABCD*J*F

3 Proposed Work

In our proposed scheme, fitness function of GA is used in order to generate the initial key or best key for encryption in cryptography algorithm. Three experiments are performed by varying different data set. In GA, only the fittest one will survive; hence it needs to define fitness function which helps us to identify the best chromosome key among the rest of the chromosome keys. Here, the key is defined as chromosome key. The selection of chromosome key is based on fitness value. In our proposed scheme we considered two-point crossover and then three-point mutation operation is performed. To select the initial key/best chromosome key among the keys, we have defined fitness function. Second, AVK is applied [3] to generate dynamic keys where new keys are always generated based on the previous key and previous block of data. In this approach, the initial chromosome key is distributed among sender and receiver through the fitness function of GA. For example, if the sender sends a data using an initial key the receiver can later generate the same initial key using fitness function. The initial key generation algorithm and the dynamic key algorithm are depicted as follows:

Statement	s/e	Frequency	Total steps
Algorithm 1: To generate initial chromosome key (P, V_1, V_2, M)			
1. {	0	-	0
2. $P \leftarrow$ Initial population	1	1	1
//selection of fitness value			1
3. $V_1 \leftarrow \frac{n}{4}$ and $V_2 \leftarrow \frac{n}{2}$	1	1	
// where n is the size of individual chromosome key			1
4. $i \leftarrow 1$;	1	1	1
5. Selection:	1	1	1
6. Repeat while $(i \leq M)$	1	$M + 1$	$M + 1$
7. {	0	-	-
8. $b_1 \leftarrow 0$;	1	M	M
9. for $i = 1$ to n do	1	$M(n + 1)$	$Mn + M$
10. if$(P_i [j] == '1')$then	1	Mn	Mn
11. $b_1 \leftarrow b_1 + 1$	1	Mn	Mn
12. if$(b_1 \geq V_1 \& b_1 \leq V_2)$then	1	M	M
13. {	0	-	0
14. $K_1 \leftarrow P_i$ Break ;	1	M	M
15. $i \leftarrow i + 1$;	1	M	M
16. }	0	-	0
17. }	0	-	0
18. $r \leftarrow i + 1$;	1	M	M
19. Repeat while $(r \leq M)$ //population$= M$	1	$M - r + 1 + 1$	$M - r + 2$
20. {	0	-	0
21. $b_2 \leftarrow 0$;	1	$M - r + 1$	$M - r + 1$
22. for $j = 1$ to n do	1	$(M - r + 1)(n + 1)$	$Mn - M - rn - r + n+1$
23. if$(P_r [j] == '1')$then	1	$(M - r + 1)n$	$Mn - rn + n$
24. $b_2 \leftarrow b_2 + 1$;	1	$(M - r + 1)n$	$Mn - rn + n$
25. if$(b_2 \geq V_1 \& b_2 \leq V_2)$then	1	$(M - r + 1)$	$(M - r + 1)$
26. {	0	-	0
27. $K_2 \leftarrow P_r$ Break;	1	$(M - r + 1)$	$M - r + 1$
28. $r \leftarrow r + 1$;	1	$(M - r + 1)$	$(M - r + 1)$
29. }	0	-	0
30. }	0	-	0
31. 2 point crossover(K_1, K_2);	1	1	C_1
32. 3 point mutation(K_1, K_2);	1	1	C_2

Statement	s/e	Frequency	Total steps
33. *Fitness Function* ← '1110';	1	1	1
34. n_1 ← $cmp(K_1, Fitness\ Function)$;	1	1	C_3
35. n_2 ← $cmp(K_2, Fitness\ Function)$;	1	1	C_3
36. if($n_1 \neq 0 \| n_2 \neq 0$)then	1	1	1
37. {	0	-	0
38. if($n_1 < n_2$)then	1	1	1
39. {	0	-	0
40. *Initial Chromosome Key* ← K_1;	1	1	1
41. else			
42. *Initial Chromosome Key* ← K_2;	1	1	1
43. }	0	1	0
44. else			
45. *goto selection*:			
46. }	1	1	1
47. }	0	-	0
	0	-	0
Total			6Mn-3rn+3n+13M-6r+18+ $C_1+C_2 + C_3$

Statement	s/e	Frequency	Total steps
Algorithm 2: To generate the dynamic chromosome key			
1. *Block Key Generation*($Dataset, Initial\ Chromosome\ Key$)			
2. {	0	-	0
3. $Key[1]$ ← *Initial Chromosome Key*;	1	1	1
4. i ← 2;	1	1	1
5. *While*($Dataset \neq \theta$)	1	K	K
6. {	0	-	0
7. $Key[i]$ ← $Key[i-1] \oplus Dataset[i-1]$;	1	K	K
8. $i \leftarrow i+1$;	1	K	K
9. }	0	-	0
10. return(key)	1	1	1
11. }	0	-	0
Total			3+3K

where: Fitness Function (FF) = '1110', $cmp = compare$.

4 Complexity Analysis of Algorithm 1 and Algorithm 2

The time complexity of the given algorithm is analysed for the initial chromosome key and to generate the chromosome keys for rest session using AVK [3, 4]. Let, individual chromosome key is of size "n", size of the population "M", Dataset as "K" and "r" is a variable used for initialization. To compute the fitness function, we need input "n", and "M". The complexity required to generate the initial key is O(Mn). The time complexity of crossover point and mutation point function is

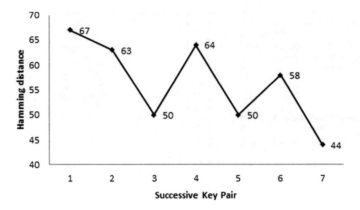

Fig. 2 Hamming distance of successive keys of experiment 1

$O(1) + 0(1)$ **or** $C_1 + C_2$ is considered as a constant time. "D" defined as the variable for all constant. Calculating all these, total number of execution of initial key generation is given below.

$$=> \quad 6Mn \ - \ 3rn \ + \ 3n \ + \ 13M \ - \ 6r \ + \ C_1 \ + \ C_2 \ + \ C_3 \ + \ 18 \ =$$
$6Mn \ - \ 3n(r-1) \ + \ 13M \ - \ 6r \ + \ D.$ thus, the overall time complexity *of the algorithm to find initial chromosome key* $= \ O(Mn)$ and to generate a dynamic key for each session is $3 + 3$ K than time complexity $= O(K)$.

5 Experiment Results

The proposed algorithm has been implemented and analyzed using Dev-C++. In this work, three different experiments have been checked for different datasets in order to get higher performance. We use the length of the chromosome keys 128 bits and population size of 24 from which 7 dynamic keys are generated.

Experiment 1: In this experiment, we have used the Dataset *"Private key cryptography is also known as symmetric key cryptography where, for encryption and decryption use the same key."* The auto-generated successive keys are shown in Fig. 2.

Experiment 2: The dataset for this experiment is *"Artificial intelligence is to study how to use systems to mimic human brain to conduct thinking and activities"* (Fig. 3).

Experiment 3: *"The encryption and decryption is done using two different key is known as asymmetric cryptosystem or public key cryptography."* is considered as dataset for the experiment 3 and auto-generated graphs are shown in Fig. 4.

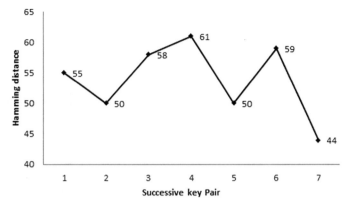

Fig. 3 Hamming distance of successive keys of experiment 2

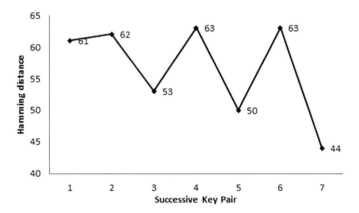

Fig. 4 Hamming distance of successive keys of experiment 3

6 Performance Analysis

The method of proposed technique is demonstrated and we have compared the average Hamming distance of some other schemes proposed earlier, such as Automatic Variable Key (AVK) [3], Alternating and Shifting Automatic Variable Key (ASAVK) and Computing and Shifting Automatic Variable Key (CSAVK) [6, 7] with our scheme named as Genetic Algorithm Key (GAK). We compared to observed that our scheme has at par Average Hamming distance and higher Standard Deviation (SD) as the other schemes. The standard deviation gives an idea that how close the entire set of data is to the average value [9, 10].

$$AHD = \sum \left(\frac{HDKP}{NHDKP} \right) \qquad (1)$$

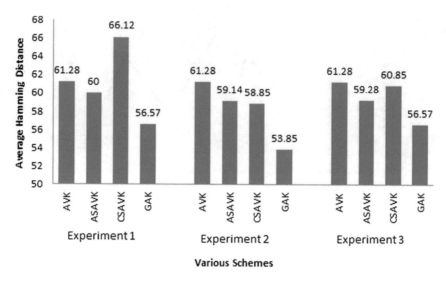

Fig. 5 Average hamming distance of GAK and various schemes of AVK

$$SD = \sqrt{\left(\frac{\sum (HDKP - AHD)2}{NHDKP}\right)}, \qquad (2)$$

where, AHD = Average Hamming Distance, SD = Standard Deviation, HDKP = Hamming Distance of Key Pair, NHDKP = Nos. of Hamming Distance of Key Pair. Results of experiments 1, 2, and 3 have been shown in Figs. 5 and 6. In graph, the x-axis indicates various schemes and y-axis represents average randomness and standard deviation. In Figs. 5 and 6 respectively.

7 Randomness Verification

For the Experiments 1, 2, and 3 we have performed NIST test to prove that our scheme is a truly random key generator technique. The NIST test suite is a statistical package consisting of 15 different tests to check the random and pseudorandom number generators for cryptographic application [22]. We have carried out four different tests for our proposed technique namely, Frequency Test, Block frequency, Run Test, Cumulative Sum Test, as per the bits length required for the test. The NIST test suite is used to calculate the P-Value. If the P-Value is >0.01, then the sequence is random otherwise, non-random.

From Tables 1, 2 and 3, we can derive that our method has P-Value >0.01, which means the keys generated using GA are truly random keys with confidence more than 90%.

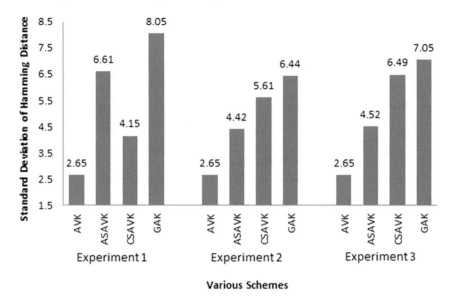

Fig. 6 Standard deviation of GAK and various schemes of AVK

Table 1 Statistical test results for experiment 1

Experiment 1			
Sl No.	Test	P-Values	Remarks
a.	Frequency test	0.122325	Random
b.	Block frequency test	0.035174	Random
c.	Runs test	0.350485	Random
d.	Cumulative sums forward	0.534146	Random
e.	Cumulative sums backward	0.911413	Random

8 Security Analysis

We have analyzed the security of our proposed algorithm for Frequency Attack, Brute Force Attack, and Differential Cryptanalysis.

8.1 Frequency Attack or Cipher Only Attack

Frequency of letters in cipher text is directed towards alphabet of the message rather than data. In our proposed scheme, dataset and key size are large. Thus, the attacker

Table 2 Statistical test results for experiment 2

Experiment 2

Sl No.	Test	P-Values	Remarks
a.	Frequency test	0.350485	Random
b.	Block Frequency test	0.035174	Random
c.	Runs test	0.739918	Random
d.	Cumulative sums forward	0.213309	Random
e.	Cumulative sums backward	0.122325	Random

Table 3 Statistical test results for experiment 3

Experiment 3

Sl No.	Test	P-Values	Remarks
a.	Frequency test	0.035174	Random
b.	Block frequency test	0.350485	Random
c.	Runs test	0.911413	Random
d.	Cumulative sums forward	0.213309	Random
e.	Cumulative sums backward	0.739918	Random

would not be able to guess an alphabet correctly and for the attacker, it would difficult to get correct alphabet.

8.2 Brute Force Attack

Attacker tries every possible combination of the key to get secret information. If size of the key is of n bits then possible combination is 2^n bits. If the attacker, somehow extracts the secret information like plaintext, ciphertext by combining all possible keys. Then also it would be infeasible to compute as the length of the key size is long. The larger the population size impossible for the attacker to get exact bits size of the population and difficult to compute correct key from the population. Therefore, the proposed scheme is secured against brute force attack.

8.3 Differential Cryptanalysis

Output depends on input and key size. In our scheme, we have used 128-bit size of the key. In GAK, in each session, "n" unique new key is generated because of that it is difficult to adopt a Differential Cryptanalysis.

9 Conclusion

A new approach of key generation technique has been proposed using GA and the concept of AVK for more secure data communication. The initial key is generated using the GA. Later, dynamic key is generated using the principle of AVK. The uniqueness of this new approach, GAK is that it does not require identical key synchronization by exchanging it between the sender and receiver. Secondly, the three experiments have been performed to randomness of our proposed scheme. The following results are illustrated after the experimental study:

- Initial chromosome key or best fit chromosome key is generated using the GA fitness function.
- AHD and SD of experiment 1, 2, and 3 have shown that our proposed algorithm has the higher unpredictable occurrence of keys.
- NIST: we have done the various statistical tests namely, Frequency Test, Block Frequency Test, Cumulative Sum Test and Runs Test, as per the bits length required for the test. NIST test suite is used to calculate the P-value. Since, P-value >0.01 for each experiment, the sequence can be accepted as random.

Considering all the experiments we can conclude that our scheme has higher unpredictability in terms of occurrence of bits as compared to the other existing techniques of AVK. Therefore, we can say that our scheme is truly random and infeasible to predict. For this reason, it is suitable for practical implement in cryptosystem.

References

1. Shannon, C.E.: A mathematical theory of communication, Part I, Part II. Bell Syst. Tech. J. **27**, 623–656 (1948)
2. Shannon, C.E.: Communication theory of secrecy systems. Bell Lab. Tech. J. **28**(4), 656–715 (1949)
3. Bhunia, C. T.: Application of AVK and selective encryption in improving performance of quantum cryptography and networks. United Nations Educational Scientific and Cultural Organization and International Atomic Energy Agency, retrieved, 10(12), p. 20010 (2006)
4. Bhunia, C.T., Mondal, G., Samaddar, S.: Theory and application of time variant key in RSA and that with selective encryption in AES. In: Proceedings of EAIT, pp. 219–221. Elsevier Publications, Calcutta CSI (2006)

5. Kumar, A., Chatterjee, K.: An efficient stream cipher using genetic algorithm. In: International Conference on Wireless Communications, Signal Processing and Networking (WiSPNET), IEEE, pp. 2322–2326 (2016)

6. Goswami, R.S., Chakraborty, S.K., Bhunia, A., Bhunia, C.T.: New approach towards generation of automatic variable key to achieve perfect security. In: 2013 Tenth International Conference on Information Technology: New Generations (ITNG), IEEE, pp. 489–491 (2013)

7. Goswami, R.S., Chakraborty, S.K., Bhunia, A., Bhunia, C.T.: New techniques for generating of automatic variable key in achieving perfect security. J. Inst. Eng. (India) Ser. B **95**(3), 197–201 (2014)

8. Singh, B.K., Banerjee, S., Dutta, M.P., Bhunia, C.T.: Generation of automatic variable key to make secure communication. In: Proceedings of the International Conference on Recent Cognizance in Wireless Communication and Image Processing. Springer, New Delhi, pp. 317–323 (2016)

9. Dutta, M.P., Banerjee, S., Bhunia, C.T.: An approach to generate 2-dimensional AVK to enhance security of shared information. Int. J. Secur. Appl. **9**(10), 147–154 (2015)

10. Banerjee, S., Dutta, M.P., Bhunia, C.T.: A novel approach to achieve the perfect security through AVK over insecure communication channel. J. Inst. Eng. (India) Ser. B **98**(2), 155–159 (2017). https://doi.org/10.1007/s40031-016-0264-2

11. Prajapat, S., Thakur, R.S.: Cryptic mining for automatic variable key based cryptosystem. In: 1st International Conference on Information Security and Privacy. Procedia Comput. Sci. **78**, 199–209 (2016). https://doi.org/10.1016/j.procs.2016.02.034

12. Xu, P., Cumanan, K., Dai, X.: Group secret key generation in wireless networks: algorithms and rate optimization. IEEE Trans. Inf. Forensics Secur. **11**(8), 1831–1846 (2016)

13. Eli B.: A fast new DES implementation in software. In: Proceedings the of International Symposium on Foundations of Software Engineering, pp. 260–273 (1997)

14. Mondal, S., Mollah, T.K., Samanta, A., Paul, S.: A survey on network security using genetic algorithm. Int. J. Innov. Res. Sci. Eng. Technol. **5**(1), 319–8753 (2016)

15. Subhrani, S., Niladri, S.C., Mandal, J.K.: Key based level genetic technique. In: 7th International Conference on Information Assurance and Security (IAS). IEEE (2012). ISBN: 978-1-4577-2154-0, https://doi.org/10.1109/isias.2011.6122826

16. Jawaid, S., Jamal, A.: Generating the best fit key in cryptography using genetic algorithm. Int. J. Comput. Appl. (0975–8887) **98**(20), 3339 (2014)

17. Kumar, A., Ghose, M.K.: Overview of information security using genetic algorithm and chaos. Inf. Secur. J. Glob. Perspect. **18**(6), 306–315 (2009). https://doi.org/10.1080/19393550903327558

18. Soni, A., Agrawal, S.: Key generation using genetic algorithm for image encryption. Int. J. Comput. Sci. Mob. Comput. **2**(6), 376383 (2013)

19. Sindhuja, K., Devi, S.P.: A Symmetric key encryption technique using genetic algorithm. J. Comput. Sci. Inf. Technol. **5**(1), 414–416 (2014). ISSN: 0975-964

20. Bhowmik, S., Acharyya, S.: Image cryptography: The genetic algorithm approach. In: 2011 IEEE International Conference on Computer Science and Automation Engineering (CSAE), vol. 2, pp. 223–227 (2011)

21. Elaine, R., Kevin, K., Shivashankar, B.N.: Artificial Intelligence, 3rd edn. McGraw Hill Publication, India, (2008)

22. Rukhin, A., Soto, J., Nechvatal, J., Smid, M., Barker, E.: A statistical test suite for random and pseudorandom number generators for cryptographic applications. Booz-Allen and Hamilton Inc., Mclean Va (2010)

A Low-Cost, High-Performance Implementation of RSA Algorithm Using GPGPU

P. S. Sasaank Srivatsa and P. V. R. R. Bhogendra Rao

Abstract Information security has been of great use in various civilian and defence domains. However, the arduous procedures involved in securely establishing a common key lead to asymmetric-key encryption. The high computational needs of asymmetric-key encryption process made them impractical to be used for real-world applications. The proliferation of low-cost, high-performance computing through GPGPUs made many applications with high-performance requirements feasible to implement and use in real-world applications. A successful attempt has been made to implement RSA algorithm using GPGPU and the results are presented in this paper.

Keywords Parallel and distributed computing · Heterogeneous computing
GPGPU computing · OpenCL · Asymmetric-key encryption

1 Introduction

Encryption is used by militaries and government for secure communication. Currently, it is extensively being used to protect the information and communication not only between government agencies but also citizens—directly or indirectly. One of the well-known implementations of encryption is protecting static data which is stored in computers and servers. In simple terms, using cryptography a simple text is encrypted which would seem cryptic to anyone else other than the authorized person. Encryption can be done in two ways, symmetric- and asymmetric-key encryptions [4]. In symmetric-key encryption, a secret key is known to both the sender and receiver, which will be used for encryption and decryption.

P. S. Sasaank Srivatsa (✉)
Keshav Memorial Institute of Technology, Hyderabad, India
e-mail: sasaank.sss@gmail.com

P. V. R. R. Bhogendra Rao
Defence Research and Development Laboratory, Hyderabad, Kanchanbagh, India
e-mail: bhogendra@drdl.drdo.in

© Springer Nature Singapore Pte Ltd. 2019
A. Abraham et al. (eds.), *Emerging Technologies in Data Mining and Information Security*, Advances in Intelligent Systems and Computing 755,
https://doi.org/10.1007/978-981-13-1951-8_65

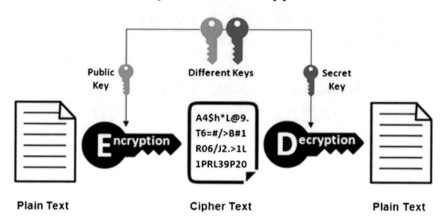

Fig. 1 Philosophy of asymmetric-key encryption

In contrast, asymmetric encryption, also known as public-key encryption, uses two different keys, namely, public key and private key, as shown in Fig. 1, for encryption and decryption [3], respectively. It is important to note that without private key, it is not possible to decrypt the ciphered text. This is why asymmetric encryption uses two related keys, to boost the security.

In asymmetric encryption, the keys must be many times longer than those in secret cryptography in order to boost equivalent security [16]. Using keys of small lengths makes the algorithm vulnerable to brute force attacks. So, generating longer keys will usually prevent brute force attacks from succeeding, thus providing more security. The lengths of the keys are longer than those of integers supported by the basic instruction set of underlying CPU. The use of long keys necessitates long integer arithmetic libraries which are not supported by compilers by default. The programmer needs to develop proprietary long integer arithmetic libraries or use open-source libraries such as GMP. But increasing the key length is counterproductive; security of the algorithm is improved but computational time of the algorithm will increase exponentially and make the algorithm non-useful for real-world problems.

As a solution to these big integer computations and to reduce the complexity, High-Performance Computing (HPC) can be used [21]. However, not only the cost of HPC systems is prohibitively expensive but also their maintenance is arduous. The General-Purpose Graphic Processing Units (GPGPUs), on the other hand, are relatively cheaper. GPUs are abundantly available in today's market, and it can safely be said that every system today is equipped with a GPU, making it a low-cost solution to high-performance computing.

Organization of this paper is as follows: Sect. 2 provides required technical background on general asymmetric-key encryptions to understand rest of the paper, Sect. 3 introduces issues involved in GPGPU programming, Sect. 4 covers proposed use of OpenCL and implementation of RSA algorithm with OpenCL, Sect. 5 presents CPU vs. GPU execution results, performance analysis, and the paper concludes with Sect. 6.

2 General Asymmetric-Key Encryption

Public-key encryption [17], also known as asymmetric-key encryption, has a pair of keys: a public key which can be shared with everyone, and a private key which is known only to the authorized person, i.e., the owner. The use of two separate keys for encryption and decryption serves two purposes: one, authentication, which is used to verify if the sender is authentic and two, authorization where only the person having knowledge of public key can decrypt and read the message. One of the many fields in which Public-key encryption finds its implementations is information security, i.e., protection against security threats [11]. Information security deals with assuring confidentiality, authentication, authorization, and non-repudiation of information and communication.

Among many best-known uses of public-key cryptography are public-key encryption and digital signatures. Few classic examples of public-key encryption are DSA, elliptic-curve cryptography, and RSA.

2.1 Simple RSA Implementation

RSA algorithm is an asymmetric-key cryptography [4]. It was first proposed in 1977 by Ron Rivest, Adi Shamir, and Leonard Adleman. It employs two different, yet mathematically linked keys—public and private keys. As their names suggest, public key can be shared and private key must be kept secret. Either of these keys can be used to encrypt. When the public key is used to encrypt, the private key must be used to decrypt and vice versa. Anyone with the knowledge of public key can encrypt a message. If this key is not large enough, someone acquainted with basic math can easily take advantage of it and decrypt the message [19].

The first step in RSA algorithm is to choose two large and unique prime numbers p and q. Find product, say n, of p and q. This product n serves as the mathematical link between public and private keys. To encrypt a character, we use its ASCII value as plain text.

Key Generation:

1. Choose two dissimilar and large prime numbers p and q, $p\ != q$
2. Compute $n = p \times q$
3. Compute $\phi(n) = (p-1)(q-1)$
4. Choose e such that $gcd(e, \phi(n)) = 1, 1 < e < \phi(n)$

5. Choose d, such that $d.e \bmod \phi(n) = 1$, i.e., d is multiplicative inverse of e in $\bmod \phi(n)$
6. Get the public key as $E = \{e, n\}$
7. Get the private key as $D = \{d, n\}$

Encryption: Cipher text, $C = P^e \bmod n$.

Decryption: Plain text, $P = C^d \bmod n$.

2.2 Sequential Implementation on CPU

Following are the steps in implementation of sequential version of RSA algorithm:

Step 1: Compute the keys as described in Sect. 2.1
Step 2: Input the plain text that should be encrypted
Step 3: Perform encryption using public key over this data through iterating over the for loop resulting in ciphertext.

3 GPGPU Programming

General-Purpose Graphic Processing Unit (GPGPU) is a graphic processing unit that performs nonspecialized calculations that would be conducted by the CPU. GPGPU pipeline, in short, can be defined as parallel processing between various combinations of GPUs and CPUs, which can be used to analyze data presuming it as an image or other graphic form. The use of multiple GPUs in one computer will render more parallelism to the already parallel nature of graphic processes [7]. The basic purpose of GPU is to provide data-parallel computing, that is, each processor performing the same task on different pieces of distributed data, which reduces the requirement for sophisticated control statements and speeds up the algorithm [10]. This implementation of GPGPU is provided by OpenCL, developed by Khronos [20] and CUDA developed by NVIDIA.

3.1 Programming with OpenCL

Open computing language is an open-sourced cross-platform for parallel programming on manifold processors found in computers, laptops, servers, and mobiles. OpenCL uses C++ as well as extended C (C99) [1] as kernel programming language. The stock library of C is superseded by a custom set of functions developed by Khronos which are aimed toward math programming. The use of C99 simplifies the

Fig. 2 OpenCL
programming model

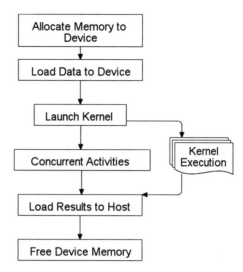

use of parallelism with vector types, vector operations, and synchronization with work items and work groups [12].

OpenCL follows load–launch–read programming model, as shown in Fig. 2. Initially, the required space is allotted in the device and data is loaded from host to device. The parallel computing program called kernel [8] is launched and after the kernel is executed, the results are loaded from device memory to host. The allotted space on the device is cleared after the data is loaded back to host.

Following are the predefined functions needed to perform the required operations:

- clCreateBuffer()—to allocate device memory
- clEnqueueWriteBuffer()—to load the data to device
- clSetKernelArgs()—to pass parameters to kernel program
- clEnqueueNDRangeKernel()—to launch the kernel
- clEnqueueReadBuffer()—to load the results to host
- clRelease()—to free device memory

The functionality which needs to execute in parallel on the dataset needs to be declared as a __global__ function and can be specified to the compiler. The kernel should be created using the keywords __kernel; it shall run in multiple threads. Each thread can be identified by its thread index, which can be obtained using function get_global_id(0). There can be multiple kernels running at the same time and the user gets to decide how many of them should be created

OpenCL defines four-level memory hierarchy for computer devices [21]. The functionality which needs to be executed in parallel on the dataset is to be declared as __global, or __local, or __constant, or __private based on the need. Global memory is shared by all the processing elements but has high latency. Local memory is shared by a group of processing elements. Constant or read-only memory is small (int, float) and has low latency, writable by the host CPU, but not by the computing devices. Private memory is stored in registers, and it is to be used only for variables declared as pointers. Sometimes, devices do not share any memory with the host [1]. These are the names reserved for use of address space qualifiers and shall not be used otherwise. At the time of launching the kernel, it is required to specify size of grid and size of block. The size of block should be a multiple of warp and should be less than max block size. However, the configuration depends highly on the current load of the device and the resources required by the kernel.

3.2 Issues in Programming with OpenCL

The following are some of the factors that influence the performance of parallel program:

- Load balancing,
- Race conditions, and
- Essential sequential computations.

The effective computation time per node is the time taken by the node with maximum load. Thus, the total time for computation will be effectively more than the time taken with the balanced load. Race condition is the condition when more than one thread tries to access the same variable and at least one of them attempts to write. In such case, it is required to synchronize the threads that are trying to access the same variable. There are certain parts of computations that cannot be computed in parallel [13]. Such essential sequential computation part should be minimized, in order to achieve higher performance of the parallel application [2].

In addition to the factors discussed in previous subsection, the following are the major issues to be addressed while programming with CUDA:

- Coalesced access to data structure memory.

The size of grid and size of block are to be correctly chosen. Otherwise, the result would be erroneous. Coalesced access to memory increases the run-time performance as the data will be cached.

4 Implementation of RSA Using OpenCL

The following are standard methods available in literature for parallelizing any problem:

4.1 Types of Parallelization

Bit-level parallelism is a form of parallel computing based on increasing processor word size.

Instruction-level parallelism Pipelining can overlap the execution of instructions when they are independent of one another.

Task-level parallelism is based on logical tasks that can be concurrently executed by processes or threads over a distributed system of processors and also called function parallelism or control parallelism.

Data parallelism emphasizes data decomposition over different computing elements and performs identical function on the data elements concurrently.

However, bit-level parallelism, instruction-level parallelism, and Task-level parallelism do not provide a solution to the asymmetric- key encryption problem using GPGPU, as GPGPU is inherently data parallel. Asymmetric-key encryption is a block encryption process and does not scale well with data parallelism and may be inferior to sequential implementation on CPU. Hence, a novel method, namely, operation-level parallelism, is adopted in the current implementation.

The proposed variant implementation supports variable key size. The main bottleneck of the RSA-based encryption method is its requirement for large integer arithmetic. In order to provide parallel implementation to RSA, the basic arithmetic operations need to be parallelized. We observed that multiplication operation, in particular, is to be parallelized.

4.2 Parallelizing RSA Algorithm

In Sect. 2.1, we can see that plaintext is raised to the power of public key for encryption and ciphertext is raised to the power of private key for decryption. This calculation gets more complex as the size of plaintext increases and also consumes more time. The maximum size of an int can store is 4 bytes. It can further be improved to 8 bytes using long long int. Any number beyond that size cannot be stored in C. To overcome this problem, we store the number as an array of digits. To perform large integer multiplication, as shown in Fig. 3, we store the multiplicand and multiplier in two different arrays and multiply each element in the multiplicand with each element in multiplier, in succession. The product is stored in a third array and it is added to obtain the final result.

In threading structure for GPU, each thread will run on a grid, and the number of blocks and thread created is decided by the user. A thread block can hold up to 1024 threads [14]. Every thread has a private memory. All the threads have access to the global memory, where they access the array element assigned to them.

| 3 | 4 | 5 | 6 | Multiplicand |
| X | 1 | 2 | 3 | Multiplier |

	18	15	12	9	0	0	Product of 1st array and last element of 2nd array
	0	12	10	8	6	0	Product of 1st array and second element of 2nd array
+	0	0	6	5	4	3	Product of 1st array and 1st element of 2nd array
	8	8	0	5	2	4	column wise sum
	4	2	5	0	8	8	Result

Fig. 3 Parallel multiplication

4.3 OpenCL Implementation

Following are the steps in parallel implementation of RSA algorithm on GPGPU using OpenCL:

Step 1: Compute the keys as mentioned in Sect. 2.1.
Step 2: Input the plaintext that needs to be encrypted.
Step 3: Allocate space on the device (GPU) and load the data.
Step 4: Set kernel arguments, and pass them into the kernel.
Step 5: Call kernel method (GPU kernel).
Step 6: Read the result back to host (CPU).

5 Performance Analysis

We employed GPGPU for parallelizing RSA algorithm in multicore CPU and GPU and to compute the time gained by parallelizing RSA. Thus, we find the speedup factor. Speedup factor [13] is the measure of comparative benefit gained by implementing a sequential algorithm in parallel. Here, the speedup factor for running "m" processes in parallel is derived as a ratio between sequential execution time on CPU, T_s, and parallel execution time on GPU, T_p.

$$\textbf{Speedup, } S = \frac{T_s}{T_p}$$

The test groups are defined as follows:

Group 1: The key size is fixed at 128 bits, and the CPU and GPU execution time is noted for encryption of plaintext with varied message size from 1 byte to 1700 bytes. The execution time is shown in Table 1, time is measured in seconds.

Table 1 Execution time for encryption on CPU and GPU

S.No.	Message size	Encryption time (Sec)		Decryption time (Sec)	
	(Bytes)	CPU	GPU	CPU	GPU
1.	50	0.000442	0.000645	0.091044	0.153431
2.	100	0.001519	0.000575	0.305869	0.298786
3.	500	0.002901	0.001025	4.252353	3.495753
4.	1000	0.005704	0.005704	9.507196	6.617569
5.	1700	0.009185	0.001443	28.634290	12.913890

Table 2 Speedup factor

S.No.	Message size	Speedup factor	
	(Bytes)	Encryption	Decryption
1.	50	0.6	0.5
2.	100	2.6	1.1
3.	500	2.8	1.2
4.	1000	4.6	1.4
5.	1700	6.3	2.3

Group 2: The key size is fixed at 128 bits, and the CPU and GPU execution time is noted for decryption of plaintext with varied message size from 1 byte to 1700 bytes. The execution time is shown in Table 2, time is measured in seconds.

The experiment was carried out on a laptop with following specifications:

- CPU: Intel® Core™ i5-4210U 1.70GHz
- Memory: 4GB
- GPU: Intel® HD Graphics Family, 1GB
- OS: Ubuntu 16.04.3 LTS Xenial Xerus

It is evident from Fig. 4 that for small data sizes, CPU version is faster than GPU version, but as the message size increases CPUs execution time increases exponentially, whereas GPUs execution time increases linearly. Similar behavior can be noticed for decryption also. It can be noted that GPGPU version of RSA algorithm produces significantly better performance for larger data sizes.

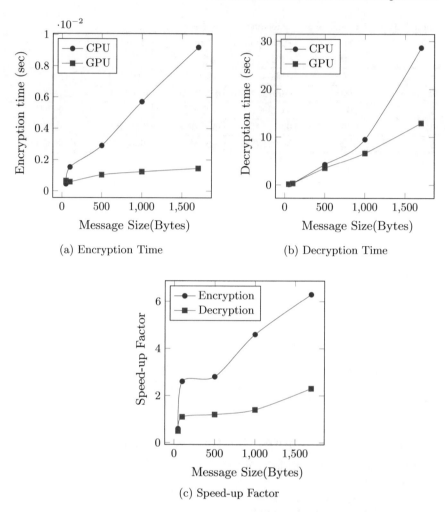

Fig. 4 Performance of GPGPU application

6 Conclusions

The GPGPU variant of RSA algorithm is successfully developed using the novel operation-level parallelism approach and its performance *vis-á-vis* sequential version is presented in this paper with their results shown in Fig. 4a, b. A speedup of 6x for encryption and 2x for decryption was obtained using OpenCL over low-end GPU of desktop computer as shown in Fig. 4c. These results show that GPU is appropriate to speed up the RSA algorithm at low cost. This RSA program can further be optimized to produce better results by making use of shared memory and registers, thus showing a promising future.

References

1. Aaftab M.: The OpenCL C Specification, Version 2.0. (2014)
2. Amdahl, G.M.: Validity of the Single-Processor Approach to Achieving Large-Scale Computing Capabilities. In: Proceedings of the American Federation of Information Processing Societies Conference, AFIPS Press, pp. 483–485 (1967)
3. Bellare, M.: Public-Key Encryption in a Multi-user Setting: Security Proofs and Improvements. Springer, Berlin Heidelberg (2000)
4. Christof, P., Jan, P.: Introduction to Public-Key Cryptography, Chapter 6 of Understanding Cryptography. A Textbook for Students and Practitioners. Springer (2009)
5. https://computing.llnl.gov/tutorials/parallel_comp/
6. Delfs, H., Knebl, H.: Symmetric-key Encryption. Introduction to Cryptography: Principles and Applications. Springer (2007)
7. Fung J., Mann S.: Using multiple graphics cards as a general purpose parallel computer: applications to computer vision. In: Proceedings of the 17th International Conference on Pattern Recognition (ICPR2004), vol. 1, pp. 805–808. Cambridge, United Kingdom (2004)
8. Howes, L.: The OpenCL Specification Version:2.1 Document Revision:23. Khronos OpenCL Working Group (2015)
9. Hoffman, A.R., et al.: Supercomputers: Directions in Technology and Applications, pp. 35–47. National Academies (1990)
10. Hwu, W., Keutzer, K., Mattson, T.G.: The concurrency challenge. IEEE Design Test Comput. **25**(4), 312–320 (2008)
11. Information Security Resources. SANS Institute (2014). https://www.sans.org/security-resources/
12. Introduction to OpenCL Programming 201005, pp. 89–90. AMD (2017)
13. Srinivas, J.V.S., Bhogendra Rao, P.V.R.R., Kamakshi Prasad, V.: Parallel implementation of back propagation on master slave architecture. In: Proceedings of the International Conference on Computational Intelligence and Multimedia Applications (ICCIMA 2007)
14. OpenCL C PROGRAMMING GUIDE, Khronos Group (2013)
15. Conformant Companies. Khronos Group (2015)
16. Kchlin W.: Public key encryption, ACM SIGSAM Bulletin, pp. 69–73 (1987)
17. Menezes, A.J., van Oorschot, P.C., Vanstone, S.: Handbook of Applied Cryptography (1997)
18. Reinders, J.: Understanding task and data parallelism, ZDNet (2017)
19. Rivest, R., Shamir, A., Adleman, L.: A method for obtaining digital signatures and public-key cryptosystems. Commun. ACM **21**(2), 120–126 (1978)
20. https://www.ssl2buy.com/wiki/symmetric-vs-asymmetric-encryption-what-are-differences/
21. Stone, J.E., David, G., Guochin, S.: OpenCL: a parallel programming standard for heterogeneous computing systems. Comput. Sci. Eng. **12**, 66–73 (2010)

Attacks and Threats on RSA

Sreemoyee Biswas and Namita Tiwari

Abstract This paper in a nutshell introduces the conventional RSA algorithm and its application in small devices. It also states that there are numerous attacks possible when RSA is used with small parameter values to reduce the time required for encryption and decryption. The paper also includes a description of each such attack. It is also stated that most of the attacks mentioned in the paper are not enough to question the security and credibility of RSA when used with large values of its parameters such as value of primes, product of primes, value of modulus, encryption exponent, and decryption exponent.

1 Introduction

1.1 Importance of Data Security

Modern Digital Technology has proved its usefulness in every sphere of human life. But with the advent of newer and modern technology, there rose a demand to ensure its security in this computer world. Introducing a new technology for the welfare of the mankind is indeed a noble deed, but at the same time, it is just equivalent to building a house at second floor with no staircase. Facilitating interactive means to use the technology with ease is equivalent to providing it with a staircase and other basic amenities to make it a place worth living. But after all this, if proper security measures are not ensured, then it is like leaving the house without any lock and key facility where anyone can breach and harm the confidentiality and personal space of the house. Thus, any advanced technology is incomplete without a proper means to ensure high level security. Thus to ensure security, different algorithms

S. Biswas (✉) · N. Tiwari
Department of Computer Science and Engineering, Maulana Azad
National Institute of Technology, Bhopal 462003, India
e-mail: shonai.biswas@yahoo.in

N. Tiwari
e-mail: namita_tiwari21@rediffmail.com

© Springer Nature Singapore Pte Ltd. 2019
A. Abraham et al. (eds.), *Emerging Technologies in Data Mining and Information Security*, Advances in Intelligent Systems and Computing 755,
https://doi.org/10.1007/978-981-13-1951-8_66

were proposed, some of which have become obsolete whereas some of them are still in rigorous use.

1.2   Cryptography

Cryptography is a branch of computer science that deals with the science of ensuring secure communication by using encryption and decryption process. Encryption—It is the process by which the original message to be sent is converted into a form which is different from the original message. It is done with the help of various encryption algorithms. Decryption—It is the process by which the encrypted text is converted back into the original message. It is done using various available decryption algorithms. Plaintext—The message to be transmitted is called plaintext. Ciphertext—The encrypted text is called ciphertext which is transmitted over the communication channel. Key—It is a secret bit of strings used to establish secure communication by transforming the plaintext into ciphertext and vice versa [7].

To perform cryptography, we use some cryptographic algorithms. They are broadly of three types, Symmetric Key Cryptography, Asymmetric Key Cryptography, and Hashing Cryptography. Symmetric Key Cryptography—In this, an indistinguishable key is used both for Encryption and Decryption process. Examples are DES, 3DES, AES, etc. Asymmetric Key Cryptography—In this, separate and divergent keys are used for encryption and decryption process. It is also coined as Public Key Cryptography [9]. Most common example of which is RSA algorithm. Hashing—It is quite different from the above mentioned types. The symmetric and asymmetric key cryptography provides both encryption and decryption processes. But unlike them, Hashing condenses the original message into an irreversible form. It can be used only to assure confidentiality and authenticity of data and cannot be used to retrieve the original message. Some of the examples are MD5 and SHA-1.

2   RSA Algorithm

2.1   Overview

RSA algorithm is a race of nonsymmetric cryptographic algorithm that relies on the complexity of factorization of large numbers proposed by Rivest et al. [10]. The algorithm has undergone through rigorous test for variety of attacks for more than 20 years. And it received gradual acceptance and appreciation for being best option present by users worldwide. RSA algorithm can not only be used for data encryption but it has its application for authentication also. It is safe, and the associated ease of understanding and implementing makes it a good choice [12].

2.2 The Basic Theories of Mathematics and Principles of RSA Algorithm

2.2.1 Foundations of Mathematics of RSA Algorithm [6]

Theorem 1 *Euler's theorem: If a and m are integers and are relatively prime, i.e.,* $\gcd(a, m) = 1$, *then Eq. (1) holds.*

$$a^{\phi(m)} = 1 (\text{mod } m) \tag{1}$$

Chiefly when p is a prime number, for any a, there exists

$$a^p = 1 (\text{mod } p) \tag{2}$$

Theorem 2 *If* $m \geq 1$, *and* $\gcd(a, m) = 1$, *then c surely exists, such that*

$$c * a = 1 (\text{mod } m) \tag{3}$$

where c is the inverse of a modulo m value, and it is represented by $a^{-1} (\text{mod } m)$, *or* a^{-1}.

Theorem 3 *If* $a \equiv b (\text{mod } m_1)$, $a \equiv b (\text{mod } m_2)$, ... , $a \equiv b (\text{mod } m_k)$, *then*

$$a = b \text{ mod}(m_1 * m_2 * \cdots * m_k) \tag{4}$$

2.2.2 The Chief Principles of RSA Algorithm

RSA algorithm elucidation [10]

1. Pick two large prime numbers: p, q
2. Calculate: $n = p * q$, $\phi(n) = (p - 1) * (q - 1)$
3. Arbitrarily select d: $1 < d < \phi(n)$, $\gcd(d, \phi(n)) = 1$
4. Compute e: $e * d = 1 (\text{mod } \phi(n))$
5. Encryption algorithm for any plaintext M, $M \in \mathbb{Z}_n = \{0, 1, ..., n - 1\}$, generated ciphertext is C:
$$C = M^e (\text{mod } n) \tag{5}$$

6. Decryption of the generated ciphertext C is

$$M = C^d (\text{mod } n) \tag{6}$$

Fig. 1 Categories of attacks
possible on RSA algorithm

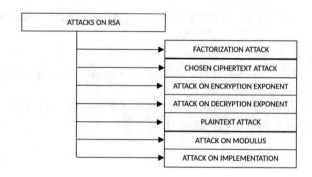

3   Basic Attacks on RSA Algorithm

RSA algorithm is one among the most popular and most widely used Public Key
Cryptographic Algorithm. Due to its popularity and simplicity, as it is easy to use by
users, similarly it is also very vulnerable to attacks by attackers/intruders. Some of
the famous attacks are shown in Fig. 1.

3.1   Factorization Attack

Factorization attack refers to an attempt to factorize the product of primes, i.e., n
which is $p * q$, where p and q are two randomly selected large prime numbers. If one
succeeds in factorizing n, then he directly reaches the two prime numbers p and q
and consequently finds d (the decryption exponent) and e (the encryption exponent)
where e is public. There are many factorization algorithms but none of them can
factor a large integer in polynomial time. Hence, it is suggested that n should be of
more than 300 decimal digits and hence modulus value ought to be at least 1024
bits. But when using RSA for small devices such as smart cards, etc., and to reduce
execution time of encryption and decryption, small values of p and q are selected
and hence it becomes feasible to factor n.

It is interesting to scrutinize some rational restricted models of computation and
establish that in such model, factoring is synonymous to the RSA problem [3]. Boneh
and Venkatesan [3] showed that any SLP that factors n by formulating queries for
at most logarithmic number to an oracle solving the Low-Exponent RSA (LE-RSA)
problem can be successfully adapted into a real-time polynomial algorithm for factoring n.

Fig. 2 Working of chosen
ciphertext attack

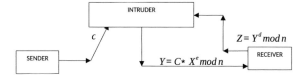

3.2 Chosen Ciphertext Attack

The algorithm followed to establish the chosen ciphertext attack is as follows: Sender
(S) sends the ciphertext (C) for transmission. Attacker (A) gets C and chooses random
X, such that $X \in \mathbb{Z}_n^*$ A calculates $Y = (C * X)^e (\mathrm{mod}\ n)$ and send Y to receiver (R).
A sends Y for decryption to R and gets Z, where $Z = Y^d (\mathrm{mod}\ n)$. This step is an
instance of chosen ciphertext attack. A can now easily find P by [5]:

$$Z \quad = Y^d (\mathrm{mod}\ n) \tag{7}$$
$$= (C * X^e)^d (\mathrm{mod}\ n) \tag{8}$$
$$= (C^d * X^{ed})(\mathrm{mod}\ n) \tag{9}$$
$$= (C^d * X)(\mathrm{mod}\ n) \tag{10}$$
$$= (P * X)(\mathrm{mod}\ n) \tag{11}$$
$$i.e.\ P \quad = Z * X^{-1} \mathrm{mod}\ n \tag{12}$$

X^{-1} can be fetched using extended euclidean algorithm and eventually the value of
P is found (Fig. 2).

3.3 Attack on Encryption Exponent

The small values of prime numbers p and q make the algorithm's execution faster
but at the same time, it also increases the vulnerability of the algorithm to attacks.
Smaller values of p and q also result in the small value of the encryption exponent
e. This poses a potential threat to the algorithm (Fig. 3).

Fig. 3 Categories of attack
on encryption exponent

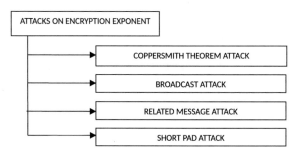

3.3.1 Coppersmith Theorem Attack

This theorem states that in a modulo-n polynomial $f(x)$ of degree e, an algorithm can be utilized of the complexity equal to log n to fetch the roots if one of the roots is more minimal than $n^{1/e}$ [5].

For RSA cryptosystem, $C = f(P) = P^e \bmod n$ where C is the formed ciphertext, P is the plaintext to be encrypted, e is the encryption exponent, and n is the product of primes.

Here, if $e = 3$ and only 2/3rd of plaintext bits are known, then algorithm can find all the bits of plaintext [5].

3.3.2 Broadcast Attack

It is a type of attack which is possible only when a message is broadcasted to multiple number of recipients [2].

When the same message is sent to multiple receivers but with same value of e but with different values of moduli say n_1, n_2, n_3, respectively. For example [5], let us consider $e = 3$, then

$$C_1 = P^3 (\bmod n_1) \tag{13}$$
$$C_2 = P^3 (\bmod n_2) \tag{14}$$
$$C_3 = P^3 (\bmod n_3) \tag{15}$$

Using Chinese Remainder Theorem, one can easily formulate and find an equation of the form:

$$C' = P^3 (\bmod n_1 * n_2 * n_3) \tag{16}$$

This means $P^3 < n_1 * n_2 * n_3$, that is $C' = P^3$ is in regular arithmetic. Hence, $C' = P^3$. Therefore, $P = C'^{1/3}$.

3.3.3 Related Message Attack

This type of attack persists when two conditions are present:

1. Sender encrypts two plaintexts P_1 and P_2 using same value of e and sends C_1 and C_2 to the receiver.
2. P_1 is linearly related to P_2. If these two conditions persist, then it is feasible to recover the plaintexts P_1 and P_2 using the values of C_1 and C_2.

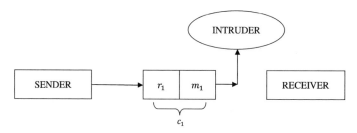

Fig. 4 Padding message m_1 with r_1

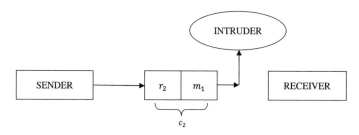

Fig. 5 Padding message m_1 with r_2 pad value and resending it

3.3.4 Short Pad Attack

Sender pads r_1 to plaintext P_1 and then encrypts it to form C_1 and sends it to the receiver. But the attacker intercepts the message. The receiver informs the sender that he has not received the message. Then the sender again sends the message by padding it with r_2 and encrypts it to form C_2. But again the attacker intercepts this C_2.

With short r_1 and r_2 values and small value of e (encryption exponent), the attacker can recover the original plaintext. This is called Short Pad Attack (Figs. 4 and 5).

3.4 Attack on Decryption Exponent

3.4.1 Revealed Decryption Exponent Attack

This is the type of attack which is possible when decryption exponent (d) is known to the attacker. Using the value of d, it is practically feasible to fetch n (the product of primes) and to factorize n to find p and q which are the two prime numbers . Once d is revealed to the attacker, one may think that altering the value of d may be a solution to the problem. But it is not so because as we have seen earlier that by using the value of d, one can reach the value of prime numbers p and q, hence to prevent further attacks, the values of p, q, n, e, and d, all must be changed (Fig. 6).

Fig. 6 Categories of attacks
on decryption exponent

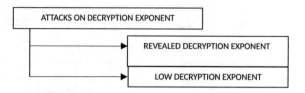

3.4.2 Low Decryption Exponent Attack

This attack prevails when the value of decryption exponent (d) chosen is small, so as to reduce the time required for decryption.

Wiener showed that if $d < 1/3n^{1/4}$, a special type of attack based on continuous fraction can jeopardize the security of RSA [5].

There exists another condition $-q < p < 2q$. With these two conditions existing simultaneously, it is easy and practically feasible to factor n in polynomial time [2].

3.5 Plaintext Attacks

3.5.1 Short Message Attack

As the message is short, one can guess it with more ease by using the ciphertext. This weakness can be reduced to a great extent by padding it with random bits and hence prevent its exposure [11]. Padding using random bits is done by using a standard recommended procedure called Optical Asymmetric Encryption Padding (OAEP) (Fig. 7).

OAEP encryption begins by converting data into code of a seed, a hash, padding octets, and the secret session key into an octet [8]. Masking operations efficaciously randomize these octets before they are subjected to become as the unsigned binary representation of an integer [8]. The number of padding octets is selected such that the encoding occupies one less octet than needed for a unsigned binary representation of the value of modulus. This ensures that the integer is less than the modulus as required in RSA [8]. In another way, the encoded messages can be regarded as an octet string with the same length as the modulus value, but with the most significant octet set to '00'h [8].

Fig. 7 Categories of
plaintext attacks

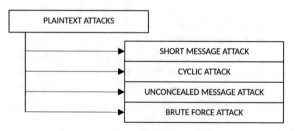

3.5.2 Cyclic Attack

As we already know, that the plaintext is a permutation of the ciphertext, we can also conclude that if we continuously perform encryption on the ciphertext, then after some number of iterations, we will get the plaintext.

This is called the Cyclic Attack on RSA. The algorithm which is used to establish this attack is

$$C_1 = C^e (\text{mod } n)$$
$$C_2 = C_1^e (\text{mod } n)$$
$$\vdots$$
$$C_k = C_{k-1}^e (\text{mod } n)$$

If $C_k = C_{k-1}^e (\text{mod } n)$ then stop and plaintext is C_{k-1}.

This is one of such attacks which has great potential to challenge the security and credibility of the conventional RSA algorithm. But on a positive note, just like the factoring attack, if n has a large value, then it becomes practically infeasible to carry out the huge number of iterations and hence chances to find the plaintext become negligible.

3.5.3 Unconcealed Message Attack

Sometimes, though very rarely but it does happen that plaintexts encrypt to themselves such as plaintext $= 0$ or plaintext $= 1$ [4]. This usually happens with odd values of the encryption exponent, e. Such messages are called Unconcealed Messages. And the attack which exploits this property is called Unconcealed Message Attack.

3.5.4 Brute Force Attack

This attack is when the attacker tries to break confidentiality of message by repetitively trying to use the permutations of the available ciphertext.

There lies another situation which increases the risk of brute force attack in RSA, which is the presence of repetitive numbers or characters in plaintext. Due to same value of p, q, e, n, and d used for encryption, the same characters of plaintext get encrypted to same ciphertext. And repetitive values in ciphertext, further exposes the plaintext to attacker.

For example, let us consider some values of p, q, e, n, and d, and original plaintext "aabba" and let the ciphertext corresponding to a and b be 121 and 692. Then, the resulting ciphertext will be 121121692692121.

Hence, we can conclude that the repetition is clearly seen in the ciphertext.

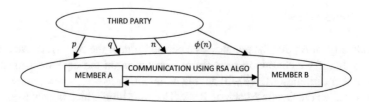

Fig. 8 Illustration of attack on modulus value

3.6   Attack on Modulus

It usually happens that if a community or organization decides to use RSA algorithm for secure communication within themselves, then they hire a third party and the responsibility of providing security lies with third party. In such circumstances, the third force decides the values of p, q, n, and modulus $\phi(n)$. Then, any member of the third party can exploit the value of $\phi(n)$ to reach to plaintext (Fig. 8).

3.7   Attack on Implementation

3.7.1   Timing Attack

Timing attacks are a class of side channel attacks where an intruder gets access to information from the execution of the cryptosystem rather than from any ingrained weakness in the mathematical properties of the system. This way of attack makes it different from other attacks as the other attacks are contingent on the loopholes of mathematical properties of the algorithm unlike this one.

4   Conclusion

In this paper, a detailed description of working of RSA algorithm is described and its associated attacks. Along with it, some common attacks on it are described which are possible when using RSA with small parameter values. Using RSA with larger parameter values has no feasible attacks because such attacks cannot be done in polynomial time. But using RSA with small parameter values has gained importance due to its application in small devices. In such devices, to reduce encryption and decryption time, small parameter values are chosen for RSA algorithm but there lies risk of security breach. To avoid such flaws, one must first understand the risks associated and thoroughly understand the attacks which are possible. For future work, one can understand the risks and incorporate minor changes in the existing RSA algorithm to enhance its security further during its application with small parameter values. Attacks such as Brute force attack, Cyclic Attack, Unconcealed Message Attack, and others usually exploit the repetitive nature of the conventional algorithm. If irregularities can be introduced, then these can be prevented to a great extent.

References

1. Aggarwal, D., Maurer, U.: Breaking RSA generically is equivalent to factoring. In: Joux, A. (eds.) Advances in Cryptology—EUROCRYPT 2009 (2009). https://doi.org/10.1007/978-3-642-01001-9_2
2. Bellare, M., Rogaway, P.: Optimal asymmetric encryption padding how to encrypt with RSA. In: Advances in Cryptology—EUROCRYPT'94 (1994)
3. Boneh, D., DeMillo, R.A., Lipton, R.J.: On the importance of checking cryptographic protocols for faults. In: Fumy, W. (eds.) Advances in Cryptology—EUROCRYPT'97 (1997). https://doi.org/10.1007/3-540-69053-0_4
4. Chaun-ling, J.: Application of RSA asymmetrical encryption algorithm in digital signature. Commun. Tech. **42**, 192–196 (2008)
5. Forouzan, B.A., Mukhopadhyay, D.: Cryptography and Network Security (SIE), 2nd edn. McGraw Hill Education, New Delhi, India (2010)
6. Kai-cheng, L.: Computer Cryptography—The Confidentiality and Security of Data in the Computer Network, 3rd edn. Tsinghua University Press, Beijing, China (2003)
7. Lamprecht, C., van Moorsel, A., Tomlinson, P., Thomas, N.: Investigating the efficiency of cryptographic algorithms in online transactions. Int. J. Simul. Syst. Sci. Technol. **7**, 63–75 (2006)
8. Manger, J.: A chosen ciphertext attack on RSA optimal asymmetric encryption padding (OAEP) as standardized in PKCS #1 v2.0. In: Kilian, J. (eds.) Advances in Cryptology—CRYPTO 2001 (2001). https://doi.org/10.1007/3-540-44647-8_14
9. Nguyen, T.D., Nguyen, T.D., Tran, L.D.: Attacks on low private exponent RSA: an experimental study. In: 13th International Conference on Computational Science and Its Applications (2013). https://doi.org/10.1109/ICCSA.2013.32
10. Rivest, R.L., Shamir, A., Adleman, L.: A method for obtaining digital signatures and public-key cryptosystem. Commun. ACM **21**, 120–126 (1978)
11. Sun, H.M., Wu, M.E., Steinfeld, R., Guo, J., Wang, H.: Cryptanalysis of short exponent RSA with primes sharing least significant bits. In: Franklin M.K., Hui L.C.K., Wong D.S. (eds.) Cryptology and Network Security, CANS 2008 (2008). https://doi.org/10.1007/978-3-540-89641-8_4
12. Xiaolin, Y., Nanzhong, C., Zhigang, J., Xiaobo, C.: Trusted communication system based on RSA authentication. In: 2nd International Workshop on Education Technology and Computer Science (2010). https://doi.org/10.1109/ETCS.2010.460

Private Communication Based on Hierarchical Identity-Based Cryptography

D. Kalyani and R. Sridevi

Abstract Public Key Infrastructure (PKI) is an important tool for securing information in the communication. Presently, a PKI framework demonstrates a pattern toward an emerging worldwide PKI which turns out to be more complicated. In this paper, we address the issue of a contact acquiring a message that it missed, from different contacts of the client while keeping up the secrecy of all gatherings required in the networks. Along these lines in this paper, we built up a specific peer-to-peer public key framework model realizing efficient hierarchical identity-based encryption a modification to the proposed by Boneh HIBE. The proposed scheme need not set up a maximum potential recipients set ahead of time, and it has steady size of the general public key, private key, and header of cipher content. We have utilized our proposed key issuing method to distribute private keys of the users to avoid escrow issue problem. We have presented performance analysis with comparison results assuming a PKI network environment.

Keywords Cryptography · IBE · HIBE · Pairings · Key issuing protocol · PKI

1 Introduction

Identity-Based Encryption (IBE) [1, 2] is a public key encryption scheme, where one's public key can be used as any unique identity such as email address, IP address, or any unique string. Introductory proposition of IBE was first proposed by Shamir [1] in 1984 and later feasible practical identity-based encryption was proposed by D. Boneh et al. in 2001. IBE is an augmentation of public key encryption to give a more freedom to sender by choosing a unique identity of the receiver that avoid

D. Kalyani (✉)
Department of IT, VNRVJIET, Bachupally, Hyderabad 500090, India
e-mail: kalyani_d@vnrvjiet.in

R. Sridevi
Department of CSE, JNTU Hyderabad, Kukatpally, Hyderabad 500085, India
e-mail: sridevirangu@jntuh.ac.in

© Springer Nature Singapore Pte Ltd. 2019   749
A. Abraham et al. (eds.), *Emerging Technologies in Data Mining and Information Security*, Advances in Intelligent Systems and Computing 755,
https://doi.org/10.1007/978-981-13-1951-8_67

generation of two keys in which the public key is no need to be generated whereas private key is generated by Trusted Authority (TA). After Boneh and Franklin's proposal and implementation, other IBE systems are based on the Bilinear Diffie–Hellmann (BDH) assumption. The security of the identity-based cryptosystems is proved using BDH computational hardness assumptions. Later, different schemes similar to IBE include a Certificate-Based Encryption (CBE) scheme, where a client needs both a private key and a forward endorsement from a Certified Authority. Also, the client requires the Public Key Encryption with Keyword Search (PEKS). We might want to have an answer where every authority can designate keys to its sub-authorities, who can keep delegating keys additionally down the hierarchy of importance to the clients. An IBE framework that permits delegation as above mentioned is called Hierarchical Id-Based Encryption (HIBE). Initially, Horwitz and Lynn [8] proposed this idea, who also presented a partial solution for this problem, and later Gentry and Silverberg [5] first described fully functional HIBE.

1.1   ID-Based PKI

In various hierarchical structures, the public key infrastructures generally exist, whereby there may be some intermediate CAs between a client and a root CA to which all clients are relied upon to trust. The major specialized distinction between a certificate-based PKI and a identity-based PKI is the official between people in general/private keys and the person. This can be accomplished by utilizing an endorsement in the conventional PKI. In the character- based setting, people in general key is bound to the transmitted information while the authoritative between the private key and the individual is overseen by the Trusted Authority (TA). Boneh and Franklin [4] recommended in that key escrow can be evaded by utilizing different TAs and limit cryptography. On the other hand, due to this inherent component, the client dependably needs to set up a free secure channel with his TA for recovering private key material.

1.2   Motivation and Contribution

In this paper, we address the problem where there is only a subset of contacts available to a user to directly communicate a message and the remaining users who are not online need to receive their message from those who already received it without compromising their privacy. We present a hybrid cryptographic approach as a solution to this problem. In this paper, we propose a hierarchical identity-based PKI for private communication between peers, that is a modification to the Boneh et al. [4] HIBE. We also compare the performance of our scheme with existing schemes. The proposed scheme is efficient and practical for large receivers.

1.3 Organization of the Paper

The paper is organized as follows. In Sect. 2, we present related work and in Sect. 3, we presented PKI for private messaging based on modified hierarchical identity-based encryption scheme for public key infrastructure. In Sect. 4, we presented our scheme security analysis in the standard model and also discussed the performance of our scheme and conclusions are in Sect. 5.

2 Related Work

The extension of public key encryption and to overcome the key management problem as the public key of the receiver, in 1984, Shamir [1] introduced an Identity-Based Cryptography (IBC). Although the initial concept and idea of Identity-Based Encryption (IBE) was introduced by Shamir [1], the first realization of IBE using bilinear pairings was given by Boneh [3] and Franklin [2] in 2001. After that many IBE schemes such as Gentry's [6] and Brent Waters [10] practical IBE without random oracles, Alexandra et al. [9] IBE with efficient revocation and Amit Sahai [12] fuzzy IBE were developed. To avoid key escrow problem in IBE [7, 13], in 2002, Craig Gentry et al. [5] and Horwitz [8] proposed Hierarchical Identity-Based Encryption (HIBE) scheme and the an extensive survey is given in [27]. In 2005, Boneh [4] proposed hierarchical identity-based encryption with constant size ciphertext. In 2006, Boyen et al. [11] proposed anonymous hierarchical identity-based encryption without random oracles. Distributed Key Generation (DKG) [26] was proposed by Pedersen [14] to qualify a gathering of elements to cooperatively set up a secret-sharing [23–25] environment over a public channel. Lee et al. [15] and others [16, 17] showed and given a safe key issuing convention for IBC. Due to more development and popularity in online communication and the importance of privacy of the information, Guha et al. [20] proposed a solution that replaces the personal details of users by fake information. Luo et al. [21] and Gentry [22] proposed schemes for broadcast encryptions to enforce privacy to the published information. Later, Baden [18] and Gunther et al. [19] make use of applications of an attribute-based encryption method for social network domains.

3 Proposed Hierarchical Id-Based Key Infrastructure

In this section, we present our proposed hierarchical Id-based cryptosystem and discuss our proposed PKI structure using this proposed method.

3.1   Design of HIB PKI

We discuss the points of interest of setting up the parameters of a setup, enlisting contacts, contacts creating refresh solicitations to be prepared by different contacts, refresh reaction to such a demand, and re-key of the framework at a client and how contacts refresh themselves.

1. **Setup** A user P will have a two level HIBE system parameters. This is by calling setup (2). This will generate generates the public parameters and the master key of the user as follows:

   - Select a generator $g \in G$ and a random $\alpha \in Z_p$
   - Set $g_1 = g$
   - Pick random $g_2, g_3, h_1, h_2 \in G$
   - params $= (g, g_1, g_2, g_3, h_1, h_2)$
   - master-key $= g_2$

2. **Private key securing and Issuing**
   Generate the private key of the given kth level ID using a $k - 1$ level private key using Root and other KPA's including user partial key presented in our paper [17] as follows:

   a. Registered user ID chooses a $r \in Z_q$ which is random secret and compute $Q_{ID} \in \mathbb{G}_1$, $R = rP$ and $D_{ID} = r.Q_{ID}$, verify D_{ID} and store in the data base.
   b. SD $(\mathscr{S}, \{ID_i\}_{i=1}^n, \{SK_{ID_i}\}_{i=1}^n, PP)$
   c. Reconstructs partial secret key of the KPA's.
   d. User computes private key as $\mathbb{S} = s_0.s_p.H_1(ID)$.
   e. Finally, checks $e(D_{ID}, P) = e(Q_{ID}, Y)$ which provides the correctness of the private key.

3. **Registering a Contact** The main idea is to setup a two level ($l = 2$) HIBE system at each user. When a user P registers a CP_i it will create a new random first level identifier $I_{r_i} \in Z_p$ and corresponding private key $(d_{I_{r_i}})$. The private key and the identifier will be communicated to CP_i using a private channel. $d_{I_{r_i}}$ is of the form $(g_2(h_1^{I_{r_i}} g_3)^r, g^r, h_2^r)$, where $r \in G$ is random.

   - CP_i keeps both I_{r_i} and $(d_{I_{r_i}})$ private along with the public parameters of P
   - P stores the tuple $< I_{r_i}, r >$.

4. **Contact request and update** The interesting case is when a contact C_{Preq} needs to obtain the latest update of P and P is no longer available online. In such a situation, as highlighted by in the requirements, C_{Preq} will be able to generate a request for $P's$ update, Q_p. This is generated as follows:
   Suppose the identifier assigned to C_{Preq} by P is I_{r_1}.

- Select a random $I_{r_2} \in Z_p$
- Set $ID_{req} = h_1 I_{r_1} h_2 I_{r_2}$
- Update Request to be published: $QP = <PID, ID_{req}>$, here PID is an string of P known to all $P's$ contacts.
  $C_{P_{req}}$ publishes $<PID, ID_{req}>$ and any of $P's$ other contacts will be able to respond to this request. This request information can be made publicly available using a common medium. The steps in creating the response is described next.

5. **Encryption and Update Response** The contact of P observes the tuple $< PID, ID_{req} >$ and runs the below encryption.
   $Encrypt(params\ P, ID_{req}, M_P)$:

   - Select a random $s \in Z_p$
   - $CT_{resp} = (e(g_1, g_2)^s.M, g^s, (ID_{req}g_3)^s) = (A, B, C)$
     The contact publishes the tuple $< PID, ID_{req}, CT_{resp} >$ as the response S_P.

6. **Decryption of the Update**

   - Private key for $ID_{req} : d_{ID_{req}} = (a_0.b_2.(h_1^{I_{r_1}}.h_2^{I_{r_2}}.g_3), a_1.g^t) = (a'_0, a'_1)$
   - Finally to decrypt $CT_{resp} = (A, B, C) : (Ae(a'_1, C))/(e(B, a'_0)) = M_P$

7. **Re-key** The set of contacts at a user C can change in two ways:

   - When a new contact joins
   - When an existing contact is removed.

In user setup, the generated HIBE configuration if of the form params $= (g, g_1, g_2, g_3, h_1, h_2)$ and master-key $= g_2^\alpha$, where $g_1 = g^\alpha$ and $\alpha \in Z_p$ is random. In the case of re-key a user :

   - Generates a new random $\alpha' \in Z_p$
   - Sets master-key $= g_2^{\alpha'}$
   - Set $g_1 = g^{\alpha'}$

With this change P will have to update the private keys of the contacts. Note that in contact registration process P stored the tuple $< I_{r_i}, r >$ for each contact C_{P_i}. To update contacts:
First generate a random $u \in Z_p$
Initialize a list $< id'i, A_i >$ and for each contact $C_{P_i} \in C$:

   - generate the first component of the private keys of the contacts as $g_2^{\alpha'}(h_1^{I_{r_i}} g_3)^{r_i} = A$. This r value is from the stored $< I_{r_i}, r >$.
   - Add $< u^{I_{r_i}}, A >$ to the $< id'_i, A_i >$ list. Finally the complete re-key information to be published is $< PID, g_1, u, [< id'_1, A_1 >, \ldots, < id'_n, A_n >] >$ where $n = |C|$, Note that id'_i is the identifier of the C_{P_i} blinded using u where $id'_i = u^{I_{r_i}}$.

4 Security and Performance Analysis

We present a high-level theoretical evaluation and analysis. Formal proof of security is also included in this paper.

A. Update Request

A contact of user P generates a random identifier for any other party to use in encryption of an update message (which is included in the update request QP). As described in the previous section, C this takes the form: $ID_{req} = h_1^{I_{r_1}} h_2^{I_{r_2}}$ Here, h_1 and h_2 are public values but I_{r_1} and I_{r_2} values are only known to the contact who generates the request.

B. Update response

When a contact of P responds to a Q_P with a response S_P which is of the form $< P, ID_{req}, CT_{resp} >$. Here, CT_{resp} is the original HIBE encryption of M_P using the identity I_{r_1}, I_{r_2}.

C. Re-key

When a user P is re-keyed the information published is $< PID, g_1, u, [< id_1', A_1 > , \ldots, < id_n', A_n >] >$. Here, $g_1 = g^{\alpha'}$ is a public parameter of P in the original HIBE scheme and α' value is safe due to the discrete logarithm problem.

Comparisons and Performance analysis

Table 1 compares key differences between a certificate-based PKI and an identity-based approach.

Table 1 Comparison with existing methods

Feature	Certificate-based PKI	Id-based PKI	H-Id-based PKI (proposed)
Public key generation	Using random information	Using an explicit identifier	Using an explicit root and PKG's
Private key generation	By a user or the CA	By the PKG	By the PKG by level PKG's (MSK by root)
Key certification	Yes	No	No
Key distribution	Requiring an integrity protected Channel	Requiring an integrity protected Channel	Distributing nature
Public key retrieval	Key owner	Public directory	Public directory
Escrow facility	No	Yes	No

5 Conclusions

In this paper, we propose a new public key infrastructure for private messing using efficient hierarchical identity-based encryption scheme, a modification to the proposed by Boneh that can be used for public key infrastructure. We addressed the problem of distributing a message from a common user using cryptographic primitives, where the clients requesting the message can request messages anonymously. Security analysis of the proposed method is presented along with performance analysis, comparison results assuming a public key infrastructure.

References

1. Shamir, A.: Identity-based cryptosystems and signature schemes. In: Advances in Cryptology Crypto'84, LNCS, vol. 196, pp. 47–53. Springer (1984)
2. Boneh, D., Franklin, M.K.: Identity-based encryption from the Weil pairing. In: Kilian, J. (ed.) Advances in Cryptology—CRYPTO 2001, volume 2139 of Lecture Notes in Computer Science, pp. 213–229. Springer (2001)
3. Boneh, D., Boyen, X.: Efficient selective id secure identity-based encryption without random oracles. In: EUROCRYPT 2004, vol. 3027, pp. 223–238. Springer (2004)
4. Boneh, D., Boyen, X., Goh, E.: Hierarchical identity based encryption with constant size ciphertext. In: EUROCRYPT 2005, vol. 3494, pp. 440–456. Springer (2005)
5. Gentry, C., Silverberg, A.: Hierarchical id-based cryptography. In: Zheng, Y. (ed.) Advances in Cryptology—ASIACRYPT 2002, vol. 2501, pp. 548–566. Springer (2002)
6. Gentry, C.: Practical identity based encryption without random oracles. In: Vaudenay, S. (ed.) Advances in Cryptology—EUROCRYPT 2006, vol. 4004, pp. 445–464. Springer (2006)
7. Gentry, C., Halevi, S.: Hierarchical identity based encryption with polynomially many levels. In: Reingold, O. (ed.) Theory of Cryptography—TCC 2009, vol. 5444, pp. 437–456. Springer (2009)
8. Horwitz, J., Lynn, B.: Toward hierarchical identity-based encryption. In: Knudsen, L.R. (ed.) Advances in Cryptology—EUROCRYPT 2002, vol. 2332, pp. 466–481. Springer (2002)
9. Boldyreva, A., Goyal, V., Kumar, V.: Identity-based encryption with efficient revocation. In: Ning, P., Syverson, P.F., Jha, S. (eds.) ACM Conference on Computer and Communications Security, pp. 417–426. ACM (2008)
10. Waters, B.: Efficient identity-based encryption without random oracles. In: EUROCRYPT 2005, vol. 3494, pp. 114–127. Springer (2005)
11. Boyen, X., Waters, B.: Anonymous hierarchical identity based encryption (without random oracles). In: CRYPTO 2006, vol. 4117, pp. 290–307. Springer (2006)
12. Sahai, A., Waters, B.: Fuzzy identity-based encryption. In: Cramer, R. (ed.) Advances in Cryptology—EUROCRYPT 2005, vol. 3494, pp. 457–473. Springer (2005)
13. Seo, J.H., Emura, K.: Efficient delegation of key generation and revocation functionalities in identity-based encryption. In: Dawson, Ed. (ed.) Topics in Cryptology—CT-RSA 2013, vol. 7779, pp. 343–358 (2013)
14. Pedersen, T.P.: Non-interactive and information-theoretic secure verifiable secret sharing. In: Proceedings of the 11th Annual International Cryptology Conference on Advances in Cryptology, CRYPTO91, pp. 129–140. Springer, London, UK (1992)
15. Lee, B., Boyd, E., Daeson, E., Kim, K., Yang, J., Yoo, S.: Secure key issuing in ID-based cryptography. In: Proceedings of the Second Australian Information Security Workshop-AISW 2004, pp. 69–74

16. Gangishetti, R., Gorantla, M.C., Das, M.L., Saxena, A., Gulati, V.P.: An efficient secure key issuing protocol in ID-based cryptosystems. In: Proceedings of the International Conference on Information Technology: Coding and Computing (ITCC'05), vol. 1, pp. 674–678. IEEE Computer Society (2005)

17. Kalyani, D., Sridevi, R.: Robust distributed key issuing protocol for identity based cryptography. In: 2016 International Conference on Advances in Computing, Communications and Informatics (ICACCI), Jaipur, 2016, pp. 821–825. https://doi.org/10.1109/ICACCI.2016.7732147

18. Baden, R., Bender, A., Spring, N., Bhattacharjee, B., Starin, D.: Persona: an online social network with user-defined privacy. SIGCOMM Comput. Commun. Rev. **39**(4), 135146 (2009). https://doi.org/10.1145/1594977.1592585

19. Gunther, F., Manulis, M., Strufe, T.: Cryptographic treatment of private user profiles. In: Danezis, G., Dietrich, S., Sako, K. (eds.) Proceedings of the RLCPS FC 2011 Workshops, LNCS, vol. 7126, pp. 40–54. Springer (2011)

20. Guha, S., Tang, K., Francis, P.: NOYB: privacy in online social networks. Proceedings of the WOSN, p. 4954. ACM, New York, NY, USA (2008)

21. Luo, W., Xie, Q., Hengartner, U.: Facecloak: an architecture for user privacy on social networking sites. In: Proceedings of the IEEE CSE, pp. 26–33. IEEE, Washington, DC, USA (2009)

22. Gentry, C.: Practical identity-based encryption without random oracles. In: Vaudenay, S. (ed.) Proceedings of the Advances in Cryptology—EUROCRYPT 2006, vol. 4004, pp. 445–464 (2006)

23. Tentu, A.N., Mahapatra, B., Venkaiah, V.Ch., Kamakshi Prasad, V.: New secret sharing scheme for multipartite access structures with threshold changeability. In: ICACCI 2015, Kochi, India, 10–13 August 2015

24. Tentu, A.N., Rao, A.A.: Efficient Verifiable Multi-secret Sharing Based on Y.C.H. CSS 2014, vol. 448, pp. 100–109

25. Tentu, A.N., Paul, P., Venkaiah, V.Ch.: Conjunctive hierarchical secret sharing scheme based on MDS codes. In: IWOCA 2013, vol. 8288, pp. 463–467

26. Jahid, S., Mittal, P., Borisov, N.: EASiER: encryption-based access control in social networks with efficient revocation, pp. 411–415. ACM (2011)

27. Kalyani, D., Sridevi, R.: Survey on identity based and hierarchical identity based encryption schemes. Int. J. Comput. Appl. **134**(14), 32–37 (2016)

Part VII
Expert System

An Approach to the Parameter Based Doctor Ranking

Mohammed Mahmudur Rahman, Md. Aman Ullah, Zinnia Sultana and Md. Rashedul Islam

Abstract The term "ranking" can be defined as the comparative relationship among the set of items that makes it possible to evaluate complex information according to certain criteria. It makes the list of items by reducing detailed measures to a sequence of ordinal numbers. Ranking methods vary depending on items or objects. In this research work, present an approach to the parameter based doctor ranking. The main aim of this paper is to develop a web-based system that will rate the registered doctors by calculating their some special quality. This research work has considered here nine different standards and applying the mathematical process of calculation such as min-max Normalization and Bayesian ranking etc. On the basis of the final calculating result, the system will have ranked the doctors that will help the users to find out their desired doctors.

Keywords Bayesian · Min-Max normalization · Parameter based · Ranking

1 Introduction

Internet browsing has become an inevitable part of our daily life for different purposes such as information sharing, online services, and business purposes and so on. The popularity of the web has grown rapidly since last decade as because of the internet has become the sources of information, but the huge number of information make it more difficult for end users to retrieve information. Therefore the importance

M. M. Rahman (✉) · Md. A. Ullah · Z. Sultana · Md. R. Islam
International Islamic University Chittagong, Chittagong, Bangladesh
e-mail: provaiiuc@raudah.usim.edu.my

Md. A. Ullah
e-mail: ullah047@yahoo.com

Z. Sultana
e-mail: zinniaiiuc@yahoo.com

Md. R. Islam
e-mail: rashed_maths@yahoo.com

© Springer Nature Singapore Pte Ltd. 2019
A. Abraham et al. (eds.), *Emerging Technologies in Data Mining and Information Security*, Advances in Intelligent Systems and Computing 755,
https://doi.org/10.1007/978-981-13-1951-8_68

search engines have become flourishing in terms of information retrieval (IR). They help users providing accurate location of the relevant content according to their information needs. The IR problem can also be formulated as a ranking problem, which means that, given a query, the documents are sorted by the ranking function, and then the ranked list is returned to the user.

There are many sites for finding and taking the appointment of doctors at online. But there is no such sites are available that can provide doctors ranking based on their qualities. So there is a huge possibility of fraud doctor's appointment by the patients. So, this work approach path is an initiative for trying to make a list of top doctors according to their ranking value based on their specialties and details to resolve the problem in terms of Bangladesh. It has been a great work using different algorithms and methods for ranking in different fields. Our approach is parameter based ranking of doctors calculating on their special qualities so that patients can easily identify the best option to get the desired doctor and can take appointment according to the doctor's schedule. We have considered nine parameters and weights has given for each of the individual parameters and applied different mathematical calculations on them. After completing calculations the list of registered doctors is stored into the database according to their ranking value and the users will find the sorted list of doctors based on their search category.

2 Background

The Ranking is becoming a key issue for various information retrievals (IR) applications, such as document retrieval and collaborative filtering. Different methods, such as RankNet, RankSVM, and RankBoost, try to create ranking functions automatically by using some training data [1]. Related algorithms, such as SVMmap [2] and AdaRank [3], have proved effective for IR applications. Search engines rank their web pages using the combination of query-dependent and query-independent methods. Query-independent methods try to measure the importance of a web page based on link analysis rather than considering how well it matches to the specific query; such examples like the HITS algorithm, PageRank and TrustRank [4]. Query-dependent methods try to measure the degree to which a web page matches a specific query rather than considering the importance of a web page. Query-dependent ranking consider the number of matches of the various query words on the page itself, in the URL or in any anchor text referring to the page. The Google search engine [5] use PageRank algorithm to accomplish their web page ranking, which exploits the global, rather than local, hyperlink structure of the web [6].

It is very difficult to compare with others. Although we try to find some related work in this fields. There are some site and algorithm ranking and rating. Our system works with a number of parameters. If we look at the ranking system the most popular and used algorithm is Elo rating system [7]. Basically, Elo rating system is used to rating into two things like two players of a game (e.g. Chess). Elo rating system would be more accurate to all of the ratings, and none of them as the Elo rating. Especially it is very efficient to rank between two players calculating their number of wins, losses and score. Another Rating is "The Glicko rating system" which is the improved version of Elo rating system. It was invented by Mark Glickman and initially it used for only chess rating system. Then gradually it use for another rating. Both are works for rating game and especially they are not parameter based algorithm. Hence there is also some parameter based rating algorithm those for university ranking, but those algorithms are not given. They only take the algorithm and their parameter.

3   Methodology

In our research implement a system for the doctors ranking according to parameter based approach. Through which all registered doctors will be stored in accordance with their ranking value calculating by the system. As a result it becomes very easy to find out the well-ranked doctors by the patients according to their requirement. This work also implement the appointment and scheduling part, so that it becomes quite easy for the users to find right doctor and get appointment and scheduling in a single moment. The overall system design flowchart has given below (Fig. 1).

This algorithm is work with nine parameter/criteria which are Patient review, Research paper, Degree, University, Membership of any organization, Website, Social network, BMDC Registration, Achievement. Based on those parameter/criteria we calculate Bayesian method, Normalization and other calculations to rate the doctor. It takes a doctor one by one and then calculates its parameters with given weights, after calculating parameters it sums all parameters. Finally it gives a

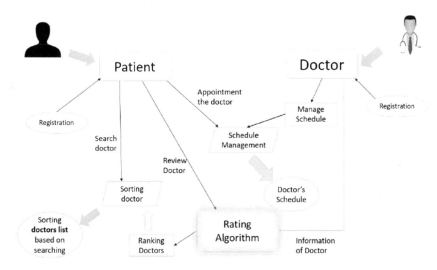

Fig. 1 System design

rating value against the input doctor and saves the rating value. Similarly, we calculate for all doctors.

3.1 Patient Review (P1)

Bayesian Method is very useful to rating someone and it is also most usable rating method to find absolute rank value [8]. For this method, we have to find out the four things.

Overall Number ← Summation of total number of reviews

For the overall average, firstly we have to find the individual number of each rating (e.g., five star). Then multiply the individual number with its rating (e.g. for five star: 152 * 5)

*Overall Average ← (totalRatingNum * 5   + … + totalRatingNum * 1)/(Overall Number)*

Similarly, *Individual Number ← Summation of all reviews to a single item*

*Individual Average ← (individualRatingNum * 5   + … + individualRatingNum * 1)/(Individual Number)*

Using the four values, we will get the Bayesian ranking value:-

*Bayesian ← (Overall Number * Overall Number + individual Number * Individual Average)/(Overall Number + Individual Number)*

This Bayesian value consists 1–5. Then we moderate the value by divide with 5 to consist the value into 0–1. Finally, we multiply with the percentage which passes in a parameter.

3.2 Degree (P2)

This function gives a value by calculating the number of degrees using normalization. In this work, simply categories the medical degrees into six-tier [9]. From the Fig. 2, just neglect the First-tier and it may cover the others category. Our work fixed a value for each category. A doctor may have more than one degree. So, this work sums the all degree's values for an individual doctor. Then find the overall maximum degree value and minimum degree value.

function Normalization (value, max, min) {return (value-min)/(max-min)}

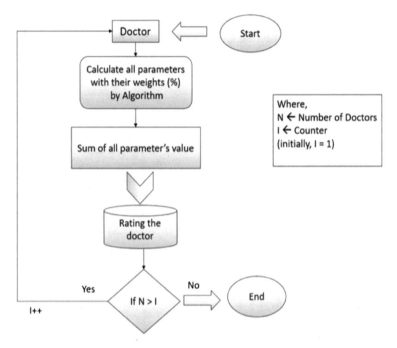

Fig. 2 Flow chart of the ranking algorithm

3.3 University (P3)

University is related to the degree because a number of degrees may equal to the number of University. Here we just only take the best university ranking value. For example, we take the minimum value of the particular doctor's University ranking values. As a degree we also take maximum and minimum value from all doctors. Then we get the normalization value by those values as a degree. Now this work needs to modify this normalization value. Because as the system of University ranking the good are serialize in ascending order. So we reverse the value by subtracting from 1 (e.g., 1–0.78). Now our work get the deserve normalization value. Now, as usual, we multiply with the percentage value and return: (1-normaVal) * percentage.

3.4 Research Paper (P4)

A research paper may contain a deep knowledge of a particular topic. An established research paper might publish in the journal. Here our work focus on the journal because a good journal must maintain and concern their quality and stability to publish new research paper. There are some organizations that rate those publications.

A doctor may have several numbers of the research paper. So, this work takes all paper of his/her and sums those SJR values which take from "www.scimagojr.com". Then normalized the value and multiply by the percentage then return it.

3.5 Achievement (P5)

A good doctor may have single or more than one achievement/reward. It represents the doctor's specialist, proportional success, popularity, experiment skill, something better than others etc. But that achievement may gain from different place and level. So, we categorized them into three levels as International, National and Local. And also put the value against each category. Finally, we calculate all values of the category to get individual value, maximum value and minimum value. Similarly, we find normalize value and multiply with its percentage.

3.6 Membership (P6)

Basically, Membership means he/she engaged with which and how many Organization or Group. Thus we simply named it as "membership". The Organization or Group also may have different types like Global (worldwide), National (Government) and Local (normal). For each type, we also fixed a value to separation by its area. Then this work calculates all membership values to get individual value, the maximum value and minimum value. Finally, we also calculate normalization value and return the multiplication of normalization value and its percentage value.

3.7 Bmdc (P7)

BMDC stands for Bangladesh Medical & Dental Council [10]. It says the recognition of Medical and Dental Qualifications granted by Medical and Dental Institutions in Bangladesh and Institutes outside Bangladesh. Basically BMDC authenticates a doctor. Generally, it may vary in different nation. So it is very important to identify a doctor. First, we check does he/she is registered of BMDC then only return true and false.

3.8 Website (P8)

A doctor may have a personal website. There doctor may write an article, give tips and suggestion and others information. There are many organization and website to rate the website. These works select the popular ranking site "Alexa" to get the ranking value [11]. Our work only take one website from and find the maximum value and minimum value to get normalization value, so they rating the website in ascending order. So this works need to modify the normalization by subtraction the normalization form 1. Finally, return the multiplication of normalization value and its percentage value.

3.9   Social Network (P9)

Nowadays social network is an essential factor in all sectors. Most of the time people connect to the social network. Doctor and Patient also may connect to the social network. There are several social networks like Facebook, Twitter, Google Plus, LinkedIn, Instagram, etc [12]. The popular doctor may have a good number of followers/fans. Basically our work takes only the follower of the doctor from his/her Profile.

3.10   Algorithm

int **function**RatingAlgo(doctor)

    review ← patientReview(p)
    degree ← degree(p)
    varsity ← university(p)
    paper ← researchPaper(p)
    achieve ← achievement(p)
    member ← membership(p)
    bmdc ← bmdc(p)
    site ← website(p)
    socialNet ← socialNetworking(p)

return   review + degree + varsity + paper + achieve + member + bmdc + site + socialNet

4   Experimentation and Results

The method of min-max normalization has been used for seven parameters from our nine parameters. In case of BMDC registration the value is either one or zero as explained before the only different calculation method that is Bayesian Ranking has been used for patient rating particulars. In every case, each value of a parameter

Fig. 3 Bayesian review rating and algorithm rating

was multiplied by its given percentage. After getting the value of each parameter of a doctor, the system will sum up the values for all registered doctors and make the sorted in descending order.

Our research work cannot find any free dataset according to the requirement, and it was quite impossible for us to collect real data. Finding no solution we have collected almost 100 doctor's information from different sources on internet (for example: emedicalpoint.com, bmdc.org.bd, etc.) [13, 14]. So, there are not in a position that all collected data are true and real. Our research works have tried our level best to collect true data as it can be possible. As there is no source of taking the patient rating value, this work have given randomly patient rating values so that our work can make sure that calculation was perfect and the system works properly. Finally, those data given as input in our system, and then the system returns the rating of the doctor. According to our system, it takes nine parameter's value.

For example, 15 doctors data taken for testing evaluations. Here doctors data table:

After processing all data, moreover, the calculation is proceeds with all over doctor's information (not only ten doctors). Whatever the output of the ten doctors is given in the table below with their individual parameters value:

Now we find all parameter's values of each doctor. The system calculates all parameters with given weights and returns the rating against the doctor. Then the system modified the rating value which is consists at 1–10. Gradually it finishes for all doctors. As a result, we find a list of rating and ranking among the 10 doctors. So, in this part, we take another rating value for the same doctors only based on the patient review and here we obviously use Bayesian Ranking to compare the two values.

If we look at the two tables of ranking values, we can easily compare the two ranking and detect the difference. The two ranking of two tables is totally different. There is low difference among the rating value of the review rating, but in the algorithm rating, there is difference between the rating values and lots of variations among all values.

If we look at the Chart (Fig. 3) we can realize the difference that the rating value is change according to another parameter. Those parameters create the difference and influence the all of the rating values. It also remarkable that most of the value is lower than review rating because it only judge by the review (which is not an academic qualification or knowledge), but on another hand, the algorithm calculates several

Table 1 Doctors data (input)

D.N	P1	P2	P3	P4	P5	P6	P7	P8	P9
D1	16094	6	15509	0.1041	0	7	1	505329	7412
D2	32188	13	10512	0.0967	0	5	0	505329	5154
D3	48282	6	28500	0	0	4	1	505329	7412
D4	64376	6	21850	0.0654	0	2	1	501025	12000
D5	80470	6	18542	0	3	7	1	502547	12000
D6	96564	6	14067	0	0	2	1	501028	11245
D7	112658	6	18452	0	3	4	1	507512	8523
D8	128752	10	19532	0.0443	0	13	1	457895	315201
D9	144846	14	15555	0	3	2	1	495875	14585
D10	160940	6	25784	0.102	0	8	1	3214564	12547

Table 2 Output of all parameters

D.N	P1	P2	P3	P4	P5	P6	P7	P8	P9
D1	23.32	0	7.921	0.443	0	2	10	3.637	0.008
D2	23.283	9.545	8.591	0.407	0	1.2	0	3.637	0
D3	23.333	0	6.179	0	0	0.8	10	3.637	0.008
D4	23.324	0	7.071	0.256	0	0	10	3.643	0.024
D5	23.332	0	7.514	0	0.6	2	10	3.641	0.024
D6	23.364	0	8.114	0	0	0	10	3.643	0.021
D7	23.292	0	7.526	0	0.6	0.8	10	3.634	0.012
D8	23.307	5.455	7.381	0.155	0	4.4	10	3.701	1.093
D9	23.359	10.909	7.915	0	0.6	0	10	3.65	0.033
D10	23.351	0	6.543	0.436	0	2.4	10	0	0.026

Table 3 Rating on patient reviews

D.N	Review rating	Ranking	Algo. rating	Ranking
D1	3.3314	7	2.36643	3
D2	3.32614	10	2.33319	5
D3	3.33335	4	2.19789	9
D4	3.33203	6	2.21592	8
D5	3.33311	5	2.35555	4
D6	3.33765	1	2.25711	7
D7	3.32743	9	2.29323	6
D8	3.3296	8	2.77458	2
D9	3.33703	2	2.82331	1
D10	3.3358	3	2.13778	10

numbers of qualifications then rate. It is rare that all have good qualifications of all criteria. That is why the algorithm rate value is lower (Table 1).

5 Conclusion

Employing nine different indicator data and calculating the obtained values by the system using min-max normalization, Bayesian ranking and so on we have ranked some doctors in terms of their detail information. According to our results, there are 10 doctors ranking are shown in Table 2 and Table 3 respectively. There is in Fig. 3 we have shown the graphical representation of the difference between our proposed approach and Bayesian ranking. Here some other indicators might be considered to get the more sophisticated result and weight of the parameters must be changed after added furthermore indicators in future. In future, we will try to implement an expert search system and also build up an expert system for appointment scheduling process to make our system more effective so that user can get the factual result in an efficient manner.

References

1. Jen-Wei, K., Pu-Jen, C., Hsin-Min, W.: Learning to Rank from Bayesian Decision Inference, pp. 827–835. ACM (2009)
2. Yue, Y., Finley, T., Radlinski, F., Joachims, T.: A support vector method for optimizing average precision. In: SIGIR '07: Proceedings of the 30th Annual International ACM SIGIR Conference on Research and Development in Information Retrieval, pp. 271–278 (2007)
3. Xu, J., Li, H.: Adarank: a boosting algorithm for information retrieval. In: SIGIR '07: Proceedings of the 30th Annual International ACM SIGIR Conference on Research and Development in Information Retrieval, pp. 391–398. ACM (2007). ISBN: 978-1-59593-597-7

4. Query-independent Ranking (n.d.).: Wikipedia. Retrieved from https://en.wikipedia.org/wiki/Ranking
5. Brin, S., Page, L.: The anatomy of a large scale hypertextual web search engine. In: Proceedings of the 7th International World Wide Web Conference (2009)
6. Albert, R., Jeong, H., Barabsi, A.: Diameter of the world wide web. Nature **401**, 130–131 (1999)
7. Smokovec, S., Tatras, T.H., Slovakia.: World university rankings qualify teaching and primarily research. In: 11th IEEE International Conference on Emerging eLearning Technologies and Applications, 978-1-4799-2162-1(13) (2013)
8. Shaw-Pin, M., Joon, J.S.: Bayesian ranking of sites for engineering safety improvements: decision parameter, treatability concept, statistical criterion, and spatial dependence. In: Accident Analysis and Prevention. Elsevier (2005)
9. Shu-Ming, H., Ssu-An, L., Chiun-Chieh, H., Wei-Chao, L.: Novel Cloud Service for Improving World Universities Ranking, pp. 127–131 (2012)
10. Paul, S.: How to Find BMDC Registered Doctors in Bangladesh Online, Sujonhera. Retrieved 22 Sept 2016
11. Kathiresan, G., Mahendran, S.: Cluster Analysis of Top 200 Universities in Mathematics, pp. 408-413. IEEE (2015)
12. Faiz, M.M.U., Al-Mutairi, M.S.: Engineering Education for a Resilient Society: A Case Study of the Kingdom of Saudiarabia, pp. 82–88, 20–24. IEEE (2015)
13. Frey, S.: Top Ten Doctors Review Sites, Quora. Retrieved 20 Dec 2011, From https://www.quora.com/What-are-the-best-doctor-review-sites (2011)
14. The Best Doctors Review Sites, liveClinic healthcare blog. Retrieved from (2017)

Land Use/Land Cover Modeling of Sagar Island, India Using Remote Sensing and GIS Techniques

Ismail Mondal, Sandeep Thakur, Phanibhusan Ghosh, Tarun Kumar De and Jatisankar Bandyopadhyay

Abstract Image classification is an important process of land use and land cover mapping. For effective image classification, quite a few aspects have to be considered including the accessibility of quality of satellite imagery, ground control points, a accurate classification method and the skill and proficiency of the user in the processes involved. This study classifies and maps the land use/land cover (LULC) of the Sagar Island using two images (1975 and 2015) and additionally verifies the precision of the classification method used. The study has been divided into two sections (1) Landuse/Land cover (LULC) classification and (2) accuracy assessment. Unsupervised classification was performed using Non-Parametric Rule and change detection was done for the 40 years study period. It was observed that 7.60% of mangrove vegetation's were converted to cropland. Similarly, 40.26% of agricultural (mono-crop) land was converted to agriculture land, 1.48% of mud flat was converted to mangrove swamps, 1.87% wetland area converted to aquaculture land, and 22.54% agricultural (mono-crop) land converted to the settlement with homestead orchard respectively. Other LULC conversions are agricultural (mono-crop) land to cropland (40.26%), mud flat to shallow water (1.36%), wetlands to cropland (0.055%). The study had an overall classification accuracy of 79.53% and Kappa coefficient (K) 0.7465. This overall classification accuracy of the LULC maps is quite significant in terms of their potential use for land use change modeling of Sagar Island.

Keywords Land use land cover · Image processing · Anthropogenic activities Sagar Island

I. Mondal (✉) · S. Thakur · T. K. De
Department of Marine Science, University of Calcutta, 35 Bc Road, Kolkata 700019, India
e-mail: ismailmondal58@gmail.com

P. Ghosh
Institute of Engineering & Management, Saltlake, Sector V, Kolkata 700091, India

J. Bandyopadhyay
Department of Remote Sensing and GIS, Vidyasagar University, Midnapore 721102, India

© Springer Nature Singapore Pte Ltd. 2019
A. Abraham et al. (eds.), *Emerging Technologies in Data Mining and Information Security*, Advances in Intelligent Systems and Computing 755,
https://doi.org/10.1007/978-981-13-1951-8_69

771

1   Introduction

Coastal zones around the planet are hotspots of developmental activities and therefore are under escalating pressure. The coasts highly productive zones and thus form a significant constituent of the global life support system [1]. The need and dependency of human on the coasts have resulted in unplanned changes of the land use land cover (LULC) of the zone Thu, making it very imperative to study and map the changes from time to time in order to sustainably manage the resources of the region.

LULC information are crucial for ecological defence and spatial planning and hence are a prerequisite for strategy making, commerce, and organizational rationale [2]. LULC categorization is imperative as it provides information that can be utilized as input for modeling, particularly when addressing topics regarding the environment, especially in models that involve climate change and policies making [3]. Therefore, the collective LULC thematic maps allow a inclusive way of comprehending the interactions of geo-biophysical, socioeconomic behaviors [4]. In order to supplement this objective to an extent that more useful information in LULC can be extracted, Remote Sensing is frequently paired with Geographic Information System (GIS) tools.

The present study of geomorphological and LULC changes of Sagar Island were produced using satellite-derived remote sensing techniques by taking into account the natural and anthropogenic activities in the island. Various coastal geomorphological landforms regarding shoreline changes affect the LULC of the island and their detailed study is suggested for remedial measures pertaining to erosion control in Sagar Island. Coastal problems and brisk attrition and crucial coastal themes of Sagar Island were investigated by quite a few researchers by multi-temporal satellite data [5, 6]. Accuracy estimation is a major action in LULC studies for the validation of the thematic maps created. It measures the errors of commission that signify the probability of a classified pixel matching the LULC type of its equivalent real-world position [7–9]. The error matrix and kappa coefficient are considered as standard means of appraisal of image classification precision.

The main objective of this study is to classify the land use/land cover map using Geospatial techniques and to carry out the accuracy assessment in order to find out if the hypothesis of using ancillary data could lead to improvement of land use classification.

2   Materials and Methods

2.1   Study Area

Sagar Island is situated at the mouth of the Hooghly estuary in the western most part of Indian Sundarban. Sundarban is one of the largest mangrove forests, which is situated in both India and Bangladesh. Sundarban is situated between 21°37′21″N to

21°52′28″N and 88°01′46″E to 88°9′25″E, spread over Southeastern part of North 24 Parganas and south 24 Parganas districts (Fig. 1). The island is only 6.5 m above sea level [10] and is connected to the mainland by ferry service across the Muriganga River. From the 102 islands situated in Indian Sundarban, 54 islands are inhabited and the oddment of 48 islands is covered with mangrove forest [11]. The maximum width and length of this island are 12 km and 30 km respectively and it stretches from north to south direction. In 1951 the Sagar Island covered an area of 286 km^2 but in 2015 due to an extensive rate of erosion it has dwindled down to 239 km^2. Being a tide-dominated deltaic island it is influenced daily by tidal fluctuations and ravaged by tropical cyclones from time to time. The island has a very active geomorphic history influenced by macro-tidal environment. The geomorphology of this island is affected by natural coastal processes and anthropogenic activities [12].

2.2 Data and Methodology

Multispectral satellite images (MSS and OLI) of two different time series (1975 and 2015) having Path/Row-148/45 for Landsat MSS and Path/row-138/45 for OLI has been taken for the study. At first, the geometric correction was done for the two images. Then radiometric correction was carried out for each of the bands of satellite images. It includes two steps: (1) converting the DN values into radiance values (2) converting the radiance values into reflectance values. The 1975 MSS image was taken as the base layer map after it was resampled to 30 m in order to compare it with the spatial resolution of the OLI image.

2.3 Geomorphological Condition of the Study Area

The generalized geomorphological map given is suitable for a regional understanding of the existing as well as an earlier process. However, in this study prominence has been given for active coastal zone and their temporal changes. Visual image interpretation was used to identify sub-features under broad categories. Some essential interpretation keys, such as shape, size, pattern, tone, texture, shadow, site, situation, association and spectral properties corresponding to coastal units [13].

The high resolutions google earth image scene of the study area was used to confirm the understanding of geomorphological influences. The major classes of geomorphic units of the study area are sand dunes, tidal flat, mangrove marsh, tidal inlets and tidal creeks, aquaculture, beach face and inactive tidal flat with soil profile (Figs. 2 and 3).

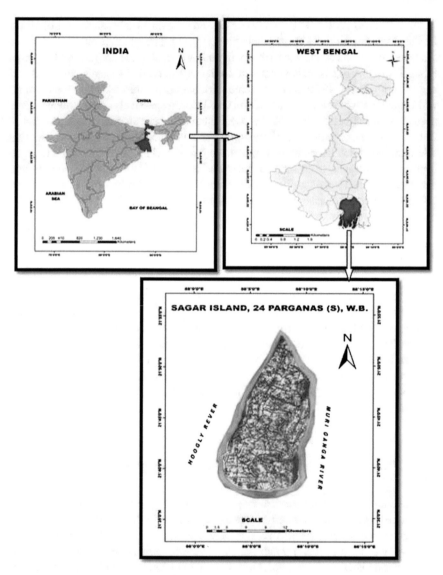

Fig. 1 Location map of the study area

3 Results and Discussion

3.1 *Land Use/Land Cover Classification*

Unsupervised classification technique was performed using all spectral bands in each satellite image. This is a widely adopted classification algorithm. The MSS images

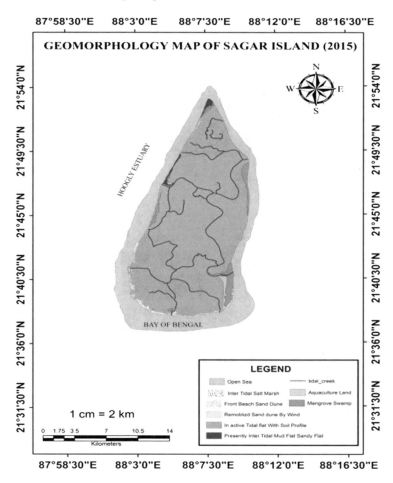

Fig. 2 Geomorphology map of the Sagar Island

were subjected to a three step process to produce the land use land cover classes. These include feature extraction; and identifying and mapping of the following eleven land cover and land use classes: water body, cropland, and agricultural (mono-crop) land, the settlement with homestead orchard, sandbar, mud flat, aquaculture, mangrove vegetation, mangrove swamp, and wetlands. The image classification was guided by information gathered from the ground control points of the study area. The results are shown in (Figs. 4, 5, 6 and 7) with ten classes for 1975 and twelve for 2015 satellite images respectively. The classified images are further reclassified into different forest and non-forest areas.

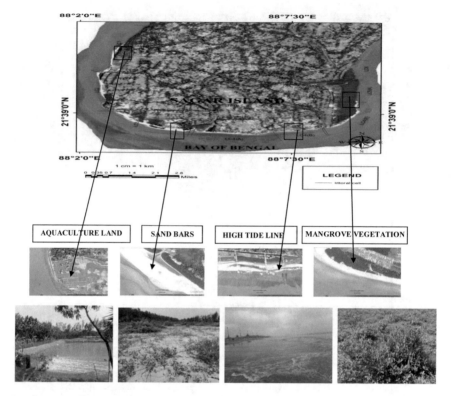

Fig. 3 Google earth images and satellite images with geomorphological photographs from field data

3.2 LULC Map of 1975 Landsat MSS Imagery

The unsupervised classification of the 1975 Landsat MSS image yielded the LULC classes shown in (Figs. 5 and 6). The classification of land area of deep mangrove vegetation is 1516.63 ha, representing (4.31%) of the total area. This is basically found in the southern eastern part with the highest concentration. Likewise, cropland and mono-crop land covers 4280.56 ha (22.18%) and 3453.69 ha (17.89%) respectively, and are mostly scattered throughout the island. The settlement with homestead orchard covers an area of 2377.62 ha (12.32%) and is located mainly within patches around the tidal creeks. Wetland area is 155.952 ha (i.e., 0.81%) and is situated in the southern portion of the island. Mudflat has an area of 377.505.22 ha (2.62%) with small patches scattered in the area. The sandbar and tidal creek have respective land areas of 428.218 ha (i.e., 2.22%) and 831.094 ha (4.31%). They are situated at the Northern, Southern and Western parts respectively while the latter has very small patches scattered across the map. The land area of water body is 5706.87 ha representing 29.57% and is mainly concentrated around the whole island.

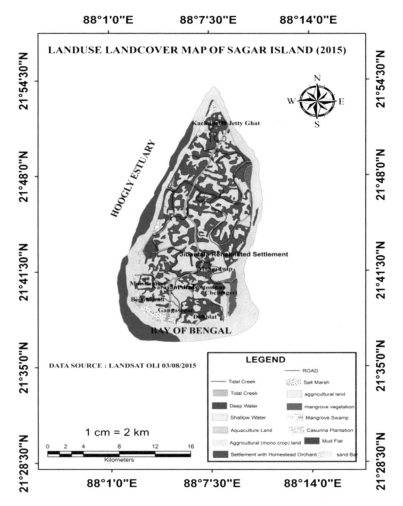

Fig. 4 Land use/Land cover map GIS based of 2015

3.3   LULC Map of 2015 Landsat 8 OLI Image

12 LULC classes were obtained from the Landsat 8 OLI (2015) data after running the unsupervised classification on the image (Figs. 7 and 8). The land area of high-density mangrove vegetation in 2015 is 630 ha (3.26%) which is concentrated around the southland Middle Western and eastern portion of Sagar Island. Mangrove swamp and salt marsh covers an area of 330.39 ha (1.71%) and (0.58%) and mainly around the southeastern and lower portion of the island. The agricultural cropland occupies 4387.68 ha (22.73%), which is the most dominant LULC class. This is scattered around the entire map mainly; Northern, Southern, Eastern and small patches around the western part and middle portion of the map. Agricultural (mono-crop) land occu-

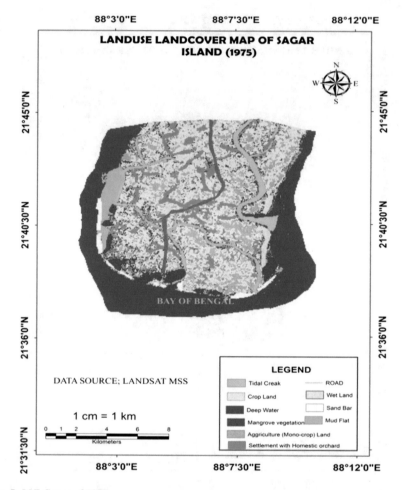

Fig. 5 LULC map of 1975

pies 12.23% of the area. Aquaculture land could be mainly found along south-eastern part and marshy areas cover 233.46 ha (1.21%).

The settlement with homestead orchard area occupies of 4131.54 ha (21.42%) with big patches centered around North, North East, South East, Western and on the middle portions of the island. Sandbar has an area of 418.5 ha (2.68%) with very small patches at the southern part of the island. Water body, having 4752 ha (20.82%) is the least area coverage and mainly concentrated around the area. The LULC maps indicate that some changes of the landscape have occurred over a 40 year period from 1975 through 2015 (Figs. 5 and 7). But the number of classes increased from 10 in 1975 to 12 in 2015. There were many LULC which converted into another class viz. Mangrove vegetation to mangrove swamp (0.33%), agricultural mono-crop to (40.26%) cropland. Factors, which could influence this trend, is the population

PIE GRAPH 1975

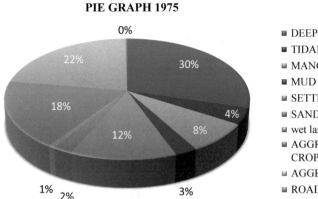

Fig. 6 Represents the LULC classes and their percentages 1975

growth, rapid urbanization, sand removal activities, but these are not investigated. Further analysis of these factors is needed to better explain the impact of these factors on forest cover change.

3.4 Accuracy Assessment of LULC Classes

Accuracy assessment is a significant component of image classification to determine the precision of the LULC map. The higher level of accurateness defines the prominence of the thematic map. An accuracy appraisal was executed on the Landsat 8 OLI (2015) classified image which generated an assessment statement containing an error matrix, overall accuracy, and kappa statistics. An overall classification accuracy of 79.53% and a Kappa coefficient (overall kappa statistics) of 0.7465 was achieved. None of the LULC classes recorded accuracy below 50%.

Overall Classification Accuracy = 79.53%
Overall Kappa Statistics = 0.7465.

3.5 LULC Change Trend from 1975 to 2015

The change trend analysis of the LULC of Sagar Island reveals a change in the size of the ten LULC over the 40 year period of the study (Table 1). Agricultural (mono-crop) land has the most positive change while cropland experienced the most negative change. The results also indicate that from 1975 to 2015 major LULC conversions have taken place within the Sagar Island. The most evident in the results is that agricultural (mono-crop) land made the negative change. Conversely, settlement

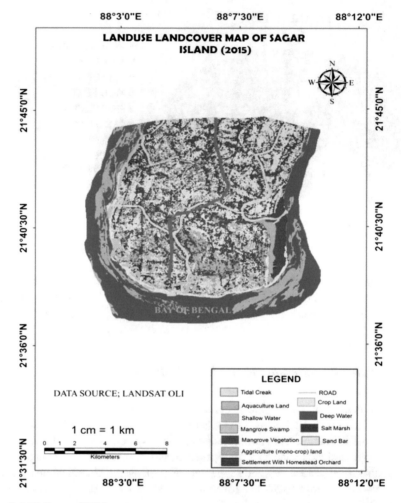

Fig. 7 LULC map of 2015

with homestead orchard (9.08%) and wetland to aquaculture have made the positive change in 2015. Other LULC classes such as cropland have increased (0.55%) from 1975, mangrove vegetation has decreased (−1.04%) and agricultural (mono-crop) land also reduced and have been converted to settlement and cropland respectively.

Some mangrove and casuarinas vegetation's were also subjected to erosion at Gobindapur, Sumatinagar, and Dublat. The status of mangroves was critical in the vicinity of Dublat and Basantpur on the east coast due anthropogenic pressure where as near Beguakhali and Ganga Sagar on the southwest it was due to a combination of both anthropogenic pressure and wave actions. The extensive deforestation mangrove forest was found near Chemagari (Fig. 9), [14].

PIE GRAPH 2015

- ■ DEEP WATER
- ■ MANGRIVE VEGETATION
- ■ SALT MARSH
- ■ SAND BAR
- ■ AGGRICULTURAL LAND
- ■ ROAD

- ■ SHLLOW WATER
- ■ MANGROVE SWAMP
- ■ SETTLEMENT
- ■ TIDAL CREEK
- ■ AGGRICULTURAL (mono crop) LAND
- ■ AQUACULTURE LAND

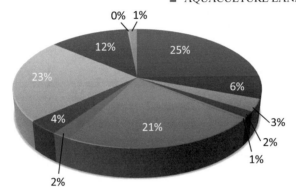

Fig. 8 Represents the LULC classes and their percentages of 2015

Table 1 Trend analysis of LULC of 1975–2015

LULC	1975	2015	(1975–2015)	%
	Area (ha)	Area (ha)		
Deep water	5706.87	4752.72	5711.814664	−4.94466
Mangrove vegetation	1516.63	630	2377.62	−1.04203
Settlement	2377.62	4131.54	2377.62	9.085973
Sand bar	428.218	418.5	4280.56	−0.05046
Tidal creek	831.094	716.85	3453.69	−0.59208
Agricultural land	4280.56	4387.68	45.486	0.553857
Agricultural (mono-crop) land	3453.69	2359.89	155.952	−5.66757
Road	45.486	58.86	505.22	0.069275
Wet land	155.952	233.46	18907.96266	0.401508
Mud flat	505.22			

Fig. 9 Deforestation mangrove forest in Chemagari

3.6   LULC Conversion Analysis Within the Sagar Island

The summary of the major LULC conversions that have been taken place from 1975 to 2015 within the Sagar Island is in (Fig. 10, Table 2). The most evident result is that mangrove vegetation decreased by 7.60% from its area in 1975. Similarly, 40.26% of agricultural (mono-crop) lands was converted to agriculture land, 1.48% of mud flat to mangrove swamp, 1.87% wetland converted to aquaculture land and 22.54% agricultural (mono-crop) land converted to the settlement with homestead orchard respectively. Other LULC conversions are agricultural (mono-crop) land to cropland (40.26%), mudflat to shallow water (1.36%), and wetlands to cropland (0.055%).

4   Conclusions

The present study of Sagar Island in Indian Sundarban using Geospatial technology can provide accurate, cost effective, and instantaneous information; but unless these are applied in "real time" monitoring and assessment of mangrove dynamics from appropriate public platforms, optimum success cannot be achieved [15]. From 1975 to 2015 in the context of Sagar Island, continual deforestation of mangrove forest has been done. Erosion also played a very active role in this deforestation as some mangrove and casuarinas vegetation's were eroded at Gobindapur, Sumatinagar, and Dublat. The status of mangroves in the vicinity of Dublat and Basantpur along the east coast; Beguakhali and Ganga Sagar on the southwest coast was critical due to anthropogenic pressure and wave actions. Although extensive deforestation of mangrove forest was found near Chemaguri it was not critical. Accuracy of Overall

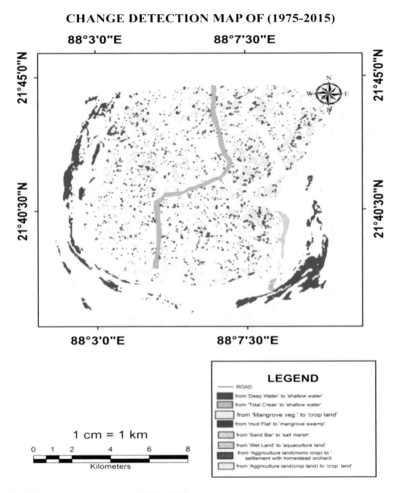

Fig. 10 Change detection map of 1975–2015

Classification was 79.53% and Kappa coefficient (K) was 0.7465. This accuracy improvement in classification of the LULC maps is significant in terms of their potential use for land change modeling of the region and thus substantiates the claim that Geospatial techniques are very crucial role for the land use and land cover mapping.

Table 2 Shows LULC conversion analysis (1975–2015)

Sl No	Area over changed	Area (ha)	Percentage (%)
1	Deep water to shallow water	8619300	24.0416719
2	Tidal creek to shallow water	162900	0.454374294
3	Tidal creek to mangrove swamp	117900	0.328856533
4	Mangrove vegetation to agriculture land	2727000	7.606376302
5	Mud flat to shallow water	486900	1.358102171
6	Mud flat to mangrove swamp	532800	1.486130287
7	Wet land to crop land	19800	0.055227815
8	Wet land to aquaculture land	671400	1.872724991
9	Agriculture mono-crop to settlement	8081100	22.54047948
10	Agriculture mono-crop to agriculture land	14432400	40.25605623
	Total area	35851500	100

References

1. Mondal, I., Bandyopadhyay, J.: Coastal Zone Mapping through Geospatial Technology for Resource Management of Indian Sundarban, West Bengal, India (2014)
2. Rwanga, S.S., Ndambuki, J.M.: Accuracy assessment of land use/land cover classification using remote sensing and GIS. Int. J. Geosci. **8**, 611–622 (2017). https://doi.org/10.4236/ijg.2017.8 4033
3. Disperati, L., Gonario, S., Virdis, P.: Assessment of land-use and landcover changes from 1965 to 2014 in Tam Giang-Cau Hai Lagoon, Central Vietnam. Appl. Geogr. **58**, 48–64 (2015)
4. Moran, E.F., Skole, D.L., Turner, B.L.: The Development of the International Land-Use and Land-Cover Change (LUCC) Research Program and Its Links to NASA's Landcover and Land-Use Change (LCLUC) Initiatives. Kluwer Academic Publication, Netherlands (2004)
5. Gopinath, G., Seralathan, P.: Rapid erosion of the coast of Sagar Island, West Bengal, India. Environ. Geol. **48**, 1058–1067 (2005)
6. Purkait, B.: Coastal erosion in response to wave dynamics operative in Sagar Island, Sundarbans delta, India. Front. Earth Sci. China **3**, 21–33 (2009)
7. Congalton, R.G.: A review of assessing the accuracy of classifications of remotely sensed data. Remote Sens. Environ. **37**, 35–46 (1991). https://doi.org/10.1016/0034-4257(91)90048-B
8. Campbell, J.B.: Introduction to Remote Sensing, 4th edn. The Guilford Press, New York (2007)
9. Jensen, J.R.: Introductory Digital Image Processing: A Remote Sensing Perspective, 3rd edn. Pearson Prentice Hall, Upper Saddle River, NJ (2005)
10. Mukherjee, S.: Sundarbans biosphere reserve (in) workshop on Sundarbans Day, 3rd June 2001, organised by Sundarbans Biosphere Reserve and Zilla Parishad—24 Parganas (South)in collaboration with Department of Sundarbans Affairs, Zoological Survey of India, Botanical Survey of India and Calcutta Wildlife Society (2001)
11. Manjunath, K.R., Tanumi, K., Kundu, N., Panigrahy, S.: Discrimination of mangrove species and mudflat classes using in situ hyperspectral data: a case study of Indian Sundarbans (2013)
12. Mondal, I., Bandyopadhyay, J., Dhara, S.: Detecting shoreline changing trends using principle component analysis in Sagar Island, West Bengal, India. J. Spat. Inf. Res. **25**(1), 67–73 (2016). https://doi.org/10.1007/s41324-016-0076-0 (Springer Nature)
13. Jensen, John R.: Introductory Digital Image Processing. Prentice-Hall, New Jersey (1986)

14. Jayappa, K.S., Mitra, D., Mishra, A.K.: Coastal geomorphological and land-use and land-cover study of Sagar Island, Bay of Bengal (India) using remotely sensed data. Int. J. Remote Sens. **27**(17), 3671–3682 (2006). https://doi.org/10.1080/01431160500500375
15. Datta, D., Deb, S.: Analysis of coastal land use/land cover changes in the Indian Sunderbans using remotely sensed data. Geo-spat. Inf. Sci. **15**(4), 241–250 (2012). https://doi.org/10.108 0/10095020.2012.714104

A Hybrid Approach to Improve Recommendation System in E-Tourism

Mohammed Mahmudur Rahman, Zulkifly Bin Mohd Zaki, Najwa Hayaati Binti Mohd Alwi and Md. Monirul Islam

Abstract Recommendation Systems help users search large amounts of digital contents and identify more effectively the items—products or services—that are likely to be more attractive or useful. As such, it can be characterized as tools that help people making decisions, i.e., make a choice across a vast set of alternatives. This research work has explored decision-making processes in the wide application domain of online services, specifically, hotel booking. This research work is a combination of collaborative filtering (Item-based) recommendation and knowledge-based recommendation system. In which collaborative filtering recommendation will work for user searching and knowledge-based recommendation will work as default recommendation system. In knowledge-based recommendation system it reads the user profile along with his activity of certain last time period as our main knowledge base where this work define the fact of user's activity. Then this research work applies sorting and counting algorithm. Contextual data are temporarily stored in the knowledge base as the time user stay logged in. Each login will take an updated contextual database. In searching, using item-based k-nearest neighbor algorithm for prediction by collaborative filtering. This work proposed a new rating system which based on hotels performance.

Keywords Personalized recommendation · Rating system · E-tourism
Hybrid recommendation · Collaborative filtering
Knowledge-based recommendation

M. M. Rahman (✉) · Z. B. M. Zaki · N. H. B. M. Alwi
Universiti Sains Islam Malaysia, Nilai, Negeri Sembilan, Malaysia
e-mail: mmr@cse.iiuc.ac.bd

Z. B. M. Zaki
e-mail: zulkifly@usim.edu.my

N. H. B. M. Alwi
e-mail: najwa@usim.edu.my

Md. Monirul Islam
International Islamic University Chittagong, Chittagong, Bangladesh
e-mail: monirliton@yahoo.com

© Springer Nature Singapore Pte Ltd. 2019
A. Abraham et al. (eds.), *Emerging Technologies in Data Mining and Information Security*, Advances in Intelligent Systems and Computing 755,
https://doi.org/10.1007/978-981-13-1951-8_70

1 Introduction

E-Tourism is a growing industry and it is very important in modern economy includes different scopes and functions. Now a day tourists use e-tourism sites for their trip planning such as hotel booking, ticket reservation and so on. This research work is focused on hotel booking in e-tourism. Because there are thousands and millions of hotel in the e-tourism site and that makes confuse tourist to choose. The approach of this work is to help them by recommending hotels by matching their interest. There is two type of technique with the recommendation:

1. Personalized recommendations.
2. Non-personalized recommendation.

This research work proposed a recommendation system which is a combination of collaborative filtering and knowledge-based recommendation systems. And this system will contain a contextual database which will contain browsing records of each user. By learning from the user profile, Hotel details and context recommender system will recommend the hotel for a specific user and also proposed a rating system that based on hotels performance. After learning about the current recommendation systems which are running by many renowned organizations, realize that they are suggesting user's similar products. For this reason, sometimes users get confused about choosing a product. And also there are some recommendation systems which are tried to fulfill this problem by browsers cookies. But it is not the ultimate solution for that problem. After that traditional rating system for rate, the products do not give us perfect rating about the product. Because it's based on user review and a big part of the user are not interested in reviewing products. The objectives of this work is to improve recommendation system by a new combination of recommendation systems that can satisfy user's expectation by recommending products and come over the major drawbacks of traditional recommendation systems. And also to build a new rating system which will be based on items performance.

2 Background

According to Beel et al. [1] and Aldhahri et al. [2] in the past few years, there are lots of researches done on a recommendation for travel and tourism sector. Most of them used traditional collaborative filtering recommendation and some recommendation system combine it with some other recommendation system like content based and case based filtering recommendation. The knowledge-based recommendation also used several times. In 2008 an application was proposed by Sebastia et al. [3] named 'E-tourism' which is a vacationer proposal and arranging application to help clients on the association of a relaxation and traveler motivation. Initial, a recommender framework offers the client a rundown of the city puts that are likely important to the client. This rundown considers the client statistic order, the client enjoys in previous outings and the inclinations for the present visit. Second, an arranging module plans the rundown of prescribed spots as per their worldly attributes and the client limitations; that is the arranging framework decides how and when to play out the suggested exercises. This is an extremely pertinent element that most recommender frameworks need as it permits the client to have the rundown of suggested exercises sorted out as a motivation, i.e., to have an absolutely executable arrangement.

In [4] they proposed a hybrid recommendation system which includes item-based collaborative filtering, genetic algorithm, and hybrid recommendation algorithm. And they have described mobile recommendation, web recommendation and collaborative filtering approach that are used previously. The proposed hybrid tourism recommendation system adopts item-based collaborative filtering to offer high preference recommendation list for tourists. A genetic algorithm is used for an optimal solution in presence of various constraint conditions. Just applying item-based recommendation can process preference offers only. Recommendation system based on web mining discussed in Zhao and Ji [5]. They have analyzed some major recommendation methods based on web mining, such as collaborative filtering and association rules mining. In Aldhahri et al. [2] they discuss crowdsourcing recommendation. Crowdsourcing is an approach where requesters can call for workers with various capacities to handle an errand for money related reward.

Some recommendation systems are based on user action online. Those system record user action and then analyze data to user's interest. This work described in Elkhelifi et al. [6] and proposes a new recommender algorithm based on multi-dimensional users behavior and new measurements. It is used in the framework of a recommender system that uses knowledge discovery techniques to the problem of making product recommendations during a live user interaction. Rating is the most important part of the recommendation. All the recommendation works with the rating. But nowadays the rating system of hotels in e-tourism based on user review. And most of the users of e-tourism sector are not interested in review hotels [7]. For this reason, hotel rating cannot tell perfectly about hotels condition and it's very bad for the e-tourism industry. That is why recommendation based on rating is not reliable. If the rating measure hotel performance along with user review then it will be a perfect solution.

3 Our Approach

3.1 Proposed System Architecture

In this recommendation model has two part one is knowledge-based recommendation system and another is collaborative filtering recommender system. In knowledge-based recommendation it has two parts:

1. Inference Engine: In this, it holds the facts, rules, and queries for applying on the data in the knowledge base. The algorithm also applied in inference engine.

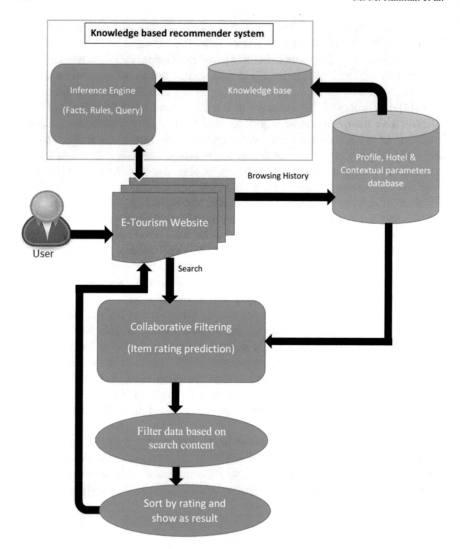

Fig. 1 The architecture of a recommendation which is a combination of collaborative filtering recommender and knowledge-based recommender system

2. Knowledge Base: It holds data about the fact for being processed by the inference engine.

And in collaborative filtering recommendation, have applied item-based nearest neighbor algorithm. This recommendation system use k-nearest neighbor algorithm for predicting the unrated hotels by individual's collaborative filtering recommender (Fig. 1).

Every interaction of user will be recorded as a browsing history and it will be recorded in the contextual database [8, 9]. This data will use as experienced data for knowledge. This research proposes a new rating system for hotel, based on its performance along with user review or user rating.

3.2 Knowledge-Based Recommender System

In this recommendation system, knowledge-based recommender system will work as a default recommendation system. And this system will take data from the main database to the knowledge base and inference engine will find the facts from those data and then apply rules on data based on the query. Knowledge will work in two different part, (1) knowledge base and (2) Inference engine.

Knowledge base:
A knowledge base (KB) [10, 11] is a part of the knowledge-based system used to store complexly organized and unstructured data utilized by a computer system. The underlying utilization of the term was regarding expert systems which were the principal knowledge-based frameworks. The first utilization of the term knowledge base was to portray one of the two sub-frameworks of a knowledge-based system (Fig. 2).

Inference Engine:
The inference engine is the most important part of a knowledge-based system. It's called the brain of a knowledge-based system [12, 13]. Knowledge base stores data about facts and inference engine applies rules on it. And it has three components:

• An interpreter: The interpreter executes the chosen agenda items by applying the corresponding base rules.

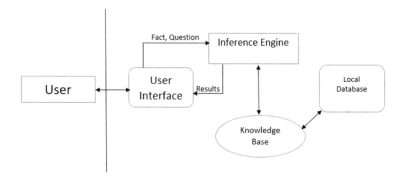

Fig. 2 Knowledge-based system

- A scheduler: The scheduler maintains control over the agenda by estimating the effects of applying inference rules in light of item priorities or other criteria on the agenda.
- A consistency enforcer: The consistency enforcer attempts to maintain a consistent representation of the emerging solution.

For inference engine, in this research paper have followed 'RETE' algorithm, which is an inference rule algorithm. This algorithm is a method for comparing a set of patterns to a set of objects in order to determine all the possible matches. And it's an efficient pattern matching algorithm deals with a Production Memory (PM) and a Working Memory (WM) which each of them may change over time. Working Memory is a set of items that represent facts about the system's current situation or state. Each item in the WM is called Working Memory Element (WME).

All the rules are made and work by following the example pattern. Inference works with two different methods:

Forward chaining:
It takes a rule and if its conditions are true ads its conclusion to working memory until no more rules can be applied. If the conditions of the rule if A and B then C are true then C is added to working memory. In forwarding chaining, the system simply tests the rules in the order that occurs, therefore, rule order is important.

Backward chaining:
The backward chaining inference engines try to prove a goal by establishing the truth of its conditions. The rule is if A and B then C the backward chaining engine will try to prove C by first proving A and then proving B. Proving these conditions to be true may well invoke further calls to the engine and so on.

From those methods, in this research work choose forward chaining method for the inference engine. First inference engine will apply the rules on the data that knowledge base stored. By the way, this research has used the ranking method in inference rules. And it picks the highest counted value for a particular fact. Then collect all the results and merge all of them. After merging those result system will show the result as a recommendation [14, 15].

The knowledge base is a database that stores fact data from the local database. So this research work just need to retrieve the specific data will recommend. To retrieve data we have used sorting algorithm as rule and taking some fact like last visited hotel, most visited hotel, most visited category and etc. for the rule. Here two facts described with algorithm firstly, this work considered for sorting are the last attempt of the user context that he booked. Then secondly all of his most associated hotels where he booked more or rated more. And finally, most rated and viewed hotels will be referred to him/her.

General sorting algorithm for the date:
    Step 1—Set MAX to location date
    Step 2—Search the MAXIMUM element in the Date_list
    Step 3—Swap with value at location MIN_date
    Step 4—Increment MAX_Date to point to next element

Step 5—Repeat until list is sorted

Then the system will peak the last visited hotel by a particular user. Sorting algorithm for count hotel that is booked maximum time by the user.

Counting algorithm:

Step 1—If unsorted hotels are stored in an array h [ ].

Step 2—Sorted hotels will store in sh [ ] array.

Step 3—Algorithm uses an additional array to put the temporary value which we can say th [ ].

Step 4—In this th[ ] possible values are inserted.

After that system will peak the highest visited hotel by particular user. Like above all the rules based on facts will come with a result and then inference engine will make a recommendation of a set of hotels to the logged in user.

3.3 Collaborative Filtering Recommender System

The collaborative filtering has two types of approach, first one is user-based collaborative recommendation and secondly item-based collaborative recommendation. User-based collaborative recommendation has some drawbacks like if there are not enough users in a system it will be difficult to recommend for the system. In the meantime, the item-based collaborative recommendation doesn't need much similar user to target user because it only read the target users profile then search related item from item list matching with user's interest. So this research works peak item-based recommendation for this system. It also has two part (1) item recommendation and (2) item rating prediction. This work will use item rating prediction as a search recommendation system. When user search for a content then system will start doing prediction about the hotels that target user did not rate. Then next step take that content as a condition for filter hotel data. After that, it will give the filtered data set to the next step for sorting by their rating. And show it to the user as a search result. For this collaborative filtering, the recommendation will use the algorithms of item-based filtering. And it uses k nearest neighbor algorithm. Unlike the user-based collaborative filtering algorithm, the item-based approach looks into the set of items the target user has searched for and takes filtered data from the previous step and computes how similar they are to the target rating and then selects k most similar items $\{i_1, i_2 \ldots \ldots i_k\}$. Once the most similar items are found, the prediction is then computed by taking an outweighed average of the target ratings on these similar items. This research work describes these two aspects namely, the similarity computation and prediction computation.

Similarity computation

One critical step in the item-based collaborative filtering algorithm is to compute the similarity between items and then to select the most similar items. The basic idea in similarity computation between two items i and j are to first isolate the users who have rated both of these items and then to apply a similarity computation technique to determine the similarity $s_{i,j}$.

Prediction Computation

Based on similarity computed prediction will be calculated. There are the techniques we use for this calculation:-

Weighted sum: As the name implies, this method computes the prediction on item i for a user u by computing the sum of the ratings given by the user on the items similar to i. Each rating is weighted by the corresponding similarity $s_{i,j}$ between items i and j. we can denote the prediction $P_{u,i}$ as

$$P_{u,i} = \frac{\sum_{all\ similar\ item,N} (S_{i,N} * R_{u,N})}{\sum_{all\ similar\ items,N} (|S_{i,N}|)}$$

Basically, this approach tries to capture how the active user rates the similar items. The weighted sum is scaled by the sum of the similarity terms to make sure the prediction is within the predefined range.

Proposed rating system

Nowadays rating system for the hotel is to take a review about it from users then take the mean as a rating of that hotel. This is the traditional system of rating a hotel. It cannot measure the performance of the hotel currently. Because most of the visitor of a hotel are not similarly interested in a review. In this case if a hotel have more customer but not expected review from their visitor then they will have less rating in website and on the other site if we consider another hotel that they have not that much visitor but most of them are review their hotel, then this hotel will get most rating than the first one, although they are providing the same facility. So this research work came up with a new idea for rating system which will help to make the exact rating for a hotel. In this work proposed system considers about four cases:

1. User rating
2. Hotel Category
3. No. of hotel's booking (of last 10 days in E-Tourism site)
4. No. of hotel's visitor (of last 10 days in E-Tourism site)

This work will measure this performance in 10 out of 100%. For that we divided the 100% of the four cases, user rating 20%, hotel category 30%, No. of booking 30% and No. visitors 20%. How the cases will be calculated, it is given below:

1. User rating (20%): We will define user rating as U and from user rating, 20% will be taken for new rating system,

$$U = \frac{user\ rating}{100} \times 20$$

2. Hotel Category (30%): All over the world hotels are categorized by the facility which they have given in their hotels and it measured by stars. And till now the

7-star hotel is the highest luxurious in the world. So seven stars are available for categorizing hotels. So this research work fixed a value of '4.2857' for each star of a hotel, and that will be the score of hotel category out of 30%. Hotel category will be defined as C,

$$C = Hotels\ category\ stars \times 4.2857$$

3. No. of hotel's booking (30%): This part will work with hotels booking data of last 10 days. Initially, these works have fixed a minimum value for highest performance and it is 200 booking. If a hotel gets 200 or above booking confirmed by the record of the e-tourism website then its booking performance will be 100%. And for every 10 booking, it counts 5% out of 100%. Then this work will take 30% of that value. Hotel booking will be defined as B,

$$B = \frac{Booking Performance}{100} \times 30$$

4. No. of hotel's visitor (20%): This part will work with a number of the visitor of a hotel on an e-tourism site of last ten days. For this also fixed a value and the minimum highest value is 1000 visitors. So if a hotel gets 1000 or above number of visitors that hotels visitor hit performance will be 100% and every 10 visitors will be counted as 1%. Then this system will take 20% of that. Hotels visitor's performance will define as V,

$$V = \frac{Visito'rs\ hit\ performance}{100} \times 20$$

After all those cases value taken final equation will summarize the values and take 10% of those values. Because these researches work are counting hotel rating out of 10. The final equation is,

$$Hotel\ rating = \frac{U + C + B + V}{100} \times 10$$

Initially, these research works have calculated fifteen hotels data with the proposed system. The performance of the new approach is shown in Fig. 3 and red line in the chart representing rating with performance and blue bars are representing traditional rating. The workflow algorithm of rating system is given below:

If user does any activity on the system then the workflow algorithm will start,

Step-1: Start
Step-2: Update activity parameters value for hotels.
Step-3: Calculate the changed performance parameters.
Step-4: Update rating of that hotel.
Step-5: End.

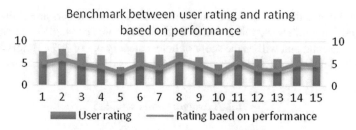

Fig. 3 Benchmarking of user review based rating and rating with performance

4 Experimentation and Results

To evaluate the given approach, this research work has created an e-tourism website for a knowledge-based recommendation in asp.net platform and used rapid miner studio for collaborative filtering recommendation. For that platform have collected data from a renowned hotel booking website 'Booking.com' and an American data set catalog 'https://catalog.data.gov/dataset'. In website for a knowledge-based recommendation we have used five facts and they are (1) last visited the hotel, (2) most visited the hotel, (3) average price, (4) hotel category and (5) hotel rating. Then this work made rules and apply them on the specified site. The site which created, online for the limited time and has taken 50 users feedback about recommendation and it was successful as expected. In collaborative filtering recommendation, these works have used a huge data set with missing rating values in rapid miner studio.

It predicted missing ratings for particular users. Measuring the performance of this process and in this RMSE: 1.439, MAE: 1.135 and NMAE: 0.284. Finally, for the new rating system, this work proposed applied on fifteen hotels performance and user ratings. In Fig. 3 the benchmark between current rating and rating based on performance. It shows that how current rating system giving wrong information about a hotel. If a hotel, fortunately, get good customer review then it will get good rating although its service performance is medium. But on the other hand, if a hotel, unfortunately, gets less review than a hotel which not good at service performance as this one, it will get a bad rating. That is shown in the benchmark.

5 Conclusion

In this research work, proposed a new rating system combined with knowledge-based recommender system and collaborative filtering recommender system which is based on users contextual parameters or browsing history where it is a hybrid approach and here knowledge-based recommender worked as the default recommended and collaborative work as searching recommender. The proposed model has two part—first part is implemented by .NET framework and second part is implemented by rapid miner. Both parts worked properly and benchmark of this work proposed new rating system based on performance and traditional rating system shows the difference between existing and this work where this result is better than others on a different perspective. In future work, this research work proposes to apply the recommendation to an e-tourism site and in collaborative filtering, if the item rating prediction can combine with user-based collaborative filtering it will be a good recommendation and enriching the contextual information for knowledge base recommendation system. It will help knowledge-based to increase its performance.

References

1. Beel, J., Gipp, B., Langer, S., Breitinger, C.: Research-paper recommender systems: a literature survey **17**(4), 305–338 (2016)
2. Aldhahri, E., Shandilya, V., Shiva, S.: Towards an effective crowdsourcing recommendation system a survey of the state-of-the-art. In: IEEE Symposium on Service-Oriented System Engineering, ISBN: 978-1-4799-8356-8 (2015)
3. Sebastia, L., Garcia, I., Onaindia, E., Guzman, C.: e-Tourism: a tourist recommendation and planning application. In: 20th IEEE International Conference on Tools with Artificial Intelligence (2008)
4. Chen, J.-H., Chao, K., Shah, N.: Hybrid recommendation system for tourism. In: IEEE 10th International Conference on e-Business Engineering (2013)
5. Zhao, X., Ji, K.: Tourism e-commerce recommender system based on web data mining. In: The 8th International Conference on Computer Science & Education (2013)
6. Elkhelifi, A., Ben Kharrat F., Faiz, R.: Recommendation systems based on online user's action. In: IEEE International Conference on Computer and Information Technology; Ubiquitous Computing and Communications; Dependable, Autonomic and Secure Computing; Pervasive Intelligence and Computing, ISBN: 978-1-5090-0154-5 (2015)
7. Yifan, Y., Junping, D., Dan, F., Lee, J.M.: Design and implementation of tourism activity recognition and discovery system. In: 12th World Congress on Intelligent Control and Automation (WCICA) (2016)
8. Zhou, X., Xu, Y., Li, Y., Josang, A., Cox, C.: The state-of-the-art in personalized recommender systems for social networking. Springer Link, Vol. 37, Issue 2, pp 119–132, ISSN- 1573-7462 (2012)
9. Antonio, D.: Exploring recommender system for decision making in e-tourism. Politecnico Di Milano. http://hdl.handle.net/10589/59342 (2012)
10. Towle, B., Quinn, C.: Knowledge-based recommender systems using explicit user models. AAAI Technical Report WS-00-04. (2000)
11. Burke, R.: Knowledge-based recommender systems. Researchgate/publication/2378325 (2000)
12. Thiengburanathum, P., Cang, S., Yu, H.: Overview of personalized travel recommendation systems. In: 22th International Conference on Automation & Computing, University of Essex, Colchester CO4 3SQ (2016)
13. Tang, Z., Wen, Z.: Recommendation system based on collaborative filtering in rapid miner. Comput. Model. New Technol. **18**(11), 1004–1008 (2014)
14. Zhang, L., Hu, C., Chen, Q., Chen, Y., Shi, Y.: Domain knowledge based personalized recommendation model and its application in cross-selling. In: Proceedings of the International Conference on Computational Science, ICCS, Vol. 9, pp. 1314–1323. Elsevier (2012)
15. Burke, R.: Integrating knowledge-based and collaborative-filtering recommender systems. AAAI Technical Report WS-99-01 (1999)

Developing a Vision-Based Driving Assistance System

Ashfak Md. Shibli, Mohammed Moshiul Hoque and Lamia Alam

Abstract Driver's inattention and distraction are the most prominent reasons for road accidents. Since a fraction of second distraction may cause a severe accident and hence active attention of driver is mandatory while driving a car. Intelligent driving assistance system may reduce accident rate that are mostly occur due to inattentiveness and in turn improve the efficiency in driving. It is quite challenging task for computer vision to monitor the driver's level of attention continuously and assist him/her when level of attention is low due to distractions. This paper presents a driving assistance system that can compute the driver's attention and determine his/her level of attention while driving using a simple webcam and computer vision technique. The driver's attention level is determined by estimating his/her face direction, gaze direction, mouth movement, and head pose from video stream captured by the camera. If the level of attention crosses a predetermined value, the system initiates an audio sound to create the alertness of the driver. Experimental results show the system is functioning well and successful to generate alarms for 89.34% cases of inattention.

Keywords Computer vision · Driving assistance · Human–computer interaction
Level of attention · Intelligent vehicle

A. Md. Shibli · M. M. Hoque (✉) · L. Alam
Department of Computer Science & Engineering, Chittagong University
of Engineering and Technology, Chittagong 4349, Bangladesh
e-mail: mmoshiulh@gmail.com

A. Md. Shibli
e-mail: shibli.emon@gmail.com

L. Alam
e-mail: lamiacse09@gmail.com

© Springer Nature Singapore Pte Ltd. 2019
A. Abraham et al. (eds.), *Emerging Technologies in Data Mining and Information
Security*, Advances in Intelligent Systems and Computing 755,
https://doi.org/10.1007/978-981-13-1951-8_71

1 Introduction

Driving assistance system is an assistive technology for the drivers to help them drive safely and efficiently without risking their lives, passengers or around the vehicles. As like other assistive technology driving assistance system gives real-time feedback to be attentive in the driving environment avoiding human errors. It may be designed in two approaches: sensor-base and computer vision-base. In the sensor-based system, several sensor need to be embedded in the driver's body to estimate attention level. Embedding sensors in human body is very complex, uncomfortable and sometimes provides noisy data. In computer vision system there is no disturbance made to driver's body parts. Rather facial or visual data of the driver is taken by camera and estimates driver's attention or other behaviors to determine a driver is attentive or not. In this work, we have built a computer vision-based assistance system. Sometimes a very high-skilled driver fall victim to a severe injury because of accidents that are happened for lack of attention.

Driving is a sensitive task as it requires concentration and coordination of the driver and the driving environment at the same time. It is a simultaneous application of multiple sensory, physiological and cognitive abilities. Though these activities are involved, we often see drivers do non-driving activities like looking to other direction, talking to the passengers facing towards them. For some young minds there are more entertainment facilities and embedding more electronic sophisticated communication devices with the internet is common in new vehicles coming now and near future. Another wireless device engagement that is mobile phone is entirely a natural phenomenon while driving nowadays. Some of the activities that race for acquiring driver's attention while driving can demean driving operation and have a severe effect on traffic security. The distraction of driver from essential tasks while driving is one of the forms of driver inattention and is arrogated to be a contributing factor in over half of crash events due to distraction [1]. Moreover, as more entertainment devices and vehicle backing systems hit the vehicle industries, it is probable that the rate of perplexity-related accidents will grow.

It is one of the vital issues to keep the level of attention of the driver is a high value while he/she is in driving to ensure safety of himself/herself and the passengers. In this work, the level of attention is refers to the measure of concentration of the driver during driving in terms of physiological parameters. A high level of attention of drivers is very much vital because it retains safety of the vehicle and the humans in or around the vehicle. If the driver's attention level decreases to a shallow level because of the physiological etiquette, severe accident may occur within a moment.

Therefore, it is observed that the distraction is caused by activities inside the vehicle such as looking to other direction than required for driving, mobile phones, talking to the passengers facing them, etc. Our primary focus is the adaptation of the driver while even more integration of such distractive technologies whether the adaptation failed or attention is affected. Therefore, maintaining a sufficient amount of attention level during driving plays a noteworthy role in demeaning the rate of accident and assures secure travel. Therefore, continuous monitoring of driver's attention is also a prominent factor to improve performance of the driver. In our system, we will take the visual data of the driver and extract facial features to evaluate three parameters such as, eye aspect ratio (EAR), facial angle (FA), and lip motions (LM). We estimate the level of attention of the driver using three parameters and initiating an audio sound when his/her attention level falls in a low value.

2 Related Work

In a recent work, a system is developed to measure the driver's level of attention with two parameters such as lip motion and face angle using the Microsoft Kinect sensors [2]. Though it was efficient in lip motion and head movement detection and attention estimation, eye detection or eye closure weren't detected which is a vital parameter to make more error free. Garcia et al. [3], the proposed system detected drowsiness then they have detected distraction with the camera, NIR illuminator sensors and to measure head orientation an electromagnetic sensor used which is to be attached to head. Yawning [4] is identified upon mouth opening, mouth covering and facial feature distortion. Here only yawning is used as a parameter of tiredness. In [5], proposes a system to decide the drowsiness drive by Hidden Markov Model (HMM) using the head pose-estimation and eye blink. HMM predicts the driver's tiredness condition using yaw, pitch and eye blink. Smith and his colleagues presented a system to track a person's head robustly and facial features and the system is stable that it can auto reinitialize itself [6]. A Hybrid system is introduced in another work [7] that is a combination of template matching of open eyes and closed and template of yawing. But it was not tested in a real environment instead tested in a simulation environment. Using IR beam differentiating bright and dark eye and background removal techniques are applied in another work [8] that efficiently detects eye either with or without glasses. In [4] yawning analysis was done to determine facial fatigue. Here to distinguish yawning while the mouth is covered local binary pattern features and the neural network is used. There is also a twofold expert system [9] proposed to detect drowsiness by detecting a black region of a binary image like previous one is used. Using Template matching of open eyes and closed and template of yawing and techniques based on computer vision computed gaze, head and body posture to recognize VFOA in [10].

Most of the previous works had given good results only in the laboratory or simulated environments and use very limited parameters to estimate attention. Our propose framework is design to work in real environment, and estimate driver's level of attention using four parameters such as face direction, gaze direction, lip motion, and drowsiness. Moreover, a small webcam is used in the system that is not blocking the front view for the human eye and does not affect driver's efficiency.

3 Proposed Framework

The primary focus of our work is to develop a driving assistance system that can estimate the different attention level of the driver in terms of three parameters: face angle (FA), eye aspect ratio (EAR), and lip movement (LM). With the video stream from the driver's facial region, the system detects his/her face. Then the system compute EAR to calculate closeness of eyes [11]. The system calculates the mouth height which is our lip motion parameter. According to the height, we calculate the talking or yawning of a driver. Finally, the system estimate the face angle by solving perspective n point problem with the n 3D points of the face and corresponding 2D projections in the image. We use OpenCV's Solve-PnP feature to estimate face pose robustly. Figure 1 illustrates the schematic view of proposed framework.

3.1 Capturing Driver's Visual Data

In order to collect the driver's frontal face, we have used a simple USB webcam, Logitech C525, auto focusing HD 720p. As it is USB plug and play device, we don't need any software dependencies to connect or get it work. Just placing on the dashboard of the car and positioning with the field of view of driver the video stream is captured by system and sent to application interface.

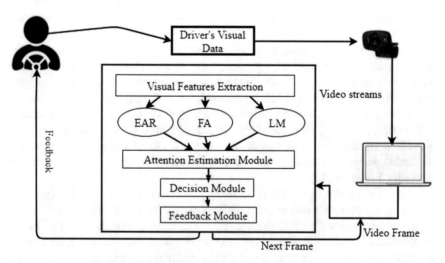

Fig. 1 Schematic framework of the driving assistance system

3.2 *Visual Feature Extraction*

A shape predictor is used to detect the critical region of interest from the visual features of input image frame. The prediction method will detect the prominent formation on the face image. Determination of facial landmarks can be done in two steps: (i) localize the face in the image and (ii) detect the vital facial structures.

For detecting face, we used dlib's pre-trained face detector which is a combined with histogram oriented gradient and linear svm. The facial landmark detector is used to identify landmarks with better accuracy. The landmark detector provides the $68(x, y)$ locations that depict the facial shape on the face image. In pre-trained image set the (x, y) coordinates of p facial landmarks are denoted by following Eq. 1,

$$S = (x_1^T, x_2^T, \ldots\ldots, x_p^T)^{.T} \in \Re^{2p} \tag{1}$$

The current estimate is denoted by $S(t)$ and each regressor is an update vector of an image. S(t) is included to the shape to enhance the estimation.

$$\hat{S}^{(t+1)} = \hat{S}^{(t)} + r_t(I, \hat{S}^{(t)}) \tag{2}$$

Ensemble of regression trees are trained using training sets and estimate the locations of facial landmark directly from intensity values. The annotation in Fig. 2 indicates the 68 points that are found by using iBUG 300-W dataset. The dlib facial landmark predictor is trained using these points and dataset.

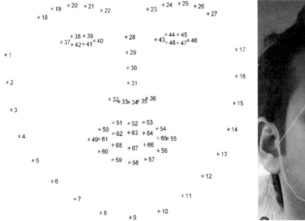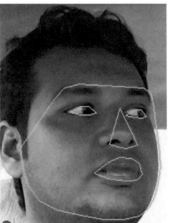

Fig. 2 Sixty eight facial landmarks coordinates by the dlib face detector and detected landmarks and connected in a video frame in our system

3.3   Attention Estimation

In order to determine the attention level of the driver, Eye region (to estimate EAR) head movement (to estimate FA) and yawning (to estimate lip movements) are used. Cumulated data from all parameters are combined to estimate overall attention level.

Eye Region Extraction and EAR Calculation. In Fig. 3 eye points skeleton is shown which are used to calculate EAR. These points are used to determine the Euclidean distance between points P and Q which is the distance of the connecting segments. The distance, D between two points $P(p_1, p_2, p_3, \ldots \ldots, p_n)$ and $Q(q_1, q_2, q_3, \ldots \ldots, q_n)$ can be determined by Eq. (3) [10].

$$D(P, Q) = \sqrt{(p_1 - q_1)^2 + (p_2 - q_2)^2 + \cdots + (p_3 - q_3)^2}$$

$$= \sqrt{\sum_{i=1}^{n} (p_i - q_i)^2} \tag{3}$$

As our coordinates are 2D coordinates, then the distance between two points is evaluated from Eq. 3. The EAR is then calculated with the following Eq. 4 by summation of distance between the selected eye landmark points,

$$EAR = \frac{\sum_{n=1}^{2} \sqrt{\sum_{i=1}^{2} (p_{ni} - q_{ni})^2}}{2\sqrt{\sum_{i=1}^{2} (p_{3i} - q_{3i})^2}} \tag{4}$$

Here the 37, 38, 39, 40, 41, 42 and 43, 44, 45, 46, 47, 48 points are the left and right eye points $p_3, p_1, p_2, q_3, q_2, q_1$ respectively in Fig. 4.

With this coordinates, the Eq. 4 can be simplified as in Eq. 5.

$$EAR = \frac{\|p_1 - q_1\| + \|p_2 - q_2\|}{2\|p_3 - q_3\|} \tag{5}$$

Two eye's EAR is then averaged to get the average EAR of both eyes. After calculating the EAR, we determine eye blinks according to some threshold EAR_T. EAR_T is evaluated by empirical data by our experiments. To check whether there is blink happened $EAR_T = 0.25$ is prescribed. We are assuming that in a car driver's distance is fixed.

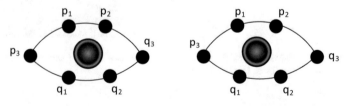

Fig. 3 The landmark points of left and right eyes

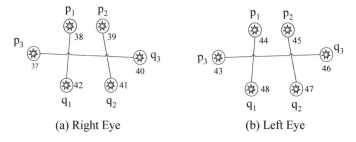

Fig. 4 The points to calculate distance and EAR

According to Caffier et al. [10], three drowsiness states are classified according to the blink duration (BD). Such as BD is less than 400 ms classify as awake, BD is greater than 400 ms and less than 800 ms classify as drowsy and BD is greater than 800 ms is classify as Sleeping respectively.

We have not used the speed of car but the frame rate of the camera is used to determine the time in driving. The mentioned algorithm acquired more than 94% accuracy to detect eye blink. We can detect every frame for blink detection if the blinks are occurred for few frames, then we can call the driver is drowsy and not attentive. If video frame rate is 30 fps, then in a 800 ms video, there will be $30 \times 0.8 = 24$ frames. So if anyone blinks up to 24 frames continuously we can conclude that he/she is in drowsy.

Facial Angle (FA) Estimation. To determine the head pose angles, we used anthropometric constant values of the human head [12]. We also got intrinsic parameters of the camera, approximated the optical center by center of the image frames. We have approximated the focal length by the width of images in pixels and assumed that deformation does not exist. The points from facial landmarks among them we will use following points in Fig. 5 to estimate head pose.

We have used Perspective N Point problem-solving algorithm here. Calculating the camera matrix, 2D, and 3D landmarks points we have solved perspective N point problem to get the rotation and translation vectors. OpenCV's Solve-pnp function uses the following formula gives the rotation and translation vectors.

$$sm' = A[R|t]M' \qquad (6)$$

$$s\begin{bmatrix} u \\ v \\ 1 \end{bmatrix} = \begin{bmatrix} f_x & 0 & c_x \\ 0 & f_y & c_y \\ 0 & 0 & 1 \end{bmatrix} \begin{bmatrix} r_{11} & r_{12} & r_{13} & t_1 \\ r_{21} & r_{22} & r_{23} & t_2 \\ r_{31} & r_{32} & r_{33} & t_3 \end{bmatrix} \begin{bmatrix} X \\ Y \\ Z \\ 1 \end{bmatrix} \qquad (7)$$

We get translation (t) and rotation (r) vectors from Eq. (7) and used Rodrigues' rotation function as in Eq. (8) to get the rotation matrix of the head.

$$v_{rot} = v \cos \theta + (k \times v) \sin \theta + k(k \cdot v)(1 - \cos \theta) \qquad (8)$$

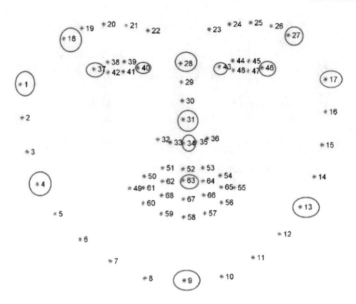

Fig. 5 Landmarks points to estimate head pose

We have converted the rotation matrix to Euler angles to get the yaw, pitch and roll angles by python's function in Eq. (9).

$$\theta = math \cdot a \tan 2(x, y) \tag{9}$$

Then from the pitch angle, we have determined if the driver is feeling drowsy if the angle is greater than facial angle threshold $EAR_T = 35°$.

Lip Motion (LM) Estimation. Same procedure is used to calculate mouth area. Leveraged the Euclidean distance measurement function to calculate the average distance between the two lips. The facial landmark points 51, 52, 53, 57, 58, 59 are used to determine the average Euclidian distance between two lips. Points are shown in Fig. 6.

A similar formula is used to calculate the distances. 51, 52, 53, 57, 58, 59 points are $p_1, p_2, p_3, q_3, q_2, q_1$. The distances are calculated according to the following Eq. 10 where MH is the mouth height measurement.

$$MH = \frac{1}{3} \sum_{n=1}^{3} \sqrt{\sum_{i=1}^{2} (p_{ni} - q_{ni})^2} \tag{10}$$

Simplifying the equation, we yield Eq. 11,

$$MH = \frac{1}{3}(\|p_1 - q_1\| + \|p_2 - q_2 + \|p_3 - q_3\|) \tag{11}$$

Fig. 6 Landmark points to calculate lip motion

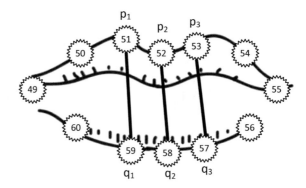

If a mouth is open at the highest level for 5 s [13], then it can be called a yawn. If the mouth is open more than calculated mouth height threshold MH_T, then it $LM = 1$ other than $LM = 0$. MH_T will vary with distance from the camera to the driver. We have taken MH_T value around 20 from a range where yawn may happen.

Level of Attention Estimation (EAR + FA + LM). The attention level is measured using the EAR, FA, and LM. In order to determine the different attention level of the driver Eq. (12) is used. This equation if formed by combining the Eqs. 4, 7, and 9 respectively.

$$LA = f(EAR, FA, LM)$$

$$= \begin{cases} EAR^{<EAR_T}; \ if \ t \geq 800 \ ms, \ 0.20 < EAR_T \leq 0.30 \\ FA^{>FA_T}; \ if \ FA_T \geq 35°, \ FA = FA \times (-1) if \ FA < 0 \\ LM; \ 1 \ if \ MH \geq MH_T, t > 5s, \ 15 < MH_T < 25 \\ 0 \ if \ MH < MH_T \end{cases} \quad (12)$$

The parameters' values are fed through the Eq. (12) and with the any of the conditions are true we decrement attention level by 3. We have taken the attention level as a range from 0 to 100. The initial attention level is 100. Here, EAR threshold EAR_T and mouth height threshold MH_T are changed in range from 0.20 to 0.30 and 15 to 25 according to distance from the camera. We typically used average 0.25 as EAR_T and 20 as MH_T. In order to classify the level of attention we used a set of rules in which all the values are determined empirically. For examples, if the attention range is less than 30 then it is classify to low attention and in that case alarm will activate to create the awareness of the driver.

4   Experimental Results

To evaluate the system, we run an experiment in real driving environment with various aged, skin colored and having different behavior participants. 5 participants eagerly were willing to test on our system and give time for the different type of experiments. The participants' average age was 25, and they have mixed skin colored. We have tested our system at various lighting conditions and other aspects of a driver behavior along with our system objectives.

4.1   Procedure

An USB camera is connected with the laptop and installs this camera in a location of car frontal so that it can capture the frontal face of the driver. During testing, the participant is asked to seat in a fixed position. The camera is fixed at a distance of 3 in. above from the driving wheel of the car. An experimenter is seated beside the percipient to run the interface and collecting data.

4.2   Visualization in Decision Module

We observed the charts starting from the program runs. We distinguish among different output curves according to the variation in the input. For understanding the changes, we have separated five regions to depict relative changes with three parameters (EAR, mouth height, pitch angle). For example, we have got the EAR and attention curve comparison shown in Fig. 4 for the experiments. If the human closes his eye randomly concerning the system, the result changes in region 1 of Fig. 4, (second plot from top) blue colored EAR value on the y-axis. The yellow colored plot is the Mouth Height changes over the period our experiment. In the last subplot of the interface shows the variation of pitch angle of the driver. The overall attention level changes are combined calculation and shown in the first subplot which is a green colored line in Fig. 7. This figure shows the GUI with corresponding measures of attention level. On right of figure the real-time video window is shown where the attention level and three dependent parameters are given. If there is any sound alert, there will be a flashing text also in this window.

4.3   Results

We test the proposed framework with five drivers having different skin tones, different age and skills in diverse lighting conditions.

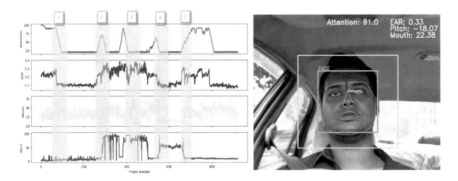

Fig. 7 Complete interface of graphical representation of LA and output window with real-time changing view of driver and parameters

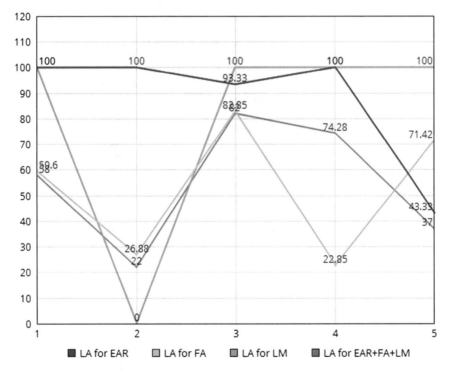

Fig. 8 Graphical representation of attention level for individual's values of EAR, FA, LM and their combinations (EAR + FA + LM)

Overall Attention Level Estimation. We have tested the parameters individually for attention level and combined them using the Eq. 12. Figure 8 shows the level of attention with single parameters and combined parameters. Results revealed that the graph with combined parameters is changed moderately than the graph with

Table 1 Level of attention estimation for individual and combined parameters

Participant	Eye Aspect Ratio (EAR)	Face Angle (FA)	Lip Motion (LM)	LA for Single parameter (EAR) (%)	LA for Single parameter (FA) (%)	LA for Single parameter (LM) (%)	LA for all three (EAR+FA+LM) (%)	Attention Level
1	0.32	20.86	No	100	59.6	100	58.0	HMA
2	0.36	44.41	Yes	100	26.88	0	22.0	LWA
3	0.28	29.31	No	93.3	82.85	100	82.0	VHA
4	0.30	27.29	No	100	22.85	100	74.3	HA
5	0.13	10.48	No	43.3	71.42	100	37.0	LMA
Average				87.3	52.72	80	54.7	MA

Fig. 9 Output of the feedback module to generates audio sound

individual parameter. Although the level of attention 100% for LM the combined parameters shows the fair attention level.

Table 1 shows for the participant 1, when the face angle is 20.86°, there is no lip motion as the mouth remains closed, so we can determine the level of attention of the driver using Eqs. (6), (11) and (12) respectively. For example, when consider the all three parameters such as EAR, FA and LM the level of attention is found 58.0% that is categorized as higher medium attention (HMA).

Feedback Module. It is certain that acute car crashed may happen if the level of attention of the driver is very low. Therefore, it will be the highest priority to create the awareness of the driver especially when his/her attention level is poor. In our current implementation, the system initiates an audio sound to create the awareness of the driver when attention level fewer than 30%. The system checks the driver's attention level and generates audio sound constantly until the high level of attention is resumed. Feedback module flashed red alert in the window and also generates beep sound as shown in Fig. 9. The total 122 trials are observed and the system can generate audio sound 89.34% of cases while the driver is in low attention category.

As an example, the red-colored text in Fig. 9 indicates that the system is producing an audio sound when the level of attention of a driver is less than 30%.

5 Conclusion

The work presented in this paper illustrates a driving assistance system using computer vision technique. This system continuously monitors the attention level of the driver while driving a car and categorized it in terms of high, very high, low, very low, low medium etc. In the current implementation, face angle, eye aspect ratio and lip movements are used to determine the various attention level. Experimental evaluation shows that the proposed assistant system can estimate the level of attention of a driver with reasonable accuracy and quite successful in generating the audio sound as an awareness signal. This system may be used as a driving assistance that will help to reduce the accident rate. The performance of the system may be improved by including drowsiness detector, talking/non-talking detector, alcohol detector, and night vision camera. These are left as future extensions.

Acknowledgements All procedures performed in studies involving human participants were in accordance with the ethical standards of the institutional and/or national research committee and the 1964 Helsinki declaration and its later amendments or comparable ethical standards.

References

1. Stutts, J., Reinfurt, D., Staplin, L., Rodgman, E.: The role of driver distraction in traffic crashes. Technical Report, AAA Foundation for Traffic Safety (2001)
2. Chowdhury, P., Alam, L., Hoque, M.M.: Designing an empirical framework to estimate the driver's attention. In: 5th International Conference on Informatics, Electronics and Vision, Dhaka, pp. 513–518 (2016)
3. García, I., Bronte, S., Bergasa, L.M., Almazán, J., Yebes, J.: Vision-based drowsiness detector for real driving conditions. In: IEEE Intelligent Vehicles Symposium, Alcala de Henares, pp. 618–623 (2012)
4. Ibrahim, M., Soraghan, J., Petropoulakis, L., Di Caterina, G.: Yawn analysis with mouth occlusion detection. Bio. Sig. Proc. & Con. **18**, 360–369 (2015)
5. Choi, I., Kim, Y.: Driver's drowsiness inference based on hidden Markov model using head pose and eye-blink tracking. In: Proceedings of HCI, Korea, pp. 61–65 (2014)
6. Smith, P., Shah, M., Lobo, N., da, V.: Determining driver visual attention with one camera. In: IEEE Transactions on Intelligent Transportation Systems, Vol. 4, no. 4, pp. 205–218 (2003)
7. Devane, P., Chavan, S.: Improved method for driver drowsiness detection using hybrid template matching algorithm. In: International Journal of Science and Research (2015)
8. Ji, Q.: Real-time eye, gaze, and face pose tracking for monitoring driver vigilance. Real-Time Imag. **8**(5), 357–377 (2002)
9. Anitha, C., Venkatesha, M., Adiga, B.: A two fold expert system for yawning detection. Proc. Com. Sci. **92**, 63–71 (2016)
10. Caffier, P., Erdmann, U., Ullsperger, P.: The spontaneous eye-blink as sleepiness indicator in patients with obstructive sleep apnoea syndrome-a pilot study. Sleep Med. **6**(2), 155–162 (2005)

11. Soukupova, T., Cech, J.: Real-time eye blink detection using facial landmarks. In: 21st Computer Vision Winter Workshop (2016)
12. Patacchiola, M., Cangelosi, A.: Head pose estimation in the wild using convolutional neural networks and adaptive gradient methods. Patt. Rec. **71**, 132–143 (2017)
13. Gupta S., Mittal S.: Yawning and its physiological significance. Int. J. App. Basic Med. Res. **3**(1), 11 (2013)

A Proposed Web-Based Architecture for Diabetes Awareness, Prevention, and Management

Md. Ariful Islam, Syed Akhter Hossain and Khondaker Abdullah Al Mamun

Abstract Diabetes is a growing concern and number of diabetes patient is increasing worldwide. Diabetes causes complication which leads to death in a long term. Proper lifestyle management is the key to control this disease. Treatment of diabetes is costly and complication caused by diabetes requires additional treatment. In developing countries, the situation is even worse. To address this problem, an innovative and cost-effective solution is required which will prevent diabetes and helps patient to modify their lifestyle. Web-based architecture can be very fruitful in this regard as it can be accessed from anywhere with IT enabled devices. With the existing infrastructure it can be used as a tool to create awareness, share knowledge and it can also help people to manage their lifestyle effectively. Interactive web-based platform can be used to show tips, diet measurement information and generate preventive measures. In this paper we have reviewed ICT enabled services and articles for diabetes awareness and prevention and proposed a web-based framework that can be used for screening diabetes and educating people about it. We are optimistic that, our proposed framework can improve the healthcare services and reduce the cost of healthcare making life easier.

Md. Ariful Islam (✉) · K. A. A. Mamun
AIMS Lab, Department of Computer Science and Engineering,
United International University, 1212 Dhaka, Bangladesh
e-mail: arifmdislam@outlook.com

K. A. A. Mamun
e-mail: mamun@cse.uiu.ac.bd

S. A. Hossain
Department of Computer Science and Engineering, Daffodil International University,
1207 Dhaka, Bangladesh
e-mail: akhterhossain@daffodilvarsity.edu.bd

© Springer Nature Singapore Pte Ltd. 2019   813
A. Abraham et al. (eds.), *Emerging Technologies in Data Mining and Information Security*, Advances in Intelligent Systems and Computing 755,
https://doi.org/10.1007/978-981-13-1951-8_72

1 Introduction

Diabetes Mellitus (DM) is a chronic disease and occurs if the pancreas cannot produce required amount of insulin or the body cannot use insulin effectively. Type 1 and type 2 are two different types of diabetes among which type 1 diabetes is more severe.

Type 2 diabetes leads to type 1 diabetes if left untreated and not properly controlled. Gestational diabetes is another kind of diabetes that occurs during the time of pregnancy [1].

According to World Health Organization (WHO), 8.5% of adults who are at least 18 years old had diabetes in the year 2014. In the year 2012, 1.5 million people died because of diabetes and 2.2 million people died because of high distribution of fasting plasma glucose in blood. It is the projection of WHO that, in 2030 diabetes will become the seventh major cause of death [2].

To reduce the occurrence of this disease we need to create awareness and take preventive steps as much as possible. To reduce the deaths caused by the diabetes, proper lifestyle management is required. Present day healthcare system is not well-equipped to face this crisis of diabetes even in developed countries [3]. The situation is worse in developing countries. According to WHO, the diabetes increase rate is alarming in South-East Asian Region (SEAR) and total of all diabetes patient, more than 20% will be found in this region. Currently, around 7 million cases of diabetes are available in Bangladesh and the number of diabetes patients is increasing by the rate of 5–6% in a year [4]. To overcome this crisis self-care activity is required in addition to existing healthcare programs and medical interventions. AADE (American Association of Diabetes Education) has a guideline supported by American Diabetes Association for self-care educations. A set of activities are defined in that guideline. Most important of them are performing exercise, taking healthy foods, medication and self-monitoring. All these activities are hard to follow for a diabetes patient. However, the use of ICT can be very helpful for a patient in this context [5]. The use of Information and Communication technology can enhance patient's self-care activity. Moreover, patient can find it interesting and easy to maintain diabetes through modern technologies like mobile apps and websites.

According to Global Report on diabetes, Geneva, 2016 by WHO, the number of diabetes patient has increased dramatically 108 million to 422 million from the year 1980 to 2014 which is almost four times higher. For both Men and Women highest number of deaths occurred in the age of 50–69. The prevalence of diabetes is also increasing rapidly. In 1980 worldwide prevalence of diabetes was 4.7%. However, it increased to the number 8.5% in the year 2014 which is almost double compared to the 1980 [6]. The details of diabetes prevalence categorized by countries with different income group are shown in Fig. 1.

The objective of this article is to analyze ICT enabled services, articles to promote ICT in an effective way for diabetes awareness, prevention and management. To promote ICT in this context, we have discussed possible solutions and we proposed a framework for screening diabetes patient and help them to manage diabetes effectively.

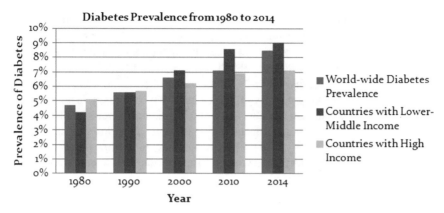

Fig. 1 Prevalence of diabetes from 1980 to 2014 categorized by countries with different income group

2 ICT for Diabetes Awareness, Prevention and Management

Diabetes is non-curable but preventable disease. Creating awareness is the first step to prevent diabetes. To create awareness, we need to spread the knowledge about preventive measures of this disease. World Diabetes Day (WDD) is celebrated yearly on November 14 to aware people about diabetes and to show them the means of prevention. This campaign is supported by International Diabetes Federation (IDF) and corresponding member associations. Other ways those are followed to create awareness are organizing flash mob, speaking in schools and colleges, spreading leaf lets, handbooks, banner, making news on newspapers [7]. However, as we live in an age of technology, we can aware people more effectively by using information and communication technologies. Nowadays, almost all kinds' of people including children and age old peoples use some kind of communication devices [8].

Smartphone, Personal computers (PC), Notebooks or tablets are found even with peoples in developing countries. With the help of Internet all these devices can be used as a tool or communication media to share news about the prevention of diabetes which will actually aware people [9]. Another important sector that can be considered is Social Media. Social media has become an important part of our life. More people are engaging in social media every day. Social Media can be used as an easy, cost-effective solution to reach millions of people across worldwide. This platform can be used to share online resources. Web pages, statistics, medication and a lot of information can be shared in social media to raise awareness. Health promotion is also possible through internet based social media which include Facebook, Twitter, Myspace, blogs, messaging application like Viber and whatsApp [10].

Mobile applications are becoming more popular. More mobile apps can be created to aware people about this disease. SMS-based reminder, motivational messages can also be used to modify people's lifestyle. This process has already proved its impact [11, 12]. In the end we can say that, ICT has become integral part of our life. If we use ICT in an effective way, creation of awareness will be easier and we will be able to reach more people in shortest possible time.

Diabetes is a lifelong disease and once occurred, there is no escape from this. However, prevention is big deal for type 2 diabetes which actually leads to type 1 diabetes. Type 2 diabetes can be prevented by taking regular exercise, eating healthy food and by checking risk overall, by modifying lifestyle. This lifestyle modification can prevent type 2 diabetes by 50% [13]. However, these preventing measures can be performed effectively by the people who are properly educated with the knowledge of diabetes. This diabetes education can be performed in various ways. ICT can be the right tool and also a media to perform knowledge education and share knowledge about diabetes.

Steps which can be taken to prevent and manage diabetes with the help of ICT are SMS based mobile awareness, internet based health monitoring systems, Information Technology for creating awareness, Use of Mobile Health (mHealth) technology and Using application services (Mobile app.).

The number of mobile phone has increased dramatically and a related study showed that, the number of messages sent tripled globally between the year 2007 and 2010. Being low cost and instant transmission mobile messaging has become well accepted among people of all status. For this reason, mobile phone messaging (SMS) can be used to send educational advice as well as motivational messages to change lifestyle for preventing diabetes [13].

Websites can be created to spread knowledge as well as interactive website can keep track of a diabetes patient. It can also be used as a tool distant learning, communication media and as a platform for Telemedicine where patients can have medication information through online [14]. A recent study showed that, in rural china a smart web aid helped people to identify and diagnose diabetes a way that was never been possible before. They gave that system a name "Smart Web Aid for Preventing Type 2 Diabetes (SWAP-DM2)". Using SWAP-DM2, around 2219 patient identified with diabetes risk and within 6-month 113 diabetes diagnosis was possible [15]. For this reason, it is now feasible to take help from internet based websites to prevent diabetes and also control this disease by educating people through online.

3 Related Work

There are lots of websites available for diabetes education and support. Some of these websites also offer diabetes risk test for type 2 diabetes. American Diabetes Association has its own website. They provide important features which includes risk test for type 2 diabetes, medication guideline, diet plan and BMI check. However, there is no option available for Blood glucose monitoring, SMS/Email notifications [16]. There are other websites also available with similar features. However, in an extensive search there is no web-based system found which supports diabetes prevention, management and screening in a single platform. Our proposed framework can screen diabetes based on the input of user, suggest medication, provide diabetes education for all kinds of user, and provide support by assigning a doctor. Table 1 shows the summary of some mostly used websites for diabetes management and prevention.

In this summary we have added basic features for diabetes awareness, prevention and management. Limitation is added to visualize the strength of a particular web. The country in which the website is developed also added. It is matter of sorrow that no interactive website is found developed and used in Bangladesh. Bangladesh Diabetes Association has its website only provide diabetes information.

4 Diabetes Screening and Management: A Proposed System

To identify diabetes patient and manage the disease we propose a web-based architecture where diabetes screening will be performed automatically based on user input.

The process follows given steps;

1. At first, user has to create his account by giving necessary information to identify a person. Creation of account will be done through email/cellphone verification.
2. Then, system will generate some questions regarding diabetes symptoms for rapid assessment which will include BMI (Body Mass Index), diet and exercise pattern, obesity, previous diabetes record in any other family members and other assessments.
3. If diabetes risk is detected, then system will ask for blood glucose test.
4. If there is no risk detected then, system will provide (electronic media) some basic knowledge and preventive steps of diabetes. After that, system will update its database and end the procedure.
5. User with diabetes risk need to perform blood glucose test. Then, user need to input his result in the system. System will then identify whether the user has diabetes, prediabetes or normal condition.
6. If no diabetes found (fasting glucose < 5.6 mmol/L) then system will follow steps designed for user without diabetes risk. If prediabetes found (fasting glucose is 5.6–6.9 mmol/L), then system will generate basic treatment guideline and preventive steps, provide diet plan and required exercise pattern. It will also set a reminder for checking in next 30 days. The reminder message can be sent to registered email address/cellphone number. In the end system will update its database for this particular patient.
7. However, if the value of fasting glucose is greater than 7.0 mmol/L, then system will undergo a weeklong observation for the user. In this time, the system will check user's BMI level, diet pattern, blood glucose level, obesity and some other necessary parameters.

Table 1 Summary of some mostly used websites for diabetes management and prevention

Name	Address	Features											Developed in	Limitations
		Diabetes education	Test for diabetes	Glucose monitoring	Medication guideline	Diet plan	Exercise	Blood pressure Monitoring	BMI checking	Communication (Online/Offline)	Support	SMS/Email Notifications		
American diabetes Association	http://www.diabetes.org/	✓	DM2[a]		✓	✓	✓	×	✓	✓	✓	×	USA	No active blood glucose, Blood pressure monitoring options available. Diabetes screening is not also available
Diabetes Self-Management	http://www.diabetesselfmanagement.com/	✓	×	×	✓	✓	✓	×	×	✓	✓	✓	Not found	No active blood glucose, Blood pressure monitoring options available. No BMI checking and Diabetes testing options available either. Diabetes screening is not also available
Sanofi Diabetes India	http://www.sanofidiabetes.in/	✓	×	×	✓	✓	✓	×	×	×	×	×	India	No active blood glucose, Blood pressure monitoring options available. No BMI checking and Diabetes testing options available either. Diabetes screening is not also available
Diabetes Australia	https://www.diabetesaustralia.com.au/	✓	DM2[a]	×	✓	✓	✓	×	×	✓	✓	×	Australia	No active blood glucose, Blood pressure monitoring options available. Diabetes screening is not also available

(continued)

Table 1 (continued)

Name	Address	Features											Developed in	Limitations
		Diabetes education	Test for diabetes	Glucose monitoring	Medication guideline	Diet plan	Exercise	Blood pressure Monitoring	BMI checking	Communication (Online/Offline)	Support	SMS/Email Notifications		
The National Institute of Diabetes and Digestive and Kidney Diseases	http://www.niddk.nih.gov/Pages/default.aspx	✓	✗	✗	✓	✓	✓	✗	✗	✓	✓	✓	USA	No active blood glucose, Blood pressure monitoring options available. No BMI checking and Diabetes testing options available either. Diabetes screening is not also available
Lilly Diabetes	http://www.lillydiabetes.com/	✓	✗	✗	✓	✓	✓	✗	✗	✓	✗	✗	USA	No active blood glucose, Blood pressure monitoring options available. No BMI checking and Diabetes testing options available either. Diabetes screening is not also available
Fit2Me	http://www.fit2me.com/	✓	✗	✗	✓	✓	✓	✗	✗	✓	✓	✗	USA	No active blood glucose, Blood pressure monitoring options available. No BMI checking and Diabetes testing options available either. Diabetes screening is not also available
Canadian Diabetes Association	http://www.diabetes.ca/	✓	✗	✗	✓	✓	✓	✗	✓	✓	✓	✓	Canada	No active blood glucose, Blood pressure monitoring options available. No BMI checking and Diabetes testing options available either. Diabetes screening is not also available

[a]DM2 refers to diabetes mellitus type 2 (Type 2 Diabetes)

8. If the user proved as normal or in prediabetes state, the system will follow steps of a user who have prediabetes.
9. If diabetes is confirmed, then system will engage the user as patient. It will also subscribe the user to daily feed which is email/SMS based. This feed will provide necessary information about controlling diabetes and also motivational messages to change lifestyle. System will also generate a diet plan according to patient's condition, BMI level and required exercise pattern.
10. Then, a doctor will be assigned for the patient depending his/her geographical area.
11. At the end system will update the database with new information about the patient.

These steps are for new users. If the user is not new and also is not a patient then system will start it steps from rapid assessment. If the user is a patient then system will generate his/her statistics considering his/her diabetes condition. Then system will start steps from measuring blood glucose level. Figure 2 shows the flowchart of the proposed framework. Figure 3 shows the architecture of the framework.

5   Features of the Proposed System

ICT is already an important part of our life and the use of ICT in the healthcare sector is promising. Web-based services are the most widespread sector of ICT. With almost all kinds of IT-enabled devices a website can be viewed and service can be taken. Considering this fact, we proposed this our frame with required feature for diabetes awareness, prevention and management. With these features it will be easier to screen the disease, monitor health and spread awareness as well. Features of proposed system summarized below:

1. SMS-based reminder and Knowledge sharing.
2. Engagement of more people and aware them on Social Media.
3. Sharing knowledge about preventive steps on social media in a regular basis.
4. Interactive website to keep track on diabetes patient as well as to aware them about the disease.
5. Use of Telemedicine/mHealth/eHealth facilities (if exist) to spread knowledge, awareness and lifestyle management information which can be very fruitful in preventing the disease.
6. It can also be used as a media for remote diagnosis of the disease and also it can help to identify diabetes patient by providing them the knowledge of diabetes symptoms and measurement procedures.
7. With the integration of GPS (Global Positioning System), it can be also used to track down nearest healthcare center in case of emergency.
8. Communication is possible with the assigned doctor in case of emergencies.

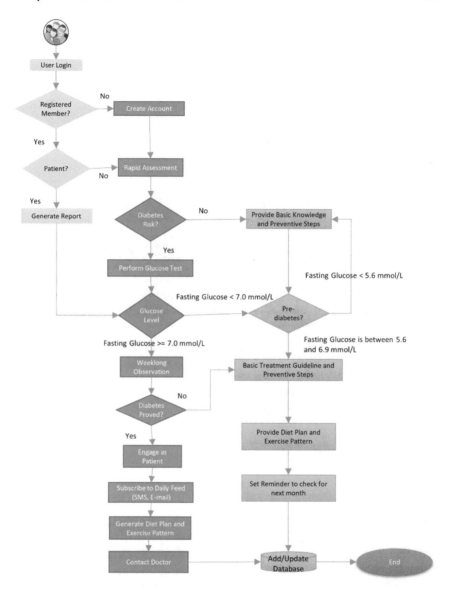

Fig. 2 Flowchart of web-based smart diabetes screening and management system

9. It can also help doctors to keep track on the patient which will reduce the need of physical meeting; a cost-effective solution when patient and doctor live far away, especially in rural areas.

Overall, the use of ICT can be very fruitful for the management and prevention of diabetes disease. It is a disease that needs personal awareness and lifestyle medication

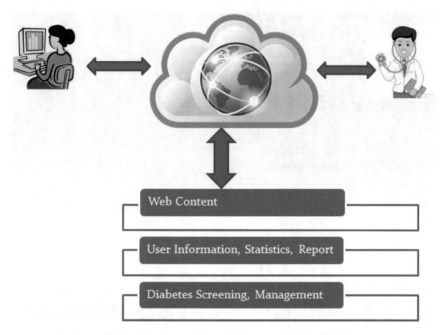

Fig. 3 Architecture of web-based smart diabetes screening and management system

to control it. The use of ICT is a big deal in this context. As people use ICT regularly, it will be easy for them to take necessary steps through ICT.

6 Conclusion and Future Work

The number of diabetes patient is increasing rapidly and it is definitely a challenge to reduce the rate. To address this issue, preventive measures are needed to be taken. Creating awareness, sharing knowledge are the ways of preventing diabetes. There are also some challenges to resolve which include financial challenges. Implementation of web-based platform can be very fruitful addressing all of these issues. Internet and web have its own infrastructure and it is already is an active part of our life. So, if we use this proposed web-based infrastructure in a proper way it will be cost-effective, easy and user friendly solution for diabetes awareness, prevention and management.

To reach the goal and to successfully implement our research work and its features, there are still some steps needed to be taken. We need to develop the web-based framework and to experiment it for further study and effectiveness of this type of services. We need to collect user feedback more to make it more perfect and user friendly. We need to consult with physicians as it is a service that concerns about human health. However, it is good news that, to build a more reliable, smart cloud architecture we are collaboratively working with hospitals, doctors and professionals.

Acknowledgements This is to acknowledge that, this research has been carried out with the funding of ICT Division, Bangladesh and help and support from AIMS Lab, United International University and Daffodil International University.

References

1. Anand, A., Muthukrishnan, C., Akella, A., Ramjee, R.: Redundancy in network traffic: findings and implications. In: Proceedings of the SIGMETRICS/Performance'09, pp. 37–48 (2009)
2. Mohan, P., Padmanabhan, V.N., Ramjee, R.: Nericell: rich monitoring of road and traffic conditions using mobile smartphones. In: Proceedings of the 6th ACM Conference on Embedded Network Sensor Systems—SenSys'08, p. 323 (2008)
3. Ahmad, F.S., Tsang, T.: Diabetes prevention, health information technology, and meaningful use: challenges and opportunities. Am. J. Prev. Med., **44**(4) Suppl. 4, S357–S363 (2013)
4. Basar, M.A., Alvi, H.N., Bokul, G.N., Khan, M.S., Anowar, F., Huda, M.N., Al Mamun, K.A.: A review on diabetes patient lifestyle management using mobile application. In: 2015 18th International Conference on Computer Information Technology, pp. 379–385, September 2015
5. Guo, S.H.-M., Chang, H.-K., Lin, C.-Y.: Impact of mobile diabetes self-care system on patients' knowledge, behavior and efficacy. Comput. Ind. **69**, 22–29 (2015)
6. Global Report on Diabetes: World Health Organization. http://www.who.int/diabetes/global-report/en/ (2016)
7. Mohan, P., Padmanabhan, V.N., Ramjee, R.: TrafficSense: rich monitoring of road and traffic conditions using mobile smartphones. In: 6th ACM Conference on Embedded Network Sensor Systems, pp. 1–29 (2008)
8. Boulos, M.N.K., Wheeler, S., Tavares, C., Jones, R.: How smartphones are changing the face of mobile and participatory healthcare: an overview, with example from eCAALYX. Biomed. Eng. **10**(1), 24 (2011)
9. Almarabeh, T., Rajab, L., Majdalawi, Y.K.: Awareness and usage of computer and internet among medical faculties' students at the University of Jordan. J. Softw. Eng. Appl. **9**(May), 147–154 (2016)
10. Korda, H., Itani, Z.: Harnessing social media for health promotion and behavior change. Health Promot. Pract. **14**(1), 15–23 (2013)
11. Khurshid, A., Brown, L., Mukherjee, S., Abebe, N., Kulick, D.: Texting for health: an evaluation of a population approach to type 2 diabetes risk reduction with a personalized message. Diabetes Spectr. **28**(4), 268–275 (2015)
12. Rai, A., Chintalapudi, K. K., Padmanabhan, V.N., Sen, R.: Zee: zero-effort crowdsourcing for indoor localization. In: Proceedings of the 18th Annual International Conference on Mobile Computing Netwoking—Mobicom'12, p. 293 (2012)
13. Johnston, D.G.: Effectiveness of mobile phone messaging in prevention of type 2 diabetes by lifestyle modification in men in India: a prospective, parallel-group, randomised controlled trial. Lancet Diabetes Endocrinol. **1**(March 2016), 191–198 (2013)
14. Kwon, H.S., Cho, J.H., Kim, H.S., Song, B.R., Ko, S.H., Lee, J.M., Kim, S.R., Chang, S.A., Kim, H.S., Cha, B.Y.: Establishment of blood glucose monitoring system using the internet. Diabetes Care, **27**(2), 478–483 (2004)

15. Chen, P., Chai, J., Cheng, J., Li, K., Xie, S., Liang, H., Shen, X., Feng, R., Wang, D.: A smart web aid for preventing diabetes in rural China: preliminary findings and lessons. J. Med. Internet Res. **16**(4), e98 (2014)
16. American diabetes Association. http://www.diabetes.org/?referrer=https://www.google.com.bd/

Rule Languages for the Semantic Web

Sonia Mehla and Sarika Jain

Abstract Ontology plays a major role in Semantic Web to describe the meaning about data on the web. Some inferences can be gleaned from the ontology model itself, but others may not be expressible in the ontology language (usually OWL) and require a more functional representation. Semantic Rules are required to infer implicit inference. It is a way of expressing additional things that can be inferred from your dataset. Rules layer is on top of the OWL in semantic web-layered architecture. This layer is less developed and active area of research. Various Rule languages have been developed by the authors for the Semantic web such as RuleML (Rule Markup Language), SWRL (Semantic Web Rule Language), RIF (Rule Interchange Format), R2ML (REWERSE Rule Markup Language) and many more. This paper aims to discuss the state of the art with respect to semantic rule-based technologies. It gives an overview of the rules and rule languages that are currently available to support rule-based- and ontology-based reasoning, and it also reviews some of the limitations of these technologies in terms of their inability to deal with uncertain or imprecise data, incomplete knowledge, decidability and their poor performance in some reasoning contexts.

Keywords Semantic web · SWRL · Ontology · Non-monotonicity · Decidability

1 Introduction

The vision of the future potential of the World Wide Web (WWW) is the Semantic Web which provides a global infrastructure for the dissemination, exploitation, and representation of human knowledge. Representational formalisms make the Seman-

S. Mehla (✉) · S. Jain
Department of Computer Applications, National Institute of Technology Kurukshetra,
Kurukshetra 136119, Haryana, India
e-mail: soniamehla_6160014@nitkkr.ac.in

S. Jain
e-mail: jasarika@nitkkr.ac.in

© Springer Nature Singapore Pte Ltd. 2019
A. Abraham et al. (eds.), *Emerging Technologies in Data Mining and Information Security*, Advances in Intelligent Systems and Computing 755,
https://doi.org/10.1007/978-981-13-1951-8_73

tic Web different from the conventional Web that makes the meaning of information content explicit. The availability of knowledge representation language is the first step in the development of the semantic web which is able to express human knowledge in a form that is amenable to machine processing. OWL is built on the top of XML and RDF in a layered architecture of semantic web. It is based on the description logic (DLs) to represent the concepts, properties, and individuals. A concept is described as a set of individuals and properties are described as a relation between individuals. These knowledge representation languages have expressive limitation. Rules can improve the expressive power and helps to infer implicit information from existing knowledge. The Semantic Web stack accentuates the demand of rule languages for the Web. These languages can enhance the knowledge representation by allowing complex relations that cannot be represented in Owl. Rules layer is on the top of the OWL in a layered architecture of the Semantic web. Rules describe the complex relationships between individuals and it is capable to extend the expressiveness of ontology languages. The set of inference rules can be used to conduct automated reasoning over the ontology. The researchers have proposed some semantic web rule languages to fulfill these requirements such as: Rule ML, SWRL, RIF, R2ML, WORL and many more. Prerequisites of rule language include rule integration, interoperability between rule languages, expressiveness and compatibility with other standards of the semantic web.

The review and comparison of rule languages for the Semantic Web has been done in this paper based on some comparison criteria. It reviews some of the limitations of these rule languages in terms of their inability to deal with uncertain or imprecise data, incomplete knowledge, decidability and their poor performance in some reasoning contexts. The major goal is to determine the comparison parameters and unique features of different rule languages. It may provide some research guideline in choosing the rule languages based on the application requirements.

The remaining paper is formulated as follows. The various rule languages have been described with one sample use case in Sect. 2. Comparative study of the rule languages based on the discussed parameters is described in Sect. 3 and then concludes this paper in Sect. 4.

2 Review of Rule Languages

A properly designed rule-based Inferencing engine is an interactive system used to support the decision makers in acquiring useful knowledge from raw data, documents, and personal knowledge to solve complex problems and concrete tasks. There are various rule languages proposed by the authors for the rule-based reasoning. Each of them supports different logics and functions. Let discuss them one by one and after a discussion on a few of the proposed rule languages, we provide a sample use case and discuss the rest in the light of this use case.

- **RuleML 0.9**: It stands for rule markup language 0.9 version introduced by Hirtle et al. [1] in 2006. It is an integration of <degree> with RuleML. It is a suitable form for rule interchange.
- **Vague-SWRL**: Still fuzzy SWRL is unable to represent imprecise knowledge accurately using single membership degree in fuzzy sets. Wang et al. [2] proposed a vague-SWRL based on the vague set in 2008. In vague-SWRL membership degrees have been modified by using second degree weight of vague classes and properties.
- **F-SW-if-then-unless-RL**: It stands for fuzzy semantic web if-then unless rule language which is proposed by Wang et al. [3] in 2009. It can be applied to perform reasoning on incomplete and uncertain knowledge. It is capable of reasoning and representing both non-monotonicity and fuzziness which can enhance the expressiveness of rule languages. Two types of negation (negation and negation as failure) are introduced for non-monotonic rules.
- **f-NSWRL**: In 2009 Wang et al. [4] suggested a fuzzy Non-monotonic Semantic Web Rule Language (f-NSWRL) based on negation as failure and negation to reason and represent non-monotonic fuzzy knowledge. It is the integration of two forms of negation and f-SWRL to represent the exceptions in default rules. It can assign the priorities between two similar rules to resolve the confliction.
- **f-R2ML**: It stands for fuzzy reverse rule markup language. In 2011 Wang et al. [5] enhanced the expressive power of reverse rule markup language (R2ML) using fuzzy rules and proposed a f-R2ML language. It is able to markup the crisp and fuzzy rules.
- **ESWRL**: Calero et al. [6] provides an extension over the SWRL in 2012 to enhance the expressiveness of the rules with adding four operators: not exists used to ask for remove and missing facts, Not operator to represent negative facts, dominance to make priorities between rules and mutex to create exclusions during rule executions.
- **Bayes-SWRL**: Liu et al. [7] proposed an extension to SWRL using Bayesian logic programs. This language is mainly used to predict the event from partial knowledge. Probability theory is more suitable to predict the event from a partial knowledge. Little research has been done on probability theory with semantic web, which is important for real-world problems like post disaster risk evaluations.
- **WORL**: It stands for web ontology rule language that is developed by Cao et al. [8] in 2014. It is an integration of OWL2 RL with eDatalog language. It has data complexity PTime with respect to standard semantics. This language is able to represent negation, unary external checkable predicates and minimal number restrictions in concept inclusion axioms at the left hand side. However, this language can also allow to use the syntax of description logic.
- **C-SWRL**: In 2017 Jajaga et al. [9] introduced C-SWRL to perform reasoning over stream data. It is an aggregation of RDF streams to enable time aware and closed-world reasoning with SWRL rules with the use of C-SPARQL filtering. This language is able to support non-monotonic behavior with the use of OWLAPI constructs.

2.1 Sample Use Case

We discuss a motivating use case from the earthquake disaster management. A knowledge base of earthquake management has been created [16]. Here, we will try to express post evolution rule for earthquake management by using rule languages. When the ground shakes strongly, buildings can be destroyed or damaged and their occupants may be injured or killed. Instances of earthquake or ground shaking created to store the parameters and resources have been stored based on those parameters in the ontology. The action depends on the intensity level, location like country and city. One sample rule is described to estimate the casualties in particular location due to earthquake. Further provide the number of resources based on the estimated casualties. It varies from place to place and mainly depends on the location and intensity of the earthquake.

Rule: *If the earthquake occurred in location x with intensity y then there is possibility of z Casualties.*

Earthquake (?x), has location (?x, ?y), has intensity (?x, ?z) ->Casualties (y?, ?a)

- **Rule ML**: It stands for the Rule Markup Language introduced by W3C. In 2000 RuleML Markup Initiative [10] designed the RuleML (Rule Markup Language) to encompass different ways of rules. It expresses the rules in XML syntax for the Web. Set of statements in the rules should be written within Atom. It's top element is 'Imp' (Implies) that represents the rule implication which is in the If-Then form.

This can be marked up as the following RuleML Datalog rule (an implication):

Example:

```
<Implies>
<head>
< Atom>
<Rel> Casualties </Rel>
<Var>z</Var></Atom>
< /head>
<body>
< Atom>
<Ind> earthquake <Ind>
<Rel> occurred < /Rel>
<Rel> location< /Rel>
<Var>x</Var>
<Rel> intensity </Rel>
<Var>y</Var>
< /Atom>
< /body>
</Implies>
```

- **SWRL**: It stands for semantic web rule language proposed by Horrocks et al. [11] in 2004 that enables horn-like rules to be integrated with an OWL Knowledge base. It is an integration of OWL DL and OWL-Lite sublanguages of the OWL with the

Unary/Binary Datalog RuleML sublanguages of the Rule Markup Language. This language enhanced the description logics with rules to overcome the limitations of ontology languages. It is a homogeneous approach in which SWRL rules are integrated with an OWL ontology and represented in a common logical language.

Example:

XML concrete Syntax:

```
<ruleml:imp xml:base>
<ruleml:_rlab ruleml:href="#example2"/>
< ruleml:_body>
<swrlx:classAtom>
< owlx:Class owlx:name="Earthquake" />
< ruleml:var>y</ruleml:var>
< /swrlx:classAtom>
< swrlx:individualPropertyAtom swrlx:property="haslocation">
<ruleml:var>y</ruleml:var>
< ruleml:var>x</ruleml:var>
< /swrlx:individualPropertyAtom>
<swrlx:individualPropertyAtom swrlx:property="hasintensity">
<ruleml:var>y</ruleml:var>
<ruleml:var>z</ruleml:var>
< /swrlx:individualPropertyAtom>
< /ruleml:_body>
<ruleml:_head>
< swrlx:individualPropertyAtom swrlx:property="Casualties">
<ruleml:var>x</ruleml:var>
<ruleml:var>a</ruleml:var>
</swrlx:individualPropertyAtom>
< /ruleml:_head>
< /ruleml:imp>
```

- **RIF**: It stands for rule interchange format which includes formal and precise specification of the semantics, syntax, and XML serialization. It was introduced in 2005 [12] within the World Wide Web Consortium (W3C). It's main motive is to develop a standard exchange format for rules on the Semantic Web. It is designed to enable interoperability among the rule languages. It is used to facilitate the sharing and exchanging of rules on the web.

Example:

```
Document (Prefix(<#>)
Group(ForAll ?x ?y ?z ?a
(:Earthquake(?x) :has_location(?x ?y) :has_intensity(?x ?z)) :-
:casualties(?y ?a)))
```

- **f-SWRL**: SWRL is unable to represent imprecise and vague knowledge and information. In 2006 Pan et al. [13] proposed an f-SWRL, extension to SWRL with integration of fuzzy assertions and fuzzy rules. Using fuzzy logic weight factor has been included in SWRL atoms to define the truth value between 0 and 1 that represents the importance of the atom in a rule.

Example:

> *Earthquake(?x)∧has_location(?x,?y)∧has_intensity(?x,?z)−>Casualties(y?,*
> *?a)∗0.6*

- **R2ML**: It stands for REWERSE Rule Markup Language which is an interchange format for rules. Wagner [14] presents a reverse rule markup language (R2ML) in 2006 combining the Rule Markup Language (RuleML), Object Constraint Language (OCL) and the Semantic Web Rule Language (SWRL). It facilitates to interchange the rules on different systems. The main motive of R2ML is to accede with semantic web standards like OWL and RDF (s). It supports integrity, derivation and production rules.

Sample described is based on the production rule as given below:

Example:

```
<r2ml:RuleBase Prefix<#>>
<r2ml:ProductionRuleSet....>
<r2ml:ProductionRule r2ml:ruleID="...">
<r2ml:Documentation>...</r2ml:Documentation>
<r2ml:conditions>
 <r2ml:AttributionAtom
r2ml:attribute=":Ground_Shaking.has_intensity">
<r2ml:subject>
 <r2ml:ObjectVariable r2ml:name="Earthquake"
r2ml:class="Ground_Shaking">
</r2ml:subject>
<r2ml:dataValue>
<r2ml:TypedLiteral r2ml:lexicalValue="x" r2ml:datatype="xs:integer"/>
</r2ml:dataValue>
</r2ml:AttributionAtom>
<r2ml:AttributionAtom r2ml:attribute=":Ground_Shaking.has_location">
<r2ml:subject>
 <r2ml:ObjectVariable r2ml:name="Earthquake"
r2ml:class=":Ground_Shaking">
</r2ml:subject>
<r2ml:dataValue>
<r2ml:TypedLiteral r2ml:lexicalValue="y" r2ml:datatype="xs:string">
```

```
        </r2ml:dataValue>
        </r2ml:AttributionAtom>
        <r2ml:producedActionExpr>
        <r2ml:UpdateActionExpr r2ml:property=":Location.casualties">
        <r2ml:contextArgument>
        <r2ml:ObjectVariable r2ml:name="y " r2ml:class=":Location"/>
        </r2ml:contextArgument>
        <r2ml:TypedLiteral r2ml:lexicalValue="z" r2ml:datatype="xs:integer">
        </r2ml:UpdateActionExpr>
        </r2ml:producedActionExpr>
        </r2ml:ProductionRule>
        </r2ml:ProductionRuleSet>
        </r2ml:RuleBase>
```

- **Jena rules**: It is Java-based framework [15] to develop the Semantic Web applications. Jena API's are used to extract data from and write to OWL ontologies and
  RDF graphs.

  Example

  @prefix:<#>
  [Rule1: (?x :haslocation ?y)(?x :hasintensity ?z) -> (?y :Casualties ?a)

3   Result and Discussion

Some parameters are discussed in this section to review and compare the rule languages: handling uncertainty, exception handling, non-monotonicity, decidability
and rule priorities. Table 1 presents a comparative study of rule languages based on
the discussed parameters.

- **Handling Uncertainty**: Knowledge is imperfect or imprecise in real-world scenarios. Some applications handle the random information and events, others cope
  with fuzzy and imprecise knowledge, and still others deal with distorted or missing
  information resulting in uncertainty.
- **Exception Handling**: It is important to handle the exceptions to perform reasoning
  on incomplete knowledge. Exceptions occur rarely in the rules. Unless condition
  is used to represent the exceptions in rules. Exclusive operator can be used to
  represent the unless part of the condition.

  For example "*if X then Y unless Z*" where Z is the exception condition.

- **Non-monotonicity**: Our knowledge about the world is always incomplete and
  therefore we need to perform reasoning in the absence of complete knowledge.
  When new information is added then truthness of prepositions does not change
  in monotonic logic. Human reasoning is non-monotonic in nature which revises
  their conclusions, when new information becomes available.

Table 1 Comparative study of rule languages

Year	Rule languages	Logical foundation	Handling uncertainty	Exception handling	Non-monotonicity	Decidability	Rules priority
2000	Rule ML	–	✗	✗	✗	–	✗
2004	SWRL	OWL-DL/OWL-Lite+ Rule ML	✗	✗	✗	✗	✗
2005	RIF	–	✗	✗	✗	–	✗
2006	f-SWRL	SWRL+ Fuzzy Set	✓	✗	✗	✗	✗
2006	R2ML	OCL+ SWRL+ Rule ML	✗	✗	✗	–	✗
2006	Rule ML 0.9	Rule ML+ <degree>	✓	✗	✗	–	✗
2008	Vague-SWRL	Vague set+ SWRL	✓	✗	✗	✗	✗
2009	F-SW-if-then unless-RL	If-then-rules + unless+ fuzzy set+ OWL DL	✓	✓	✓	–	✗
2009	f-NSWRL	F-SWRL + (negation + negation as failure)	✓	✓	✓	✗	✓
2011	f-R2ML	Fuzzy set+ R2ML	✓	✗	✗	–	✗
2012	ESWRL	SWRL+ four operators	✗	–	✓	–	✓
2013	Bayes-SWRL	SWRL+ BLP	✓	✗	✗	✗	✗
2013	Jena rule	–	✗	✗	✗	✗	✗
2014	WORL	OWL2 RL+ eDatalog	✗	✓	✓	✓	–
2017	C-SWRL	RDF streams+ SWRL	✗	✗	✓	✗	✗

(✓ indicates yes, ✗ indicates no and – indicates yes/no)

- **Decidability**: It is a crucial issue of reasoning in those systems which combining DL KBs and rules. The rule languages are decidable which can be solving the problems algorithmically.
- **Rule Priority**: Priority relationship between rules is important to model the realistic scenarios. Standardized rule languages are not able to define an exclusion relationship or priority order among rules. Conflict resolution strategy is required when more than one similar rule can be fired at the same time to increase the expressiveness of rule languages.

Table 1 summarizes the features of the previously discussed rule languages. It is no wonder that almost all languages have the capability to support all the parameters. Since 2000 many languages have been proposed, which have been shown in the first two columns of the table. Logical foundation column described the integrated form of different sub languages or operators to develop the proposed languages. Handling uncertainty column shows that Fuzzy SWRL, Bayes-SWRL, Vague-SWRL, F-SW-if-then-unless-RL, f-NSWRL, f-R2ML, Rule ML 0.9 are able to handle uncertain information. In next two columns F-SW-if-then-unless-RL, WORL and f-NSWRL can support both exception handling and non-monotonicity but ESWRL and C-SWRL can only support non-monotonicity. In the next section reveals languages with decidability support, which can be found in only WORL. ESWRL and f-SWRL can provide the priorities between the rules as shown in the last column.

4 Conclusion

Review and comparison of rule languages for the semantic web have been done in this paper. One of the main goal is to determine some criteria that can be used to compare the rule languages. Additionally, some key features are compared in existing rule languages based on the criteria. These language comparisons should give some research guideline in choosing the rule languages. Day by day expressiveness of the rule languages are improving. Priority between rules are important to resolve the confliction in application. Little research has been done on this issue in the semantic web. Our future work is to propose a conflict resolution method which will provide rule priority to perform the rule base reasoning.

Acknowledgements This work was financially supported by the Government of India, Ministry of Defense, DRDO. This paper is done in NIT Kurukshetra where the authors are affiliated to the Department of Computer Applications.

References

1. Hirtle, D., Boley, H., Grosof, B., Kifer, M., Sintek, M., Tabet, S., Wagner, G.: Schema Specification of RuleML 0.9. http://www.ruleml.org/0.9/
2. Wang, X., Ma, Z.M., Yan, L., Meng, X.: Vague-SWRL: a fuzzy extension of SWRL. In: International Conference on Web Reasoning and Rule Systems, pp. 232–233. Springer, Berlin, Heidelberg (2008)
3. Wang, X., Ma, Z.M., Yan, L., Cheng, J.: If-then and if-then-unless rules in the semantic web. In: Proceedings of the 2009 IEEE/WIC/ACM International Joint Conference on Web Intelligence and Intelligent Agent Technology, vol. 01, pp. 357–360. IEEE Computer Society (2009)
4. Wang, X., Ma, Z.M., Xu, C., Cheng, J.: Nonmonotonic fuzzy rules in the semantic web. In: 6th International Conference on Fuzzy Systems and Knowledge Discovery, vol. 2, pp. 275–279. IEEE (2009)
5. Wang, X., Meng, X., Sun, J., Chen, J.: f-R2ML: a fuzzy rule markup language. In: 8th International Conference on Fuzzy Systems and Knowledge Discovery, vol. 2, pp. 1275–1279. IEEE (2011)
6. Calero, J.M.A., Ortega, A.M., Perez, G.M., Botía, J.A., Gómez-Skarmeta, A.F.: A nonmonotonic expressiveness extension on the semantic web rule language. J. Web Eng. **11**(2), 93–118 (2012)
7. Liu, Y., Chen, S., Li, S., Wang, Y.: Bayes-SWRL: a probabilistic extension of SWRL. In: 9th International Conference on Computational Intelligence and Security, pp. 702–706. IEEE (2013)
8. Cao, S.T., Nguyen, L.A., Szałas, A.: WORL: a nonmonotonic rule language for the semantic web. Vietnam J. Comput. Sci. **1**(1), 57–69 (2014)
9. Jajaga, E., Ahmedi, L.: C-SWRL: SWRL for reasoning over stream data. In: 11th International Conference on Semantic Computing, pp. 395–400. IEEE (2017)
10. The Rule Markup Initiative. http://www.ruleml.org/
11. Horrocks, I., Patel-Schneider, P.F., Boley, H., Tabet, S., Grosof, B., Dean, M.: SWRL: a semantic web rule language combining OWL and RuleML. W3C Member submission. 21, 79 (2004)
12. Hawke, S.: Rule interchange format working group charter. W3C Semantic Web Activity (2005)
13. Pan, J.Z., Stoilos, G., Stamou, G., Tzouvaras, V., Horrocks, I.: f-SWRL: a fuzzy extension of SWRL. Lect. Notes Comput. Sci. **4090**, 28 (2006)
14. Wagner, G., Giurca, A., Lukichev, S.: A usable interchange format for rich syntax rules integrating OCL, RuleML and SWRL. In: Proceedings of Workshop Reasoning on the Web (2006)
15. Jena, A.: Reasoners and rule engines: Jena inference support. The Apache Software Foundation (2013)
16. Jain S., Mehla S., Mishra S.: An ontology for natural disasters with exceptions. In: International Conference System Modeling & Advancement in Research Trends IEEE-Explore, pp. 232–237 (2016)

Towards a Semantic Knowledge Treasure for Military Intelligence

Sanju Mishra and Sarika Jain

Abstract Information integration is essentially important for military operations because the range of relevant information sources is significantly distinct and dynamic. This work aims to develop a semantic knowledge treasure comprising of military resource ontology and procedures, as a model for better interoperability of heterogeneous resources of Indian Military. This model can interpret and govern the context of military information automatically, thereby facilitating the military commanders with decision making in several operations, such as command and control, weapon selection, situation awareness and many more. To design the military resource ontology, we specify the key concepts of the ontology. These concepts are based on terms extracted from heterogeneous resources. We develop an intelligent tool "QueryOnto" as an interface to the military resource ontology that provides a commander's decision support service and demonstrates how to apply the military ontology in practice. Web Ontology Language (OWL) and SPARQL query are used to implement the complete task.

Keywords Military resource ontology · Semantic knowledge treasure · OWL
SPARQL

1 Introduction

Current military has converted itself into an information-based fighting force and the demand for semantically rich data has become easily apparent. Two main factors contribute to the difficulties encountered in incorporating intelligence into military domain. First is the varying sources of information presenting information in various heterogeneous formats and the other is its highly dynamic nature. Military informa-

S. Mishra (✉) · S. Jain
National Institute of Technology Kurukshetra, Kurukshetra, Haryana, India
e-mail: sanju.tiwari.2007@gmail.com

S. Jain
e-mail: jasarika@nitkkr.ac.in

© Springer Nature Singapore Pte Ltd. 2019
A. Abraham et al. (eds.), *Emerging Technologies in Data Mining and Information Security*, Advances in Intelligent Systems and Computing 755,
https://doi.org/10.1007/978-981-13-1951-8_74

tion is scattered in various forms, incomplete, dynamic and publicly not available many times. Military forces are required to prepare with vigorous capabilities to support the response to threats in Battle Space Management, personnel, systems and facilities. It is required to collect the information in an integrated form as a Knowledge Treasure to automate and support a multitude of tasks. There is a need of a Resource Manager (machine) which will display the location, status, specifications, remaining quantity, etc., for any kind of resource: equipment, personnel or supplies. This resource manager should be able to handle critical scenarios and helpful to execute the prepared plan in a timely manner. Military intelligence demands information integration and gathering of all scattered information at one place which requires the use of semantic web technologies. At a growing phase of ontology development various government organizations and military applications are using ontologies from the last few decades.

This paper presents an ontological approach to develop a Semantic Knowledge Treasure (for Indian Army) by integrating the information from heterogeneous sources and gathering it in one place. The semantic knowledge treasure contains the ontology of resources with concepts, their properties and instances; and procedures of visualization, navigation, reasoning, and querying. The Military Resource Ontology thus developed offers to construct such a model and to specify certain concepts related to information in the domain of military operations and define them to the degree that they can be made "machine readable" through the knowledge based systems. Web Ontology Language (OWL), is used to express the semantic annotation of several significant military information resources. We present a distributed reasoning approach over large military resources. Various competency questions are discussed in this paper to evaluate the ontology. These competency questions are helpful to take the decisions in various military domains such as Decision Support, Situation Awareness and Battle Space Management. A QueryOnto Tool is designed to execute the competency questions to retrieve the information from Ontology. This paper is organized in different sections. Section 2 includes related work. A methodology is discussed in the Sect. 3 to construct the Military Resource Ontology. Section 4 represents the intelligent system 'QueryOnto'. In Sect. 5 conclusions and future work has been drawn.

2 Related Work

Several data models and formats are available to manage the large volumes of information and data from heterogeneous information sources (structured and unstructured) which needs to be integrated and interpreted. For unstructured data content extraction is required. So Natural Language Processing techniques are required for this. For structured data, various data models and formats have to be considered. In all these available models, reasoning processes are not very efficient, so ontological models are required which could be automatically supported information fusion process. Ontology is considered as a technology that serves a method to interchange semantic information between machines and people [1].

Several military applications that use ontology to execute valid military affairs has been growing. Military organizations in the United States have begun to recognize the potential for semantic markup of information. The U.S. Army has initiated research with the "Defense Advanced Research Projects Agency" (DARPA) to design domain ontologies in different areas such as leading acquiring weapons, identifying automated targets and geopolitical crises. Some research on upper ontology and its military application has been conducted [2]. Matheus et al. proposed Situation Awareness Ontology to describe battlefield situations, that dramatically update according to space or time [3]. The ontology contains object, relation and event as core concepts of the ontology. They have presented, how to describes the battlefield situations into formal languages based on the ontology. The U.S. Army designed a C4ISR ("Command, Control, Communications, Computers, Intelligence, Surveillance and Reconnaissance") Data Ontology in order to incorporate the legacy military systems and support interoperability between them [4]. Dianic presented a common method to combine heterogeneous military information systems using the domain ontology [5]. Here we provide an overview of various military ontologies developed so far.

- *Command and Control for Military*

Stoutenberg et al. [6] designed rules and ontologies to address emerging mission needs. For developing concise, extensible and modular core Command and Control ontology, Smith et al. [7] express a process to build a core ontology for supporting interoperability in a military environment. Nguyen et al. [8] conferred the designing of a group of ontologies for the messaging systems within the military and first responder Command and Control applications. ONTO-CIF [9] is a core ontology of intelligence analysis for command and control. THOR [10] ontology is developed as an extended work of existing command and control ontologies.

- *Ontologies for Situation Awareness*

SAW (Situation Awareness) [11] was the first model to support the situation awareness. SAW manages a situation as being comprised of entities, interacting in order to accomplish one or more objectives. Entities are abstract or physical objects and semantic relations modeling their connections. CONON (CONtext ONtology) [12] ontology expresses the conceptual description for situation awareness. Some features, person, local coordinates and objects with their activities are considered as collection of context. In civil-military operational areas, AKTiveSA [13] focused on the development of technology to enhance situation awareness and information integration. As a core element for military ontology, ATCIS [14] was implemented which focused on analyzing the target, reporting the battlefield situation, decision making and operational order.

- *Ontologies for Military Coalitions*

Some missions are directed jointly by various armies against common adversaries. These missions are named as Military Coalitions. Dorion et al. [14] created military coalition ontology. The model of this ontology was based on the C2IEDM ("Joint Consultation, Command and Control Information Exchange Data Model") [15] and useful to develop the automatic procedures to execute textual messages. Five main concepts of this ontology are highlighted, corresponding to: organizations, individuals, equipments, features and facilities.

- *Planning Ontology*

DSO National Laboratories (DSO) Singapore designed a DAML+OIL military plan ontology [16], describing concepts of the military domains, along with geographic features, specialities, military organizations and many more. The military plan ontology comprises 98 concepts, 26 relations and 34 instances. Network Attack Planning Ontology (NAPO) [17] intended to give support on a tactical level during the network operation planning.

Proposal

After studying several works on military ontologies we find some areas where work needs to be done. There are some research gaps which need to incorporate into this research. These gaps are:

- Models that are existing, prove to be either too common to fit our predicted task, or tailored for various applicative scenarios, and therefore irrelevant.
- The current work focuses upon developing a resource manager with a special focus on Equipments, Militant_Group, Attack, Operations, Locations and their sub domains. Up to our knowledge, no such ontology exists along with the reasoning procedures to visualize, search and query the ontology.
- Evaluation of some ontologies (Onto-CIF, PLAN) has not been done.

After the analysis of literature, we have planned a few objectives for this work. It is required to create a generic ontology for Indian Military domains which will integrate all entities of the domain at one place and reasoning will infer new knowledge.

- *Construct a Military Resource Ontology for Indian Defense* This ontology will provide the integrated information of Indian Forces (Army, Air Force, Naval...) and also facilitate other information like what, when, where, which mission, why the data was generated, to the domain experts with proper access.
- *Reasoning with Military Resource Ontology* Reasoning means inference of useful facts, decisions and automated initiated actions by the processing of structured knowledge in ontology. We have developed a tool which has features such as Querying, Searching and Visualizing the ontology. For example, if any person wants to know about the militant group which was responsible in Pathankot Attack 2016. By writing a query, essential information can be extracted from the Military Resource Ontology. SPARQL is used to write high level queries based on concepts. Military Resource Ontology is constructed as a doctrine-base military domain

hierarchical description which models a prospect of fundamental concepts and relationships described in military domains and scenarios designed by subject experts. This ontology provides the entities, concepts, and relationships essential to describe military resource's content, thus enabling intelligent reasoning over military resources in order to increase information integration and command and control for commanders.

3 Construction of Military Resource Ontology

Military information is scattered over the web in various forms (text, audio, and video) which is also incomplete information and publicly not available many times. It is required to collect the information in an integrated form to automate and support the task, such as visualization, reasoning, navigation, translation, mapping, merging and constraint checking. The structure of military organizations is in hierarchical form. It is classified in Corps and Divisions which are composed in brigades, battalions, companies, sections, platoons, etc. The Military Resource Ontology is constructed for keeping precise and unique knowledge of Indian Army. It is defined based on terms extracted from different sources. It is useful as a semantic knowledge treasure to store the military information and able to express what user wants to say. It is a domain ontology with a light-weight level of expressiveness which represent essential concepts of a military domain and relations among them.

To construct the Military Resource Ontology, we have proposed the following methodology. There are four major phases to develop an ontology:

(i)   Conceptual Identification
(ii) Concept Analysis and Organization
(iii) Encoding
(iv) Evaluation

3.1 Concept Identification

To develop an ontology [18, 24, 25], one must identify the concepts of the particular domain. Existing ontologies can be reused for collecting the concepts. For representing the ontology scope we have designed a series of queries that we would expect Subject Matter Experts (SMEs) to respond. These competency questions are the questions which should be answered by the knowledge treasure built from the domain ontology. Example of Competency Questions for Military Resource Ontology is presented in Table 1.

The primary task, to construct the Military Resource Ontology is to extract terms from scratch such as Google, Wikipedia, Survey Paper, Magazines and by interviewed many military commanders. We find out the core concepts, the relationship

Table 1 List of competency questions

S. No.	Competency question
1.	Which branch of Indian Army handles the Army Aviation Corpus?
2.	Which Helicopters are used only for attacks?
3.	Name the operation which handles the Pathankot attack.
4.	Display the name of the workshop which was used in Operation_Parakram.
5.	Which terrains are performed by Artillery Regiments?

between these concepts and review the existing ontologies such as Military_Ontology [20], OpenCyc [21], SUMO [22], DOLCE [23] etc. Concepts are organized into various classes and subclasses.

3.2   Concept Analysis and Organization

This section involves constructing the conceptual model by following the identification of terms in the previous section. Various terms are collected in the form of concepts and bound them into the hierarchy. These concepts are written top-down as well as bottom-up to create the ontology. Then the top-down approach, bottom-up approach, incremental approach applied to extend the ontology horizontally as well as vertically. The concept hierarchy will be created vertically in the knowledge treasure and horizontally in the database. All concepts of Military Resource Ontology are stored as a class hierarchy and related to the data properties and object properties. Concepts are expressed by the classes and relationships are expressed by the properties. There are two types of properties, Data and Object properties. Maximum work has been done by following the steps.

- *Describe the classes and hierarchy*

For constructing ontologies, the amalgam of bottom-up and top-down approaches is followed. We have started from the top level concept 'Mil_Res_Ontology'. Noy and McGuinness [1] had suggested various instructions to be follow while designing a hierarchy of concepts. These instructions are considered and used to analyze against the concept hierarchy which are constructed for own ontology. These instructions had fixed several rules that supports to design the concept hierarchy.

Subclasses of a class generally

(1) have extra properties that the super class does not have, or
(2) associated within various relationships than the super classes, or
(3) restrictions vary from those of the super class.

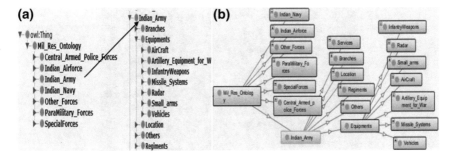

Fig. 1 Top level concepts. **a** Hierarchical structure. **b** Graph structure

- *Analyze Relationships*

In hierarchical relationships, there is only one relation, namely "hasSubclass". These relationship belongs to the similar hierarchies. Associative relationships are appointed among concepts that belong to dissimilar hierarchies and expressed by analyzing verbs related among concepts and appointing relation name that would construct a relevant statement with the concept name.

3.3 Encoding

In this section, we interpret the concept hierarchy into OWL documents representing the ontology. The ontology describes a rich description of military content using concepts and relationships acquired from the subject experts and military scenarios precisely to empower this sort of inference. Various ontology implementation languages are available. OWL ("Web Ontology Language") is used to develop Military Resource Ontology. OWL provides stronger machine interpretability of Web content than that supported by XML, RDF, and RDF Schema (RDFS) by providing additional vocabulary including a formal semantics. There are several ontology development tools are existing such as Apollo, Protégé and many more. To construct Military_Resource_Ontology, we choose Protégé 5.1 [19]. It is available as an open source [8] and having a plug-and-play environment for significant application development.

- *Hierarchical Structure of Ontology*

Military Resource Ontology displayed in a hierarchical form. Ontology can be represented in a tree structure like in Fig. 1a.

- **Graph Structure of Ontology**

It is also possible to represent the Military Resource Ontology in a graph structure like in Fig. 1b.

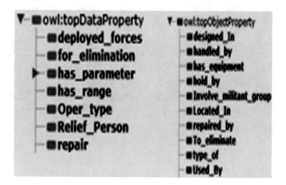

Fig. 2 Properties of military resource ontology

Fig. 3 Instances of concepts **a** Attack, **b** Militant_Group, **c** Operations

- *Properties of Classes*

An ontology must express the properties of the classes to response the competency questions expressed in Table 1. Properties are categorized in two types, Data Properties are used to relate instances to a user-defined value and Object Properties are used to relate the instances of a class to individuals of another class (Fig. 2).

- *Instances*

The last step is creating the domain database, i.e., creating instances of classes in the concept hierarchy. Describe an instance, or member of a class need (1) selecting a class, (2) forming an instance of that class, and (3) putting the attribute values. For example, we can create an individual instance, such as 'Jash-e-Mohammad' to represent a peculiar type of militant group. So, Jash-e-Mohammad is an instance of the class Militant_Group which represents all militants. Some instances are expressed in Fig. 3.

3.4 Evaluation

In order to evaluate the ontology, there are two types of evaluations have been achieved. External evaluation achieved by writing competency questions to inquire

Fig. 4 Visualization of
ontology

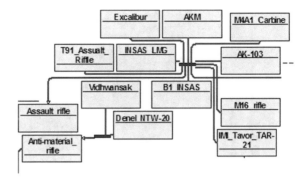

whether the Knowledge Treasure is efficient to response them or not. In Internal
Evaluation, by using HermitT 1.3.8 and Pellet reasoners, consistency has been verified and they provide the same result. SPARQL language and Pellet reasoner are used
to structure the competency questions and to execute them. SPARQL is a SQL-like
language for querying with the Knowledge Treasure and used to querying answering from the Knowledge Treasure. For evaluating the Military Resource Ontology,
there are competency questions discussed in Table 1 and specific use cases using the
ontology to automated reasoning, describing a group of queries the ontology must
be ready to respond. These questions are written in natural language and developed
by the discussion with experts and studying motivating scenarios.

4 QueryOnto Tool

Military Resource Ontology can used for reasoning in various ways, such as browsing, keyword searching, querying/answering etc. In this paper, we have designed a
tool "QueryOnto" which serve reasoning with ontology. There are three *main features of this tool, to visualize the complete ontology (KB-Browser), to search* any
concept of Military Resources in ontology (KeyWordSearch), to query various competency questions with Military Resource Ontology (Question/Answering) (Figs. 4
and 5).

5 Conclusion

We have developed a Semantic Knowledge Treasure for Military Intelligence, which
could be used by various Military Organizations. We have designed our own tool
"QueryOnto" for a wide range of military applications like visualizing, searching
and querying the Military Resource Ontology. Several Competency Questions are
prepared for the concept formalization and implementation of Knowledge Treasure.

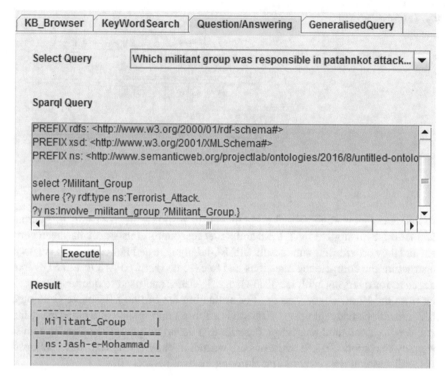

Fig. 5 Querying-answering with ontology

All competency questions are successfully executed on this tool. For future work, our goal is to develop an online decision support system that uses the Knowledge Treasure to guide the Military Organizations through inferences of military actions, which are set through the attributes of the Military Resources. Moreover, it is developed for future evolution and expansion. Other application and domain specializations may be incorporated in the future without updating the core ontology.

Acknowledgements This work was supported by Defence Research and Development Organisation (DRDO, New Delhi) followed by the grant no ERIP/ER/1506048/M/01/1611. The work presented in this paper is done at NIT Kurukshetra, India, where the authors are affiliated to the Department of Computer Applications.

References

1. Noy, N.F., McGuinness, D.L.: Ontology development 101: a guide to creating your first ontology, Stanford knowledge systems. Laboratory technical report KSL-01-05 and Stanford medical informatics technical report SMI-2001-0880 (2001)
2. Ra, M., Yoo, D., No, S.: Construction and applicability of military ontology for semantic data processing. In: Proceedings of the Third International Conference on Web Intelligence Mining and Semantics (WIMS'13), 12–14 June 2013, Madrid, Spain (2013)

3. Matheus, C.J., Kokar, M.M., Baclawski, K.: A core ontology for situation awareness. In: Proceedings of the Sixth International Conference on Information Fusion, Australia (2003)
4. Army Net-Centric Data Strategy: C4ISR Data Ontology Supports Army Net-Centric Data Strategy. http://data.army.mil
5. Dianic, A.: The need for scalability in network centric warfare—it's all in the semantics. SEC BSSD Bull. (2006)
6. Stoutenberg, S., Obrst, L., McCandless, D., Nichols, D., Franklin, P., Prausa, R., Sward, R.: Ontologies for rapid integration of heterogeneous data for command, control and intelligence. In: Ontology for the Intelligence Community (2007)
7. Smith, B., Mietinen, K., Mandrivk, W.: The ontology of command and control. In: Proceedings of the 14th International Command and Control Research and Technology Symposium (2009)
8. Nguyen, D.N., Kopena, J.B., Loo, B.T., Regli, W.C.: Ontologies for distributed command and control messaging. In: Proceedings of International Conference on Formal Ontology in Information Systems (FOIS) (2010)
9. Valentina, D.: Developing a core ontology to improve military intelligence analysis. Int. J. Knowl.-Based Intell. Eng. Syst. **17**, 29–36 (2013)
10. Emily, L., Nguyen, D.N., Marcello, B., William, C.R., Joseph, B.K., Thomas, W.: Military ontologies for information dissemination at the tactical edge. JOWO@ IJCAI (2015)
11. Wang, X.H., Zhang, D.Q., Gu, T., Pung, H.K.: Ontology based context modeling and reasoning using OWL. In: Proceedings of the 2nd IEEE Annual Conference on Pervasive Computing and Communication Workshops, USA (2004)
12. Smart, P.R., Russell, A., Shadbolt, N., Schraefel, M.C., Carr, L.A.: AKTiveSA: a technical demonstrator system for enhanced situation awareness in military operations other than war. Comput. J. **50**, 703–716 (2007)
13. Yoo, D., No, S., Ra, M.: A practical military ontology construction for the intelligent army tactical command information system. Int. J. Comput. Commun. Control **9**(1), 93–100 (2014)
14. Dorion, E., Matheus, C.J., Kokar, M.M.: Towards a formal ontology for military coalition operations. In: Proceedings of the 10th International Command and Control Research and Technology Symposium, USA (2005)
15. MIP: Multilateral interoperability programme, overview of the c2 information exchange data model (C2IEDM) (2003)
16. Lee, C.H.: Phase I Report for Plan Ontology. DSO National Labs, Singapore (2002)
17. Heerden, R.V., Chan, P., Leenen, L., Theron, J.: Using an ontology for network attack planning. Int. J. Cyber Warfare Terror. **6**(3) (2016)
18. Mishra, S., Malik, S., Jain, N.K., Jain, S.: A realist framework for ontologies and the semantic web. J. Procedia Comput. Sci. **70**(2), 483–490 (2015)
19. Knublauch, H., Fergerson, R., Noy, N.F., Musen, M.: The prot´eg´e OWL plugin: an open development environment for semantic web applications. In: Proceedings of ISWC 2004, number 3298 in LNCS, pp. 229–243 (2004)
20. http://rdf.muninn-project.org/ontologies/military.html
21. Matuszek, C., Cabral, J., Witbrock, M., DeOliveira, J.: An introduction to the syntax and content of Cyc. In: AAAI Spring Symposium (2006)
22. Niles, I., Pease, A.: Towards a standard upper ontology. In: Proceedings of the 2nd International Conference on Formal Ontology in Information Systems (FOIS) (2001)
23. Gangemi, A., Guarino, N., Masolo, C., Oltramari, A., Schneider, L.: Sweetening ontologies with dolce. In: Proceedings of the European Workshop on Knowledge Acquisition, Modeling, and Management (2002)
24. Jain, S., Mehla, S., Mishra, S.: An ontology for natural disasters with exceptions. In: International Conference System Modeling & Advancement in Research Trends IEEE-Explore, pp. 232–237 (2016)
25. Mishra, S., Jain, S.: A study of various approaches and tools on ontology. In: IEEE Conference CICT 2015 in ABES College of Engineering, pp. 57–61 (2015)

A Fuzzy AHP Approach to IT-Based Stream Selection for Admission in Technical Institutions in India

Oindreela Saha, Arpita Chakraborty and Jyoti Sekhar Banerjee

Abstract We encounter various problems and abundance of decisions, making our world intricate with the passage of time. Some decisions are relatively simple and insignificant, whereas others are complicated and have a notable effect on our life. In order to prevent imprecise decision-making, we need a proper decision-making framework. One of the most important decisions of an engineering aspirant's life is the selection of a stream for admission in a technical institute. It is often associated with lots of arguments, confusion, and vagueness. To overcome such situation, this paper provides a Fuzzy Analytical Hierarchy Process (FAHP) based stream selection scheme that prioritizes the vagueness of the decision makers during the stream selection procedure. Here authors considered entrance examination rank, class XII or equivalent exam marks, science related projects, and coding proficiency as the selection parameters for Information Technology (IT)-based stream selection during admission in engineering or technical colleges in India. A similar approach may be adapted to select colleges also.

1 Introduction

Technical Education and learning are the most significant processes in today's world. The development of any nation strongly depends on Technical education. Technical education has a strong influence on the socio-economical standard of a nation [20]. A sharp change has been noticed in the field of technical education marking a beginning of new era in the world of technology. Earlier, traditional engineering streams like civil, mechanical and electrical were the only available choice. The emergence

O. Saha · A. Chakraborty · J. S. Banerjee (✉)
Department of ECE, Bengal Institute of Technology, Kolkata 700150, India
e-mail: tojyoti2001@yahoo.co.in

O. Saha
e-mail: oindreelasaha19@gmail.com

A. Chakraborty
e-mail: chakraborty_arpita2006@yahoo.com

© Springer Nature Singapore Pte Ltd. 2019
A. Abraham et al. (eds.), *Emerging Technologies in Data Mining and Information Security*, Advances in Intelligent Systems and Computing 755,
https://doi.org/10.1007/978-981-13-1951-8_75

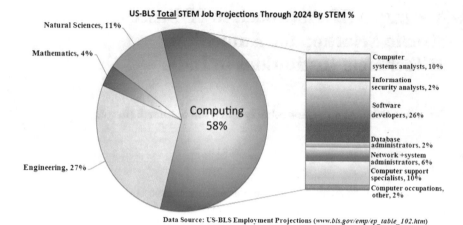

Fig. 1 Percentage of new STEM (Science, Technology, Engineering and Mathematics) Jobs by area through 2024 [16]

of Information Technology (IT) companies have brought the opportunity to study of various IT-based engineering streams, as a result Computer Science & Engineering (CSE), Electronics and Communication Engineering (ECE), Information Technology (IT) and Electronics and Instrumentation Engineering (EIE) came into limelight. In Fig. 1, the comparative study of IT-based streams versus core engineering is shown. The figure clearly portrays the advent of IT-based engineering streams in the recent time with the evidence of better job opportunities [16].

Recently, a lot of researches aimed on decision-making problems in various fields like relay node selection in cognitive radios [3–9, 11, 13] cooperative communications and sensor networks, personnel selection, etc., are carried on. The students are getting puzzled while selecting their streams for the betterment of their career. We often interrogate ourselves which stream to select? Often we take decisions based on our intuitions. Our every action is based on our decisions. It is always wise to understand the situation before drawing any conclusion. Relevant information will always help us to take the appropriate decision. But often we do not consider the vagueness in our decisions. For accurate decision-making, identification of the problem is the major step followed by evaluation of factors affecting the concerned problem and choosing of alternatives. The application of MADM (Multiple Attribute Decision Making) tools like GRA, TOPSIS, AHP, etc., has made the study of MADM problems more accurate and logical. In this piece of work we have proposed Fuzzy Analytical Hierarchy Process (FAHP) based stream selection scheme that prioritizes the vagueness of the decision makers during the stream selection procedure, thereby making our decision-making more logical and realistic. We have considered entrance exam rank, class XII or equivalent exam marks, science projects, and coding or programming proficiency as our decision-making criteria. In this paper we have made ourselves confined to only IT-based stream for the insight study of the concerned problem using Fuzzy AHP.

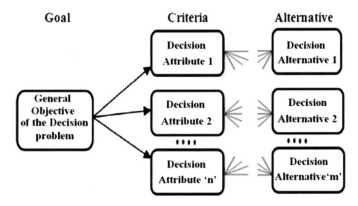

Fig. 2 MADM structure

The paper has been organized as follows. The system model is defined in Sect. 2. Section 3 contains the IT-Based stream selection parameters. FAHP-based stream selection scheme is presented in Sect. 4. Section 5 shows the results and discussions followed by the conclusion in Sect. 6.

2 System Model

Decision-making becomes tangled with the conflicting criteria. For eliminating such heterogeneity a logical framework is needed. Multiple attribute decision-making (MADM) is an approach to solve a problems with conflicting decision parameters and selecting the alternative which will meet the requirement in an efficient way. It involves disintegration of a problem into three levels, i.e., goal, criteria and alternatives. The disintegrated structure of a MADM process [15] has been shown in Fig. 2.

We have used Fuzzy AHP to solve a real-life MADM problem in this paper. Our aim is to choose the appropriate IT-based stream for admission in technical institute in India for a student where we have considered rank, class XII or equivalent examination marks, science projects, and coding or programming proficiency as our influencing factors or decision parameters. The alternatives are as follows: CSE, ECE, IT and EIE.

Table 1 Descriptions of selection parameters

Selection parameters	Description
1. Rank	Rank scored by candidates in entrance examination
2. Class XII or equivalent marks	Marks obtained in class XII or equivalent examination
3. Science related projects	Science projects, modeling or designing
4. Coding proficiency	Language known like C, C++, Java, etc.

3   IT-Based Stream Selection (IT-BSS) Parameters

The most critical concern in any MADM process is the parameters selection. Thus, in Fuzzy AHP scheme, construction of the hierarchy tree is the crucial factor to solve the problem. Like other MADM processes, Fuzzy AHP [18, 19] also provides a framework to the decision maker and helps to evaluate the complex problem. The decision-making problem is disintegrated into intermediate stages or hierarchy by keeping the goal at the top, the selection parameters in the middle followed by the alternatives at the bottom. In this paper, authors have analyzed the IT-based stream selection problem using Fuzzy AHP method. The level 1 of the hierarchy tree comprises the goal, i.e., selection of IT-based stream [20]. The criteria (shown in Table 1) are presented in level 2 and the alternatives, i.e., CSE, ECE, IT and EIE are shown in level 3 (shown in Fig. 3).

4   Fuzzy AHP-Based Stream Selection Scheme

FAHP is a decision-making tool used in various applications where multi criteria-based decisions are necessary. As the basic AHP [2, 18, 19, 21] does not take into account the vagueness of the decision makers, it has been modified by incorporating the fuzzy logic [1, 12]. In Fuzzy pairwise comparison of both the criteria and alternatives are carried out through the linguistic variables which are generally represented by the triangular numbers. Researchers proposed various methods where pairwise comparison matrices are used with fuzzy triangular numbers (TFN). In this paper we are employing Buckley's technique [10] to determine the relative importance weights of both the criteria and alternatives.

Stage 1: From the Linguistic terms (see Table 2) Decision maker compares the criteria or alternatives. With the help of the above Triangular Fuzzy Number (TFN) table, we can assign Triangular numbers depending on the linguistic terms. The comparison matrix is shown in Eq. 1, where $\tilde{d}_{ij}^{k}$ denotes the kth Decision maker's preference of ith criterion over jth criterion. To cite an example $\tilde{d}_{12}^{1}$ represents the first decision maker's preference of 1st criteria over the second criteria.

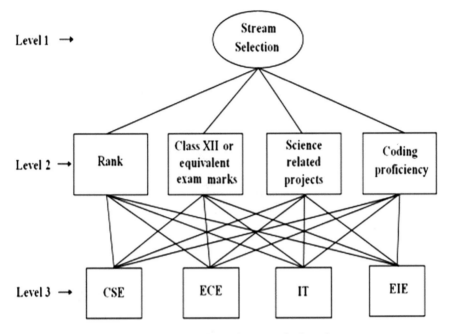

Fig. 3 Fuzzy AHP hierarchy for proposed IT-based stream selection scheme

$$A = \begin{bmatrix} \tilde{d}^k_{11} & \tilde{d}^k_{12} & \cdots & \tilde{d}^k_{1n} \\ \tilde{d}^k_{21} & \cdots & \cdots & \tilde{d}^k_{2n} \\ \cdots & \cdots & \cdots & \cdots \\ \tilde{d}^k_{n1} & \tilde{d}^k_{n2} & \cdots & \tilde{d}^k_{nn} \end{bmatrix} \tag{1}$$

Stage 2: Preferences of each decision maker are averaged and calculated as in the Eq. 2 for more than one decision maker.

$$\tilde{d}_{ij} = \frac{\sum_{k=1}^{K} \tilde{d}^k_{ij}}{K} \tag{2}$$

Stage 3: Pairwise comparison matrices are updated in Eq. 3 according to averaged preference calculated above.

$$\tilde{A} = \begin{bmatrix} \tilde{d}_{11} & \cdots & \tilde{d}_{1n} \\ \vdots & \ddots & \vdots \\ \tilde{d}_{n1} & \cdots & \tilde{d}_{nn} \end{bmatrix} \tag{3}$$

Table 2 Linguistic terms and corresponding fuzzy numbers

Intensity of importance in a numeric scale (Saaty scale)	Definition	Triangular fuzzy scale	Triangular fuzzy reciprocal scale
1	Equal significance	(1, 1, 1)	(1, 1, 1)
3	Moderate significance	(2, 3, 4)	(1/4, 1/3, 1/2)
5	Essential or strong significance	(4, 5, 6)	(1/6, 1/5, 1/4)
7	Very strong significance	(6, 7, 3)	(1/8, 1/7, 1/6)
9	Extreme significance	(9, 9, 9)	(1/9, 1/9, 1/9)
2	In between values of the two adjacent judgments	(1, 2, 3)	(1/3, 1/2, 1)
4		(3, 4, 5)	(1/5, 1/4, 1/3)
6		(5, 6, 7)	(1/7, 1/6, 1/5)
8		(7, 8, 9)	(1/9, 1/8, 1/7)

Stage 4: According to the Geometric Means method of Buckley, the geometric mean of Fuzzy comparison values of every criterion gets calculated is shown in Eq. 4 Where $\tilde{r}_i$ still denotes triangular values.

$$\tilde{r}_i = \left(\prod_{j=1}^{n} \tilde{d}_{ij} \right)^{1/n} , \quad i = 1, 2, \ldots, n \tag{4}$$

Stage 5: The fuzzy weights of each criterion can be found from Eq. 5 by following the subsequent three sub steps.

Stage 5a: The vector summation of each $\tilde{r}_i$ is to be found.

Stage 5b: Find the (-1) power of summation vector. Substitute the triangular Fuzzy numbers to arrange it in an increasing order.

Stage 5c: To get the fuzzy weight of criterion i ($\tilde{w}_i$), multiply each $\tilde{r}_i$ with the reverse vector as shown below:

$$\tilde{w}_i = \tilde{r}_i \otimes (\tilde{r}_1 \oplus \tilde{r}_2 \oplus \cdots \oplus \tilde{r}_n)^{-1}$$
$$= (lw_i, mw_i, uw_i) \tag{5}$$

Stage 6: Since $\tilde{w}_i$ are still fuzzy triangular numbers, they are to be defuzzified by Centre of Area method proposed by Chou and Chang [14], by applying the Eq. 6.

$$M_i = \frac{lw_i + mw_i + uw_i}{3} \tag{6}$$

Stage 7: As M_i is a non-fuzzy number we need to normalize it in Eq. 7.

$$N_i = \frac{M_i}{\sum_{i=1}^{n} M_i} \tag{7}$$

This whole process consisting of above seven stages need to perform to normalize the weights of criteria and alternatives. Then the product of each alternative with the related criteria gives us the score of each alternative. The alternative with the highest score is selected as best decision. Now we will use this process in our IT-based stream selection problem.

5 Results and Discussion

In this process selection of suitable IT-based stream, four most significant criteria are considered; rank as the first criteria, class XII or equivalent examination marks as the second, Science projects as the third and coding proficiency as the fourth criteria. The Authors also have considered that rank is the most important decision parameter followed by class XII or equivalent examination marks, science projects and coding proficiency. The average pairwise comparison of the criteria is represented in the Table 3.

A. *Determining the weights of criteria*

After completing the first three steps of methodology, the geometric means of fuzzy comparison matrix are tabulated in the Table 4.

Then the weights are calculated with the help of Eq. 5 and shown in the Table 5.

Finally, the non-fuzzy normalized weights are calculated using the Eqs. 6 and 7 and shown in the Table 6.

Now let us assume that with equal number of seats for each stream, i.e., CSE (500), ECE (500), & IT (500), EIE (500) there are altogether 2000 number of seats available for the 3000 candidates who have cracked the degree engineering entrance examination in a particular geographic region. Here the number of students, in the published list of candidates who have cracked the entrance examination, is more considering a few drop outs during admission. Now the authors have considered a scenario where the priority of stream selection is as follows: (i) CSE, (ii) ECE, (iii)

Table 3 Comparison matrix for criteria

Criteria	A	B	C	D
A	(1, 1, 1)	(2, 3, 4)	(4, 5, 6)	(4, 5, 6)
B	(1/4, 1/3, 1/2)	(1, 1, 1)	(2, 3, 4)	(2, 3, 4)
C	(1/6, 1/5, 1/4)	(1/4, 1/3, 1/2)	(1, 1, 1)	(1, 1, 1)
D	(1/6, 1/5, 1/4)	(1/4, 1/3, 1/2)	(1, 1, 1)	(1, 1, 1)

Table 4 Geometric means of fuzzy comparison values

Criteria	$\tilde{r}_i$		
A	2.37	2.94	3.46
B	1.00	1.31	1.68
C	0.449	0.506	0.594
D	0.449	0.506	0.594
Total	4.268	5.262	6.328
Reverse	0.234	0.190	0.158
Increasing order	0.158	0.190	0.234

Table 5 Relative fuzzy weights of each criterion

Criteria	$\tilde{w}_i$		
A	0.374	0.558	0.809
B	0.158	0.248	0.393
C	0.070	0.096	0.138
D	0.070	0.096	0.138

Table 6 Averaged and normalized relative weights of criteria

Criteria	M_i	N_i
A	0.580	0.553
B	0.266	0.253
C	0.101	0.096
D	0.101	0.096

IT, (iv) EIE. The scenario may vary with time and location. The streams get selected on the basis of rank as shown in Table 7 (A). Considering that fact that there may be some dropouts of students, the range of rank for each category is intentionally kept more than the seat availability. Table 7 (B) basically depicts the belongingness to a stream of a candidate with respect to Coding Proficiency and interest in Science Projects.

Now for an example, a student **X** has scored **370** rank in entrance examination and rated himself **6** in Coding Proficiency and **3** in Science project. Relative importance of streams determines the weight matrixes of alternatives under the four decision factors. Finally, each alternative stream's weight is calculated in the same way.

B. *Determining the weights of streams with respect to criteria*

Using the above mentioned Eqs. 1–7, the normalized weights for each alternative with respect to each Criterion are calculated and shown in the Table 8.

By using the Tables 6 and 8, the individual aggregated scores of each alternative with respect to criteria are calculated and are given in the Table 9.

Table 7 (A) Interpretation of Stream on the basis of rank. (B) Stream wise efficiency measuring scale

(A)

Stream	Rank range
CSE	1–1500
ECE	1–2000
IT	1–2500
EIE	1–3000

(B)

Streams	Coding proficiency	Science projects
CSE	9–10	5–6
ECE	1–4	9–10
IT	7–8	1–4
EIE	5–6	7–8

Table 8 Normalized non-fuzzy weights of each alternative with respect to each criterion

Alternatives	A	B	C	D
	N_i	N_i	N_i	N_i
CSE	0.467	0.25	0.278	0.96
ECE	0.278	0.25	0.095	0.253
IT	0.16	0.25	0.467	0.179
EIE	0.095	0.25	0.16	0.471

Table 9 Aggregated results for each alternative

Criteria		Scores of alternatives with respect to criteria			
	Weights	CSE	ECE	IT	EIE
A	0.553	0.467	0.278	0.16	0.095
B	0.253	0.25	0.25	0.25	0.25
C	0.096	0.278	0.095	0.467	0.16
D	0.096	0.096	0.253	0.179	0.471
	Total	0.356	0.249	0.212	0.175

Thus, the stream with the highest weight (stream CSE with 0.356) is selected as the best stream for this particular candidate X. In this letter, the results are verified through an online technical computing software [17]. The weights of the criteria have been calculated using three different methods like geometric mean, eigenvector & triangular fuzzy number and are shown in Figs. 4a, b and 5a respectively. Finally, the stream with the highest weight (CSE with weight 0.356) gets selected as the best stream for this particular candidate X (see Fig. 5b).

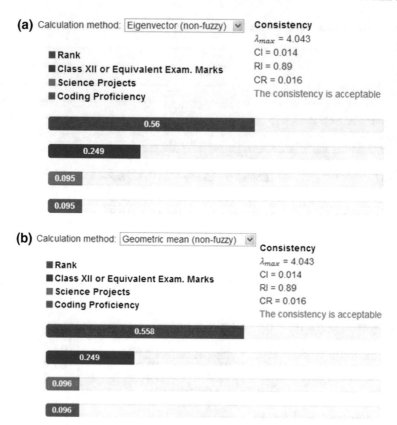

Fig. 4 **a**, **b** The results of the pairwise comparison matrix for criteria Table 3

6 Conclusion

The selection of an appropriate stream is one of the vital decisions of an engineering aspirant's life. To solve such problems, authors previously used other MADM tool like AHP, but it does not take into account the vagueness of the decision makers. Again, the application of simulation software programs in Fuzzy AHP is very simple and fast. Keeping the above criterion in mind, our proposed Fuzzy AHP approach to IT-based stream selection scheme can successfully cope up with the problem of vagueness among the decision makers and the other above discussed problems.

(a)

Calculation method: | Triangular fuzzy elements ▾ | **Consistency**

NI = 0.076

■ **Rank: 0.507 0.558 0.561**
■ **Class XII or Equivalent Exam. Marks: 0.213 0.249 0.272**
■ **Science Projects : 0.096 0.096 0.096**
■ **Coding Proficiency: 0.096 0.096 0.096**

(b)

■ CSE: 0.296 **0,359** 0,371
■ ECE: 0.206 **0,251** 0.279
■ IT: 0.18 **0,214** 0.239
■ EIE: 0.156 **0,176** 0.192

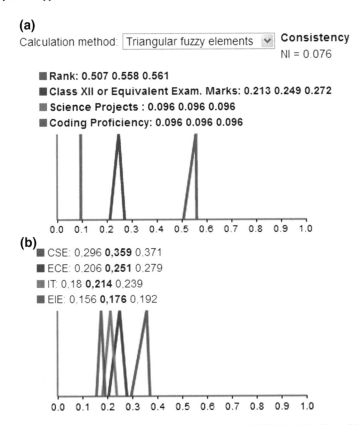

Fig. 5 **a** The results of the pairwise comparison matrix for criteria Table 3. **b** Ranking of the streams

References

1. Banerjee, J.S., Chakraborty, A., Chattopadhyay, A.: Fuzzy based relay selection for secondary transmission in cooperative cognitive radio networks. In: Proceedings of OPTRONIX, pp. 279–287. Springer (2017)
2. Banerjee, J.S., et al.: Relay node selection using analytical hierarchy process (AHP) for secondary transmission in multi-user cooperative cognitive radio systems. In: Proceedings of ETAEERE, pp. 745–754. Springer (2018)
3. Banerjee J.S., Chakraborty A., Chattopadhyay, A.: A novel best relay selection protocol for cooperative cognitive radio systems using fuzzy AHP. J. Mech. Continua Math. Sci. 13(2), 72–87 (2018)
4. Banerjee, J.S., Chakraborty, A., Chattopadhyay, A.: Reliable best-relay selection for secondary transmission in co-operation based cognitive radio systems: a multi-criteria approach. J. Mech. Continua Math. Sci. 13(2), 24–42 (2018)
5. Banerjee, J.S., Goswami, D., Nandi, S.: OPNET: a new paradigm for simulation of advanced communication systems. In: Proceedings of International Conference on Contemporary Challenges in Management, Technology & Social Sciences, SEMS, pp. 319–328, Lucknow, India (2014)

6. Banerjee, J.S., Chakraborty, A.: Fundamentals of software defined radio and cooperative spectrum sensing: a step ahead of cognitive radio networks. In: Kaabouch, N., Hu, W. (eds.) Handbook of Research on Software-Defined and Cognitive Radio Technologies for Dynamic Spectrum Management, pp. 499–543. IGI Global, USA (2015)
7. Banerjee, J.S., Chakraborty, A.: Modeling of software defined radio architecture and cognitive radio, the next generation dynamic and smart spectrum access technology. In: Rehmani, M.H., Faheem, Y. (eds.) Cognitive Radio Sensor Networks: Applications, Architectures, and Challenges, pp. 127–158. IGI Global, USA (2014)
8. Banerjee, J.S., Chakraborty, A., Karmakar, K.: Architecture of cognitive radio networks. In: Meghanathan, N., Reddy, Y.B. (eds.) Cognitive Radio Technology Applications for Wireless and Mobile Ad Hoc Networks, pp. 125–152. IGI Global, USA (2013)
9. Banerjee, J.S., Karmakar, K.: A comparative study on cognitive radio implementation issues. Int. J. Comput. Appl. **45**(15), 15, 44–51 (2012)
10. Buckley, J.J.: Fuzzy hierarchical analysis. Fuzzy Sets Syst. **17**(3), 233–247 (1985)
11. Chakraborty, A., Banerjee, J.S., Chattopadhyay, A.: Non-uniform quantized data fusion rule alleviating control channel overhead for cooperative spectrum sensing in cognitive radio networks. In: Advance Computing Conference (IACC), pp. 210–215. IEEE (2017)
12. Chakraborty, A., et al.: An Advance Q Learning (AQL) approach for path planning and obstacle avoidance of a mobile robot. Int. J. Intell. Mechatron. Robot. **3**(1), 53–73 (2013)
13. Chakraborty, A., Banerjee, J.S.: Effects of different digital modulation schemes for image transmission through wireless fading channels. LAP Lambert Academic Publishing (2017)
14. Chou, S.W., Chang, Y.C.: The implementation factors that influence the ERP (enterprise resource planning) benefits. Decis. Support Syst. **46**(1), 149–157 (2008)
15. Dey, P.K.: Integrated project evaluation and selection using multiple-attribute decision-making technique. Int. J. Prod. Econ. **103**(1), 90–103 (2006)
16. Henderson, R.: Industry employment and output projections to 2024. Monthly Lab. Rev. **138**, 1 (2015)
17. Holecek, P., Talašová, J.: A free software tool implementing the fuzzy AHP method
18. Paul, S., et al.: A fuzzy AHP-based relay node selection protocol for wireless body area networks (WBAN). In: Proceedings of OPTRONIX 2017 (Press), pp. 1–6. IEEE (2018)
19. Paul, S., et. al.: The extent analysis based fuzzy AHP approach for relay selection in WBAN. In: Proceedings of CISC 2018 (Press). AISC-Springer (2018)
20. Saha, O., Chakraborty, A., Banerjee, J.S.: A decision framework of IT-based stream selection using analytical hierarchy process (AHP) for admission in technical institutions. In: Proceedings of OPTRONIX 2018 (Press), pp. 1–6. IEEE (2018)
21. Saaty, T.L.: How to make a decision: the analytic hierarchy process. Eur. J. Oper. Res. **48**(1), 9–26 (1990)

Author Index

© Springer Nature Singapore Pte Ltd. 2019
A. Abraham et al. (eds.), *Emerging Technologies in Data Mining and Information
Security*, Advances in Intelligent Systems and Computing 755,
https://doi.org/10.1007/978-981-13-1951-8